刘伯里 院士

20世纪80年代,刘伯里院士在美国进行合作交流

20世纪90年代,刘伯里院士在国内做学术报告

2010 年，刘伯里院士从教 58 周年纪念

刘伯里院士和夫人陈淑华

中国工程院 院士文集

Collections from Members of the
Chinese Academy of Engineering

刘伯量文集

A Collection from Liu Boli

北 京
冶金工业出版社
2016

内 容 提 要

《刘伯里文集》收录了刘伯里院士团队自1964年至2014年间发表的研究报告和学术论文,内容主要涉及裂变产物分离和裂变废液处理,锝、铼和卤素等放射性药物化学。本书对我国从事放射化学和放射性药物化学行业研究的科研工作者有很好的参考价值。

图书在版编目(CIP)数据

刘伯里文集/刘伯里著. —北京:冶金工业出版社,2016.7
(中国工程院院士文集)
ISBN 978-7-5024-7108-8

Ⅰ.①刘… Ⅱ.①刘… Ⅲ.①放射化学—文集 Ⅳ.①O615-53

中国版本图书馆 CIP 数据核字(2016)第 208269 号

出 版 人　谭学余
地　　址　北京市东城区嵩祝院北巷39号　邮编　100009　电话　(010)64027926
网　　址　www.cnmip.com.cn　电子信箱　yjcbs@cnmip.com.cn
策　　划　任静波　责任编辑　李　臻　于昕蕾　美术编辑　彭子赫
版式设计　孙跃红　责任校对　石　静　责任印制　牛晓波
ISBN 978-7-5024-7108-8
冶金工业出版社出版发行;各地新华书店经销;三河市双峰印刷装订有限公司印刷
2016年7月第1版,2016年7月第1次印刷
787mm×1092mm 1/16;39印张;2彩页;901千字;606页
179.00元

冶金工业出版社　投稿电话　(010)64027932　投稿信箱　tougao@cnmip.com.cn
冶金工业出版社营销中心　电话　(010)64044283　传真　(010)64027893
冶金书店　地址　北京市东四西大街46号(100010)　电话　(010)65289081(兼传真)
冶金工业出版社天猫旗舰店　yjgycbs.tmall.com
(本书如有印装质量问题,本社营销中心负责退换)

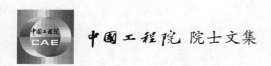

《中国工程院院士文集》总序

2012年暮秋，中国工程院开始组织并陆续出版《中国工程院院士文集》系列丛书。《中国工程院院士文集》收录了院士的传略、学术论著、中外论文及其目录、讲话文稿与科普作品等。其中，既有院士们早年初涉工程科技领域的学术论文，亦有其成为学科领军人物后，学术观点日趋成熟的思想硕果。卷卷文集在手，众多院士数十载辛勤耕耘的学术人生跃然纸上，透过严谨的工程科技论文，院士笑谈宏论的生动形象历历在目。

中国工程院是中国工程科学技术界的最高荣誉性、咨询性学术机构，由院士组成，致力于促进工程科学技术事业的发展。作为工程科学技术方面的领军人物，院士们在各自的研究领域具有极高的学术造诣，为我国工程科技事业发展做出了重大的、创造性的成就和贡献。《中国工程院院士文集》既是院士们一生事业成果的凝炼，也是他们高尚人格情操的写照。工程院出版史上能够留下这样丰富深刻的一笔，余有荣焉。

我向来认为，为中国工程院院士们组织出版院士文集之意义，贵在"真、善、美"三字。他们脚踏实地，放眼未来，自朴实的工程技术升华至引领学术前沿的至高境界，此谓其"真"；他们热爱祖国，提携后进，具有坚定的理想信念和高尚的人格魅力，此谓其"善"；他们治学严谨，著作等身，求真务实，科学创新，此谓其"美"。《中国工程院院士文集》集真、善、美于一体，辩而不华，质而不俚，既有"居高声自远"之澹泊意蕴，又有"大济于苍生"之战略胸怀，斯人斯事，斯情斯志，令人阅后难忘。

读一本文集，犹如阅读一段院士的"攀登"高峰的人生。让我们翻开《中国工程院院士文集》，进入院士们的学术世界。愿后之览者，亦有感于斯文，体味院士们的学术历程。

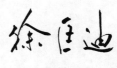

2012年7月

刘伯里院士简介

刘伯里（1931~），江苏常州人。放射化学和放射性药物化学家，中国放射性药物领域的主要开拓者。1997年当选为中国工程院院士。1953年毕业于华东师范大学化学系。他和合作者早期承担并完成了中国核燃料后处理工程低放裂变废液的任务，参加了核爆炸现场裂变产物污染苦咸水的去污，核潜艇原子反应堆第一回路水放射性的净化和从高放裂变废液中提取 ^{137}Cs、^{90}Sr 等裂变产物的研究，为完成有关军工任务做出了重要贡献。20世纪70年代初，他开始致力于核能的和平利用，从事放射性药物的研究，研究了15种核素的放射性药物，在放射性药物分子设计和应用两方面取得了系列创新成果。他和合作者首次提出并用实验证实了锝配合物的稳定性规律，系统研究了心、脑、肾上腺和有关受体放射性药物的构效关系；指导并参加研制的 ^{99m}Tc-ECD、^{99m}Tc-MIBI 获国家一类新药证书和生产证书，取得了较大的经济和社会效益。还研制出有望成为具有中国自主知识产权的新的脑斑块显像剂。此外，还研制出 ^{67}Ga 和 ^{111}In 两类新的心肌显像剂。首创了湿热熔融标记法和冠醚介质催化交换法，前者用于放射性药物标记的新工艺，后者实现了低温快速同位素交换。合作著有《锝药物化学及其应用》，在国内外主要专业刊物上发表论文240余篇。曾获全国科技大会奖（1979年）、国家教委甲类科技进步奖二等奖（1993年）、国家教委科技进步奖二等奖（1998年）、国家科技进步奖二等奖（1999年）、国防科工委以及省部级科技进步奖等9项奖励。

一、成长历程

刘伯里1931年3月17日出生于江苏常州。他在省立常州中学读高中时期，就接触到了一些进步书籍，新中国成立后通过对马列主义和中国革命史的学习及社会活动，确立了自己的信仰和一生奋斗目标。从此，他一直在各方面严格要求自己。

1953 年，刘伯里毕业于华东师范大学化学系，同年分配到北京师范大学工作，师从胡志彬教授。在 5 年的助教工作中，他埋头业务，读了两三遍研究生课程，打下了良好的理论基础和实验技能，并开展了物理化学方面的科学研究。1958 年是刘伯里学术生涯的转折点，学校要选调一批人才转向原子能科学研究，刘伯里被选送到中国科学院原子能研究所，师从留美归来的冯锡璋教授学习放射化学。在启蒙导师冯锡璋的指导下，他精读了不少专著并开始从事放射性废液处理的研究。冯锡璋不仅把他引入了原子能科学的殿堂，而且使他认识到"科研工作一定要走在生产需要的前面"。经过细致深入的调研，刘伯里根据中国核燃料后处理工业发展的需要，确定了裂变废液的处理和裂变核素的分离回收作为他的研究方向。从 20 世纪 50 年代末开始，他利用中国的天然无机矿物，如高岭土、蒙脱石、蛭石、沸石等，对主要铀裂变产物进行了交换吸附的研究，并取得了一些有意义的结果。60 年代中期，根据"备战、备荒、为人民"和三线建设的需要，为了确保长江上游不被核污染，要求将核工厂排放的放射性废液安全降低到国家规定的标准。刘伯里当时正在山西武乡革命老区参加农村四清运动，中途被调回北京，随即开展这项研究。由于时间紧、任务重，他和同事们从接受任务开始，根本顾不上放射性的危险，五年多的时间中，他们几乎没有节假日，全身心地投入研究。此后，他又接着从事了裂变核素的电迁移行为研究。由于经常接触毒性极大的核素 ^{239}Pu 和接受很大的辐射剂量，在十多年的工作中，刘伯里等接受的辐射总剂量是很大的。因此，他不到 40 岁时，头发已经脱落和变白，从外形看上去宛如个老人。"文化大革命"时期他由于有海外关系，因此各种到现场试验的机会都没有，虽然他也感到委屈，但他总是想到自己所做的工作只要对祖国和人民有益，能够使国家真正强大起来，不再受帝国主义列强的凌辱和宰割，个人受一点委屈不算什么。

20 世纪 70 年代中期，随着原子能和平利用的发展，刘伯里认识到放射性药物领域是放射性核素应用方面极为活跃的一个分支，又能直接为人民健康服务，造福人类，因此研究方向又转向了放射性药物。1974 年，北京市科技局的主要领导白介夫（后任北京市政协主席）和郑林（原核工业部原子能所党委书记）两位局长亲临北京师范大学放射化学与辐射化学研究室，听取了该室的汇报后，当场拍板给40个科研编制，成立北京市回旋加速器放射性药物实验室，刘伯里任该室董事长。40多年来，刘伯里和他的合作者在这块园地辛勤耕耘，取得了不少令人瞩目的成果。迄今为止，刘伯里和他的合作者著有《锝药物化学及其应用》一书，在国内外主要专业刊物上发表论文240余篇。

20世纪80年代以来，刘伯里除了担任繁重的教学和科研工作外，还担任许多行政和社会工作。历任北京师范大学化学系副主任、应用化学研究所所长、应用科学与技术学院院长，曾任中国核学会核化学和放射化学学会副理事长、中国核学会同位素学会常务理事、北京回旋加速器放射性药物实验室董事长、中国《大百科全书·核化学与放射化学卷》副主编、中国放射化学教材委员会主任、国家同位素工程技术研究中心工程技术委员会委员以及核化学与放射化学杂志常务编委等职。他积极组建了中国第一个放射性药物教育部重点实验室，并担任该重点实验室筹建期的学术委员会主任。他作为主要负责人之一参与了985非动力核技术创新平台的建立（2004年11月），并被任命为985非动力核技术创新平台学术委员会主任，促进了中国放射性药物的研发和青年人才的培养。由于他对我国放射化学事业的卓越贡献以及献身祖国的赤胆忠心，1997年当选为中国工程院院士，2006年被评为北京市优秀共产党员。

二、主要研究领域和学术成就

1. 开拓放射性药物的研究

刘伯里在放射性药物领域，研究了15种核素的放射性药物，尤其在锝化学以及锝药物的理论设计和应用方面取得了系列成果。

（1）锝化学与^{99m}Tc-放射性药物的研究。1980年以来，由于^{99m}Tc具有理想的核性质，锝药物得到了很大的发展。刘伯里及其合作者将堆积模型应用于锝化学，首次定量研究了锝配合物的结构稳定性规律，得出了所有稳定锝配合物，其空间立体角系数之和（SAS值）为0.97 ± 0.13。这解释了为什么世界各国过去在合成锝放射性药物时发现的某些锝配合物不稳定，提出了各种不同配位原子可能形成的稳定锝放射性药物的方案,这些设想后来均为实验所证实。这一成果使稳定锝配合物的设计、合成摆脱了过去经验的束缚,缩短了筛选时间,为锝放射性药物的设计提供了理论和实验依据。此外,刘伯里等还系统地研究了心、脑、肾上腺和有关受体锝放射性药物的构效关系,提出了脑显像剂的吸收机理和滞留机理,设计合成了对中枢神经系统特定受体具有高亲和性、选择性和高的脑摄取的 ^{99m}Tc 标记的分子探针。研制出有望成为中国自主知识产权的新放射性药物有:脑灌注显像剂 ^{99m}Tc-MPBDA、脑 Aβ 斑块显像剂 ^{99m}Tc-TZPH2-4-BAT、心肌灌注显像剂 ^{99m}Tc(CO)$_3$(MIBI)$_3$ 以及肾功能显像剂 ^{99m}Tc-BPHA 等。其中, ^{99m}Tc-MPBDA 具有良好的脑吸收和滞留性质, ^{99m}Tc(CO)$_3$(MIBI)$_3$ 的性质优于目前临床上广泛使用的心肌灌注显像剂 ^{99m}Tc-MIBI, ^{99m}Tc-BPHA 的性质与临床上使用的肾功能显像剂 ^{99m}Tc-DTPA 相当,脑 Aβ 斑块显像剂 ^{99m}Tc-TZPH2-4-BAT 具有较高的脑摄取。为了降低诊断成本,为广大人民的健康服务,他和合作者积极研制了脑灌注显像剂 ^{99m}Tc-ECD 和心肌灌注显像剂 ^{99m}Tc-MIBI,目前已通过卫生部批准,在全国临床推广应用,产生了较大的社会和经济效益。

(2) 放射性卤素药物。在放射性肾上腺显像剂的研制中,在王世真首次标记 6 位碘代胆固醇的基础上,刘伯里和他的合作者系统地研究了 ^{82}Br、^{131}I、^{211}At 标记的 6 位胆固醇和 ^{131}I 标记的 6 位甲基胆固醇和 6 位乙基胆固醇。其中,放射性核素溴、砹标记的 6 位胆固醇和 ^{131}I 标记的 6 位乙基胆固醇是独创的。此外,他还系统地研究了放射性卤素同位素交换动力学规律。

在放射性卤素药物标记方面,刘伯里首创了湿热熔融交换法和冠醚介质催化交换法。前者已用于放射性药物标记的新工艺,后者实现了低温快速同位素交换,为生物活性物质的标记开辟了新途径。另外,为了有目的地设计合成新型的放射性药物,刘伯里领导的研究组应用量子化学、分子力学以及分子图形学等计算机辅助药物设计方法研究了一系列脑受体与其配体相互作用的模型,得到了一些性质优良的放射性卤素示踪剂。如在系统研究新型 Aβ 斑块分子显像剂的基础上,发现 ^{18}F 标记的苯并恶唑类衍生物 [^{18}F]ZB-35a 在小鼠体内外的综合性质和美国食品与药品管理委员会(FDA)批准用于临床诊断阿尔茨海默病的 ^{18}F 标记的 Aβ 斑块显像剂 [^{18}F]AV45 的性质相当,目前正在进行临床前研究。此外,放射性碘标记的苯基苄基醚类 Aβ 斑块显像剂具有全新的化学结构,且各项性质优于已报道的[^{123}I]IMPY。再如在系统研究 sigma-1 受体显像剂的基础上,设计合成了一种新型的^{18}F 标记的 sigma-1 受体显像剂,其体内外综合性质均优于目前国际上唯一用于人体研究的^{18}F 标记的 sigma-1 受体显像剂[^{18}F]FPS。这些成果都处于放射性药物的研究前沿,受到各国同行的瞩目。

(3) 其他核素的放射性药物。在美国加州大学 Crocker 核实验室,刘伯里参加了对生产医用核素^{123}I 重要的^{127}I(p,xn)核反应的激发函数、薄靶产额及累计产额的测定,其中包括^{127}I(p,n)^{127}Xe、^{127}I(p,3n)^{125}Xe、^{127}I(p,5n)^{123}Xe、^{127}I(p,6n)^{122}Xe 和^{127}I(p,7n)^{121}Xe 等核反应,并大规模生产高纯度的^{123}I。在国内,他指导并参加了用 CS-30 回旋加速器研制^{67}Ga、^{111}In、^{201}Tl 和^{57}Co 等医用核素。其中,^{67}Ga 自 1987 年开始经卫生部批准,首次在国内大批量生产,供全国临床应用,^{57}Co 也批量生产,供制作穆斯鲍尔源,改变了依靠进口的局面,取得了很好的社会和经济效益。另外,成功地研制出 +1 价的^{111}In-BAT-TE 和^{68}Ga-BAT-TECH 配合物用作新的心肌显像剂,并定量地研究了上述两类药物的构效关系,为放射性^{111}In 和正电子发射核素^{68}Ga 的应用,以及为镓、铟放射性药物的发展开辟了广阔的前景。

2. 研究裂变产物分离和裂变废液处理

在中国核工业发展的初期,刘伯里根据中国核燃料后处理工业发展的需要,积极从事裂变废液处理的研究。自 1958 年以来,他系统地研究了中国蒙脱石、蛭石、斜发沸石等对裂变产物^{137}Cs、^{90}Sr、^{147}Pm、^{144}Ce 和^{106}Ru 的交换吸附行为,并取得了一些有意义的结果。

1965 年,根据三线建设的需要,刘伯里和他的同事们承担了核燃料后处理工厂(821 工程)中裂变废液处理的任务。刘伯里是该项目的主要技术负责人,组织并参加了该项目主要的实验研究。该实验包括几段流程,仅离子交换段就采集了全国上千种不同产地的蛭石、沸石矿样,并研究了它们对^{137}Cs、^{90}Sr、^{144}Ce 等裂变产物的交换吸附行为。为了确保数据的可靠性,他们建立了全套主要裂变核素的分析方法和低本底总 β 的测量方法,筛选了各类净化^{106}Ru 的阴离子树脂,设计了快速混凝沉降器,选择了各种快速混凝剂,试验了各种沉凝方案以及进行了热实验等。经过五年的艰苦奋斗,经过几百次小型流程试验和三次扩大试验,他们终于成功地确定了一套符合国家标准要求的处理流程，采用混凝沉降和离子交换两段流程处理大量低放裂变废水，达到了国家安全排放指标。该项成果填补了国内空白，达到了当时国际先进水平，并在三线核工业建设的有关工程设计中被采用。

1970~1972年，刘伯里作为主要负责人之一，参加了国家海洋局下达的海水淡化器脱盐除放射性裂变产物污染的研究项目。以核爆炸污染的苦咸水为体系，用冷却5天的浓缩 ^{235}U、^{239}Pu，经反应堆辐照后的裂片为指示剂，研究裂变产物在电渗析脱盐过程中的迁移行为和规律。此项研究取得了满意结果，为净化核爆炸现场的污染水源提供了有效手段，具有实际应用价值。此后，他还参加了中国核潜艇原子反应堆第一回路水中泄漏放射性核素的净化研究，为完成军工任务做出了重要贡献。

1974年，刘伯里参加了用人工合成的无机离子交换材料从高放裂变废液1AW中，分离和提取 ^{137}Cs 和 ^{90}Sr 的研究并进行了热试验，该项目经验收，达到了国际先进水平。

三、重视教育事业，积极开展国际学术交流

刘伯里长期从事教学和科研工作，在承担研究项目的同时，积极从事人才的培养，为研究生开设了原子核物理、放射化学、放射性药物化学和锝化学及其应用等基础和专业课程。1960~1964年培养放射化学研究生4名，1978年以来培养硕士生45名，博士生32名，博士后1名。其中，博士生崔孟超获得了2012年北京市优秀博士论文。20世纪60年代初培养研究生时，他还是一名讲师，在此特殊环境下，他勇敢地探索培养了中国第一批放射化学专业研究生。当时被聘任为答辩委员会主席学部委员的徐光宪和答辩委员会委员的刘元方都对刘伯里给予了很高的评价。1980年以来培养的研究生中，不少学生在国外深造或工作，国外的同事对他培养的学生同样给出了很高的评价。例如，2011年，在荷兰阿姆斯特丹举办的第19届国际放射性药物科学会议（19th International Symposium on Radiopharmaceutical Sciences, Amsterdam, The Netherlands, August 28—September 2, 2011）大会开幕式上，SRS会议之父、美国医学科学院院士、华盛顿大学的Welch教授在题为《The birth of the International Symposia of Radiopharmaceutical Chemistry and ISRS and SRS today》的大会报告中，把刘伯里作为放射性药物领域最有世界

影响力的重量级人物进行介绍，认为他推动了 ISRS 的发展，很多杰出的放射性药物工作者都出自他的团队。再如美国密苏里大学的 Lewis 教授来信感谢他送来了得到全面良好训练、创新性和动手能力都非常强的好学生。目前，他的学生大多在工作中做出了出色的成绩，有些已经成为了某些大学里放射性药物领域的学术负责人。刘伯里鼓励自己的学生学成回国服务，他常常对自己的学生说："出去可以开阔眼界，学到知识和技能，但是不能忘记我们最终的根是在中国。"受他的影响，他的一些学生学成归来后，做了大量促进中外放射性药物"产学研，技工贸"的交流合作活动。但是，有的学生觉得留在国外更合适，他也很理解他们的选择。他认为："要从更长远的角度来看人才外流的问题。现在国际学术交流活动很多，但是事实上最有收获的还是跑到国外跟他们一起做项目。为什么？因为即便是大型国际学术交流，展示的几乎都是过去一两年的成果，也就是说你了解到的都是过去式，如果你跟着他们一块儿做项目，那么你了解的就是进行时，而且如果你保持敏感的话，你还可以捕捉到将来时。作为一名科研工作者，我更深刻地体会到改革开放给国家带来的巨大变化。报效祖国有各种方式，而且最优秀的人才留学，不可能一个都不回来。哪怕是有 10% 的人回来，对国家而言都是宝贵的财富。"

　　刘伯里重视因材施教，要求学生结合自身的实际条件，掌握一定的理论和方法。他认为，知识的掌握，贵在熟练，只有反复熟练地理解和掌握了，才有可能谈得上发展和创新，所谓"熟能生巧"，这个"巧"字，就是科学发明创新的契机。许多新理论、新方法都是从这个"巧"字上开始萌发衍生而发展起来的。不仅如此，每当从事新的规律的探索时，自己最愿意应用的理论工具或实验方法，常常是自己掌握最熟练的那一部分。刘伯里不仅在学术上对学生指导有方，而且对学生生活也给予了很多关心和支持。当得知 2004 级硕士研究生张秋艳的父亲患脑瘤需要做手术而其家境又困难的消息时，他立即尽力解囊相助，使她的父亲的手术及时进行，病情及时得到控制。该学生非常感动，在学业上更加勤奋、努力。

　　为了发展中国的放射化学与放射性药物化学事业并扩大影响，刘伯里积极从事国际学术交流活动，先后 7 次被邀赴美国讲学和进行合作研究。1983 年，他担任美国纽约州立大学布法罗分校访问学者。1985 年，他担任美国加利福尼亚大学戴维斯分校访问教授。1988~1990年，他担任美国宾夕法尼亚大学放射学系的访问教授和脑放射性药物研究项目的顾问。三次应邀赴日本讲学，在九州大学、东北大学进行合作研究。1980~1981年，他担任日本九州大学访问研究员。作为中日双边会议的主要发起人和倡导者之一，他四次作为中国代表团团长出席在日本召开的第一次（福冈）、第三次（福冈）、第五次（仙台）和第七次（京都）中日双边会议，并且担任在中国召开的第二次（北京）、第四次（上海）和第八次（北京）中日双边会议的主席。1986年在北京召开的国际核化学和放射化学会议中，他担任国际组织委员会委员并做了大会特邀报告。他还是国际原子能机构放射性同位素生产培训班的特邀讲课人（1988年，1992年，1999年），受到亚太地区各国学生及机构官员的好评。另外，他还以中国代表身份两次参加了1993年和1995年在曼谷和悉尼举行的亚太地区核合作计划中有关放射性药物的国际会议。

作为中国放射性药物领域的主要开拓者，刘伯里认为，不仅应当争取成为一个具有创新精神的科学家，而且还力争成为一个懂得研究成果开发和转化的精明的企业家。如今每门学科正在向自己邻近的学科迅速扩充和渗透，同时研究成果转化为商品的周期也越来越短，对于应用研究来说，其成果用论文的形式发表，只是一个开端，决不应当是终结。他和他的合作者研制的心、脑放射性药物 ^{99m}Tc-MIBI 和 ^{99m}Tc-ECD 已由开发团队商品化，正在为全国人民的健康服务。他们不停地在放射性药物领域积极探索，争取早日将放射性药物的科研成果，转化为具有中国自主知识产权的具有国际竞争力的商品。

目　录

裂变产物分离和裂变废液处理

» 裂变产物和裂变废液中 Cs^{137} 的分离和回收 ························· 3
» 蛭石在低放裂变废水处理中的应用 ································· 19
» 电渗析器对长寿命裂片元素去除的基本研究（一） ····················· 32
» 利用无机离子交换剂——多聚锑酸从1AW中提取裂变核素 ^{90}Sr 的研究 ······ 39
» 用磷酸锆-磷钼酸铵复合交换剂从动力堆元件废液中提取 ^{137}Cs 的研究 ······· 45

放射性药物化学

锝、铼放射性药物

» Quantitative Study of the Structure-stability Relationship of Tc Complexes ············ 55
» Quantitative Study of the Structure-stability Relationship of $Tc^V O(Ⅲ)$ Complexes ··· 70
» Stability of Isomers of ^{99m}Tc-HMPAO and Complex of ^{99m}Tc-CBPAO Based on CNDO/2 Method ·· 84
» 铼的配位化学研究 Ⅰ. 铼化合物的堆积饱和规律 ························ 90
» Synthesis of New N_2S Ligands, Preparation of ^{99m}Tc Complexes and Their Preliminary Biodistribution in Mice ···································· 96
» 一种新的 ^{99m}Tc 标记的 $[Tc(CO)_3(MIBI)_3]^+$ 络合物作为心肌显像剂 ············ 103
» Solvation Effects on Brain Uptakes of Isomers of ^{99m}Tc Brain Imaging Agents ······ 109
» Preparation of New Technetium-99m NNS/X Complexes and Selection for Brain Imaging Agent ·· 118
» Synthesis, Separation and Biodistribution of ^{99m}Tc-CO-MIBI Complex ············· 129
» 新的 ^{99m}Tc 标记 σ 受体肿瘤显像剂 ···································· 137
» Synthesis and Biological Evaluation of Technetium-99m-labeled Deoxyglucose

- Derivatives as Imaging Agents for Tumor 146
- Preparation and Biodistribution of Novel ^{99m}Tc(CO)$_3$-CNR Complexes for Myocardial Imaging 153
- Caution to HPLC Analysis of Tricarbonyl Technetium Radiopharmaceuticals: An Example of Changing Constitution of Complexes in Column 162
- Preparation and Biological Evaluation of ^{99m}Tc-CO-MIBI as Myocardial Perfusion Imaging Agent 168
- Synthesis and Biological Evaluation of ^{99m}Tc, Re-monoamine-monoamide Conjugated to 2-(4-aminophenyl)Benzothiazole as Potential Probes for β-amyloid Plaques in the Brain 178
- Copolymer-based Hepatocyte Asialoglycoprotein Receptor Targeting Agent for SPECT 184
- ^{99m}Tc-and Re-labeled 6-dialkylamino-2-naphthylethylidene Derivatives as Imaging Probes for β-amyloid Plaques 199
- Synthesis and Biological Evaluation of Novel Technetium-99m Labeled Phenylbenzoxazole Derivatives as Potential Imaging Probes for β-amyloid Plaques in Brain 207
- ^{99m}Tc-labeled Dibenzylideneacetone Derivatives as Potential SPECT Probes for *in vivo* Imaging of β-amyloid Plaque 215
- Novel Cyclopentadienyl Tricarbonyl Complexes of ^{99m}Tc Mimicking Chalcone as Potential Single-Photon Emission Computed Tomography Imaging Probes for β-amyloid Plaques in Brain 234

卤素放射性药物

- Labelling of 6-Iodocholesterol in Melt 257
- Radioiodination of Alkyl Halides with Na^{125}I Catalyzed by Dicyclohexyl-18-crown-6 260
- 卤代烷烃与^{125}I,^{77}Br快速交换的新方法 263
- Radiobromine Labeled Cholesterol Analogs Synthesis and Tissue Distribution of Bromine-82 Labeled 6-Bromocholesterol 268
- 无载体^{18}F的制备及快速标记 273
- Halogen Exchanges Using Crown Ethers: Synthesis and Preliminary Biodistribution of 6-[^{211}At]-astatomethyl-19-norcholest-5(10)-en-3β-ol 277
- 复相体系同位素交换的研究热液熔融快速标记法 282
- A Fast, Kit-type Radioiodination Procedure for ^{127}I-for-^{123}I Exchange 289
- Cyclotron Production of High-purity ^{123}I Ⅰ. A Revision of Excitation Functions,

- Thin-target and Cumulative Yields for ^{127}I(p, xn) Reactions ·········· 299
- Comparison of [^{82}Br] 4-Bromoantipyrine and [^{125}I] 4-iodoantipyrine: the Kinetics of Exchange Reaction and Biodistribution in Rats ·········· 316
- Radioactive Iodine Exchange Reaction of HIPDM: Kinetics and Mechanism ·········· 326
- A Kit Formulation for Preparation of Iodine-123-IBZM: a New CNS D-2 Dopamine Receptor Imaging Agent ·········· 335
- 新型肾上腺显影剂——6-(^{82}Br)溴甲基胆固醇的合成及其在动物组织内的分布 ·········· 342
- Novel Anilinophthalimide Derivatives as Potential Probes for β-amyloid Plaque in the Brain ·········· 347
- Synthesis and Evaluation of Novel Benzothiazole Derivatives Based on the Bithiophene Structure as Potential Radiotracers for β-amyloid Plaques in Alzheimer's Disease ·········· 361
- Novel Imaging Agents for β-amyloid Plaque Based on the N-benzoylindole Core ·········· 379
- Novel (E)-5-styryl-2,2′-bithiophene Derivatives as Ligands for β-amyloid Plaques ·········· 386
- Synthesis and Evaluation of Novel ^{18}F Labeled 2-pyridinylbenzoxazole and 2-pyridinylbenzothiazole Derivatives as Ligands for Positron Emission Tomography (PET) Imaging of β-amyloid Plaques ·········· 405
- ^{18}F-labeled 2-phenylquinoxaline Derivatives as Potential Positron Emission Tomography Probes for in vivo Imaging of β-amyloid Plaques ·········· 433
- (E)-5-styryl-1H-indole and (E)-6-styrylquinoline Derivatives Serve as Probes for β-amyloid Plaques ·········· 449
- Synthesis and Preliminary Evaluation of ^{18}F-labeled Pyridaben Analogues for Myocardial Perfusion Imaging with PET ·········· 463
- Radioiodinated Benzyloxybenzene Derivatives: a Class of Flexible Ligands Target to β-amyloid Plaques in Alzheimer's Brains ·········· 478

其 他

- 放射性同位素半衰期测定中可能遇见的问题 ·········· 509
- (n,γ)核反冲法浓集^{24}Na 的研究 ·········· 513
- WilzBach 气曝法标记 H^3-三尖杉酯碱 ·········· 522
- H^3-L 门冬酰胺的标记 ·········· 525
- 用中子活化分析法测定 ^{238}U/^{235}U 同位素丰度比 ·········· 529

- 关于不同卤原子之间的交换反应 ……………………………………………… 537
- Synthesis, Characterization, and Biodistribution of [^{113m}In]TE-BAT: a New Myocardial Imaging Agent …………………………………………… 544
- A New Myocardial Imaging Agent: Synthesis, Characterization, and Biodistribution of Gallium-68-BAT-TECH ……………………………………… 552
- Carbon-11 Labeled Stilbene Derivatives from Natural Products for the Imaging of Aβ Plaques in the Brain ………………………………………… 563
- Evaluation of Molecules Based on the Electron Donor-acceptor Architecture as Near-infrared β-amyloidal-targeting Probes ………………………… 575
- Radiopharmaceuticals in China: Current Status and Prospects …………… 583

裂变产物分离和裂变废液处理

裂变产物和裂变废液中 Cs^{137} 的分离和回收*

本文利用蒙脱石，经过 630℃ 的热处理，改善了它的机械过滤性能，作为柱式无机离子交换剂，从裂变产物和裂变废液中分离和回收无载体放射性同位素 Cs^{137}。产品经过鉴定认为是放射性纯的。没有发现其他元素对 γ 能谱的显著干扰。

本工作首先研究了蒙脱石在不同条件下对几种重要的裂变元素（Cs^{137}、Sr^{90}、Pm^{147}、Ce^{144}、Ru^{106} 及 Zr^{95}）的交换吸附性能，同时对这几种裂变元素分别进行了柱式交换实验，然后从冷却一年和三年的裂变产物和裂变废液中分离和回收了 Cs^{137}。最后确定了 Cs^{137} 的回收率在 92% 以上。并探讨了蒙脱石对 Cs^{137} 的交换机理。

1 引言

目前从裂变产物和裂变废液中分离和回收 Cs^{137} 的方法多半采用共沉淀法[1~10]、离子交换法[11~15]、萃取法[16,17] 以及某些其他方法[18,19]。在这些方法中，共沉淀法是大规模的从裂变产物和裂变废液中分离和回收 Cs^{137} 的主要方法。有机离子交换剂在实际应用技术上虽比较简单，但由于它的热稳定性和辐射稳定性差，限制了它的应用范围[20]。由于无机离子交换剂的热稳定性和辐射稳定性较好，因此近年来应用无机离子交换剂分离裂变元素的工作也有一定的进展[21~25]。

目前各国利用天然黏土矿（蒙脱石、蛭石、海绿石和伊利石等）对裂变废液的处理和对某些裂变产物（Cs^{137}，Sr^{90}）的吸附已有不少报道。在这些研究中发现蒙脱石[26,42]、蛭石[27~29]、海绿石[30]、伊利石[31] 以及其他黏土矿[32,33] 对 Cs^{137} 有较高的吸附能力。

本工作利用蒙脱石对 Cs^{137} 的分离和回收进行了比较系统的研究。结果发现蒙脱石对 Cs^{137} 具有高度的选择性，交换量也较大。为了利用它作为柱式离子交换剂，采用热处理的方法，改善了它的机械过滤性能。并对最重要的几种裂变元素 Cs^{137}、Sr^{90}、Pm^{147}、Ce^{144}、Ru^{106} 和 Zr^{95} 进行了柱状的交换试验。结果表明，经过 630℃ 热处理以后的蒙脱石，交换量虽然有所降低，但却进一步增加了对 Cs^{137} 的选择性。为此，根据试验所确定的分离条件，对冷却一年和三年的裂变产物和裂变废液进行了 Cs^{137} 的分离和回收实验。产品经过物理和化学鉴定，认为是无载体，放射性纯的。最后确定了 Cs^{137} 的回收率，并探讨了蒙脱石对 Cs^{137} 的交换机理。

2 实验结果和讨论

本实验所采用的蒙脱石，在试验前都经过纯化、烘烤（105℃）和筛选。测量所用的钟罩形计数管的直径为 2.2cm，云母窗的厚度为 3.5mg/cm²。标准样品盘离云母窗的

---

* 本文合作者：赵忠顺。原发表于《原子能》，1964(10)：940~951。

距离为 2.5cm。所用的放射性同位素是无载体的 $Cs^{137}Cl$、$Sr^{90}Cl_2$、$Ru^{106}Cl_3$、$Zr^{95}Cl_4$、$Ce^{144}(NO_3)_3$、$Pm^{147}(NO_3)_3$ 以及经过一年和三年冷却的裂变产物溶液。其他稳定的化学试剂是一级或二级纯的。试液用红外线灯烘干后测量。计数装置均符合相对测量的各项要求。

关于蒙脱石在不同条件下对 Cs^{137} 吸附性能的试验,蒙脱石对 Cs^{137} 静态平衡交换量的测定,以及杂质离子(Al^{3+}、Fe^{3+}、Co^{2+})对 Cs^{137} 交换吸附的影响,作者在有关文献中已加以报道[42],此处不再赘述。

兹将实验步骤和结果分述于后。

2.1 钠离子对 Cs^{137} 吸附率的影响

由于裂变废液中含有一定量的钠离子(0.23mol/L),而且钠和铯是同一族元素,因此钠离子对 Cs^{137} 交换吸附的影响是值得考虑的问题。为此,配制了不同浓度的 $NaNO_3$ 溶液,研究对 Cs^{137} 吸附的影响。实验结果列于表 1。

表 1 钠离子对 Cs^{137} 吸附率的影响

$NaNO_3$ 浓度/N❶	放射性①/脉冲·min^{-1}		吸附率/%
	吸附前	吸附后	
0	2612	24	99.1
0.005	3511	44	98.7
0.01	3920	80	97.9
0.05	5104	145	97.1
0.1	5646	216	96.2
0.5	5082	841	83.4
1.0	5282	1358	74.3
2.0	4790	2302	51.9

①每次测量取 0.2mL。

从表 1 中数据可以看出,只有当 $NaNO_3$ 浓度达 0.5N 时,才产生比较显著的影响。但这个浓度已超过一般废液中钠离子的浓度。因此上述离子的存在,对 Cs^{137} 的分离不会产生显著的影响。

2.2 改进蒙脱石过滤性能的试验

为了改进蒙脱石的机械过滤性能,以便作为柱式无机交换剂使用,采用了热处理的方法[26]。将蒙脱石在 100~1000℃ 的温度下,分别烘烤 2h,然后分别测定对 Cs^{137}、Sr^{90}、Ce^{144}、Pm^{147}、Ru^{106} 和 Zr^{95} 的吸附率。硝酸溶液的 pH 值为 3~4。实验结果见图 1。

从图 1 中曲线可以看出,在任何烘烤温度下,蒙脱石对 Ru^{106} 几乎不吸附。经 600℃ 烘烤的蒙脱石对 Pm^{147}、Ce^{144}、Sr^{90} 的吸附率迅速下降。Zr^{95} 可能由于微量胶体的形成,而使蒙脱石在 600℃ 以前对 Zr^{95} 的吸附率稍低(胶体影响离子交换),而在 600℃ 以后,蒙脱石对 Zr^{95} 还有一定量的吸附。但必须指出,经过 600℃ 烘烤过的蒙脱石对

---

❶ N 表示浓度,1N=1mol/L。

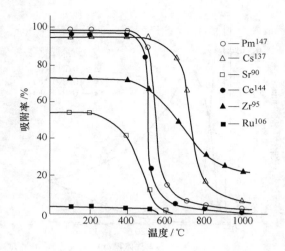

图 1　吸附率与温度的关系曲线

Cs^{137} 的吸附几乎没有显著的变化，吸附率仍在 90% 以上。而经过 600℃ 烘烤过的蒙脱土，它的机械过滤性能已完全能满足柱式交换的试验要求。温度超过 700℃ 以后，吸附率迅速下降。因此可以认为，为了从裂变产物或裂变废液中分离和回收 Cs^{137}，600～630℃ 是蒙脱石热处理的适宜温度。

2.3　裂变产物和裂变废液中 Cs^{137} 的分离和回收

首先将蒙脱土用标准筛筛选❶（12 孔），然后在 630℃ 烘烤 2h，即可作为柱式交换剂使用。在装柱前用蒸馏水浸泡 20～30min。

我们选择了在裂变产物中既是寿命长，又是产额大的几个具有代表性的裂变元素，进行了个别的淋洗实验。在以下每一柱式交换实验中，通过柱的原始溶液的流速为 1mL/6min。而淋洗剂的流速为 1mL/4min。柱高均为 14cm，柱的内径为 0.4cm，每次装入柱中的蒙脱石的质量均为 1.27g。

2.3.1　用不同硝酸浓度淋洗 Cs^{137} 的淋洗曲线

每次先将含有 Cs^{137} 的一定量溶液通过交换柱，用水洗去没有发生交换的多余的 Cs^{137}，然后用 0.4N、0.6N、0.8N、8.0N 的硝酸溶液分别淋洗，每次所取流出液的体积为 2 滴（约 0.1mL），实验结果见图 2。

由图 2 可以看出，酸度越大，淋洗曲线的峰越尖。在用 0.4N HNO_3 淋洗时，当通过 30mL 时，Cs^{137} 还几乎不能被淋洗下来（用 8.0N HNO_3 淋洗时，由于峰值过高，因而在图中用虚线表示）。

2.3.2　Sr^{90}、Ru^{106}、Pm^{147}、Ce^{144}、Zr^{95} 的单独淋洗曲线

从图 2 的淋洗曲线可以看出，0.4N HNO_3 几乎完全不能把 Cs^{137} 淋洗下来，因此选用 0.4N HNO_3 对 Sr^{90}、Ru^{106}、Pm^{147}、Ce^{144}、Zr^{95} 分别进行单独淋洗实验，实验结果见图 3 和图 4。

---

❶ 孔径的数目是以每厘米长所具有孔的数目计算得到的。

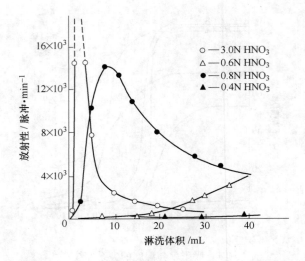

图 2 不同硝酸浓度淋洗 Cs^{137} 的淋洗曲线

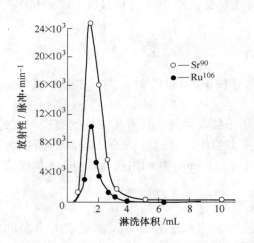 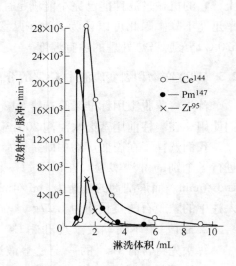

图 3 用 0.4N HNO_3 淋洗 Sr^{90}、 　　图 4 用 0.4N HNO_3 淋洗 Ce^{144}、Pm^{147}、
　　　Ru^{106} 的淋洗曲线 　　　　　　　　　　　Zr^{95} 的淋洗曲线

从图 3 和图 4 可以看出，用 0.4N HNO_3 分别淋洗 Sr^{90}、Ru^{106}、Pm^{147}、Ce^{144}、Zr^{95} 时，均在 14mL 左右就已接近本底。而从图 2 又可以看到，当用 0.4N HNO_3 淋洗 Cs^{137} 时，淋洗体积为 14mL 时，Cs^{137} 还没有下来。因此可以采用 0.4N HNO_3 作为 Cs^{137} 和其他裂变元素分离的淋洗浓度。

2.3.3 裂变产物中 Cs^{137} 的分离

实验是用经 580 天照射、冷却三年的裂变产物。在反应堆中铀棒经过 580 天照射后，裂变形成的长寿命放射性同位素列于表 2[35]。

表2 经580天照射后的裂变产物

同位素	半衰期	裂片产额	射线类别
Kr^{35}	10.27 年	0.293	β^-, γ
Sr^{90}	28 年	5.77	β^-
Tc^{99}	2.1×10^5 年	0.06	β^-
Ru^{106}	1.01 年	0.38	β^-, γ
Sb^{195}	2.0 年	0.021	β^-, γ
Te^{127m}	105 天	0.035	β^-, γ
Cs^{137}	30 年	6.15	β^-, γ
Ce^{144}	285 天	6.0	β^-, γ
Pm^{147}	2.6 年	2.7	β^-
Sm^{151}	80 年	0.45	β^-, γ
Eu^{155}	1.7 年	0.033	β^-, γ

将裂变产物溶液通过交换柱后，用0.4N HNO_3淋洗，淋洗曲线见图5。

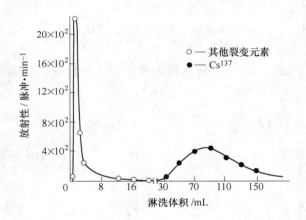

图5 裂变元素的淋洗曲线

从图5可以看出Cs^{137}和其他裂变元素的分离是比较好的。但是由于0.4N HNO_3对Cs^{137}淋洗酸度太小，故淋洗曲线峰低，宽度太大。为此，用0.4N HNO_3淋洗其他裂变元素时，当淋洗液的放射性强度接近本底时，再改换8.0N HNO_3淋洗Cs^{137}，淋洗曲线见图6。

比较图5和图6可以清楚地看出，用8.0N HNO_3淋洗Cs^{137}，可以获得较高比度的产品。应该指出，用6N、4N HNO_3淋洗时也可以获得较好的结果，只是淋洗峰稍宽一些。也曾用8.0N、6N HCl 淋洗Cs^{137}，淋洗效果与相同浓度的HNO_3相比较，大致相似。

2.3.4 裂变废液中Cs^{137}的回收

不同萃取过程所产生的裂变废液，其中含有不同量的外加杂质。按照杂质含量较高的雷多克斯（Redox）[19]流程废液中所含的量，在冷却三年的裂变产物溶液中加入外

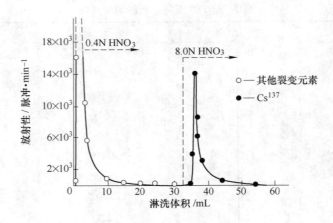

图 6　从裂变产物中分离 Cs^{137} 的淋洗曲线

加杂质，配成实际的裂变废液，进行以下的分离实验。所配的废液中，外加杂质的含量如下：

Al(NO$_3$)$_3$·9H$_2$O　　　143.3g/L　　NaNO$_3$　　　　　　　19.5g/L

Fe(NO$_3$)$_3$·6H$_2$O　　　5g/L　　　　Ni(NO$_3$)$_2$·6H$_2$O　　　4.5g/L

Cr(NO$_3$)$_3$　　　　　　　小于1g/L

此外用磷酸三丁酯和煤油使溶液达到饱和。溶液 pH = 1。从裂变废液中回收 Cs^{137} 的淋洗曲线见图7。

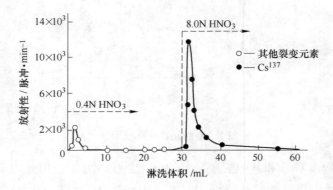

图 7　从裂变废液中回收 Cs^{137} 的淋洗曲线

图6和图7均为在淋洗之前从柱中流过的原始放射性溶液的总体积和总强度相同条件下的实验结果。只是图6是从裂变产物中分离 Cs^{137} 的淋洗曲线。而图7是从裂变废液中回收 Cs^{137} 的淋洗曲线。虽然两图中的其他裂变元素淋洗峰的高低和宽度相差较大，但 Cs^{137} 淋洗峰的高低和宽度则相差不大。这意味着，在一定的酸度范围内，废液中的外来杂质，对 Cs^{137} 在蒙脱石上的交换吸附，不产生显著的影响。而对其他裂变元素的影响则较为明显。因此可以认为，利用蒙脱石从裂变废液中回收 Cs^{137} 要比从裂变产物中分离 Cs^{137} 更为有利。这也进一步显示了蒙脱石对 Cs^{137} 具有高度的选择性。

2.3.5 冷却一年的裂变废液中 Cs^{137} 的回收

为了进一步了解此法是否适用于从冷却时间较短的裂变产物中分离 Cs^{137}，利用已确定的实验条件，进行了从冷却一年的裂变产物中分离 Cs^{137} 的研究。

将冷却一年的裂变产物溶液通过交换柱，分别用 0.4N HNO_3 和 8.0N HNO_3 淋洗，淋洗结果见图 8。

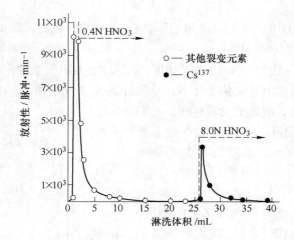

图 8　从冷却一年的裂变废液中回收 Cs^{137} 的淋洗曲线

从图 8 可以看出，Cs^{137} 和其他裂变元素的分离是比较理想的，因此可以认为所确定的分离条件，不只适用于从老的裂变产物或裂变废液中分离和回收 Cs^{137}，也适用于从冷却一年的或冷却时间较短的裂变产物或裂变废液中分离和回收 Cs^{137}。

2.3.6 含有载体铯的裂变废液中 Cs^{137} 的回收

为了进一步了解已确定的实验条件是否适用于从高强度裂变废液中回收 Cs^{137}，在实际裂变废液中加入一定量的载体铯，浓度为 3.6×10^{-6} mol/L。溶液 pH = 1。

利用前述分离条件将所配的溶液，通过交换柱，然后用 0.4N HNO_3 和 8.0N HNO_3 分别淋洗，结果见图 9。

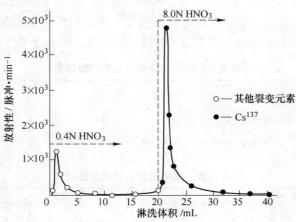

图 9　从含有载体铯的裂变废液中回收铯的淋洗曲线

从图9的曲线可以看出：Cs^{137}峰的宽度和图6及图7中Cs^{137}淋洗峰的宽度完全相同。由此可以认为，所确定的分离条件，对于从较强的裂变废液中回收Cs^{137}也是适用的。

图9中其他裂变元素的淋洗峰较图6和图8为低，这可能是因为当含有稳定铯的载体溶液在通过交换柱时，由于蒙脱石对Cs^{137}有高度的选择性，致使其他裂变元素在蒙脱石上的交换几率减小。

2.3.7　产品鉴定

（1）淋洗法。从图5～图9的淋洗曲线可以看出：其他裂变元素和Cs^{137}淋洗曲线峰的位置与图2～图4中用0.4N HNO_3单独对Cs^{137}、Ce^{144}、Ru^{106}、Sr^{90}、Pm^{147}、Zr^{95}进行淋洗的淋洗峰的位置是完全一致的，这可以初步表明Cs^{137}和其他裂变元素的分离是比较理想的。

为了进一步了解在上述实验条件下，从实际裂变废液中回收Cs^{137}时，在Cs^{137}的淋洗曲线位置上是否有其他裂变元素存在，我们把不含Cs^{137}的裂变废液（22mL，放射性总强度为1513600脉冲/min，酸度为0.1N HNO_3溶液）通过交换柱，然后分别用0.4N HNO_3和8.0N HNO_3进行淋洗，结果如图10所示。从Cs^{137}的淋洗曲线位置上（即8.0N HNO_3的淋洗部分）所收集的流出液，其放射性总强度为576脉冲/min。因此可以认为，在所得的Cs^{137}产品中，其中全部放射性杂质不超过千分之三。总的去污因素为2.6×10^3。

（2）载体法。在强硝酸溶液中磷钨酸是铯的有效沉淀剂[19,40]。为此，取3mL Cs^{137}产品溶液（放射性总强度为173500脉冲/min，酸度为8N HNO_3溶液），加入13mg稳定铯作为载体。然后逐滴缓慢地加入浓度为10%的磷钨酸。经过搅拌和离心分离再测量上层溶液中的放射性。为了使Cs^{137}沉淀得更加完全，另加入沉淀剂进行多次沉淀，直至溶液中的放射性强度保持恒定为止。小心将上层溶液和沉淀分离。同时每次用5mL 0.4N HNO_3溶液洗涤沉淀并测量全部上层溶液和洗涤液中的放射性强度，实验结果分别列入表3及表4。

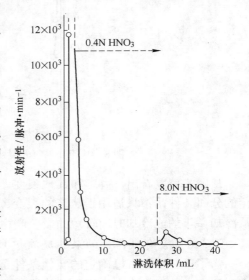

图10　不含Cs^{137}的裂变废液的淋洗曲线

表3　上层液中放射性强度测量结果

原始液的总强度/脉冲·min^{-1}	173500			
沉淀次数	1	2	3	4
每次沉淀后上层溶液中的放射性强度①/脉冲·min^{-1}	21	11	3	4

① 每次测量取0.1mL。

表4　洗涤液中放射性强度测量结果

沉淀的总放射性强度/脉冲·min^{-1}	173322				
洗涤次数	1	2	3	4	5
每次洗涤液中的总放射性强度/脉冲·min^{-1}	118	69	42	30	14

经过四次沉淀，上层溶液的总体积为 6mL，放射性总强度为 178 脉冲/min。从表3和表4分别可以看出，经过四次沉淀，溶液中的放射性强度已保持恒定。同时沉淀经过五次洗涤，洗涤液中的放射性强度从第三次起就已接近本底。如果将沉淀后上层溶液中的放射性强度的总和与全部洗涤液中的放射性强度的总和加在一起，也不超过原始总强度的千分之五。

（3）能谱鉴定。利用上述的分离条件，将从裂变产物和裂变废液中分离出的 Cs^{137} 产品和 Cs^{137} 标准源分别进行 γ 能谱的测定❶，并相互进行比较，结果见图11～图13。

从图11～图13的 γ 能谱分析，可以认为产品的 γ 能谱的单能性很好。同时产品经过六次能谱鉴定，重复性和单能性均很好，没有发现其他 γ 放射性同位素的干扰。结果与文献［36］一致。此外，从表2也可以看出，冷却三年的裂变产物，除 Sr^{90}、Tc^{99} 和 Pm^{147} 在衰变过程中只放出 β 射线以外，其他裂变产物在衰变过程中既放出 β 射线也放出 γ 射线。产品经过 γ 能谱鉴定没有发现其他 γ 杂质的干扰，这就意味着这些元素在产品中是不存在的。而 Sr^{90} 和 Pm^{147} 虽然只是纯 β 裂变产物，但对于上述两种裂变元素已作了单独的淋洗实验，从图2～图4可以看出，用 0.4N HNO_3 淋洗时，Sr^{90}、Pm^{147} 和 Cs^{137} 是可以分离的。因此可以估计产品也不含有 β 放射性杂质，可以认为产品是放射性纯的。此外，对图8和图9所

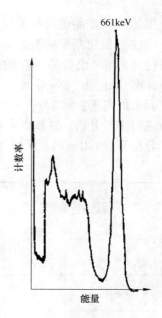

图11　Cs^{137} 标准 γ 能谱图
（条件：2.5cm×2.5cm 的 NaI（Tl）晶体；道宽1V）

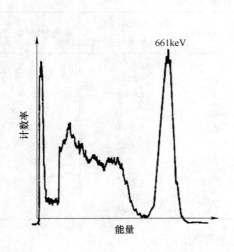

图12　Cs^{137} 产品 γ 能谱图
（条件：2.5cm×2.5cm 的 NaI(Tl) 晶体；道宽1V）

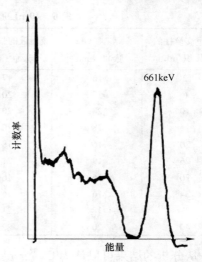

图13　Cs^{137} 产品 γ 能谱图
（条件：2.5cm×2.5cm 的 NaI(Tl) 晶体；道宽1V）

---

❶ 能谱的测定工作是委托其他单位完成的。

获得的 Cs^{137} 产品也进行了 γ 能谱鉴定，同样没有发现其他 γ 杂质的干扰（能谱图略）。

（4）稳定杂质的鉴定。经过 600~630℃ 热处理的蒙脱石，不仅改善了它的机械过滤性能，同时也提高了它的抗酸性。蒙脱石经过化学全分析，已知含有 Fe、Al、Ca、Ti、Mg、Si、K、Na 等元素（百分含量见表8）。为了检查 Cs^{137} 产品中是否存在以上稳定元素的离子，将 15mL 8.0N HNO_3 溶液通过装有 1.27g 蒙脱石的交换柱，然后对流出液进行定性分析，在试剂灵敏度的范围内，没有发现上述离子的存在（其中 Si 和 Ti 未作分析）。所用试剂及灵敏度列入表5。

表5 分析试剂及灵敏度

离子	分析试剂	试剂灵敏度①
K^+	钴亚硝酸钠	$m = 4\gamma$ $1 : G = 1 : 12500$
Na^+	醋酸铀酰锌	$m = 12.5\gamma$ $1 : G = 1 : 4000$
Mg^{3+}	镁试剂	$m = 0.5\gamma$ $1 : G = 1 : 100000$
Fe^{3+}	亚铁氰化钾	$m = 0.05\gamma$ $1 : G = 1 : 1000000$
Al^{3+}	铝试剂	$m = 0.1\gamma$ $1 : G = 1 : 2 \times 10^4$
Ca^{3+}	乙二醛-2-羟基胺	$m = 0.05\gamma$ $1 : G = 1 : 1000000$

① m 为检出限量；G 为限介稀度。

表6 40mL 原始液通过交换柱时的吸附率

通过交换柱的体积/mL	本底（b）/脉冲·min^{-1}	流出液的强度（a）/脉冲·min^{-1}	$a-b$	原始液强度/脉冲·min^{-1}	吸附率/%
2	16	17	1	1969	99.9
6	17	20	3	1969	99.9
10	20	20	0	1969	100
14	18	20	2	1969	99.9
20	16	18	2	1969	99.9
26	19	21	2	1969	99.9
30	16	16	0	1969	100
34	16	16	0	1969	100
38	18	21	3	1969	99.9
40	20	22	2	1969	99.9

从分析的结果可以认为 Cs^{137} 产品中不含稳定杂质的离子。

2.4 裂变废液中 Cs^{137} 回收率的确定

按照雷多克斯流程废液中的杂质含量配成试液，其中加入已知量的 Cs^{137}，总强度为 896000 脉冲/min。通过交换柱的总体积为 40mL。然后用 20mL 0.4N HNO_3 淋洗（为了和前面从裂变废液中回收 Cs^{137} 的实验条件保持完全相同），再用 8.0N HNO_3 淋洗

Cs^{137}，回收的总强度为827225脉冲/min，所以：

$$回收百分率 = \frac{827225}{896000} \times 100\% = 92.3\%$$

应该指出，当40mL原始溶液全部通过交换柱时，其吸附率还在99%以上，这表明原始液中的Cs^{137}有99%都吸附在交换柱上，当用20mL 0.4N HNO_3淋洗时，根据图2的实验可以认为，Cs^{137}是不会被淋洗下来的，但是在淋洗过程中我们仍随时测量流出液的放射性，没有发现Cs^{137}被淋洗下来。那么以回收率和吸附率相比较，回收率又为什么只有92.3%呢？这主要是由于用8.0N HNO_3淋洗Cs^{137}时，淋洗曲线的尾部拖得较长，没有全部把Cs^{137}淋洗下来。

2.5 酸度对柱式交换的影响

裂变产物和裂变废液的酸度往往很大，考虑到酸度对无机离子交换剂的交换性能影响较大，我们配制了不同浓度的HNO_3溶液，其中分别加入相同强度的Cs^{137}作为指示剂进行实验。结果见图14。

图中每一次试验所用的原始液比度均为67000脉冲/(min·mL)。

从图中曲线可以看出，从裂变废液中分离或回收Cs^{137}时，通过交换柱的原始液酸度越高，吸附率逐渐降低。

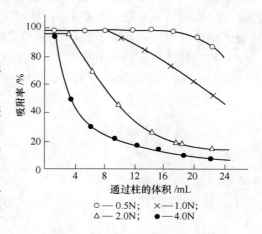

图14 酸度影响实验曲线

2.6 蒙脱石交换后的再生试验

从经济角度考虑，蒙脱石作为离子交换剂，能否再生不是主要问题，但是从实际应用上讲，在分离中如能尽量减少操作手续和提高工作效率，有必要对交换剂再生问题进行试验。为此，将放射性同位素Cs^{137}配制的原始溶液（比度为44980脉冲/(min·mL)）通过交换柱，用8.0N HNO_3淋洗下Cs^{137}，然后再用蒸馏水洗去柱中的酸，再用Cs^{137}原始溶液进行交换试验，结果见表7。

表7 蒙脱石交换后的再生实验[①]

第一次（再生前）			第二次（再生后）		
原始液通过柱的体积/mL	放射性/脉冲·min⁻¹	吸附率/%	原始液通过柱的体积/mL	放射性/脉冲·min⁻¹	吸附率/%
2	3	99.9	2	5	99.9
6	5	99.9	6	6	99.9
10	8	99.8	10	8	99.8
14	7	99.8	14	6	99.8
20	9	99.7	20	5	99.8
22	14	99.8	24	7	99.8
26	10	99.8	28	5	99.9
30	11	99.8	30	7	99.8

① 每次测量取0.2mL。

从表7中数据可看出，第一次和第二次通过柱的放射性溶液，其吸附率均在99%以上。因此可以认为蒙脱石作为无机离子交换剂是可以再生使用的。其实，根据前面的分离和回收 Cs^{137} 的实验条件来看，蒙脱石是可以连续使用的，因为分离 Cs^{137} 时所用的淋洗剂本身就是 0.4N HNO_3 和 8.0N HNO_3。所以只要在每次分离手续操作完毕后，用蒸馏水洗去柱中的酸即可再次使用，从而省去了再生的手续。一次装柱大约可以使用15次。

2.7 击穿和饱和交换容量的测定

前文中[42]蒙脱石静态平衡交换量的测定，是用未经热处理的蒙脱石，因而不能反映柱式交换的情况。另一方面采用载体静态法测得的交换量也不够准确。因此进一步用柱式载体动态法，测定了经 600~630℃ 热处理后的蒙脱石的击穿和饱和交换容量。

溶液的载体浓度为 0.0040N $CsNO_3$（以 Cs^{137} 作指示剂），蒙脱石重 1.1g。溶液的 pH 值为 4~5，柱高 12cm，内径为 0.4cm，流速为 1mL/8min，实验结果见图15。

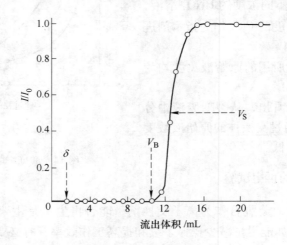

图15 击穿和饱和容量测定曲线

根据图15的击穿曲线，并利用下述公式：

$$击穿容量 = \frac{(V_B - \delta) \times C}{m}$$

$$饱和容量 = \frac{(V_S - \delta) \times C}{m}$$

式中，V_B——在击穿时流出液的总体积，mL；
V_S——当流出液的强度为原始液强度一半时（即 $I/I_0 = 0.5$ 时）流出液的总体积，mL；
C——载体溶液浓度，N；
m——交换剂质量，g；
δ——交换柱内自由体积，mL。

求得蒙脱石的击穿容量为 0.033mmol/g，饱和容量为 0.041mmol/g。

图 15 中，I/I_0 为流出液与原始液放射性强度之比。从所得数据来看，蒙脱石经过 630℃ 热处理后，交换量降低很大。但从分离无载体 Cs^{137} 来考虑，并不影响它的使用。因为重要的是经热处理后的蒙脱石提高了对 Cs^{137} 的选择性，并改善了它的机械过滤性能。

3 交换机理

采用 X 射线分析、差热分析和化学全分析法❶进行了研究。差热分析曲线和化学全分析数据分别见图 16 和表 8（X 射线分析数据从略）。

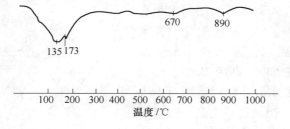

图 16 差热分析实验曲线

表 8 蒙脱石化学全分析结果

成 分	SiO_2	Fe_2O_3	TiO_2	Al_2O_3	CaO	MgO	Na_2O	K_2O	灼烧减量
含量/%	71.43	0.84	0.47	7.54	2.00	2.31	0.24	0.15	13.93

目前一般都认为蒙脱石的结构是层状的[38]，在蒙脱石的晶格结构中，四面体晶层内三价的铝离子对四价的硅离子的置换以及八面体晶层内低价离子，特别是镁对三价铝的置换，出现了电荷的不平衡。电荷的不平衡被晶架单元层与层之间的交换性阳离子所补偿，认为这是阳离子交换吸附的主要可能原因[38]。

为了探讨蒙脱石对 Cs^{137} 交换的高度选择性，选择了几种不同价态的阳离子对 Cs^{137} 在蒙脱石上分配系数的影响进行了研究。实验结果见表 9。蒙脱石构造示于图 17。

表 9 不同价态阳离子对 Cs^{137} 在蒙脱石上分配系数的影响

化合物	烘烤温度/℃			
	105		630	
离子浓度/N	0.05	0.5	0.05	0.5
KNO_3	61.1	22.9	23.4	21.6
NH_4NO_3	64.9	27.3	26.6	25.1
HNO_3	122.5	59.3	47.4	30.2
$Co(NO_3)_2$	215.0	59.2	128.0	41.6
$Al(NO_3)_3$	270.7	73.0	137.5	56.3
$NaNO_3$	444.2	98.3	240.7	65.3

---

❶ 由河北省地质中心实验室代为分析鉴定。

由表 9 可看出，不同价态的阳离子，对 Cs^{137} 在蒙脱石上的分配系数的影响顺序为：

$$K^+ > NH_4^+ > H^+ > Co^{2+} > Al^{3+} > Na^+$$

上述顺序与该离子水化半径的顺序大致相同。K^+、NH_4^+ 离子的水化半径与铯离子的水化半径最为接近，对分配系数的影响也较显著。因此离子水化半径的大小，在交换过程中起着重要的作用。考虑到离子交换是不带水的吸附，若要进入蒙脱石的层间发生交换，需经过脱水过程，Cs^{137} 离子水化半径最小，比其他离子易脱水，因此容易发生交换。另外，蒙脱石在 630℃ 烘烤 2h 后，硅氧层间就失去了全部的水分，同时蒙脱石也失去了膨胀的性能[39]。当蒙脱石完全失去硅氧层间的水以后，硅酸盐单位晶胞间空间的宽度约为 0.32nm，而这个宽度与 Cs^{137} 脱水后的晶体离子直径相近似，因此我们认为当蒙脱石在 630℃ 烘烤后，其之所以对 Cs^{137} 具有一定的选择性，一方面是由于 Cs^{137} 容易脱

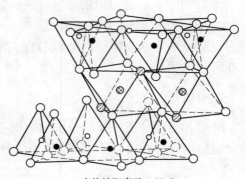

交换性阳离子，nH_2O

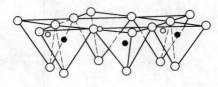

○—氧；　　⊗—铝、铁或镁；
◐—氢氧；　○,●—硅或铝

图 17　蒙脱石构造示意图

水，另一方面，Cs^{137} 脱水后离子的大小与这个空穴的大小相当，因此 Cs^{137} 不但比其他离子容易发生交换，而且比其他离子保持得较为牢固（脱水后的铯离子半径数值是最大的）。近年来有人研究 Cs^{137} 在蛭石[28]、海绿石[30] 以及其他黏土矿[32] 上的交换吸附时，发现这些黏土矿对 Cs^{137} 也有较好的选择性，并提出了类似的看法。

4　结论

（1）一定孔径的蒙脱石经过 600~630℃ 烘烤 2h，完全可以作为离子交换剂从冷却一年到三年的裂变产物或裂变废液中分离和回收 Cs^{137}，回收率在 92.3% 以上，产品经过化学和物理鉴定，认为是无载体和放射性纯的，纯度为 99.5%。

（2）本方法的优点是简便、经济，产品是高比度和无载体的，是从裂变产物或裂变废液中分离 Cs^{137} 的新方法。

（3）利用蒙脱石从裂变废液中回收 Cs^{137} 要比从裂变产物中分离 Cs^{137} 更为有利，废液的其他成分不会有显著的变化，因而也就不会影响废液进一步的加工和处理。

（4）蒙脱石对 Cs^{137} 有较高的选择性，经过热处理以后，交换量虽有所降低，但重要的是改善了它的机械过滤性能，进一步增加了 Cs^{137} 的选择性，因此，可以进行 Cs^{137} 的分离和回收。蒙脱石可以再生使用。

（5）初步探讨了 Cs^{137} 的交换机理及其高度的选择性。

本文承杨承宗、胡志彬、冯锡璋、徐光宪教授和刘元方副教授的审阅，特此表示衷心的感谢。

参 考 文 献

[1] G. B. Barton et al., Ind. Eng. Chem., 50, 212(1958).
[2] G. B. Barton et al., Ind. Eng. Chem., 51, 741(1959).
[3] А. Картрайт, Атомная техника за рубежом, вып. 6, 34(1962).
[4] P. R. Gray, J. Inorg. Nucl. Chem., 12, 304(1960).
[5] С. Е. Брелер, Радиоактивные элементы стр. 476, 1957.
[6] R. E. Burns et al., HW-31444(1954).
[7] T. T. MoKenzie, HW-47761(1957).
[8] R. L. Moore and R. E. Burns, Proc. 2nd Intern. Conf. Peaceful Uses of Atomic Energy, 1958, Vol. 18, P/1678.
[9] A. E. Rupp, IDO-14363, 1956; Progress in Nuclear Energy, Ser. III, Vol. 1, p. 345, Pergamon Press Ltd London.
[10] Р. Л. Мур, Р. И. Бернс, Труты второй международной конференции по мирному использованию атомной энергии, Женева, 1958, Избранные доклады иностранных ученых, Том 5, стр, 376, P/1768.
[11] 重松恒信, 大盐敏树, Isotopes and Radiation, 4, No. 2, 105(1961).
[12] J. L. Woodhead, A. T. Fudge, E. N. Jenknis, Analyst, 81, 570(1956).
[13] M. V. Šušic et al., Proc. 2nd Intern. Conf. Peaceful Uses of Atomic Energy, 1958, Vol. 18, P/488, p. 82.
[14] J. C. Dalton et al., Anal. Chem. Acta, 15, 317(1956).
[15] W. G. Mathers et al., Anal. Chem., 35, 2064(1963).
[16] V. Kouřim et al., Chem. Listy, 52, 262(1958); Collection, 24, 1474(1958).
[17] J. Van R. Smit et al., AERE-R-4039(1962).
[18] E. Lazzarini, Energia Nucleare (Milan), 9, 40(1962).
[19] C. B. Amphlett, Treatment and Disposal of Radioactive Wastes, 1961.
[20] I. R. Higgins, ORNL-1325(1955).
[21] K. A. Kraus and H. O. Phillips, J. Amer. Chem. Soc., 78, 694(1956).
[22] H. L. Garon and T. T. Sugihara, Anal. Chem., 34, 1082(1962).
[23] J. Van R. Smit, AERE-R-3884(1961).
[24] C. B. Amphlett et al., J. Inorg. Nucl. Chem., 6, 220(1958).
[25] H. Buchwald et al., J. Inorg. Nucl. Chem., 5, 341(1958).
[26] G. Loanid, Proc. 2nd Intern. Conf. Peaceful Uses of Atomic Energy, 1958, Vol. 18, P/1315.
[27] Ю. А. Кокотов и др., Радиохимия, 2, вып. 6, 1242(1962).
[28] В. М. Николаев и др., Радиохимия, вып. 1, 32(1963).
[29] Бернс, Глюкауф, Труды второй международной конференции по мирному использованию атомной энергии, Женева, 1958, Избранные доклады иностранных ученых, Том 5, стр. 491.
[30] С. З. Рогинский и др., Радиохимия, Том 2, вып. 4, 431(1960).
[31] W. J. Lacy et al., ORNL-TM-127(1962).
[32] Ю. А. Кокотов и др., Радиохимия, вып. 2, 199(1961).
[33] N. Oi and E. Ohashi, Bull. Chem. Soc. Japan, Vol. 35, No. 8, 1306(1962).
[34] W. J. Kaufman, NYO-1571(1951).

[35] М. П. Грецушкина, Таблица состова продуктов мгновенного деления U^{335}, U^{388}, Pu^{330}, Москва, 1964.
[36] C. E. Crouthamnl, Applied γ-ray Spectrometer, 1960.
[37] G. W. Brindley, X-ray Identification and Crystal Structure of Clay Mioneral, 1951.
[38] R. E. 格里姆, 粘土矿物学, 地质出版社, 1960.
[39] Ф. В. Чухуов, Коллоидны в земнойкоре, Москва, 1955.
[40] V. Kouřim, J. Inorg. Nucl, Chem., 12, 370(1960).
[41] 刘伯里、王德昭等, 未发表, 1960.
[42] 赵忠顺、唐天才、刘伯里, 科技, 第4期, 296(1963).

蛭石在低放裂变废水处理中的应用*

1 前言

从当前世界各国以工业规模处理低放裂变废水的现状来看，根据废水的不同比放射性和稳定组分的含量，一般仍采用凝聚、蒸发、离子交换等单级或联合手段[1,2]。对于某些类型的废水，也以中间工厂的规模采用泡沫分离[1]和电渗析等[3]方法。经处理后的废液，在排放到周围环境以前，为了尽可能降低流出液中 Cs^{137}、Sr^{90} 等核素的含量，大都在最后一级应用或准备采用离子交换法。

天然无机交换剂蛭石，由于具有较高的阳离子交换容量，对主要裂变核素，特别是对 Cs^{137} 具有较高的交换选择性，在水中溶胀小，且在较大的 pH 值范围内稳定，以及价格低廉等优点，在裂变废水处理中作为阳离子交换剂加以应用，已引起世界各国的广泛注意。

西特霍斯（Seedhouse）[4]等人最先利用磷酸盐凝聚——蛭石交换流程，以工业规模处理哈威尔的低放裂变废水。其后克拉克（Clarke）[5]等人以中间工厂规模开展了蛭石离心床的试验。后来克拉克[6]等人对凝聚——蛭石流程又进行了一些改进。近年来汤姆斯（Thomas）等人[7,8]报道了利用凝聚——蛭石流程处理低放裂变废水的转运经验。

由于蛭石对主要长寿命裂变核素具有良好的交换性能，不少国家[9~19]报道了有关蛭石对 Cs^{137}、Sr^{90}、R. E.、Ru^{106} 等核素交换吸附的研究。特别是蛭石对 Cs^{137} 具有高度选择性，有关的交换机构和动力学的研究[20~24]也报道了有关蛭石的预处理[25~27]和后处理[28]，以及蛭石在处理低放废水中应用的综合评论[29]。

本工作的目的在于研究我国蛭石在低放裂变废水处理中应用的性能。我们到现场收集了 1000 余种蛭石样品，测定了蛭石对 Cs^{137}、Sr^{90} 的分配系数。初步选出了以 M-11 为代表的一批性能良好的蛭石。在此基础上，首先研究了 M-11 蛭石的转型条件，并着重研究了 M-11 蛭石对 Cs^{137} 的高度选择性。其次，进一步观察 M-11 蛭石对其他裂变核素的交换性能以及影响交换的主要因素。经过小型工艺试验和扩大实验的验证，确定了一套利用凝聚——蛭石两段流程处理低放裂变废水的工艺参数。初步探讨了 M-11 蛭石在蒸发冷凝液体系和含盐量很高的去污废液中的除铯效果。

另外，对蛭石的前处理，蛭石的装料、卸料，特别是蛭石应用中的技术难关——蛭石的水力学过滤性能，也进行了工业规模的大型试验。结合工业使用的实际情况和蛭石本身的特性，确定了一组使蛭石的过滤性能满足实际应用需要的运转参数。最后，根据 M-11 蛭石在低放裂变废水处理中应用的效果，提出了应用的范围和初步结论。

---

* 本文合作者：金昱泰、孙兆祥、翁皓珉、陈文琇。原发于《北京师范大学学报（自然科学版）》，1976（增刊1）：76~90。

2 实验结果

2.1 我国蛭石质量的概况

为了全面了解我国蛭石的分布和质量情况,并使蛭石能符合工业使用上的各项要求,不仅要注意到蛭石的质量,同时也要考虑到产区、储量和运输等条件。此外,由于天然蛭石地质形成条件的不同,各地区中存在着差异性,因此不仅考察了不同地区间蛭石质量的差异情况,同时更着重研究了同一地区蛭石质量的差异性。经过现场工作和实验室分配系数的测定,以及小型工艺试验的相互配合,初步摸索出了一些鉴定蛭石质量的经验规律。

2.1.1 蛭石分配系数的测定

分配系数 K_d 是指在一定实验条件下,单位质量 1g 蛭石吸附的放射性核素强度与单位体积 1mL 溶液中余下的放射性核素强度比值。可见分配系数的大小反映了蛭石交换放射性核素能力的大小。分配系数的量值,可按下式计算:

$$K_d = -\frac{(C_0 - C_t)V}{wC_t}$$

式中,C_0 为交换液的原始放射性强度;C_t 为交换 t 时间后交换液的放射性强度;V 为交换液的总体积,mL;w 为蛭石的质量,g。

测定分配系数的实验条件是:交换液 25mL,Cs^{137}、Sr^{90} 的比放射性约为 0.5μCi/mL,稳定盐含量,$NaNO_3$ 为 1500×10^{-6},Ca 离子为 3×10^{-6},交换液 pH 值为 8~9,蛭石质量 250mg,粒度 1~2mm,振荡时间 1h,振荡前后测量交换液的比放射性。交换液的量大约为蛭石的 100 倍。在计算 Sr^{90} 分配系数时对 Y^{90} 用吸收法做了校正。

实验证明,蛭石样品在 Cs^{137} 的分配系数和蛭石外观物理特征之间有以下一些经验规律:

(1) 外观金黄色、黄褐色大片蛭石,油脂光泽,膨胀率❶较高。K_d 在 1000 以上。

(2) 外观黄色、黄褐色碎片蛭石,油脂光泽,膨胀率较差,K_d 在 1000 左右。

(3) 外观土黄色、块状蛭石,K_d 在 800~900。

(4) 外观鳞片状蛭石,直径一般在 0.5~5mm,其中浅黄色蛭石有油脂光泽,K_d 在 400~600;浅黑色蛭石有珍珠光泽,K_d 在 200~300。

(5) 外观黄绿色蛭石,焙烧后变白色,膨胀率差,$K_d < 200$。

(6) 黑色蛭石粉末,$K_d < 200$。

实验结果表明:我国不少地区蛭石的平均质量较好,储量也大。有些地区的某些蛭石,也呈现良好的交换性能。必须指出:有些蛭石的分配系数较大,但水力学过滤性能较差,在实际上仍不能使用。因此在利用蛭石分配系数的大小来判断蛭石质量的好坏时,必须同时兼顾蛭石的水力学过滤性能。此外,在分配系数测定中,蛭石的加工破碎方法、蛭石的粒度、蛭石的清洗程度、交换液的 pH 值和含盐量等,都应保持相同的实验条件。

---

❶ 膨胀率是指蛭石受热后,沿蛭石平面垂直方向膨胀大小的量度。

利用蛭石的外观特征，可以作为定性的初步鉴别，而分配系数的测定是一种较为简便的定量方法。

2.1.2 蛭石的差异性

如上所述，不同地区的蛭石样品，差异性较大，其外观特征也存在区别。同一地区的蛭石样品，由于矿脉部位的不同，仍然具有差异性。现将 G 地区不同矿脉部位的蛭石对 Cs^{137}、Sr^{90} 的分配系数列于表 1。

表 1 G 地区蛭石对 Cs^{137}、Sr^{90} 的分配系数

矿脉部位编号	K_d（Cs^{137}）	K_d（Sr^{90}）
1	80	410
2	120	350
3	240	400
4	360	580
5	440	500
6	600	540
7	880	1620
8	910	740
9	800	1100

由此可见，蛭石的差异性与蛭石质量和储量的估计有着密切的关系。为此在蛭石普查的基础上进一步测定了 M 地区蛭石质量的差异情况，实验结果见表 2。

表 2 M 地区蛭石对 Cs^{137}、Sr^{90} 的分配系数

矿脉部位编号	K_d（Cs^{137}）	K_d（Sr^{90}）
1	780	990
2	910	910
3	790	760
4	820	1140
5	1400	1400
6	1070	870
7	990	1600
8	930	1100
11	870	1540

从表 2 可以看出，M 地区的蛭石质量虽呈现一定的差异，但并不显著，而且蛭石的平均质量和水力学过滤性能均较好。从 M-11 矿点来看，不同部位的蛭石样品也比较均一，见表 3。

表 3 M-11 矿点蛭石对 Cs^{137} 的分配系数

蛭石编号	K_d（Cs^{137}）[①]
M-11A	920
M-11B	870
M-11C	1200

① 实验值为 6~8 个平行样品的平均值。

对 M-11 矿点共取样 76 个，绝大部分数据表明，该矿区蛭石样品的分配系数较高，过滤性能良好。为此，确定以 M-11 蛭石为代表，进行了以下的全部实验。

2.2 蛭石对主要裂片核素的交换性能

2.2.1 蛭石的转型

一般蛭石的可交换离子为 Ca^{2+}、Mg^{2+} 等离子。原型天然蛭石在高 pH 值体系交换过程中，逐步产生 $Ca(OH)_2$ 及 $Mg(OH)_2$ 沉淀，覆盖于蛭石表面以致影响蛭石的进一步交换[30]。为此，事先应对蛭石进行化学预处理（转型）。现将用 $0.1N\ HNO_3$ 和 $1N\ NaNO_3$ 进行预处理的实验结果列于表 4。

表 4　M-11 蛭石经转型后对 Cs^{137}、Sr^{90} 分配系数的影响

转型条件	分配系数	K_d（Cs^{137}）	K_d（Sr^{90}）
M-11 原型蛭石		870	1540
$0.1N\ HNO_3$ 浸泡 24h $1N\ NaNO_3$		990	1930

实验结果表明：转成钠型后的蛭石，对 Cs^{137}、Sr^{90} 的交换性能均略有提高。用 $1N\ NaNO_3$ 预处理，也能得到类似的结果。

2.2.2 蛭石的动态转型

工业上蛭石的转型，一般采用动态转型比较方便。为了提供可靠的动态转型数据，测定了转型过程中 Na 与蛭石中 Ca 或 Mg 的置换平衡过程，实验结果如图 1 所示（图中 Na^+、Ca^{2+} 数据均用 Lange 6-型火焰光度计测得）。

M-11 蛭石动态转型的实验条件为：M-11 蛭石用量为 75g；转型溶液为 $0.1N\ HNO_3$ 和 $1N\ NaNO_3$ 混合液，转型速度为 1 床体积/6h，共 4 床体积。实验结果表明，以每 6h 一个床体积的转型速度，共用四个床体积的混合液转型，蛭石钠型的转换是相对完全的。转型时，转型液所消耗的 $NaNO_3$ 含量与转型床体积的关系如图 2 所示。

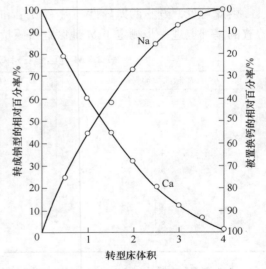

图 1　M-11 蛭石 Ca(Mg)-Na 置换平衡曲线

转型液在使用后，按图 2 的数据，适当补充 $NaNO_3$ 的盐量，可多次使用，以下的动态交换实验，均采用上述转型条件。

2.2.3 蛭石对 Cs^{137} 的高度选择性

（1）蛭石对 Cs^{137} 的穿透床体积。蛭石对 Cs^{137} 交换过程中，干扰显著的非放组分是同族元素 Na^+（此外尚有 NH_4^+）。在低放裂变废水处理中，蛭石对 Cs^{137} 选择交换能力的大小，主要表现在铯与钠的竞争上。

实验所用进料液的 Cs^{137} 比放射性为 1.67×10^{-8} Ci/L，其中含 $NaNO_3$ 1550×10^{-6}，Ca^{2+} 3×10^{-6}，pH 值为 8~9，通过 M-11 钠型蛭石，床体积 15mL，流速 4 床体积/h。实验结果如图 3 所示。

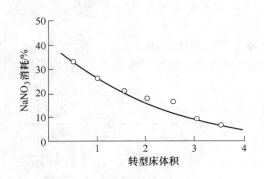

图 2　转型液 $NaNO_3$ 消耗与床体积的关系

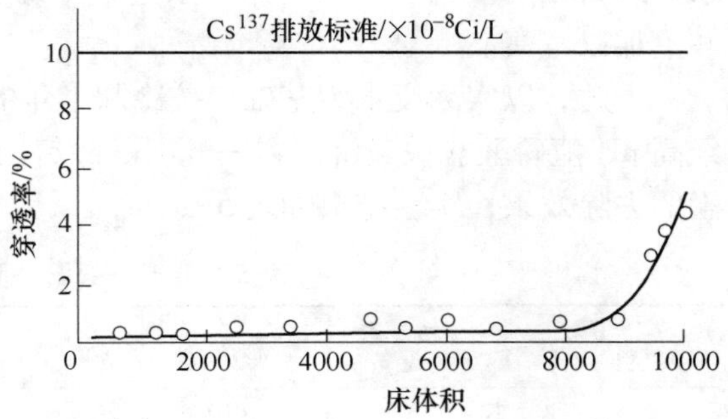

图 3　M-11 蛭石对 Cs^{137} 的穿透曲线

图 3 表明，当将流出液强度为进料液强度的 1% 定为穿透时，则蛭石对 Cs^{137} 的穿透体积为 8800 床体积，可见 M-11 钠型蛭石对 Cs^{137} 有很高的选择交换性能。同样在用实际废水试验的结果表明，经过 M-11 钠型蛭石柱 Cs^{137} 的穿透床体积在 8000~11000 之间。英国哈威尔实验室在 0.01N Na^+ 溶液中痕量铯的穿透体积为 3000L/kg 蛭石，约为 2000 床体积后来发展成为离心蛭石床[8]，Cs^{137} 的穿透床体积为 2800。将图 3 的结果与之比较，可以看出 M-11 蛭石的质量是优良的。

（2）蛭石对 Cs^{137} 的饱和交换量。由于蛭石对 Cs^{137} 具有高度的选择性，因此从穿透到交换饱和还有不少的容量可以利用，因为流出液中 Cs^{137} 的增加是比较缓慢的。考虑到蛭石柱在实际应用中是串级使用，因此用载体加指示剂的方法，求出蛭石的击穿容量和饱和交换量。

实验条件：稳定 Cs 载体浓度为 0.05N $CsNO_3$；
指示剂：强度为 10000 计数/mL；
$NaNO_3$ 含量为 1000×10^{-6}；
M-11 钠型蛭石 2.177g，粒度为 0.5~2mm；
流速为 4 床体积/h。

实验结果如图 4 所示。

用下式计算蛭石的击穿交换量和饱和交换量：

$$击穿交换量 = \frac{(V_B - \delta)C}{m} = 0.346 \text{mEq/g}$$

$$饱和交换量 = \frac{(V_S - \delta)C}{m} = 0.493 \text{mEq/g}$$

式中，m 为蛭石质量，g；C 为载体浓度，mEq/g。

图 4 的结果表明，在上述实验条件下，M-11 蛭石的饱和交换量较击穿交换量略大三分之一。

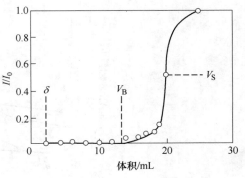

图 4　M-11 蛭石对 Cs^{137} 的击穿和饱和交换曲线
V_B—击穿体积；V_S—击穿 50% 体积；
δ—交换柱的自由体积

2.2.4 在两段流程中蛭石对各主要裂变核素的交换性能

对于原水比放射性为 1.0×10^{-7} Ci/L，冷却时间为半年的低放裂变废水，一般采用凝聚-离子交换法加以处理。以使废水的总 β 达到 1.0×10^{-9} Ci/L，即降低两个量级的要求。蛭石在两段流程处理实际废水过程中对各核素的交换究竟如何，必须从流程的总体上加以考察。

为此，以实际废水为基础，按冷却半年的裂变产物的比例加以调整，配制成原水，经高 pH 值磷酸盐快速沉凝器—蛭石—除钌阴柱两段流程，着重考察蛭石对主要裂变核素的去除效果，实验结果如表 5 所示。

表 5　凝聚—蛭石—除钌树脂处理效果　　　　　　　　（Ci/L）

项　目	总 β	Sr^{90}	Cs^{137}	Ru^{106}	总稀土	Zr^{95}-Nb^{95}
原　水	2.29×10^{-7}	1.94×10^{-8}	6.40×10^{-8}	9.5×10^{-9}	7.60×10^{-8}	1.61×10^{-8}
凝聚砂滤出水	4.00×10^{-8}	1.05×10^{-11}	3.00×10^{-8}	2.95×10^{-9}	$(1.0\sim 3.0)\times 10^{-10}$	1.51×10^{-11}
蛭石除钌柱出水 2800 床体积	$(2.1\sim 2.6)\times 10^{-9}$	1.0×10^{-12}	$(1.4\sim 7.0)\times 10^{-11}$	$(1.0\sim 1.3)\times 10^{-9}$	2.5×10^{-12}	3.0×10^{-12}

实验结果表明，M-11 蛭石对 Cs^{137} 的去除保持了良好的选择性，同时对 Sr^{90} 有较高的去污能力，总稀土 Zr^{95}-Nb^{95} 得到进一步的净化。对于冷却半年放射性为 1.0×10^{-7} Ci/L 的低放裂变废水来说，凝聚过程已将 Sr^{90}、总稀土、Zr^{95}-Nb^{95} 净化到满意的要求。凝聚清液中残留的主要核素是 Cs^{137} 和 Ru^{106}。Ru^{106} 由除钌树脂解决，而这里阳离子交换剂的主要任务是除 Cs^{137}，因此采用高 pH 值磷酸盐凝聚—蛭石—除钌树脂两段流程。由于各段都发挥了自己的特点，互相配合，蛭石总的处理周期比较长，达到 9300 床体积，而这时 Cs^{137} 的流出液强度仍在 1.4×10^{-9} Ci/L。与美国的凝聚—羧酚[31]两段流程比较，羧酚除 Cs^{137} 的周期为 2000 床体积，而且在凝聚过程中特别增加了除 Cs^{137} 的添加剂。与蛭石平行的高 pH 值磷酸盐凝聚—蛭石—除钌树脂流程的实验结果表明，羧酚除 Cs^{137} 的周期约为 1000 床体积。图 5 绘出了两者对比的实验结果。

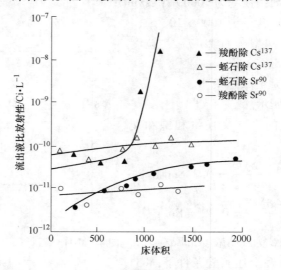

图 5　蛭石—羧酚流出液中 Cs^{137}、Sr^{90} 强度与床体积的关系

2.2.5 影响蛭石和其他核素交换的主要因素

低放裂变废水的活性组分和非放组分随生产运行情况的不同而有一定的波动。特别是非放组分的种类和含量变动较大，因此进一步研究这些条件的变化对凝聚的效果和蛭石交换性能的影响是十分必要的。

以实际废水为基础，分别研究不同石油磺酸含量、肥皂、洗衣粉对蛭石交换的影响，同时也研究了蛭石粒度变化，流速变化的影响。实验结果见表6～表9。

表6 原水中各核素的放化组分

核素	总β	Sr^{90}	Cs^{137}	总稀土	Zr^{95}-Nb^{95}	Ru^{106}
比放射性/Ci·L^{-1}	1.85×10^{-7}	6.29×10^{-8}	3.87×10^{-8}	5.69×10^{-8}	1.48×10^{-9}	4.09×10^{-9}

表7 凝聚方案

试剂	NaOH	Fe	$KMnO_4$	PO_4^{3-}
投料次序	1	2	3	4
投量	调至 pH = 11.5	15×10^{-6}	1×10^{-6}	120×10^{-6}

表8 不同石油磺酸含量对蛭石交换 Cs^{137} 的影响

原水中石油磺酸含量	1000×10^{-6}	500×10^{-6}	250×10^{-6}	100×10^{-6}	0
原水中 Cs^{137} 比放射性/Ci·L^{-1}	1.8×10^{-8}	1.7×10^{-8}	1.7×10^{-8}	1.7×10^{-8}	1.97×10^{-8}
流出液在该床体积时的比放射性/Ci·L^{-1}	2000 床体积 17×10^{-11}	2134 床体积 6.9×10^{-11}	2000 床体积 5.3×10^{-11}	2100 床体积 6.4×10^{-11}	2000 床体积 4.2×10^{-11}
去污因数	135	245	321	267	469

表9 肥皂及蛭石粒度、流速变化对 M-11 蛭石交换 Cs^{137}、Sr^{90} 的影响

实验编号	条件变化	比较项目	流出液组分		
			总β	Sr^{90}	Cs^{137}
1	粒度 0.5～2mm 流速 4 床体积/h	床体积	2000	2000	2000
		比放射性/Ci·L^{-1}	2.81×10^{-9}	2.32×10^{-11}	6.04×10^{-11}
		去污因数	64.8	2710	604
2	粒度 0.3～2mm	床体积	2100	2100	2100
		比放射性/Ci·L^{-1}	2.63×10^{-9}	1.26×10^{-11}	4.18×10^{-11}
		去污因数	69.2	4992	926
3	流速 2 床体积/h	床体积	1100	1100	1100
		比放射性/Ci·L^{-1}	2.46×10^{-9}	1.43×10^{-11}	4.21×10^{-11}
		去污因数	74.0	4400	920
4	原水中加入 100×10^{-6} 肥皂	床体积	1600	1600	1600
		比放射性/Ci·L^{-1}	3.54×10^{-9}	1.16×10^{-11}	1.8×10^{-10}
		去污因数	51.4	5420	215

注：所列比放射性、去污因数值均为相应床体积之前各取样点分析结果的平均值。

实验条件：原水经高 pH 值磷酸盐凝聚、砂滤清液上柱，原水总盐量为 1000×10^{-6}，其中含 Ca^{2+} 70×10^{-6}，pH 值为 8.5。蛭石为 M-11 钠型，1N $NaNO_3$ 转型。蛭石粒度 0.5～2mm。流速 4 床体积/h，床体积 15mL。图 6 和表 9 的实验结果表明，肥皂、洗衣粉、不同含量的石油磺酸虽然对蛭石交换 Cs^{137} 有一定影响，但并不显著。而上述因素对有机树脂交换的影响是十分明显的，例如：洗衣粉对羧酚树脂交换 Cs^{137} 的穿透床体积可降低一半左右。蛭石粒度由 0.5～2mm 降至 0.3～2mm 范围，对 Cs^{137}、Sr^{90} 的去除效果有所改进。流速由 4 床体积/h 改为 2 床体积/h 时，对上述核素的去除也有所提高。

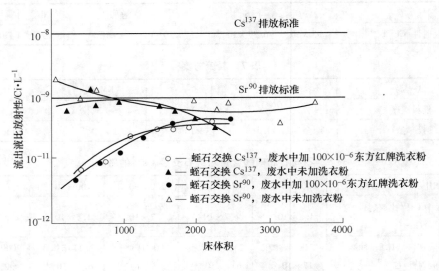

图 6 洗衣粉对蛭石交换 Sr^{90}、Cs^{137} 的影响

2.2.6 综合条件的扩大试验

根据上述的条件试验，最后进行了快速沉凝器蛭石除钌树脂的扩大试验。原水由 1AW 萃取水相余液冲稀配制，着重考察了石油磺酸，洗衣粉，NH_4^+，Na^+ 和其他有机、无机杂质对蛭石交换和各段流程的综合影响。原水非放组分和活性组分见表 10 和表 11。

表 10 综合条件扩大实验的原水水质 pH = 8

	东方红洗衣粉	300×10^{-6}	石油磺酸	250×10^{-6}
	Na_2CO_3	300×10^{-6}	$(NaPO_3)_6$	60×10^{-6}
非放组分	$H_2C_2O_4$	17×10^{-6}	$(NH_4)_2CO_3$	110×10^{-6}
	$KMnO_4$	10×10^{-6}	$NaHSO_4$	20×10^{-6}
	柠檬酸	1×10^{-6}	CCl_4	1×10^{-6}

实验证明，在综合条件下，两段流程的总效果是良好的。总 β 的去污因数接近 100。蛭石经 4700 床体积后，Cs^{137} 的流出液比放射性仍在 4.69×10^{-7} Ci/L 水平，远未穿透。除 Ru^{106} 外，其他核素在流出液中的水平也低于排放标准。总 β 主要由流出液中 Ru^{106} 的含量所决定，因此进一步提高 Ru^{106} 的去除效果，改进除钌树脂的交换性能，对两段流程总 β 去污因数的提高起主要的作用。

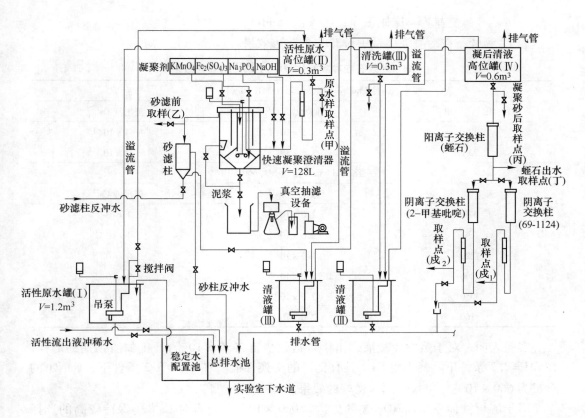

图 7 综合条件扩大试验流程示意图

（所有原水罐、清液罐均用普通钢板内衬塑料板制成；稳定水配制池及总排水池内衬耐酸磁板；
全部管道系不锈钢管；砂滤柱为两个，倒换使用）

表 11 凝聚—蛭石—除钌树脂处理效果 （Ci/L）

核 素	总 β	Sr^{90}	Cs^{137}	Ru^{106}	总稀土	Zr^{95}-Nb^{95}
床体积	4700	4700	4700	1100	4700	4700
原水比放射性	1.36×10^{-7}	6.75×10^{-9}	5.47×10^{-9}	5.92×10^{-9}	9.14×10^{-8}	2.65×10^{-8}
凝聚砂滤出水的比放射性	5.35×10^{-9}	2.27×10^{-11}	2.59×10^{-9}	2.35×10^{-9}	3.13×10^{-10}	9.38×10^{-11}
凝聚—蛭石—除钌树脂出水的比放射性	1.37×10^{-9}	6.96×10^{-12}	4.69×10^{-11}	1.37×10^{-9}	5.71×10^{-11}	6.3×10^{-11}

2.2.7 蛭石在蒸馏冷凝液体系和去污废液中除 Cs^{137} 的性能

由于 M-11 蛭石对 Cs^{137} 有高度选择性，因此除了研究蛭石在两段流程中处理一般低放裂变废水外，对蒸馏冷凝液和去污废液中的应用也作了初步的探索。兹将实验结果列于表12。

实验条件：M-11 钠型蛭石粒度为 0.5～2mm；

流速为 4 床体积/h；床体积 15mL。

表 12 为蛭石在冷凝液体系中对 Cs^{137} 的除去效果。

表 12　蛭石在冷凝液体系中对 Cs^{137} 的除去效果　　　　　　（Ci/L）

冷凝液 pH = 2.2，含 $NaNO_3$ 30×10^{-6}		
床 体 积	原水中 Cs^{137}	流出液 Cs^{137}
142	3.5×10^{-9}	4.44×10^{-12}
338	3.5×10^{-9}	8.34×10^{-11}
冷凝液 pH = 6.2 含 $NaNO_3$ 660×10^{-6}		
床 体 积	原水中 Cs^{137}	流出液 Cs^{137}
310	2.0×10^{-8}	2.7×10^{-11}
830	2.0×10^{-9}	3.4×10^{-11}
1220	2.0×10^{-8}	2.0×10^{-11}
2400	3.0×10^{-8}	4.9×10^{-11}
3050	2.0×10^{-8}	2.4×10^{-11}
4100	2.0×10^{-8}	6.6×10^{-11}

实验表明，对于蒸发冷凝液，由于含盐量极低（30×10^{-6}），正如所预期的那样，在 pH = 2.2 条件下，蛭石对 Cs^{137} 具有良好的交换性能。在 pH = 6.2 条件下，$NaNO_3$ 的含量为 660×10^{-6}，也得到同样良好的效果。

对于解析液体系，$NaNO_3$ 含量高达 25000×10^{-6}，蛭石对 Cs^{137} 仍保持较高的选择性。实验结果表明，当床体积达 1000 时，蛭石对 Cs^{137} 仍保持 90% 左右的去除率。实验结果列于表 13。

表 13　蛭石在去污废液中对 Cs^{137} 的去除效果

床 体 积	原水中 $Cs^{137}/Ci \cdot L^{-1}$	流出液 $Cs^{137}/Ci \cdot L^{-1}$	去除效果/%
144	2.2×10^{-8}	1.1×10^{-10}	99.5
230	2.2×10^{-8}	5.5×10^{-10}	97.5
280	2.2×10^{-8}	5.3×10^{-10}	97.6
579	2.2×10^{-8}	1.1×10^{-10}	99.5
717	7.0×10^{-8}	1.8×10^{-9}	97.2
1269	7.0×10^{-8}	9.9×10^{-9}	87.3

2.3　蛭石的前处理和水力学实验

天然蛭石在使用以前，必须经过适当的加工处理，这部分工作没有详细的资料，而这些参数在工业规模处理废水时是十分重要的。为此我们对蛭石破碎、洗涤、装料、卸料以及水力学过滤性能进行了工业规模的试验。对于柱高为 1.4m，内径为 0.8m，粒度为 0.8~2.5mm 的蛭石柱，其流速达到每小时 10 床体积，从而满足了工业上每小时 4 床体积的要求。此外还试验成功了用反冲压出的卸料方法，为蛭石卸料提供了一种新的有效途径。

（1）蛭石的前处理。根据蛭石的特性和选用的粒度范围，采用锤式破碎较好。滚式破碎易黏合，球磨破碎粒度太细。

将筛选后 1~3mm 粒径的蛭石进行水洗，用自来水反冲洗涤，直至泥沙全部洗净为止。在实际使用时，这一步骤十分重要，不能疏忽，否则蛭石上柱后，阻力会逐步增加，影响使用。

（2）蛭石的水力学实验。蛭石水力学实验的目的是要探讨蛭石的阻力、片径、柱高等对流速的影响。由于大型水力学实验的条件变化比较困难，本实验只在要求的床高、粒径和实际的水质条件下，观察其流速的变化。

实验条件：

M-11 蛭石粒径 0.8~2.5mm，床高 1.4m；内径 0.8m，原水由自来水配制，其中含洗衣粉 100×10^{-6}，Na_2CO_3 500×10^{-6}，石油磺酸 1000×10^{-6}，pH 值为 9~10。实验结果列于表 14。

表 14 大型水力学实验流速和压力降的关系

测量次数编号（每隔12h）	流速/床体积·h^{-1}	压力降/MPa	石油磺酸
1	10.0~10.1	0.166	无
2	10.1~10.3	0.164	无
3	10.0~10.3	0.167	500×10^{-6}
4	10.1~10.3	0.167	500×10^{-6}
5	10.3	0.165	500×10^{-6}
6	10.1~10.3	0.165	500×10^{-6}
7	9.9~10.0	0.170	1000×10^{-6}
8	9.6~9.7	0.178	1000×10^{-6}
9	9.3~9.4	0.183	1000×10^{-6}
10	9.9~10.0	—	1000×10^{-6}

（3）蛭石柱的装卸料。

装料：打开交换柱上盖，使柱内充满水，将已洗净的蛭石倒入柱内，装料完毕，适量反冲。蛭石柱体积为 $0.7m^3$。

卸料：光由下口反冲，使蛭石松动，然后用正压反冲法，上口加压使蛭石从卸料口流出。需用水量 5t，流速 20t/h，卸料时间 15min。

3 结论

（1）M-11 地区蛭石对 Cs^{137} 具有较高的选择性。废水体系在下列条件范围内变化均可适用：

1）pH 值范围：2.2，6，8~9，11.3~11.6。

2）含盐量范围：

$NaNO_3$：5×10^{-6}，600×10^{-6}，1000×10^{-6}，1500×10^{-6}，25000×10^{-6}；

$(NH_4)_2CO_3$：100×10^{-6}。

3）有机杂质含量范围：

石油磺酸：100×10^{-6}，250×10^{-6}，500×10^{-6}，1000×10^{-6}；

洗衣粉：100×10^{-6}，300×10^{-6}；

肥皂：100×10^{-6}。

4）废水中 Cs^{137} 比放射性范围：1×10^{-7} Ci/L，5×10^{-8} Ci/L，1×10^{-8} Ci/L，2×10^{-9} Ci/L，因此蛭石与高 pH 值磷酸盐凝聚配合，可以比较合理地处理原水比放射性为 1×10^{-7} Ci/L，冷却时间为半年左右的低放裂变废水。

对于不同类型的废水，蛭石作为阳离子交换剂对 Cs^{137} 同样保持了较高的选择性，可以在其他类型的废水处理中单独使用，或和其他处理手段联合使用。

（2）M-11 地区蛭石在两段流程中作为阳离子交换剂对 Sr^{90} 有较高的去污能力，但选择性较 Cs^{137} 为差。

总稀土对 Zr^{95}-Nb^{95} 和 Ru^{106}（阳离子部分）也有一定的净化效果。

（3）确定了两段流程中，利用 M-11 蛭石的工艺参数。

1）粒度：$1 \sim 3$ mm。

2）转型条件：1N $NaNO_3$；4 床体积/24h。

3）交换速度：4 床体积/h。

4）卸料方法：反冲压出法。

（4）利用天然蛭石作为阳离子交换剂，前处理工作和放射性固体废物较有机树脂的使用为多，对 Sr^{90} 等核素的选择性也不如有机树脂为好。

参 考 文 献

[1] Symposium on Practices in the treatment of low and intermediate level radioactive wastes, STI-PUB-116 (1965).

[2] Management of Low and intermediate Level radioactive wastes, STI-PUB-264(1970).

[3] Ф. В. Раузен и др., Proc. 4th Inter. Conf. on Peaceful Uses of Atomic Energy A/Conf 11, 427(1972).

[4] K. G. Seedhouse et al., AERE-ES/R-2089.

[5] J. H. Clarke et al., AERE-R-4314(1963).

[6] J. H. Clarke et al., AERE-R-6153(1969).

[7] K. T. Thomas et al., IAEA-SM-137/38, STI-PUB-264(1970).

[8] K. T. Thomas et al., BARC-340(1968).

[9] D. C. Sammon et al., AERE-R3-3274(1960).

[10] Werner Hoffman, Nukleonik, 3, 195(1961).

[11] K. Ueno, JAERI-5003(1961).

[12] В. М. Николаев и др., Радиохимия, 5, No. 1, 32(1963).

[13] C. R. Frost., AAEC/TM-205(1963).

[14] Ю. В. ЕгороВ и др., Радиохимия, 8, 8(1966).

[15] Ю. В. ЕгороВ и др., Радиохимия, 8, 397(1966).

[16] K. T. Thomas et al., BARC-315(1967).

[17] K. T. Thomas et al., BARC-349(1968).

[18] Chandra Umesh, BARC-454(1970).

[19] R. C. Reynolds, NYO 3912-4(1971).

[20] G. R. Frysinger, TID-7613(1961).

[21] D. G. Jacobs et. al., Health phys., 2, 391(1960).
[22] D. G. Jacobs, TID-7644(1963).
[23] B. L. Sawhney, Soil Sci. Soc. Amer., Proc., 30, 565(1966).
[24] W. D. Klobe et al., Soil Sci. Soc. Amer., Proc., 34, 746(1970).
[25] T. Tammura et al., Health phys., 9, 697(1963).
[26] H. W. Levi et al., Allgen. Chem., 337, 105(1965).
[27] Brat, Satya et al., BARC-534(1971).
[28] R. C. Rastogi, TID-21882(1965).
[29] 大盐敏树，原子能译丛，第四期，303(1966).
[30] T. D. Wright et al., AERE-E/R-2707.
[31] R. E. Blanco et al., STI-PUB-116(1965).
[32] R. H. Burns et al., Proc. 2nd Intern, Conf, on Peaceful Uses of Atomic Energy, Conf/P/308(1953).

电渗析器对长寿命裂片元素去除的基本研究（一）[*]

20世纪50年代末期，美国开始利用电渗析处理弱放废水的研究工作[1]，一些国家利用电渗析处理中低放废水亦有报道[2,3]。处于极低浓度下的放射性元素在废水中的状态和行为对电渗析处理废水的效果影响很大。在一般电渗析中不太考虑浓水和淡水的出水比例，而放射性废水处理中对电渗析出水的浓、淡比例则必须满足一定的要求。尽管电渗析处理放射性废水在工艺操作上还存在一些问题，但它目前仍是发展中的处理中低放废水的一种手段，国内有些单位也正在进行研究[4,5]。

Walters等人[6]最先利用电渗析器浓缩放射性裂变废水，初步指出了除盐和除去放射性（总放）的关系。Nishidoi等人[2]也开展了利用电渗析器处理放射性裂变废水的研究，并利用浓水二次淡化的流程使浓缩比达到100。

Sammon等人[7]还讨论了浓缩比达到100的流程。

Coleman等人[8]利用填充树脂床电渗析从冷凝裂变废液中进行了除钌的研究。Bub[9~11]和Facchinl等人[12]利用电渗析器研究了裂变产物的分离。

Yoshihide Honda等人[13]用电渗析器研究了海水中某些裂变产物的状态。

综上所述，利用电渗析处理放射性废水时放射性元素的状态对除放效果影响很大，对这些问题的研究，将有助于电渗析处理放射性废水在工业上的应用。

1 利用连续式电渗析器研究除盐除放的关系

1.1 组装参数和实验设备

隔板：外形尺寸为 100mm×270mm×2mm 的聚氯乙烯板，每块隔板有五条来回折流串联的水道，流长100cm，流水道总面积为146cm^2，有效面积为49.2%，流水道内不黏结菱形格网。

电极：阳极用石墨电极，阴极用不锈钢作电极。

极框板：由14mm硬塑料板制成，尺寸和内形与隔板相同，极水道出口在极框两边横侧。

离子交换膜：上海化工厂生产的异相膜根据电渗析淡化苦咸水同类型工作，选用以下参数组装：

v(流速) = 4cm/s；

i(电流密度) = 3mA/cm^2；

---

[*] 本文合作者：翁皓珉，金昱泰，孙兆祥，陈文琇。原发表于《北京师范大学学报（自然科学版）》，1979(2)：47~53，62。

η(电流效率) = 60%；

I(总电流) = $\frac{3 \times 146}{60\%}$ = 0.736A；

C_0(泵水含盐量) = 87mmol；

C(淡水含盐量) = 0.87mmol；

根据以上参数组装为一级22对串联，一次脱盐为90%～99%。

1.2 流程简图

电渗析流程示意图如图1所示。

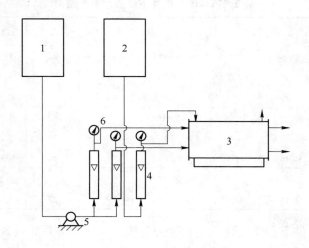

图1 电渗析流程示意图

1,2—原水和极水贮槽；3—电渗析器；4—流量计；5—不锈钢泵；6—压力表

1.3 实验水质

原水为含盐5000ppm的苦咸水，加入放射性同位素配制模拟裂片核素污染水，具体组分见表1。

表1 放射性原水成分

稳定离子	含量/ppm	放射性元素	强度/Ci·L^{-1}	含量/%
总 盐	4679	总 放	1×10^{-5}	
Cl$^-$	910	^{137}Cs	8×10^{-7}	8
SO$_4^{2-}$	2152	^{90}Sr-^{90}Y	1×10^{-6}	10
Mg^{2+}	172	^{144}Ce	5.6×10^{-6}	56
Ca^{2+}	225	^{106}Ru	1×10^{-7}	1
Na$^+$	1010	^{131}I	2.6×10^{-6}	26

加入的放射性核素^{137}Cs、^{90}Sr、^{144}Ce、^{131}I均系离子状态Cs$^+$、Sr^{2+}、Ce^{3+}、I$^-$，指示剂^{106}Ru是硝酸体系。

放射性原水配制100L，在实验中循环使用。

1.4 实验条件

立式安装，极水进出口在上，浓淡水流量比为1∶1，均为4L/h，极水流量为6L/h，实验间隙运转，按不同除盐要求分三组进行。

1.5 实验结果

不同除盐条件下除去总放的结果，见表2。

表2 不同除盐条件下除去总放的结果

组	编号	操作条件		含盐量/ppm	除盐/%	水温/℃	总放/Ci·L^{-1}	除放/%
一	原-1			4500		22	1.3×10^{-5}	
	淡1-1	0.38A	30V	240	95	27	8.3×10^{-7}	94.3
	淡1-2	0.39A	35V	160	96.5	27	6.9×10^{-7}	95.0
二	原-2			5000			7.3×10^{-6}	
	淡2-1	0.38A	40V	29	99.4	27	2.4×10^{-7}	95.5
	淡2-2	0.45A	60V	<10	99.7	28	1.34×10^{-7}	98.0
三	原-3			5000			4.5×10^{-6}	
	淡3-1	0.45A	60V	<10	99.7	27	5.6×10^{-8}	98.5

在上述除盐条件下，除盐除总放与除去各核素之间的关系如表3所示。

表3 除盐过程中除去各核素的实验结果

编号	除盐/%	除放/%	^{106}Ru		^{144}Ce		^{137}Cs		^{90}Sr		^{131}I		总放/Ci·L^{-1}	
			强度/Ci·L^{-1}	除去/%	强度/Ci·L^{-1}	除去/%	强度/Ci·L^{-1}	除去/%	强度/Ci·L^{-1}	除去/%	强度/Ci·L^{-1}	除去/%	实测	加和
原-1			7.8×10^{-8}		4.6×10^{-6}		1.2×10^{-8}		6.7×10^{-7}		5.4×10^{-6}		1.3×10^{-5}	1.1×10^{-5}
淡1-1	94.6	94.3	1.3×10^{-5}	82.9	4.3×10^{-7}	91	4.7×10^{-8}	96.4	3.0×10^{-8}	95.5	9.6×10^{-6}	98	8.3×10^{-7}	6.6×10^{-7}
浓1-1			5.6×10^{-5}		2.8×10^{-6}		2.0×10^{-6}		6.1×10^{-7}		3.9×10^{-6}		1.05×10^{-5}	9.4×10^{-6}
淡1-2	96.5	95	1.2×10^{-5}	85	4.8×10^{-7}	90	4.8×10^{-8}	96.4	2.1×10^{-8}	96.4	2.8×10^{-7}	95	6.9×10^{-7}	8.3×10^{-7}
浓1-2			7.6×10^{-8}		2.7×10^{-6}		2.3×10^{-6}		6.1×10^{-7}		2.5×10^{-6}		1.06×10^{-5}	8.5×10^{-6}
原-2			3.5×10^{-8}		2.3×10^{-6}		1.0×10^{-6}		4.3×10^{-7}		4.5×10^{-6}		7.4×10^{-6}	9.1×10^{-6}
淡2-1	99.4	95.5	8.2×10^{-9}	76.5	2.7×10^{-7}	89	1.3×10^{-8}	98.7	5.5×10^{-9}	99.3	5.6×10^{-8}	99	3.4×10^{-7}	3.5×10^{-7}
浓2-1			5.2×10^{-8}		1.5×10^{-6}		1.5×10^{-6}		6.2×10^{-7}		7.2×10^{-6}		8.3×10^{-5}	1.1×10^{-5}
淡2-2	99.7	98	6.6×10^{-9}	81	1.9×10^{-7}	92	1.6×10^{-9}	99.9	2.7×10^{-9}	99.7	3.4×10^{-8}	99.3	1.5×10^{-7}	2.0×10^{-7}

续表3

编号	除盐/%	除放/%	¹⁰⁶Ru 强度/Ci·L⁻¹	¹⁰⁶Ru 除去/%	¹⁴⁴Ce 强度/Ci·L⁻¹	¹⁴⁴Ce 除去/%	¹³⁷Cs 强度/Ci·L⁻¹	¹³⁷Cs 除去/%	⁹⁰Sr 强度/Ci·L⁻¹	⁹⁰Sr 除去/%	¹³¹I 强度/Ci·L⁻¹	¹³¹I 除去/%	总放/Ci·L⁻¹ 实测	总放/Ci·L⁻¹ 加和
浓2-2			5.1×10^{-6}		1.6×10^{-6}		3.0×10^{-6}		1.3×10^{-6}		1.3×10^{-5}		1.5×10^{-5}	1.9×10^{-5}
原-3			4×10^{-5}		9.6×10^{-7}		9.7×10^{-7}		3.2×10^{-7}		2.4×10^{-6}		4.5×10^{-6}	5.2×10^{-6}
淡3-1	99.7	98.5	4.1×10^{-9}	90	2.5×10^{-6}	96.3	4.4×10^{-9}	99.5	2.4×10^{-9}	99.7	1.0×10^{-8}	99.6	5.6×10^{-8}	6.7×10^{-8}
浓3-1			4.5×10^{-8}		2.2×10^{-6}		1.8×10^{-7}		7.2×10^{-7}		6.0×10^{-6}		1.1×10^{-5}	1.1×10^{-5}

原-1是原始配制的放射性苦咸水，原-2是原-1的一次循环水，原-3是原-2的循环水。在实验中原-1的浓淡室出水合并就是原-2的实验用水，原-2的浓淡室出水合并就是原-3的实验用水。实验中发现放射性原水的强度，每循环一次就会降低，发生"丢失"现象。这种"丢失"主要是由电渗析系统中膜对放射性核素的吸附所造成的。三组实验中各放射性核素的吸附情况，如表4所示。

表4 三组实验中各放射性核素的吸附情况

编号	总放 强度/Ci·L⁻¹	总放 吸附/%	¹⁰⁶Ru 强度/Ci·L⁻¹	¹⁰⁶Ru 吸附/%	¹⁴⁴Ce 强度/Ci·L⁻¹	¹⁴⁴Ce 吸附/%	¹³⁷Cs 强度/Ci·L⁻¹	¹³⁷Cs 吸附/%	⁹⁰Sr 强度/Ci·L⁻¹	⁹⁰Sr 吸附/%	¹³¹I 强度/Ci·L⁻¹	¹³¹I 吸附/%
原-1	1.3×10^{-5}		7.8×10^{-8}		4.5×10^{-6}		1.2×10^{-6}		6.1×10^{-7}		5.4×10^{-6}	
原-2	7.5×10^{-6}	42	3.5×10^{-8}	68	2.3×10^{-6}	55	1.0×10^{-6}	18	4.3×10^{-7}	30	4.5×10^{-6}	17
原-3	4.5×10^{-6}	40	3.5×10^{-8}	≈0	9.6×10^{-7}	60	9.7×10^{-7}	≈0	3.2×10^{-7}	21	2.4×10^{-6}	48
总吸附①		65		68		80		18		47		57

① 总吸附为第三组原水各核素对原-1的相对吸附百分数。

实验后拆开电渗析器，对膜上吸附的放射性进行了测量，并用0.5N HNO₃浸泡膜24h，取样分析，结果如表5所示。

表5 膜上吸附的放射性

项目\核素	¹⁰⁶Ru	¹⁴⁴Ce	¹³⁷Cs	⁹⁰Sr	¹³¹I
强度/Ci·L⁻¹	—	2.3×10^{-6}	3.7×10^{-7}	2.5×10^{-7}	4.5×10^{-7}
解吸/%		65.8	30	86	15

实验数据表明：用0.5N HNO₃解吸膜上吸附的放射性核素是很不完全的，但可以看出，¹⁴⁴Ce吸附最多。为了弄清¹⁴⁴Ce在膜上吸附情况和对膜性能的影响，我们进一步研究了¹⁴⁴Ce的吸附和迁移过程。

2 连续式电渗析器对¹⁴⁴Ce的饱和吸附实验

在稳定的苦咸水中，单独加入¹⁴⁴Ce放射性同位素，配水量为50L，在实验中循环使用。所用离子交换膜为经过 0.5N HNO_3 浸饱 24h 的旧膜，整个流程装置和操作参数与上述电渗析除盐除放实验相同。

实验共进行 100h，每隔 6h 取样分析。并补加 ^{144}Ce 核素，使放射性强度保持 10^{-6} c/l，结果如表 6 所示。

表6 对 ^{144}Ce 的饱和吸附实验结果

编号	操作条件		电阻/Ω		^{144}Ce			吸附淡化情况	
	U/V	I/A	原水	淡水	原水/Ci·L^{-1}	淡水/Ci·L^{-1}	浓水/Ci·L^{-1}	吸附/%	除去/%
1	62	0.36	280	9.3×10^4	3.3×10^{-6}	1.8×10^{-7}	2.5×10^{-6}	58	94.4
2	60	0.34	280	9×10^4	1.3×10^{-6}	3.1×10^{-7}	1.1×10^{-6}	46.9	76.1
3	63	0.34	280	6.1×10^4	1.5×10^{-6}	2.4×10^{-7}	1.2×10^{-6}	50.5	84.7
4	64	0.35	280	3.8×10^4	2.0×10^{-6}	3.3×10^{-7}	1.2×10^{-6}	59.4	84
5	63	0.34	280	2.3×10^4	2.3×10^{-6}	5.1×10^{-7}	1.6×10^{-6}	54.7	78
6	61	0.38	270	1.4×10^4	3.1×10^{-6}	4.7×10^{-7}	2.6×10^{-6}	50.1	85
7	62	0.33	275	2.1×10^4	2.7×10^{-6}	4.0×10^{-7}	2.8×10^{-6}	37.8	84.5
8	61	0.36	275	1.65×10^4	3.0×10^{-6}	4.4×10^{-7}	3.2×10^{-6}	38.3	85.6
9	70	0.39	280	1.5×10^4	4.4×10^{-6}	2.4×10^{-7}	6.2×10^{-6}	25.1	94.6
10	60	0.37		1.8×10^4	3.9×10^{-6}	1.2×10^{-7}	5.7×10^{-6}	25.9	97
11	62	0.35	220	8×10^3	4.1×10^{-6}	2.0×10^{-7}	5.3×10^{-6}	31.9	95.2
12	70	0.40	240	1.0×10^4	5.6×10^{-6}	1.5×10^{-7}	7.8×10^{-6}	29.4	97.4
13	65	0.38	240	5.1×10^4	5.1×10^{-6}	0.6×10^{-7}	7.7×10^{-6}	22.6	98.8
14	70	0.42	240	6×10^3	4.9×10^{-6}	1.12×10^{-7}	7.3×10^{-6}	24.3	97.7
15	63	0.42	260	4.5×10^4	3.7×10^{-6}	1.33×10^{-7}	6.7×10^{-6}	8.4	96.4
16	65	0.49		8×10^3	4.6×10^{-6}	7.6×10^{-6}	9.5×10^{-6}	0	98.3

注：吸附率 = $\dfrac{\text{原水强度} - \dfrac{\text{浓淡水强度之和}}{2}}{\text{原水强度}} \times 100\%$。

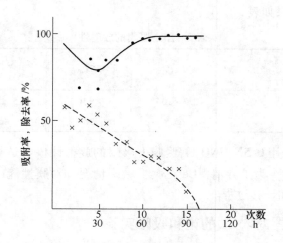

图2 ^{144}Ce 在膜上吸附（虚线）及除去（实线）的关系

3 电渗析器离子交换膜达到饱和吸附后对各核素除去的影响

原水配制：在单一 ^{144}Ce 放射性原水中加入 ^{137}Cs、^{90}Sr、^{106}Ru 和 ^{131}I 等核素，实验条件同前，各放射性核素除去的情况如表 7 所示。

表 7 各放射性核素除去情况

实验编号	水质	含盐	除盐/%	总放 强度/Ci·L^{-1}	除去/%	^{90}Sr 强度/Ci·L^{-1}	除去/%	^{137}Cs 强度/Ci·L^{-1}	除去/%	^{144}Ce 强度/Ci·L^{-1}	除去/%	^{106}Ru 强度/Ci·L^{-1}	除去/%	^{131}I 强度/Ci·L^{-1}	除去/%
21	原水	4200×10^{-6}		6.4×10^{-6}		2.7×10^{-7}		9.5×10^{-7}		6.57×10^{-6}		6.4×10^{-8}		1.4×10^{-8}	
	淡水	100×10^{-6}	97.6	1.3×10^{-7}	98.0	2.0×10^{-9}	99.2	6.6×10^{-9}	99.2	1.1×10^{-7}	98.2	2.0×10^{-9}	70.0	1.7×10^{-9}	87.8
22	原水	4200×10^{-6}		5.1×10^{-6}		3.4×10^{-7}		6.6×10^{-7}		4.7×10^{-6}		4.3×10^{-8}		7.7×10^{-9}	
	淡水	115×10^{-6}	97	1.2×10^{-7}	97.6	2.7×10^{-9}	99.1	1.2×10^{-8}	98.2	1.3×10^{-7}	97.3	1.5×10^{-8}	64.7	5.4×10^{-10}	93.1
23	原水	4100×10^{-6}		4.7×10^{-6}		3.7×10^{-7}		6.1×10^{-7}		3.6×10^{-6}		3.6×10^{-8}		4.7×10^{-9}	
	淡水	100×10^{-6}	97.6	6.3×10^{-8}	98.7	9.9×10^{-10}	99.6	6.9×10^{-9}	98.4	3.1×10^{-8}	99.1	1.3×10^{-8}	64.2	2.1×10^{-10}	95.4
24	原水	4100×10^{-6}		5.1×10^{-6}		3.8×10^{-7}		6.6×10^{-7}		3.8×10^{-6}		3.3×10^{-8}		4.8×10^{-7}	
	淡水	110×10^{-6}	97.4	1.2×10^{-7}	97.7	1.6×10^{-9}	99.7	1.4×10^{-8}	98.0	3.5×10^{-8}	99.2	1.3×10^{-8}	61.3	7.0×10^{-8}	85.5
25	原水	4200×10^{-6}		5.4×10^{-6}		5.6×10^{-6}		9.2×10^{-7}		2.83×10^{-6}		3.6×10^{-8}		1.5×10^{-7}	
	淡水	100×10^{-6}	97.6	3.7×10^{-7}	93.3	2.6×10^{-7}	99.5	1.7×10^{-8}	98.0	2.8×10^{-8}	99	1.1×10^{-8}	70.3	2.1×10^{-8}	85.9
26	原水	4500×10^{-6}		6.0×10^{-6}		5.9×10^{-7}		9.2×10^{-7}		3.8×10^{-6}		4.3×10^{-8}		1.2×10^{-7}	
	淡水	120×10^{-6}	95.1	3.6×10^{-7}	94.0	2.0×10^{-9}	99.6	1.9×10^{-8}	98	2.1×10^{-8}	99.5	1.2×10^{-8}	71.4	2.0×10^{-8}	83.3

4 实验结果讨论

（1）表 2 和表 3 的数据表明：电渗析除盐过程中相应地除去了离子状态的放射性核素。在电渗析过程中，首先是溶液中的离子吸附在膜上，然后再迁移到浓区。所以实验中测得的放射性离子在淡区的减少，应该是在膜上的吸附和电迁移的总和。当微量放射性离子和常量离子是同一元素时，它们在电渗析中淡化的效果是完全一致的。

否则，由于它们在溶液中的行为不一致，膜对微量和常量离子的吸附有一定的差异，淡化效果就不同。在本实验苦咸水中常量离子和微量放射性离子不是同一元素，所以在电渗析中除盐和除总放是不一致的，除盐和各核素的除去效果也不相同。

（2）表3表明：^{106}Ru的除去效果不好，因为^{106}Ru在溶液中价态复杂，以多种离子状态存在，而且在某些条件下容易生成配合离子和胶体。这些原因使^{106}Ru在电渗析淡化过程中的除放效果较差。

（3）表4~表6的数据表明：^{144}Ce在交换膜上的吸附较多，^{144}Ce的除去百分数是电迁移加交换膜吸附的总和。在电渗析中^{144}Ce先吸附在交换膜上然后向浓区迁移，^{144}Ce在交换膜上饱和吸附后，它的除去百分数基本上没有变化，这表明，在膜上^{144}Ce吸附饱和后，电迁移是主要过程。

（4）^{144}Ce饱和吸附后的离子交换膜能继续对^{137}Cs、^{90}Sr、^{131}I、^{144}Ce和^{106}Ru进行迁移。实验结果表明，在相同的实验条件下，各核素的除去百分数与交换膜的饱和吸附无关。交换膜的饱和吸附^{144}Ce后不但可以继续淡化^{144}Ce，而且也可以继续淡化^{137}Cs、^{90}Sr、^{131}I和^{106}Ru等核素。

（5）电渗析淡化器在除盐过程中除去微量放射性元素，还与电渗析器组装是否合理，离子交换膜的选择等因素有关。为了使除放有较好的效果，必须综合考虑这些因素。

参 考 文 献

［1］ Robert A.，ORNL-2557(1959).
［2］ M. Nishidoi et al.，J. At. Energy Soc. of Jap. 2. 460(1960).
［3］ Ф. В. Раузен и др.，Proceedings of the 4th international conference on the peaceful uses of atomic Energy Vol. 11 387~398 Vienna(1972).
［4］ 北京原子能研究所：电渗析法处理弱放废水实验报告，内部资料(1976).
［5］ 上海原子核研究所：用电渗析法处理低水平放射性废水，内部资料(1977).
［6］ W. R. Walters et al.，Ind. Eng. chem. 47，61(1955).
［7］ D. C. Sammon et al.，AERE-R-3137(1960).
［8］ L. F. Coleman et al.，BNWL-72(1965).
［9］ Bub. G. T. et al.，TID-11612(1960).
［10］ Bub. G. T. et al.，Rev. Sci. Instr. 32 857(1961).
［11］ Bub. G. T. et al.，TID-13423(1961).
［12］ A. Facchinl et al.，ORNL-TM-1035(1965).
［13］ Yoshihide Honda et al.，Radioisotope 21(1972).

利用无机离子交换剂——多聚锑酸从 1AW 中提取裂变核素 ^{90}Sr 的研究*

摘 要 本文叙述了许多改进聚锑酸的制备方法，得到含 5 个结晶水的 PAA 晶体，从模拟的 1AW 废液中直接回收 ^{90}Sr。对温度、流速、酸度在回收 ^{90}Sr 的影响，以及淋洗条件等进行了一系列的实验，并用裂变产物进行了模拟回收 ^{90}Sr 的实验。说明这种改进的 PAA 无机离子交换剂，对锶具有较大的选择交换容量，对杂质离子具有较好的去污效果和较快的交换速度。能够从 1AW 模拟液中直接回收 ^{90}Sr，并初步确定了回收 ^{90}Sr 的流程方案。

1 前言

反应堆元件后处理产生的 1AW 高放废液中，^{90}Sr 占有很高的产额。近年来 ^{90}Sr 作为热源而引人注目。从 1AW 中先将 ^{90}Sr 提取出来，这不仅可以大幅度地降低 β 辐射强度，而且为从 1AW 中提取其他有用核素创造了有利条件。

在用 D_2EHPA 萃取，有机离子交换和硫酸盐沉淀载带法从裂变废液中提取 ^{90}Sr 时[1~3]，由于 D_2EHPA 的辐射乳化现象，有机离子交换剂容易发生辐射分解，以及沉淀法工艺陈旧复杂等，近年来用无机离子交换剂直接从裂变废液中分离回收 ^{90}Sr 的研究，引起了广泛的注意。

鉴于无机离子交换剂对某些核素具有高度的选择性，以及良好的热和辐射稳定性，人们对许多不溶性水合氧化物、杂多酸盐、多价金属酸式盐进行了广泛的研究，以便用来从酸性裂变废液中提取 ^{90}Sr[4~7]。其中对水合五氧化二锑的研究取得了较大的进展[7~11]。Baetsle 等人[12]用多聚五氧化二锑从 100Ci 的裂变废液中回收了 ^{90}Sr。Davis 等人[13]设计了利用 PAA 从商用反应堆废料中回收 ^{90}Sr 的工艺流程。虽然目前在工艺上使用 PAA 从裂变废液中大规模回收 ^{90}Sr 的工作尚未见报道，但一般认为从 1AW 中直接回收 ^{90}Sr，PAA 是最有希望在工艺上得到应用的无机离子交换剂。

本工作改进了 PAA 的合成方法。获得了颗粒状的 PAA 晶体，对锶的选择交换容量比国外文献报道的大[11,12,14]，并具有良好的动力学过滤性能。对模拟的 1AW 废液分离 ^{90}Sr 的研究，表明利用 PAA 是有效的。根据实验数据，初步确定了用 PAA 从 1AW 中分离 ^{90}Sr 的流程。

2 实验部分

2.1 试剂及仪器

化学试剂均为化学纯和分析纯。同位素指示剂 ^{85}Sr、^{137}Cs、^{144}Ce 及不同冷却时间的裂

---

* 本文合作者：翁皓珉、李太华、韩俊、孙兆祥、刘正浩。原发表于《核材料与工程》，1982，2(3)：238~244。

变产物，由原子能研究所提供，放射性测量使用美国 CANBERRA 公司的 Jupiter 程控能谱分析系统及 FH-408 自动 γ 测量仪。离子交换柱的大小为 38cm×25cm，交换床为 38cm×12cm，带有保温套，用 CS-50 型超级恒温水浴调节温度。

2.2 无机离子交换剂 PAA 的制备及结构测定

（1）制备[14]方程式为：

$$SbCl_5 + nH_2O \longrightarrow Sb_2O_5 \cdot nH_2O + HCl$$

将 $SbCl_5$ 水解，得白色沉淀。经过陈化、过滤，用 2N HNO_3 洗涤，再在低温下干燥。然后在酸性溶液中裂解，室温下干燥、过筛。取 40～400 目（0.425～0.0374mm）颗粒做实验样品。

（2）结构测定。取上述样品用 X 衍射仪测定，结果与美国 ASTM 卡片 20-111 的数据一致。PAA 的结构式为 $Sb_2O_5 \cdot (4H_2O)_n$。

（3）PAA 结合水的测定。用 1000mg 样品放入马弗炉中，在不同温度下烘烤。每 3h 称重一次，痕重后即为该温度下样品的质量。由此算出不同温度下样品的结晶水。PAA 样品的通式为 $Sb_2O_5 \cdot 5.1H_2O$。

2.3 分配系数、交换容量测定及钠、钡、铁、铈和锆离子对锶交换容量的影响

（1）分配系数的测定。取三份 250mg PAA 样品，分别放在 100mL 塑料瓶中，分别加入 20mL 以 ^{85}Sr、^{137}Cs、^{144}Ce 为指示剂的 0.012N 的 Sr^{2+}、Cs^+、Ce^{3+} 之 1N HNO_3 溶液。在 25℃ 恒温下作振荡交换。测得 Sr^{2+}、Cs^+、Ce^{3+} 的分配系数 K_d 分别为 4690、30、30。

（2）锶交换容量测定。取 1g PAA 样品，在 2N HNO_3 中浸泡 2～3h，然后转入带有保温套的玻璃柱中，保持 64℃。以 ^{85}Sr 为指示剂的 1N HNO_3 + 0.10N $Sr(NO_3)_2$ 溶液，在 2BV/h 的流速下通过交换柱，直到 $C/C_0 = 0.5$ 为止（C_0、C 分别为料液和流出液的比放（cpm/mL））。测得 PAA 对锶的交换容量为 2.03mmol/g。

（3）钠、钡、铁、铈和锆离子对锶交换容量的影响。实验条件同上。在用 ^{85}Sr 作指示剂的 0.10N $Sr(NO_3)_2$ + 1N HNO_3 料液中，分别按比例加入各种杂质离子，测定杂质离子的影响。结果见表 1。

表 1 钠、钡、铁、铈和锆对锶交换容量影响

加入离子的浓度	料液	1.97N Na^+	0.03N Ba^{2+}	1.05N Fe^{3+}	0.68N Ce^{3+}	0.92N Zr^{4+}
锶的交换容量/mmol·g^{-1}	2.03	1.54	2.03	1.70	1.52	1.71

2.4 用 PAA 从 1AW 中分离 ^{90}Sr

2.4.1 模拟 1AW 溶液的组分

以冷却四个月的混合裂变产物为指示剂，加入示踪剂的 ^{85}Sr，配成模拟液，组成见表 2。

表2 模拟液成分

离子	Na^+	Cs^+	Rb^+	Zr^{4+}	U^{4+}	Sr^{2+}	Ba^{2+}	Fe^{3+}	RE（以Ce^{3+}代）	HNO_3
浓度 /g·L^{-1}	2.300	0.750	0.075	2.000	0.250	0.500	0.250	2.000	3.750	1N

2.4.2 锶的选择交换容量

取1g PAA样品，湿法装柱，保持在13℃，模拟液以0.5BV/h的流速通过交换柱。流出液每隔1mL取一次样，测定放射性强度，直到$C/C_0=0.5$为止。测得锶的穿透曲线，见图1。当$C/C_0=0.01$时，对锶的选择交换容量0.1mmol/g。当$C/C_0=0.50$时，则为0.33mmol/g。

2.4.3 温度对锶的选择交换容量的影响

升高温度，交换离子的淌度增高，溶液的黏度降低，交换离子在PAA颗粒内的扩散加速。锶的选择交换容量也就提高。实验结果证实了这一点。结果见表3及图2。

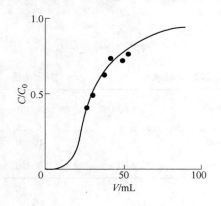
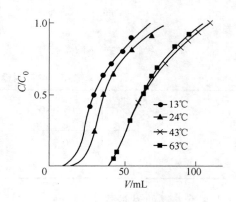

图1 锶的穿透曲线　　　　　图2 温度对锶穿透的影响

表3 流速为0.5BV/h时温度对锶选择交换容量的影响　　　　（mmol/g）

穿透/% \ 温度/℃	13	24	43	63
1	0.100	0.165	0.440	0.440
50	0.330	0.390	0.700	0.690

2.4.4 流速对锶选择交换容量的影响

其他条件相同，流速影响的结果见表4及图3。

表4 温度为63℃时流速对锶选择交换容量的影响　　　　（mmol/g）

穿透/% \ 流速/BV·h^{-1}	0.5	1	2	3	5
1	0.440	0.440	0.385	0.330	0.297
50	0.700	0.740	0.700	0.780	0.770

2.4.5 酸度对锶选择交换容量的影响

配制了 1N、2N、3N 和 4N 的 HNO_3 的模拟液（其他组分浓度均相同）。结果见表5及图4。酸度的影响不明显。2N HNO_3 的结果稍高，是由 H^+ 对吸附的杂质离子的洗脱作用造成的。

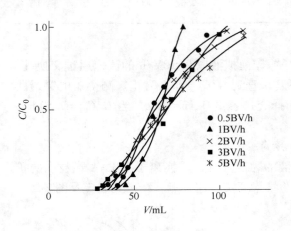

图 3　流速对锶穿透的影响

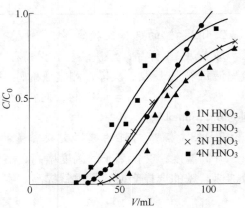

图 4　酸度对锶穿透的影响

表 5　酸度对锶选择交换容量的影响　　　　　　　　　　（mmol/g）

穿透/% \ 酸度	1N HNO_3	2N HNO_3	3N HNO_3	4N HNO_3
1	0.330	0.360	0.330	0.280
2	0.780	0.850	0.780	0.720

2.4.6 锶的淋洗条件

（1）淋洗剂的选择。Murthy 等人[11]和 Baetsle 等人[12]指出：$AgNO_3$、$Pb(NO_3)_2$ 和 $NaNO_3$ 对吸附在 PAA 上的锶都有解吸作用。首先用静态法进行试验。取吸附了锶的 PAA 样品三份（各 1g），分装在三支 10mL 的试管中，分别加入 8N HNO_3 + 0.5N $AgNO_3$、1N HNO_3 +4N $Pb(NO_3)_2$、2N HNO_3 +4N$NaNO_3$ 3mL，置于水浴中加热至 90℃ 不断搅拌，半小时更换一次解吸液，直到解吸不下放射性锶为止。最后测定解吸液的总放射性和 PAA 上的残留放射性，求得解吸率。上述三种解吸液的解吸率分别为 98.2%、92.5% 和 35.5%。故选 8N HNO_3 + 0.5N $AgNO_3$ 作淋洗剂。

（2）锶的淋洗。将吸附了锶的 PAA 柱，用 5BV 的 2N HNO_3 洗涤残留料液和杂质后，在 64℃ 下用上述淋洗剂，在 0.5BV/h 下淋洗。结果见图5。25BV 的淋洗剂能洗脱 99.8%。

2.4.7 PAA 柱重复使用效果

在淋洗 PAA 柱上的 Sr^{2+} 被 Ag^+ 取代。用 8N HNO_3 作再生液。64℃ 时，以 1BV/h 的流速进行再

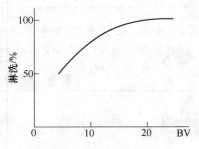

图 5　锶的淋洗曲线

生，直到流出液中无 Ag^+ 为止。再生液用量 80BV 即可。然后用再生 PAA 柱吸附、淋洗、再生循环使用。结果见表 6。

表 6　循环次数与交换容量的关系　　　　　　　　　　　　　（mmol/g）

穿透/%　循环次数	1	2	3	4
1	0.440	0.253	0.220	0.110
50	0.700	0.465	0.468	0.490

2.5　从 1AW 中分离回收先将 ^{90}Sr 产品质量鉴定

将模拟液在 64℃下，以 3BV/h 的流速通过交换柱。用 2N HNO_3、5BV 洗涤杂质和残留液，再用 1N HNO_3 + 1N $H_2C_2O_4$ 溶液 5BV 洗涤 ^{95}Zr-^{95}Nb，然后淋洗。淋洗液中含 $^{141,144}Ce$ 19.4%，^{85}Sr 80.6%。$^{141,144}Ce$ 的去污因数为 33❶。

2.6　利用 PAA 从 1AW 中分离回收 ^{90}Sr 的流程方案

根据实验数据，初步设计了用 PAA 从 1AW 中分离回收 ^{90}Sr 的工艺流程，见图 7。操作步骤如下：

（1）1AW 料液在 64℃下，以 3BV/h 的流速上柱。穿透 1% 用料液 33BV，穿透 50% 用 66BV。

（2）以 1~2BV/h 的流速洗涤残留液及杂质。

（3）洗涤 ^{95}Zr-^{95}Nb。

（4）以 1~2BV/h 的流速淋洗 ^{90}Sr。

图 7 为分离回收 ^{90}Sr 的流程示意图。

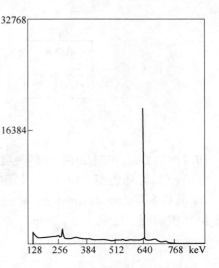

图 6　淋洗液的 γ 能谱

3　讨论

（1）由于改进了 PAA 的制备方法，获得了动力学性能好、对锶的选择交换容量大的颗粒状结晶样品。利用它可以有效地从 1AW 中分离裂变核素 ^{90}Sr。

（2）流速从 0.5BV/h 到 5BV/h 变化，选择交换容量变化不显著。锶穿透 1% 时，从 0.440mmol/g 降为 0.300mmol/g；穿透 50% 时从 0.700mmol/g 上升至 0.780mmol/g。温度对选择交换容量影响较大，它随温度上升而增大。温度升高到 63℃，选择交换容量与 43℃时相近，趋于恒定。酸度对锶的选择交换容量影响不大。

（3）用 8N HNO_3 + 0.5N $AgNO_3$ 淋洗锶，效果良好，用 19BV 可洗脱锶 97%，用 25BV 可洗脱 99.8%。

（4）用 PAA 分离回收 ^{90}Sr 流程简单。其他 γ 放射性核素沾污少。淋洗液中仅有

---

❶ 去污因数 $F = \dfrac{(^{141}Ce + ^{144}Ce)\text{ 在总料液内的放射性}}{(^{141}Ce + ^{144}Ce)\text{ 在淋洗液内的放射性}}$。

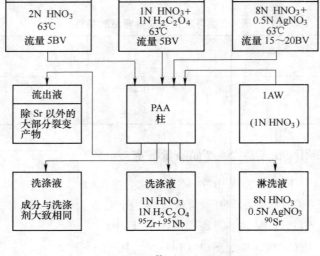

图7 分离回收^{90}Sr的流程示意图

19.4%的Ce。粗制品中^{90}Sr的纯度与回收率均比沉淀法高。

（5）循环使用PAA柱，对锶的选择交换容量明显下降。经三次循环使用，穿透1%时的容量为0.25mmol/g；穿透50%时为0.47mmol/g，尚可使用。一次使用PAA柱则比较简便、适宜。

参 考 文 献

[1] L. A. Bray and H. H. Van Tuyl，HW-69534（1961）.

[2] G. L. Richardson，BNWL-4（1965）.

[3] G. L. Richardson，HW-79762 Part 2.（1964）.

[4] V. Vesley and V. Pekarek，Talanta 19，1245（1972）.

[5] V. Vesley and V. Pekarek，Talanta 19，219（1972）.

[6] C. B. Amhplett，Inoganic Ion Exechanger（1964）.

[7] L. H. Baetsle，D. Van Deyck，D. Huys and A. Guery. EUR-2497（e）（1965）.

[8] M. Abe and T. Ito，Bull Chem. Society of Japan 41，333（1968）.

[9] M. Abe and T. Ito，Journal of Chem. Society of Japan Pure Chem.，Sec.，87，1174（1966）.

[10] I. H. Baetsle，D. Huys，J. Inorg. And Nuel. Chem.，30，639（1968）.

[11] T. S. Murthy，K. R. Balesubramanian，B. A. R. C. -894（1976）.

[12] L. H. Baetsle，D. Huys，BLG-487（1973）.

[13] D. K. Davis，J. A. Partsidge，O. H. Korki，BNWL-1063（1971）.

[14] 翁皓珉，李太华等．无机离子交换剂聚锑酸的制备及其性能测定（待发表）.

用磷酸锆-磷钼酸铵复合交换剂从动力堆元件废液中提取 ^{137}Cs 的研究*

1 前言

从20世纪60年代中期开始,采用无机离子交换剂提取铯已成为重要的分离手段,它与有机树脂和有机溶剂相比,具有良好的辐照稳定性、热稳定性及高度的选择性,因此,近些年以来,许多国家对无机离子交换剂曾进行了广泛的研究。在无机离子交换剂中,对磷酸锆(ZrP)研究得比较多,它对碱金属的分离比有机树脂具有更好的效果。但它受到酸度的限制,比较适合于中等酸度的介质。如用 ZrP 从 1AW 废液中提取 ^{137}Cs 时,须对废液进行稀释或脱硝至 0.5N 硝酸以下才能使用。

另外,杂多酸盐例如磷钼酸铵(AMP)、磷钨酸铵(APW)等,都能在强酸高放废液中优先于其他任何金属离子吸附铯,它们是早已被人们注意的一种无机离子交换剂。但据报道,应用 AMP 分离 ^{137}Cs 还存在两个问题:一是 AMP 颗粒成型比较困难,采用多种方法制备的 AMP 均是微晶沉淀,过滤性能差,不适于装柱;二是淋洗效果差,浓缩倍数低[1~4]。

上述不溶性磷酸盐和杂多酸盐各有其优缺点,在单独使用时都受到一定的限制,而复合交换剂磷酸锆-磷钼酸铵(ZrP-AMP)比 ZrP 的耐酸性能好,比 AMP 的淋洗效果好,且容易成型。

2 实验

2.1 ZrP-AMP 的制备[5]

将 1mol 的硝酸氧锆加热溶解在 4mol 的硝酸中,过滤得到硝酸氧锆溶液,在不断搅拌的条件下,加入过量的磷酸,生成白色磷酸锆胶体,抽滤成胶冻,再向其中加入钼酸铵,充分搅拌均匀,胶冻逐渐变成黄色,直至颜色不再变深为止。然后将胶冻铺在玻璃板上,厚度约 0.5cm,在室温下放置一昼夜,之后在 50℃烘干,再放入蒸馏水中"炸裂",经过分离、干燥和过筛,则得到颗粒状产品。

2.2 动态实验

(1) 试剂。均为分析纯或化学纯。

示踪剂:^{137}Cs$^+$、^{22}Na$^+$、^{86}Rb$^+$、$^{85+89}$Sr^{2+}、^{141}Ce^{3+}、^{241}Am^{3+} 及冷却 135 天的混合裂变产

---

* 本文合作者:韩俊、孙兆祥、刘正浩、翁浩珉、李太华、唐志刚。原发表于《北京师范大学学报(自然科学版)》,1983(3):47~54。

物。离子交换柱：带夹套的玻璃交换柱，内径0.4cm，柱高30cm，床高7cm。

测量装置：FH-408自动定标器；FT-603阱型γ闪烁探头；美国CANBERRA公司的Jupiter程控能谱分析系统。料液：1N HNO$_3$模拟液加^{137}Cs。交换剂粒度：50～100目（270～150μm）。交换剂质量：1g。

（2）模拟液的配制。在1N HNO$_3$溶液中，按模拟液的组成加入各成分的质量，模拟液成分见表1，边加边搅拌，直至全部溶解，过滤去掉少许悬浮物，溶液稍带浅绿色。

表1 模拟液成分

元素	浓度/g·L^{-1}	元素	浓度/g·L^{-1}	元素	浓度/g·L^{-1}
Na$^+$	2.5	Sr^{2+}	0.5	Fe^{3+}	2.0
Cs$^+$	0.75	Ba^{2+}	0.25	Zr^{4+}	2.0
Rb$^+$	0.075	Ce^{3+}	3.75	U^{4+}	0.25

模拟液中硝酸的酸度为1N。

（3）实验方法。称取1g 50～100目（0.3～0.15mm）交换剂，经过1N HNO$_3$浸泡，再转入带夹套的交换柱中，然后经过吸附、洗残液、洗涤、淋洗和再生等步骤，构成一次动态循环。

（4）ZrP-AMP从模拟的1AW废液中提取^{137}Cs的性能试验：

1）铯的吸附性能实验，如图1所示。

2）铯的淋洗性能实验，如图2所示。

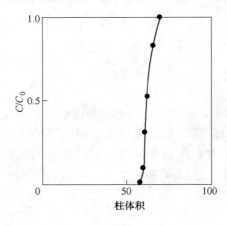

图1 铯的穿透曲线

（C为流出液中铯的浓度；C$_0$为进料液中铯的浓度；柱体积：0.9mL；流速：3BV/h；温度：10℃）

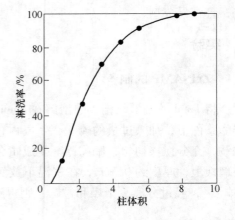

图2 铯的淋洗曲线

（淋洗液：0.1N HNO$_3$+5N NH$_4$NO$_3$；柱体积：0.9mL；流速：3BV/h；温度：60℃）

3）流速对铯吸附的影响，如图3所示。

4）循环次数对交换容量和淋洗效率的影响，见表2。

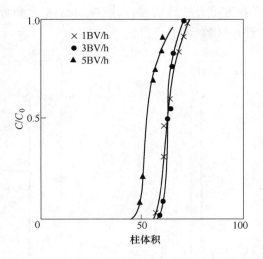

图 3　不同流速时铯的穿透曲线

（柱体积：9mL；温度：10℃）

表 2　循环次数对吸附和淋洗的影响

循环次数	$C/C_0=0.01$ 时的交换容量/mEq·g^{-1}	洗至 11 柱体积时的淋洗效率/%
1	0.28	97
2	0.29	99
3	0.29	99

连续循环实验的吸附速度和淋洗速度均为 3 柱体积/h，吸附温度 10℃，淋洗温度 60℃，淋洗液为 0.1N HNO_3 和 5N NH_4NO_3。

5) 去污实验。Na、Rb、Sr、Ce、Am 五种核素的去污因数，是以它们相应的放射性核素作指示剂单个测定的，而 Zr、Nb、Ru 等核素，由于它们在溶液中的化学状态比较复杂，很难反映真实料液的情况，所以采用冷却了 135 天的混合裂变产物作指示剂进行测定。实验结果见表 3 和图 4～图 8。

表 3　^{86}Rb 等核素的去污因数

核　素	放射性总强度计数	0.5N HNO_3 洗下的强度计数	0.1N HNO_3 + 5N NH_4NO_3 洗下的强度计数	交换剂滞留强度计数	去污因数
^{86}Rb	493606	30534	45734	41618	11
^{22}Na	510760	10790	262	75	1956
$^{85+89}$Sr	301568	4228	300	136	1005
^{141}Ce	674680	1620	427	426	1580
^{241}Am	403080	1535	200	—	2015

表 3 中去污因数的计算方法如下：

$$去污因数 = A_1/A$$

式中，A_1 为原始含有的沾污元素的放射性，计数/min；A 为经分离后仍与欲测核素相存在的沾污元素的放射性，计数/min。

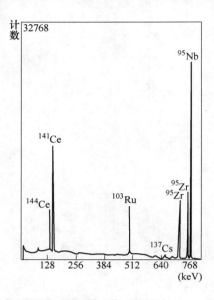

图 4　1AW 能谱

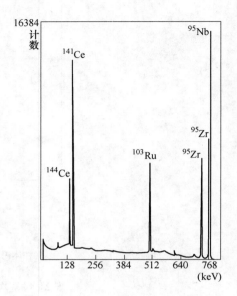

图 5　流出液能谱

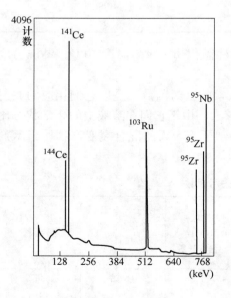

图 6　0.5N HNO_3 洗涤液能谱

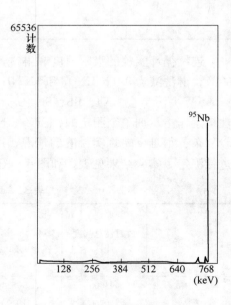

图 7　0.5N $H_2C_2O_4$ 洗涤液能谱

表 4 中各项数据的计算方法：通入的料液按 40mL 计，^{137}Cs 的浓度为 0.75mg/mL，淋洗效率为 99%，则淋洗液中 ^{137}Cs 的总质量为 29.7mg。同样，根据料液中各种核素的含量及其去污因数，分别计算出每种核素在淋洗液中的质量，就可得出各种核素在产品中的质量分数。

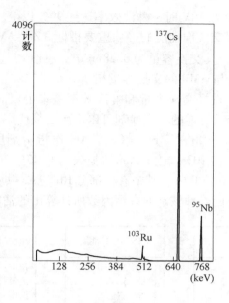

图 8 淋洗液能谱

表 4 淋洗液中各核素的质量分数

核素	去污因数	淋洗液中^{137}Cs的总质量/mg	淋洗液中去污元素的总质量/mg	淋洗液中各核素的质量之比/%
^{86}Rb	11		0.27	0.92
^{22}Na	1956		0.051	0.17
$^{85+89}$Sr	1005	29.7	0.020	0.067
^{141}Ce	1580		0.095	0.32
^{241}Am	2015		0.0005	0.0017

从 Am 去污因数的测定，以计算出 Am 的损失率，它包括在洗涤液和淋洗液中的损失。如果把经过吸附之后，留在交换柱上的 Am 看做是这两部分的总损失，并从表 3 可得 Am 在吸附段的去污因数为 232，由此得 Am 的损失率为 0.43%。

3 结果与讨论

（1）ZrP-AMP 复合交换剂通过生成胶体而后炸裂成型的方法，是一种比较新的制备途径，克服了 AMP 不易成型的缺点。AMP 虽然具有较好的离子交换性能，但由于颗粒小，过滤性能差，在应用上受到限制，使用时通常使用不溶解的惰性介质作载体。而 ZrP-AMP 具有良好的力学性能和过滤性能，实验证明，经过 4h 的强烈振荡，没有发现破碎现象；连续六天的动态实验，流速保持不变，也曾用 1N HNO$_3$ 浸泡半年之后，再连续进行动态实验，流速仍然稳定。

（2）ZrP-AMP 复合交换剂改善了 AMP 的淋洗性能。AMP 的另一个缺点是淋洗效果差，因而南非的流程曾采用 NaOH 溶解 AMP，然后再通过 ZrP 交换柱进行分离，而 ZrP-AMP 复合交换剂仍保留了 ZrP 的淋洗性能，采用 0.1N HNO$_3$ 和 5N NH$_4$NO$_3$ 作淋洗剂，当温度为60℃、流量为9BV时，淋洗效率达到99%以上。

（3）交换容量是表示交换剂性能的一项重要指标。ZrP-AMP 在 1N HNO$_3$ 模拟液中吸附^{137}Cs，当穿透1%时，其交换容量为0.28mEq/g，此值明显地超过了 ZrP 在相同条件下的交换容量，基本上与 AMP 的交换容量相当。

（4）循环三个周期，选择交换容量和淋洗效率基本上没有变化，这说明 ZrP-AMP 与铯的离子交换过程完全是可逆的，交换剂可以多次反复使用。

（5）以单个核素 ^{22}Na、^{86}Rb、$^{85+89}$Sr、^{141}Ce、^{241}Am 作指示剂所进行的去污实验表明，除 ^{86}Rb 以外，去污因数都在 10^3 以上，Am 的损失为 0.43%。但 ^{86}Rb 的去污因数只有11，表明 Rb 的性质十分类似于 Cs，其中有一部分 Rb 与 Cs 共吸附在交换柱上，还须对两者进一步分离。根据前面五种核素的去污因数所计算出的沾污百分数为 1.5%。

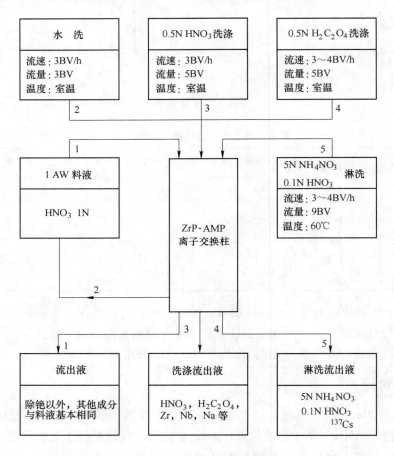

图9 用 ZrP-AMP 从 1AW 中回收^{137}Cs 流程图

4 结论

ZrP-AMP 是一种性能良好的复合型无机离子交换材料。它在制备方法上采取了生成胶体、烘干、炸裂的途径，克服了磷钼酸铵不易成型的缺点，为这类交换剂进行柱式操作提供了条件。ZrP-AMP 具有对^{137}Cs 选择交换容量大、力学性能好、杂质的去污因数高和比较容易淋洗等特点，能从 1N HNO$_3$ 的 1AW 废液中，不经预处理而直接回收^{131}Cs。

参 考 文 献

[1] V. Vesley ami V. Pekarek, Talanta, 19(1972), 219, 1245.
[2] B. J. Mehta and D. K. Baxi., Radiochimica Acta, 23(1976), 104~106.
[3] J. Dolezal, et al., J. Radioanal, Chem., 21(1974), 381~387.
[4] T. S. Murthv, et al., RARC-893 1977.
[5] A, Raggenbrass, DP-1066, Vol, Ⅱ, 1966, 33.
[6] J. Van and K. Smit, AERE-3884, 1961.

Studies on Recovery of ^{137}Cs from Waste Solution of Power Reactor Fuels with Zirconium Phosphate—Ammonium Phosphomolybdate Complex Inorganic Ion Exchanger

Han Jun, Sun Zhaoxiang, Liu Zhenghao, Weng Haomin,
Li Taihua, Tang Zhigang, Liu Boli

Abstract In this paper the preparation of zirconium phosphate-ammonium phos-phomolybdate is studied. The compound has very high selectivity for Cs$^+$. Its break-through capacities at $C/C_0 = 0.01$ were found to be 0.28mEq/g for caesium in simulation solutions of 1N HNO$_3$. An overall elution yield of 99% is attained in a total of 9 bed volume of the elutrient with 0.1N HNO$_3$ +5N NH$_4$NO$_3$ at 60℃. The contamination percent was 1.5 in eluant. Meantime, the loss of ^{241}Am is less than 0.5%. It can be used as a recovery material for ^{137}Cs from 1AW waste solution.

放射性药物化学

Quantitative Study of the Structure-stability Relationship of Tc Complexes*

Abstract By studying structural parameters (van der Waals radius and bond length from X-ray crystallography) of more than 100 known Tc compounds and using the cone packing model, a stability indicator of the Tc compounds was derived. The indicator is based on the solid angle factor sum (SAS), which is calculated from van der Waals radii and bond lengths between all coordinating atoms and Tc. The average SAS value is 0.97 ± 0.13. The SAS value reflects the limits of the packing of ligand around Tc. The quantitative calculation of the SAS is useful for predicting stability and designing new Tc radiopharmaceuticals.

Abbreviations

acac	acetylacetone	HBPz$_3$	hydrotris(pyrazolyl)borate
AcO	acetate	HM-PAO	hexamethyl PnAO
bipy	2,2′-bipyridine	Im	imidazole
Bu	butyl	MDP	methylenediphosphonate
CN	coordination number	Me	methyl
cyclam	1,4,8,11-tetraazocyclotetradecane	meim	methylimidazole
depe	1,2-bis(diethylphosphino)ethane	MP	mesoporphyrin IX dimethylester
diars	o-phenylenebis(dimethylarsine)	NTA	nitrilotriacetic acid
diphos	o-phenylenebis(dimethylphosphine)	OHphsal	N-(O-hydroxyphenyl) salicylideneiminate
dmpe	1,2-bis(dimethylphosphino)ethane		
dpa	N,N′-diphenylacetamidinate	OS	oxidation state
dppb	1,2-bis(diphenylphosphino)benzenedppe	OXMe	2-methyl-8-quinolinate
	1,2-bis(diphenylphosphino)ethane	pen	D-penicillamine
dtt	N,N′-di-p-tolyltriazenide	Ph	phenyl
EDTA	N,N,N′,N′-ethylenediaminetetraacetate	phsal	N-phenylsalicylideneiminate
		pic	picrate
ema	N,N′-ethylenebis(2-mercaptoacetamide)	PM-P′AO	propylene-PnAO
		PnAO	[3,3′-(1,3-propanediyldiimino)bis(3-methyl-2-butanone oximato)$^{-3}$]
EN	ethylenediamine		
Et	ethyl	py	pyridine

---

* Copartner: Y. Wei, H. F. Kung. Reprinted from *Appl. Radiat. Isot.*, 1990, 41(8):763-771.

quin	quinoline		dithiolate
salH	salicylaldehyde	tbp	4-t-butylpyridine
Salpd	N, N'-propane-1,3-diylbis(salicyli-deiminate)	tmbt	2,3,5,6-tetramethyl-thiophenyl
		TM-PAO	tetramethyl PnAO
SAS	solid angle factor sum	TPP	meso-tetraphenylporphyrin
SC	1,2-di(carbomethoxy)ethane-1,2-		

1 Introduction

Due to favorable cost effectiveness, convenience and superior physical characteristics, complexes of ^{99m}Tc ($t_{1/2}$ = 6h, 140keV) are the major radiopharmaceuticals for routine nuclear medicine procedures. However, because of the complexity of Tc chemistry, it is often difficult to predict the stability and physicochemical properties based on the chemical structures of potential ligands for Tc compounds. In designing new ^{99m}Tc radiopharmaceuticals, the first criterion is the stability of ^{99m}Tc complexes. The design parameters and rationale for targeting the formation of stable ^{99m}Tc radiopharmaceuticals generally are not well-defined. A large number of ^{99m}Tc compounds have been reported in the literature, and several systematic analyses of *in vitro* stability of the complexes have also been reported (Deutsch and Libson, 1984; Deutsch et al., 1983; Clarke and Podbielski, 1987; Davison and Jones 1982a, b; Kruichkov et al., 1986; Melnik and Van Lier, 1987; Kennedy and Pinkerton, 1988 a, b, c). However, no systematic studies based on the cone packing model (Fischer and Li, 1985) were attempted. In order to refine the design parameters for new ^{99m}Tc complexes, and to improve the success rate for preparation of stable complexes, which may lead to useful radiopharmaceuticals, the quantitative structure-stability relationship of Tc compounds using the cone packing model were investigated. This paper will concentrate on the first order stability effects, which are mainly derived from the coordinating ligand atoms of Tc complexes. The second order effects, the steric or electronic factors, from atoms not directly involved in coordinating the Tc, are also important and will be the subject of future reports.

2 Theory and Methods

Solid angle factor sum (SAS)

In order to investigate the structure-stability relationship of Tc compounds, X-ray crystallography data (van der Waals radius, R, bond length, ML) of more than 100 known Tc compounds (containing various ligand systems: S_4, N_2S_2, N_4, N_2O_2, et al.) were collected and analyzed. Using the cone packing model (Fischer and Li, 1985) and the above mentioned structural parameters, a general indicator for the stability of the Tc compounds was derived. The indicator is based on the solid angle factor sum (SAS), which is calculated from van der Waals radii and bond lengths between all coordinating atoms and Tc (Fig. 1). The van der Waals radii for coordinate atoms are averaged and normalized as listed in Table 1.

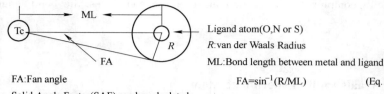

FA: Fan angle
Solid Angle Factor(SAF) can be calculated:
Solid Angle Factor sum (SAS):

Ligand atom(O,N or S)
R: van der Waals Radius
ML: Bond length between metal and ligand

$$FA = \sin^{-1}(R/ML) \quad \text{(Eq.1)}$$
$$SAF = (1-\cos FA)/2 \quad \text{(Eq.2)}$$
$$SAS = \sum_i SAF_i \quad \text{(Eq.3)}$$

Fig. 1 Definition of fan angle (FA), solid angle factor (SAF) and solid angle factor sum (SAS) for a ^{99}Tc complex

Table 1 Comparison of reported and normalized van der Waals radii of coordinate atoms

Coordinate atoms	Reported	Normalized	Coordinate atoms	Reported	Normalized
—H	1.2	1.2	—P	1.90	1.78
—C	—	1.5	—S	1.85	1.80
—N	1.5	1.5	—Cl	1.80	1.75
=N	—	1.3	—As	2.0	1.87
—O	1.4	1.4	—Br	1.95	1.86
=O	—	1.29	—Tc	—	1.57
—F	1.345	1.35	—I	2.15	2.02

Two methods can be used to obtain ML (metal to ligand bond length) values, with which the SAS values are calculated:

(1) Direct method. ML values obtained directly from X-ray crystallography data.

(2) Indirect method. ML values derived from average of reported crystallography data of the same ligand atom. For example, averaged bond length of Tc-S, Tc-N and Tc-O can be obtained based on reported data of various Tc complexes containing such bonding.

The initial paper on the cone packing model (Fischer and Li, 1985), describes the application of this technique to the quantitative-stability relationship of complexes for lanthanides and actinides. Since these two series of elements consist of f orbital electrons, the coordinate bonds between these outer shell electrons and ligand atoms are generally weak. Therefore, the van der Waals radii of coordinating atoms are indistinguishable before and after the complex formation. However, the Tc contains d electrons which from much tighter coordinating bonds between the metal and the ligand atoms. The van der Waals radii become smaller after complexation. This effect is more pronounced for soft ligand atoms such as sulfur, phosphorus, chlorine, bromine, etc. In order to compensate for this effect, the van der Waals radii for these atoms are normalized by the following technique: the average bond lengths and bond angles for TcL_n complexes (complexes with n equivalents of same ligand atoms) are collected from the X-ray crystallography data reported in the literature. By setting $\sum_i SAF_i = 1$, the normalized van der Waals radii for each atom are calculated (Table 1). These normalized values are used for calculating SAS values for each Tc complex (indirect method; for those complexes, the bond lengths are unknown).

3 Results

3.1 Van der Waals radii of coordinate atoms

The reported van der Waals radii and the values used for SAS calculation in this paper are not always the same. After normalization, as expected, the soft ligand atoms, such as S, O, N, etc. usually show a lower van der Waals radius value (Table 1).

3.2 ^{99m}Tc complexes and their SAS values

More than 100 ^{99m}Tc complexes, with detailed X-ray crystallography data, have been reported in the literature (as at the end of 1987). Using the measured bond length and the normalized van der Waals radii, SAS values were calculated (Table 2) and summarized in Fig. 2.

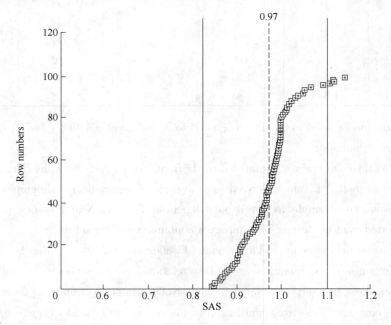

Fig. 2 Distribution of SAS values of all of the Tc compounds
(The average value is 0.97 ± 0.13)

4 Discussion

4.1 SAS values

The average SAS value for all of the Tc complexes is 0.97 ± 0.13 (Fig. 2). The standard deviation, 0.13, is smaller than the SAF value of any one of the possible coordinate atoms; therefore, it is reasonable to assume that the coordination number based on the predicted SAS values is within the range of standard deviation. It is possible to accurately determine the number of atoms (and the types of atoms) required to form a stable Tc complex. Despite the fact that the Tc complexes differ in valence state, net charge, coordinating number and coordinating atoms, the majority of them exist in a specific "saturation package range". The saturation packing range is a reflection of the limits on packing of coordinating atoms around the Tc atom. The most stable structures of Tc complexes are the result of interplay between several factors such as bond length, bond angle and steric hindrance, etc. These factors dictate the SAS value for each Tc compound, moving the value closer to 0.97, the average value. Theoretically, the optimal SAS value is 1.00, when a "perfect" cone packing is achieved. When packing of the coordinate atoms deviates from the optimum, either over or under saturation, the stability of the complex will be decreased. The stability is quantitatively related to the SAS value.

Recently, qualitative structural analysis of Tc complexes has been reported (Melnik and Van Lier, 1987). The basic conclusions are: (1) Tc-L bond lengths increase with higher coordination numbers; and (2) Tc-L bond lengths increase with larger van der Waals radii of the ligands. All of these qualitative factors are consistent with the quantitative analysis demonstrated in this paper. The quantitative method reported in this paper takes one step further in calculating the factors influencing stability of Tc complexes.

4.2 Application of SAS values in designing new radiopharmaceuticals

It is possible to use packing parameters (SAS values) to optimize the design of Tc complexes, including the selection of coordinating atoms, number of coordinating ligands and potential 3-dimensional arrangements. All of the factors will determine the stability of Tc complexes and also define the physico-chemical parameters, such as size, shape, net charge and lipid-solubility, etc. The properties of Tc complexes will then determine the biodistribution of these complexes and the potential diagnostic properties.

Based on the data reported in this paper, it is possible to estimate the stability of various $Tc^V O$ complexes. Using S, N, O as potential ligands and pyramidal basic core structure (four coordinating atoms) as one example, the SAS values and sequences are listed in Table 2. The SAS values for $Tc^V OS_4$ and $TcO^V N_2S_2$ are 0.993 and 0.9298, respectively. These types of TcO (Ⅲ) complexes are very stable and suitable for further evaluation as potential radiopharmaceuticals. On the other hand, the SAS value for $Tc^V ON_4$ is 0.905, which is not very stable; therefore the N_4 ligands generally prefer to form $Tc^V O_2 N_4$ complexes (Troutner et al., 1984). The best example in this series of Tc complexes is the formation of the $Tc^V O_2$(cyclam) complex, SAS = 0.98, which is much closer to 0.97, the average value, than that of $Tc^V O_4$, SAS = 0.905, However, when the N_4 ligand, cyclam, was changed to an open ring N_4 system, such as PnAO, the complex configuration changed to $Tc^V ON_4$ (SAS = 0.905). Of course, the intra-molecular H bonding does increase the stability (Jurisson et al., 1986); This is a secondary effect (effects of atoms not directly linked with Tc) which will be the subject for future reports. Generally, $Tc^V OX_4$ complexes (X: Cl or Br) are not stable (SAS values 0.89 and 0.86 for Cl and Br, respectively) but they are good starting materials for other $Tc^V O$(Ⅲ) complexes

(Cotton et al., 1979). The calculated value for $Tc^V OO_4$ complexes is 0.84, a very low value, and there is only one report in the literature of this type of Tc complex. Another type of Tc complex is $Tc^V OS_2O_2$, SAS = 0.905, which may have comparable stability to that of $Tc^V ON_4$. Both of these complexes are more stable than $Tc^V ON_2O_2$ (SAS = 0.87). Ligand exchange reactions starting with the $Tc^V ON_2O_2$ and stronger ligands, such as the N_2S_2 ligand, have been reported (Epps, 1984) and the results support the prediction by the cone-packing model (SAS value). One may argue that the SAS values based on solidstate X-ray crystallography data may not be a true reflection of the stability of complexes in solution. The SAS value may be the first step towards understanding the structure-stability relationship, other methods will have to be developed to evaluate the stability in solution.

In conclusion, the packing model can be applied to predict the stability of Tc complexes. The solid angle factor sum (SAS) of the Tc complex is quantitatively related to the stability. The average SAS value for the Tc complexes is 0.97 ± 0.13, which is very close to the theoretically optimal value of 1.00. This quantitative structure-stability relationship of Tc complexes is potentially useful for designing new radiopharmaceuticals used in diagnostic nuclear medicine.

Acknowledgements

This work is supported by the Chinese National Scientific Foundation and by a Grant from the National Institute of Health (NS-15809). The authors thank Mrs Fiona Chapman, Jean Posthauer, and Heather Cullen for their assistance in preparing this manuscript.

References

Clarke M. J. and Podbielski L. (1987) Medical diagnostic imaging with complexes of ^{99m}Tc. Coord. Chem. Rev. 78, 253.

Cotton F. A., Davison A., Day V. W., Gage L. D. and Trop H. S. (1979) Preparation and structural characterization of salts of oxotetrachloro technetium (V). Inorg Chem. 18, 3024.

Davison A. and Jones A. G. (1982a) The chemistry of technetium (V) Int. J. Appl. Radiat. Isot. 33, 875.

Davison A. and Jones A. G. (1982b) The chemistry of technetium Ⅰ, Ⅱ, Ⅲ and Ⅳ. Int. J. Appl. Radiat. Isot. 33, 867.

Deutsch E. and Libson K. (1984) Recent advances in technetium chemistry: bridging inorganic chemistry and nuclear medicine. Comments Inorg. Chem. 3, 83.

Deutsch E., Libson K., Jurrisson S. and Lindoy L. F. (1983) Technetium chemistry and technetium radiopharmaceuticals. Prog. Inorg. Chem. 30, 75.

Epps L. A. (1984) The chemistry of neutral, lipid soluble technetium (V) complexes of aminoalcohols and aminothiols. Ph. D. dissertation, p. 74.

Fischer R. D. and Li X. -F. (1985) The solid angle sum rule: A new ligand cone packing model of optimal applicability in organolanthanoid chemistry. J. Less-Common Metals 112, 303.

Jurisson S., Schlemper E. O. and Troutner D. E. (1986) Synthesis, characterization, and X-ray structural determinations of technetium (V)-oxo-tetradentate amine oxime complexes. Inorg. Chem. 25, 543-549.

Kennedy C. M. and Pinkerton T. C. (1988a) Technetium carboxylate complexes- Ⅰ. A review of Tc, Re and Mo carboxylate chemistry. Appl. Radiat. Isot. 39, 1159-1165.

Kennedy C. M. and Pinkerton T. C. (1988b) Technetium carboxylate complexes-II. Structural and chemical studies. Appl. Radiat. Isot. 39, 1167-1177.

Kennedy C. M. and Pinkerton T. C. (1988c) Technetium carboxylate complexes-III. A new synthetic route to hexakis (isonitrile) technetium (I) salts. Appl. Radiat. Isot. 39, 1179-1186.

Kruichkov S. V., Grigoriev M. S. Kuzina A. F., Gulev B. F. and Spitsyn V. I. (1986) Synthesis and structure of a new octanuclear bromide cluster of technetium with quadruple metal-metal bonds. Dokl. Akad. Nauk SSSR 288, 893.

Melnik M. and Van Lier J. (1987) Analyses of structural data of technetium compounds. Coord. Chem. Rev. 77, 275.

Troutner D. E., Volkert W. A., Hoffman T. J. and Holmes R. A. (1984) A neutral lipophilic complex of ^{99m}Tc with a multidentate amine oxime. Int. J. Appl. Radiat. Isot. 35, 467.

Appendix

SAS Values of Tc Complexes

	Ligand	OS	CN	—L × n	Bond length (Å)	SAS*	Ref.
1	Tc_2O_7	VII	4	—Oμ	1.84	0.849	a
				=O ×3	1.678		
2	$[NMe_4][TcO_4]$	VII	4	=O ×4	1.676	0.9	b
3	$(NH_4)[TcO_4]$	VII	4	=O ×4 (298K)	1.702	0.863	c
				208K	1.709	0.853	
				141K	1.711	0.85	
4	$KTcO_4$	VII	4	=O ×4	1.711	0.85	d
5	$H_2(TcCl_6) \cdot 9H_2O$	IV	6	—Cl ×6	2.33	1.019	e
6	$(NH_4)_2[TcCl_6]$	IV	6	—Cl ×6	2.353	0.994	f
7	$TcCl_6^{2-}$	IV	6	—Cl ×6	2.37	0.977	g
8	$TcOCl_4^-$	V	5	=O	1.65	0.89	g
				—Cl ×4	2.3		
9	$TcOCl_5^{2-}$	V	6	=O	1.65	0.989	g
				—Cl cis ×4	2.36		
				—Cl trans	2.5		
10	$[(Ph_3P)_2N][TcOCl_4]$	V	5	=O	1.61	0.895	h
				—Cl ×4	2.31		
11	$[AsPh_4]_2[TcCl_6]$	IV	6	—Cl ×6	2.3515	0.99	i
12	$[TcOF_4]_3$	VI	6	—F ×3	1.837	0.912	j
				—Fμ ×2	2.26		
				=O	1.66		
13	$(NH_4)_2[TcBr_6]$	IV	6	—Br ×6	2.5	0.995	k

	Ligand	OS	CN	—L × n	Bond length (Å)	SAS*	Ref.
14	(TcCl$_4$)$_n$ chain	IV	6	—Cl × 2	2.242	0.985	l
				—Cl × 2	2.383		
				—Cl × 2	2.492		
15	[TcBr$_6$]$^{2-}$	IV	6	—Br × 6	2.52	0.976	g
16	[TcOB$_4$]$^-$	V	5	=O	1.66	0.862	g
				—Br × 4	2.48		
17	[TcOBr$_5$]$^{2-}$	V	6	=O	1.66	0.956	g
				—Br cis × 4	2.54		
				—Br trans	2.74		
18	TcCl$_6^{2-}$	IV	6	—I × 6	2.72	0.991	g
19	[Tc(dmg)$_3$(SnCl$_3$)OH]·3H$_2$O	V	7	—Oμ	2.03	1.05	m
				—N × 6	2.089		
20	[(TcO(salpd))$_2$O]	V	6	—O × 2	2.01	0.959	n
				—N × 2	2.123		
				—O	1.685		
				=O	1.9		
21	[TcOCl(salpd)]	V	6	—Cl	2.44	0.971	n
				=O	1.66		
				—N × 2	2.115		
				—O × 2	1.98		
22	[TcO(OEt)Br$_2$L$_2$]	V	6	=O	1.684	0.994	o
				—OEt	1.855		
				—Br × 2	2.56		
				—N × 2	2.14		
23	(PPh$_4$)[TcCl$_4$(sal)]	IV	6	—Cl × 4	2.34	0.945	p
				—O × 2	2.01		
24	[Cl(pic)$_4$Tc'—O—TcCl$_4$(pic)]·H$_2$O	3.5	6	Tc—Cl × 4	2.369	0.963	q
				—N	2.2		
				—O	1.835		
			6	Tc'—O	1.795	0.916	
				—N × 4	2.146		
				—Cl	2.39		
25	(n-Bu$_4$N)$_3$[Tc(NCS)$_6$]	III	6	—N × 6	2.045	0.961	r
26	[(H$_2$EDTA)Tc(μ-O)]$_2$·5H$_2$O	IV	7	—Oμ × 2	1.913	1.014	s
				—Tc'	2.331		
				—O × 2	2.011		
				—N × 2	2.207		

Continued

	Ligand	OS	CN	—L × n	Bond length (Å)	SAS*	Ref.
27	trans-[TcO(OH$_2$) · ((acac)$_2$en)]$^+$	V	6	—N × 2	2.002	0.936	t
				—O × 2	2.016		
				—OH$_2$	2.282		
				=O	1.648		
28	trans-[TcO((sal)$_2$en)Cl]	V	6	—N × 2	2.045	0.967	t
				—O × 2	1.99		
				—Cl	2.526		
				=O	1.626		
29	[Tc(OH)(MDP)]$^-$	V	6	—O × 6	1.989	0.869	u
30	[TcO(HBPz$_3$)Cl$_2$]	V	6	—Cl × 2	2.328	0.986	v
				=O	1.656		
				—N × 2	2.087		
				—N trans	2.259		
31	trans-[TcOCl(phsal)$_2$]	V	6	—O × 2	1.965	0.969	w
				=O	1.67		
				—N × 2	2.155		
				—Cl	2.38		
32	[TcO(C$_{13}$H$_9$NO$_2$)Cl]	V	5	—O × 2	1.948	0.879	x
				=O	1.634		
				—N	2.055		
				—Cl	2.302		
33	trans-[TcO$_2$(Im)$_4$]$^+$	V	6	=O × 2	1.71	0.993	y
				—N × 4	2.15		
34	trans-[TcO$_2$(1-meim)$_4$]$^+$	V	6	=O × 2	1.71	0.996	y
				—N × 4	2.145		
35	trans-[TcO$_2$(Tbp)$_4$]$^+$	V	6	=O × 2	1.743	0.972	z
				—N × 4	2.15		
36	trans-[TcO$_2$(EN)$_2$]$^+$	V	6	=O × 2	1.747	0.964	aa
				—N × 4	2.158		
37	trans-[TcO$_2$(cyclam)]$^+$	V	6	=O × 2	1.752	0.982	ba
				—N × 4	2.126		
38	Ba[TcO(EDTA)]$_2$	V	7	=O	1.657	0.978	ca
				—O × 2	2.16		
				—O × 2	2.03		
				—N trans × 2	2.35		
39	[TcO(OHphsal)(quin)]	V	6	=O	1.659	0.95	da
				—O × 3	2.009		
				—N trans	2.19		
				—N	2.055		

Continued

	Ligand	OS	CN	—L×n	Bond length (Å)	SAS*	Ref.
40	meso-[TcO(TM-PAO)]	V	5	=O	1.677	0.903	ea
				—N×4	1.998		
41	[TcO(PM-PAO)]	V	5	=O	1.671	0.908	ea
				—N×4	1.996		
42	DL-[TcO(HM-PAO)]	V	5	=O	1.682	0.908	ea
				—N×4	1.99		
43	meso-[TcO(HM-PAO)]	V	5	=O	1.6735	0.907	ea
				—N×4	1.995		
44	[TcO(PnAO)]	V	5	=O	1.679	0.9	fa
				—N×4	2.001		
45	trans-[Tc(NO)(NH_3)$_4$ · (OH_2)]$^{2+}$	2.5	6	—NO	1.708	0.94	ga
				—N×4	2.161		
				—O	2.163		
46	[Tc_2O_2(SCH_2CH_2S)$_3$]	V	5	Tc-S×4	2.335	0.955	ha
				=O	1.665		
				Tc'-S×4	2.324	0.966	
				=O	1.661		
47	[Tc(SC(NH_2)$_2$)$_6$]$^{3+}$	III	6	—S×6	2.427	0.987	ia
48	[TcO(meso-SC)]$^-$	V	5	=O	1.672	0.968	ja
				—S×4	2.316		
49	[Tc(NHC_6H_4S)$_3$]	VI	6	—N×3	1.994	1.047	ka
				—S×3	2.351		
50	[TcO(SCH_2CHO)$_2$]$^-$	V	5	=O	1.662	0.916	la
				—S×2	2.29		
				—O×2	1.95		
51	[TcO($SCH_2C(O)S$)$_2$]$^-$	V	5	=O	1.672	0.965	ma
				—S×4	2.32		
52	[TcO(SCH_2CH_2S)$_2$]$^-$	V	5	=O	1.64	0.994	na
				—S×4	2.3		
53	[TcO(pen)(pen H)]	V	6	—O	1.64	0.997	oa
				—O	2.214		
				—N×2	2.197		
				—S×2	2.29		
54	cis-[TcO(OXMe)$_2$Cl]	V	6	=O	1.649	0.966	pa
				—Cl	2.36		
				—O×2	1.971		
				—N×2	2.197		
55	trans[Tc(tmbt)$_3$(NCMe)$_2$]	II	5	—S×3	2.249	0.922	qa
				—N×2	2.043		

Continued

	Ligand	OS	CN	—L × n	Bond length (Å)	SAS*	Ref.
56	(AsPh$_4$)[TcNBr$_4$]	VI	5	≡N —Br × 4	1.596 2.482	0.887	ra
57	(AsPh$_4$)[TcNCl$_4$]	VI	5	≡N —Cl × 4	1.581 2.322	0.901	sa
58	[TcN(C$_9$H$_6$NS)$_2$]	V	5	≡N —S × 2 —N × 2	1.623 2.356 2.135	0.844	ta
59	[Tc(NS)(S$_2$CNEt$_2$)$_2$Cl$_2$]	III	7	—NS —S × 4 —Cl × 2	1.75 2.468 2.421	1.106	ua
60	[TcN(S$_2$CNEt$_2$)$_2$]	V	5	≡N —S × 4	1.604 2.401	0.884	va
61	[Tc(S$_2$CNEt$_2$)$_3$(PMe$_2$Ph)]	III	7	—S × 6 —P	2.481 2.33	1.112	wa
62	[Tc(S$_2$CNEt$_2$)$_3$(CO)]	III	7	—S × 6 —C	2.482 1.861	1.139	xa
63	trans, trans.-[TcN(NCS)$_2$(NCMe)(PPh$_3$)$_2$]	V	6	≡N —N × 2 —P × 2 —NCMe	1.629 2.057 2.509 2.491	0.911	ya
64	trans-[Tc(dppe)$_2$(NCS)$_2$]	II	6	—P × 4 —P × 2	2.44 2.04	0.954	za
65	trans-[Tc(dppe)$_2$Br$_2$]$^+$	III	6	—P × 4 —Br × 4	2.5 2.44	0.948	ab
66	[Tc(dmpe)$_3$]$^+$	I	6	—P × 6	2.4	0.988	bb
67	trans-[Tc(dmpe)$_2$Cl$_2$]$^+$	III	6	—P × 4 —Cl × 2	2.436 2.324	0.977	bb
68	trans-[TcO(dmpe)$_2$(OH)](F$_3$CSO$_3$)$_2$	V	6	=O —OH —P × 4	1.66 1.96 2.477	0.991	bb
69	trans-[TcHN$_2$(diphos)$_2$]	I	6	—H —N2 —P × 4	1.7 2.05 2.359	0.993	cb
70	trans-[TcCl$_2$(PPh(OEt)$_2$)$_4$]$^+$	III	6	—Cl × 2 —P × 4	2.415 2.41	0.962	db
71	[TcCl$_5$(PPh$_3$)]$^-$	IV	6	—Cl × 5 —P	2.336 2.57	0.983	eb
72	trans-[Tc(acac)$_2$Cl(PPh$_3$)]	III	6	—Cl —P —O × 4	2.42 2.465 2.013	0.872	fb

Continued

	Ligand	OS	CN	—L×n	Bond length (Å)	SAS*	Ref.
73	TcP$_4$	IV	7	—P×6	2.398	1.064	gb
				—Tc′	3.002		
74	TcP$_3$	III	6	—P×6	2.386	1.002	hb
75	Tc$_2$As$_3$	V	6	—As×6	2.517	0.992	ib
76	trans-[Tc(diars)$_2$Cl$_2$]$^+$	III	6	—As×4	2.5085	1.007	jb
				—Cl×2	2.329		
77	trans-[Tc(diars)$_2$Cl$_2$]ClO$_4$	III	6	—As×4	2.5145	1.007	kb
				—Cl×2	2.318		
78	[Tc(diars)$_2$Cl$_4$]·PF$_6$	V	8	—As×4	2.442	1.228	kb
				—Cl×4	2.578		
79	mer-[TcCl$_3$(PMe$_2$Ph)$_3$]	III	6	—Cl×3	2.33	0.956	
				—P×3	2.455		
80	[TcCl$_3$(CO)(PMe$_2$Ph)$_3$]	III	7	—C	1.86	1.115	lb
				—Cl×3	2.48		
				—P×3	2.44		
81	[(TPP)(Tc(CO)$_3$)$_2$]	I	6	—C×3	1.892	0.941	mb
				—N×3	2.322		
82	[Tc(NO)Br$_2$(CNCMe$_3$)$_3$]	I	6	—N	1.726	1.015	nb
				—C×3	2.08		
				—Br×2	2.582		
83	[Tc(PMe$_2$Ph)$_2$(CO)$_2$(dtt)]	I	6	—P×2	2.4115	0.999	ob
				—C×2	1.874		
				—N×2	2.184		
84	[Tc(PMe$_2$Ph)$_2$(CO)$_2$(dpa)]	I	6	—P×2	2.409	0.994	ob
				—C×2	1.875		
				—N×2	2.2025		
85	cis-[Tc(CO)$_2$·(P(OEt)$_2$Ph)$_4$]ClO$_4$	I	6	—C×2	1.9	1.033	pb
				—P×4	2.4175		
86	[Tc$_2$(CO)$_{10}$]	0	6	—C×5	1.979	0.944	qb
				—Tc′	3.036		
87	Na$_2$[((NTA)Tc(μ-O))$_2$]	IV	6	—N	2.148	0.99	rb
				—O×3	2.047		
				—Oμ×2	1.919		
				—Tc′	2.363		
88	YTc$_2$Cl$_8$·9H$_2$O	2.5	5	—Cl×4	2.364	0.985	sb
				—Tc′	2.105		
89	K$_3$Tc$_2$Cl$_8$·nH$_2$O	2.5	5	—Cl×4	2.364	0.98	tb
				—Tc′	2.117		

Continued

	Ligand	OS	CN	—L×n	Bond length (Å)	SAS*	Ref.
90	(NH$_4$)$_3$[Tc$_2$Cl$_8$]	2.5	5	—Cl×4 —Tc′	2.3575 2.13	0.981	ub
91	(n-Bu$_4$N)$_2$[Tc$_2$Cl$_8$]	III	5	—Cl×4 —Tc′	2.329 2.147	1.009	vb
92	[Tc$_2$(MeCOO)$_4$Br]	2.5	6	—Br —Tc′ —O×4	2.843 2.112 2.06	0.996	wb
93	K[Tc$_2$(MeCOO)$_4$Cl$_2$]	2.5	6	—Cl —Tc′ —O×4	2.589 2.126 2.074	0.996	xb
94	[Tc$_2$(MeCOO)$_4$Cl]	2.5	6	—Cl —Tc′ —O×4	2.656 2.117 2.0645	0.995	yb
95	[Tc$_2$(Me$_3$CCOO)$_4$Cl$_2$]	III	6	—Cl —Tc′ —O×4	2.408 2.192 2.032	1.03	zb
96	[Tc$_2$(OC$_5$H$_4$N)$_4$Cl]	2.5	6	—Cl —Tc′ —O×2 —N×2	2.679 2.095 2.087 2.087	1.05	ac
97	(Me$_4$N)$_3$[Tc$_6$Cl$_{14}$]	11/6	6	—Tc′ —Tc′×2 —Cl×3 —Cl	2.163 2.685 2.37 2.954	0.93	bc
98	(Me$_4$N)$_2$[Tc$_6$Cl$_{12}$]	5/3	7	Tc1–Tc′ —Tc′×2 —Cl×3 —Cl	2.218 2.5675 2.376 3.213	0.923	bc
			6	Tc2-Tc′ —Tc′×2 —Cl×3	2.218 2.568 2.365	0.874	
99	K$_8$(H$_3$O)[Tc$_2$Cl$_8$]$_3$·3H$_2$O	2.5	5	—Tc′ —Cl×4	2.14 2.38	0.97	cc
100	[H$_3$O(H$_2$O)$_3$]$_2$[Tc$_6$Br$_6\mu_3$(Br·OH)$_6$]	5/3	9	—Tc′×4 —Br —μ_3(OH)×2 —μ_3(Br)×2	2.59 2.55 2.46 2.46	1.091	dc

	Ligand	OS	CN	—L × n	Bond length (Å)	SAS*	Ref.
101	[Tc$_8$Br$_4\mu$-Br$_8$]Br	13/8	9	Tc1-Br × 3	2.5	0.987	ec
				—Tc' × 3	2.513		
				—Tc' × 3	3.479		
			10	Tc2-Br × 2	2.52	1.046	
				—Br	2.929		
				—Tc' × 4	2.513		
				—Tc' × 3	3.411		
102	[TcO(O$_2$C$_6$H$_4$)$_2$]$^-$	V	5	=O	1.648	0.840	x
				—O × 4	1.957		
103	trans-[Tc((acac)$_2$en)(PPh$_3$)$_2$]PF$_6$	III	6	—P × 2	2.5075	0.890	fc
				—O × 2	2.0195		
				—N × 2	2.058		

* SAS values are calculated by equations(1), (2) and(3) in Fig. 1.

References in Appendix

a—Krebs B. (1971) *Z. Anorg. Allg. Chem.* **380**, 146.

b—Gilman K. Z. et al. (1986) *Dokl. Akad. Nauk. SSSR* **287**, 650.

c—Faggiani R. et al. (1980) *Acta Cryst.* **B36**, 231.

d—Krebs B. and Hasse K. (1976) *Acta Cryst.* **B32**, 1334.

e—Koz' min P. A. et al. (1975) *Koord. Khim.* **2**, 473.

f—Elder R. C. et al. (1979) *Acta Cryst.* **B35**, 136.

g—Thomas R. W. et al. (1985) *Inorg. Chem.* **24**, 1472.

h—Cotton F. A. et al. (1979) *Inorg. Chem.* **18**, 3024.

i—Baldas J. et al. (1984) *Acta Cryst.* **C40**, 1343.

j—Edwards A. J. et al. *J. Chem. Soc. A* **1970**, 252.

k—Shiplicof S. F. et al. (1975) *Zh. Neorg. Khim.* **20**, 330.

l—Elder M. and Penfold B. R. (1966) *Inorg. Chem.* **5**, 1197.

m—Deutsh E. et al. (1976) *Proc. Natl. Acad. Sci. USA.* **73**, 4287.

n—Bandoli G. et al. (1984) *J. C. S. Dalton.* **1984**, 2505.

o—Fackler P. H. et al. (1984) *Inorg. Chem.* **23**, 3968.

p—Mazzi U. et al. (1982) *Transition Met. Chem.* **7**, 63.

q—Kastner M. E. et al. (1986) *Inorg. Chim. Acta*, **114**, L11.

r—Trop H. S. et al. (1980) *Inorg. Chem.* **19**, 1105.

s—Burgi H. B. et al. (1981) *Inorg. Chem.* **20**, 3829.

t—Jurisson S. et al. (1984) *Inorg. Chem.* **23**, 227.

u—Libson K. et al. (1980) *J. Am. Chem. Soc.* **102**, 2476.

v—Thomas R. W. et al. (1979) *J. Am. Chem. Soc.* **101**, 458.

w—Bandoli G. et al. (1982) *J. C. S. Dalton.* 2455.

x—Bandoli G. et al. (1984) *Inorg. Chim. Acta* **95**. 217.

y—Fackler P. H. et al. (1985) *Inorg. Chim. Acta* **109**, 39.

z—Kastner M. E. et al. (1984) *Inorg. Chim.* **23**, 4683.

aa—Kastner M. E. et al. (1982) *Inorg. Chim.* **21**, 2039.

ba—Zuckman S. A. et al. (1981) *Inorg. Chim.* **20**, 2386.

ca—Deutsch E. et al. (1982) *Coord. Chem. Rev.* **44**, 191.

da—Mazzi U. et al. (1986) *J. C. S. Dalton* 1623.

ea—Jurisson S. et al. (1986) *Inorg. Chem.* **25**, 543.

fa—Fair C. K. et al. (1984) *Acta Cryst.* **C40**. 1544.

ga—Radonovich L. J. and Hoard J. L. (1984) *J. Phys. Chem.* **88**, 6711.

ha—Davison A. et al. (1985) *Can. J. Chem.* **63**, 319.

ia—Abrams M. J. et al. (1984) *Inorg. Chem.* **23**, 3284.

ja—Bandoli G. et al. (1984) *Transition Met Chem.* **9**, 127.

ka—Baldas J. et al. (1982) *Aust. J. Chem.* **35**, 2413.

la—Jones A. G. et al. (1981) *Inorg. Chem.* **20**, 1617.

ma—DePamphilis B. V. et al. (1978) *J. Am. Chem. Soc.* **100**, 5570.

na—Smith J. R. et al. (1978) *J. Am. Chem. Soc.* **100**, 5571.

oa—Franklin K. J. et al. (1982) *Inorg. Chem.* **21**, 1941.

pa—Wilcox B. E. et al. (1984) *Inorg. Chem.* **23**, 2962.

qa—Davison A. et al. (1986) *Inorg. Chim. Acta* **120**, L15.

ra—Baldas J. et al. (1985) *Aust. J. Chem.* **38**, 215.

sa—Baldas J. et al. (1984) *J. C. S. Dalton* 2395.

ta—Baldas J. et al. (1986) Inorg. Chem. **25**, 150.
ua—Baldas J. et al. (1984) Aust. J. Chem. **37**, 751.
va—Baldas J. et al. (1981) J. C. S. Dalton 1798.
wa—Batsanov A. S. et al. (1984) Z. Anorg. Allg. Chem. **510**, 117.
xa—Baldas J. et al. (1982) J. C. S. Dalton 451.
ya—Baldas J. et al. (1984) J. C. S. Dalton 833.
za—Bandoli G. et al. (1984) Inorg. Chem. **23**, 2898.
ab—Libson K. et al. (1983) Inorg. Chem. **22**, 1695.
bb—Vanderheyden J. L. et al. (1984) Inorg. Chem. **23**, 3184.
cb—Struchkov V. Y. T. et al. (1982) Z. Anorg. Allg. Chem. **494**, 91.
db—Mazzi U. et al. (1977) Inorg. Chem. **16**, 1042.
eb—Bandoli G. et al. (1982) J. C. S. Dalton 1381.
fb—Bandoli G. et al. (1977) J. C. S. Dalton 1837.
gb—Ruhl R. et al. (1982) J. Solid State Chem. **44**, 134.
hb—Ruhl R. and Jeitscho W. (1982) Acta Cryst. **B38**, 2784.
ib—Jeitschko W. et al. (1985) J. Solid State Chem. **57**, 59.
jb—Elder R. C. et al. (1980) Acta Cryst. **B36**, 1662.
kb—Glavan K. A. et al. (1980) J. Am. Chem. Soc. **102**, 2103.
lb—Bandoli G. et al. (1976) J. C. S. Dalton 125.
mb—Tsutsui M. et al. (1975) J. Am. Chem. Soc. **97**, 3952.
nb—Linder K. E. et al. Inorg. Chem. **25**, 2085.
ob—Marchi A. et al. (1985) Inorg. Chem. **24**, 4744.
pb—Cingi M. B. et al. (1975) Inorg. Chim. Acta. **13**, 47.
qb—Bailey M. F. and Dahl L. F. (1965) Inorg. Chem. **4**, 1140.
rb—Anderegg G. et al. (1983) Helv. Chim. Acta **66**, 1593.
sb—Cotton F. A. and Shive L. W. (1975) Inorg. Chem. **14**, 2032.
tb—Cotton F. A. et al. (1982) Inorg. Chem. **21**, 1211.
ub—Bratton W. K. and Cotton F. A. (1970) Inorg. Chem. **9**, 789.
vb—Cotton F. A. et al. (1981) Inorg. Chem. **20**, 3051.
wb—Koz' min P. A. et al. (1983) Koord. Khim. **9**, 1114.
xb—Koz' min P. A. et al. (1982) Koord. Khim. **8**, 851.
yb—Koz' min P. A. et al. Koord. Khim. **7**, 1719.
zb—Cotton F. A. and Gage L. D. (1977) Nouv. J. Chim. **1**, 441.
ac—Cotton F. A. et al. (1980) J. Am. Chem. Soc. **102**, 1570.
bc—Koz' min P. A. et al. (1985) Koord. Khim. **11**, 1559.
cc—Koz' min P. A. et al. (1975) Koord. Khim. **1**, 248.
dc—Koz' min P. A. et al. Usp. Khim. **54**, 637.
ec—Kruichkov (1986) Dokl. Akad. Nauk. SSSR **288**, 893.
fc—Jurisson S. et al. (1984) Inorg. Chem. **23**, 4743.

Quantitative Study of the Structure-stability Relationship of $Tc^V O(\text{III})$ Complexes*

Abstract This paper describes the quantitative structure-stability relationship of $Tc^V ON_2 S_2$ compounds using a stability indicator based on the solid angle factor sum (SAS) for predicting the *in vitro* stability. The SAS values of six $Tc^V ON_2 S_2$ and one $Tc^V ON_4$ (d, l-HMPAO) complexes were calculated from their X-ray crystallography data. The rank order of *in vitro* stability of these compounds, as measured by ligand exchange reaction, is directly related to that of the SAS values. The SAS values are potentially useful for predicting stability and designing new ^{99m}Tc radiopharmaceuticals.

1 Introduction

In the preceding paper the basic concept of using solid angle factor sum (SAS) values, which are derived from the cone packing model, is described (Wei et al., 1990). Based on this method the SAS values of more than 100 Tc compounds were analyzed and the average SAS value was found to be within the region of 0.97 ± 0.13. The majority of stable Tc compounds fall within this region with one standard deviation. Statistically, half of the total Tc compounds are at Tc(V) valence state, and the majority of these Tc(V) complexes contain a $Tc^V O$ center core.

The Tc=O bond is a strong coordination bond due to the pairing of two $d_{x,y}$ electrons of Tc(IV) outer electron shell. The $Tc^V O$ is equivalent to stable metal ions with d^0 inert electronic configuration; therefore the $Tc^V O$ center core is amenable to various coordinating ligand atoms, ligand size, and atoms or functional groups attached to the complexing atoms. Many $Tc^V O$ complexes can be synthesized with different stability (SAS values), as in the examples shown below. The diversity of $Tc^V O$ complexes is suitable for designing new radiopharmaceuticals with a specific range of stability and biological characteristics.

Several series of $Tc^V O$ complexes, such as $Tc^V ON_2 S_2$ and $Tc^V ON_4$, have received a lot of attention in the past few years, and radiopharmaceuticals based on these ligand systems have been developed or are being investigated for human applications as brain perfusion or kidney functional imaging agents. The ^{99m}Tc complexes of diamidedithiol (DADS) and CO_2-DADS are anionic and chemically stable. They have been shown to be superior kidney imaging agents (Davison et al., 1981a; Fritzberg et al., 1981, 1982, 1984; Kasina et al., 1986; Klingensmith et al., 1982, 1984; McAfee et al., 1985; Davison et al., 1981b). In developing ^{99m}Tc labeled brain perfusion imaging agents, significant interest has been placed on the formation and stability of neutral and lipid-soluble Tc complexes. Several neutral and lipid-soluble $Tc^V ON_2 S_2$ complexes have been re-

---

* Copartner: H. F. KUNG, Y. WEI, S. PAN. Reprinted from *Appl Radiat. Isot.*, 1990, 41(8):773-781.

ported as potential brain perfusion imaging agents(Kung et al., 1984, 1985; Efange et al., 1987, 1988; Kung et al., 1989a, b; Lever et al., 1985; Scheffel et al., 1988; Lever et al., 1988).

The first clinically useful ^{99m}Tc brain perfusion imaging agent, ^{99m}Tc-d, l-HMPAO, a Tc^{V}ON$_4$ complex has reached nuclear medicine clinics for human use. It represents a milestone in the development of new ^{99m}Tc radiopharmaceuticals for brain imaging. Not only is the agent lipid-soluble and able to penetrate the intact blood-brain barrier(BBB), it also displays very high *in vivo* instability (it decomposes in seconds after i. v. injection, probably by reacting with intracellular glutathione) by which it is trapped inside the brain. The regional uptake and retention are generally related to the regional perfusion(Hung et al., 1988; Nechvatal et al., 1984; Jurisson et al., 1987; Leonard et al., 1986; Neirinckx et al., 1987; Podreka et al., 1987). HMPAO, an optically active ligand, was developed by modifying a neutral Tc^{V}ON$_4$ complex, [^{99m}Tc]PnAO (Volkert et al., 1984; Troutner et al., 1984). The d, l-isomer of [^{99m}Tc]HMPAO shows prolonged brain retention, while the meso-isomer displays higher *in vitro* stability and little *in vivo* brain retention (Leonard et al., 1986; Neirinckx et al., 1987).

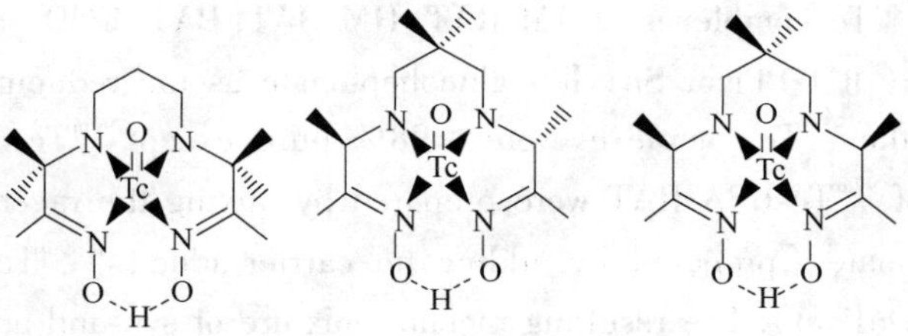

[^{99m}Tc]PnAO   [^{99m}Tc]d,l-HMPAO   [^{99m}Tc]meso-HMPAO

Scheme 1

A second neutral and lipid-soluble ^{99m}Tc brain perfusion imaging agent, [^{99m}Tc]l, l-ECD (ethylene cysteine dimer), is based on an N$_2$S$_2$ ligand. Interestingly, the agent is also an optically active ^{99m}Tc complex. The l, l-isomer shows the desired brain uptake and prolonged retention. The D, D-isomer displays high initial uptake but the brain retention is disappointing (L'eveille et al., 1988; Walovitch et al., 1988; Cheesman 1988, 1989; Holman et al., 1989).

This paper presents the application of SAS value calculation to a series of Tc^{V}ON$_2$S$_2$ compounds and Tc^{V}O-d, l-HMPAO. SAS values of several Tc^{V}ON$_2$S$_2$ complexes (based on published and unpublished X-ray crystallography data) are calculated, and the order of stability is confirmed by experimental results from exchange reactions between ligands, by which TcO complexes with various measurable differences in stability were formed. The application of this factor, SAS value, on predicting the sequence of the ligand exchange reaction and on designing new Tc radiopharmaceuticals is also discussed.

2 Theory and Methods

2.1 Calculation of solid angle factor sum(SAS)

As described in a previous paper(Wei et al., unpublished) the following equations are employed for the calculation of the SAS values for various Tc complexes:

FA: fan angle

$$FA = \sin^{-1}(R/ML) \tag{1}$$

Where, R is radii of coordinating atom; ML is metal to ligand bond length.

SAF: solid angle factor

$$SAF = (1 - \cos FA)/2 \tag{2}$$

SAS: solid angle factor sum

$$SAS = \sum_i SAF_i \tag{3}$$

2.2 Ligand exchange reaction

Ligands were prepared by methods reported previously (Kung et al., 1984, 1985; Efange et al. 1987, 1988; Kung et al., 1988a, b; Troutner et al., 1984), except ECD, which was a gift from Dr Cheesman (NEN/DuPont). All of the chemicals used in this report were of chemical grade.

No carrier-added ^{99m}Tc complexes of TM-BAT, HM-BAT, PAT, ECD and d-l-HMPAO were prepared by using Sn(II)-PPi or Sn(II)-glucoheptonate as the reducing agent for [^{99m}Tc] pertechnetate. All of the ^{99m}Tc complexes were ≥98% pure, except ^{99m}Tc-d,l-HMPAO (~90% pure). The isomers of ^{99m}Tc-BPA-BAT were prepared by mixing the racemic ligand BPA-BAT with ^{99m}Tc-glucoheptonate (prepared by adding no carrier-added [^{99m}Tc] pertechnetate to a Glucoscan; (NEN)/DuPont). The resulting racemic mixture of syn-and anti-isomers was separated by HPLC (reverse phase PRP-1 column, acetonitrile: 3,3-dimethyl-glutaric acid buffer, pH 7.0, 80 : 20) (Kung et al., 1989a, b).

Ligand exchange reactions between no carrier-added ^{99m}Tc complex (0.5mL) and competitive ligand (1mg) were performed (pH 6.8, 80°C). The pH of the solution was adjusted to 6.8 by adding the appropriate amount of 0.1N sodium hydroxide solution. At different time intervals, samples were removed and analyzed by an HPLC method (reverse phase PRP-1 column, acetonitrile: 3,3-dimethylglutaric acid buffer, 0.5mM, pH 7.0).

For the ligand exchange reaction between the specific ^{99m}Tc-BPA-BAT isomer and the ligand, purified no-carrier-added isomer (5mCi/0.5mL), either ^{99m}Tc-BPA-BAT(syn) or ^{99m}Tc-BPA-BAT(anti), was mixed with racemic BPA-BAT ligand at a concentration of 0.25mg/mL of the respective isomer. The exchange reaction was allowed to proceed at 80°C for 180min. Samples from the resulting mixture were injected directly into the HPLC and eluted in accordance with the system described above.

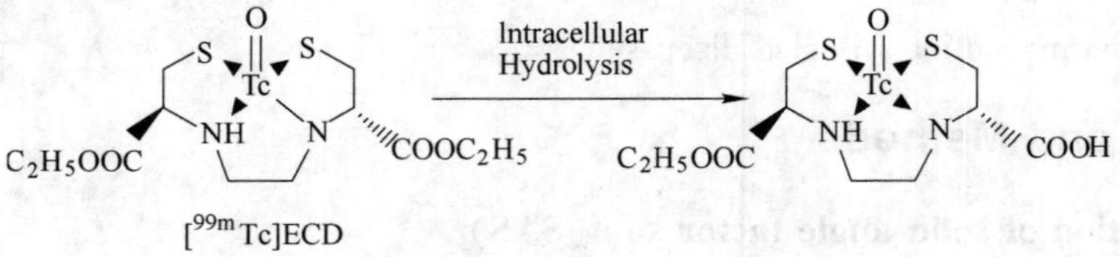

Scheme 2

3 Results

3.1 Calculation of SAS values for $Tc^V ON_2S_2$ complexes and $Tc^V O$-d, l-HMPAO

Calculations of the SAS values based on X-ray crystallography data are presented in Table 1. The chemical structures and the rank order of SAS values are shown in Fig. 1. The bond lengths and van der Waals radii of each complex are listed and the FA, SAF and SAS values were calculated with the equations listed above.

Table 1 SAS values for seven $Tc^V O$ complexes

Name	ML	R	R/ML	FA	SAF	SAS
Tc-TM-BAT[①]						
Tc(1)-S(1)	2.265A	1.80A	0.7946	52.62	0.1965	
Tc(1)-S(2)	2.281A	1.80A	0.7891	52.10	0.1929	
Tc(1)-O(1)	1.692A	1.40A	0.8276	55.86	0.2194	
Tc(1)-N(1)	1.899A	1.50A	0.7900	52.18	0.1935	
Tc(1)-N(2)	2.154A	1.50A	0.6965	44.15	0.1413	0.9436
Tc-PBA-BAT(syn)[②]						
Tc(1)-S(1)	2.297A	1.80A	0.7836	51.59	0.1894	
Tc(1)-S(2)	2.283A	1.80A	0.7884	52.04	0.1925	
Tc(1)-O(1)	1.671A	1.40A	0.8378	56.91	0.2270	
Tc(1)-N(1)	2.127A	1.50A	0.7052	44.85	0.1456	
Tc(1)-N(2)	1.920A	1.50A	0.7813	51.38	0.1879	0.9424
Tc-ECD[④]						
Tc(1)-S(1)	2.293A	1.80A	0.7850	51.72	0.1903	
Tc(1)-S(2)	2.273A	1.80A	0.7919	52.36	0.1947	
Tc(1)-O(1)	1.666A	1.40A	0.8403	57.18	0.2290	
Tc(1)-N(1)	2.168A	1.50A	0.6919	43.78	0.1390	
Tc(1)-N(2)	1.924A	1.50A	0.7796	51.23	0.1869	0.9399
Tc-PBA-BAT(anti)[③]						
Tc(1)-S(1)	2.284A	1.80A	0.788	51.998	0.19215	
Tc(1)-S(2)	2.276A	1.80A	0.791	52.28	0.1941	
Tc(1)-O(1)	1.681A	1.40A	0.8328	56.39	0.2233	
Tc(1)-N(1)	2.158A	1.50A	0.695	44.03	0.1405	
Tc(1)-N(2)	1.919A	1.50A	0.782	51.44	0.1884	0.9384
Tc-HM-BAT[①]						
Tc(1)-S(1)	2.283A	1.80A	0.7883	52.03	0.1924	
Tc(1)-S(2)	2.280A	1.80A	0.7894	52.13	0.1931	
Tc(1)-O(1)	1.688A	1.40A	0.8296	56.06	0.2208	
Tc(1)-N(1)	2.166A	1.50A	0.6926	43.83	0.1394	

Continued 1

Name	ML	R	R/ML	FA	SAF	SAS
Tc(1)-N(2)	1.938A	1.50A	0.7741	50.72	0.1835	0.9292
Tc-PAT[3]						
Tc(1)-S(1)	2.304A	1.80A	0.7813	51.38	0.1922	
Tc(1)-S(2)	2.276A	1.80A	0.7908	52.26	0.1940	
Tc(1)-O(1)	1.672A	1.40A	0.8373	56.86	0.2270	
Tc(1)-N(1)	2.213A	1.50A	0.6778	42.67	0.1324	
Tc(1)-N(2)	1.939A	1.50A	0.7736	50.68	0.1832	0.9287
Tc-d,l-HMPAO[5]						
Tc(1)-N(1)	2.075A	1.50A	0.7229	46.29	0.1545	
Tc(1)-N(2)	2.067A	1.50A	0.7257	46.53	0.1560	
Tc(1)-O(1)	1.682A	1.40A	0.8323	56.34	0.2229	
Tc(1)-N(3)	1.910A	1.50A	0.7853	51.75	0.1905	
Tc(1)-N(4)	1.910A	1.50A	0.7853	51.75	0.1905	0.9144

① Tulip T. New England Nuclear Inc./DuPont, personal communication.
② Kung et al. (1988a, b).
③ Mach (1989).
④ Cheesman et al. (1988).
⑤ Jurisson et al. (1987).

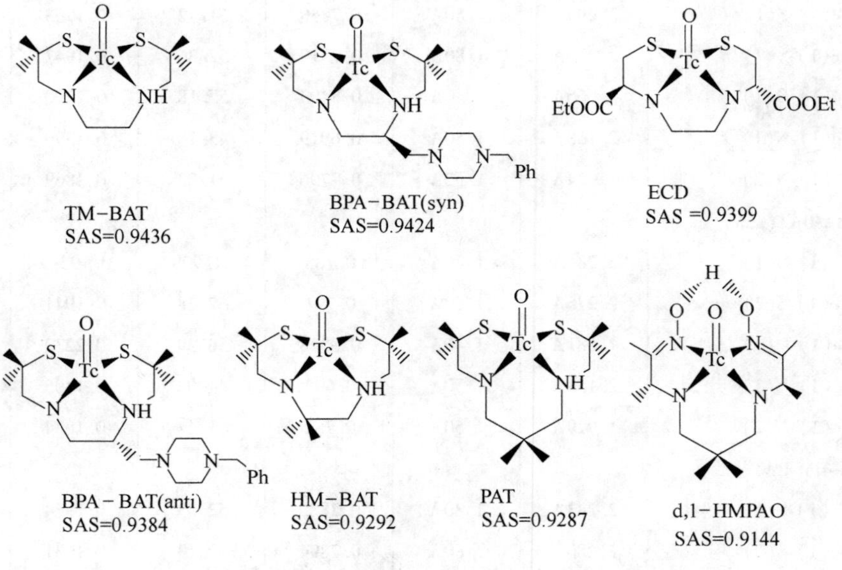

TM-BAT
SAS=0.9436

BPA-BAT(syn)
SAS=0.9424

ECD
SAS=0.9399

BPA-BAT(anti)
SAS=0.9384

HM-BAT
SAS=0.9292

PAT
SAS=0.9287

d,l-HMPAO
SAS=0.9144

Fig. 1 Chemical structures and SAS values of seven $Tc^V=O$ complexes
(The structures are arranged based on rank order of expected stability (SAS values))

3.2 Results of ligand exchange experiments

In order to verify that the SAS value is an indicator of *in vitro* stability of Tc complexes, relative *in vitro* stability of several Tc complexes was evaluated by a ligand exchange method. Stronger ligands which form Tc complexes of higher SAS values will be able to strip the Tc=O core from a weaker Tc complex (lower SAS value), while the same process will not occur when using a weaker ligand to compete with a stronger Tc complex. Since Tc-TM-BAT is the most stable compound in this group of compounds this is the ligand of choice for competitive ligand exchange reaction. The following exchange reactions were studied:

(1) ^{99m}Tc-Hm-BAT (SAS = 0.9292) vs TM-BAT (SAS = 0.9436) and vice-versa (Fig. 2).

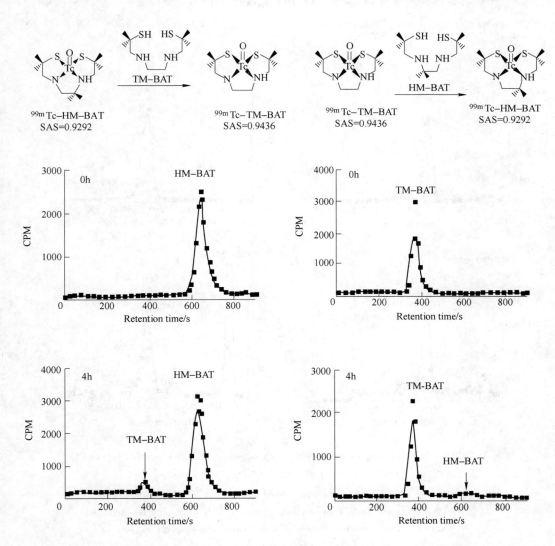

Fig. 2　HPLC profiles of samples before and after competitive exchange reaction between HM-BAT (SAS = 0.9292) and TM-BAT (SAS = 0.9436)

(Formation of ^{99m}Tc-TM-BAT is more pronounced than that for ^{99m}Tc-HM-BAT, indicating that ^{99m}Tc-TM-BAT is more stable. The arrow indicates the new peak after competitive exchange reaction)

(2) ^{99m}Tc-ECD (SAS = 0.9399) vs TM-BAT (SAS = 0.9436) and vice-versa (Fig. 3).
(3) ^{99m}Tc-PAT (SAS = 0.9287) vs TM-BAT (SAS = 0.9436) (Fig. 4).
(4) ^{99m}Tc-HMPAO (SAS = 0.9144) vs TM-BAT (SAS = 0.9436) (Fig. 5).
(5) ^{99m}Tc-BPA-BAT (syn) (SAS = 0.9424) vs BPA-BAT (racemic) and ^{99m}Tc-BPA-BAT (anti) (SAS = 0.9384) vs BPA-BAT (racemic) (Fig. 6).

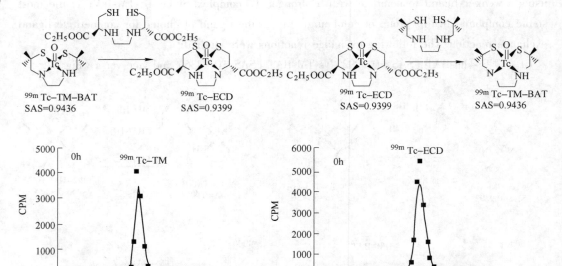

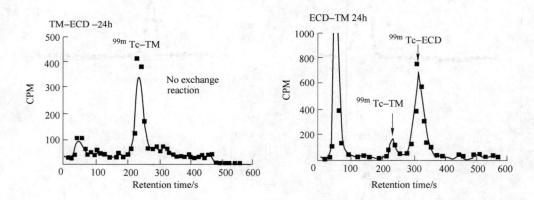

Fig. 3 HPLC profiles of samples before and after competitive exchange reaction between ECD (SAS = 0.9399) and TM-BAT (SAS = 0.9436)

(Formation of ^{99m}Tc-TM-BAT is more pronounced than that for ^{99m}Tc-ECD, indicating that ^{99m}Tc-TM-BAT is more stable. There was decomposition of ^{99m}Tc-ECD during the exchange reaction. The arrow indicates the new peak after competitive exchange reaction)

4 Discussion

The order of stability, based on SAS values (see Fig. 1), for the Tc complexes evaluated in this

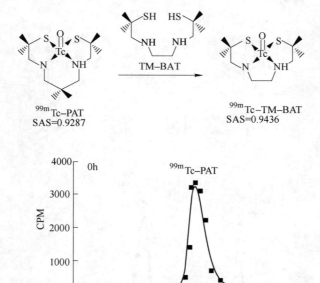

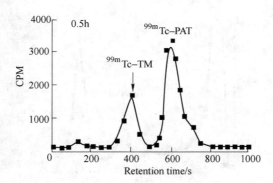

Fig. 4 HPLC profiles of samples before and after competitive exchange reaction between ^{99m}Tc-PAT (SAS = 0.9287) and TM-BAT (SAS = 0.9436), indicating that ^{99m}Tc-TM-BAT is more stable

(The arrow indicates the new peak after competitive exchange reaction)

paper is:

TM-BAT > BPA-BAT (syn) > ECD > BPA-BAT (anti) > HM-BAT > PAT > HMPAO
0.9436   0.9424   0.9399   0.9384   0.9292   0.9287   0.9144

The order of stability of Tc complexes can be verified by the ligand exchange reaction. The ligands which form stronger complexes with higher SAS values can successfully compete with weaker Tc complexes, resulting in the transfer of the Tc=O core from the weaker to the stronger ligand.

The HPLC profiles in Fig. 2 indicate that competitive ligand exchange between ^{99m}Tc-HM-BAT (SAS = 0.9292), a weaker complex, and a stronger ligand, TM-BAT (SAS = 0.9436),

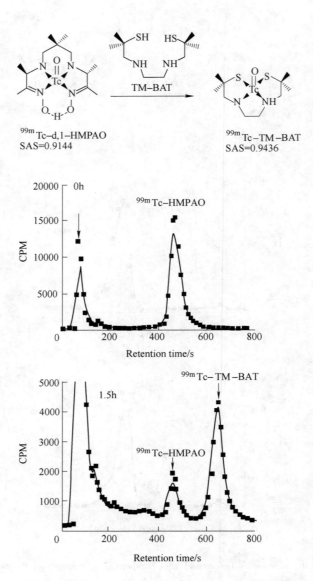

Fig. 5 HPLC profiles of samples before and after competitive exchange reaction between
^{99m}Tc-HMPAO (SAS = 0.9144) and TM-BAT (SAS = 0.9436),
indicating that ^{99m}Tc-TM-BAT is more stable

(There was significant decomposition during the exchange reaction. The decomposition may be catalyzed by the free mercapto group of N_2S_2 ligand. The arrow indicates the new peak after competitive exchange reaction)

can take place. At 4h, a new peak representing the formation of ^{99m}Tc-TM-BAT was observed on the profile (Fig. 2). The reverse is not true. The weaker ligand, HM-BAT, cannot compete with ^{99m}Tc-TM-BAT (a stronger complex); therefore the HPLC profile essentially showed the absence of formation of ^{99m}Tc-HM-BAT.

The exchange reaction between ECD and TM-BAT confirmed the same observation: the more stable complex, ^{99m}Tc-TM-BAT, will form, at the expense of the less stable complex, ^{99m}Tc-ECD

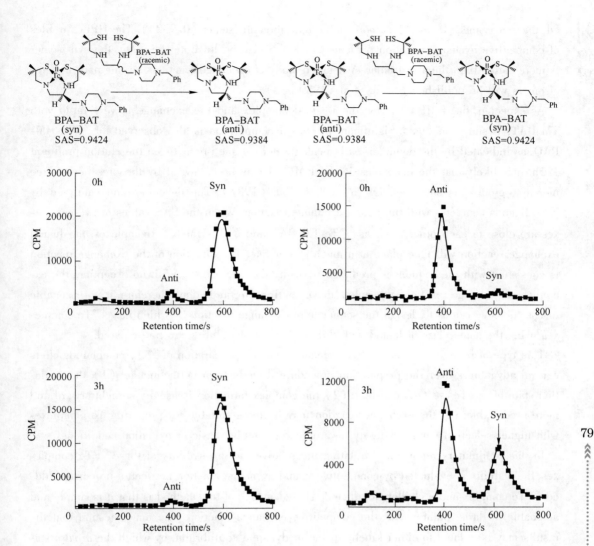

Fig. 6 HPLC profiles of samples before and after competitive exchange reaction between ^{99m}Tc-BPA-BAT (syn) (SAS = 0.9424) and ^{99m}Tc-BPA-BAT (anti) (SAS = 0.9384)
(The arrow indicates the new peak after competitive exchange reaction)

(Fig. 3), while the reverse reaction showed no exchange reaction. In this experiment, ^{99m}Tc-ECD displayed *in vitro* decomposition upon prolonged heating. The water soluble component with a retention time <100s probably results from hydrolysis of the ester groups. The acid derivatives are unlikely to compete with N_2S_2 ligands. The decomposition did not affect the conclusion of the rank order of *in vitro* stability based on SAS values. Similarly, the exchange reaction between ^{99m}Tc-PAT (SAS = 0.9287) and the stronger ligand TM-BAT is also demonstrated (Fig. 4). The exchange reaction between the syn-and anti-isomers of BPA-BAT proves to be very intriguing, since these two isomers can be formed simultaneously when a racemic ligand is used for competing with the individual ^{99m}Tc labeled isomer. Based on the SAS values calculated, the syn-isomer (0.9424) is more stable than the anti-isomer (0.9384). The HPLC profiles of competitive exchange reaction presented in Fig. 6 clearly illustrate the point: the syn-isomer is more satble than the anti-isomer. A detailed study of exchange reactions on this pair of stereoisomers will be published (Kung et al., 1989 a, b).

As expected, the HMPAO forms the weakest complex and the exchange is very rapid with TM-BAT, as shown in Fig. 5. Significant *in vitro* decomposition is also observed for ^{99m}Tc-HMPAO, as indicated by the dominant early peak (water soluble product) on the elution profile at 1.5h. It is likely that the decomposition of HMPAO may be catalyzed by the presence of free mercapto groups, as suggested by Neirinckx et al. (1987). The decomposition catalyzed by N_2S_2 ligands competes with the ligand exchange reaction. When the SAS values of two complexes are close to each other, such as ^{99m}Tc-TM-BAT and ^{99m}Tc-HM-BAT complexes, the ligand exchange reaction will take place at a much slower rate. The direction of the exchange reaction is consistent with the sequences predicted by using the SAS values. With no exception, the examples of exchange reactions presented above confirm the prediction. Therefore, it is reasonable to use the SAS values to design the sequence of exchange reaction by which new ^{99m}Tc complexes using the same class of ligand with different levels of stability can be prepared.

This type of properly chosen exchange reaction for the preparation of ^{99m}Tc compounds offers several advantages over the preparation procedure directly from pertechnetate: (1) the oxidation state of the $Tc^V=O$ is constant; (2) the complex formation is usually straightforward and more predictable; (3) the exchange reaction may be more suitable for preparing Tc complexes with higher selectivity and sensitivity towards certain acidic, basic or oxidation conditions.

In this preliminary report on correlating the relative *in vitro* stability of these ^{99m}Tc complexes, the stability is evaluated by competitive ligand exchange reaction between a no-carrier-added ^{99m}Tc complex and a challenging ligand. The advantage of this method is that it is simple and amenable for large numbers for the competitive exchange reaction. One may question that the results may be related to either kinetic or thermodynamic equilibrium, by which the *in vitro* stability of these ^{99m}Tc complexes is determined. A more precise measurement of *in vitro*- stability is the formation constant (K_d). All of the ligands studied in this paper form a neutral complex with $Tc^V O$. During the complex formation three equivalents of hydrogen ions were released. By titrating the pH change, it may be possible to determine the formation constants for each complex. The formation constant is probably more reliable as an indicator for *in vitro* stability. However, technically it is much more difficult to achieve. This work is currently ongoing in our laboratory. Since our ligand exchange experiments are carried out under the same conditions, it is reasonable to assume that the relative stability is reflected by the direction exchange reaction. Using this type of study, the relative order of stability can be determined.

The preliminary results of exchange reactions between ligands of different strengths presented here reaffirm the validity of the cone packing model and its application to predicting the stability of Tc complexes. Furthermore, the SAS values are potentially useful in assisting the design of new radiopharmaceuticals based on ^{99m}Tc. It may be appropriate to estimate the SAS values and estimate the stability of various Tc complexes before an extensive effort is made on the synthesis and characterization of a new series of Tc complexes. On the other hand, it is also feasible to re-evaluate the unsuccessful "old" Tc complexes reported previously and the potential effects of instability on the unsuccessful radiopharmaceuticals.

There are other factors that can be applied for determining the stability and relative potency of a ligand exchange reaction, such as fan angle at the transvacancy position (opposite to the Tc=O bond). This angle is a reflection of the space available for the sixth ligand to form a coordinate covalent bond with Tc=O complexes. This will be discussed in a future paper. In addition, the primary effect of the stability of Tc complexes is due to the coordinating ligand atoms. The atoms or functional groups surrounding the complexing atoms also have a significant effect on stability and, more importantly, the biological distribution of the Tc compounds. Examples shown above, such as the different SAS values for the same type of $Tc^V ON_2S_2$ complexes, clearly demonstrated that even though the primary complexing atoms are the same—N_2S_2 atoms—the SAS values and the *in vitro* stability are not equal. As a consequence, the *in vivo* biodistribution properties of these N_2S_2 complexes were not equal. The secondary effects on the stability by the atoms or functional groups not directly involved in coordinate covalent bond formation are currently under study.

In conclusion, the stability indicators (SAS values) of six known $Tc^V ON_2S_2$ complexes and one $Tc^V ON_4$ complex were calculated based on their X-ray crystallography data. The rank order of *in vitro* stability is confirmed by competitive ligand exchange reaction. The SAS values are potentially useful for predicting *in vitro* stability and designing new ^{99m}Tc radiopharmaceuticals.

Acknowledgements

This work is supported by a grant awarded by the National Institute of Health (NS-18509). The authors thank Dr T. Tulip of NEN/DuPont and Dr S. Jurisson of Squibb & Sons for providing X-ray crystallography data. The authors also thank Mrs Fiona Chapman and Ms Heather Cullen for their assistance in preparing this manuscript.

References

Cheesman E. H., Blanchette M. A., Ganey M. V., Mehen L. J., Miller S. J. and Watson A. D. (1988a) Technetium-99m ECD: ester-derivativized diamine-dithiol Tc complexes for imaging brain perfusion. J. Nucl. Med. 29, 788 (Abstract 197).

Cheesman E. H., Blanchette M. A., Calabrese J. C., Ganey M. V., Mehen L. J., Morgan R. A., Walovitch R. C., Watson A. D., Williams S. J. and Miller S. J. (1989) J. Labelled Compd. Radiopharm. 421 (Abstract 180).

Davison J., Jones A., Orvig C. et al. (1981a) A new class of oxotechnetium (5+) chelate complexes containing a $TcON_2S_2$ core. Inorg. Chem. 20, 1629.

Davison A., Jones A. G., Orvig C. et al. (1981b) A new class of oxotechnetium (5+) chelate complexes containing a $TcON_2S_2$ core. Inorg. Chem. 20, 1632.

Efange S. M. N., Kung H. F., Billings J., Guo Y.-Z. and Blau M. (1987) [^{99m}Tc] Bis (aminoethanethiol)

(BAT) complexes with amine sidechains-Potential brain perfusion imaging agents for SPECT. J. Nucl. Med. 28, 1012.

Efange S. M. N., Kung H. F., Billings J. and Blau M. (1988) The synthesis and biodistribution of [^{99m}Tc] piperidinyl bis(aminoethanethiol) complexes: potential brain perfusion imaging agents for SPECT. J. Med. Chem. 31, 1043.

Fritzberg A. R. Klingensmith W. C., Whitney W. P. et al. (1981) Chemical and biological studies of Tc-99m N, N-bis(mercaptoacetamido) ethylenediamene: a potential replacement for I-131 iodohippurate. J. Nucl. Med. 22, 258.

Fritzberg A. R., Kuni C. C., Klingensmith W. C. et al. (1982) Synthesis and biological evaluation of Tc-99m N, N-bis (mercaptoacetyl)-2, 3-diaminopropanoate: a potential replacement for [131I] o-iodohippurate. J. Nucl. Med. 23, 592.

Fritzberg A. R., Kasina S., Eshima S. et al. (1984) Synthesis and evaluation of N_2S_2 complexes of Tc-99m as renal function agent. J. Nucl. Med. 25, 16 (Abstract).

Holman B. L., Hellman R. S., Goldsmith S. J. et al. (1989) Biodistribution, dosimetry, and clinical evaluation of technetium-99m ethyl cysteinate dimer in normal subjects and in patients with chronic cerebral infarction. J. Nucl. Med. 30, 1018.

Hung J. C., Corlija M., Volkert W. A. and Holmes R. A. (1988) Kinetic analysis of technetium-99m d, l-HMPAO decomposition in aqueous media. J. Nucl. Med. 29, 1568.

Jurisson S., Aston K. Fair C. K. et al. (1987) Effect of ring size on properties of technetium amine oxime complexes. Inorg. Chem. 26, 3576.

Kasina S., Fritzberg A. R., Johnson D. et al. (1986) Tissue distribution properties of technetium-99m-diamide-dimercaptide complexes and potential use as renal radiopharmaceuticals. J. Med. Chem. 29, 1933.

Klingensmith W, C., Gerhold J. P., Fritzberg A. R. et al. (1982) Clinical comparison of Tc-99m N, N'-bis (mercaptoacetamido) ethylenediamine and [131I] ortho-iodohippurate for evaluation of renal tubular function: concise communication. J. Nucl. Med. 23, 377.

Klingensmith W. C., Fritzberg A. R., Spitzer V. M. et al. (1984) Clinical evaluation of Tc-99m N, N'-bis(mercaptoacetyl)-2,3-diaminopropanoate as a replacement for I-131 hippurate: concise communication. J. Nucl. Med. 25, 42.

Kung H. F., Molnar M., Billings J., Wicks R. and Blau M. (1984) Synthesis and biodistribution of neutral lipid-soluble Tc-99m complexes which cross the blood-brain barrier. J. Nucl. Med. 25, 326.

Kung H. F., Yu. C-C., Billings J., Molnar M. and Blau M. (1985) Synthesis of new bis-aminoethanethiol (BAT) derivatives: possible ligands for Tc-99m brain imaging agents. J. Med. Chem. 28, 1280.

Kung H. F., Liu B. L. and Pan S. (1989a) Kinetic study of ligand exchange reaction between ^{99m}Tc-glucoheptonate and N-benzyl-N-methyl-piperazinyl-bis(aminoethanethiol) (BPA-BAT). Appl. Radial. Isot. 40, 677.

Kung H. F., Guo Y. -Z., Yu C. -C., Billings J., Subramanyam V. and Calabrese J. (1989b) New brain perfusion imaging agents based on Tc-99m bis-aminoethanethiol (BAT) complexes: stereoisomers and biodistribution. J. Med. Chem. 32, 433.

Leonard J. P., Nowotnik D. P. and Neirinckx R. D. (1986) Tc-99m d, l HMPAO: A new radiopharmaceutical for imaging regional brain perfusion using SPECT—A comparison with iodine-123 HIPDM. J. Nucl. Med. 27, 1819.

L' eveille J., Demonceau G., Rigo P., De Roo M., Taillefer R., Burgess B. A., Morgan R. A. and Walovitch R. C. (1988) Brain tomographic imaging with Tc-99m-ethyl cysteinate dimer (Tc-ECD): a new stable brain perfusion agent. J. Nucl. Med. 29, 758 (Abstract 73).

Lever S. Z., Burns H. D., Kervitzky T. M., Goldfarb H. W., Woo D. V., Wong D. F. and Epp L. A. (1985) Design, preparation and biodistribution of a Technetium-99m triaminodithiol complex to access regional cerebral blood flow. J. Nucl. Med. 26, 1287.

Lever S. Z., Baidoo K. E., Kramer A. V. and Burns H. D. (1988) Synthesis of a novel bifunctional chelate designed for labeling proteins with technetium-99m. Tetrahedron Lett. 29, 3219.

McAfee J. G., Subramanian G., Schneider R. et al. (1985) Technetium-99m DADS complexes as renal function and imaging agents: II. Biological components with I-131 hippuran. J. Nucl. Med. 26, 375.

Nechvatal G., Canning L. R., Cummings S. et al. (1984) In Technetium in Chemistry and Nuclear Medicine (Eds Micolini M., Bandoli G. and Mazzi U.) Vol. 2. Raven Press, New York.

Neirinckx R. D., Canning L. R., Piper I. M. et al. (1987) Technetium-99m-d-l HM-PAO: A new radiopharmaceutical for SPECT imaging of regional cerebral blood perfusion. J. Nucl. Med. 28, 191.

Podreka I., Suess E., Goldenberg G. et al. (1987) Initial experience with Technetium-99m-HM-PAO brain SPECT. J. Nucl. Med. 28, 1657.

Scheffel U., Goldfarb H. W., Lever S. Z., Gungon R. L., Burns H. D. and Wagner H. N. Jr (1988) Comparison of technetium-99m aminoalkyl diaminodithiol analogs as potential brain blood flow imaging agents. J. Nucl. Med. 29, 73.

Troutner D. E., Volkert W. A., Hoffman T. J. and Holmes R. A. (1984) A neutral lipophilic complex of ^{99m}Tc with a multidentate amine oxime. Int. J. Appl. Radiot. Isot. 35, 467.

Volkert W. A., Hoffman T. J., Seger R. M. and Holmes R. A. (1984) ^{99m}Tc-propylene amine oxime (^{99m}Tc-PnAO); a potential brain radiopharmaceutical. Eur. J. Nucl. Med. 9, 511.

Walovitch R. C., Makuch J., Knapik G., Watson A. D. and Williams S. J. (1988) Brain retention of Tc99m-ECD is related to in vivo metabolism. J. Nucl. Med. 29, 747 (Abstract 26).

Wei Y., Liu B.-L. and Kung H. F. Quantitative study of the structure-stability relationship of Tc complexes. Appl. Radiat. Isot. 41, 763.

Stability of Isomers of ^{99m}Tc-HMPAO and Complex of ^{99m}Tc-CBPAO Based on CNDO/2 Method*

Abstract Formation energy, total energy of system and net charge of isomers of ^{99m}Tc-HMPAO have been calculated by CNDO/2 method. The computed results indicate that the order of stability is ^{99m}Tc-meso 2-HMPAO > ^{99m}Tc-d. l-HMPAO > ^{99m}Tc-meso 1-HMPAO. After cyclization of d. l-HMPAO, the calculated results show that stability of ^{99m}Tc-d. l-CBPAO is increased rapidly. This study reveals structural features that may be employed to account for disparate pharmacokinetics of the meso and d. l-HMPAO ^{99m}Tc-radiopharmaceuticals.

Key words stability of isomers, ^{99m}Tc-HMPAO structural configuration, formation energy

1 Introduction

It is well known that 3, 6, 6, 9-tetramethyl 4, 8-diazaundecane-2, 10-dione dioxime (HMPAO) is now widely used in nuclear medicine as a potential brain imaging agent. HMPAO, with two asymmetric carbon, produces d. l and meso isomers as shown in Fig. 1.

The isomers of ^{99m}Tc-HMPAO reveal dramatic differences of biodistribution in brain uptake and retention that may be caused by structural configuration and various stability of isomers. In this paper, we use three indicators to describe the stability of isomers and complex of ^{99m}Tc-CBPAO based on corn packing model[1] or CNDO/2 method[2].

The computed results indicated that the order of stability of isomers is as follows: ^{99m}Tc-meso 2-HMPAO > ^{99m}Tc-d. l-HMPAO > ^{99m}Tc-meso 1-HMPAO. The difference of stability between ^{99m}Tc-d. l-HMPAO and ^{99m}Tc-meso-HMPAO may play an important role in the understanding of mechanistic information about the in vivo different localization properties. d. l-HMPAO can also be stabilized in vitro by substituting a cyclo-butyl group on the PAO backbone. After cyclization of d. l-HMPAO, the stability of ^{99m}Tc-d. l-CBPAO is increased rapidly. This study reveals structural features that may be useful to account for disparate pharmacokinetic of the meso and d, l-HMPAO ^{99m}Tc-radiopharmaceuticals.

2 Outline of Calculation Method

2.1 Cone packing model[1]

To facilitate discussion by this model the relation between the technetium atom and its ligand atom is illustrated in Fig. 2.

In order to investigate the stability of ^{99m}Tc-HMPAO isomers, the SAS value is calculated

---

* Copartner: Meng Zhaoxing. Reprinted from *Radiochimica Acta*, 1993, 63: 217-220.

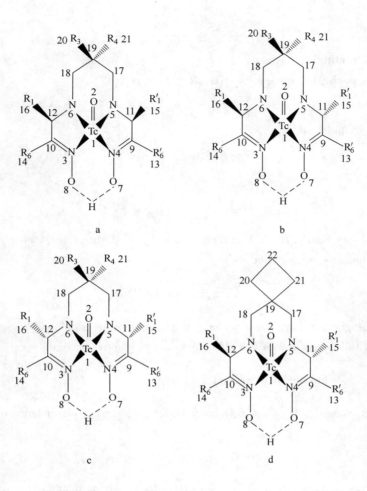

Fig. 1 ^{99m}Tc-meso 1-HMPAO(a), ^{99m}Tc-d,l-HMPAO(b), ^{99m}Tc-meso 2-HMPAO(c), ^{99m}Tc-d,l-CBPAO(d)

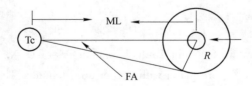

Fig. 2   Definition of fan angle (FA) for a ^{99m}Tc-complex
(Ligand atom (N). R—Van der Waals Radius; ML—Bond Length between Metal and Ligand)

from Van der Waals radii and bond lengths between coordinating atom and Tc as described in a previous paper[1]. The definition of fan angle (FA), solid angle factor (SAF) and solid angle factor sum (SAS) are given as following Eq. 1-Eq. 3. ML(metal to ligand bond length) values obtained directly from X-ray crystallography[3].

FA: fan angle $\qquad$ $FA = \sin^{-1}(R/ML)$ $\qquad$ (1)

SAF: solid angle factor $\qquad$ $SAF = (1 - \cos FA)/2$ $\qquad$ (2)

SAS: solid angle factor sum $\quad\quad \text{SAS} = \sum_i \text{SAF}_i \quad\quad\quad (3)$

2.2 CNDO/2 method[2]

In CNDO/2 approach, HF equation is given as:

$$\sum_\nu (F_{\mu\nu} - \varepsilon_i \delta_{\mu\nu}) c_{v_i} = 0$$

Total energy of the system is:

$$E = E_{\text{elec}} + \sum_{A<B} Z_A Z_B R_{AB}^{-1}$$

where

$$E_{\text{elec}} = \frac{1}{2} \sum_{\mu\nu} P_{\mu\nu}(H_{\mu\nu} + F_{\mu\nu})$$

where, $P_{\mu\nu}$ is density matrix; $H_{\mu\nu}$ is Hamilton matrix; $F_{\mu\nu}$ is Hartree-Fock matrix.

Precision of iterative computation:

$$\frac{|E_{\text{elec}}^{K-1} - E_{\text{elec}}^{K}|}{|E_{\text{elec}}^{K+1}|} \leqslant 10^{-6}$$

The computer programs for quantum mechanical calculations are validated by comparing calculation results of this program with the published data of known compounds[6].

$$\text{Formation energy of Tc complexes} = \sum_{i=1}^{n} \frac{e_{\text{Tc}}^+ \cdot e_i^-}{\Gamma_i} \quad\quad (4)$$

where, e_{Tc}^+ is net charge of Tc; e_i^- is net charge of ligand atom i; Γ_i is distance between Tc and ligand atom i.

$$\text{Net Charge } \Delta e = Z_A - P_{AA} \quad\quad (5)$$

where, Z_A is atomic core number; $P_{AA} = \sum_\mu^A (P_{\mu\mu}^\alpha + P_{\mu\mu}^\beta)$, $P_{\mu\mu}^\alpha$ is Population number of φ_μ of A atom of α electron, $P_{\mu\mu}^\beta$ is Population number of φ_μ of A atom of β electron.

3 Results

Selected CNDO parameters are listed in Table 1, parameters of C, N, O and H are taken from literature, the parameters of Tc are taken with proper adjustments. The structure parameters of ^{99m}Tc-d, l-HMPAO and ^{99m}Tc-meso-HMPAO are taken from Ref. [3] and listed in Table 2. The comparison of SAS values, formation energy and total energy of system are listed in Table 3. The net charge distribution of methyl group and the interaction energy between different group within ^{99m}Tc-HMPAO and ^{99m}Tc-CBPAO are listed in Table 4 and Table 5.

Table 1 CNDO/2 parameters

Atom	n	ξ	$-(I_\mu + A_\mu)/2$	$-\beta_A^0$
H	1s	1.2000	7.1710	4.5000
O	2s	2.2000	24.8175	15.5000
	2p	1.9750	7.7955	15.5000

Continued 1

Atom	n	ξ	$-(I_\mu + A_\mu)/2$	$-\beta_A^0$
N	2s	1.9237	19.2070	12.5000
	2p	1.9170	6.9620	12.5000
C	2s	1.6250	14.0510	21.0000
	2p	1.6250	5.5720	21.0000
Tc	5s	1.2460	3.6500	20.1600
	5p	0.1370	0.9060	3.3530
	4d	2.3230	4.4270	21.4610

Table 2 Bond length between Tc and coordination atoms of ^{99m}Tc-HMPAO

Type of HMPAO	Coordination atom				
	O_2	N_3	N_4	N_5	N_6
	Bond length/Å				
d,l-HMPAO	1.6825	2.0736	2.0676	1.9115	1.9106
meso 1-HMPAO	1.6773	2.0824	2.0724	1.9124	1.9164
meso 2-HMPAO	1.6704	2.0804	2.0794	1.9184	1.9044

Note: 1Å = 0.1nm.

Table 3 Comparison of SAS values formation energy and total energy of system

	d,l-HMPAO	meso 1-HMPAO	meso 2-HMPAO	d,l-CBPAO
Netcharge distribution				
Tc:	0.978	0.875	1.081	1.577
O:	-0.114	-0.162	-0.163	-0.469
N_1:	-0.236	-0.440	-0.241	-0.176
N_2:	-0.176	-0.205	-0.301	-0.007
N_3:	-0.059	-0.080	-0.405	-0.441
N_4:	-0.321	-0.060	-0.008	-0.200
Formation energy	0.4575	0.4209	0.6312	0.7563
Total energy of system	-179.44320	-178.5840	-180.3409	—
SAS values	0.9140	0.9114	0.9147	—

Table 4 Net charge distribution of methyl group of ^{99m}Tc-HMPAO and ^{99m}Tc-CBPAO

Name	$C_{13}(R_5')$	$C_{14}(R_5)$	$C_{15}(R_1')$	$C_{16}(R_1)$	$C_{21}(R_4)$	$C_{20}(R_3)$	$C_{22}(C)$
d,l	-0.280	1.907	0.106	0.341	-0.261	0.303	—
meso 1	1.519	0.811	-0.307	-0.744	-0.555	-0.234	—
meso 2	-1.523	-0.792	-0.507	-0.226	-0.936	1.082	—
C-d,l	-0.210	1.909	0.108	-0.481	0.152	0.655	-1.051

Table 5 Energy of interaction between R_3 and $R_4(R_1'O)$

Site of interaction	d,l-HMPAO	meso 1-HMPAO	meso 2-HMPAO	d,l-CBPAO
R_3 action with O(TcO)	-0.010100	0.010708	-0.049820	-0.089823
R_3 action with R_1	0.023063	0.038085	-0.054446	-0.070324
R_3 action with R_1'	0.007169	0.015719	-0.120038	0.015790

4 Discussion

It is clearly shown in Table 3, the order of stability of isomers, the SAS values as calculated by the conepacking model, is parallel to that obtained by quantum mechanical calculation based on the CNDO/2 method, although the difference of SAS values among isomers is quite small. The order of stability is ^{99m}Tc-meso 2-HMPAO > ^{99m}Tc-d, l-HMPAO > ^{99m}Tc-meso 1-HMPAO. This fact also can be explained by their different configuration and conformation[4]. From the structural data for the isomeric TcO(HMPAO) by Jurisson[3], the stick drawings of TcO(HMPAO) isomers were shown in Fig. 1. It is obvious from the figures that ^{99m}Tc-meso 1-HMPAO is a "boat" conformation and ^{99m}Tc-meso 2-HMPAO is a "chair" conformation. In the former conformation, the central carbon atom of the propyl backbone is directed up and towards the technetium oxo group and in the later conformation that is directed down and away from the TcO group. In the case of Fig. 3, the interaction between $R_3(R_4)$ and TcO is more stronger than that in Fig. 4.

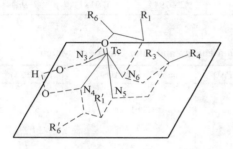
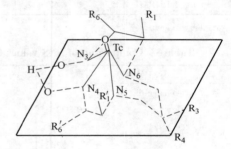

Fig. 3 ^{99m}Tc-meso 1-HMPAO   Fig. 4 ^{99m}Tc-meso 2-HMPAO

The total energy of ^{99m}Tc-meso 2-HMPAO is more higher than that of ^{99m}Tc-meso 1-HMPAO, which indicates that meso-2 is more stable as compared with meso-1. ^{99m}Tc-d, l-HMPAO is a "boat" conformation in Fig. 5 similar with ^{99m}Tc-meso 1-HMPAO. But the group of R'_1 is directed up and towards the TcO group. The total energy of ^{99m}Tc-d, l-HMPAO is higher than that of ^{99m}Tc-meso 1-HMPAO. ^{99m}Tc-d, l-HMPAO is more stable than ^{99m}Tc-meso 1-HMPAO and unstable as compared with ^{99m}Tc-meso 2-HMPAO. From the vacant technetium coordination point of view, the stable form of meso-2 provides much more hindered than that of unstable form of d, l-HMPAO and thus alter the efficiency of intracellular trapping[4].

It is also clearly shown from our calculation results that after cyclization of d, l-HMPAO. The formation energy of ^{99m}Tc-d, l-CBPAO is much higher than the isomers of ^{99m}Tc-HMPAO. The stability of ^{99m}Tc-d, l-CBPAO is increased rapidly in Fig. 6. The both two agents exhibit qualitatively similar biodistribution, but the brain uptake and retention of the CBPAO agent is about 30%-50% less than that of the HMPAO agent. This result can be understood in terms of the inherently greater stability of ^{99m}Tc-d, l-CBPAO relative to ^{99m}Tc-d, l-HMPAO.

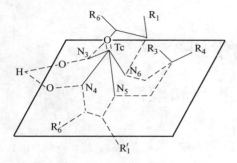
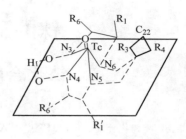

Fig. 5 ^{99m}Tc-d, l-HMPAO 　　　　Fig. 6 ^{99m}Tc-d, l-CBPAO

This research project was supported by The P. R. C. National Natural Science Foundation.

References

[1] Yi Wei, Liu Bo-Li. Kung, H. F. : Quantitative study of the structure-stability relationship of Tc-complexes. Int. J. Appl. Radiat. Isot. 41. 763-771 (1990).

[2] Liu Bo-Li. Meng Zhao-Xing, Kung, H. F. : Quantitative study of the structure-stability relationship of TcO (V) complexes. 1Xth International Symposium on Radiopharmaceutical Chemistry. 6-10th. April (1992). Abstracts: 21-23.

[3] Jurisson, S. , Schlemper, E. O. , Troutner, D. E. , et al. : Synthesis. Characterization and X-ray Structural Determinations of Technetium (V)-OXO-Tetradentate Amine Oxime Complexes, Inorg. Chem. 25, 543-549 (1986).

[4] Willson, G. , Flor. Mark. , Green, A. : Comparison of the isomers of Tc-HMPAO, Purdue University (1987).

[5] Meng Zhao-Xing, Liu Bo-Li: A study on the location of Tritium-labelling of Harringtonine by Wilzbach method. J. Nucl. Radiochem. 9(4) ,224-231 (1987) (in Chinese).

[6] Liu Bo-Li, Meng Zhao-Xing, Zhang Xiao-Xing, Li Ling: A study on the structure-activity relationship of cholesterol analogues as adrenal imaging agent with CNDO/2 method, J. Isot. 5 (1) ,1-7 (1992) (in Chinese).

铼的配位化学研究

Ⅰ. 铼化合物的堆积饱和规律*

摘　要　将锝化合物稳定性研究中引入的改进的堆积模型用于铼化合物的稳定性研究。在查得406种不含金属—金属键的化合物结构基础上，对其立体角系数和（SAS）进行了计算和分析，发现铼化合物的稳定中心（平均立体角系数和$\overline{SAS}$）为0.951，标准偏差（σ）为0.084，其2σ不大于一个常见配位原子的立体角系数（SAF），这反映了铼周围配位原子堆积的限度。

关键词　堆积模型　铼化合物　稳定性　归一化半径

近十年来，^{186,188}Re越来越引起核医学界的重视，它们在肿瘤、骨癌骨痛、风湿和类风湿关节炎的治疗上有着广阔的应用前景[1]，骨痛的治疗已发展到临床应用阶段。铼的放射性药物多数是铼的配合物，有关铼药物的设计基本上处于仿锝阶段。由于锝、铼的化学性质相似，锝化学的研究方法也适用于铼，但铼有稳定核素，因此，有关铼化合物的实验数据资料更为丰富，为铼化学研究提供了方便，其研究成果反过来又可指导锝化学的研究。20世纪60年代以来，单晶四元衍射技术已经普及，从而获得了大量铼化合物的结构参数。本文在收集1992年以前406种铼化合物的晶体数据[2]基础上，应用锝化合物稳定性研究中的改进的堆积模型[3~7]，进行铼配合物的稳定性研究。由于绝大多数放射性药物不含金属—金属键，因而有关双核配合物及原子簇配合物将另外处理。

1　归一化半径的计算

根据李醒夫[8,9]等建立的空间堆积模型，一个配位原子L的立体角系数SAF为该原子在中心原子M的单位球面上投影面积的百分数，所有配位原子的立体角系数和（SAS）应小于单位球的总面积除以4π，即小于1。计算公式为：

$$\theta_{FA} = \sin^{-1}\left(\frac{R}{l}\right) \tag{1}$$

$$SAF = \frac{1}{2}(1 - \cos\theta_{FA}) \tag{2}$$

$$SAS = \Sigma SAF \tag{3}$$

在李醒夫的模型中，配位原子半径采用范氏半径R，文献［3］和［6］在处理锝化合物时，采用了归一化半径R'。由于锝、铼的化学性质非常相似，因此在铼的化合物堆积模型中也应采用归一化半径R'。铼化合物的归一化半径R'的计算结果列入表1，由表1看出，用铼化合物求出的归一化半径R'(Re)与锝化合物的归一化半径R'(Tc)极

---

* 本文合作者：刘国正。原发表于《核化学与放射化学》，1995，17(3)：164~169。

其接近[3]。

根据不同情况，归一化半径可用三种方法计算：(1) 选择 ReL_6 体系，按范氏半径计算 SAS，若 SAS≤1，则范氏半径 R 即为归一化半径 R'；(2) 按范氏半径计算 SAS，若 SAS>1，则令 SAS=1，SAF=1/6，$R'=0.7454l$，由平均键长 $\bar{l}$ 即可求得 R'；(3) 若找不到 ReL_6 体系，可用 $n>6$ 的 RL_n 体系。氢原子的归一化半径 1.07 即是用此法求得的。表 1 中 Se 的 R' 用已有方法虽无法求得，但 As、Se、Br 位于同一周期，As 和 Br 的 R' 很接近，所以 Se 的 R' 一定介于其间。

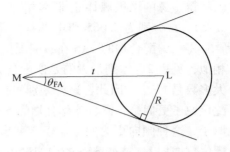

图 1　空间堆积模型示意图

表 1　锝、铼化合物中归一化半径及与范氏半径的比较　　（nm）

配位原子	H	C	N	O	F	P	S	Cl	As	Se	Br	I
R	0.12	—	0.15	0.14	0.135	0.19	0.185	0.180	0.20	0.20	0.195	0.215
R'(Re)	0.107	0.150	0.149	0.140	0.135	0.180	0.178	0.175	0.186	0.185	0.185	0.202
R'(Tc)	0.12	0.150	0.15	0.140	0.135	0.178	0.180	0.175	0.187	—	0.186	0.202

2　SAS 的计算及其堆积饱和规律

求出 R' 后，依次按式(1)～式(3)求出各配位体系的 SAS（参见文献[2]）。将 SAS 进行统计运算，结果为：$n=406$，$\overline{SAS}=0.951$，$\sigma=0.084$。统计分布情况示于图 2。由统计结果可以看出，铼化合物也存在一个饱和堆积的稳定区间，其中心 $\overline{SAS}$ 为 0.951，区间宽度 2σ 为 0.168，它不大于一个正常配体的 SAF 值。

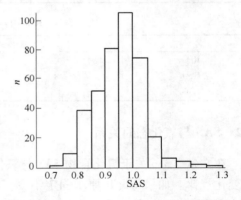

图 2　SAS 的分布状况

3　讨论

3.1　锝铼配合物的相似性及 SAS 的变化规律

表 1 所示的锝、铼配合物对应的归一化半径 R' 非常接近。H、P、S 的归一化半径 R' 略有差别。由于在锝化合物中未找到 TcH_n 体系，因而采用了范氏半径。在铼的化合物中，有已知结构 ReH_9^{2-}（K_2ReH_9）[10]，因而未采用范氏半径。P 的范氏半径 R 大于 S 的，与对应 Re 的归一化半径 R' 大小顺序相同，而与对应锝的归一化半径 R' 正好相反。由于在锝化合物中，计算 P、S 归一化半径的化合物数目太少，出现 0.002nm 的偏差是可能的。因此，相比之下，对应的 Re 数据可能更为可靠些。锝、铼化合物中，配位原子的归一化半径相似，是由于它们的化合物常常具有相同的构型和极其相似的键长，这也是锝、铼药物生物分布性质常常相似的原因。

锝、铼的堆积规律也非常相似。锝、铼的 SAS 平均值（$\overline{SAS}$）分别为 0.966 和 0.951，σ 分别为 0.065 和 0.084。

锝、铼配合物的 SAS 在不同配位数下的统计平均值 $\overline{SAS}$ 随配位数的变化规律也很相似，见表 2。差别最大的是四配位体系，$\overline{SAS}$ 分别为 0.861 和 0.838，四配位体系绝大多数是 MO_4^-，因此，这种差别实际上反映了 TcO_4^- 和 ReO_4^- 的差别，前者键长更短，配位原子更拥挤，锝、铼化合物的 SAS 随配位数增大而增大。如果固定氧化态和配位原子，这种倾向更为明显。表 3 列出了 Re(Ⅶ)-O 配位体系的 $\overline{SAS}$。从表 3 可以看出，配位数增加 1，往往 $\overline{SAS}$ 的增加量小于一个正常配体的 SAF（0.168），体系通常通过键长微微增加以缓冲 SAS 的增大。

表 2　SAS 随配位数的变化关系

配位数	Tc			Re		
	n	$\overline{SAS}$	σ	n	$\overline{SAS}$	σ
4	6	0.861	0.018	70	0.838	0.040
5	26	0.924	0.047	37	0.910	0.038
6	64	0.972	0.038	287	0.979	0.068

表 3　Re(Ⅶ)-O 配位体系的 $\overline{SAS}$

名　称	n	$\overline{SAS}$	σ
ReO_4	45	0.838	0.035
ReO_5	8	0.905	0.017
ReO_6	10	0.991	0.030

3.2　SAS 与氧化态的关系

在文献［3］关于锝化合物 SAS 与氧化态的变化关系中，未能观察到两者间有明显的变化规律。如果将配位原子固定不变，在给定配位数的条件下，将 $\overline{SAS}$ 随氧化态的变化列入表 4。由表 4 看出，电负性较强的氧易与高价金属结合，随中心离子氧化态的降低，结合力越来越弱，键长变长，SAS 变小。然而 Deutsch[11] 曾观察到 Tc—P 键长随氧化态升高而增大。在 Re_mP_n 类化合物中，氧化态分别为 2、3、4 时，$\overline{SAS}$ 没有明显的增大趋势，反而存在着降低趋势（$\overline{SAS}$ 分别为 1.022（$n=2$），0.994（$n=3$），1.003（$n=6$）），配位原子相同的含磷化合物，SAS 也是降低的。如 trans-$ReCl_2$（dppbe）的 Re(Ⅱ)—P 键长为 0.2415mm；$[ReCl_2(dpcp)_2]^{2+}$ 的 Re(Ⅳ)—P 键长为 0.2508nm。SAS 分别为 0.977 和 0.952。由于氧化态与 SAS 的关系还受到原子或离子软硬配合规律的制约，因而泛泛讨论 SAS 与氧化态的关系意义不大。

表 4　ReO_6 的 SAS 随氧化态的变化

氧化态	n	$\overline{SAS}$	σ
7	10	0.991	0.030
6	7	0.933	0.061
5	45	0.917	0.033
4	3	0.835	0.005

3.3 堆积饱和规律与电中性原则及十八电子规则

堆积饱和规律反映了以中心原子为球心的球面上，配位原子间的相互排斥有着一定的限度。排斥力太大，体系不稳定；排斥力太小，体系会进一步与其他配体结合。SAS 小于 0.783（0.951 − 2σ）的体系有三个：第一个是 Nd_4ReO_4，它在温度为 100K 时，SAS 为 0.776，但常温下为 0.838；第二个是 $Dy(ReO_4)_3$，它有三个 SAS，分别为 0.754、0.867、0.771；第三个是 $[NaCH_3CN][ReO_2(CH_2\text{-}t\text{-}Bu)_2]$，其 SAS 有两个值 0.705 和 0.751，此化合物很不稳定，晶体必须处于氮气氛围中，否则将被氧化分解。SAS 大于 1.119（0.951 + 2σ）的六配位体系有 5 个：一个是 ReF_5NCl，它的 SAS 为 1.164，很不稳定，极易与空气中的水发生水解反应；与之相似的 ReF_5NF，SAS 为 1.060，也很不稳定；另一个化合物 $ReNF_4 \cdot ReF_5NCl$ 有两个 SAS，分别为 1.266 和 1.165，这个化合物也很活泼；第四个是 $ReNCl_4$，SAS 为 1.160，活泼性不如 $ReNF_4 \cdot ReF_5NCl$，但也很容易与其他配体发生取代反应；第五个是 $Re_2(CO)_6(S_4)_2^{2-}$，其 SAS 为 1.119，它和其相似物 $Re_2(CO)_6(Se_4)_2^{2-}$（SAS 为 1.114）也都不太稳定。

除了可以用 SAS 与统计平均值 $\overline{SAS}$（0.951）之差粗略地判断一个化合物的稳定性之外，其他因素也影响着化合物的稳定性。电中性原则即是其中的一个。当配位体系有较高的负电荷时，即使 SAS 较小也很难再吸收一个配体。四配位体系绝大多数是 ReO_4^-，$\overline{SAS}$ 为 0.838，在水溶液中很稳定，由于 ReO_4^- 已有一个负电荷，倘若再吸收一个 O^{2-}，除了增加配位原子间固有斥力外，又增加了很大的静电斥力，因此配位数不易增加。但在酸性水溶液中，当 ReO_4^- 与质子结合后就可以与水分子继续结合。在固体中，由于配位原子 O_2^- 与反离子紧密接触，不同条件下就出现了 ReO_5^{3-}、ReO_6^{5-} 体系。这些现象说明，若按 SAS 的规律，ReO_6^{5-} 会更稳定些。由于存在较强的静电斥力，ReO_4^- 才稳定存在。近几年来，已制备出含 ReO_6 独立结构的配合物，其中氧原子与其他基团相连，如 $NEt_4[ReO_2(O_2C_6H_2\text{-}t\text{-}Bu_2)_2]$，$ORe(OTeF_5)_5$，$N\text{-}t\text{-}Bu_4[ReO(OPph_3)(tccat)_2]$，$N\text{-}t\text{-}Bu_4[ReO(MeOH)(tccat)_2]Re_2(OMe)_{10}$。

当中心离子的氧化态较高时，甚至整个配位离子带正电荷，SAS 往往偏大。但是除去一个负性配体会使体系带有更高的正电荷，静电作用会把负性配体再次吸引回来。在铼化合物中，带正电荷的配位离子的配位数皆≥6。

表 5 列出了 SAS 大于 1.05 的化合物（从图 2 可以看出这些化合物占有很小的比例）。表 5 中第一组 SAS 偏大即为电中性要求造成的，其中个别含 F^-、Cl^- 化合物的配位离子已带一个负电荷，这是由于此配位原子给电子能力太差，使中心原子对配体的要求很强，甚至出现了第四组中的化合物 K_2ReF_8。

十八电子规则是另一个重要影响因素。对于低价铼的化合物，中心离子与配位原子间通常存在着反馈键以分散中心离子上的电荷，要保证有效的键相互作用，就要尽量使 d 轨道充满电子。有时 SAS 值较大，仍然不易失去一个配体。第二组化合物基本上属于这种情况，甚至出现了第四组中的化合物，$K_4[Re(CN)_7] \cdot 2H_2O$，它有如此高的负电荷和如此高的配位数。第三组化合物具有三棱柱结构，三棱柱结构存在的原因在化学上仍是一个疑问，文献 [12] 曾对三棱柱结构及研究中尚存在的疑问进行过详细探讨。

表5　SAS大于1.05的配合物

序号	配合物	SAS	配合物	SAS
第一组	ReF_5NF	1.060	$ReNCl_4$	1.160
	ReF_5NCl	1.164	$ReCl_4O \cdot ReO_3Cl$	1.073
	$ReNF_4 \cdot ReF_5NCl$	1.266	$[ReOBr_4(H_2O)]^+$	1.089
		1.165	$[S_4N_3]^+[ReCl_4(NSCl_2)]^-$	1.085
	$[Re_2F_9O_2]^+[Sb_2F_{11}]^-$	1.062	$Cs_2[ReCl_3O_3]$	1.075
	$[Asph_4][ReCl_4(NSCl)_2]$	1.084	$Re(NC-P_r^n)Cl_3(dppbe)$	1.063
	$[Pph_4][ReCl_4(N_2S_2)]$	1.095	$X_yl-NH_3^+[ReCl_3Me(N-X_yl)_2]^-$	1.061
	$[Na15冠-5][ReCl_4(N_2S_2)]$	1.079	$[ReBr_3(N-P_r^n)(dppbe)]$	1.054
	$[Pph_4][ReF_2Cl_2(N_2S_2)]$	1.087	$[ReO_3(C_9S_3)]^+BF_4^-$	1.077
第二组	$[R—SRe(CO)_5]BF_4$	1.05	$[Re_2(CO)_6(S_4)_2][NBu_4]_2$	1.119
	$Re_3(CN)_3(CO)_{12}$	1.074	$[Re_2(CO)_6(Se_4)_2][Pph_4]_2$	1.114
第三组	$[Re(S_2C_2ph_2)_3]$	1.076	$[Pph_4][Re[(SCH_2)_3CCH_3]_2]$	1.063
第四组	K_2ReK_8	1.185	Re_3As_7	1.238
	$K_4[Re(CN)_7] \cdot 2H_2O$	1.068	$[ReCl_2(CN-t-Bu_3(PMeph_2)_2]^+$	1.119

注：未列入含氢配体的化合物。

4　结论

SAS值的分布存在一个稳定区间，说明了配位原子的堆积必须满足空间几何条件。配位原子间的相互排斥力随距离很快地变化是SAS存在稳定区间的根本原因。决定一个化合物能否稳定存在，配位数是多少，除了受配位原子相互斥力的影响外，还受到电荷因素、十八电子规则的制约。简单地用SAS与统计平均值$\overline{SAS}$之差的大小判定配合物的稳定性往往不太准确，但是对于同一类化合物，SAS与稳定性的确存在着确定的关系，这在铼化学研究中已得到充分证明。

参　考　文　献

[1] 刘国正，刘伯里. 铼放射性药物的现状和展望. 同位素，1995，8(1)：53.

[2] 刘国正. 络合物键长及其SAS数据表. 北京师范大学化学系内部资料. 1995.

[3] 魏毅，刘伯里. 锝化学研究 I. 锝化合物的结构稳定性规律. 核化学与放射化学. 1988，10(2)：65~67.

[4] 魏毅，刘伯里. 锝化学研究 II. 空间堆积模型在锝化学的应用. 核化学与放射化学，1989，11(1)：7~12.

[5] 魏毅，刘伯里. 锝化学研究 III. 与TcO核有关事实的阐释. 核化学与放射化学，1989，11(2)：78~83.

[6] Wei Yi, Liu Boli, Kung H F. Quantitative Study of the Structrue-Stability Relationship of Tc Complexes. Appl Radiat Isot, 1990, 41: 763~771.

[7] Kung H F, Liu Boli, Wei Yi, et al. Quantitative Study of the Structrue-Stability Relationship of $Tc^V O$ (III) Complexes. Appl Radiat Isot, 1990, 41: 773~781.

[8] Fischer R O, Li Xingfu. The "Solid Angle Sum Rule": A New Ligand Cone Packing Model of Optimal Applicability in Organolanthanoid Chemistry, Part Ⅰ. J Less-Common Met, 1985, 112: 303~325.

[9] 李醒夫,冯锡章,徐英庭,等. 堆积饱和规律与堆积均匀规律——镧系元素配合物的结构特征. 化学学报, 1985, 43: 502~506.

[10] Abrahams S C, Ginsberg A P, Knox K. Transition Metal-Hydrogen Compounds Ⅱ. The Crystal and Molecular Structrue of Potassium Rhenium Hydride K_2ReH. Inorg Chem. 1964, 3: 558~567.

[11] Deutsch E. Aspects of the Chemistry of Technetium Phosphine Complexes. In: Topical Symposium on the Behavior and Utilization of Technetium'93. Sendi, Japa. 1993.

[12] 刘国正. 三棱柱配位结构稳定性理论的评价. 化学通报, 1995, 3: 52~55.

Study of Coordination Chemistry of Rhenium

Ⅰ. Packing Saturation Rule

Liu Guozheng Liu Boli

(Department of Chemistry, Beijing Normal University, Beijing, 100875, China)

Abstract The stability of rhenium compounds in crystal is studied based on the reformed packing model. The structural data of 406 rhenium compounds containing no metal-metal bonds are collected. The sums of their solid angle factor (SAF) are calculated and statistically analysed. It turns out that the stability center ($\overline{SAS}$) is 0.951 ± 0.084. The value of 2σ is comparable to the SAF of a common coordinating atom. This result means that the packing of the coordinating atoms arround the central atom is restricted by geometrical factors.

Key words packing model, rhenium compounds, stability, normaliged "Van der waals" radii

Synthesis of New N_2S Ligands, Preparation of ^{99m}Tc Complexes and Their Preliminary Biodistribution in Mice[*]

Abstract Two new N_2S ligands of MPBDA (N-(2-mercapto-propyl)-1,2-benzenediamine) and MEBDA (N-(2-mercapto-ethyl)-1,2-benzenediamine) were synthesized, and their lipophilic complexes of ^{99m}Tc-MPBDA and ^{99m}Tc-MEBDA were prepared in high yield (>92%) by stannous reduction of ^{99m}TcO$_4$. Effects of pH, reaction temperature and time on the formation of ^{99m}Tc complexes were investigated. Potential use of ^{99m}Tc-MPBDA and ^{99m}Tc-MEBDA in cerebral and myocardial imaging were evaluated in mice. High uptakes are demonstrated in both brain and heart, and residence times within heart are long. Two minutes following i. v. Administration, 1.85% ID of ^{99m}Tc-MPBDA and 1.49% ID of ^{99m}Tc-MEBDA appear in brain, and 1.64% ID of ^{99m}Tc-MPBDA and 1.63% ID of ^{99m}Tc-MEBDA are in heart. At 1h after injection, 1.17% ID and 1.26% ID of ^{99m}Tc-MPBDA remain respectively in brain and heart, and 0.67% ID and 1.18% ID of ^{99m}Tc-MEBDA remain in brain and heart respectively. The activity in blood clears rapidly and the clearance half-life is less than 15 minutes. The properties of ^{99m}Tc-MPBDA make it a potential useful agent for cerebral and myocardial perfusion imaging.

Key words N_2S ligands, organic synthesis, technetium complexes, biodistribution

1 Introduction

A useful radiopharmaceutical for SPECT imaging rCBF must have a high cerebral uptake and a long residence time in the brain without redistribution during imaging acquisition.

Technetium-99m-labelled neutral lipophilic agents can cross the intact blood-brain barrier (BBB). In 1984, Volkert et al. showed that ^{99m}Tc-PnAO {propylene amino oxime, 3,3'-(1,3-propanediyldiimino)bis-(3-methyl-2-butanone)-dioxime} can cross the intact BBB[1,2]. However, the residence time within the brain is not long enough to allow single-headed rotating SPECT imaging. Subsequently, Nowotnik et al. found that ^{99m}Tc-d, l-hexamethylpropylene amine oxime (HMPAO) has a high cerebral uptake and long residence time to allow SPECT imaging[3], However, the complex of ^{99m}Tc-d, l-HMPAO is not stable in vitro, the brain/blood ratio is low[4], and quantitative SPECT images underestimate blood flow at high flow rates[5].

The neutral, lipid soluble Tc(V) oxo complexes based on the 3,6-diazaoctane-1,8-dithiol (N_2S_2) ligand system have shown significant uptake in the brain[6-8]. However, most of them were rapidly cleared from brain. In 1989, ^{99m}Tc-l, l-ethyl cysteinate dimmer (ECD) was shown to cross the BBB in several species with significant retention only in higher species[9,10]. The

---

[*] Copartner: Miao Yubin. Reprinted from *Journal of Labelled Compounds and Radiopharmaceuticals*, 1999, 42: 629-640.

cerebral retention of ^{99m}Tc-l, l-ECD is caused by hydrolysis from a diethyl ester to the monoethyl ester by an enzyme in the brain[11]. The complex of ^{99m}Tc-l, l-ECD is stable in vitro and has a high brain/blood ratio, but the SPECT images may not accurately reflect rCBF at high flow rates[12].

Neutral, lipophilic ^{99m}TcO-MRP20 (N-(2(1H pyrolylmethyl))(N'-(4-pentene-3-one-2))ethane-1, 2-diamine) has shown significant cerebral uptake and long residence time to allow SPECT imaging[13]. The retention of ^{99m}TcO-MRP20 in brain is caused by its decomposition.

Liu Boli et al.[14] obtained the order of stability between ^{99m}Tc-d, l-CBPAO (4,8-diaza-3, 9-dimethyl-6,6-(trimethylene)-undecane-2, 10-dionebisoxime) and isomers of ^{99m}Tc-d, l-HMPAO by CNDO/2 method in 1993. They suggested that the brain retention ability of these complexes is relate to their stability in vitro. In 1995, based on the computed results of the solid angle factor (SAFv) of the vacancy trans to Tc(V)O-chelates, Liu Boli et al.[15] inferred that the proper instability of Tc(V)O-chelates for brain imaging agent may be valuable to their high brain retention. In attempt to further increase stability in vitro, C. Cutler et al.[16] have reported on three new PAO derivatives ^{99m}Tc-OCBPAO (4,8-diaza-3, 9-dimethyl-6, 6-(3-oxacyclobutylene)-undecane-2, 10-dionebisoxime), ^{99m}Tc-EOCBPAO (4, 8-diaza-3, 9-diethyl-6, 6-(3-oxacyclobutylene)-undecane-2, 10-dionebisoxime), ^{99m}Tc-IPOCBPAO (5, 9-diaza-2, 4, 10, 12-tetramethyl-7, 7-(3-oxacyclobutylene)-tridecane-2, 11-dionebisoxime). After animal distribution investigations and in vitro stability determinations of ^{99m}Tc-d, l-PAO derivatives, they have suggested that the retention of these complexes within the brain depends on the in vitro instability. The in vitro stability of PAO complexes appears to determine the extent to which these complexes will be retained in the brain. In our previous work[17], based on crystal data of ^{99m}Tc-d, l-HMPAO, the structures of ^{99m}Tc-d, l-PAO complexes have been optimized by method of INDO/1. The computed results indicate that brain retention ability of ^{99m}Tc-d, l-PAO complexes depends on the instability in vitro of complexes.

In this paper, based on the idea above, two new N_2S ligands of MPBDA (N-(2-mercaptopropyl)-1, 2-benzenediamine) and MEBDA (N-(2-mercapto-ethyl)-1, 2-benzenediamine) have been synthesized and their lipophilic complexes of ^{99m}Tc-MPBDA and ^{99m}Tc-MEBDA have been prepared. Preliminary biodistribution of ^{99m}Tc-MPBDA and ^{99m}Tc-MEBDA have been investigated in mice. Details on synthesis of MPBDA and MEBDA have been described in this paper.

2 Materials and Methods

2.1 Synthesis of MPBDA and MEBDA

Organic compounds were characterized by melting point, ^{1}HNMR, IR spectroscopy, elemental analysis and Mass spectroscopy. All chemicals were of reagent grade and were used as received. The ligands of MPBDA and MEBDA were synthesized as following procedure:

4.5g (0.06mol) of propylene sulfide was added dropwise into a solution of 6.6g (0.06mol)

1,2-phenylenediamine in 40mL of absolute ethanol. The reaction mixture was refluxed under nitrogen for 8h. The solvent was removed under reduced pressure and the residue was purified though silica gel (40% EtOAc/hexane). Dry hydrogen chloride gas was bubbled into the solution to give 3.6g (26.7% yield) of white solid. The precipitated solid was collected under suction. Recrystallization from ethanol-ether provided MPBDA sample for analysis. Mp: 158-159℃. IR (KBr): 3372cm^{-1}, 3284cm^{-1}, 2955cm^{-1}, 2921cm^{-1}, 2862cm^{-1}, 2608cm^{-1}, 2536cm^{-1}, 1624cm^{-1}, 1583cm^{-1}, 1520cm^{-1}, 1460cm^{-1}, 1379cm^{-1}, 755cm^{-1}. MS (FAB$^+$, m/z): 183 (M$^+$), 150, 136, 121 (100), 109, 92. ^{1}HNMR (d$_6$-DMSO): δ 1.34 (s, 3H), 3.02-3.20 (m, 5H), 6.60-6.80 (m, 2H), 7.10-7.30 (m, 2H). Anal. Calc. for C$_9$H$_{14}$N$_2$S · HCl: C, 49.53; H, 6.93; N, 12.84. Found: C, 49.54; H, 7.06; N, 12.52.

2.6g (21.2% yield) of white MEBDA sample was obtained in similar manner. Mp: 148-149℃. IR (KBr): 3310cm^{-1}, 3120cm^{-1}, 2832cm^{-1}, 2536cm^{-1}, 1624cm^{-1}, 1530cm^{-1}, 1480cm^{-1}, 1454cm^{-1}, 1325cm^{-1}, 755cm^{-1}; MS (FAB$^+$, m/z): 169; ^{1}HNMR (d$_6$-DMSO): δ 2.50 (s, 1H), 2.70-2.90 (m, 3H), 3.30-3.40 (m, 2H), 6.70-6.90 (m, 2H), 7.10-7.30 (m, 2H), 9.50 (w, 2H); Anal. Calc. for C$_8$H$_{12}$N$_2$S · HCl: C, 46.93; H, 6.41; N, 13.69. Found: C, 46.80; H, 6.55; N, 14.08.

2.2 Technetium complexation

Sodium pertechnetate-^{99m}Tc was obtained from ^{99}Mo-^{99m}Tc generator (China Institute of Atomic Energy, Beijing) by eluting with normal saline. All other chemicals were of reagent grade and were used as received. The ^{99m}Tc complexes of MPBDA and MEBDA were prepared as following procedure:

4mg of MPBDA (or MEBDA) was dissolved in 0.2mL absolute ethyl alcohol in a 10mL glass vial. One drop of Tween-80, 1mL of distilled water and 0.1mL of stannous chloride solution (formed by 1mg of stannous chloride dihydrate in 1mL of 2mol/L hydrochloride acid solution) were injected into the vial. After adjusting pH of mixture, 0.6mL of sodium pertechnetate (about 1-1.2mCi) was injected into the vial. The mixture reacted at room temperature (17℃) for 15min. Complex radiochemical purity (RCP) was assayed by thin layer chromatography.

2.3 Analysis of ^{99m}Tc complexes

The distribution of radioactivity was determined by method of thin layer chromatography. A 1-2μL sample was applied to Xinghua No.1 chromatography strip (1cm × 10cm, Beijing, China). The chromatogram was developed by ascending chromatography in tanks containing methanol/chloroform (V:V = 1:9) to a depth of 1cm. The chromatography separated ^{99m}Tc-MPBDA or ^{99m}Tc-MEBDA (R_f = 0.9) from ^{99m}TcO$_4^-$ (R_f = 0.1) and other impurities (R_f = 0.1). The radiochemical purity of ^{99m}Tc-MPBDA and ^{99m}Tc-MEBDA were respectively more than 95% and 92%. After adjusting pH, the solution was used for animal studying.

2.4 Biodistribution of ^{99m}Tc-MPBDA and ^{99m}Tc-MEBDA in mice

Biodistribution studies were performed in Kunming mice of either sex weighing between

18-20g. Each animal was administered 100μL of the saline solution containing ^{99m}Tc complex of MPBDA (or MEBDA) through the lateral tail vein. The radiochemical purity of ^{99m}Tc-MPBDA and ^{99m}Tc-MEBDA were respectively 96.2% and 94.1%. Mice were killed separately at 2min, 5min, 15min, 30min and 60min after injection. The organs and tissues of interest were removed and assayed for radioactivity in a gamma counter. The various organs studied were perfused prior to measurement of ^{99m}Tc. Suitable standards, representing 1/100 of the injected dose, were prepared from the injection material. Blood values were taken as 7% of the total body weight.

3 Results and Discussion

3.1 Synthesis of MPBDA and MEBDA, preparation and analysis of ^{99m}Tc complexes

The route of synthesis of MPBDA and MEBDA are shown in Fig. 1. One molecular equivalent of propylene sulfide (ethylene sulfide) with equal molecular equivalent of 1,2-phenylenediamine to provide MPBDA in 26.7% yield (MPBDA in 21.2% yield).

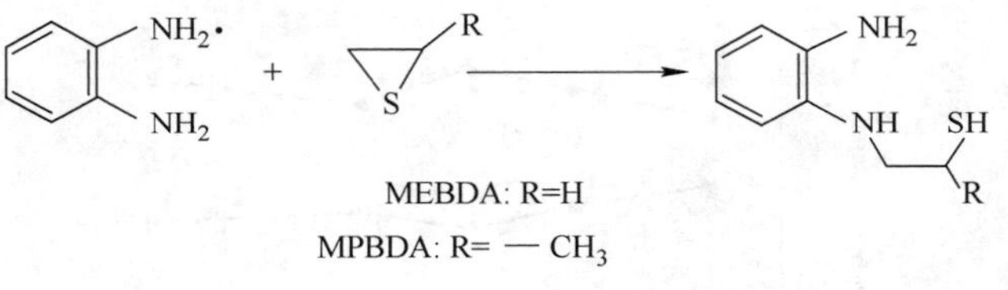

Fig. 1 Synthesis of MEBDA and MPBDA

The solution of MPBDA (MEBDA) and stannous chloride as reductant allow the ^{99m}Tc complex to be prepared simply by adding generator eluate to the vial. Thin layer chromatography permits the quantitative determination of radioactive components following complex formation. The chromatography separated ^{99m}Tc-MPBDA or ^{99m}Tc-MEBDA ($R_f = 0.9$) from ^{99m}TcO$_4^-$ ($R_f = 0.1$) and other impurities ($R_f = 0.1$). Addition of ^{99m}Tc-pertechnetate to the vial contains MPBDA and stannous chloride provides lipophilic complex of ^{99m}Tc-MPBDA in more than 95% labelling yield immediately after complex formation. The radiochemical purity of ^{99m}Tc-MEBDA is more than 92%.

3.2 Effects of pH, reaction temperature and time on the formation of ^{99m}Tc complexes

The ligands of MPBDA and MEBDA readily form neutral, lipophilic complexes with ^{99m}Tc to provide new radiopharmaceuticals for cerebral perfusion imaging. Effects of pH, reaction temperature and time were investigated, and the results are shown respectively in Fig. 2-Fig. 4. The complex reaction may be performed only on the condition of acid media. If pH is raised to 6-7, precipitate is occurred in the solution. It is shown that this reaction is not very sensitive to temperature over 17℃. Therefore, it is convenient to choose room temperature as optimal reaction temperature. At pH 2.5 and room temperature (17℃), the radiochemical purity of ^{99m}Tc-MPBDA is more than 96% and the radiochemical purity of ^{99m}Tc-MEBDA is more than 92% only when mixture have reacted for 5min, and the radiochemical purity of complexes do not increase significantly when reaction time prolong.

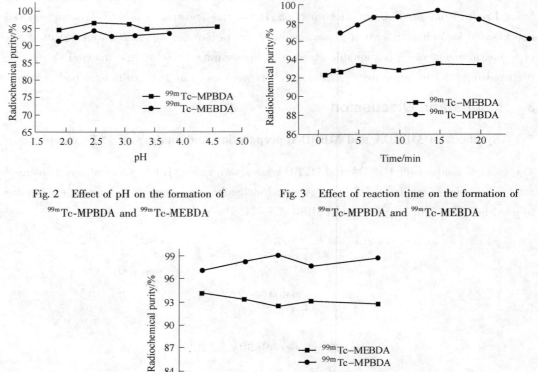

Fig. 2　Effect of pH on the formation of ^{99m}Tc-MPBDA and ^{99m}Tc-MEBDA

Fig. 3　Effect of reaction time on the formation of ^{99m}Tc-MPBDA and ^{99m}Tc-MEBDA

Fig. 4　Effect of reaction temperature on the formation of ^{99m}Tc-MPBDA and ^{99m}Tc-MEBDA

3.3　Biodistributions of ^{99m}Tc-MPBDA and ^{99m}Tc-MEBDA complexes in mice

The biodistributions of ^{99m}Tc-MPBDA and ^{99m}Tc-MEBDA in mice are shown in Table 1 and Table 2. Two minutes following i. v. administration of these complexes in mice, 1.85% ID of ^{99m}Tc-MPBDA and 1.49% ID of ^{99m}Tc-MEBDA appear in the brain. Retention of ^{99m}Tc-MPBDA is high, 1.17% ID of ^{99m}Tc-MPBDA remain in the brain at 60min after injection. It is shown that about more than 10% ID of activities remain in blood at 2min postinjection. However, the activities clearance from blood are rapid, and the clearance half-lives are less than 15min. The ratios of brain/blood of ^{99m}Tc-MPBDA and ^{99m}Tc-MEBDA in mice are 1.42 and 0.85 respectively at 1h postinjection.

1.64% of ^{99m}Tc-MPBDA and 1.69% of ^{99m}Tc-MEBDA of the injected dose are extracted into heart at 2min after injection and retention of activities are high. In comparison with the activity at 2min postinjection, more than 75% of both complexes activities remain in the heart at 1h

Table 1 Biodistribution of ^{99m}Tc-MPBDA in mice ($x \pm s.d.$, $n=3$) expressed as % injected dose/total tissue and % injected dose/g tissue

Tissues	2min		5min		15min		30min		60min	
	%ID/organ	%ID/g	%ID/organ	%ID/g	%ID/organ	%ID/g	%ID/organ	%ID/g	%ID/organ	%ID/g
Blood	14.44 ± 0.35	10.86 ± 1.09	8.72 ± 0.06	6.56 ± 0.20	6.19 ± 0.07	4.66 ± 0.13	4.07 ± 0.06	3.06 ± 0.10	2.77 ± 0.09	2.08 ± 0.28
Brain	1.85 ± 0.38	4.80 ± 0.99	1.80 ± 0.02	4.95 ± 0.30	1.32 ± 0.12	3.67 ± 0.29	1.16 ± 0.03	3.08 ± 0.12	1.17 ± 0.05	2.95 ± 0.26
Heart	1.64 ± 0.33	22.69 ± 4.16	1.49 ± 0.03	22.71 ± 3.35	1.40 ± 0.12	19.76 ± 2.05	1.30 ± 0.06	18.72 ± 0.35	1.26 ± 0.06	15.21 ± 1.31
Liver	14.78 ± 1.72	21.10 ± 2.71	15.77 ± 0.41	23.68 ± 2.41	14.43 ± 0.11	22.46 ± 0.75	14.90 ± 0.32	19.51 ± 1.09	10.17 ± 1.08	14.50 ± 1.91
Kidneys	3.79 ± 0.29	17.83 ± 2.92	3.53 ± 0.16	17.88 ± 1.96	3.65 ± 0.22	18.73 ± 0.43	3.38 ± 0.36	16.24 ± 0.60	3.01 ± 0.16	12.80 ± 0.61
Spleen	0.42 ± 0.05	5.79 ± 1.59	0.49 ± 0.05	5.74 ± 0.24	0.50 ± 0.13	6.16 ± 0.39	0.36 ± 0.05	5.15 ± 0.15	0.37 ± 0.05	4.31 ± 0.62
Lungs	3.71 ± 0.37	20.74 ± 3.66	2.99 ± 0.09	23.27 ± 3.93	2.33 ± 0.31	18.31 ± 1.15	1.71 ± 0.23	11.37 ± 0.44	1.30 ± 0.06	8.53 ± 1.22

Table 2 Biodistribution of ^{99m}Tc-MEBDA in mice ($x \pm s.d.$, $n=3$) expressed as % injected dose/total tissue and % injected dose/g tissue

Tissues	2min		5min		15min		30min		60min	
	%ID/organ	%ID/g	%ID/organ	%ID/g	%ID/organ	%ID/g	%ID/organ	%ID/g	%ID/organ	%ID/g
Blood	11.01 ± 0.88	8.74 ± 0.70	6.62 ± 1.26	5.26 ± 1.00	4.77 ± 1.09	3.78 ± 0.87	4.65 ± 0.17	3.69 ± 0.14	2.86 ± 0.05	2.27 ± 0.04
Brain	1.49 ± 0.27	4.40 ± 0.75	1.22 ± 0.38	3.69 ± 1.18	0.82 ± 0.08	2.50 ± 0.21	0.69 ± 0.01	2.20 ± 0.03	0.67 ± 0.02	1.93 ± 0.02
Heart	1.63 ± 0.19	26.42 ± 1.63	1.38 ± 0.16	18.77 ± 2.30	1.26 ± 0.07	20.07 ± 0.61	1.16 ± 0.03	24.02 ± 0.64	1.18 ± 0.07	22.92 ± 4.15
Liver	32.62 ± 1.76	75.45 ± 13.4	43.31 ± 5.58	69.42 ± 7.92	47.54 ± 7.36	67.17 ± 6.07	37.97 ± 3.81	64.72 ± 5.16	23.05 ± 1.24	55.82 ± 3.24
Kidneys	3.02 ± 0.26	16.71 ± 1.18	2.73 ± 0.19	12.74 ± 0.12	3.25 ± 0.31	15.07 ± 1.79	3.29 ± 0.25	17.98 ± 1.85	5.89 ± 0.13	23.75 ± 0.53
Spleen	0.79 ± 0.03	26.93 ± 0.46	1.67 ± 0.05	18.48 ± 0.64	1.53 ± 0.13	20.16 ± 0.08	1.48 ± 0.15	22.46 ± 1.57	1.29 ± 0.08	18.02 ± 2.48
Lungs	10.67 ± 0.38	51.25 ± 4.12	5.29 ± 0.18	38.03 ± 1.26	5.24 ± 0.11	36.78 ± 0.14	5.88 ± 0.12	35.62 ± 0.17	3.03 ± 0.18	31.56 ± 0.24

Table 3 The brain/blood and heart/tissue ratios of ^{99m}Tc-MPBDA and ^{99m}Tc-MEBDA

Item	^{99m}Tc-MPBDA					^{99m}Tc-MEBDA				
	2min	5min	15min	30min	60min	2min	5min	15min	30min	60min
Brain/Blood	0.44	0.75	0.79	1.01	1.42	0.51	0.70	0.66	0.60	0.85
Heart/Blood	2.09	3.46	4.24	6.12	7.31	3.02	3.57	5.31	6.51	10.01
Heart/Liver	1.08	0.96	0.88	0.96	1.05	0.35	0.27	0.30	0.37	0.41
Heart/Lungs	1.09	0.98	1.08	1.65	1.78	0.52	0.49	0.55	0.67	0.73

after injection. The heart/blood, heart/liver, heart/lungs ratios of ^{99m}Tc-MPBDA and ^{99m}Tc-MEBDA respectively are 7.31 and 10.01, 1.05 and 0.41, 1.78 and 0.73 at 1h postinjection. 5min following i. v. Administration of ^{99m}Tc-MPBDA, the liver contains its peak activity of more than 15% ID. 15 minutes following i. v. Administration of ^{99m}Tc-MEBDA, the liver contains its peak activity of more than 45% ID. The structure and retention mechanism in brain and heart of these ^{99m}Tc complexes need to be determined by further research work.

4 Conclusions

The rapid blood clearance and significant cerebral and myocardial uptake coupled with a long clearance time in mice brain and heart of ^{99m}Tc-MPBDA suggest that it will be a potential cerebral and myocardial perfusion imaging agent in high species animals.

Acknowledgment

This work was supported financially by National Natural Science Foundation and State Committee of Science and Technology of P. R. China.

References

[1] Volkert W A, Mckenzie E H, Hoffman T J, et al. Int. J. Nucl. Med. Biol. 11:243(1984).
[2] Troutner D E, Volkert W A, Hoffman T J, et al. Int. J. Nucl. Med. Biol. 11:467(1984).
[3] Nowotnik D P, Canning L R, Cumming S A, et al. Nucl. Med. Comm. 6:499(1985).
[4] Neirinckx R D, Canning L R, Piper I M, et al. J. Nucl. Med. 28:191(1987).
[5] Mathias C J, Welch M J, Lich L, et al. J. Nucl. Med. 29:747(1988).
[6] Lever S Z, Burns H D, Kervitsky T M, et al. J. Nucl. Med. 26:1287(1985).
[7] Kung H F, Molnar M, Billings J, et al. J. Nucl. Med. 25:326(1984).
[8] Kung H F, Yu C C, Billiings J, et al. J. Med. Chem. 28:1280(1985).
[9] Demonceau G, Leveille J, De Roo M, et al. J. Nucl. Med. 29:747(1988).
[10] Vallabhajosula S, Zimmerman R E, Picard M, et al. J. Nucl. Med. 30:599(1989).
[11] Walovitch R C, Hill T C, Garrity S T, et al. J. Nucl. Med. 30:1892(1989).
[12] Walovitch R C, Platts S H, Walsh R A, et al. J. Nucl. Med. 30:792(1989).
[13] Morgan G F, Deblaton M, Clemens P, et al. J. Nucl. Med. 32:500(1991).
[14] Liu Boli, Meng Zhaoxing. Radiochimica. Acta. 63:217 (1993).
[15] Liu Boli, Jia Hongmei. J. Label. Comps. Radiopharm. 37:788(1995).
[16] Cutler C, Jiang Z, Chen B, et al. J. Label. Comps. Radiopharm. 37:785(1995).
[17] Miao Yubin, Zhang Huabei, Liu Boli. Isotopes. 10:103(1997).

一种新的 ^{99m}Tc 标记的 [Tc(CO)$_3$(MIBI)$_3$]$^+$ 络合物作为心肌显像剂*

摘　要　最近，Alberto 等人报道了一种在低压（约 10^5 Pa）条件下制备水溶性的有机金属络合物 [^{99m}Tc(CO)$_3$(OH$_2$)$_3$]$^+$ 的方法。该络合物在水和空气中均比较稳定，且水配体很容易被其他的络合能力较强的配体所取代，这使得羰基络合物可作为放射性药物应用于核医学。考虑到 [^{99m}Tc(CO)$_3$(OH$_2$)$_3$]$^+$ 是一种 +1 价的络合物，于是用六甲氧基异丁基异腈（MIBI）取代上述络合物中的 3 个水配体，制得了一种新的有机锝络合物，其放射化学产率可达 85%，且具有较好的体外稳定性。由于该络合物与已被广泛应用的心肌显像剂 [^{99m}Tc(CO)$_3$(MIBI)$_3$]$^+$ 具有相似的结构，因此，期望 [^{99m}Tc(CO)$_3$(MIBI)$_3$]$^+$ 也能在心肌中浓集，并可改善 [^{99m}Tc(MIBI)$_6$]$^+$ 心/肝比值偏低的缺点。随后进行的小鼠体内生物分布实验表明，该络合物的确能在心肌中浓集，且具有较高的心/肝比和与 [^{99m}Tc(MIBI)$_6$]$^+$ 相当的心/血比，这使得其具有成为心肌灌注显像剂的潜力。

关键词　三羰基锝络合物　心肌显像　生物体内分布

在过去的 20 年里，过渡金属的放射性核在对功能性组织的显像方面起到了重要的作用[1,2]。例如 ^{99m}Tc，由于其核性质非常适于 SPECT 显像，因此锝放射性药物化学获得了很大的发展，其中研究得最多的是含有 [Tc＝O]$^{3+}$ 核的化合物，因为在水溶液体系中很容易制得 +5 价锝的化合物。而对于低氧化态锝的有机金属化学，人们尚未进行系统的研究，这主要是由于缺乏合适的反应母体。传统的前体化合物如 Tc$_2$(CO)$_{10}$，Tc(CO)$_5$，[Tc(CO)$_6$]$^+$ 需在高压条件下制备，这对于放射性物质既不实际，又存在着潜在的危害，因此安全因素是制备低氧化态的有机金属络合物的最大障碍。

近来，一种常压下制备 fac-[M(CO)$_3$]$^+$（M = Tc 或 Re）核的方法引起了人们的关注，仅仅在硼氢化合物和一氧化碳的作用下，就可使 [MO$_4$]$^-$ 直接羰基化[3]。若四氢呋喃和氯化钠也存在于反应体系中，则 [MO$_4$]$^-$ 羰基化后就只有一种产物 [MCl$_3$(CO)$_3$]$^{2-}$，它能与多种配体发生交换反应，生成各种热力学或动力学稳定的羰基化合物[3~7]。例如，水分子就很容易取代该络合物中的氯离子，当其溶于水中时，可定量地生成羰基化的水合离子 [M(CO)$_3$(OH$_2$)$_3$]$^+$。对 [Re(CO)$_3$(OH$_2$)$_3$]$^+$ 的水解研究表明，这种羰基化合物对水和氧气均不敏感，但却可以像第一过渡系列金属的水合离子那样形成缩合产物[5,8]。同样，这些水合离子也很容易发生配体交换反应[9~12]，因此我们用 MIBI 取代 [^{99m}Tc(CO)$_3$(OH$_2$)$_3$]$^+$ 中的 3 个水配体，这样就制得了一种新的有机锝络合物 [^{99m}Tc(CO)$_3$(MIBI)$_3$]$^+$。由于该络合物与 [^{99m}Tc(MIBI)$_6$]$^+$ 这种已被广泛应用的心肌显像剂具有相似的结构，我们的主要目的就是研究其是否也能在心肌中

---

* 本文合作者：蒋燕。原发表于《科学通报》，2001, 46(9): 727~730。

浓集，且其心/肝比是否比$[^{99m}Tc(MIBI)_6]^+$有所提高。

1 材料与方法

1.1 试剂与仪器

MIBI药盒（含MIBI及添加剂L-半胱氨酸，甘露醇），由北京师宏药物研制中心赠送；硼氢化钠、碳酸钠、酒石酸钾钠，北京试剂公司产品；一氧化碳气体，北京市北氧特种气体研究所产品。仪器：FT-408自动定标器，FT-603井型闪烁探头，北京核仪器厂生产；^{99}Mo-^{99m}Tc发生器，中国原子能科学研究院生产。

1.2 实验方法

整个实验步骤如图1所示。首先合成有机锝的水溶性阳离子$[^{99m}Tc(CO)_3(OH_2)_3]^+$。先将5mg硼氢化钠、4mg碳酸钠和15mg酒石酸钾钠装入一个10mL的青霉素小瓶中，密封后通入一氧化碳气体10min，将瓶中空气排净。然后用注射器加入3mL高锝酸钠淋洗液（最高活度可达30GBq），并将溶液加热至75℃反应30 min。待溶液冷却至室温后，向其中加入PBS（0.1mol/L NaCl/0.05mol/L 磷酸盐）缓冲溶液，调节pH值至7.4[11,13]。反应产物用薄层色谱层析体系进行分析，固定相为聚酰胺薄片，展开剂为乙腈，$[^{99m}Tc(CO)_3(OH_2)_3]^+$的$R_f$ = 0.1~0.2。

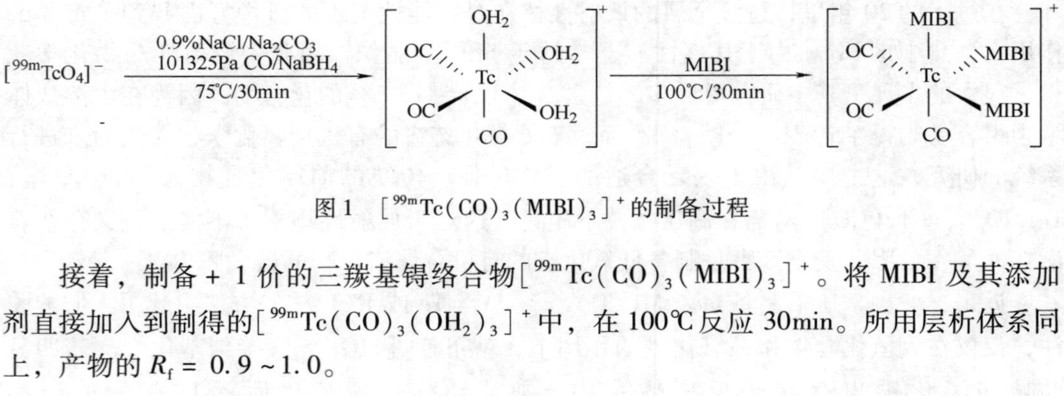

图1 $[^{99m}Tc(CO)_3(MIBI)_3]^+$的制备过程

接着，制备+1价的三羰基锝络合物$[^{99m}Tc(CO)_3(MIBI)_3]^+$。将MIBI及其添加剂直接加入到制得的$[^{99m}Tc(CO)_3(OH_2)_3]^+$中，在100℃反应30min。所用层析体系同上，产物的R_f = 0.9~1.0。

2 结果

2.1 锝-99m标记的三羰基络合物的制备

羰基锝水合离子$[^{99m}Tc(CO)_3(OH_2)_3]^+$的放射化学产率可达85%。它与MIBI发生配体交换反应制得三羰基锝络合物$[^{99m}Tc(CO)_3(MIBI)_3]^+$。由于反应进行得非常彻底，因此该络合物最终的产率仍有85%。这两种标记产物均有较好的体外稳定性，在pH值为7的生理盐水中，至少6h之内没有解离。

2.2 生物体内分布实验

上述两种标记产物在小鼠体内的分布结果列于表1和表2。由表1可见，

$[^{99m}Tc(CO)_3(OH_2)_3]^+$（放射化学产率为85%）在各种组织内无特异性浓集，但60min内的放射性保留程度高，这意味着该化合物具有很好的体内稳定性。从表2的数据中我们可看出，$[^{99m}Tc(CO)_3(MIBI)_3]^+$在小鼠的心脏中有明显浓集，且放射性保留较好，而其在肝中的浓度却比较低。在注射后5min时，产物的心/肝比为1.88，而到了15min时，心/肝比增至2.49，60min后可达4.13。这个增长主要是由肝中的放射性活度随时间的延长而逐渐降低，但心脏中的活度却基本上保持不变而造成的。

表1　$[^{99m}Tc(CO)_3(OH_2)_3]^+$在小鼠体内的生物分布[①]

组织	时间/min			
	5	15	30	60
心/%	2.68±0.88	2.39±0.66	2.39±0.56	2.08±0.31
肝/%	10.33±0.82	11.95±0.66	10.83±1.68	10.73±0.90
肺/%	6.53±1.41	5.38±0.76	5.37±0.78	5.19±1.18
血/%	9.27±1.13	6.04±0.73	5.79±0.66	4.50±0.72
肾/%	18.29±4.22	14.25±1.78	13.88±1.02	12.28±2.32
脑/%	0.31±0.09	0.27±0.03	0.29±0.06	0.22±0.03
肉/%	1.47±0.37	1.46±0.48	1.66±0.61	1.37±0.30
骨/%	2.82±0.95	1.68±0.22	2.38±0.54	1.65±0.12
胆/%	2.02±1.02	2.38±0.37	2.94±0.69	3.20±0.80

① 放射化学产率=85%，表内数据为每克组织的放射性摄取率（ID%/g），数值为平均值t标准偏差，$n=3$。

表2　+1价$[^{99m}Tc(CO)_3(MIBI)_3]^+$络合物在小鼠体内的生物分布[①]

组织	时间/min			
	5	15	30	60
心/%	21.62±2.84	20.63±3.89	20.77±1.60	19.39±0.91
肝/%	11.48±0.88	8.28±0.90	6.44±1.03	4.69±0.34
肺/%	6.16±0.48	4.14±0.90	3.89±0.54	3.20±0.85
血/%	1.03±0.21	0.32±0.05	0.24±0.07	0.13±0.04
肾/%	57.90±1.80	33.66±0.42	23.94±2.11	11.30±0.73
脑/%	0.18±0.02	0.18±0.02	0.12±0.01	0.08±0.01
肉/%	7.08±0.52	3.78±0.88	6.14±0.78	4.42±0.36
骨/%	3.72±0.84	2.10±0.10	1.94±0.29	1.42±0.31
脾/%	6.26±1.29	3.84±0.90	2.84±0.41	1.62±0.32

① 说明同表1。

此外，在注射后5min时，产物在肾中的浓集大大高于其在肝中的摄取，这说明该络合物只有很少一部分是通过肝代谢的。且作为亲水性的阳离子，其在脑中的放射性活度的确非常低。

由于制得的$[^{99m}Tc(CO)_3(MIBI)_3]^+$放射化学纯度尚不是最理想的，因此生物体内分布结果也随产率的不同而有所变化。当产物的放射化学纯度由85%降到75%时，注射后60min时的心/肝比和心/血比也随之降低到2.61和135.46，但其心/肺比却增至8.89。心脏与其邻近组织的放射性摄取比值列于表3和表4。

表3 当产物的放射化学产率为85%时心脏与其邻近组织的放射性摄取比

心脏/组织	时间/min			
	5	15	30	60
心/肝	1.88	2.49	3.23	4.13
心/肺	3.51	4.98	5.34	6.06
心/血	20.99	64.47	86.54	149.15

表4 当产物的放射化学产率为75%时心脏与其邻近组织的放射性摄取比

心脏/组织	时间/min			
	5	15	30	60
心/肝	1.45	2.39	1.82	2.61
心/肺	5.17	7.17	8.20	8.89
心/血	19.12	54.15	77.25	135.46

3 讨论

本研究的主要目的是制备一种新的+1价三羰基锝络合物$[^{99m}Tc(CO)_3(MIBI)_3]^+$，并通过动物实验探讨其生物分布规律，看其是否能在心脏中浓集。

第1个目标已较好地完成，$[^{99m}Tc(CO)_3(MIBI)_3]^+$的放射化学产率可达85%以上。但由于制备$[^{99m}Tc(CO)_3(OH_2)_3]^+$的过程中需通入一氧化碳气体，使其临床应用受到了较大限制。最近，文献[13]、[14]报道，$[^{99m}Tc(CO)_3(OH_2)_3]^+$这种反应前体可通过一步法药盒制得。这样，我们只需通过一个很简单的两步法药盒，就可方便地得到$[^{99m}Tc(CO)_3(MIBI)_3]^+$，大大增加了其应用的前景。

我们还将放射化学产率分别为85%和75%的两种产物的心/肝、心/血和心/肺比值与$[^{99m}Tc(MIBI)_6]^+$的靶器官与非靶器官比值进行了比较，如图2~图4所示。结果表明，注射后60min时三者中以$[^{99m}Tc(MIBI)_6]^+$的心/肺比为最高，而其心/肝比却是最低。此外，我们还可以看出，$[^{99m}Tc(CO)_3(MIBI)_3]^+$的放射化学产率越高，其心/肝比也越高，而心/肺比则越低。根据这些初步的结果，我们推测制得的$[^{99m}Tc(CO)_3(MIBI)_3]^+$产物中包含了少量的$[^{99m}Tc(MIBI)_6]^+$，其在生物体内的分布是两者综合作用的结果。当$[^{99m}Tc(CO)_3(MIBI)_3]^+$的放射化学产率较高时，产物中$[^{99m}Tc(MIBI)_6]^+$的量则相对较

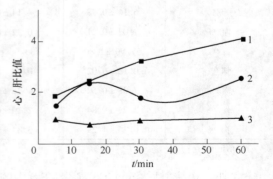

图2 不同放射化学产率的
$[^{99m}Tc(CO)_3(MIBI)_3]^+$与$[^{99m}Tc(MIBI)_6]^+$的
心/肝比值的比较

1—$[^{99m}Tc(CO)_3(MIBI)_3]^+$的放射化学产率为85%；
2—$[^{99m}Tc(CO)_3(MIBI)_3]^+$的放射化学产率为75%；
3—$[^{99m}Tc(MIBI)_6]^+$

少，其在生物体内的分布与纯的$[^{99m}Tc(CO)_3(MIBI)_3]^+$更为接近。由此看出，若产物中存在适量的$[^{99m}Tc(MIBI)_6]^+$，既可在一定程度上保留$[^{99m}Tc(CO)_3(MIBI)_3]^+$心/肝比高的特点，又可改善其心/肺比较低的缺陷。因此，比例适当的混合物表现出的生物分布特性较单独使用$[^{99m}Tc(CO)_3(MIBI)_3]^+$或$[^{99m}Tc(MIBI)_6]^+$更为理想。综上所述，羰基锝络合物$[^{99m}Tc(CO)_3(MIBI)_3]^+$的良好的心脏摄取和保留，及其高的心/肝比使得该络合物成为一种很有发展潜力的心肌显像剂。

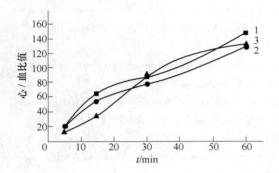
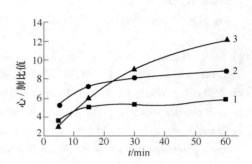

图3 不同放射化学产率的$[^{99m}Tc(CO)_3(MIBI)_3]^+$与$[^{99m}Tc(MIBI)_6]^+$的心/血比值的比较

1—$[^{99m}Tc(CO)_3(MIBI)_3]^+$的放射化学产率为85%；
2—$[^{99m}Tc(CO)_3(MIBI)_3]^+$的放射化学产率为75%；
3—$[^{99m}Tc(MIBI)_6]^+$

图4 不同放射化学产率的$[^{99m}Tc(CO)_3(MIBI)_3]^+$与$[^{99m}Tc(MIBI)_6]^+$的心/肺比值的比较

1—$[^{99m}Tc(CO)_3(MIBI)_3]^+$的放射化学产率为85%；
2—$[^{99m}Tc(CO)_3(MIBI)_3]^+$的放射化学产率为75%；
3—$[^{99m}Tc(MIBI)_6]^+$

致谢

感谢北京师范大学师宏药物研制中心提供 MIBI 药盒及唐志刚老师在实验过程中的大力协助。本工作为国家自然科学重点基金资助项目（批准号：29731020-1）。

参 考 文 献

[1] Steigman J, Eckelman W. The Chemistry of Technetium in Medicine, Vol 1. Washington D C：The National Research Council, Board on Chemical Sciences and Technology, 1992.

[2] Juris son S, Berning D, Jia W, et al. Coordination compounds in nuclear medicine. ChemRev, 1993, 93：1137～1156.

[3] Alberto R, Schibli R, Egli A, et al. Metal carbonyl synthesis：low pressure carbonylation of $[MOCl_4]^-$ and $[MO_4]^-$：the technetium (Ⅰ) and rhenium (Ⅰ) complexes $[NEt_4]_2[MCl_3(CO)_3]$. J Organomet Chem, 1995, 493：119～127.

[4] Alberto R, Schibli R, Schubiger P A, et al. Reactions with the technetium and rhenium carbonyl complexes $[NEt_4]_2[MX_3(CO)_3]$. Polyhedron, 1996, 15：1079～1089.

[5] Alberto R, Schibli R, Waibel R. Basic aqueous chemistry of $[M(OH_2)_3\text{-}(O_3)]^+$ (M = Re, Tc) directed towards radiopharmaceutical application. Coordination Chemistry Reviews, 1999, 190～192：901～909.

[6] Abram U, Abram S, Alberto R, et al. Ligand exchange reaction starting from $[Re(CO)_3Br_3]^{2-}$. Inorg Chim Acta, 1996, 248: 193~202.

[7] Schibli R, Alberto R, Abram U, et al. Structural and ^{99}Tc NMR investigations of complexes with fac-$[Tc(CO)_3]^+$ moieties and macro cyclic thio ethers of vanons ring sizes. Inorg Chem, 1998, 37: 3509~3516.

[8] Alberto R, Schibli R, Schubiger P A, et al. A simple single-step synthesis of $[^{99}Tc_3H_3(CO)_{12}]$ from $[^{99}TcO_4]^-$ and its X-ray crystal structure. Chem Commun, 1996, 61(9): 1291~1292.

[9] Schibli R, Alberto R, Schaffland A O, et al. Deniatization strategies of small biomolecules for the labeling with the organometallic $^{99m}Tc(CO)_3^-$ core. J Labelled Cpd Radiopharm, 1999, 42(Suppl 1): 147~149.

[10] Schibli R, Alberto R, Labella R, et al. Invivo and invitro evaluation of bifunctional chelators for the labeling with fac-$[Tc(OH_2)_3(CO)_3]^+$. J Labelled Cpd Radiopharm, 1999, 42(Suppl 1): 228~230.

[11] Waibel R, Alberto R, Willnda J, et al. Stable one-step technetium-99m labeling of his-tagged recombinant proteins with a novel Tc(I)-carbonyl complex. Nature Biotechnology, 1999, 17: 897~901.

[12] Marmion M E, Macdonald J R. Preparation and biodistribution of $^{99m}Tc(I)$-tricarbonyl complexes containing isonitrile and phosphine ligands. J Nucl Med, Proceedings of the 47th American Nuclear Medicine Annual Meeting. Reston: The Society of NuclearMedicine, 2000.

[13] Waibel R, Alberto R, Willnda J, et al. Method for the preparation of facial metal tricarbonyl compounds and their use in the labelling of biologically active substrates. A Patent WO, 9848848, 1998-11-05.

[14] Alberto R, Schibli R, Schubiger A P. First application offac-$[TC(OH_2)_3(CO)_3]^+$ in bioorgano metallic chemistry: design, structure, and in vitro affinity of a 5-HT1A receptor ligand labeled with ^{99m}Tc. J Am Chem Soc, 1999, 121: 6076~6077.

Solvation Effects on Brain Uptakes of Isomers of ^{99m}Tc Brain Imaging Agents*

Abstract Analysis of electrostatic hydration free energies of the isomers of the ^{99m}Tc-BAT and ^{99m}Tc-DADT complexes is carried out using the computer simulation technique. The results show that not only a correlation exists between the logarithm of the brain uptake and the electrostatic hydration free energy for the isomers of ^{99m}Tc-brain radiopharmaceuticals, but also a linear relationship exists between the logarithm of the ratio of the brain uptake of the syn isomer to that of the anti one and the difference between the electrostatic hydration free energy of the syn-isomer and that of the anti one. Furthermore, the investigation on the important factors influencing the brain uptakes of ^{99m}Tc-radiopharmaceuticals and the reasons of the different biodistribution of the isomers of the ^{99m}Tc-complexes is explored at the molecular level. The results may provide a reference for the rational drug design of brain imaging agents.

Key words ^{99m}Tc-brain imaging agent, isomer, brain uptake, electrostatic hydration free energy

In recent years, rapid development has been achieved in the field of ^{99m}Tc-labelled radiopharmaceuticals for brain perfusion imaging agents. ^{99m}Tc-d, l-HMPAO, ^{99m}Tc-L, L-ECD and ^{99m}Tc-MRP20 have been clinically used in the diagnosis of brain disease. It has been shown that it is necessary for the ^{99m}Tc-brain radio-pharmaceuticals to be neutral and lipophilic. Moreover, the properties of brain uptake and retention of ^{99m}Tc complexes depend on its stereochemistry. There are different biodistributions between the isomers of ^{99m}Tc-HMPAO and ^{99m}Tc-ECD and between ^{99m}Tc-BAT and ^{99m}Tc-DADT complexes[1-4]. Up to now, great achievements have been made in the experiments including synthesis and biodistribution of ^{99m}Tc complexes. It is well known that the investigation on the relationship between the properties of uptake and retention and the structure of ^{99m}Tc-radiopharmaceuticals on the molecular level may provide a reference for the rational drug design and synthesis experiments. Although quite a few progress in the stability of the complexes[5] and the structure-activity relationship[6-8] involved in the brain uptake and the retention mechanism have been reported, the calculation methods used in the investigation mentioned above are based on the data of X-ray crystallography. Solvent effects on the properties of ^{99m}Tc complexes have not been considered yet, while solvents have an important influence on the structure and dynamic properties of the solutes. Furthermore, biodistributions of ^{99m}Tc-brain radiopharmaceuticals depend on their properties *in vivo*. Therefore, it is necessary to study the structure-activity relationship of ^{99m}Tc complexes in solution and simulate the properties of ^{99m}Tc complexes *in vivo*. ^{99m}Tc-labelled perfusion imaging agents used in clinic nowadays have been

---

* Copartner: Jia Hongmei, Ma Xiaohui, Wang Cunxin. Reprinted from *Chinese Science Bulletin*, 2002, 47(21): 1786-1791.

extensively studied and all with TcO core. If there are substituents on the backbone, there will exist syn/anti orientation of substituents to the Tc=O bond. It has been demonstrated by the experiments that the biological properties of the syn-and anti-isomers are very different, while the essential reasons of the different biodistributions generated by the syn-and anti-stereochemistry have not been reported yet. To the best of our knowledge, the electrostatic contribution to the hydration free energy (ΔG_{el}) reflects the interaction between pharmaceutical molecule and water. And the brain uptakes of the isomers of ^{99m}Tc-BAT (C-substituents) and ^{99m}Tc-DADT (N-substituents) complexes have been reported in the literature. Therefore, in this note, the electrostatic hydration free energies of syn- and anti-isomers of ^{99m}TcO-N_2S_2 complexes are calculated based on the finite difference solution of the Poisson-Boltzmann (PB) equation. The correlation between the brain uptake and electrostatic hydration free energy is also studied. Furthermore, the investigation on the important factors influencing the brain uptakes of ^{99m}Tc-radiopharmaceuticals and the reasons of the different biodistributions of the isomers of the ^{99m}Tc-complexes is carried out. The results may provide a reference for the rational drug design of brain imaging agents.

1 Methods of Calculation

1.1 Structure optimization of ^{99m}Tc complex

The backbone structures of ^{99m}Tc complexes for brain perfusion imaging agents are shown in Fig. 1. The structures of R_1 and R_2 substituents are listed in Table 1 and Table 2, respectively. It should be noted that there is a methyl group between the BAT moiety and the amino-R_1 group for ^{99m}Tc-BAT complex, while the ethyl or hexyl group exists between the DADT moiety and the amino-R_2 group for ^{99m}Tc-DADT complex. The data of X-ray crystallography indicate that the Tc(=O)N_2S_2 core forms a distorted square pyramidal with oxygen in the apical position. The square plane defined by the nitrogen and sulfur atoms are also distorted, with the two nitrogen atoms lying above and below the plane and the two sulfur atoms lying above and below the plane by about 2.0×10^{-11} m. The Tc atom is situated above the plane by about 7.4×10^{-11} m. There are two isomers for the Tc=O core: syn-isomer in which substituent on the N_2S_2 skeleton is oriented syn to the Tc=O bond, and the anti-isomer in which it is counter to the Tc=O bond. Since the ^{99m}TcO-N_2S_2 core is the common structure of ^{99m}Tc-BAT and ^{99m}Tc-DADT complexes, geometry optimization of ^{99m}TcO-N_2S_2 skeleton is performed at first with the Hartree-Fock method and LANL2DZ basis set of G98W programs. On the basis of the optimizational geometry, the structures of 10 pairs of syn- and anti-isomers of ^{99m}TcO-BAT and ^{99m}TcO-DADT complexes are built up graphically with the Builder Module in Insight II programs. The molecular charges are taken form the esff force field. Then, all coordinates are refined by performing energy minimization and 5 ps molecular dynamics simulation in water at the temperature of 298 K using GROMACS programs[9]. Finally, the electrostatic effect is calculated from the conformation with the lowest energy.

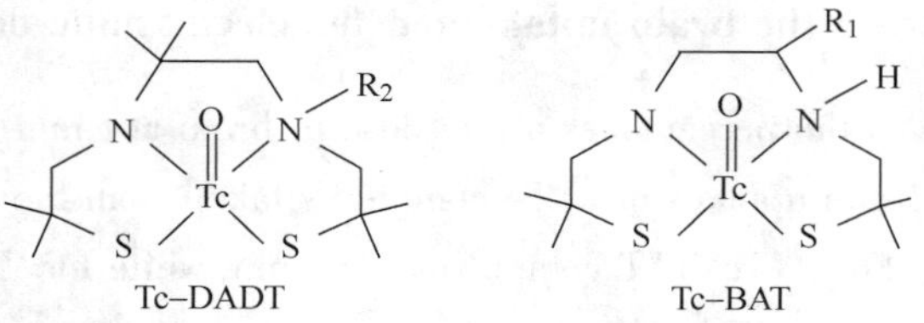

Fig. 1 Structures of ^{99m}Tc-BAT and ^{99m}Tc-DADT complexes

1.2 Calculation of electrostatic hydration free energy of ^{99m}Tc complex

Hydration free energy refers to the energy change of the complex when it is transfered from vacuum to water. It is an important physical quantity that reflects the interactions between the solute and water. Total hydration free energy includes two components: electrostatic contributions and non-electrostatic contributions. The former is the change of the electrostatic energy when the molecule is transfered from vacuum to water, while the latter is the Van der Waals' interaction related to the volume and surface area of the solute molecule. In this note, the treatment of electrostatic hydration free energy is based on the finite difference PB method[10,11]. The classical PB equation is given by:

$$\nabla \cdot [\varepsilon(r)\nabla \cdot \phi(r)] - \varepsilon(r)\kappa^2(r)\sinh[\phi(r)] + 4\pi\rho^f(r)/(k_BT) = 0 \quad (1)$$

where, $\phi(r)$ is the electrostatic potential of the position vector r; $\varepsilon(r)$ is the dielectric constant; ρ^f is the fixed charge density. The term $\kappa^2 = 1/\lambda^2 = 8\delta q^2 I/ek_BT$, where κ denotes the Debye-Huckel inverse constant, λ is the Debye-Huckel inverse length and I is the ionic strength of the bulk solution. The variables of $\phi, \varepsilon, \kappa, \rho$ are all functions of the position vector r. ϕ_o is the electrostatic potential in unit of k_BT/q (k_B is the Boltzmann constant, T is the absolute temperature and q is the charge on a proton). The value of ϕ_o is given by:

$$\phi_o = \frac{\sum_i^6 \varepsilon_i\phi_i + 4\delta q_o/\kappa_B Th}{\sum_i \varepsilon_i + \kappa^2 h^2} \quad (2)$$

where, ϕ_i and ε_i are the electrostatic potential and the relevant dielectric constant of six grid points around the point o, respectively; h is the linear grid distance. The value of ε_i for the solute is set to 2, and the dielectric constants for water and vacuum are set to 80 and 1, respectively. The number of grids of the system is set to 65. The electrostatic contribution to the solvation free energy is calculated by:

$$\Delta G_{el} = \frac{1}{2}\sum_i q_i(\phi_i^{80} - \phi_i^1) \quad (3)$$

where, q_i and ϕ_i refer to the charge and the potential at the grid i, respectively. The electrostatic desolvation free energy used in the present work is defined as ΔE, which is the electrostatic hydration free energy in the reversed process, i.e. $\Delta E = -\Delta G_{el}$.

1.3 Correlation between the brain uptake and the electrostatic desolation free energy

The brain uptake refers to the percentages of the dose in brain at t min postinjection relative to the total injected dose. Brain uptakes of ^{99m}Tc complexes taken from the literatures[1-4] are listed in Table 1 and Table 2. For ^{99m}Tc-BAT complexes, $t = 2$min, while for ^{99m}Tc-DADT complexes, $t = 5$min. When the logarithm of the brain uptake, logBU, is treated as dependent parameter and the electrostatic desolvation free energy, ΔE, as independent one, the quantitative correlation between logBU and ΔE is obtained with SPSS programs.

1.4 Correlation between $\log(BU_{syn}/BU_{anti})$ and $\Delta(\Delta E)$

In this work, $\log(BU_{syn}/BU_{anti})$ is the logarithm of the ratio of the brain uptake of the syn-isomer to that of the anti-isomer. $\Delta(\Delta E)$ denotes the difference between ΔE of syn-isomer and that of anti-isomer ($\Delta(\Delta E) = \Delta E_{syn} - \Delta E_{anti}$). The correlation between $\log(BU_{syn}/BU_{anti})$ and $\Delta(\Delta E)$ is also derived with SPSS programs.

2 Results of Calculation

The values of BU and ΔE for ^{99m}Tc-BAT and ^{99m}Tc-DADT complexes are shown in Table 1 and Table 2, respectively. The analysis results based on the data of Table 1 and Table 2 are presented in Eq. 4-Eq. 6.

Table 1 BU and ΔE of the isomers of the ^{99m}Tc-BAT complexes

R_1	Isomers	Brain uptake①/%	ΔE/kJ·mol^{-1}
	syn	2.27	42.15
	anti	0.99	48.06
	syn	0.04	54.61
	anti	0.05	57.32
	syn	2.77	53.75
	anti	0.57	62.45
	syn	2.34	43.75
	anti	1.08	47.69
	syn	1.80	44.89
	anti	0.61	49.65
	syn	1.97	44.18
	anti	1.08	47.30
	syn	1.85	45.74
	anti	1.55	49.79

Continued 1

R_1	Isomers	Brain uptake[①]/%	ΔE/kJ·mol^{-1}
(N-ethyl-3,5-dimethylpiperidine)	syn	1.88	44.44
	anti	0.88	46.53
(N-ethyl-2,6-dimethylpiperidine)	syn	1.17	46.51
	anti	0.29	52.40
(N-ethyl-4-phenylpiperidine)	syn	1.19	58.04
	anti	1.46	58.29

① % Dose/organ, $t = 2$ min.

Table 2 BU and ΔE of the isomers of the ^{99m}Tc-DADT complexes

R_2	Isomers	Brain uptake[①]/%	ΔE/kJ·mol^{-1}
(propyl-piperidine)	syn	1.95	43.52
	anti	1.11	44.88
(propyl-dimethylamine)	syn	1.94	43.24
	anti	1.43	43.48
(propyl-4-methylpiperidine)	syn	1.48	43.44
	anti	1.15	43.79
(propyl-4-propylpiperidine)	syn	0.74	44.92
	anti	1.00	43.99
(propyl-morpholine)	syn	0.69	62.09
	anti	0.61	61.13
(propyl-isopropylamine)	syn	0.47	46.33
	anti	0.88	44.97
(hexyl-piperidine)	syn	0.10	47.96
	anti	0.18	47.46
(hexyl-dimethylamine)	syn	0.12	48.30
	anti	0.18	48.32
(hexyl-morpholine)	syn	0.68	65.12
	anti	1.14	64.64
(hexyl-isopropylamine)	syn	0.05	50.19
	anti	0.06	49.92

① % Dose/organ, $t = 5$ min.

The indicator parameter I is set to 1 if there is no phenyl ring on the R_1 substituent, otherwise I is equal to 0. The correlation between BU and ΔE of the isomers of the ^{99m}Tc-BAT complexes is obtained as:

$$\log BU = -0.117(\pm 0.012)\Delta E - 1.401(\pm 0.169)I + 6.902(\pm 0.707)$$
$$n = 20, r = 0.923, s = 0.206, F_{2,17} = 48.640(\text{sig. } F = 0.000) \qquad (4)$$

If regression analysis is based on the complexes in which there is no phenyl ring, the result is given by:

$$\log BU = -0.123(\pm 0.013)\Delta E + 5.774(\pm 0.647)$$
$$n = 16, r = 0.925, s = 0.215, F_{1,14} = 82.526(\text{sig. } F = 0.000) \qquad (4')$$

As to the ^{99m}Tc-DADT complexes, another indicator parameter I' was defined as 1 when there is no ether O atom on the R_2 substituent. On the other hand, I' was equal to 0 when there is O atom. Based on these definitions, the correlation between BU and ΔE was presented as Eq. 5:

$$\log BU = -0.196(\pm 0.022)\Delta E - 3.617(\pm 0.407)I' + 12.285(\pm 1.417)$$
$$n = 20, r = 0.908, s = 0.223, F_{2,17} = 40.116(\text{sig. } F = 0.000) \qquad (5)$$

For the case that there is no O atom on the R_2 substituent, the result is shown in Eq. 5':

$$\log BU = -0.226(\pm 0.010)\Delta E + 10.024(\pm 0.475)$$
$$n = 16, r = 0.986, s = 0.097, F_{1,14} = 478.205(\text{sig. } F = 0.000) \qquad (5')$$

The correlation between $\log(BU_{syn}/BU_{anti})$ and $\Delta(\Delta E)$ is given by:

$$\log(BU_{syn}/BU_{anti}) = -8.75 \times 10^{-2}(\pm 0.012)\Delta(\Delta E) - 5.29 \times 10^{-2}(\pm 0.041)$$
$$n = 20, r = 0.863, s = 0.149, F_{1,18} = 52.399(\text{sig. } F = 0.000) \qquad (6)$$

where, n is the number of the ^{99m}Tc-complexes analysized in the regression analysis; r the multi-coefficient; s the standard error; $F_{d_{fr}, d_{fe}}$ the F value, where d_{fr} denotes the number of the independent parameters, $d_{fe} = n - d_{fr} - 1$, sig. F means the significance level of F.

3 Discussion

It is well known that the brain uptake is attributed to the lipophilicity, electrical properties as well as stereochemistry of ^{99m}Tc-complexes. As to the ^{99m}Tc-BAT complexes discussed in our work, since there is a chiral center on the BAT backbone (Fig. 1), the BAT complexes with chiral center have two isomers: the R_1 substituent is syn to the Tc=O bond and the other is anti to the Tc=O bond, following chelation with the ^{99m}Tc. Similarly, ^{99m}Tc-DADT complexes also exist as syn- and anti-isomers. The stereochemistry results in the difference between the lipophilicity, volume as well as other properties of the isomers. Therefore, the brain uptakes of the isomers are also different.

Because clinically useful brain uptake is the basis of the diagnosis of ^{99m}Tc-complexes, the complexes need to cross the intact blood brain barrier. It is required that the hydration layers of ^{99m}Tc-complex should be degenerated partly before the agent enters the brain tissue. To some extent this process approximates to the reserved process of hydration, namely desolvation. Therefore, the electrostatic desolvation free energy ΔE, is used to represent the ability that affects the process of crossing the intact blood brain barrier of ^{99m}Tc-complex. It can be deduced that the lower the value of ΔE, the less the desolvation free energy. In other words, the easier the molecule can be transfered from water to the lipid layer, the higher the initial brain uptake. As the results obtained from Eq. 4 and Eq. 5 show, the correlation between logBU and ΔE of the ^{99m}Tc-BAT and ^{99m}Tc-DADT complexes is good. It indicates that the brain uptakes are really relative to ΔE.

For the ^{99m}Tc-BAT complexes, since there is only one methyl group between the amino-R_1 substituent and the $TcON_2S_2$ backbone, the intramolecular hydrogen bond can be formed between the nitrogen atom of R_1 substituent and the amine hydrogen in the syn-isomer, which has been proved by X-ray crystallographic data[4,12] of ^{99m}Tc-BAT complexes. The formed intramolecular hydrogen bond can make the compact of the syn-isomer so that the lipophilicity will be increased as well as ΔE will be decreased, i. e. the molecule can more readily penetrate through the intact blood brain barrier. Therefore, the brain uptake of the syn-isomer could be higher than that of the anti-isomer. The conclusion obtained above also has been demonstrated by the results of the biodistribution. The brain uptakes of most syn-isomers are higher than that of the anti-ones. In the case of ^{99m}Tc-PPP complex, the brain uptake of the syn-isomer is lower than that of anti one. This is due to the relative rigidity of the phenylpiperizine pendant group. The rigidity precludes the compacting of the syn-isomer even though there exists the intramolecular hydrogen bond. So the difference between the molecular volume of both the syn-isomer and the anti-isomer is not so much. The results in our work indicate that there is only a little difference between ΔE of the syn-isomer and that of the anti one of ^{99m}Tc-PPP complex. Therefore, ΔE is the important factor influencing the brain uptake.

As for the ^{99m}Tc-DADT complexes, it is impossible to form the intramolecular hydrogen bond for both the syn- and anti-isomers. But the results of calculation also show that the brain uptake is negatively dependent on ΔE. This means that the high brain uptake will benefit from the decrease of ΔE.

It is a challenge to investigate the quantitative structure-activity relationship of drug isomers because there is only a little differences between the physicochemical parameters of the isomers such as octanol-buffer partition coefficient, while there exists much more differences in the biodistribution of isomers. Therefore, it is difficult to explain the difference of the biodistribution using apparent physicochemical properties of the drug isomers. In Eq. 4 and Eq. 5, it is demonstrated that ΔE is the important factor to affect the brain uptake. In order to explore the dependence of the brain uptake on the stereochemistry, the logarithm of the ratio of the brain uptake of the syn isomer to that of the anti one for 20 pairs of the ^{99m}Tc isomers is treated as a dependent variable and the difference of ΔE between the syn- and anti-isomer ($\Delta(\Delta E) = \Delta E_{syn} - \Delta E_{anti}$) is treated as an independent variable. Then, the linear correlation is found between both dependent and independent variables (Fig. 2). In Fig. 2 and Eq. 6, it is shown that due to the difference of ΔE between the synisomer and the anti one, the degree of degenerating the hydration layers of the isomers is also different when they cross the intact blood brain barrier. Therefore, the difference of the brain uptake is attributed to the different ΔE generated by the stereochemistry.

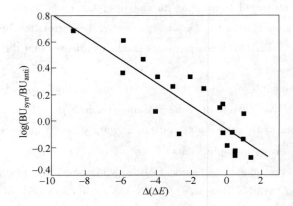

Fig. 2 The relationship between $\log(BU_{syn}/BU_{anti})$ and $\Delta(\Delta E)$

In summary, the electrostatic desolvation free energy ΔE of the ^{99m}Tc-BAT and ^{99m}Tc-DADT complexes can give reasonable explanation for the difference of the brain uptakes between syn- and anti-isomers. The value of ΔE is an important factor reflecting the brain uptake. It is the first time to investigate the quantitative structure-activity relationship of the syn- and anti-isomers of ^{99m}Tc labelled brain perfusion agents in water. The dehydration process is proposed in this note to study the ability of the ^{99m}Tc labelled brain imaging agents crossing the intact blood brain barrier and describe well the solvation effects on the brain uptakes of syn-isomer and anti-isomer in a rational way. Because the syn- and anti-stereochemistry is one of the most important characteristics of brain radiopharmaceuticals, the results herein can provide significant information for studying solvation effects on biodistribution of ^{99m}Tc complexes and designing new potential brain imaging agents.

Acknowledgements

This work was supported in part by the National Natural Science Foundation of China (Grant No. 29731020-1), the Beijing Natural Science Foundation (Grant No. 5992002), and the Development Project of Science and Technology of Beijing City Education Committee.

References

[1] Scheffel, U., Goldfarb, H. W., Lever, S. Z. et al., Comparison of technetium-99m aminoalkyl diaminodithiol (DADT) analogs as potential brain blood flow imaging agents, J. Nucl. Med., 1988, 29:73.
[2] Efange, S. M. N., Kung, H. F., Billings, J. et al. Technetium-99m bis(aminoethanethiol) complexes with amine sidechains—potential brain perfusion imaging agents for SPECT, J. Nucl. Med., 1987, 28:1012.
[3] Efange, S. M. N., Kung, H. F., Billings, J. J. et al., Synthesis and biodistribution of ^{99m}Tc-labeled piperidi-

nyl bis(aminoethanethiol) complexes: potential brain perfusion imaging agents for single photon emission computed tomography, J. Med. Chem. , 1988, 31:1043.

[4] Kung, H. F. , Guo. Y. Z. , Yu, C. C. et al. , New brain perfusion imaging agents based on ^{99m}Tc-bis(aminoethanethiol) complexes: stereoisomers and biodistribution, J. Med. Chem. , 1989, 32:433.

[5] Wei, Y. , Liu, B. L. , Kung, H. F. , Quantitative study of the structure-stability relationship of Tc complexes, Appl. Radiat. Isot. , 1990, 41:763.

[6] Meng, Z. X. , Jia, H. M. , Yang, W. et al. , Study of technetium chemistry, X , QSAR analysis of ^{99}Tc^m-labelled N_2S_2 analogues of brain imaging agents, J. Nucl. Radiochem. , 1998, 20:14.

[7] Jia, H. M. , Meng, Z. X. , Liu, B. L. , Study of technetium chemistry, XI, Study on QSAR of isomers of ^{99}Tc^m-brain radiopharmaceuticals, J. Nucl. Radiochem. , 1998, 20:20.

[8] Liu, B. L. , Jia, H. M. , Quantitative study of brain retention mechanism of ^{99m}Tc(V)O-chelates, J. Labelled Cpd Radiopharm. , 1995, 37:788.

[9] Berendsen, H. J. C. , van der Spoel, D. , van Drunen, R. , GROMACS: A message-passing parallel molecular dynamics implementation, Comp. Phys. Comm. , 1995, 91:43.

[10] Sitkoff, D. , Sharp, K. A. , Honig, B. , Accurate calculation of hydration free energy using macroscopic solvent models, J. Phys. Chem. , 1994, 98:1978.

[11] Honig, B. , Nicholls, A. , Classical electrostatics in biology and chemistry, Science, 1995, 268:1144.

[12] Francesconi, L. C. , Graczyk, G. , Wehrli, S. et al. , Synthesis and characterization of neutral M^{V}O (M = Tc, Re) amine-thiol complexes containing a pendant phenylpiperidine group, Inorg. Chem. , 1993, 32:3114.

Preparation of New Technetium-99m NNS/X Complexes and Selection for Brain Imaging Agent[*]

Abstract Based on excellent experiment results of ^{99m}TcO-MPBDA-Cl, two new ligands MPTDA and MPDAA are synthesized. Then series of ^{99m}TcO^{3+} complexes are prepared through adding different halide anions, followed by tests of physical chemistry qualities and biodistribution experiments. And results of these experiments show that complexes formed with MPTDA and MPDAA have better lipophilicity than those formed with MPBDA, still maintain the good brain retention ability of this type of compounds, but radioactivity uptake in blood is higher than that of ^{99m}TcO-MPBDA and ratios of brain/blood are reduced. Obvious affections are fetched out on brain uptake and retention if fluoride, bromide or iodide anions are added. Results of experiments can be explained in reason with theoretic computation. It is confirmed that ^{99m}TcO-MPBDA-Cl has potential to develop a new type of brain imaging agent considering integrated factors such as brain uptake, retention and toxicity.

Key words ^{99m}Tc, brain perfusion imaging agent, MPBDA, MPTDA, MPDAA

Quantification of regional cerebral blood flow (rCBF) plays an important role in the diagnosis of various cerebrovascular and neurological diseases. In nuclear medicine, brain perfusion imaging can provide situation of whole or regional cerebral blood flow perfusion (CBF or rCBF) to neurologists, help to find abnormality of cerebral blood flow before cerebrovascular and neurological diseases induce pathological changes in configuration or structure of brain. They offer important information for forepart diagnosis, therapy, curative effects observe and prognosis.

At present, ^{99m}Tc brain perfusion imaging agents developed overseas, ^{99m}TcO-D, L-HMPAO and ^{99m}TcO-L, L-ECD have been broadly applied in the domestic clinical work. But they do have certain disadvantages. It is an urgent affair to develop ^{99m}Tc radiopharmaceuticals with our own intelligent property rights whose characteristics are equal to the advanced levels abroad. In 1998 a new NNS ligand MPBDA was synthesized, then ^{99m}TcO-MPBDA was prepared and found have good initial brain uptake and retention ability[1,2]. This study (1) tries to modify MPBDA ligand, add methyl or methoxy group on phenyl moiety to gain better lipophilicity of complexes and then increase brain uptake; (2) changes anions participating in complexes formation to improve brain uptake and retention, finds out optimal complexes. In succession of biological experiments, a series of quantum chemistry computation are made to probe into structure-effect relationship of these compounds, furthermore to provide a theoretic basis for designing this kind of radiopharmaceuticals.

---

[*] Copartner: He Qiange, Chen Xiangji, Miao Yubin. Reprinted from *Science in China Ser. B Chemistry*, 2004, 47(6): 499-506.

	A	B	R
MPBDA			—H
MPTDA			—CH₃
MPTDAA			—OCH₃

1 Experimental

1.1 Synthesis of ligands

Synthesis of ligand MPBDA has been published in Ref. [1]. Steps of synthesizing MPTDA are as follows: 1.4g (0.019mol) of propylene sulfide is added dropwise into a solution of 3.0g (0.025mol) of 3,4-diaminobenzene in 20mL absolute ethanol. The reaction mixture is refluxed under nitrogen. Reaction is inspected by silica gel thin layer chromatography (TLC) with mixture V(anhydrous ether) : V(petroleum ether) = 1 : 2 as fluid phase. R_f value of the material is 0.1. The new compound with $R_f = 0.3$ can be seen after about 2.5h. Almost all solvent is removed under the reduced pressure and the residue is dissolved in 100mL of anhydrous ether. The ether solution is extracted 3 times with 20mL NaOH solution (0.5mol/L). Water solution is adjusted to pH = 7 with hydrochloric acid, then reversed extraction is processed with anhydrous ether. Effect of the extraction is checked by TLC until there is no trace of production in water. The ether is incorporated and condensed under the reduced pressure. The thick residue is purified through silica gel column (V(anhydrous ether) : V(petroleum ether) = 1 : 2). Dry hydrogen chloride gas is bubbled into the collected solution to give white deposite. The precipitated solid is collected under suction and blown to dry in nitrogen gas. 0.6g white powder was gained (16.1% yield), with melt point of 135-137℃. Synthesis of ligand MPDAA has analogy with that of MPTDA, but the yield is 12.8% and melt point is 80-83℃.

1.2 Preparation of complexes with $^{99m}TcO^{3+}$ core

1.0 mg of MPBDA and 1.0 mg of γ-cyclodextrin are dissolved in 1.0mL distilled water in a 10mL vial. 0.1mL of stannous chloride solution (formed by 1mg of stannous chloride dihydrate in 1.0mL of 1mol/L hydrochloride acid solution) is injected into the vial continuously. Shaking the vial makes the mixture uniform. After adjusting pH of mixture to 3.2, 1.0mL (74MBq) of fresh sodium pertechnetate ($Na^{99m}TcO_4$) is added into the system. The mixture reacts at room temperature (about 20℃) for 15min. Complex labeling yield is assayed by thin layer chromatography (TLC). The supporter is Xinhua No.1 chromatography strip and fluid is the mixture of methanol/chloroform (V/V = 1 : 9). The chromatography separates ^{99m}TcO-MPBDA (R_f = 0.9-1.0) from $^{99m}TcO_4^-$ (R_f = 0-0.1) and $^{99m}TcO_2 \cdot nH_2O$ (R_f = 0-0.1).

If certain halide salt is added to concentration double to [Cl$^-$] before adding Na^{99m}TcO$_4$, the complex prepared would be ^{99m}TcO-MPBDA-F, ^{99m}TcO-MPBDA-Br or ^{99m}TcO-MPBDA-I.

Labeling methods of MPTDA and MPDAA resemble that of MPBDA.

1.3 Determination of partition coefficient P

0.1mL solution of fresh ^{99m}Tc complex (approx. 3.7MBq) is put into a centrifugal tube, then add 1.9mL phosphate buffer solution (PBS, 0.025mol/L, pH = 7.4) and 2.0mL 1-octanol into it. The mixture is surged well at room temperature, then centrifuged at 4000r/min for 5min. A 0.2mL aliquot of both phases is taken to 2 tubes and counted in a well gamma counter. The partition coefficient, P, is calculated using the following equation: P = (counters in octanol-counters in background)/(counters in water-counters in background). Common logarithm of P, lg$P$, is called hydrophobic parameter.

1.4 Determination of stability *in vitro*

Complexes are placed at room temperature and labeling yields are determined after 1h, 3h and 6h.

1.5 Determination of electric charge

Complex drop is put at the central section of Xinhua No.1 chromatography strip and blown to dry. The strip is put on the surface of electrophoresis fluid(V(PBS) : V(physiological saline) : V(distilled water) : V(ethanol) = 2 : 8 : 75 : 250). Voltage and current are set to 180V, 10mA, respectively. After 6h the strip is cut into pieces. Radioactivity of two ends and the origin is counted.

1.6 Biodistribution in mice

Biodistribution analyses are performed in Kunming female mice weighing 20g. Each mouse is intravenously injected with 0.1mL solution containing 555-740kBq ^{99m}TcO complexes. The mice are killed separately 2min, 5min, 15min, 30min and 60min after injection. The organs and tissues of interest are removed, weighed and assayed for radioactivity. Blood values were taken as 7% of the total body weight.

1.7 Quantum chemistry computation

Geometry optimizations are performed with the Hartree-Fock method and LANL2DZ basis set of G98W programs[3]. The complexes of MPBDA, MPTDA and MPDAA are optimized in vacuum by setting convergence criteria as defult[4].

2 Results and Discussions

2.1 Synthesis of ligands

Compared with MPBDA, MPTDA and MPDAA are more unstable and difficult to separate from reaction mixture. The new compounds are characterized by IR, ^{1}H-MRI and MS analysis.

MPTDA: IR (cm^{-1}): 3381 (NH$_2$, NH); 1623, 1558, 1516, 1455, 823 (φ); ^{1}H-NMR: 7.69 (s, 1H, 6-C$_6$H$_3$), 7.57-7.58 (d, 1H, 3-C$_6$H$_3$), 7.21-7.23 (d, 1H, 4-C$_6$H$_3$), 3.71-3.72 (m, 2H, NCH$_2$), 3.40-3.60 (m, 1H, SCH), 2.59-2.60 (m, 1H, SH), 2.38 (s, 3H, φ-CH$_3$), 1.40-1.41 (d, 3H, CCH$_3$); MS, m/z: 196 (M$^+$), 149 (M$^+$—CH$_3$—SH), 135 (M$^+$—CH(CH$_3$)SH).

MPDAA: IR (cm^{-1}): 3445 (NH$_2$, NH), 1614, 1560, 1513, 1452, 847 (φ), 1216 (C-O-C); ^{1}H-NMR: 8.5-10.5 (NH, NH$_2$), 7.55-7.57 (d, 1H, 3-C$_6$H$_3$), 7.18 (s, 1H, 6-C$_6$H$_3$), 6.69-6.71 (d, 1H, 4-C$_6$H$_3$), 3.81 (s, 3H, OCH$_3$), 3.59-3.70 (m, 2H, NCH$_2$), 3.43 (m, 1H, SCH), 2.52 (m, 1H, SH), 1.38-1.43 (d, 3H, CCH$_3$); MS, m/z: 212 (M$^+$), 165 (M$^+$—CH$_3$—SH), 151 (M$^+$—CH(CH$_3$)SH).

Construing of ^{1}H-NMR indicates there are isomers produced in synthesis of MPTDA and MPDAA. The isomers are difficult to be separated by column chromatography because the structures of them are very similar. They may be separated by recrystallizing. But ligands MPTDA and MPDAA are unstable when they contain little amount of solvent. So recrystallizing method is not adopted.

2.2 Labeling of ^{99m}TcO^{3+} core and tests of physical chemistry *in vitro*

In form of NNS/X complexes, the chloride anions could not be avoided absolutely because fresh Na^{99m}TcO$_4$ was obtained by eluting ^{99}Mo-^{99m}Tc generators with physiological saline and stannous chloride was used as reductant. The amount of halide salt was controlled to keep the halide anion concentration times to the concentration of chloride anion, then the effects of added halide anions could exceed that of chloride to make certain complexes. It is well known that F$^-$, Br$^-$ and I$^-$ have toxicities to organism. So the amount of halides added is fixed on double to amount of Cl$^-$. The behaviors of NNS/X complexes are very similar in TLC and there is no difference in their R_f values.

Labeling yields of these 12 kinds of complexes are all higher than 95% and no obvious change can be seen even when they are placed at room temperature for 6h. Electrophoresis determinations indicate that complexes in the series are all neutral. Table 1 contains hydrophobic parameters of all these compounds.

Table 1 lg$P$ of all these NNS/X complexes

Element	MPBDA	MPTDA	MPDAA
F	2.17	2.30	2.23
Cl	1.64	1.65	1.81
Br	1.92	1.79	2.09
I	2.10	2.26	2.14

2.3 Biodistribution in mice

Table 2-Table 9 contain data of ^{99m}Tc-MPBDA-X and ^{99m}Tc-MPTDA-X.

Table 2 %ID/organ of ^{99m}TcO-MPBDA-F

Organ	%ID/organ				
	2min	5min	15min	30min	60min
Blood	12.56 ± 0.88	5.31 ± 0.75	4.85 ± 0.49	4.15 ± 0.34	3.29 ± 0.13
Brain	1.81 ± 0.22	2.63 ± 0.21	2.46 ± 0.52	2.82 ± 0.23	2.68 ± 0.05
Heart	2.21 ± 0.07	2.57 ± 0.16	2.84 ± 0.26	2.80 ± 0.21	2.68 ± 0.34
Liver	59.30 ± 6.71	49.36 ± 3.06	51.46 ± 3.25	47.02 ± 7.64	34.13 ± 5.76
Kidney	6.09 ± 1.71	6.33 ± 1.00	6.76 ± 1.03	7.49 ± 0.70	6.25 ± 1.00
Spleen	1.34 ± 0.13	1.30 ± 0.28	1.36 ± 0.06	1.36 ± 0.14	1.12 ± 0.56
Lung	7.09 ± 2.58	4.21 ± 0.40	3.87 ± 0.40	5.23 ± 1.90	2.05 ± 0.34

Table 3 %ID/organ of ^{99m}TcO-MPBDA-Cl

Organ	%ID/organ				
	2min	5min	15min	30min	60min
Blood	7.06 ± 0.12	4.93 ± 0.17	2.90 ± 0.02	2.23 ± 0.04	1.97 ± 0.05
Brain	2.49 ± 0.46	1.83 ± 0.18	1.23 ± 0.10	1.09 ± 0.32	0.93 ± 0.13
Heart	2.23 ± 0.39	1.72 ± 0.09	1.33 ± 0.14	1.23 ± 0.06	1.34 ± 0.06
Liver	15.68 ± 2.77	16.47 ± 1.29	16.28 ± 1.63	14.08 ± 1.45	8.54 ± 0.83
Kidney	3.30 ± 0.41	2.74 ± 0.24	2.93 ± 0.37	2.93 ± 0.37	3.14 ± 0.18
Spleen	0.36 ± 0.12	0.31 ± 0.04	0.36 ± 0.03	0.36 ± 0.03	0.32 ± 0.10
Lung	2.52 ± 0.39	2.03 ± 0.22	1.61 ± 0.30	0.99 ± 0.19	0.82 ± 0.12

Table 4 %ID/organ of ^{99m}TcO-MPBDA-Br

Organ	%ID/organ				
	2min	5min	15min	30min	60min
Blood	12.53 ± 0.98	6.62 ± 0.32	3.86 ± 0.25	3.02 ± 0.26	2.09 ± 0.14
Brain	4.65 ± 0.62	3.35 ± 0.51	1.78 ± 0.15	1.80 ± 0.23	1.22 ± 0.09
Heart	2.97 ± 0.47	2.79 ± 0.19	2.40 ± 0.14	1.99 ± 0.30	2.12 ± 0.38
Liver	24.76 ± 2.65	26.53 ± 4.78	20.57 ± 3.29	18.26 ± 1.94	13.51 ± 1.69
Kidney	6.51 ± 0.45	5.47 ± 0.84	4.75 ± 0.20	4.00 ± 0.57	3.91 ± 0.33
Spleen	0.62 ± 0.14	0.73 ± 0.14	0.47 ± 0.45	0.45 ± 0.12	0.34 ± 0.08
Lung	4.61 ± 0.29	2.98 ± 0.11	2.02 ± 0.22	1.65 ± 0.20	13.4 ± 0.22

Table 5 %ID/organ of ^{99m}TcO-MPBDA-I

Organ	%ID/organ				
	2min	5min	15min	30min	60min
Blood	9.16 ± 0.32	5.71 ± 0.30	4.55 ± 0.28	4.38 ± 0.16	2.38 ± 0.18
Brain	2.98 ± 0.55	1.91 ± 0.22	1.67 ± 0.16	1.55 ± 0.13	1.28 ± 0.10
Heart	2.62 ± 0.26	1.83 ± 0.24	1.75 ± 0.20	1.70 ± 0.06	1.76 ± 0.26
Liver	27.42 ± 4.73	29.00 ± 7.45	28.23 ± 5.29	26.41 ± 2.92	22.29 ± 2.34
Kidney	3.88 ± 0.49	3.93 ± 0.76	4.02 ± 4.15	4.15 ± 0.21	3.93 ± 0.31
Spleen	0.60 ± 0.10	0.74 ± 0.21	0.58 ± 0.13	0.45 ± 0.06	0.41 ± 0.07
Lung	3.17 ± 0.58	2.44 ± 0.27	2.29 ± 0.33	2.12 ± 0.40	2.19 ± 0.17

Table 6 %ID/organ of ^{99m}TcO-MPTDA-F

Organ	%ID/organ				
	2min	5min	15min	30min	60min
Blood	10.91 ± 0.32	4.77 ± 0.18	2.65 ± 0.09	1.78 ± 0.13	1.76 ± 0.03
Brain	0.41 ± 0.05	0.37 ± 0.15	0.50 ± 0.08	0.49 ± 0.11	0.66 ± 0.13
Heart	0.87 ± 0.07	0.71 ± 0.23	0.63 ± 0.06	0.67 ± 0.15	1.76 ± 0.03
Liver	39.41 ± 2.65	34.17 ± 8.61	35.03 ± 0.96	31.90 ± 1.67	36.13 ± 2.06
Kidney	3.13 ± 1.18	2.26 ± 0.50	2.52 ± 0.11	0.40 ± 0.17	2.28 ± 0.10
Spleen	1.47 ± 0.49	1.35 ± 0.22	0.80 ± 0.26	2.20 ± 0.21	3.47 ± 0.83
Lung	15.74 ± 2.04	10.53 ± 4.84	16.66 ± 1.65	13.22 ± 2.17	9.91 ± 1.81

Table 7 %ID/organ of ^{99m}TcO-MPTDA-Cl

Organ	%ID/organ				
	2min	5min	15min	30min	60min
Blood	14.35 ± 0.55	9.94 ± 0.41	8.47 ± 0.30	7.10 ± 0.27	3.77 ± 0.20
Brain	1.31 ± 0.04	1.09 ± 0.05	1.05 ± 0.05	1.07 ± 0.07	0.94 ± 0.03
Heart	1.68 ± 0.09	1.30 ± 0.13	1.09 ± 0.06	1.11 ± 0.03	0.79 ± 0.04
Liver	20.95 ± 1.56	23.18 ± 3.10	19.91 ± 3.25	17.12 ± 0.31	18.69 ± 1.30
Kidney	4.61 ± 0.87	4.28 ± 0.41	4.25 ± 0.63	3.88 ± 0.60	3.02 ± 0.22
Spleen	0.72 ± 0.01	0.65 ± 0.10	0.77 ± 0.09	0.72 ± 0.09	0.55 ± 0.08
Lung	3.77 ± 0.25	3.01 ± 0.58	2.61 ± 0.34	2.12 ± 0.15	1.65 ± 0.15

Table 8 %ID/organ of ^{99m}TcO-MPTDA-Br

Organ	%ID/organ				
	2min	5min	15min	30min	60min
Blood	27.22 ± 0.14	21.36 ± 0.51	14.56 ± 0.09	12.96 ± 0.53	8.90 ± 0.20
Brain	2.47 ± 0.02	2.42 ± 0.13	1.95 ± 0.06	1.71 ± 0.06	1.16 ± 0.11
Heart	3.35 ± 0.23	2.12 ± 0.37	1.89 ± 0.25	1.62 ± 0.05	1.28 ± 0.09
Liver	34.59 ± 1.18	29.21 ± 7.13	31.00 ± 1.02	32.52 ± 1.81	25.44 ± 1.84
Kidney	7.72 ± 0.66	6.91 ± 1.02	6.40 ± 0.42	5.79 ± 0.94	5.12 ± 0.19
Spleen	1.33 ± 0.32	1.34 ± 0.46	0.96 ± 0.10	1.03 ± 0.12	0.89 ± 0.14
Lung	7.44 ± 0.43	5.61 ± 1.21	4.43 ± 0.56	3.61 ± 0.86	2.36 ± 0.15

Table 9 %ID/organ of ^{99m}TcO-MPTDA-I

Organ	%ID/organ				
	2min	5min	15min	30min	60min
Blood	18.48 ± 0.99	11.54 ± 0.48	8.36 ± 0.68	6.02 ± 0.06	5.19 ± 0.10
Brain	1.78 ± 0.09	1.73 ± 0.20	1.34 ± 0.05	1.20 ± 0.03	0.87 ± 0.09
Heart	2.64 ± 0.10	2.00 ± 0.26	1.29 ± 0.16	1.17 ± 0.07	1.03 ± 0.14
Liver	36.98 ± 3.19	35.11 ± 4.48	32.88 ± 4.10	26.80 ± 3.24	29.39 ± 3.21
Kidney	5.79 ± 0.25	5.47 ± 0.13	5.55 ± 0.95	4.86 ± 0.25	4.03 ± 0.40
Spleen	1.80 ± 0.15	1.38 ± 0.10	1.47 ± 0.41	0.95 ± 0.09	0.77 ± 0.12
Lung	10.28 ± 1.02	6.95 ± 0.93	4.98 ± 0.11	4.01 ± 0.82	2.70 ± 0.12

It can be seen from the tables that each compound has good ability of initial brain uptake and retention. The increment of the methyl group does not help to improve brain uptake, but the radioactivity in blood increases, and then ratios of brain/blood decline. The fluoride, bromide and iodide anions add have obvious effects on brain uptake and retention, especially when Br^- is added the complexes have best ability of initial brain uptake. Complexes with F^- are lower at the level of brain uptake but they accumulate in brain. Scince toxicity induced by halide anions will restrict complexes used as pharmaceuticals, biodistribution experiments of MPDAA are carried out only with ^{99m}TcO-MPDAA-Cl and the data are listed in Table 10.

Table 10   %ID/organ of ^{99m}TcO-MPDAA-Cl

Organ	%ID/organ				
	2min	5min	15min	30min	60min
Blood	20.54 ± 2.65	11.74 ± 0.16	7.66 ± 1.46	5.45 ± 0.49	4.12 ± 0.29
Brain	1.47 ± 0.74	1.67 ± 0.29	1.91 ± 0.49	1.48 ± 0.11	1.55 ± 0.47
Heart	3.68 ± 1.39	2.57 ± 0.86	1.45 ± 0.21	1.23 ± 0.07	1.16 ± 0.08
Liver	55.3 ± 23.30	72.0 ± 13.68	77.74 ± 4.84	70.3 ± 10.97	54.92 ± 5.69
Kidney	11.55 ± 3.16	12.41 ± 0.79	11.69 ± 1.55	9.66 ± 1.91	7.05 ± 1.02
Spleen	1.09 ± 0.61	1.61 ± 0.50	1.67 ± 0.19	1.86 ± 0.35	1.75 ± 0.39
Lung	8.39 ± 1.97	7.11 ± 0.54	4.61 ± 0.65	3.73 ± 0.51	2.76 ± 0.34

Brain uptake of ^{99m}TcO-MPDAA-Cl after 15min is higher than that of ^{99m}TcO-MPBDA-Cl (Fig. 1). But the radioactivity in blood increases and ratios of brain/blood decline.

Complexes formed with MPTDA and MPDAA have better lipophilicity than those formed with MPBDA, but it does not help to improve brain uptake as anticipation. There are two reasons: (1) better lipophilicity makes more combining with serous albumen; (2) increment of methyl or methoxyl group leads to changes of metabolism routes *in vivo*, complexes are more easily to be translated into polar species.

2.4 Quantum chemistry computation

After geometry optimization, we found that: there were two kinds of configurations of MPDAA complexes formed from its A-typed ligand. There were two kinds of orientation between methoxy group and the Tc=O bond. One was that methyl group and Tc=O bond were in the same direction (Fig. 2); the other was in opposite direction (Fig. 3). Howerver, there was only one type of configuration of MPDAA complexes formed from its B-typed ligand.

The computational results (Table 11-Table 13) show that:

(1) There is good correlation between theoretically calculated dipole moment and hydrophobic parameter. The higher the calculated dipole moment, the lower the hydrophobic parameter is. The dipole moments of all isomers are in the order of F < I < Br < Cl, and the hydrophobic parameters of all isomers are in the order of F > I > Br > Cl.

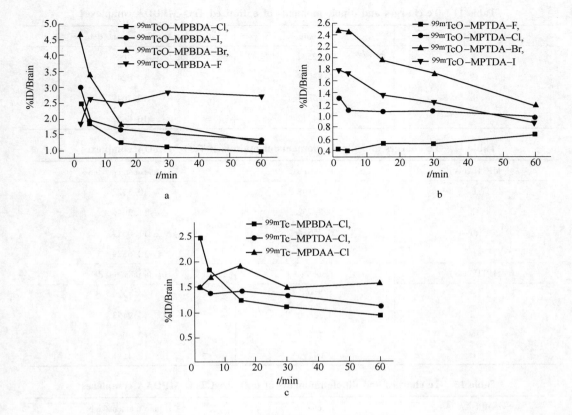

Fig. 1　Uptake-time curves in brain

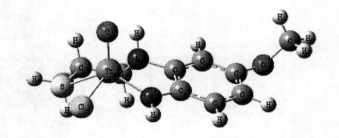

Fig. 2　Optimized syn-TcO-MPDAA-Cl

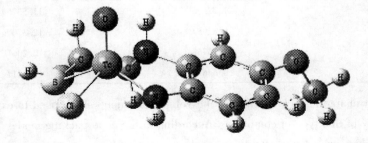

Fig. 3　Optimized anti-TcO-MPDAA-Cl

Table 11 Tc charges and dipole moments of optimized TcO-MPBDA complexes

MPBDA	Tc charge	Dipole moment/deb
F	1.22582	9.8088
Cl	0.83048	10.6240
Br	0.75837	10.4649
I	0.64712	10.0851

Table 12 Tc charges and dipole moments of optimized TcO-MPTDA complexes

MPTDA(A)	Tc charge	Dipole moment/deb
F	1.22198	10.3728
Cl	0.82752	11.2046
Br	0.75554	11.0529
I	0.64493	10.6829
MPTDA(B)	Tc charge	Dipole moment/deb
F	1.21967	10.1229
Cl	0.82440	10.9453
Br	0.75264	10.7968
I	0.64193	10.4291

Table 13 Tc charges and dipole moments of optimized TcO-MPDAA complexes

Syn-MPDAA(A)	Tc charge	Dipole moment/deb
F	1.22360	9.4449
Cl	0.82832	10.2567
Br	0.75615	10.0982
I	0.64507	9.7154
Anti-MPDAA(A)	Tc charge	Dipole moment/deb
F	1.22451	10.0398
Cl	0.82944	10.8471
Br	0.75727	10.6958
I	0.64611	10.3301
MPDAA(B)	Tc charge	Dipole moment/deb
F	1.22576	11.5375
Cl	0.82957	12.3700
Br	0.75763	12.2061
I	0.64636	11.8117

(2) The net charge of technetium and the volume of anion can be used to explain the brain retention ability of this type of complexes. According to the cone packing model[5] and glutathione attack mechanism[6,7], the glutathione in brain plays an important role in brain retention of radiopharmaceuticals. Reductive L-glutathione (GSH) is a very popular tripeptide existing in

animals and human with a concentration of 2×10^{-3} mol/L. It can provide thiol anion by losing a proton easily, then attack the technetium atom from the contrapuntal vacancy to the Tc=O bond. The complexes are then turned into polar species, thus they could not penetrate the B. B. B. and be trapped in the brain.

The volume of contrapuntal cavity and the net charge of technetium are the main factors that can influence the thiol anion's attack. The larger the cavity and the more positive technetium atom, the more easily the thiol anion attacks technetium atom, then more of this complex will be trapped in brain. Fluorine anion has the smallest radius among these four halide anions, and the technetium net charges of its complexes are also the highest. So we can predict that the complexes formed with fluorine anion will have the highest brain retention. Experiments verified this point. The radius of Cl, Br and I anion is longer than that of F anion, and the technetium net charges of their complexes decrease. So their brain retention abilities decrease accordingly.

(3) The calculated results show that there is little difference in dipole moment and technetium net charge between two isomers of ^{99m}Tc-MPTDA. However, much difference exists among the isomers of ^{99m}Tc-MPDAA, and the most component may play the major role in the animal experiments.

3 Conclusions

Two new tridentate ligands MPTDA and MPDAA are synthesized. Series of complexes are prepare. Though increment of methyl or methoxyl group does not help to improve brain uptake, halide anions added affect the biodistribution in certain degrees. It is confirmed that ^{99m}TcO-MPBDA-Cl has potential to develop a new type of brain imaging agent considering integrated factors such as brain uptake, retention, ratio of brain/blood and toxicity.

The stability *in vitro* of ^{99m}TcO-MPBDA-Cl is better than that of ^{99m}Tc-D, L-HMPAO, and it overcomes the defect that results of ^{99m}TcO-L, L-ECD are affected by enzyme distribution because the retention mechanism is enzyme hydrolysis, less time of half clearance from blood is in favor of brain imaging. The intellectual property rights are whole owned by ourselves. So ^{99m}TcO-MPBDA-Cl could be declared for new medicine as a kind of brain imaging agents (Table 14).

Table 14 Biodistribution of ^{99m}TcO-MPBDA-Cl and ^{99m}TcO-L, L-ECD, ^{99m}TcO-D, L-HMPAO[①]

Name	Brain uptake (%ID/g organ)		Retention ratio	Half clearance time from blood
	2min	60min	R[②]	$T_{1/2}$/min
^{99m}TcO-L, L-ECD	0.71[③]	0.12	0.17	>30
^{99m}TcO-MPBDA-Cl	4.80	2.95	0.61	<15
^{99m}TcO-D, L-HMPAO	6.56	4.13	0.63	47.3

① Results of this table are based on mice weighing between 18-20g;

② R = brain uptake (60min)/brain uptake (2min);

③ this is brain uptake of ^{99m}TcO-L, L-ECD at 5min. The retention mechanism of ^{99m}TcO-L, L-ECD is enzyme hydrolyzed, and the enzyme only exists in brain of primates so the retention ability is worse in brain of mice.

Acknowledgements

This work was supported by the National Natural Science Foundation of China (Grant No. 29731020).

References

[1] Yubin Miao, Boli Liu, Synthesis of new N_2S ligands, preparation of ^{99m}Tc complexes and their preliminary biodistribution in mice, J. Labelled Cpd. Radiopharm. ,1999,42(7):629-640.

[2] Yubin Miao, Boli Liu, Technetium-99m-MPBDA, a new potential cerebral perfusion imaging agent: SPECT study on monkey, Technetium, Rhenium and Other Metals in Chemistry and Nuclear Medicine (ed. Nicolini, M. , Mazzi, U.), Italy: SGE Editoriali, 1999, 819-823.

[3] Frisch, M. J. , Trucks, G. W. , Schlegel, H. B. et al. , Gaussian 98, Revision A. 7, Pittsburph PA: Gaussian Inc, 1998.

[4] Jia, H. M. , Liu, B. L. , Meng, Z. X. et al. , Studies of technetium chemistry XIII: The relationship between solvation free energies and brain uptakes of ^{99m}Tc complexes, J. Nucl. & Rad. Chem. (in Chinese), 2002, 24(1):42-49.

[5] Fisher, R. O. , Li, Xingfu, The solid angle sum rule: a new ligand cone packing model of optimal applicability in organolanthanoid chemistry, J. Less-Common. Met. , 1985, 112(10):303-325.

[6] Ballinger, J R. , Reid, R. H. et al. , Technetium-99m HM-PAO stereoisomers: differences in interaction with glutathione, J. Nucl. Med. , 1988, 29(12):1998-2000.

[7] Nock, B. A. , Maina, T. , Yannoukakos, D. et al. , Glutathione-mediated metabolism of technetium-99m SNS/S mixed ligand complexes: a proposed mechanism of brain retention, J. Med. Chem. , 1999, 42(6): 1066-1075.

Synthesis, Separation and Biodistribution of ^{99m}Tc-CO-MIBI Complex*

Abstract ^{99m}Tc-CO-MIBI was prepared by a two-step procedure involving the convenient preparation of the $[^{99m}Tc(CO)_3(OH_2)_3]^+$ precursor and followed by the substitution of the water molecules by the MIBI (2-methoxyisobutylisonitrile) ligands. In a second step, the reaction solution was adjusted to different pH values, and then the product, ^{99m}Tc-CO-MIBI, was confirmed to be a mixture of two complexes: complex A and complex B, whose labeling yields could be over 90%. The ratio of complex B to the sum of A and B could increase gradually from 0 to 1 when pH was shifted from 3.0 to 9.0. These changes were monitored by thin layer chromatography (TLC) and high performance liquid chromatography (HPLC). The two complexes were stable within 8h at room temperature *in vitro*. The partition coefficient of the two complexes indicated that there was distinct difference between them. Biodistribution in mice demonstrated that complex B showed better myocardial imaging properties than that of complex A. The heart/liver ratios of complex A, the mixture, and complex B were 1.57, 1.93, and 2.33, respectively, for 30 min post-injection. The discovery of chemical and biological properties of ^{99m}Tc-CO-MIBI would certainly promote the research on a new promising myocardial perfusion-imaging agent.

Key words biodistribution, tricarbonyl core, MIBI, myocardial perfusion-imaging agent

1 Introduction

Technetium-99m, an excellent imaging isotope, has played an essential role in the diagnostic nuclear medicine. There are some successful technetium cores such as "naked" Tc atom, $[Tc=O]^{3+}$ core, $[Tc\equiv N]^{2+}$ core, $[O=Tc=O]^+$ core and $[Tc]$-HYNIC core. However, technetium cores in its low oxidation did not receive adequate attention until the $[Tc(CO)_3]^+$ core was developed. Several years ago Alberto et al. reported a one-step synthesis of the complex $[^{99m}Tc(CO)_3(OH_2)_3]^+$ by direct reduction of $^{99m}TcO_4^-$ with $NaBH_4$ in aqueous solution in the presence of CO at 1 atm[1,2]. $[^{99m}Tc(CO)_3(OH_2)_3]^+$ is proved to be a versatile intermediate for technetium labeling.

So far, there are several radiopharmaceuticals used widely in myocardial imaging. ^{99m}Tc-MIBI is one of them though it has comparatively high liver uptake and slow liver washout. Thus we utilized the $[^{99m}Tc(CO)_3(OH_2)_3]^+$ precursor and MIBI to prepare ^{99m}Tc-CO-MIBI, which was reported by Marmion[3] and Liu[4,5] independently almost at the same time. Dyszlewski also re-

---

* Copartner: Hao Guiyang, Zang Jianying, Zhu Lin, Guo Yuzhi. Reprinted from *Journal of Labelled Compounds and Radiopharmaceuticals*, 2004, 47: 513-521.

ported it as a transport substrate of the multidrug resistance P-glycoprotein[6]. Control experiments of ^{99m}Tc-CO-MIBI vs. $[^{99m}Tc(MIBI)_6]^+$ were carried out in mice[4] and dogs[7] earlier, in which ^{99m}Tc-CO-MIBI showed to be a promising myocardial perfusion-imaging agent with good myocardial uptake and fast liver washout. Subsequently, we found that ^{99m}Tc-CO-MIBI was a mixture with two main components in different reaction conditions whose biodistribution results were distinctively different[8]. The aim of this study was to further the study of the chemical and biological properties of ^{99m}Tc-CO-MIBI.

2 Experimental

2.1 Materials

The MIBI ligand was synthesized following the previously published procedure[9-11]. Pure CO gas was purchased from NRCCRM. ^{99}Mo/^{99m}Tc generator was obtained from the Beijing Syncor Medical Corporation. All other chemicals were obtained from Beijing Chemical Reagents Company and Aldrich Co.

2.2 Preparation of $[^{99m}Tc(CO)_3(OH_2)_3]^+$

The preparation of the $[^{99m}Tc(CO)_3(OH_2)_3]^+$ precursor was carried out as follows[2]: 4mg Na$_2$CO$_3$ and 5.5mg NaBH$_4$ were added to a 10mL vial, which was tightly closed and flushed with pure CO for 15min. A 3mL quantity of generator elute containing up to 7.4GBq Na$[^{99m}TcO_4]$ in saline was added. And the solution was heated to 75℃ for 30min. The reaction mixture was flnally cooled to room temperature.

2.3 Preparation of ^{99m}Tc-CO-MIBI

^{99m}Tc-CO-MIBI (complex A and complex B) were prepared in the following general procedure: 1.0mg MIBI, dissolved in 0.5mL water, was added to 1mL $[^{99m}Tc(CO)_3(OH_2)_3]^+$ solution, which was adjusted to a certain pH value with 0.5N HCl solution. The mixture was heated to 100℃ for 15min, then cooled to room temperature.

The labeling yields of ^{99m}Tc-CO-MIBI COMPLEX were calculated by TLC. The chromatographic analyses were performed on polyamide film with acetonitrile (Sys 1) and saline: acetone: strong aqua ammonia =9: 1: 0.1 ($v/v/v$) (Sys 2) as mobile phases.

The stability of complex A and complex B was determined at room temperature by measuring their yields at different times (1h, 2h, 3h, 4h, 5h, 6h, 7h and 8h) after preparation.

2.4 HPLC analysis

Radio high-pressure liquid chromatography (HPLC) experiments were performed by using a SHIMADZU System with SCL-10Avp HPLC pump system and Park radioflow detector. The column (Waters RP C18, 250mm × 4.6mm, 5μm) was eluted at a flow rate of 1.0mL/min. Characterization of $[^{99m}Tc(CO)_3(OH_2)_3]^+$:

Mobile phase[12]:

(1) TEAP 0.05 M, pH = 2.25;

(2) MeOH 100%.

Gradient: 0-3min 100% (1), 3-6min from 100% to 75% (1), 6-9min from 75% to 66% (1), 9-20min from 34% to 100% (2), 20-27min 100% (2), 27-30min from 100% (2) to 100% (1).

Characterization of complex A and complex B:

Mobile phase: (1) : (2) = 35 : 65.

2.5 Determination of the partition coefficient for the complexes

The lipophilicity of the complex A (or B) with yields over 95% was determined as follows: 0.1mL complex A (or B) solution was mixed with 2mL 1-octanol and 1.9mL PBS (0.01M, pH = 7.4) in a centrifuge tube. The tube was vortexed at room temperature for 3min and then was centrifuged at high speed for 10min. 0.1mL samples of both phases were pipetted into other test tubes with adequate care to avoid cross contamination between the phases and were counted in a well γ-counter. The measurement was repeated for three times. The partition coefficient, P, was calculated using the following equation:

$$P = (\text{cpm in octanol} - \text{cpm background})/(\text{cpm in water} - \text{cpm background})$$

Usually the final partition coefficient value was expressed as log P.

2.6 Biodistribution studies

Samples (about 740kBq in 0.1mL solution) were injected through the tail vain into mice (18-22g, female, obtained from Animal Center of Peking University). The mice were sacrificed at 5min, 15min, 30min, and 60min post-injection. Selected organs were collected for weighing and counting. The accumulated radioactivity in the tissue of organs was calculated in terms of percentage of injected dose per gram organ (%ID/g). The biodistribution experiments were performed in three groups (Ⅰ: complex A; Ⅱ: the ^{99m}Tc-CO-MIBI mixture; Ⅲ: complex B).

3 Results and Discussion

3.1 Preparation of ^{99m}Tc complexes

Over 95% labeling yields of the $[^{99m}Tc(CO)_3(OH_2)_3]^+$ precursor evaluated by reverse phase HPLC were found (retention time: 5.5min).

^{99m}Tc-CO-MIBI was prepared by substituting the MIBI ligands for the water molecules of the $[^{99m}Tc(CO)_3(OH_2)_3]^+$ precursor. Labeling yields of the final products were evaluated by TLC and HPLC. The R_f values in TLC analysis are shown in Table 1.

Table 1 R_f values for some complexes in TLC analysis

Mobile phase[①]	$^{99m}TcO_4^-$	$^{99m}TcO_2 \cdot nH_2O$	Complex A	Complex B
Sys1	0.4-0.6	0.0	0.9-1.0	0.9-1.0
Sys2	0.1-0.2	0.0	0.6-0.8	0.3-0.4

[①] Sys1: acetonitrile; Sys2: saline: acetone: strong aqua ammonia = 9 : 1 : 0.1 ($v/v/v$).

The precursor solution was adjusted to various pH values ranging from 3.0 to 9.0, and varying final products were obtained with labeling yields over 90% evaluated by the TLC-Sys1. When the precursor solution was nearly neutral, the product was a mixture of complex A and complex B. The preparation of complex A or B was determined by the pH conditions of the solution during the substitution reaction. When pH is between 3.0 and 5.0, complex A was the main product. When pH was close to 9.0, complex B was almost the exclusive product. In both conditions mentioned above the yields of complex A and complex B calculated by the TLC-Sys2 could be over 90%. The variation of the ratios of B/(A+B) with the pH values of the reaction solution is shown in Fig. 1. The experiments demonstrated that it was feasible to obtain pure complex A, pure complex B, or a mixture of A and B with a certain ratio (Fig. 2) by adjusting the pH of the reaction mixture. The products obtained under these conditions were used in the biodistribution studies below.

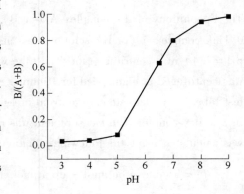

Fig. 1 The ratios of complex B to the sum of complex A and complex B

Stability experiment demonstrated that the two products A and B were stable within 8h at room temperature.

3.2 Determination of the partition coefficient for the complexes

The partition coefficient values ($\log P$) of complex A and complex B were -0.172 and 0.125, respectively.

3.3 Biodistribution studies

The results of biodistribution experiments of Ⅰ (complex A, yields 95%), Ⅱ (complex A, yields 67%; complex B, yields 30%) and Ⅲ (complex B, yields 90%) are shown in Table 2 and Fig. 3. Unlike the previous experiments in which the samples were obtained by separating the ^{99m}Tc-CO-MIBI mixtures by HPLC[8], the samples of Group Ⅰ, Ⅱ and Ⅲ were prepared directly by adjusting the pH values of the precursor solution and without further purification or separation by HPLC. The present experimental method was more convenient and more feasible for studying chemical and biological characters of complex A and complex B. Complex B

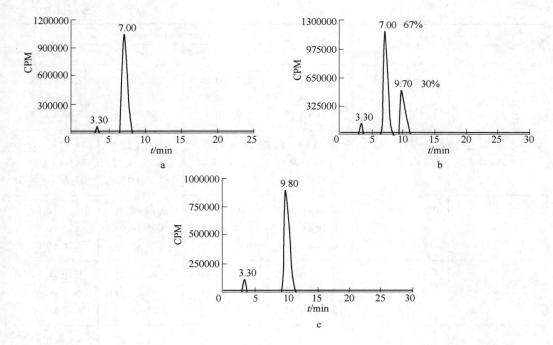

Fig. 2 HPLC chromatograms of complex A(a), the ^{99m}Tc-CO-MIBI mixture (complex A, complex B) (b) and complex B(c)

showed higher myocardial uptake and faster liver washout than complex A, and therefore complex B had considerably better heart/liver ratios in mice. For 30min post-injection the heart/liver ratios of Group Ⅰ, Ⅱ, Ⅲ and ^{99m}Tc-MIBI were 1.57, 1.93, 2.33 and 0.90[13], respectively. Since complex B had distinct improvement over ^{99m}Tc-MIBI in heart/liver ratios, it was possible to make the heart imaging at early post injection. Earlier, Liu et al. carried out the comparative pharmacology study of the ^{99m}Tc-CO-MIBI mixtures and ^{99m}Tc-MIBI in dogs[7], and they got the same results. All these showed that ^{99m}Tc-CO-MIBI could be a promising myocardial perfusion-imaging agent.

Table 2 Biodistribution of Group Ⅰ, Ⅱ and Ⅲ in mice (%ID/g, mean ± SE, n = 3)

Tissue	Groups①	Post-injection time/min			
		5	15	30	60
Blood	Ⅰ	1.39 ± 0.09	0.67 ± 0.03	0.50 ± 0.03	0.41 ± 0.02
	Ⅱ	1.69 ± 0.07	0.93 ± 0.06	0.74 ± 0.02	0.59 ± 0.03
	Ⅲ	1.22 ± 0.58	1.00 ± 0.07	0.61 ± 0.06	0.43 ± 0.00
Heart	Ⅰ	17.16 ± 1.13	15.72 ± 0.53	16.22 ± 0.58	15.61 ± 1.57
	Ⅱ	18.96 ± 1.49	22.08 ± 0.72	23.37 ± 1.12	18.57 ± 0.13
	Ⅲ	18.45 ± 4.45	24.05 ± 1.40	22.29 ± 2.86	21.28 ± 3.95
Liver	Ⅰ	8.96 ± 0.35	8.50 ± 0.45	10.42 ± 1.42	10.30 ± 0.63
	Ⅱ	13.96 ± 1.25	13.84 ± 0.89	12.26 ± 1.82	11.95 ± 1.75
	Ⅲ	11.21 ± 2.42	12.07 ± 1.64	10.02 ± 2.85	8.27 ± 1.31

Tissue	Groups①	Post-injection time/min			
		5	15	30	60
Lungs	I	4.02 ± 0.60	3.72 ± 0.24	2.79 ± 0.25	2.44 ± 0.38
	II	5.34 ± 0.47	4.56 ± 0.87	3.77 ± 1.01	2.84 ± 0.09
	III	5.91 ± 0.77	5.02 ± 0.74	4.86 ± 1.00	2.88 ± 0.66
Kidneys	I	30.59 ± 2.12	24.08 ± 5.54	17.98 ± 2.57	14.92 ± 2.62
	II	49.28 ± 4.60	39.67 ± 5.44	30.30 ± 4.12	26.14 ± 7.24
	III	71.82 ± 16.38	59.61 ± 11.13	34.78 ± 0.26	29.26 ± 4.26
Brain	I	0.20 ± 0.00	0.15 ± 0.02	0.16 ± 0.01	0.15 ± 0.01
	II	0.29 ± 0.05	0.32 ± 0.16	0.25 ± 0.03	0.25 ± 0.03
	III	0.19 ± 0.05	0.27 ± 0.09	0.18 ± 0.02	0.20 ± 0.03
Spleen	I	2.38 ± 0.54	2.48 ± 0.19	2.94 ± 0.63	1.40 ± 0.03
	II	4.78 ± 1.10	3.87 ± 0.58	2.50 ± 0.25	2.57 ± 0.05
	III	5.96 ± 0.33	5.60 ± 0.79	3.21 ± 0.70	2.07 ± 0.32
Muscle	I	7.04 ± 0.53	6.20 ± 0.62	4.91 ± 0.62	6.49 ± 0.83
	II	7.21 ± 2.75	7.25 ± 1.55	9.78 ± 1.76	6.72 ± 0.41
	III	5.72 ± 1.44	9.62 ± 5.03	8.19 ± 1.78	6.39 ± 1.15
Bone	I	2.13 ± 0.64	1.92 ± 0.58	1.73 ± 0.37	1.37 ± 0.32
	II	3.64 ± 0.46	2.26 ± 1.02	3.07 ± 0.20	2.28 ± 0.30
	III	4.76 ± 0.21	2.63 ± 1.48	2.85 ± 0.58	1.79 ± 0.47

① I: complex A, yields 95%; II: complex A, yields 67%, complex B, yields 30%; III: complex B, yields 90%.

Though there was no direct evidence to prove the structure of complex A and complex B, we presumed that complex A was $[^{99m}Tc(CO)_3(MIBI)_2(H_2O)]^+$ and complex B was $[^{99m}Tc(CO)_3(MIBI)_3]^+$ from the existing experimental results. There were differences between complex A and complex B in the retention time, the partition coefficient value and biodistribution in mice. The partition coefficient values showed that complex B had stronger lipophilicity than complex A. Furthermore, the three CO ligands of $[^{99m}Tc(CO)_3(OH_2)_3]^+$ were too stable to be substituted by any other incoming ligand generally.

4 Conclusion

^{99m}Tc-CO-MIBI was prepared by using a two-step procedure and was confirmed to be a mixture of two complexes (A and B). Furthermore, either complex A or B in the final products could be obtained with yields greater than 90% if the solution pH in the second step was adjusted to a certain value. This allows a much more convenient way to study the chemical and biologic properties of ^{99m}Tc-CO-MIBI. The biodistribution in mice demonstrates that complex B is better than complex A for myocardial imaging. However, there are still two main indefinite questions to be

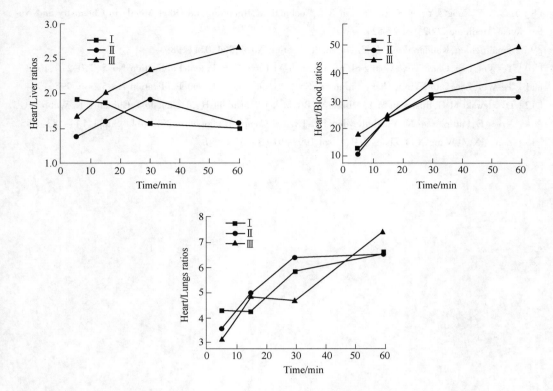

Fig. 3 Target/non-target ratios for the Group Ⅰ, Ⅱ and Ⅲ
Ⅰ—complex A, yields 95%; Ⅱ—complex A, yields 67%; complex B, yields 30%; Ⅲ—complex B, yields 90%

studied further. One is how the pH of the solution influences the synthesis of ^{99m}Tc-CO-MIBI, and the other is the structural information of complex A and complex B. Both of them are in process.

Acknowledgements

This work was supported financially by the "National 211 Engineering Project" of "Tenth-Five Year Plan of P. R. China".

References

[1] Alberto R, Schibli R, Waibel R, Abram U, Schubiger A P, Coord Chem Rev 1999;190-192:901-919.
[2] Alberto R, Schibli R, Egli A, Schubiger A P. J Am Chem Soc 1998;120:7987-7988.
[3] Marmion M E, MacDonald J R. J Nucl Med, Proceedings of the 47th Annual Meeting, 2000;124.
[4] Jiang Y, Liu B L. Chin Sci Bull 2001;46:727-730.
[5] Liu B L, Jiang Y. U. S. Patent Publication No. 20030086873, 2003.
[6] Dyszlewski M, Blake H, Dahlheimer J, Piwnica-Worms D, J Label Compd Radiopharm 2001; 44 (Suppl. 1):s63-s65.
[7] Wang J C, Liu B L, Mi H Z, Jiang Y, Tang Z G, Gui H M. Chin J Nucl Med 2002;22 (4):231-232.

[8] Hao G Y, Zang J Y, Jia F, Zhu L, Liu B L, Technetium, Rhenium and Other Metals in Chemistry and Nuclear Medicine 2002;6:527-529.

[9] Ramalingham, Kondareddiar. U. S. Patent Publication No. 4,864,051,1989.

[10] Te-Wei, Lee, Gann Ting, Chang-Shinn Su, Shyh-Yi Chyi. U. S. Patent Publication No. 5,210,270,1993.

[11] Te-Wei, Lee, Gann Ting, Chang-Shinn Su, Shyh-Yi Chyi. U. S. Patent Publication No. 5,346,995,1994.

[12] Dyszlewski M E, Bushman M J, Alberto R, Brodack J W, Knight H, Macdonald J, Chinen L K, Webb E G, Vries E, Pattipawaej M, Vanderheyden J L. J Label Compd Radiopharm 2001;44（Suppl. 1）:s483-s485.

[13] Zhang X Z, Wang X B, Zhang J B, J Isot 1997;10（3）:158-160.

新的 ^{99m}Tc 标记 σ 受体肿瘤显像剂*

摘 要 采用整体法设计合成了新的 ^{99m}Tc 标记的配合物 [N-[2-((2-oxo-2-(4-(3-phenylpropyl)piperazin-l-yl)ethyl)(2-mercaptoethyl)amino)acetyl]-2-aminoethanethiolato] technetium(V) oxide (PPPE-MAMA'-^{99m}TcO)([^{99m}Tc]-2) 及其相应的铼配合物 (PPPE-MAMA'-ReO)(Re-2)。竞争结合实验表明 Re-2 对 $σ_1$ 和 $σ_2$ 受体有中等亲和力，K_i 值分别为 (8.67 ± 0.07) μmol/L 和 (5.71 ± 1.88) μmol/L；荷 MCF-7 人乳癌裸鼠尾静脉注射 [^{99m}Tc]-2 后 0.5h、4h、20h 采集平面图像，20h 时可以看到肿瘤部位有放射性浓集，共同注射 [^{99m}Tc]-2 和抑制剂氟哌啶醇 (1mg/kg) 后显像，20h 时肿瘤部位无明显放射性浓集；体内生物分布结果显示，注射后 24h 肿瘤中的放射性摄取为 0.14% ± 0.01% ID/g，肿瘤/肌肉比为 6.02 ± 0.87。上述结果表明：虽然用整体设计法对前体化合物的结构进行了较大修饰，但得到的 ^{99m}Tc-配合物 ([^{99m}Tc]-2) 在肿瘤内仍有一定的浓集，与 $σ_1$ 和 $σ_2$ 受体仍保持一定的亲和力。在此配合物的基础上，对其进行进一步的结构修饰有可能得到对 σ 受体亲和力更高的肿瘤显像剂。

关键词 肿瘤　σ 受体　SPECT　^{99m}Tc

近年来，σ 受体被定义为一类独立的受体，目前至少有两种亚型被确认：即 $σ_1$ 型和 $σ_2$ 型受体[1]。σ 受体存在于中枢神经系统、内分泌、免疫和某些周边组织，可能在调节神经、内分泌和免疫响应中起着重要作用[2,3]。此外，σ 受体在人类许多肿瘤细胞系中有高度表达，如黑色素瘤、神经胶质瘤、乳癌、肺癌和前列腺癌等[4]。Mach 等[5] 研究发现人类增殖期的乳癌细胞中 $σ_2$ 受体的密度比静止期的高 10 倍，可作为肿瘤增殖的生物标志。因此研制具有高亲和力和选择性的 σ 受体配体，可以为 σ 受体阳性肿瘤提供敏感、特异的定位诊断方法。

^{11}C 和 ^{18}F 等正电子核素标记的 σ 受体 PET 显像剂以及 ^{123}I 标记的 SPECT 显像剂已取得了很大进展，如苯酰胺类[6~12]、二取代胍类[13]、乙二胺类[14]、哌啶[15~24] 和哌嗪[25~32] 类等。但正电子核素的半衰期较短，建造 PET 中心耗资巨大，目前在发展中国家的应用受到限制。^{123}I 是加速器生产的核素，价格昂贵，限制了其在临床上的广泛应用。^{99m}Tc 因为方便易得和适宜的核素性质成为核医学中应用最广泛的核素。但是，^{99m}Tc 为金属核素，配体经修饰与 ^{99m}Tc 偶联后，可能会影响其对受体的亲和力和选择性，因此 ^{99m}Tc 标记的受体显像剂的设计和合成一直是研究的热点和难点之一。迄今为止，有关 ^{99m}Tc 标记的 σ 受体显像剂的研究报道还很少，[^{99m}Tc]BAT-EN6[33] 对 T47D 乳癌细胞有亲和力，但没有亚型选择性。([N-[2-((3'-N'-propyl-[3,3.1] aza-bicyclononan-3α-yl)(2''-methoxy-5-methyl-phenylcarbamate)(2-mercaptoethyl)amino)acety1]-2-aminoethanethiolato] technetium(V) oxide)[34] 对 $σ_2$ 受体有高度亲和力和选择性，并在肿瘤细胞中有较高的吸收值和

---

* 本文合作者：樊彩云、贾红梅、Deuther-Conrad Winnie、Brust Peter、Steinbach Jörg。原发表于中国科学 B 辑化学，2005，35(6)：499~505。

较长的滞留时间，有希望成为乳癌肿瘤显像剂，但未见临床报道。

对于 N,N'-二取代哌嗪类化合物（见图 1），目前仅有 [11]C、[18]F 和 [123]I 标记的 σ 受体显像剂[25~32]。在构效关系研究基础上[35,36]，本文采用整体设计法用锝配位基团取代先导化合物（1）中带有 R 取代基的苯环，首次设计 [99m]Tc 标记的哌嗪类 σ 受体显像剂（PPPE- MAMA'-[99m]TcO，[^{99m}Tc]-2）（见图 1），合成标记前体，制备 [99m]Tc-配合物及相应的稳定铼配合物（Re-2），测定 Re-2 的抑制常数，并进行[^{99m}Tc]-2 在荷瘤小鼠体内的初步显像和生物分布实验。

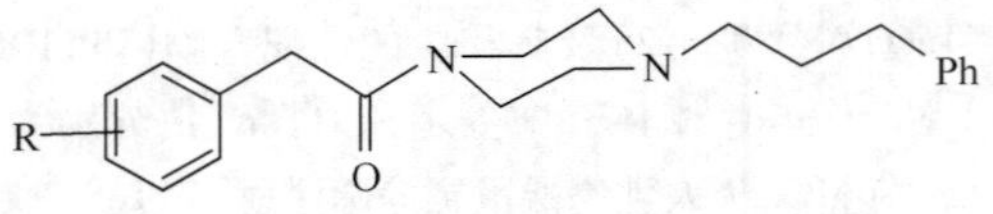

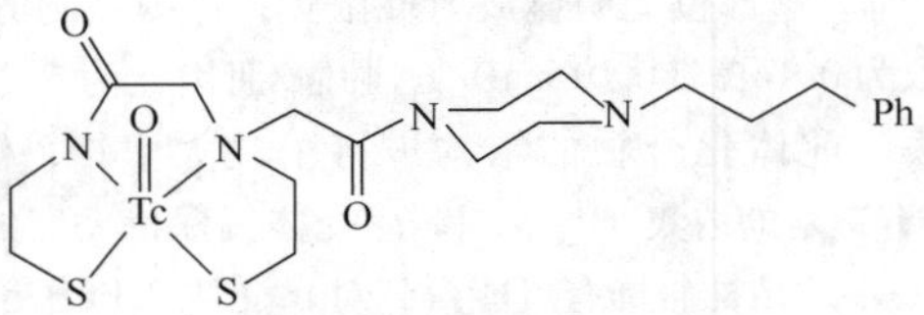

图 1　先导化合物 1 和 PPPE-MAMA'-[99m]TcO 的结构

1　实验部分

1.1　合成

1.1.1　试剂与仪器

所用化学试剂购于 Aldrich，Sigma 或北京化学试剂公司，均为分析纯。GH 药盒和 [99]Mo-[99m]Tc 发生器由中国原子能科学研究院提供。红外光谱用 AVATAR 360 型红外谱仪测定；核磁共振谱用 Bruker 500MHz 仪测定；质谱用 Trace MS 型质谱仪测定；元素分析用 Perkin-Elmer 240C 型元素分析仪测定。放射性计数用 FJ-603 井型 γ 计数器测定。化合物 3[37,38] 和 5[39] 按文献方法合成。

1.1.2　N-溴乙酰基-N'-苯丙基哌嗪（4）的合成

在 -40℃下，将 N-苯丙基哌嗪（3）（1.0g，4.9mmol）和 Et_3N（0.7mL，4.9mmol）的 CH_2Cl_2（7mL）溶液滴入搅拌下的溴乙酰溴（0.44mL，4.9mmol）的 CH_2Cl_2（5mL）溶液中，滴加完毕后，升至室温继续搅拌 30min。减压蒸馏除去溶剂得到浅黄色固体（4）。^{1}H-NMR（500MHz，$CDCl_3$）δ：7.36~7.30(m, 5H)，4.25(s, 2H)，4.17~3.66(m, 10H)，3.07~3.05(m, 2H)，1.78(m, 2H)；EIMS(m/z)，325(M)。

1.1.3　N-[2-((2-oxo-2-(4-(3-phenylpropyl)piperazin-1-yl)ethyl)(2-((triphenylmethyl)thio)ethyl)amino)-acetyl]-S-(triphenylmethyl)-2-aminoethylthiol（PPPE-MAMA'-Tr_2）(6)的合成

将化合物 5（2.9g，4.3mmol）溶解于 50mL 干燥 CH_3CN，加入化合物 4（1.4g，

4.3mmol），无水 K_2CO_3（6.0g，129mmol）和 KI（700mg，4.3mmol）后，回流8h。抽滤，向滤液中加入2g硅胶，溶剂蒸发后以乙酸乙酯：甲醇 = 20：1（体积比）为淋洗剂，经硅胶柱分离得白色泡沫0.8g（产率20.0%）。熔点：55~56℃；IR（cm^{-1}，KBr）：3357（NH），1650（C=O）；^{1}H-NMR（500MHz，$CDCl_3$）δ：7.41~7.18（m，35H），3.53~3.05（m，10H），2.69~2.66（m，4H），2.48~1.86（m，12H）；^{13}C-NMR（500MHz，$CDCl_3$）δ：170.98，168.10，144.79，144.74，129.60，129.55，128.44，128.38，127.94，127.25，126.74，126.72，126.00，66.78，66.71，58.36，57.55，54.65，54.10，52.91，52.52，41.30，38.14，33.40，32.02，30.43，28.06；EIMS（m/z），243[$(C_6H_5)_3C^+$]，437[$M+1-2(C_6H_5)_3C^+$]；ESIMS（m/z），923（$M+1$）；元素分析 $C_{59}H_{62}N_4O_2S_2$：实测值 C，76.29；H，6.81；N，5.98；理论值 C，76.75；H，6.77；N，6.07。

1.1.4　[N-[2-((2-oxo-2-(4-(3-phenylpropyl)piperazin-1-yl)ethyl)(2-mercaptoethyl)amino)acetyl]-2-aminoethanethiolato]rhenium（Ⅳ）oxide（PPPE-MAMA'-ReO）（Re-2）的合成

将化合物6（130mg，0.14mmol）溶解于5mL CF_3CO_2H 中，搅拌5min，在5℃条件下滴加 Et_3SiH 直至黄色消失，继续搅拌10min。室温减压蒸馏除去溶剂，将剩余物（7）溶于10mL甲醇，加入5mL 1mol/L醋酸钠的甲醇溶液。然后加入$(Ph_3P)_2Re(=O)Cl_3$（135mg，0.16mmol），升温至80℃回流2h，反应混合物由黄绿色变为紫色，冷至室温后，将反应混合物用30mL乙酸乙酯稀释，抽滤，滤液浓缩后以乙酸乙酯：甲醇 = 20：1（体积比）为淋洗剂，经硅胶层析柱分离得紫色固体（产率14.6%）。IR（cm^{-1}，KBr）：3441（NH），1621（C=O），963（Re=O）；^{1}H-NMR（500 MHz，$CDCl_3$）δ：7.34~7.07（m，5H），5.03~4.38（m，6H），4.04~3.52（m，6H），3.11~2.64（m，10H），1.97~1.15（m，4H）。EIMS（m/z），638（M）；元素分析 $C_{21}H_{31}N_4O_3S_2Re$：实测值 C，39.41；H，4.70；N，8.60 计算值 C，39.54；H，4.90；N，8.78。

1.1.5　[^{99m}Tc]-2 的制备

向GH（葡庚糖酸钠）药盒中加入新淋洗的$^{99m}TcO_4^-$的生理盐水溶液，摇匀，在40℃反应10min。向青霉素瓶加入0.2mL PBS（pH = 8.01），0.1mg（1mg/mL）配体（7）的乙醇溶液以及几滴吐温-80的乙醇溶液，再加入0.15mL制备好的^{99m}Tc-GH，摇匀后沸水浴加热30min。冷却后用 CH_2Cl_2 萃取，吹干后溶解于15%的乙醇/生理盐水中。用TLC和HPLC测定[^{99m}Tc]-2的放化纯分别为95%和98%。HPLC分析结果表明在相同条件下[^{99m}Tc]-2和Re-2的保留时间一致。

1.2 Sigma 受体竞争结合分析

1.2.1　材料和方法

试剂若无特殊说明，均购于 Sigma 公司。[^{3}H]镇痛新（1.354×10^{12} Bq/mmol；Perkin-Elmer Life Sciences，Boston，MA，USA）和[^{3}H]DTG（1.147×10^{12} Bq/mmol；NEN Life Science Products，Boston，MA，USA）作为σ受体配体。将配合物 Re-2 溶解于DMSO配成100mmol/L的溶液，烯丙右吗喃氢溴酸盐（获赠于 Hoffman-La Roche，Basel，Switzerland）溶解于 EtOH 配成10mmol/L的溶液，于 -25℃保存。氟哌啶醇盐酸盐溶解于DMSO配成10mmol/L的溶液，于 -25℃保存。

1.2.2 σ_1 受体和 σ_2 受体的制备

Re-2 对 σ_1 受体和 σ_2 受体的亲和力在富含各自 σ 受体亚型的组织中测定[40]：即分别在大鼠脑组织匀浆和大鼠肝组织匀浆中进行。动物的使用遵循德国实验动物法和实验室动物使用法规。

制备皮质膜蛋白[41]：雌性 Sprague-Dawley 大鼠（8 周龄）麻醉并断头处死，迅速取出脑，在冰上切开，用 10 倍体积（w/v）的冰冷 TRIS-HCl（三羟甲基氨基甲烷-盐酸）缓冲溶液（50mmol/L, pH = 7.4, 4℃）将皮质在 Teflon-glass 匀浆器中匀浆。再放入高速冷冻离心机中（15000r/min, 4℃）离心 10min。得到的膜蛋白再匀浆离心两次，然后重新悬浮于标准缓冲溶液，分装后于 -25℃ 保存。肝膜蛋白的制备和保存方法同上。

1.2.3 Re-2 与 σ_1 受体和 σ_2 受体亲和力的测定

Re-2 的体外亲和力用竞争结合分析法测定。对 σ_1 受体的亲和力用大鼠脑皮质膜蛋白和 0.8nmol/L [^{3}H]镇痛新（K_{D,σ_1} = 6.9nmol/L[32]）进行测定，对 σ_2 受体的亲和力用大鼠肝膜蛋白和 0.8nmol/L [^{3}H]DTG（K_{D,σ_2} = 29.2nmol/L[32]）在 10μmol/L 烯丙右吗喃（K_{i,σ_1} = 125nmol/L[42]）存在以封闭 σ_1 受体的条件下进行测定。膜蛋白在冰上解冻后，用培养缓冲溶液（50mmol/L Tris-HCl, pH 7.4, 21℃）稀释，再次匀浆。实验的蛋白浓度用 BCA 方法测定（167μg/mL）（Perbio Sciences, Germany）。非特异性结合试剂采用 10 μmol/L 氟哌啶醇。Re-2 的温育温度和时间分别为 21℃ 和 120min。DMSO 在稀释曲线中的最高浓度不超过 1%，这样在温育分析中 DMSO 的浓度不超过 0.1%。该条件采用氟哌啶醇（对于 σ_1 受体：从 0.01nmol/L 到 0.1μmol/L 共 7 个浓度；对于 σ_2 受体：从 0.1nmol/L 到 1μmol/L 共 13 个浓度）和 DTG（对于 σ_2 受体：从 0.1nmol/L 到 0.1μmol/L 共 6 个浓度）的竞争实验进一步验证。

所有的实验均重复 3 次。用迭代非线性曲线拟合法得出结合参数 IC_{50} 值。应用 Cheng-Prusoff 方程[43]由 IC_{50} 值计算得出抑制常数（K_i 值）。

1.3 生物评价

MCF-7 人乳癌细胞由北京师范大学生物系提供，BALB/c 裸鼠购于北京大学医学部实验动物科学部。显像用 SOPHA SPECT 仪进行。放射性计数用 LKB 1275 Minigamma 计数器测定。

选雌性裸鼠（6 周龄，13～15g），每只皮下注射 0.1mL 约 2.0×10^6 个细胞，3～4 周后肿瘤大小 1cm³ 左右。经荷瘤裸鼠尾静脉注射[^{99m}Tc]-2(0.1mL, 约 1.5×10^7Bq)，采集注射后 0.5h, 4h, 20h 荷瘤裸鼠后位平面图像（采集条件：矩阵为 128 × 128，采集计数为 5×10^5）。抑制显像研究采用[^{99m}Tc]-2 和抑制剂氟哌啶醇（1mg/kg）共同注射，20h 后显像。

经荷瘤裸鼠尾静脉注射[^{99m}Tc]-2(0.1mL, 约 3.7×10^6Bq)，24h 后处死荷瘤裸鼠（3 只），测量血液、各脏器和肿瘤的质量和放射性计数，计算肿瘤对血液、各脏器的靶和非靶比（T/NT）。

2 结果和讨论

2.1 [^{99m}Tc]-2 和 Re-2 配合物的设计制备

尽管 ^{99m}Tc 标记的受体显像剂的设计与合成一直是研究的难点，但是多巴胺转运蛋白显像剂[^{99m}Tc]-TRODAT-1[44]的研制成功和[^{99m}Tc]O-1505T[45]一期临床试验结果仍鼓舞着人们在这个领域进行着艰苦卓绝的探索。由于^{11}C、^{18}F 和 ^{123}I 标记的 N, N′-二取代哌嗪类 σ 受体显像剂取得了很大进展，因此在此基础上对该类化合物进行修饰，设计 ^{99m}Tc 标记的 σ 受体显像剂，有可能取得成功。研究表明，苯丙基胺结构可能是配体对 $σ_2$ 受体有亲和力和选择性的一个药效团[35]，而 R 部位为吸电子基时，化合物对 σ 受体的亲和力为 nmol/L 数量级，而给电子基的引入对化合物的亲和力影响不大[36]。因此采用整体设计法用锝配位基团取代前体化合物（1）中带有 R 取代基的苯环，设计了^{99m}Tc 标记的哌嗪类 σ 受体显像剂。由于^{99m}Tc 为放射性核素，而 Tc 和 Re 性质类似，一般采用测定稳定 Re 配合物的化学结构和抑制常数来推测[^{99m}Tc]-配合物的结构和对 σ 受体的亲和力等性质，因此成功制备了[^{99m}Tc]-2 及其 Re 替代物 Re-2，合成路线见图 2。

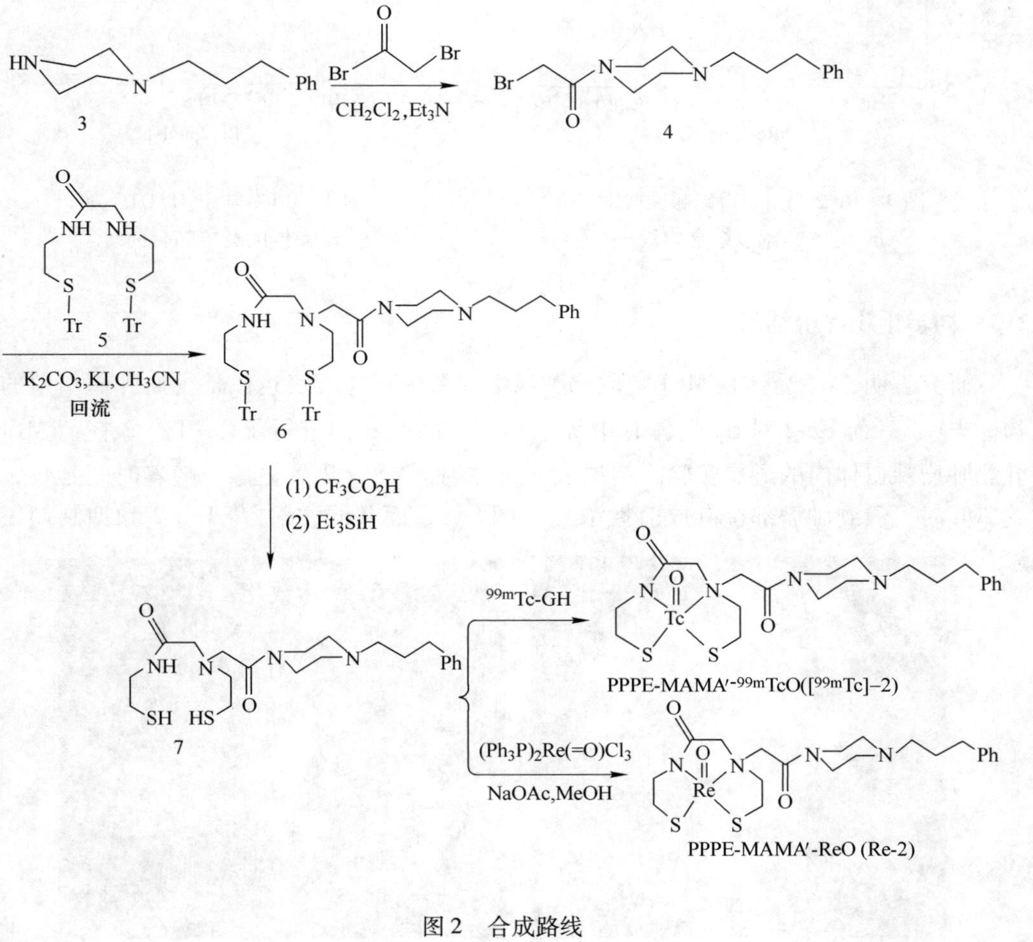

图 2　合成路线

2.2　受体结合分析

为了建立测定配体对 σ 受体亲和力的实验条件，首先进行了常用的 σ 受体配体如氟哌啶醇、DTG 对 σ_1 受体和 σ_2 受体的结合实验。结果表明，DTG 对 σ_2 受体的 K_i 值为 35.2nmol/L，氟哌啶醇对 σ_1 受体和 σ_2 受体的 K_i 值分别为 3.19nmol/L 和 24.2nmol/L，与文献报道值一致[42,46]。在此基础上，进行了 Re-2 与 σ_1 受体和 σ_2 受体的竞争结合实验。Re-2 与 σ_1 受体和 σ_2 受体的竞争结合曲线见图3和图4，K_i 值分别为 (8.67 ± 0.07) μmol/L 和 (5.71 ± 1.88) μmol/L。虽然与母体化合物相比[35]，Re-2 对 σ_1 和 σ_2 受体的亲和力均较低，但其对 σ_1 和 σ_2 受体仍保持一定的亲和力。

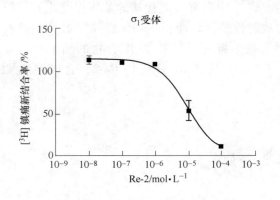 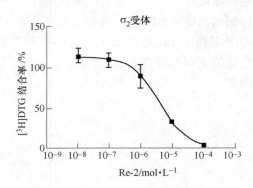

图3　Re-2 与 [^{3}H]氟哌啶醇对 σ_1 受体的竞争结合曲线

图4　Re-2 与 [^{3}H]DTG 对 σ_2 受体的竞争结合曲线

2.3　初步生物评价结果

研究表明，σ_2 受体在 MCF-7 乳癌肿瘤细胞系中有高度表达，是乳癌肿瘤增殖的生物标志[3,5]。而 Re-2 对 σ_2 受体有中等亲和力，因此我们进行了 [^{99m}Tc]-2 在荷 MCF-7 乳癌肿瘤裸鼠体内的显像实验。结果表明，注射 [^{99m}Tc]-2 后显像，随着时间延长，在约 20h 时可看到肿瘤部位有放射性浓集（图5a），而共同注射 [^{99m}Tc]-2 和抑制剂氟哌

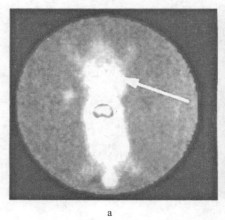 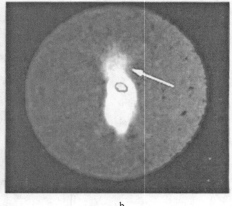

a　　　　　　　　　　　　b

图5　荷瘤裸鼠注射 [^{99m}Tc]-2 后 20h 显像可以看到肿瘤部位有放射性浓集（a）及共同注射 [^{99m}Tc]-2 和氟哌啶醇后 20h 显像肿瘤部位的放射性浓集被抑制（肿瘤位置如箭头所示）（b）

啶醇（1mg/kg）20h 后显像，可以看到肿瘤部位无明显放射性浓集（图 5b），说明 [^{99m}Tc]-2 在肿瘤部位是特异性地与 σ_2 受体结合。但是由于 [^{99m}Tc]-2 对 σ_2 受体的亲和力较低，因此需要在注射 [^{99m}Tc]-2 后较长时间才可以看到肿瘤的位置。

初步的体内分布结果表明，注射后 24h，肿瘤中的放射性摄取为 $0.14\% \pm 0.01\%$ ID/g，肿瘤/肌肉比为 6.02 ± 0.87。这与上述显像结果一致，表明 [^{99m}Tc]-2 与 σ 受体有亲和力，在肿瘤内有一定的浓集。上述初步实验结果鼓励我们对 [^{99m}Tc]-2 进行进一步的结构修饰，设计合成一系列新型的 ^{99m}Tc 标记的 σ 受体显像剂，从而筛选出对 σ 受体亲和力更高的肿瘤显像剂。相关的研究工作正在进行中。

3 结论

采用整体法设计制备了新型 ^{99m}Tc 标记的 σ 受体配体（[^{99m}Tc]-2）和相应的稳定铼配合物。Re-2 对 σ_1 和 σ_2 受体均有中等亲和力；尾静脉注射后 20h，[^{99m}Tc]-2 在荷 MCF-7 人乳癌裸鼠肿瘤部位有放射性浓集，并且注射后 24h 肿瘤/肌肉比较高。上述结果为进一步设计对 σ 受体亲和力更高的肿瘤显像剂提供了参考。

参 考 文 献

[1] Quirion R, Bowen W D, Itzhak Y, et al. A proposal for the classification of sigma binding sites. Trends Pharmacol Sci, 1992, 13: 85~86.

[2] Su T P. Sigma receptors. Putative links between nervous, endocrine and immune systems. Eur J Biochem, 1991, 200: 633~642.

[3] Guitart X, Codony X, Monroy X. Sigma receptors: biology and therapeutic potential. Psychopharmacology, 2004, 174: 301~319.

[4] Vilner B J, John C S, Bowen W D. Sigma-1 and sigma-2 receptors are expressed in a wide variety of human and rodent tumor cell lines. Cancer Res, 1995, 55: 408~413.

[5] Mach R H, Smith C R, al-Nabulsi I, et al. σ_2 receptors as potential biomarkers of proliferation in breast cancer. Cancer Res, 1997, 57: 156~161.

[6] Shiue C Y, Shiue G G, Zhang S X, et al. N-(N-benzylpiperidin-4-yl)-2-[^{18}F]fluorobenzamide: a potential ligand for PET imaging of σ receptors. Nucl Med Biol, 1997, 24: 671~676.

[7] Dence C S, John C S, Bowen W D, et al. Synthesis and evaluation of [^{18}F] labeled benzamides: high affinity sigma receptor ligands for PET imaging. Nucl Med Biol, 1997, 24: 333~340.

[8] Shiue C Y, Shiue G G, Benard F, et al. N-(N-benzylpiperidin-4-yl)-2-[^{18}F]fluorobenzamide: A potential ligand for PET imaging of breast cancer. Nucl Med Biol, 2000, 27: 763~767.

[9] Rowland D J, Tu Z, Mach R H, et al. Investigation of a new sigma-2 receptor ligand for detection of breast cancer. J Labelled Cpd Radiopharm, 2003, 46: S6.

[10] Everaert H, Flamen P, Franken P R, et al. Sigma receptor imaging by means of 123-IDAB scintigraphy: clinical application in melanoma and non-small cell lung cancer. Anticancer Res, 1997, 17: 1577~1582.

[11] John C S, Gulden M E, Li J, et al. Synthesis, in vitro binding, and tissue distribution of radioiodinated 2-[^{125}I] N-(N-benzylpiperidin-4-yl)-2-iodobenzamide, 2-[^{125}I]BP: a potential σ receptor marker for human prostate tumors. Nucl Med Biol, 1998, 25: 189~194.

[12] Staelens L, Oltenfreiter R, Dumont F, et al. *In vivo* evaluation of [^{123}I]-4-iodo-N-(4-(4-(2-methoxy-

phenyl)-piperazin-1-yl)butyl)-benzamide: a potential sigma receptor ligand for SPECT studies. Nucl Med Biol, 2005, 32: 193~200.

[13] Wilson A A, Dannals R F, Ravert H T, et al. Radiosynthesis of σ receptor ligands for positron emission tomography: ^{11}C- and ^{18}F-labeled guanidines. J Med Chem, 1991, 34: 1867~1870.

[14] Ishiwata K, Noguchi J, Ishii S-I, et al. Synthesis and preliminary evaluation of [^{11}C] NE-100 labeled in two different positions as a PET σ receptor ligand. Nucl Med Biol, 1998, 25: 195~202.

[15] Musachio J L, Mathews W B, Ravert H T, et al. Synthesis of a radiotracer for studying sigma receptors in vivo using PET: (+)-N-[^{11}C]benzyl-N-normetazocine. J Labelled Cpd Radiopharm, 1994, 34: 49~57.

[16] Collier T L, O'Brien C, Waterhouse R N. Synthesis of [^{18}F]-1-(3-fluoropropyl)-4-(4-cyanophenoxymethyl)piperidine. J Labelled Cpd Radiopharm, 1996, 38: 785~794.

[17] Waterhouse R N, Collier T L. *In vivo* evaluation of [^{18}F]1-(3-fluoropropyl)-4-(4-cyanophenoxymethyl)piperidine: a selective sigma-1 receptor radioligand for PET. Nucl Med Biol, 1997, 24: 127~134.

[18] John C S, Gulden M E, Vilner B J, et al. Synthesis, *in vitro* validation and *in vivo* pharmacokinetics of [^{125}I]N-[2-(4-iodophenyl)ethyl]-N-methyl-2-(1-piperidinyl)ethylamine: a high-affinity ligand for imaging sigma receptor positive tumors. Nucl Med Biol, 1996, 23: 761~766.

[19] Waterhouse R N, Mardon K, Giles K M, et al. Halogenated 4-(phenoxymethyl)piperidines as potential radiolabeled probes for σ-1 receptors: *in vivo* evaluation of [^{123}I]-1-(iodopropen-2-yl)-4-[(4-cyanophenoxy)methyl]piperidine. J Med Chem, 1997, 40: 1657~1667.

[20] Waterhouse R N, Mardon K, O'Brien J C. Synthesis and preliminary evaluation of [^{123}I]1-(4-cyanobenzyl)-4-[[(trans-iodopropen-2-yl)oxy]-methyl]piperidine: a novel high affinity sigma receptor radioligand for SPECT. Nucl Med Biol, 1997, 24: 45~51.

[21] Waterhouse R N, Chapman J, Izard B, et al. Examination of four ^{123}I-labeled piperidine-based sigma receptor ligands as potential melanoma imaging agents: initial studies in mouse tumor models. Nucl Med Biol, 1997, 24: 587~593.

[22] 张春丽, 刘巧平, 王荣福, 等. ^{125}I-4-(N-苄基哌啶基)-4-碘代苯磺酰胺的体内 σ 受体亲和性能研究. 同位素, 2004, 17: 198~203.

[23] Maier C A, Wünsch B. Novel spiropiperidines as highly potent and subtype selective sigma-receptor ligands. Part 1. J Med Chem, 2002, 45: 438~448.

[24] Maier C A, Wünsch B. Novel sigma receptor ligands. Part 2. SAR of spiro[[2]benzopyran-1,4′-piperidines] and spiro[[2]benzofuran-1,4′-piperidines] with carbon substituents in position 3. J Med Chem, 2002, 45: 4923~4930.

[25] Ding Y-S, Fowler J S, Dewey S L, et al. Synthesis and PET studies of fluorine-18-BMY 14802: a potential antipsychotic drug. J Nucl Med, 1993, 34: 246~254.

[26] Kiesewetter D O, de Costa B. Synthesis of N^1-3-[^{18}F]fluoropropyl-N^4-2-([3,4-dichlorophenyl]ethyl)piperazine, a high affinity ligand for sigma receptor. J Labelled Cpd Radiopharm, 1993, 33: 639~643.

[27] Van Waarde A, Buursma A R, Hospers G A P, et al. Tumor imaging with 2 σ-receptor ligands, ^{18}F-FE-SA5845 and ^{11}C-SA4503: a feasibility study. J Nucl Med, 2004, 45: 1939~1945.

[28] Elsinga P H, Kawamura K, Kobayashi T, et al. Synthesis and evaluation of [^{18}F] fluoroethyl SA4503 and SA5845 as PET ligand for the sigma receptor. J Labelled Cpd Radiopharm, 2001, 44, Suppl. 1: S4~S6.

[29] Kawamura K, Ishii S, Kobayashi T, et al. Synthesis and evaluation of ^{11}C-labeled SA4503, SA5845, and their ethyl derivatives as PET sigma receptor ligands. J Labelled Cpd Radiopharm, 2001, 44, Sup-

pl. 1: S233~S235.

[30] Kawamura K, Ishiwata K, Tajima H, et al. Synthesis and *in vivo* evaluation of [^{11}C] SA6298 as a PET sigma$_1$ receptor ligand. Nucl Med Biol, 1999, 26: 915~922.

[31] Kawamura K, Ishiwata K, Tajima H, et al. *In vivo* evaluation of [^{11}C] SA4503 as a PET ligand for mapping CNS sigma$_1$ receptor. Nucl Med Biol, 2000, 27: 255~261.

[32] Kawamura K, Elsinga P H, Kobayashi T, et al. Synthesis and evaluation of ^{11}C- and ^{18}F-labeled 1-[2-(4-alkoxy-3-methoxy-phenyl)ethyl]-4-(3-phenylpropyl)piperazines as sigma receptor ligands for positron emission tomography studies. Nucl Med Biol, 2003, 30: 273~284.

[33] John C S, Lim B B, Geyer B C, et al. ^{99m}Tc-labeled σ-receptor- binding complex: synthesis, characterization, and specific binding to human ductal breast carcinoma (T47D) cells. Bioconj Chem, 1997, 8: 304~309.

[34] Choi S-R, Yang B, Plössl K, et al. Development of a Tc-99m labeled sigma-2 receptor-specific ligand as a potential breast tumor imaging agent. Nucl Med Biol, 2001, 28: 657~666.

[35] Zhang Y, Williams W, Torrence-Campbell C, et al. Characterization of novel N, N'-disubstituted piperazines as sigma receptor ligands. J Med Chem, 1998, 41: 4950~4957.

[36] Maeda D N, Williams W, Kim W E, et al. N-arylalkylpiperidines as high-affinity sigma-1 and sigma-2 receptor ligands: phenylpropylamine as potential leads for selective sigma-2 agents. Bioorg Med Chem Lett, 2002, 12: 497~500.

[37] Moore T S, Boyle M, Thorn V M, et al. N-substituted derivatives of piperazine and ethylenediamine. The preparation of N-monosubstituted derivatives. J Chem Soc, 1929, 39: 51.

[38] Stewart H W, Turner R J, Denton J J, et al. Experimental chemotherapy of filariasis. The preparation of derivatives of piperazine. J Org Chem, 1948, 13: 134~143.

[39] O'Neil J P, Wilson S R, Katzenellenbogen J A. Preparation and structural characterization of monoamine-monoamide bis (thio) oxo complexes of technetium (O) and rhenium (O). Inorg Chem, 1994, 33: 319~323.

[40] Bowen W D. Sigma receptors: recent advances and new clinical potentials. Pharm Acta Helv, 2000, 74: 211~218.

[41] Deuther-Conrad W, Patt J T, Feuerbach D, et al. Norchloro-fluoro- homoepibatidine: specificity to neuronal nicotinic acetylcholine receptor subtypes *in vitro*. IL Farmaco, 2004, 59: 785~792.

[42] Vilner B J, Bowen W D. Modulation of cellular calcium by sigma-2 receptors: release form intracellular stores in human SK-N-SH neuroblastoma cells. J Pharmacol Exp Ther, 2000, 292: 900~911.

[43] Cheng Y, Prusoff W H. Relationship between the inhibition constant (K_i) and the concentration of inhibitor which cause 50% inhibition (IC_{50}) of an enzymatic reaction. Biochem Pharmacol, 1973, 22: 3099~3108.

[44] Kung H F, Kim H-J, Kung M-P, et al. Imaging of dopamine transporters in humans with technetium-99m TRODAT-1. Eur J Nucl Med, 1996, 23: 1527~1530.

[45] Callahan R J, Dragotakes S C, Barrow S A, et al. A phase I clinical trial of the DAT ligand ^{99m}Tc-O1505T. J Nucl Med, 2001, 42: 268P.

[46] Colabufo N A, Berardi F, Contino M, et al. Antiproliferative and cytotoxic effects of some sigma$_2$ agonists and signal antagonists in tumour cell lines. Naunyn Schmiedebergs Arch Pharmacol, 2004, 370: 106~113.

Synthesis and Biological Evaluation of Technetium-99m-labeled Deoxyglucose Derivatives as Imaging Agents for Tumor[*]

Abstract Three deoxyglucose (DG) derivatives, S-DG, MAG$_3$-DG and MAMA-BA-DG, were synthesized and labeled successfully with high labeling yields and high radio-chemical purities. Biodistribution in tumor-bearing mice demonstrated that these three new ^{99m}Tc-deoxyglucose derivatives showed accumulation in tumor and high tumor-to-muscle ratios. Among them, the ^{99m}Tc-MAG$_3$-DG showed the best characteristics as a potential tumor marker for single photon emission computed tomography (SPECT).

Key words technetium, deoxyglucose, imaging agents

Tumor is among the most common causes of death in the world. In vivo functional imaging technique can help to diagnose and stage tumors, optimize drug scheduling, and predict response to a therapeutic modality, which would be advantageous to both patient and oncologist.

Flourine-18 (^{18}F) fluorodeoxyglucose (FDG) has been used to measure normal tissue and tumor glucose utilization rates[1-5]. Although tumor metabolic imaging with [^{18}F]FDG has been studied for more than two decades, the use of this examination in clinical practice is still limited by such factors as difficult access, limited availability, and high cost[6]. In addition, positron emission tomography (PET) radio-synthesis must be performed rapidly because the half-life of F-18 is only 109 min. Thus, it would be very desirable to develop less costly imaging agents based on γ-emitter isotope, especially for developing country, where single photon emission computed tomography (SPECT) is still dominant.

Technetium-99m (^{99m}Tc) has been mostly used for labeling radiopharmaceuticals owing to its suitable physical and chemical characteristics and inexpensive isotope cost. Lots of ^{99m}Tc-labeled glucose derivatives have been synthesized in order to develop one subrogate in SPECT for [^{18}F]FDG in PET recently[7-11]. Developed by Yang, ^{99m}Tc-labeled ethylenedicysteine-deoxyglucose (ECDG) showed similarities with [^{18}F]FDG in tumor uptake[12]. This suggests that there is feasibility for ^{99m}Tc-labeled deoxyglucose as tumor metabolic imaging agents. However, [^{99m}Tc]ECDG still has some drawbacks such as slow cleanup from blood, which would cause high blood background; and large molecular weight, which would limit its penetration through blood-brain barrier (BBB).

Thus, it would be desirable to develop a smaller ^{99m}Tc- based deoxyglucose derivative with

---

[*] Copartner: Chen Xiangji, Li Liang, Liu Fei. Reprinted from *Bioorganic & Medicinal Chemistry Letters*, 2006, 16(21): 5503-5506.

rapid blood clearance and still maintaining its high tumor uptake.

The purpose of this study is to conjugate deoxyglucose with different chelating agents and to evaluate the feasibility of the ^{99m}Tc-labeled deoxyglucose derivatives as candidates for tumor-imaging agents.

The ^{99m}Tc-S-DG was synthesized according to the procedure outlined in Scheme 1. After protecting the thiol group of mercaptoacetic acid with trityl chloride, the resulting compound 3 was reacted with glucosamine using N, N′-dicyclohexyl-carbodiimide (DCC) as condensation reagent to obtain compound 4. Next, the thiol groups were deprotected in trifluoroacetic acid (TFA) to give 5. For labeling, ^{99m}Tc-S-DG was prepared by ligand-exchange reaction with ^{99m}Tc-glucoheptonate (GH).

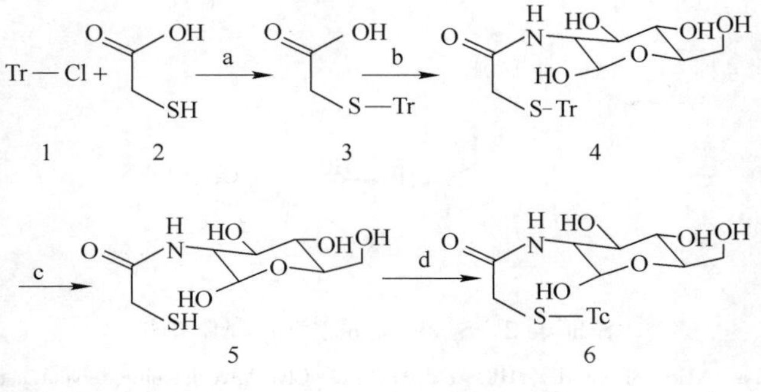

Scheme 1 Synthesis of ^{99m}Tc-S-DG

Reagents and solvents: a—Solvent: dichloromethane/acetic acid, yield: 95%; b—Glucosamine, DCC, solvent: ethanol/water, yield: 12%; c—Triethylsilane, solvent: TFA; d—^{99m}Tc-GH

Synthesis of ^{99m}Tc-MAG$_3$-DG (Scheme 2) was performed from mercaptoacetic acid. After protecting the thiol group with trityl chloride, the resulting compound 3 was reacted with N-hydroxysuccinimide (NHS) using DCC as condensation reagent to obtain the active ester 7. The active ester 7 was reacted with the amine group of glycylglycylglycine to provide the Tr-MAG$_3$ 8. Tr-MAG$_3$ was conjugated with glucosamine with DCC as condensation agent to provide the compound 9. Deprotecting and labeling were performed with the same procedure as ^{99m}Tc-S-DG.

^{99m}Tc-MAMA-BA-DG was synthesized according to the procedure in Scheme 3. After protecting the thiol group of cysteamine chloride with trityl chloride, the resulting compound 14 was reacted with bromoacetal bromide to prepare 16. The amine group of 16 was then alkylated with methyl 4-bromobutyrate to produce 17. After hydrolysis of the ester group, the resulting compound 18 was conjugated with glucosamine with DCC to obtain compound 19. Deprotecting and labeling were performed with the same procedure as ^{99m}Tc-S-DG.

The radiochemical yields of ^{99m}Tc-labeled deoxyglucose analogues were determined by TLC on three systems and the R_f values of ^{99m}Tc-species are listed in Table 1. HPLC analysis showed that the radiochemical purity is high (Fig. 1).

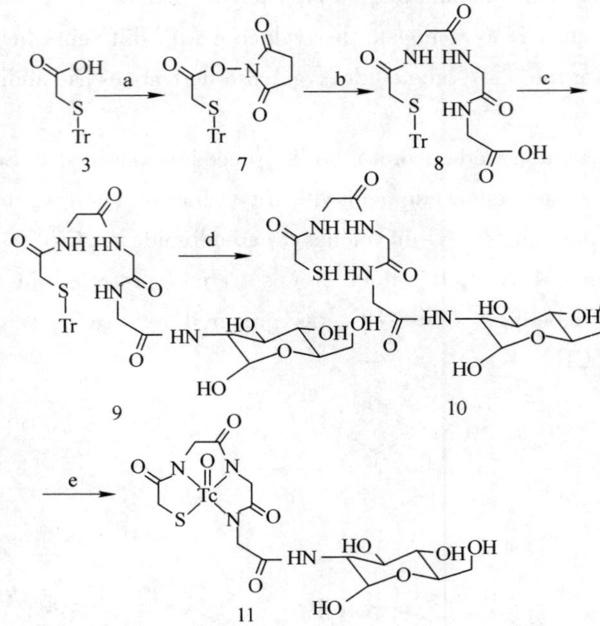

Scheme 2 Synthesis of ^{99m}Tc-MAG$_3$-DG

Reagents and solvent: a—NHS, solvent: dry THF, yield: 47%; b—Glycylglycylglycine, solvent: acetonitrile, yield: 30%; c—Glucosamine, DCC, solvent: acetonitrile/water, yield: 19%; d—Triethylsilane, solvent: TFA; e—^{99m}Tc-GH

Table 1 R_f values of ^{99m}Tc-species on TLC

^{99m}Tc-species	System 1①	System 2②	System 3③
^{99m}Tc-S-DG	0.0	0.7-0.8	0.9-1.0
^{99m}Tc-MAG$_3$-DG	0.0	0.6-0.7	0.9-1.0
^{99m}Tc-MAMA-BA-DG	0.0	0.6-0.7	0.9-1.0
^{99m}Tc-GH	0.0	0.9-1.0	0.9-1.0
^{99m}TcO$_4^-$	0.8-0.9	0.5-0.7	0.0
^{99m}TcO$_2 \cdot n$H$_2$O	0.0	0.0	0.0

① Xinhua No. 1 paper strip developed by eluent A(1mol/L of ammonium acetate/methanol (4:1)).
② Xinhua No. 1 paper strip developed by ketone.
③ Polyamide strip developed by saline.

Partition coefficients of ^{99m}Tc-DG analogues at pH 7.4 are shown in Table 2 and all of them are highly hydrophilic. ^{99m}Tc-S-DG is the highest hydrophilic one and ^{99m}Tc-MAG$_3$-DG is more hydrophilic than ^{99m}Tc-MA-MA-BA-DG.

Table 2 Partition coefficients of ^{99m}Tc-DG analogues

Name	^{99m}Tc-S-DG	^{99m}Tc-MAG$_3$-DG	^{99m}Tc-MAMA-BA-DG
log$P$	-2.92 ± 0.03	-2.19 ± 0.01	-1.71 ± 0.01

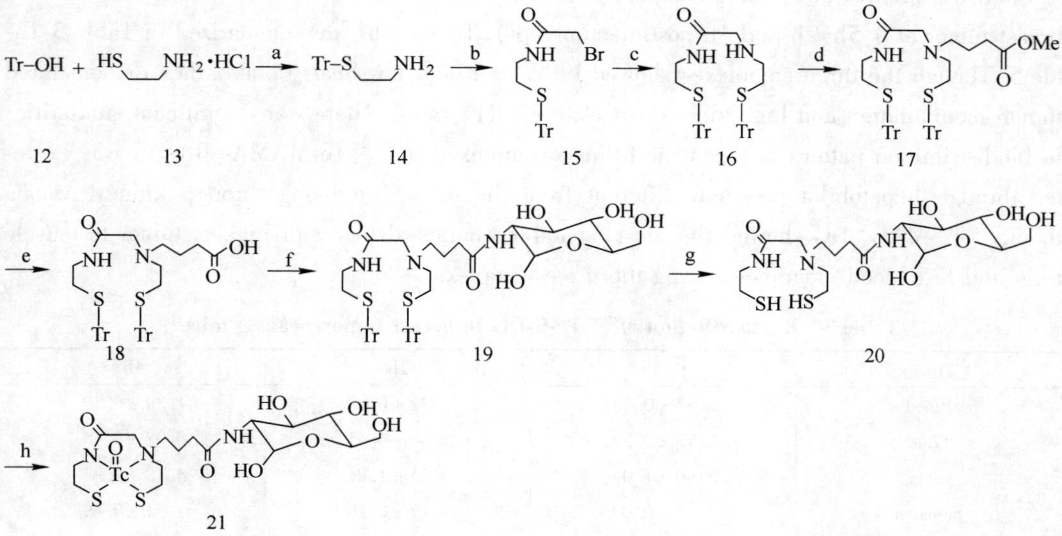

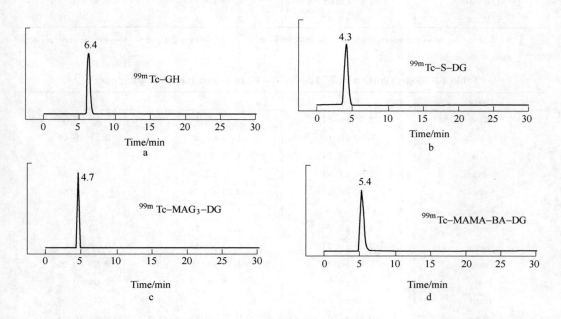

Scheme 3 Synthesis of ^{99m}Tc-MAMA-BA-DG

Reagents and solvent: a—Solvent: TFA, yield: 75%; b—Bromoacetyl bromide, triethylamine, solvent: dichloromethane, yield: 90%; c—Compound 14, triethylamine, solvent: dichloromethane, yield: 64%; d—Methyl 4-bromobutyrate, potassium iodide, potassium carbonate, solvent: acetonitrile, yield: 40%; e—5% NaOH/THF, yield: 95%; f—Glucosamine, DCC, solvent: THF/water, yield: 7.3%; g—Triethylsilane solvent: TFA; h—^{99m}Tc-GH

Fig. 1 Radio-HPLC chromatograms using an amino-column eluting with 70% acetonitrile for 30min

a—^{99m}Tc-GH, t_R = 6.4min; b—^{99m}Tc-S-DG, t_R = 4.3min;

c—^{99m}Tc-MAG$_3$-DG, t_R = 4.7min; d—^{99m}Tc-MAMA-BA-DG, t_R = 5.4min

Biodistribution of these three complexes was performed in TA-2 mice (18-22g) with MA891 breast tumor at 0.5h, 2h and 4h postinjection (pi). The results are summarized in Table 3-Table 5. Though the three complexes showed little brain and low heart uptake, they demonstrated tumor accumulation and high tumor-to-muscle (T/M) ratios. There were significant similarities in biodistribution pattern between these three complexes, but ^{99m}Tc-MAMA-BA-DG was excreted through hepatobiliary system different from the other two being through kidney. Among them, ^{99m}Tc-MAG$_3$-DG showed the most favorable characteristics with highest tumor-to-muscle ratio and fast blood clearance among the three complexes.

Table 3 Biodistribution of ^{99m}Tc-S-DG in breast tumor-bearing mice[①]

Tissue	30min	2h	4h
Blood	3.32 ± 0.50	2.22 ± 0.69	1.29 ± 0.19
Lung	2.73 ± 0.39	1.48 ± 0.45	0.98 ± 0.15
Liver	5.90 ± 1.06	4.75 ± 1.81	4.18 ± 1.42
Stomach	3.69 ± 0.09	3.17 ± 0.28	1.41 ± 0.69
Spleen	1.81 ± 0.76	1.36 ± 1.29	1.60 ± 0.40
Kidney	8.80 ± 0.61	10.81 ± 5.03	13.21 ± 2.87
Intestine	2.24 ± 0.81	4.07 ± 2.32	1.29 ± 0.47
Muscle	0.51 ± 0.18	0.45 ± 0.09	0.33 ± 0.09
Tumor	1.53 ± 0.05	1.25 ± 0.36	1.06 ± 0.18
Brain	0.11 ± 0.02	0.07 ± 0.02	0.04 ± 0.01
Heart	0.90 ± 0.18	0.57 ± 0.11	0.38 ± 0.06
T/B ratio	0.47 ± 0.05	0.59 ± 0.20	0.82 ± 0.15
T/M ratio	3.20 ± 0.87	2.85 ± 1.12	3.29 ± 0.38

① All data are the mean percentage ($n = 3$) of the injected dose of ^{99m}Tc-S-DG per gram of wet tissue ± the standard deviation of the mean.

Table 4 Biodistribution of ^{99m}Tc-MAG$_3$-DG in breast tumor-bearing mice[①]

Tissue	30min	2h	4h
Blood	2.95 ± 0.48	1.31 ± 0.17	0.88 ± 0.06
Lung	2.69 ± 0.18	1.04 ± 0.08	0.71 ± 0.06
Liver	2.87 ± 0.26	1.99 ± 0.24	1.40 ± 0.15
Stomach	2.90 ± 1.67	2.25 ± 1.24	0.64 ± 0.34
Spleen	1.02 ± 0.03	0.84 ± 0.28	1.15 ± 0.68
Kidney	10.03 ± 2.40	8.42 ± 2.37	12.80 ± 1.20
Intestine	4.80 ± 0.60	2.22 ± 0.51	1.10 ± 0.25
Muscle	0.57 ± 0.22	0.33 ± 0.13	0.20 ± 0.05
Tumor	1.61 ± 0.06	1.01 ± 0.10	0.82 ± 0.06
Brain	0.12 ± 0.01	0.06 ± 0.02	0.04 ± 0.01
Heart	0.93 ± 0.14	0.54 ± 0.15	0.36 ± 0.04
T/B ratio	0.55 ± 0.09	0.78 ± 0.13	0.94 ± 0.13
T/M ratio	3.06 ± 1.06	3.42 ± 1.33	4.35 ± 1.41

① All data are the mean percentage ($n = 3$) of the injected dose of ^{99m}Tc-MAG$_3$-DG per gram of wet tissue, ± the standard deviation of the mean.

Table 5 Biodistribution of ^{99m}Tc-MAMA-BA-DG in breast tumorbearing mice[①]

Tissue	30min	2h	4h
Blood	1.29 ± 0.09	0.36 ± 0.03	0.23 ± 0.03
Lung	1.17 ± 0.15	0.35 ± 0.07	0.24 ± 0.01
Liver	10.86 ± 1.78	5.49 ± 1.28	3.60 ± 0.94
Stomach	2.73 ± 1.08	2.83 ± 2.01	1.40 ± 0.84
Spleen	0.51 ± 0.08	0.22 ± 0.10	0.22 ± 0.07
Kidney	7.47 ± 0.53	3.76 ± 1.85	2.09 ± 0.34
Intestine	2.26 ± 1.98	2.57 ± 1.63	1.90 ± 1.18
Muscle	0.31 ± 0.03	0.14 ± 0.04	0.11 ± 0.12
Tumor	0.81 ± 0.03	0.26 ± 0.05	0.21 ± 0.03
Brain	0.06 ± 0.01	0.02 ± 0.002	0.04 ± 0.03
Heart	0.42 ± 0.01	0.12 ± 0.02	0.12 ± 0.06
T/B ratio	0.63 ± 0.02	0.74 ± 0.20	0.94 ± 0.16
T/M ratio	2.65 ± 0.15	1.91 ± 0.12	2.25 ± 0.71

① All data are the mean percentage ($n = 3$) of the injected dose of ^{99m}Tc-MAMA-BA-DG per gram of wet tissue, ± the standard deviation of the mean.

In summary, three new DG derivatives were synthesized and labeled with technetium-99m successfully. Low molecular weight accelerated clearance from blood, and different linkers and chelate cores changed excretion path. Among them, ^{99m}Tc-MAG$_3$-DG could be further studied as potential tumor imaging agents. Related work is underway and will be reported in due course.

Acknowledgement

This work was supported by the National Natural Science Foundation of China (No. 20471011).

Supplementary data

Supplementary data associated with this article can be found, in the online version, at doi: 10.1016/j.bmcl.2006.08.050.

References

[1] Bar-Shalom, R.; Valdivia, A. Y.; Blaufox, M. D. Semin. Nucl. Med. 2000, 30, 150.
[2] Delbeke, D.; Martin, W. H. Radiol. Clin. NorthAm. 2001, 39, 883.
[3] Eubank, W. B.; Mankoff, D. A. Semin. Nucl. Med. 2004, 34, 224.
[4] Zhuang, H.; Yu, J. Q.; Alavi, A. Radiol. Clin. North Am. 2005, 43, 121.
[5] Avril, N. E.; Weber, W. A. Radiol. Clin. North Am. 2005, 43, 189.
[6] Brock, C. S.; Meikle, S. R.; Price, R. Eur. J. Nucl. Med. 1997, 24, 691.
[7] Storr, T.; Obata, M.; Fisher, C. L.; Bayly, S. R.; Green, D. E.; Brudzinska, I.; Mikata, Y.; Patrick, B. O.; Adam, M. J.; Yano, S.; Orvig, C. Chem. Eur. J. 2004, 11, 195.
[8] Bayly, S. R.; Fisher, C. L.; Storr, T.; Adam, M. J.; Orvig, C. Bioconjugate Chem. 2004, 15, 923.

[9] Storr, T. ; Fisher, C. L. ; Mikata, Y. ; Yano, S. ; Adam, M. J. ; Orvig, C. Dalton Trans. 2005, 4, 654.

[10] Storr, T. ; Sugai, Y. ; Barta, C. A. ; Mikata, Y. ; Adam, M. J. ; Yano, S. ; Orvig, C. Inorg. Chem. 2005, 44, 2698.

[11] Schibli, R. ; Dumas, C. ; Petrig, J. ; Garcia-Garayoa, E. ; Schubiger, P. A. ; Spadola, L. ; Scapozza, L. Bioconjugate Chem. 2005, 16, 105.

[12] Yang, D. J. ; Kim, C. G. ; Azhdarinia, A. ; Yu, D. F. ; Oh, C. S. ; Bryant, J. L. ; Won, J. J. ; Kim, E. E. ; Podoloff, D. A. ; Schechter, N. R. Radiology 2003, 226, 465.

Preparation and Biodistribution of Novel ^{99m}Tc(CO)$_3$-CNR Complexes for Myocardial Imaging[*]

Abstract We evaluated lipophilicity and biodistribution of a series of ^{99m}Tc(CO)$_3$-ether isonitrile complexes to determine whether different lipophilicity and structure of isonitrile ligands would improve the imaging properties of the radiopharmaceutical for the heart. Novel ^{99m}Tc(CO)$_3$-MIBI analogs were prepared and analyzed by radio-HPLC, and their lipophilicity was determined. These new complexes could be bi-or tri-substituted in specified pH conditions like ^{99m}Tc(CO)$_3$-MIBI. These new complexes exhibited low liver, lungs and blood uptake compared with [^{99m}Tc(CO)$_3$(MIBI)$_3$]$^+$ though their heart uptake was not so high. Among these complexes, [^{99m}Tc(CO)$_3$(EPI)$_2$(OH$_2$)]$^+$ showed higher target to non-target ratios at 5 and 30min post-injection than that of [^{99m}Tc(CO)$_3$(MIBI)$_3$]$^+$.

Key words   Tc-99m, tricarbonyl, isonitrile, myocardial imaging

1 Introduction

Recently, Alberto[1] reported the synthesis and applications of the astonishing complex fac-[^{99m}Tc(CO)$_3$(OH$_2$)$_3$]$^+$. This simple complex combines a "lower" organometallic half with three carbonyl ligands with an "upper" Werner-type half with three water molecules as ligands[2]. It has favorable properties mainly attributed to its unusual ligand set. The σ-bound water ligands are readily displaced by other ligands, but in combination with other donor-acceptor ligands, the carbonyl groups impart a high kinetic stability on [Tc(CO)$_3$]$^+$ derivatives. In addition, a high kinetic stability results from the d^6 low-spin electron configuration of the Tc^I complex. Although it was used in hexakis(2-methoxyisobutyl isonitrile) ^{99m}Tc(I) (^{99m}Tc-MIBI), the oxidation state +1 is rather unusual in ^{99m}Tc imaging agents and difficult to stabilize with Werner-type ligands (typically with an N, O, or S donor set) alone. In this respect, organometallic complexes with CO ligands expand the available range of compounds significantly. There has been a lot of published research work utilizing this new precursor to develop new radiopharmaceuticals[3,4].

^{99m}Tc-MIBI is widely used in myocardial perfusion imaging and human tumor imaging. Therefore, preparation and preliminary evaluations of ^{99m}Tc(CO)$_3$-MIBI as a myocardial imaging agent[5-8] and a functional probe of Pgp transport activity[9] followed the introduction of the attrac-

---

[*] Copartner: Hao Guiyang, Zang Jianying, Reprinted from *Journal of Labelled Compounds and Radiopharmaceuticals*, 2007, 50(1):13-18.

tive precursor of fac-$[^{99m}Tc(CO)_3(OH_2)_3]^+$. Our group independently designed $^{99m}Tc(CO)_3$-MIBI as a myocardial imaging agent which showed high potential both in mice and dogs.

Examination of the uptake mechanism in myocardial and carcinoma cells indicates that the lipophilicity, cationic charge and the ligand of MIBI itself of ^{99m}Tc-MIBI play a significant role in its accumulation and retention. Moreover, it can be seen from our previous research that $^{99m}Tc(CO)_3$-MIBI might be bi-, tri-substituted product or both in specified pH conditions[7]. It is still unknown whether this radiolabeling reaction phenomenon is only specific for MIBI or not. Therefore, we prepared novel $^{99m}Tc(CO)_3$-ether isonitrile complexes to study this experimental phenomenon and the effects of different lipophilicity and ligand structures on biodistribution.

2 Results and Discussion

2.1 Synthesis

The general synthetic approach for the isonitrile derivatives (CNR) is shown in Scheme 1. The isonitrile derivatives could be prepared conveniently from commercial available alkoxyl amines in a two-step synthesis (Table 1)[10-12]. The use of $POCl_3$/pyridine for dehydration of formamide was more convenient and safer than other dehydration systems. The two steps were both exothermic reactions, so the good cooling conditions should be guaranteed. The Vigreux column was necessary to collect the products with high purity, though there would be a little loss in yields. All these isonitriles could be stored at -20 ℃ under N_2 for radiolabeling.

$$R\text{—}NH_2 + HCOOC_2H_5 \longrightarrow R\text{—}NHCHO + HOC_2H_5$$
$$R\text{—}NHCHO + POCl_3 + Py \longrightarrow R\text{—}N{\equiv}C + PyHCl + PyHPO_3$$

Scheme 1 Synthesis of isonitrile ligands

2.2 Radiolabeling

The $[^{99m}Tc(CO)_3(OH_2)_3]^+$ precursor was prepared in >98% yield according to the method of Alberto et al[1]. $^{99m}Tc(CO)_3$-CNR complexes were prepared as described in the experimental section (Table 1). They were analyzed by radio-TLC and radio-HPLC, and the results were shown in Table 2. Besides, there was one radio-HPLC peak corresponding to the mono-complex that appeared at the beginning of radiolabeling course and would disappear quickly with the substitution going on. When pH = 9.0-10.0, all three H_2O ligands of was shown to be readily substituted by isonitrile ligands. However, the sequential reaction would stop at bi-substitution when pH = 3.0-4.0. When pH lied between the two scopes of 3.0-4.0 and 9.0-10.0, tri-substitution only partially occurred and the ratios of bi- to tri-complex were determined by the pH values. These results above mentioned were same as that of $^{99m}Tc(CO)_3$-MIBI[7], so it should be a common characteristic for ether isonitriles reacting with $[^{99m}Tc(CO)_3(OH_2)_3]^+$.

Table 1 The isonitrile ligands and ^{99m}Tc(CO)$_3$-CNR complexes

Name	Ligand	Complex	No.
MEI	CH$_3$OCH$_2$CH$_2$NC	[^{99m}Tc(CO)$_3$(MEI)$_2$(OH$_2$)]$^+$	1a
		[^{99m}Tc(CO)$_3$(MEI)$_3$]$^+$	1b
MPI	CH$_3$OCH$_2$CH$_2$CH$_2$NC	[^{99m}Tc(CO)$_3$(MPI)$_2$(OH$_2$)]$^+$	2a
		[^{99m}Tc(CO)$_3$(MPI)$_3$]$^+$	2b
EPI	CH$_3$CH$_2$OCH$_2$CH$_2$CH$_2$NC	[^{99m}Tc(CO)$_3$(EPI)$_2$(OH$_2$)]$^+$	3a
		[^{99m}Tc(CO)$_3$(EPI)$_3$]$^+$	3b
IPPI	(CH$_3$)$_2$CHOCH$_2$CH$_2$CH$_2$NC	[^{99m}Tc(CO)$_3$(IPPI)$_2$(OH$_2$)]$^+$	4a
		[^{99m}Tc(CO)$_3$(IPPI)$_3$]$^+$	4b
THFMI	(tetrahydrofuranyl-CH$_2$-N≡C)	[^{99m}Tc(CO)$_3$(THFMI)$_2$(OH$_2$)]$^+$	5a
		[^{99m}Tc(CO)$_3$(THFMI)$_3$]$^+$	5b
MIBI	CH$_3$OC(CH$_3$)$_2$CH$_2$NC	[^{99m}Tc(CO)$_3$(MIBI)$_2$(OH$_2$)]$^+$	6a
		[^{99m}Tc(CO)$_3$(MIBI)$_3$]$^+$	6b

Table 2 Results of radiolabeling reaction in pH = 3.0-4.0 and 9.0-10.0

pH condition	Complex	HPLC elution condition TETA①:MeOH ($v:v$)	R_t/min	Yield/%
3.0-4.0	1a	50:50	7.9	>98%
	2a	40:60	9.0	
	3a	30:70	6.9	
	4a	20:80	5.6	
	5a	30:70	7.1	
9.0-10.0	1b	50:50	9.5	>98%
	2b	40:60	10.7	
	3b	30:70	10.0	
	4b	20:80	7.9	
	5b	30:70	9.2	

① TETA: triethylamine phosphate buffer, 0.05M, pH 2.25.

All complexes were stable within 6h at room temperature. Their lipophilicity was determined by partition between 1-octanol and PBS (Table 3). Most of them showed low lipophilicity because of the hydrophilicity of ^{99m}Tc(CO)$_3^+$ core. The lipophilicity turned higher when the substituent number changed from two to three for the same isonitrile ligand. The tri-substituted [^{99m}Tc(CO)$_3$(IPPI)$_3$]$^+$ had the highest lipophilicity, and the lipophilicity of [^{99m}Tc(CO)$_3$(MEI)$_2$(OH$_2$)]$^+$ was the lowest. The lipophilicity would increase with the number of carbon in ligands increasing for MEI, MPI, EPI and IPPI. At the same time, the sequences of lipophilicity of the corresponding complexes were as below: 1a < 2a < 3a < 4a, 1b < 2b < 3b < 4b. Therefore, the structure of the isonitriles strongly affects the lipophilicity of the complexes.

Table 3　Partition coefficients of $^{99m}Tc(CO)_3$-CNR complexes

Complex	1a	1b	2a	2b	3a	3b	4a	4b	5a	5b
P	0.09	0.10	0.31	0.62	0.58	1.58	2.79	10.86	0.14	0.36
logP	−1.04	−1.00	−0.51	−0.21	−0.23	0.20	0.44	1.04	−0.87	−0.44

2.3　Biodistribution

Table 4 showed the tissue distributions of the novel $^{99m}Tc(CO)_3$-CNR complexes. Tissues, including heart, blood, liver, lungs, kidneys and spleen, were collected at 5min, 30min, and 60min post-injection. Ten complexes showed distinct accumulation in heart, liver and lungs etc. The heart uptake of the new complexes except $[^{99m}Tc(CO)_3(EPI)_2(OH_2)]^+$ and $[^{99m}Tc(CO)_3(IPPI)_2(OH_2)]^+$ lay in a low level, and the activity in heart did not show a good retention from 5 to 30min post-injection. For ^{99m}Tc-MIBI and $[^{99m}Tc(CO)_3(MIBI)_3]^+$ the heart uptake at 30 min post-injection in mice could be 25% ID/g and 22% ID/g[13], respectively. However, clearance from liver, lungs and blood was very fast for the new complexes during the early stage post-injection, which resulted in apparently low uptake of liver, lungs and blood at 30min and 60min post-injection. Therefore, the new complexes exhibited lower liver, lungs and blood uptake compared with $[^{99m}Tc(CO)_3(MIBI)_3]^+$, which made the target to non-target (T/NT) ratios of several new complexes were higher than that of $[^{99m}Tc(CO)_3(MIBI)_3]^+$ (Fig. 1). The heart to liver ratio of $[^{99m}Tc(CO)_3(EPI)_2(OH_2)]^+$ was 3.37 that was about double ratio of $[^{99m}Tc(CO)_3(MIBI)_3]^+$ at 5min post injection, and for ^{99m}Tc-MIBI this ratio was usually less than 1.0 within 1h. $[^{99m}Tc(CO)_3(EPI)_2(OH_2)]^+$ exhibited almost the best performance in several important aspects for myocardial imaging among the new designed complexes. To be successful myocardial imaging agents the new complexes chiefly need to improve their heart uptake and retention to a little higher level as ^{99m}Tc-MIBI and keep their current high target to non-target ratios at the same time.

Table 4　Tissue distributions of all the complexes at 5min, 30min, and 60min post-injection

Complex	Time/min	% ID/g in tissues					
		Blood	Heart	Liver	Lungs	Kidneys	Spleen
1a	5	1.29 ± 0.16	3.04 ± 0.34	10.24 ± 1.64	2.16 ± 0.22	28.11 ± 7.05	0.85 ± 0.03
	30	0.39 ± 0.02	2.10 ± 0.02	6.80 ± 0.85	1.27 ± 0.12	9.85 ± 2.04	0.44 ± 0.12
	60	0.31 ± 0.01	1.63 ± 0.25	7.06 ± 0.63	1.34 ± 0.19	4.58 ± 0.37	0.43 ± 0.04
1b	5	2.32 ± 0.20	7.74 ± 1.31	14.00 ± 1.92	3.60 ± 0.37	34.22 ± 1.58	1.65 ± 0.23
	30	0.87 ± 0.11	4.32 ± 0.48	8.39 ± 0.38	2.10 ± 0.12	11.19 ± 1.15	0.97 ± 0.18
	60	0.39 ± 0.02	2.02 ± 0.82	5.82 ± 1.02	1.25 ± 0.20	6.03 ± 0.62	0.62 ± 0.23
2a	5	1.09 ± 0.24	11.44 ± 2.14	11.62 ± 1.24	2.64 ± 0.40	31.71 ± 4.19	1.72 ± 0.42
	30	0.36 ± 0.01	10.35 ± 1.14	5.26 ± 0.71	1.70 ± 0.36	9.19 ± 1.77	1.18 ± 0.13
	60	0.20 ± 0.02	7.54 ± 2.18	4.99 ± 2.64	1.19 ± 0.37	5.36 ± 0.25	0.81 ± 0.19

Complex	Time/min	%ID/g in tissues					
		Blood	Heart	Liver	Lungs	Kidneys	Spleen
2b	5	2.08 ± 0.27	15.39 ± 1.04	14.67 ± 1.02	4.72 ± 0.82	41.42 ± 1.74	2.93 ± 0.18
	30	0.44 ± 0.07	9.71 ± 2.24	4.34 ± 1.17	1.70 ± 0.14	10.12 ± 1.35	1.29 ± 0.09
	60	0.21 ± 0.02	5.26 ± 0.15	2.48 ± 0.05	0.97 ± 0.04	6.53 ± 0.21	1.02 ± 0.23
3a	5	1.44 ± 0.10	28.87 ± 3.19	8.60 ± 1.30	6.56 ± 0.40	62.91 ± 16.95	3.99 ± 1.34
	30	0.42 ± 0.03	19.48 ± 1.96	5.35 ± 0.45	2.98 ± 0.30	16.91 ± 1.28	2.40 ± 0.31
	60	0.26 ± 0.02	12.39 ± 0.59	5.08 ± 0.49	1.96 ± 0.26	9.44 ± 0.69	1.22 ± 0.17
3b	5	3.66 ± 0.67	15.98 ± 6.61	57.06 ± 15.74	7.12 ± 1.65	112.35 ± 26.3	7.37 ± 1.22
	30	0.75 ± 0.01	7.14 ± 1.12	13.00 ± 4.17	2.15 ± 0.37	42.14 ± 6.71	2.60 ± 0.76
	60	0.42 ± 0.02	4.71 ± 1.58	5.77 ± 1.27	0.99 ± 0.14	17.78 ± 3.98	1.40 ± 0.38
4a	5	1.63 ± 0.19	17.03 ± 4.45	14.41 ± 0.74	6.01 ± 1.86	70.61 ± 19.65	5.70 ± 1.03
	30	0.47 ± 0.07	17.08 ± 3.90	8.53 ± 0.68	2.46 ± 0.30	50.95 ± 9.11	3.99 ± 0.54
	60	0.94 ± 0.89	14.99 ± 3.09	9.77 ± 1.45	1.96 ± 0.73	46.60 ± 12.76	1.61 ± 0.04
4b	5	6.65 ± 0.74	5.77 ± 1.34	104.10 ± 8.39	9.83 ± 0.96	71.78 ± 9.51	7.74 ± 1.63
	30	1.18 ± 0.04	1.99 ± 0.27	19.07 ± 2.98	1.69 ± 0.22	26.36 ± 5.89	2.72 ± 1.39
	60	0.63 ± 0.08	1.18 ± 0.28	11.23 ± 4.66	1.01 ± 0.34	14.53 ± 3.80	0.92 ± 0.10
5a	5	0.60 ± 0.06	4.68 ± 0.53	7.67 ± 1.98	1.50 ± 0.25	18.00 ± 3.90	0.70 ± 0.15
	30	0.14 ± 0.02	2.55 ± 0.20	1.95 ± 0.12	0.59 ± 0.05	3.18 ± 0.46	0.29 ± 0.06
	60	0.08 ± 0.01	2.16 ± 0.18	1.57 ± 0.08	0.61 ± 0.25	2.53 ± 0.03	0.29 ± 0.11
5b	5	1.00 ± 0.13	5.00 ± 1.02	6.03 ± 1.30	1.78 ± 0.32	22.49 ± 6.27	0.97 ± 0.36
	30	0.23 ± 0.01	4.67 ± 0.78	2.48 ± 0.14	1.06 ± 0.16	7.75 ± 1.75	0.61 ± 0.23
	60	0.11 ± 0.06	2.52 ± 0.73	1.09 ± 0.09	0.73 ± 0.15	4.36 ± 0.60	0.45 ± 0.06

$[^{99m}Tc(CO)_3(IPPI)_3]^+$ with highest lipophilicity also had the highest liver, lungs and blood uptake at 5 min post-injection, but it had low heart uptake. The complex with the lowest lipophilicity, $[^{99m}Tc(CO)_3(MEI)_2(OH_2)]^+$, had low heart uptake though its liver, lungs and blood uptake were also low. $^{99m}Tc(CO)_3THFMI$ with the greatest difference in ligand structure did not show good biodistribution. For $^{99m}Tc(CO)_3$-MIBI, the tri-substituted product showed much better performance in myocardial imaging than bi-complex. On the contrary, for EPI and IPPI the bi-complex had much more better imaging properties. Therefore, the substituent number was not crucial factor in biodistribution for $^{99m}Tc(CO)_3$-CNR.

Usually, the biodistribution of homologous complexes of complex is affected by their lipophilicity, size and the ligand itself, etc. During the course of developing ^{99m}Tc-isonitrile complexes as myocardial imaging agents, $[^{99m}Tc(MIBI)_6]^+$ showed much better properties than others such as $[^{99m}Tc(TBI)_6]^+$ and $[^{99m}Tc(CPI)_6]^+$. It proves that the ligand plays an important role to the biological behavior of the complex, and MIBI is the best among the isonitrile analogs. It might explain from one aspect why the new complexes in this investigation did not ex-

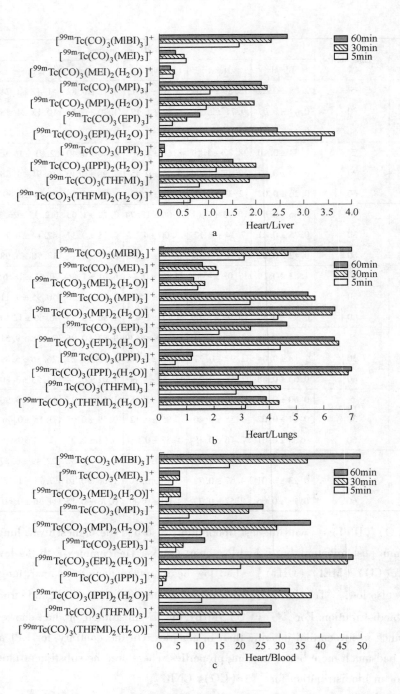

Fig. 1 The heart to non-target ratios of the ^{99m}Tc(CO)$_3$-CNR complexes
a—Heart/Liver; b—Heart/Lungs; c—Heart/Blood

hibited the same high heart uptake and retention as $[^{99m}Tc(CO)_3(MIBI)_3]^+$. In addition, the lipophilicity also showed marked influence on the biodistribution. $[^{99m}Tc(CO)_3(IPPI)_3]^+$ and $[^{99m}Tc(CO)_3(MEI)_2(OH_2)]^+$ with highest qand lowest lipophilicity, respectively, did not show good biodistribution in mice. Therefore, there is a most appropriate scope of lipophilicity to achieve the good balance of high uptake in heart and fast clearance from non-target tissues for $^{99m}Tc(CO)_3$-CNR complexes. To get this scope, more $^{99m}Tc(CO)_3$-CNR complexes should be prepared and studied in the future.

3 Conclusion

Like $^{99m}Tc(CO)_3$-MIBI, there are bi-and tri-substituted products when the five isonitriles react with fac-$[^{99m}Tc(CO)_3(OH_2)_3]^+$ in specified pH conditions. The changes from lipophilicity and ligand itself did bring great effects on the properties of the $^{99m}Tc(CO)_3$-CNR complexes for myocardial imaging. Among these new complexes, $[^{99m}Tc(CO)_3(EPI)_2(OH_2)]^+$ showed the most favorable imaging characteristics in mice.

4 Experimental

2-methoxyethaneamine, 3-ethoxypropaneamine, 3-iso propoxypropan-1-amine, tetrahydro-furfurylamine were purchased from Acros Organics Co. 3-methoxypropaneamine was purchased from Aldrich Chemical Co. Other chemicals were purchased from Beijing Chemical Reagents Company. Pure CO gas was purchased from NRCCRM, China. $^{99}Mo/^{99m}Tc$ generator was obtained from the Beijing Syncor Medical Corporation. ICR mice, 18-20g, female, were obtained from Animal Center of Peking University. Proton nuclear magnetic resonance spectroscopy was performed on Bruker Avance 500 MHz. Infrared spectrum was performed on Nicolet-170SX. The automatic gamma counts were carried out by WALLAC/WIZARD 1470, Perkin Elmer Wallac. HPLC was performed on a SHIMADZU system (SCL-10Avp pumps and SPD-10Avp UV detector) and Park Radioow detector. TLC was run on polyamide film using acetonitrile as mobile phase.

4.1 Synthesis of isonitriles (CNR)

A solution of ethyl formate (8.0mL, 100mmol), precooled to approximately 0℃, was added slowly to a stirred solution of amine analogs (95mmol) in an ice/ NaCl bath. After the slightly exothermic reaction ceased, the solution was allowed to warm slowly to room temperature and refluxed overnight. The solution was distilled through a Vigreux column to give formamide analogs determined by infrared spectroscopy. 50mmol formamide was dissolved in methylene chloride (45mL). Triethylamine (20.1mL, 0.25mol) was added and the clear solution was cooled in an ice/water bath. Phosphorus oxychloride (2.75mL, 30mmol) was added dropwise to the cooled formamide solution. The resulting suspension was stirred and allowed to slowly warm to room temperature for 1h, and at reflux temperature for 15min. 20mL cold water was added and the organic layer separated. The organic layer was washed with a saturated solution of sodium bicarbonate, water and dried with anhydrous sodium sulfate. Evaporation of the methylene chloride left a dark brown liquid. The dark brown liquid was distilled through a Vigreux column under vacuum to give the isonitrile analogs. They were determined by IR and [^1H]NMR.

4.2 Preparation of $^{99m}Tc(CO)_3$-CNR

The $[^{99m}Tc(CO)_3(OH_2)_3]^+$ precursor was prepared according to the method of Alberto et al. 1.0mg isonitrile, dissolved in 0.5mL water, was added to 1mL $[^{99m}Tc(CO)_3(OH_2)_3]^+$ solution which was adjusted to pH = 3.0-4.0 or 9.0-10.0 with 0.5N HCl solution. The solution was allowed in a boiling water bath for 15 min and examined by TLC (R_f = 0.9-1.0 for $[^{99m}Tc(CO)_3(CNR)_2(OH_2)]^+$ and $[^{99m}Tc(CO)_3(CNR)_3]^+$, R_f = 0-0.1 for $[^{99m}Tc(CO)_3(OH_2)_3]^+$ and R_f = 0.4-0.5 for $^{99m}TcO_4^-$) and radio HPLC (Alltima C18 RP column, 250mm × 4.6mm, 5μm, flow rate 1mL/min).

4.3 Stability

The labeled complexes were incubated at room temperature for up to 6h. Aliquots were taken and analyzed by radio-HPLC to assess the stability.

4.4 Determination of the partition coefficient

The lipophilicity of the complex with RCP over 95% was determined as follows: 0.1mL complex solution was mixed with 2mL 1-octanol and 1.9mL PBS (0.01M, pH = 7.4) in a centrifuge tube. The tube was vortexed at room temperature for 3min and then was centrifuged at high speed for 10min. 0.1mL samples of both phases were taken out and counted in a well-counter. The measurement was repeated for three times. The partition coefficient, P, was calculated using the following equation:

$$P = (\text{cpm in octanol} - \text{cpm background})/(\text{cpm in PBS} - \text{cpm background})$$

Usually the final partition coeffcient value was expressed as log$P$.

4.5 Biodistribution

Samples (about 740kBq in 0.1mL solution) were injected through the tail vain into ICR mice (18-20g, female). The mice were sacrificed at 5, 30, and 60min post-injection. Selected organs were collected for weighing and counting. The accumulated radioactivity in the tissues was calculated in terms of percentage of injected dose per gram organ (%ID/g).

Acknowledgements

This work was supported financially by the "National 211 Engineering Project" of "Tenth-Five Year Plan of Peoples's Republic of China". The authors would like to thank Lin Zhu for helpful assistance.

References

[1] Alberto R, Schibli R, Egli A, Schubiger A P, Abram U, Kaden T A. J Am Chem Soc 1998;120:7987-7988. DOI:10.1021/ja980745t.

[2] Metzler-Nolte N. Angew Chem Int Ed 2001;40:1040-1043. DOI:10.1002/1521-3773.

[3] Alberto R, Schibli R, Waibel R, Abram U, Schubiger AP. Coordinat Chem Rev 1999;190-192:901-919. DOI:10.1016/S0010-8545(99)00128-9.
[4] Alves S, Paulo A, Correia JDG, Gano L, Smith CJ, Hoffman TJ, Santos I. Bioconjugate Chem 2005;16:438-449. DOI:10.1021/bc0497968.
[5] Jiang Y, Liu BL. Chinese Sci Bull 2001;46:727-730 (in Chinese).
[6] Wang J, Liu B, Mi H, Jiang Y, Tang Z, Gui H. Clin J Nucl Med 2002;22:231-232 (in Chinese).
[7] Hao G Y, Zang J Y, Zhu L, Guo Y Z, Liu B L. J Label Compd Radiopharm 2004;47:513-521. DOI:10.1002/jlcr.839.
[8] Marmion M E, MacDonald J R. J Nucl Med 2000;41(Suppl.):124P.
[9] Dyszlewski M E, Blake H, Dahlheimer J, Piwnica-Worms D. J Label Compd Radiopharm 2001;44(Suppl.):s63-s65.
[10] Ramalingham K. US Patent,4,864,051,1989.
[11] Jones A G. US Patent,4,735,793,1988.
[12] Lee T W. US Patent,5,210,270,1993.
[13] Zhang X Z, Wang X B, Zhang J B. J Isot 1997;10:158-162(in Chinese).

Caution to HPLC Analysis of Tricarbonyl Technetium Radiopharmaceuticals: An Example of Changing Constitution of Complexes in Column*

Abstract Radio-HPLC is a powerful tool for analyzing radioactive species in radiopharmaceutical chemistry. In this paper, we found an example that the commonly used eluting solvent, acetonitrile, could coordinate with the popular radiopharmaceutical nuclides, technetium-99m, during chromatography. $[^{99m}Tc(CO)_3(H_2O)_3]^+$ and $[Re(CO)_3(H_2O)_3]^+$ showed quite different retention time when they were eluted using acetonitrile/water as mobile phase. However, they almost demonstrated the same retention time when they were eluted using methanol/water as mobile phase. Further analysis showed that both $[^{99m}Tc(CO)_3(H_2O)_3]^+$ and $[Re(CO)_3(H_2O)_3]^+$ could be changed into $[^{99m}Tc(CO)_3(CH_3CN)_3]^+$ and $[Re(CO)_3(CH_3CN)_x(H_2O)_{3-x}]^+$ during the separation, respectively. Some former works mistook the $[^{99m}Tc(CO)_3(CH_3CN)_3]^+$ for $[^{99m}Tc(CO)_3(H_2O)_3]^+$ when using acetonitrile and water in analysis. Quality control of the radiopharmaceuticals containing metal complex should be careful since HPLC solvent could replace some liable ligand molecules.

Key words radiopharmaceuticals, tricarbonyl technetium, HPLC

1 Introduction

Technetium-99m labeled radiopharmaceuticals are currently the main diagnostic agents used in nuclear medicine. A large amount of research on technetium-based radiopharmaceuticals is being carried out all over the world since its ideal characteristics for a nuclear medicine scan. Since the mass amount of technetium in preparation for clinical use is typically in the nanomole range, conventional techniques, such as UV, NMR, IR and elemental analysis, are invalid in analyzing such low concentrated compounds. Radio-HPLC and radio-LCMS developed recently have become powerful techniques for analyzing radio-species and quality control of radiopharmaceuticals[1-3].

Research on technetium-based radiopharmaceuticals has achieved great improvements with the advent of $[^{99m}Tc(CO)_3]^+$ core and $[^{99m}Tc(CO)_3(H_2O)_3]^+$ is a frequently used precursor for labeling small organic molecules and biomacromolecules[4-6]. Methanol/water and acetonitrile/water are the most commonly used mobile phase for reverse phase HPLC. However, organic solvent could play a role of ligand replacing the water ligand in $[^{99m}Tc(CO)_3(H_2O)_3]^+$ during the process of separation. In this paper, we explored the possibility of the liable water molecule in $[^{99m}Tc(CO)_3(H_2O)_3]^+$ being replaced by solvent molecule for HPLC.

---

* Copartner: Chen Xiangji, Guo Yunhang. Reprinted from *Journal of Pharmaceutical and Biomedical Analysis*, 2007, 43(4):1576-1579.

2 Experimental

2.1 Materials and reagents

All the reagents used were AR grade and purchased from Aldrich. HPLC solvents were obtained from Fisher Chemical. Potassium boranocarbonate was synthesized according to the literature[7]. $Na^{99m}TcO_4$ was obtained from a commercial $^{99}Mo/^{99m}Tc$ generator, Beijing Atomic Hightech Co.

2.2 Instrumentation

HPLC analyses were performed on a Shimadzu SCL-10AVP system which consisted of a binary pump with on-line degasser, a model SPD-10 Avp UV detector operating at wave length of 254nm and a Packard 500 TR series flow scintillation analyzer. The samples were separated on an RP C-18 Alltech alltima column(5μm, 250mm × 4.6mm). Electron spray ionization (ESI) mass spectra were recorded on Shimadzu LCMS-2010. Infrared spectra were recorded as KBr disks in the range of 4000-400cm^{-1} on a Nicolet Avatar 360 IR spectrophotometer. Elemental analyses were performed on an Elementar Vario EI.

2.3 Sample preparation

2.3.1 Preparation of $[^{99m}Tc(CO)_3(H_2O)_3]^+$ precursor

The $[^{99m}Tc(CO)_3(H_2O)_3]^+$ was prepared by adding 2mL of $^{99m}TcO_4^-$ from a commercial generator(10mCi) to a 10mL vial containing potassium boranocarbonate(3mg), sodium potassium tartrate tetrahydrate(6.7mg), and potassium tetraborate pentahydrate(5.5mg). The solution was heated for 15min in boiling water under N_2. $[^{99m}Tc(CO)_3(H_2O)_3]^+$ was successfully prepared with high radio-yields(>95%).

2.3.2 Synthesis of $[Re(CO)_3(H_2O)_3]^+$

Method A. The $Re(CO)_3(H_2O)_3Br$ was synthesized according to the literature[8]. Briefly, 200mg of $[Re(CO)_5]Br$ was refluxed in a 10mL round-bottom flask for 24h. Periodic rinsing of the condenser washed unreacted $[Re(CO)_5]Br$ into the reaction solution. The crude mixture was cooled and filtered. Subsequently, the solvent was removed under vacuum to give the desired product as a light grey-green powder (150mg, 75%). IR (cm^{-1}): 2018, 1939. Anal. Calcd. for $C_3H_6BrO_6Re$: C, 8.91; H, 1.50. Found: C, 9.18; H, 1.89. ESI-MS was performed with sample dissolved in methanol: m/z 385.0 $Re(CO)_3(CH_3OH)_3 \cdot H_2O$ (^{187}Re), 367.1 $Re(CO)_3(CH_3OH)_3$ (^{187}Re). Three milligrams of $Re(CO)_3(H_2O)_3$ Br was dissolved in 400μL water and 1 equiv. $AgNO_3$ was added. After the filtration of precipitate, the $[Re(CO)_3(H_2O)_3]^+$ water solution was obtained.

Method B. The $(Et_4N)_2Re(CO)_3Br_3$ was synthesized in a typical reaction[9]. Powered NEt_4Br(170mg, 0.8mmol) was slurried in 30mL of 2,3,5-trioxanoname (diglyme) and heated to 80°C under dry nitrogen. A suspension of $[Re(CO)_5]Br$(150mg, 0.37mmol) in 3mL of warm diglyme was added. The mixture was heated to 110-120℃ for 8h, during which time, some white precipitate was formed. The reaction mixture was filtered while hot and washed with several portions of cold diglyme, diethyl ether and dried in vacuo. The dried light yellow solid was the washed with ethanol to remove the unreacted NEt_4Br. Drying in vacuo yielded the product. $(Et_4N)_2Re(CO)_3Br_3$ was yielded as light-yellow solid (200mg, 70%). IR (cm^{-1}): 1999, 1868. Anal. Calcd. for $C_{19}H_{40}Br_3N_2O_3Re$: C, 29.62; H, 5.23; N, 3.64. Found: C, 29.68; H, 5.41; N, 3.59. Three milligrams of $(Et_4N)_2Re(CO)_3Br_3$ was dissolved in 400μL water and 3 equiv. $AgNO_3$ was added. After the filtration of precipitate, the $[Re(CO)_3(H_2O)_3]^+$ water solution was obtained.

2.4 HPLC methods

Two different chromatographic conditions were employed with a flow rate of 1mL/min. System 1: mobile phase consisted of H_2O containing 0.1% TFA (solvent A) and acetonitrile containing 0.1% TFA (solvent B). The HPLC gradient system started with 10% solvent B with a linear gradient to 90% solvent B from 0 to 28min. System 2: the mobile phase consisted of aqueous 0.05M TEAP (triethylammonium phosphate) buffer, pH 2.25 (solvent A) and methanol (solvent B). The HPLC system started with 100% of A from 0 to 3min. The eluent switched at 6min to 75% A and 25% B and at 9min to 66% A and 34% B followed by a linear gradient 66% A/34% B to 100% B from 9 to 20min. The test solution (2.5μL) was injected into the column and the elution was monitored by observing the UV profile at 254nm for rhenium compounds and radio-trace for technetium-99m compound.

$[Re(CO)_3(H_2O)_3]^+$ and $[^{99m}Tc(CO)_3(H_2O)_3]^+$ were examined using both system 1 (Fig. 1) and system 2 (Fig. 2). In system 1, only one radioactive peak at 16.3min could be observed as shown in Fig. 1d when $[^{99m}Tc(CO)_3(H_2O)_3]^+$ was injected. $[Re(CO)_3(H_2O)_3]^+$ showed retention time of 8.20min, and two leading peaks could be seen at time of 12.00 and 14.0min respectively as shown in Fig. 1a. When $[Re(CO)_3(H_2O)_3]^+$ was dissolved in 4℃ acetonitrile/water mixture (v/v = 50 : 50) solution for 10 min before injection, the peak at 8.20min disappeared and the two leading peaks became sharp peaks (Fig. 1b). The compound corresponding to 14.3-14.4min became the primary component rather than $[Re(CO)_3(H_2O)_3]^+$ after acetonitrile was introduced into the $[Re(CO)_3(H_2O)_3]^+$ for only 10min. If $[Re(CO)_3(H_2O)_3]^+$ was mixed with acetonitrile/water mixture for 60min before injection, the complex became the final product with retention time of 16.4-16.5min, as showed in Fig. 1c. In system 2, only one UV peak was found for $[Re(CO)_3(H_2O)_3]^+$ at 4.22min and the retention time for $[^{99m}Tc(CO)_3(H_2O)_3]^+$ was 5.3min as shown in Fig. 2.

3 Results and Discussion

Identity confirmation across ^{99m}Tc-and ^{99}Tc-or Re-complexes could be performed by co-injection on HPLC and comparison of UV and radiometric detector signal. $[^{99m}Tc(CO)_3(H_2O)_3]^+$ and

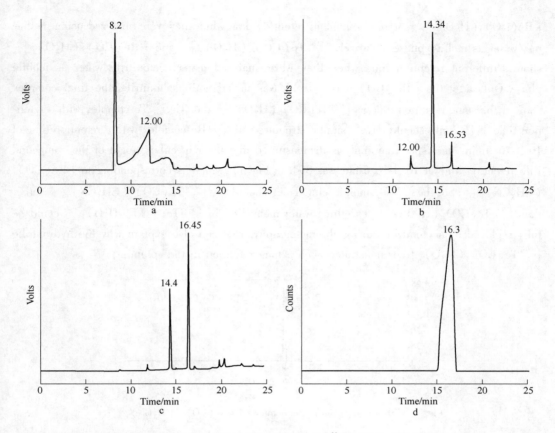

Fig. 1 HPLC chromatogram of $[Re(CO)_3(H_2O)_3]^+$ and $[^{99m}Tc(CO)_3(H_2O)_3]^+$ using system 1

a—Chromatogram of $[Re(CO)_3(H_2O)_3]^+$ in water solution; b—Chromatogram of $[Re(CO)_3(H_2O)_3]^+$ dissolved in 4 ℃ acetonitrile/water mixture (v/v = 50∶50) solution for 10min; c—Chromatogram of $[Re(CO)_3(H_2O)_3]^+$ dissolved in acetonitrile/water mixture solution for 60min; d—Chromatogram of $[^{99m}Tc(CO)_3(H_2O)_3]^+$ in water solution

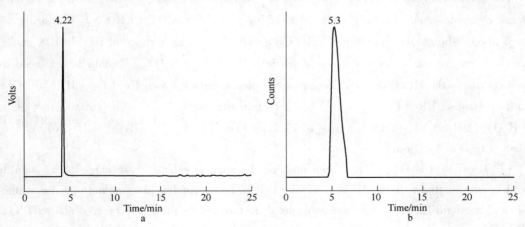

Fig. 2 HPLC chromatogram of $[Re(CO)_3(H_2O)_3]^+$ and $[^{99m}Tc(CO)_3(H_2O)_3]^+$ using system 2

a—Chromatogram of $[Re(CO)_3(H_2O)_3]^+$ in water solution;
b—Chromatogram of $[^{99m}Tc(CO)_3(H_2O)_3]^+$ in water solution

$[Re(CO)_3(H_2O)_3]^+$ showed equivalent retention time when they were analyzed using methanol/water as mobile phase. However, $[^{99m}Tc(CO)_3(H_2O)_3]^+$ and $[Re(CO)_3(H_2O)_3]^+$ showed different retention time when they were analyzed using acetonitrile/water as mobile phase. Only after the $[Re(CO)_3(H_2O)_3]^+$ was mixed with acetonitrile, the final complex showed the same retention time as $[^{99m}Tc(CO)_3(H_2O)_3]^+$ did. The ^{99m}Tc-complex with a retention time of 16.3min should have a similar structure with the Re-complex with a retention time of 16.4-16.5min. Mass spectrum analysis demonstrated that the molecular weight of the compound with retention time at 16.4-16.5min was 394.1 (Fig. 3), which is corresponding to $[Re(CO)_3(CH_3CN)_3]^+$. So, the ^{99m}Tc-complex eluted out should be $[^{99m}Tc(CO)_3(CH_3CN)_3]^+$ rather than $[^{99m}Tc(CO)_3(H_2O)_3]^+$. The three water molecules for $[^{99m}Tc(CO)_3(H_2O)_3]^+$ could be fully replaced by acetonitrile during chromatography. This can also explain why the hydrophilic $[^{99m}Tc(CO)_3(H_2O)_3]^+$ demonstrated such a long retention in the column.

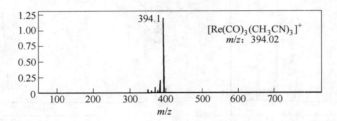

Fig. 3   ESI-MS spectrum of the compound $[Re(CO)_3(CH_3CN)_3]^+$

Some former works[10, 11] using system 1 to characterize the $[^{99m}Tc(CO)_3(H_2O)_3]^+$ showed the retention time of 13.6-13.7min for radioactivity. According to our experimental results, the eluted complex should be $[^{99m}Tc(CO)_3(CH_3CN)_3]^+$ rather than $[^{99m}Tc(CO)_3(H_2O)_3]^+$.

The mass amount of $[^{99m}Tc(CO)_3(H_2O)_3]^+$ was in range of 10^{-10} to 10^{-8} mole, so its three water molecules could be fully replaced resulting in $[^{99m}Tc(CO)_3(CH_3CN)_3]^+$ during the separation. Whereas, the quantity of $[Re(CO)_3(H_2O)_3]^+$ was in range of 10^{-6} to 10^{-3} mole, and only one or two water molecules of partial $[Re(CO)_3(H_2O)_3]^+$ could be replaced by acetonitrile in the HPLC column. The primary eluted component for $[Re(CO)_3(H_2O)_3]^+$ was still unchanged $[Re(CO)_3(H_2O)_3]^+$ (retention time of 8.2min) with partial $[Re(CO)_3(H_2O)_2(CH_3CN)]^+$ (retention time of 12.0min) and $[Re(CO)_3(H_2O)(CH_3CN)_2]^+$ (retention time of 14.4min).

$[^{99m}Tc(CO)_3(H_2O)_3]^+$ has been and will be studied extensively because of its stability and kinetic inertness. Radio-HPLC is a very helpful tool to analyze the product of radio-synthesis and acetonitrile/water was commonly used as the mobile phase for reverse phase HPLC. Though acetonitrile can not replace any strong ligand, it probably can replace the "1" weak ligand molecule in complex with a "2 + 1" coordination pattern. This will cause a change in constitution of complex after purification with HPLC, and then some important characteristics in vitro and in vivo.

4 Conclusions

In the paper, we found that $[^{99m}Tc(CO)_3(H_2O)_3]^+$ and $[Re(CO)_3(H_2O)_3]^+$ could be changed into $[^{99m}Tc(CO)_3(CH_3CN)_3]^+$ and $[Re(CO)_3(CH_3CN)_x(H_2O)_{3-x}]^+$, respectively, during the process of separation when they were eluted using acetonitrile/water mixture in HPLC. So, HPLC characterization should be carried out carefully when $[M(CO)_3(H_2O)_x]$ was eluted by acetonitrile/water. Methanol/water can be used as mobile phase since no replacement has been observed during the chromatography.

Acknowledgement

This work was supported by the National Natural Science Foundation of China (No. 20471011).

Appendix A. Supplementary data

Supplementary data associated with this article can be found, in the online version, at doi: 10.1016/j.jpba.2006.11.031.

References

[1] S. Liu, M. C. Ziegler, D. S. Edwards, Bioconjugate Chem. 11 (2000) 113-117.
[2] S. Liu, D. S. Edwards, M. C. Ziegler, A. R. Harris, S. J. Hemingway, J. A. Barrett, Bioconjugate Chem. 12 (2001) 624-629.
[3] D. Vanderghinste, M. Van Eeckhoudt, C. Terwinghe, L. Mortelmans, G. M. Bormans, A. M. Verbruggen, H. P. Vanbilloen, J. Pharm. Biomed. Anal. 32 (2003) 679-685.
[4] R. Alberto, R. Schibli, A. Egli, A. P. Schubiger, U. Abram, T. A. Kaden, J. Am. Chem. Soc. 120 (1998) 7987-7988.
[5] Y. Arano, Ann. Nucl. Med. 16 (2002) 79-93.
[6] M. A. Méndez-Rojas, B. I. Kharisov, A. Y. Tsivadze, J. Coord. Chem. 59 (2006) 1-63.
[7] R. Alberto, K. Ortner, N. Wheatley, R. Schibli, A. P. Schubiger, J. Am. Chem. Soc. 123 (2001) 3135-3136.
[8] N. Lazarova, S. James, J. Babich, J. Zubieta, Inorg. Chem. Commun. 7 (2004) 1023-1026.
[9] R. Alberto, A. Egli, R. Alberto, K. Hegetschweiler, V. Gramlich, A. P. Schubiger, J. Chem. Soc., Dalton Trans. 19 (1994) 2815-2820.
[10] D. Satpati, M. Mallia, K. Kothari, M. R. A. Pillai, J. Label. Compd. Radiopharm. 47 (2004) 657-668.
[11] D. Satpati, K. Bapat, A. Mukherjee, S. Banerjee, K. Kothari, M. Venkatesh, Appl. Radiat. Isot. 64 (2006) 888-892.

Preparation and Biological Evaluation of ^{99m}Tc-CO-MIBI as Myocardial Perfusion Imaging Agent*

Abstract ^{99m}Tc-Sestamibi has been playing an important role in the cardiac imaging for the last decades. Previously, we reported that $[^{99m}Tc(CO)_3(MIBI)_3]^+$ demonstrated a significant location in myocardium with a lower liver uptake as compared with ^{99m}Tc-Sestamibi. In this work, we found that new $[^{99m}Tc(CO)_2(MIBI)_4]^+$ could be prepared with high radiochemical purity. The intertransformations between $[^{99m}Tc(CO)_3(H_2O)(MIBI)_2]^+$, $[^{99m}Tc(CO)_3(MIBI)_3]^+$, and $[^{99m}Tc(CO)_2(MIBI)_4]^+$ were investigated and biodistribution was performed to evaluate the $[^{99m}Tc(CO)_2(MIBI)_4]^+$ as a myocardial perfusion imaging agent. The results showed that one more CO was replaced by MIBI slowing down the pharmacokinetics. The structure characterization was performed on their corresponding rhenium complexes, and the results indicated that there were differences between ^{99m}Tc-CO-MIBI and Re-CO-MIBI in preparation and hydrophobic characteristics.

Key words tricarbonyl technetium, MIBI, myocardial imaging

1 Introduction

Myocardial perfusion imaging with radiotracers plays a very important role in the evaluation of patients with coronary artery disease in clinical practice[1-10]. ^{99m}Tc(Ⅰ)-Sestamibi(shown in Fig. 1) and ^{99m}Tc(Ⅴ)-Tetrofosmin have been widely used in clinics for decades, but neither of them meets the requirements of an ideal myocardial perfusion imaging agent. Their high liver uptake and slow clearance from it make it very diffcult to interpret the heart activity in the inferior and left ventricular wall[1-8,10-13]. The complexes in lower oxidation state have been widely studied for the development of new imaging agents. One-step synthesis of $[^{99m}Tc(CO)_3(H_2O)_3]^+$ by direct reduction of $^{99m}TcO_4^-$ with sodium borohydride in aqueous solution was firstly developed by Alberto et al.[14]. The $^{99m}Tc(CO)_3$ core possesses many excellent features, such as its small volume and kinetic inertness, and the three coordinated water in this complex could be easily replaced by other ligands. Some crown-ether-containing cationic ^{99m}Tc(Ⅰ)-tricarbonyl radiotracers developed by Liu et al. showed significant localization in myocardium[15,16]. Our previous studies also showed that the water molecules in $[^{99m}Tc(CO)_3(H_2O)_3]^+$ could be replaced by two or three 2-methoxy-isobutyl-isonitrile molecules(MIBI) under different reaction conditions. Biodistribution in mice showed the proposed $[^{99m}Tc(CO)_3(MIBI)_3]^+$ could accumulate in heart rapidly post injection(p.i.) with faster liver washout than ^{99m}Tc-Sestamibi[17,18].

---

* Copartner: Chen Xiangji, Guo Yunhang, Zhang Qiuyan, Hao Guiyang, Jia Hongmei. Reprinted from *Journal of Organometallic Chemistry*, 2008, 693(10):1822-1828.

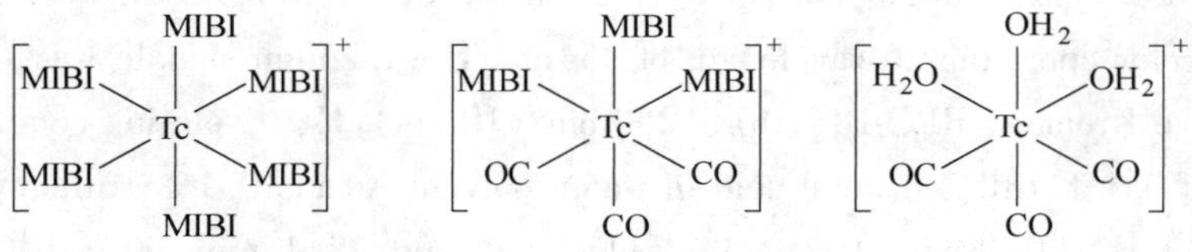

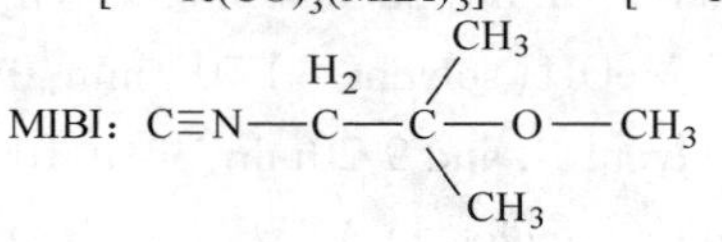

Fig. 1 The structure of ^{99m}Tc-Sestamibi, $[^{99m}Tc(CO)_3(MIBI)_3]^+$, and $[^{99m}Tc(CO)_3(H_2O)_3]^+$

In this work, we studied the inter-transformation of ^{99m}Tc(I)-CO-MIBI complexes, and found that one CO molecule in $[^{99m}Tc(CO)_3(MIBI)_3]^+$ could be further substituted by one MIBI molecule forming $[^{99m}Tc(CO)_2(MIBI)_4]^+$. Previous works indicated that $[^{99m}Tc(CO)_3(MIBI)_3]^+$ had a significant location in myocardium and lower liver uptake as compared with $[^{99m}Tc(CO)_3(H_2O)(MIBI)_2]^{+[18]}$. To evaluate the newly prepared $[^{99m}Tc(CO)_2(MIBI)_4]^+$ which possesses more MIBI molecules, the biodistribution of $[^{99m}Tc(CO)_2(MIBI)_4]^+$, and $[^{99m}Tc(CO)_3(MIBI)_3]^+$ in normal mice was performed simultaneously.

The development of technetium complexes as potential radiopharmaceuticals is facilitated by the use of rhenium, the group ⅦB congener of technetium. Rhenium generally produces complexes with similar physical and biodistribution properties to those of technetium and is often used as a non-radioactive alternative to technetium for large-scale synthesis and structural characterization[19,20]. In this work, the corresponding Re-CO-MIBI complexes were also prepared and characterized.

2 Experimental

2.1 Materials and methods

All chemicals were purchased from Aldrich and Acros. Pure CO gas was purchased from National Research Center for CRM' S. $K_2[BH_3CO_2]$ and $[Re(CO)_3(H_2O)_3]Br$ were prepared according to the literature[21,22]. $Na^{99m}TcO_4$ was obtained from a commercial $^{99}Mo/^{99m}Tc$ generator, Beijing Atomic High-tech Co. MIBI kit vials (containing 1.0mg of $Cu(MIBI)_4BF_4$, 0.15mg of $SnCl_2 \cdot 2H_2O$, 1.0mg of L-cysteine hydrochloride monohydrate, 2.6mg of sodium citrate and 20.0mg of mannitol) and $Cu(MIBI)_4BF_4$ were obtained as gifts from Beijing Shihong Pharmaceutical Center. $Cu(MIBI)_4BF_4$ was synthesized according to the literature[23]. HPLC analyses were performed on a Shimadzu SCL-10AVP system which consisted of a binary pump with on-line degasser, a model SPD-10 Avp UV detector operating at a wavelength of 254nm and a Packard 500 TR series flow scintillation analyzer. The samples were analyzed on a C-18 Alltech alltima column (5μm, 250mm × 4.6mm). HPLC separation was performed on Alltech semipreparative HPLC system which consisted of a binary pump with on-line degasser, a linear UVIS 201 detector operating at wavelength of 254nm. The rhenium sample was separated on a semi-preparative Kromasil RPC-18(10μm, 250mm × 10mm). HPLC eluting condition A: gradient mixtures of 0.1% trifluoroacetic acid in water(solvent A) and 0.1% trifluoroacetic acid in CH_3CN(solvent B) (0-28min, 10%-90% solvent B, and 28-40min, 90%-10% solvent B, 1mL/min); HPLC eluting condition B: aqueous 0.05M TEAP(triethylammonium phosphate) buffer, pH 2.25(solvent A), MeOH(solvent B) (0-3min, 0% solvent B, 3-6min, 0-25% solvent B, 6-9min, 25%-34% solvent B, and 9-20min, 34%-100% solvent B, 1mL/min). Infrared (IR) spectra were recorded on a Nicolet AVATAR 360 IR spectrometer. Mass spectra were collected on a Micromass QTOF mass spectrometer. Proton nuclear magnetic resonance (NMR) spectrum was performed on an Avance 500MHz NMR apparatus(Bruker).

2.2 Preparation of $[^{99m}Tc(CO)_3(H_2O)_3]^+$

$[^{99m}Tc(CO)_3(H_2O)_3]^+$ intermediate was prepared as previously described[24]. Briefly, 2mL of $^{99m}TcO_4^-$ from a commercial generator(10mCi) was added to a 10mL vial containing potassium boranocarbonate(3mg), sodium potassium tartrate tetrahydrate(6.7mg), and potassium tetraborate pentahydrate(5.5mg). The solution was heated for 15min in boiling water under N_2. After cooling down to room temperature, the sample was analyzed on TLC and HPLC.

2.3 Preparation of $[^{99m}Tc(CO)_3(H_2O)(MIBI)_2]^+$, $[^{99m}Tc(CO)_3(MIBI)_3]^+$, and $[^{99m}Tc(CO)_2(MIBI)_4]^+$

A MIBI kit was added to $[^{99m}Tc(CO)_3(H_2O)_3]^+$ intermediate reaction vial and the pH value was adjusted to definite pH value 1.0 for $[^{99m}Tc(CO)_3(H_2O)(MIBI)_2]^+$, 10.0 for $[^{99m}Tc(CO)_3(MIBI)_3]^+$ and 13.0 for $[^{99m}Tc(CO)_2(MIBI)_4]^+$ using 0.1N hydrochloride solution or sodium hydroxide solution. Then the reaction mixtures were heated in boiling water(15min for $[^{99m}Tc(CO)_3(H_2O)(MIBI)_2]^+$ and $[^{99m}Tc(CO)_3(MIBI)_3]^+$, and 60min for $[^{99m}Tc(CO)_2(MIBI)_4]^+$). After cooling down to room temperature, samples were filtered (2μm) and analyzed on HPLC. The radiochemical purity was examined by HPLC with gradient mixtures of 0.1% trifluoroacetic acid in water (A) and 0.1% trifluoroacetic acid in CH_3CN(B) (0-28min, 10%-90% B, and 28-40min, 90%-90% B, 1mL/min).

2.4 Biodistribution of $[^{99m}Tc(CO)_3(MIBI)_3]^+$ and $[^{99m}Tc(CO)_2(MIBI)_4]^+$

The pH values of freshly prepared $[^{99m}Tc(CO)_3(MIBI)_3]^+$ and $[^{99m}Tc(CO)_2(MIBI)_4]^+$ were adjusted to 7.4, and 0.1mL of the resulting solution(20-25μCi) was injected into the tail vein of male ICR mice (18-22g). The mice were sacrificed by decapitation at specific time points post injection. The organs of interest were excised and weighed, and the radioactivity was measured using a Wallac WIZARD 1470 automatic Gamma counter(PerkinElmer, USA). The biodistribution of radiotracer in each tissue sample was expressed as a percentage of the injec-

2.5 Synthesis of $[Re(CO)_3(MIBI)_3]^+$ and $[Re(CO)_2(MIBI)_4]^+$

$[Re(CO)_3(H_2O)_3]Br(51mg, 130\mu mol)$ was dissolved in water(6mL). One equivalent of AgNO$_3$ was added and the precipitate(AgBr) was filtered. $Cu(MIBI)_4BF_4(57mg, 95\mu mol)$ was added to the rhenium solution and the pH value was adjusted to 1.0 for $[Re(CO)_3(MIBI)_3]^+$ and 10.0-13.0 for $[Re(CO)_2(MIBI)_4]^+$. The mixtures were heated for 35min in boiling water and neutralized. After the samples were filtered(0.2μm), they were analyzed on HPLC and mass spectroscopy. $[Re(CO)_3(MIBI)_3]^+$ was purified by HPLC and the collected fraction was evaporated in vacuo. IR(KBr, cm^{-1}): 2982(m), 2242(m), 2214(s), 2059(vs), 1995(vs), 1198(m). ^{1}H NMR(500MHz, D$_2$O): δ 1.18(s, 18H), 3.19(s, 9H), 3.83(s, 6H). MS(ESI) m/z: 610.

3 Results and Discussion

3.1 Preparation of $[^{99m}Tc\text{-}CO\text{-}MIBI]^+$

The preparations of three cationic complexes are shown in Scheme 1. First, $[^{99m}Tc(CO)_3(H_2O)_3]^+$ was prepared with high labeling yield and radiochemical purity (RCP >95%) measured by TLC and HPLC. Only two water ligands were replaced by two MIBI ligands forming $[^{99m}Tc(CO)_3(H_2O)(MIBI)_2]^+$ at pH 1.0 as shown in Fig. 2a. The third water ligand could not be replaced even with longer heating time. $[^{99m}Tc(CO)_3(H_2O)(MIBI)_2]^+$ was totally converted to $[^{99m}Tc(CO)_3(MIBI)_3]^+$ by adjusting the pH value to 10.0 and heating for another 15min. $[^{99m}Tc(CO)_3(MIBI)_3]^+$ could also be directly prepared by exchanging three water ligands with three MIBI ligands when pH value was 10.0. Heating for 15min in boiling water was enough for all the three water ligands to be exchanged and the RCP was higher than 98% as shown in Fig. 2b. Some tetra-substituted product, $[^{99m}Tc(CO)_2(MIBI)_4]^+$ (ca. 5%-10%), was determined when extending the heating time from 15min to 30min. Once $[^{99m}Tc(CO)_3(MIBI)_3]^+$ was formed, it could not be changed back to $[^{99m}Tc(CO)_3(H_2O)(MIBI)_2]^+$ by adjusting the pH value to 1.0 and heating for 30min at 100℃.

$[^{99m}Tc(CO)_2(MIBI)_4]^+$ was prepared under more basic condition (pH 13.0) with extended heating time(60min) from $[^{99m}Tc(CO)_3(H_2O)_3]^+$ as shown in Fig. 2c. $[^{99m}Tc(CO)_2(MIBI)_4]^+$ could also be prepared from $[^{99m}Tc(CO)_3(MIBI)_3]^+$ with less heating time(30min) at the pH value of 13.0.

The pH value played an important role in the ligand-exchange reactions. The first and second water molecules could be easily exchanged by MIBI. However, the third water molecule could not be fully replaced by MIBI until pH value in reaction was higher than 9.0. One CO molecule could even be replaced by one MIBI when the hydroxide ion concentration was high enough. Dyszlewski et al. have also observed that some tetra-substituted product,

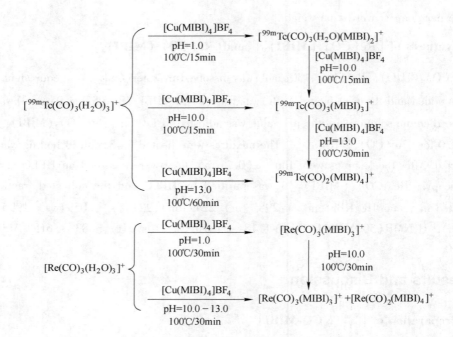

Scheme 1　Preparations and inter-transformations of the ^{99m}Tc(Ⅰ), Re(Ⅰ)

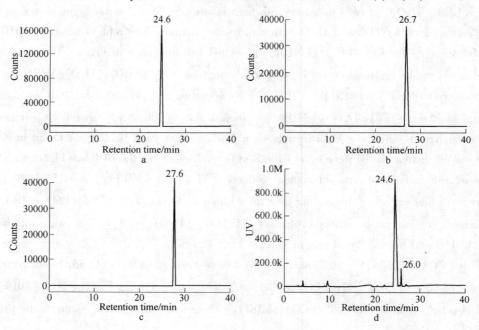

Fig. 2　The radioactivity chromatograms of ^{99m}Tc-CO-MIBI complexes and Re-CO-MIBI complexes under the eluting condition A

a—$[^{99m}Tc(CO)_3(H_2O)(MIBI)_2]^+$ with retention time of 24.6min; b—$[^{99m}Tc(CO)_3(MIBI)_3]^+$ with retention time of 26.7min; c—$[^{99m}Tc(CO)_2(MIBI)_4]^+$ with retention time of 27.6min, HPLC conditions see Section 2; d—$[Re(CO)_3(MIBI)_3]^+$ and $[Re(CO)_2(MIBI)_4]^+$ with retention times of 24.6min and 26.0min, respectively

$[^{99m}Tc(CO)_2(MIBI)_4]^+$ or $[Re(CO)_2(MIBI)_4]^+$ (ca. 5%-10%), formed during the preparation of $[^{99m}Tc(CO)_3(MIBI)_3]^+$ or $[Re(CO)_3(MIBI)_3]^{+[25]}$. Usually, the CO can not be substituted by other ligand molecule owing to its strong π-back-bonding interaction. MIBI ligand, also possessing such strong π-back-bonding interaction with technetium, labilized the bond between CO and rhenium or technetium at the opposite site. So, one of the CO ligand in $[^{99m}Tc(CO)_3(MIBI)_3]^+$ or $[Re(CO)_3(MIBI)_3]^+$ was replaced by one MIBI ligand. Till now, no further replacement of CO molecule was observed under current reaction conditions in this work.

3.2 Biodistribution of $[^{99m}Tc(CO)_3(MIBI)_3]^+$ and $[^{99m}Tc(CO)_2(MIBI)_4]^+$

The results of biodistribution studies of $[^{99m}Tc(CO)_3(MIBI)_3]^+$ and $[^{99m}Tc(CO)_2(MIBI)_4]^+$ in normal mice are shown in Table 1 and Fig. 3. $[^{99m}Tc(CO)_2(MIBI)_4]^+$ demon-strated lower initial heart uptake than $[^{99m}Tc(CO)_3(MIBI)_3]^+$ in the 0-60min after injection and both of them showed slow washout from heart. At 120min after injection, there was almost the same heart uptake of radioactivity for $[^{99m}Tc(CO)_2(MIBI)_4]^+$ and $[^{99m}Tc(CO)_3(MIBI)_3]^+$. For $[^{99m}Tc(CO)_2(MIBI)_4]^+$, the slower clearance from blood and liver caused inferior heart to blood and heart to liver ratios. However, $[^{99m}Tc(CO)_2(MIBI)_4]^+$ washed out quickly from the lungs and demonstrated less accumulation in muscle and skeleton, which is in favor of heartimaging. $[^{99m}Tc(CO)_2(MIBI)_4]^+$ with one more MIBI molecule was more hydrophobic than $[^{99m}Tc(CO)_3(MIBI)_3]^+$ as could be seen from the longer retention time on the C-18 column. But it did not accumulate in the heart quickly, and its extra MIBI molecule slowed its pharmacokinetics.

Table 1 Biodistribution data of $[^{99m}Tc(CO)_3(MIBI)_3]^+$ and $[^{99m}Tc(CO)_2(MIBI)_4]^+$ in IRC mice at 5min, 15min, 30min, 60min, 120min p.i. ($n=3$)

Time/min	Heart	Blood	Lung	Liver	Kidney	Muscle	Spleen	Bone	Brain
$[^{99m}Tc(CO)_3(MIBI)_3]^+$									
5	15.13±6.76	1.16±0.09	4.43±1.90	11.01±1.42	51.3±12.93	4.80±0.70	4.73±0.86	3.95±1.08	0.16±0.05
15	14.52±1.30	0.45±0.14	2.68±0.83	9.86±2.15	38.23±2.01	4.80±0.38	3.69±1.10	3.08±0.48	0.16±0.02
30	13.09±2.31	0.25±0.07	2.14±0.66	8.21±2.79	31.57±5.50	4.27±0.58	3.37±1.19	2.28±0.37	0.16±0.01
60	12.14±2.08	0.15±0.02	1.65±0.52	7.95±1.71	16.55±2.14	4.88±0.81	2.48±0.80	2.07±0.96	0.16±0.01
120	12.40±2.50	0.11±0.01	1.46±0.11	5.47±0.92	16.46±3.02	4.65±0.43	1.24±0.21	1.86±0.61	0.09±0.02
$[^{99m}Tc(CO)_2(MIBI)_4]^+$									
5	9.08±2.70	1.95±0.94	3.51±0.56	13.91±1.06	43.65±9.46	2.64±0.69	3.97±0.98	2.65±0.96	0.14±0.03
15	8.11±3.05	1.29±1.13	2.23±0.70	13.70±3.04	36.19±11.51	2.64±0.95	3.48±1.08	2.00±0.59	0.12±0.03
30	8.27±0.70	0.35±0.04	1.40±0.26	12.06±4.03	30.03±11.08	3.04±0.60	2.75±0.52	1.99±0.64	0.14±0.02
60	8.61±1.54	0.23±0.02	1.16±0.13	6.61±1.38	23.64±8.78	3.20±0.63	1.37±0.06	1.52±0.26	0.10±0.01
120	12.02±1.66	0.24±0.08	1.11±0.45	4.33±0.72	14.52±2.98	3.85±0.28	0.68±0.20	0.88±0.31	0.06±0.01

Note: Organ uptake is expressed as the percentage of injected dose per gram (%ID/g) of wet tissue mass.

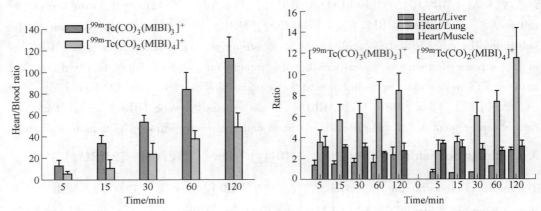

Fig. 3 Histogram representation of ratios of heart to other tissues at different time points after injection

3.3 Synthesis of $[Re(CO)_3(MIBI)_3]^+$ and $[Re(CO)_2(MIBI)_4]^+$

Only $[Re(CO)_3(MOBI)_3]^+$ was obtained at pH 1.0. After purification on HPLC, its structure was confirmed by IR, ^{1}H NMR, and ESI-MS(Fig. 5). $[Re(CO)_3(MIBI)_3]^+$ showed a retention time of 21.4min, and $[^{99m}Tc(CO)_3(MIBI)_3]^+$ showed a rerntion time of 22.3min under the same conditions as shown in Fig. 4. However, a mixture of $[Re(CO)_3(MIBI)_3]^+$ with some $[Re(CO)_2(MIBI)_4]^+$ was obtained even at the pH 14.0 as shown in Fig. 2d and Fig. 6. The mixture was analyzed by mass spectroscopy without purification as shown in Fig. 7 and species with molecular weights of 695 and 610, which were corresponding to $[Re(CO)_2(MIBI)_4]^+$ and $[Re(CO)_3(MIBI)_3]^+$, respectively, were derected. $[Re(CO)_3(MIBI)_3]^+$ was the primary product, and $[Re(CO)_2(MIBI)_4]^+$ was only in percentage of 5%-10%. We tried different HPLC eluting conditions in this work and neither $[^{99m}Tc(CO)_3(MIBI)_3^+]$ nor $[^{99m}Tc(CO)_2(MIBI)_4]^+$ demonstrated matchable retention time with its corresponding rhenium complex. Technetium complexes were always washed out with longer retention times than

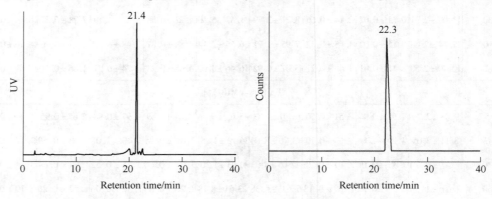

Fig. 4 The chromatograms of $[Re(CO)_3(MIBI)_3]^+$ (UV) and $[^{99m}Tc(CO)_3(MIBI)_3]^+$
(radioactivity) under the eluting condition B

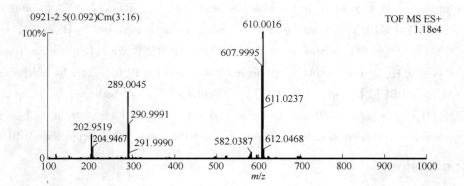

Fig. 5　ESI-MS of $[Re(CO)_3(MIBI)_3]^+$

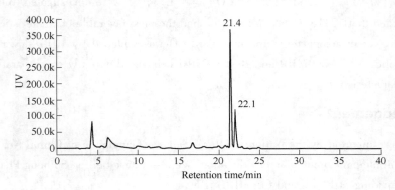

Fig. 6　HPLC chromatograms for mixture of $[Re(CO)_3(MIBI)_3]^+$ and $[Re(CO)_2(MIBI)_4]^+$ under the eluting condition B

(The retention time for $[Re(CO)_3(MIBI)_3]^+$ is 21.4 min and $[Re(CO)_2(MIBI)_4]^+$ is 221 min, respectively)

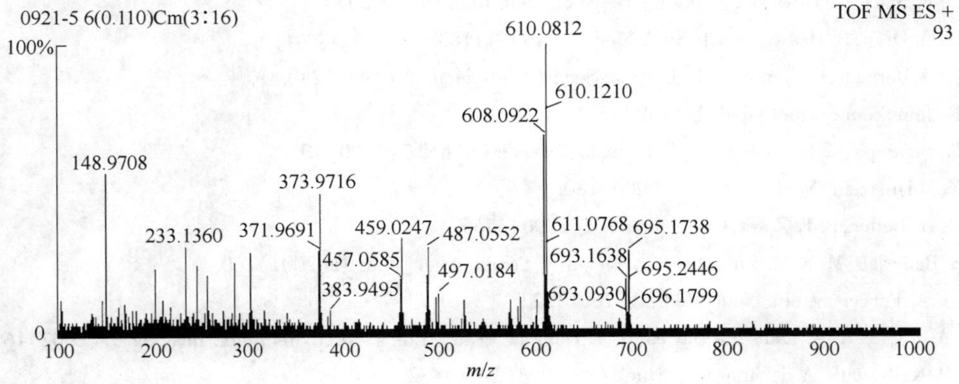

Fig. 7　ESI-MS of mixture of $[Re(CO)_3(MIBI)_3]^+$ and $[Re(CO)_2(MIBI)_4]^+$

the corresponding rhenium complexes, which indicated that technetium complexes were more hydrophobic that the corresponding rhenium complexes. These differences may be due to the different atomic radii of technetium and rhenium. Similar unmatched retention times have also been observed previously[26].

Also, unlike ^{99m}Tc-CO-MIBI, in the reaction between $[Re(CO)_3(H_2O)_3]^+$ and MIBI, only $[Re(CO)_3(MIBI)_3]^+$ was obtained under acidic condition (pH 1.0), and no $[Re(CO)_3(H_2O)(MIBI)_2]^+$ was detected. $[Re(CO)_3(MIBI)_3]^+$ could be partly converted into $[Re(CO)_2(MIBI)_4]^+$ when the pH value was readjusted to about 10.0, heating for another 30min. $[Re(CO)_3(MIBI)_3]^+$ and $[Re(CO)_2(MIBI)_4]^+$ were both observed by heating $[Re(CO)_3(H_2O)_3]^+$ with MIBI for 35min at the pH value of 10.0-14.0 in boiling water. $Re(CO)_3$ and $Tc(CO)_3$ demonstrated different characteristics when they reacted with MIBI.

4 Conclusion

The preparations and inter-transformations between $[^{99m}Tc(CO)_3(H_2O)(MIBI)_2]^+$, $[^{99m}Tc(CO)_3(MIBI)_3]^+$, and $[^{99m}Tc(CO)_2(MIBI)_4]^+$ were investigated. Biodistribution studies indicated that $[^{99m}Tc(CO)_3(MIBI)_3]^+$ had the most favorable characteristics as a myocardial imaging agent among them and one more CO was replaced by MIBI slowing down the pharmacokinetics. ^{99m}Tc-CO-MIBI and Re-CO-MIBI behaved differently in preparation and hydrophobic characteristics.

Acknowledgements

This work was supported by the National Natural Science Foundation of China (No. 20471011) and Beijing Municipal Commission of Education. We also thank Beijing Shihong Pharmaceutical Center for providing MIBI kit and $Cu(MIBI)_4BF_4$.

References

[1] A. D. Nunn, Semin. Nucl. Med. 20(1990)111.
[2] G. B. Saha, R. T. Go, W. J. MacIntyre, Nucl. Med. Biol. 19(1992)1.
[3] L. H. Opie, B. Hesse, Eur. J. Nucl. Med. 24(1997)1183.
[4] D. S. Berman, G. Germano, L. J. Shaw, Semin. Nucl. Med. 29(1999)280.
[5] D. Jain, Semin. Nucl. Med. 29(1999)221.
[6] W. Acampa, C. Di Benedetto, A. Cuocolo, J. Nucl. Cardiol. 7(2000)701.
[7] V. J. Dilsizian, Nucl. Cardiol. 7(2000)180.
[8] G. A. Beller, B. L. Zaret, Circulation 101(2000)1465.
[9] S. Banerjee, M. R. A. Pillai, N. Ramamoorthy, Semin. Nucl. Med. 31(2001)260.
[10] J. A. Parker, Semin. Nucl. Med. 31(2001)223.
[11] A. Kapur, K. A. Latus, G. Davies, R. T. Dhawan, et al., Eur. J. Nucl. Med. Mol. Imaging 29(2002)1608.
[12] P. Kailasnath, A. J. Sinusas, J. Nucl. Cardiol. 8(2001)482.
[13] J. G. Llaurado, J. Nucl. Med. 42(2001)282.
[14] R. Alberto, R. Schibli, A. Egli, A. P. Schubiger, et al., J. Am. Chem. Soc. 120(1998)7987.
[15] Y. S. Kim, Z. J. Hea, R. Schibli, S. Liu, Inorg. Chim. Acta 359(2006)2479.
[16] Z. He, W. -Y. Hsieh, Y. -S. Kim, S. Liu, Nucl. Med. Biol. 33(2006)1045.
[17] G. Y. Hao, L. Zhu, B. L. Liu, J. Label. Compd. Radiopharm. 46(2003)S321.
[18] G. Y. Hao, J. Y. Zang, L. Zhu, Y. Z. Guo, B. L. Liu, J. Label. Compd. Radiopharm. 47(2004)513.

[19] E. Deutsch, K. Libson, J.-L. Vanderheyden, A. R. Ketring, H. R. Maxon, Nucl. Med. Biol. 13 (1986) 465.
[20] J. Dilworth, S. Parrot, Chem. Soc. Rev. 27 (1998) 43.
[21] R. Alberto, K. Ortner, N. Wheatley, R. Schibli, A. P. Schubiger, J. Am. Chem. Soc. 123 (2001) 3135.
[22] N. Lazarova, S. James, J. Babich, J. Zubieta, Inorg. Chem. Commun. 7 (2004) 1023.
[23] G. Ferro-Flores, L. Ramirez-Garcia, J. Lezama, P. Garcia-Garcia, J. I. Tendilla, J. Radioanal. Nucl. Chem. 188 (1994) 409.
[24] X. Chen, Y. Guo, B. Liu, J. Pharm. Biomed. Anal. 43 (2007) 1576.
[25] M. Dyszlewski, H. M. Blake, J. L. Dahlheimer, C. M. Pica, D. Piwnica-Worms, Mol. Imaging 1 (2002) 24.
[26] Z. Zhuang, M. Kung, C. Hou, K. Ploessl, H. Kung, Nucl. Med. Biol. 32 (2005) 171.

Synthesis and Biological Evaluation of ^{99m}Tc, Re-monoamine-monoamide Conjugated to 2-(4-aminophenyl) Benzothiazole as Potential Probes for β-amyloid Plaques in the Brain*

Abstract The benzothiazole aniline (BTA) conjugated with monoamine-monoamide (MAMA) was synthesized and then labeled with ^{99m}Tc. Its corresponding rhenium analogue was synthesized, and the fluorescent staining was performed in brain sections of both Tg mouse and Alzheimer's disease (AD) patient. The fluorescent rhenium complex Re-MAMA-BTA selectively bound to the amyloid aggregates in the brain sections of both APP Tg mouse and AD patient. The analogous ^{99m}Tc-MAMA-BTA complex could enter the normal mouse brain with high initial uptake. These results are encouraging for further exploration of their derivatives as imaging agents for Aβ plaques in the brain.

Key words β-Amyloid, technetium-99m, 2-(4-Aminophenyl) benzothiazole

Alzheimer's disease (AD) is a progressive neurodegenerative disorder and a leading cause of dementia. AD is characterized by abundant senile plaques composed of β-amyloid (Aβ) peptide and numerous neurofibrillary tangles formed from filaments of highly phosphorylated tau proteins[1,2]. Formation of Aβ plaques in the brain is a pivotal event in the pathology of AD. Detection of deposited Aβ with non-invasive techniques such as positron emission tomography (PET) and single photon emission computed tomography (SPECT) could enable the diagnosis of AD in its pre-symptomatic stages[3-6]. Many specific ligands have been developed for imaging of Aβ aggregates. Out of the many agents that have been reported to target Aβ aggregates, some have so far moved forward into further evaluation studies using cohorts of AD patients: ^{18}F-FDDNP[7,8], ^{11}C-PIB[9,10], ^{11}C-SB-13[11,12], and ^{11}C-BF-207[13]. The results from the initial studies are encouraging; and all of them are in clinical evaluation as PET imaging agents for AD.

However, the development of imaging agents for SPECT was lagging far behind, especially for the ^{99m}Tc-labeled radioactive probes, which were hindered by the low brain uptake. The most widely used radionuclide for diagnostic imaging with SPECT is the metastable isotope of technetium, ^{99m}Tc, because of its favorable physical properties ($t_{1/2} = 6h$, $E_\gamma = 140 keV$), low cost, and widespread availability[14]. The development of technetium complexes as potential radiopharmaceuticals is facilitated by the use of rhenium, the group ⅦB congener of technetium. Rhenium generally produces complexes with similar physical and biodistribution properties to those of technetium and is often used as a non-radioactive alternative to technetium for large-scale syn-

---

* Copartner: Chen Xiangji, Yu Pingrong, Zhang Lianfeng. Reprinted from *Bioorganic & Medicinal Chemistry Letters*, 2008, 18 (4):1442-1445.

thesis and structural characterization[15,16].

The ^{99m}Tc-labeled Congo red cannot penetrate the blood-brain barrier (BBB), probably because of its charge and high molecular weight[17,18]. The more lipophilic chrysamine-G has been labeled with ^{99m}Tc-monoamine-monoamide bisthiol(MAMA), but studies also demonstrated too low brain uptake[19,20].

More lipophilic ligands with smaller molecular weight such as benzothiazole aniline (BTA)[21-24], stilbene, and biphenyl have been labeled with ^{99m}Tc, among which ^{99m}Tc-labeled stilbene[25] and biphenyl[26] showed high brain uptake.

In our previous works, we investigated the possible binding mechanism and site of the ligands using molecular modeling techniques[27]. In this work, the chelator MAMA was conjugated with 2-(4-aminophenyl)benzothiazole through a 5-carbon alkyl chain as linker based on the structure of thioflavin-T and thioflavin-S, aiming at the development of analogous technetium and rhenium complexes with affinity for Aβ aggregates. ^{99m}Tc-MAMA complexes 6a(^{99m}Tc) and 6b(Re) were prepared, respectively, and biological characterization was also performed to evaluate their potential as tracer for visualization of β-amyloid plaques.

The compound 2-(4-nitrophenyl)-6-hydroxybenzothiazole (1) was conjugated with 1,5-dibromopentane to afford the compound 2 (yield 79%). Then the compound 2 was joined to the monoamine-monoamide bisthiol protected with p-methoxy benzyl(MAMA-MBz) to generate the compound 3 (yield 62%). After reduction of the nitro group using stannous chloride, the compound 4 (yield 71%) was obtained. Compound 5 (yield 93%) was obtained by deprotection of the thiol groups in compound 4. The complexes 6a and 6b (yield 53%) were prepared by the reaction of compound 5 with corresponding technetium and rhenium precursors, as outlined in Scheme 1[28].

Scheme 1　^{a}MBz = p-methoxy benzyl

Biological evaluation of complex 6b was accomplished by fluorescent staining of brain sections of a Tg C57(APP, 12-month-old) mouse and AD patient(male, 57 y) according to the reported method[28,29]. The complex 6b could bind the Aβ aggregates in Tg C57(APP) mouse as shown in Fig. 1a. Complex 6b also demonstrated binding affinity to different sizes of Aβ plaques in postmortem brain section of AD patient as shown in Fig. 1b-d.

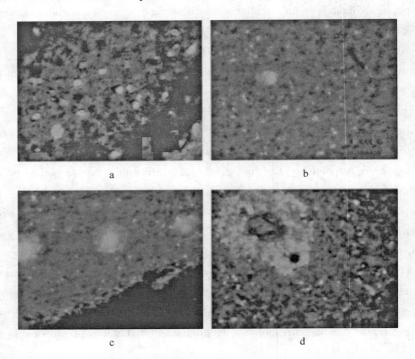

Fig. 1 Fluorescence microscopy photographs showing the complex 6b selectively stained Aβ aggregates in the Tg C57(APP) mouse and AD patient brain sections

a—Tg C57(APP) mouse brain section with 3.5mM of 6b in 40% EtOH solution for 10min;
b-d—Postmortem AD patient brain section with 3.5mM of 6b in 40% EtOH solution for 10min

BTA is a kind of small molecule which has been proved to bind to amyloid with high affinity. According to our previous molecular modeling, the analogues of stilbene and thioflavin-T should be embedded in the amyloid when they bind to it[27]. In order to keep its binding affinity, the chelating moiety should be apart from benzothiazole moiety. The linker we chose is a 5-carbon alkyl chain, which is flexible and with the length of about 0.8nm. So the interference on the interaction between BTA ligand and the amyloid by the chelating moiety is diminished. In addition, Re-MAMA-BTA bears similar structure as thioflavin-S, so it can be considered as a design of 'integrated method' based on thioflavin-S.

The corresponding ^{99m}Tc complex of 6b, complex 6a, was also prepared by ligand exchange reaction employing the precursor ^{99m}Tc-glucoheptonate(GH). The labeling yield detected by radio-TLC was higher than 95%. The resulting mixture was analyzed by reversed phase HPLC, showing that a single radioactive complex was formed with radio-chemical purity higher than

95% and the radioactive complex was stable for more than 12h in PBS at room temperature. The identity of the radioactive complex was established by comparative HPLC studies using the corresponding rhenium complex 6b as reference. The retention times for 6b(UV) and 6a(radioactivity) on HPLC were 30.35 and 32.0min, respectively. The complex 6a was rather hydrophobic with log P value 2.90 ± 0.13 measured using partition method between n-octanol and PBS[28].

The biodistribution of complex 6a was carried out in IRC normal mice(18-22g), which were sacrificed at the specific time after intravenous administration of the purified complex 1b(5-10μCi)[28]. The biodistribution data are shown in Table 1. The complex 6a showed high initial brain uptake((1.34 ± 0.16)% ID/g organ at 2min) and medium wash-out. The radioactivity was mainly excreted via the hepatobiliary system. The wash-out is to be expected since there is no amyloid in the normal mice brain. The high blood background, which may be due to the high hydrophobicity, is unfavorable for imaging application.

Table 1 Biodistribution of radioactivity measured after the administration of complex 6a in normal mice①

Organ	2min	30min	60min	120min
Blood	6.14 ± 0.47	3.88 ± 0.19	4.43 ± 0.56	4.71 ± 1.24
Heart	17.49 ± 0.70	5.31 ± 0.30	5.09 ± 0.75	3.75 ± 0.77
Lung	10.38 ± 0.87	5.03 ± 0.40	4.38 ± 0.39	3.34 ± 0.62
Liver	25.94 ± 2.24	34.51 ± 4.28	34.62 ± 5.98	32.45 ± 4.27
Spleen	4.60 ± 1.23	4.50 ± 0.86	3.51 ± 0.66	2.89 ± 0.49
Kidney	15.88 ± 1.58	8.92 ± 0.44	6.98 ± 1.33	5.10 ± 0.73
Brain	1.34 ± 0.16	0.99 ± 0.08	0.65 ± 0.10	0.44 ± 0.04

① All data are the mean percentage($n=3$) of the injected dose per gram of wet tissue(% ID/g) ± the standard deviation of the mean.

The lipophilicity plays an important role in BBB penetration. Hansch and Leo found that blood-brain barrier penetration is optimal when the log P values are in the range of 1.5-2.7, with the mean value of 2.1[30]. The ligands that can bind to Aβ aggregates are very hydrophobic and the hydrophobic interaction contributes to the binding. However, compounds with high log P value may bind to the proteins in the blood leading to high blood background. This contradiction seems not easy to be compromised. The less lipophilic MAMA ligand was chosen to adjust the lipophilicity of final compound. Former studies conjugating BTA with bisaminoethanethiol ligand showed lower brain uptake of the Tc-99m labeled compound[21,24]. And 6-Me-BTA lost its binding affinity after chelation with $^{99m}Tc(CO)_3$[22]. The choice of suitable conjugation method is critical for the binding affinity and brain uptake. Further experiments will be optimizing the pharmacokinetics while keeping its binding affinity. Introducing hydroxyl group to increase its solubility may improve the pharmacokinetics.

In conclusion, the isostructural ^{99m}Tc and Re complexes 6a and 6b were successfully synthesized and their preliminary evaluation results demonstrated the binding of 6b to Aβ aggregates in the brain slices of transgenic mouse and AD patient. In addition, complex 6a can penetrate BBB with high initial brain uptake and medium wash-out. These results are encouraging for further exploration of their derivatives as imaging agents for Aβ plaques in the brain.

Acknowledgement

The work was financially supported by NSFC(20471011).

Supplementary data

Synthetic procedures and HPLC analysis for 6a and 6b, fluorescent staining of brain sections with 6b, partition coefficient determination and biodistribution of 6a in normal mice are available in supporting information. Supplementary data associated with this article can be found, in the online version, at doi: 10.1016/j.bmcl.2007.12.071.

References

[1] Selkoe, D. J. Physiol. Rev. 2001, 81, 741.

[2] Hardy, J.; Selkoe, D. J. Science 2002, 297, 353.

[3] Selkoe, D. J. Nat. Biotechnol. 2000, 18, 823.

[4] Mathis, C. A.; Wang, Y.; Klunk, W. E. Curr. Pharm. Des. 2004, 10, 1469.

[5] Nordberg, A. Lancet Neurol. 2004, 3, 519.

[6] Lockhart, A. Drug Discov. Today 2006, 11, 1093.

[7] Shoghi-Jadid, K.; Small, G. W.; Agdeppa, E. D.; Kepe, V.; Ercoli, L. M.; Siddarth, P.; Read, S.; Satyamurthy, N.; Petric, A.; Huang, S. -C.; Barrio, J. R. Am. J. Geriatr. Psychiatry 2002, 10, 24.

[8] Small, G. W.; Kepe, V.; Ercoli, L. M.; Siddarth, P.; Bookheimer, S. Y.; Miller, K. J.; Lavretsky, H.; Burggren, A. C.; Cole, G. M.; Vinters, H. V.; Thompson, P. M.; Huang, S. C.; Satyamurthy, N.; Phelps, M. E.; Barrio, J. R. N. Engl. J. Med. 2006, 355, 2652.

[9] Klunk, W. E.; Engler, H.; Nordberg, A.; Wang, Y. M.; Blomqvist, G.; Holt, D. P.; Bergstrom, M.; Savitcheva, I.; Huang, G. F.; Estrada, S.; Ausen, B.; Debnath, M. L.; Barletta, J.; Price, J. C.; Sandell, J.; Lopresti, B. J.; Wall, A.; Koivisto, P.; Antoni, G.; Mathis, C. A.; Langstrom, B. Ann. Neurol. 2004, 55, 306.

[10] Engler, H.; Forsberg, A.; Almkvist, O.; Blomquist, G.; Larsson, E.; Savitcheva, I.; Wall, A.; Ringheim, A.; Langstrom, B.; Nordberg, A. Brain 2006, 129, 2856.

[11] Verhoeff, N. P. L. G.; Wilson, A. A.; Takeshita, S.; Trop, L.; Hussey, D.; Singh, K.; Kung, H. F.; Kung, M. -P.; Houle, S. Am. J. Geriatr. Psychiatry 2004, 12, 584.

[12] Ono, M.; Wilson, A.; Nobrega, J.; Westaway, D.; Verhoeff, P.; Zhuang, Z. -P.; Kung, M. -P.; Kung, H. F. Nucl. Med. Biol. 2003, 30, 565.

[13] Kudo, Y.; Okamura, N.; Furumoto, S.; Tashiro, M.; Furukawa, K.; Maruyama, M.; Itoh, M.; Iwata, R.; Yanai, K.; Arai, H. J. Nucl. med. 2007, 48, 553.

[14] Jurisson, S. S.; Lydon, J. D. Chem. Rev. 1999, 99, 2205.

[15] Deutsch, E.; Libson, K.; Vanderheyden, J. -L.; Ketring, A. R.; Maxon, H. R. Nucl. Med. Biol. 1986, 13, 465.

[16] Dilworth, J.; Parrot, S. Chem. Soc. Rev. 1998, 27, 43.

[17] Han, H. ; Cho, C. G. ; Lansbury, P. T. J. Am. Chem. Soc. 1996, 118, 4506.

[18] Zhen, W. ; Han, H. ; Anguiano, M. ; Lemere, C. A. ; Cho, C. G. ; Lansbury, P. T. J. Med. Chem. 1999, 42, 2805.

[19] Dezutter, N. A. ; De Groot, T. J. ; Busson, R. H. ; Janssen, G. A. ; Verbruggen, A. M. J. Labelled Compd. Radiopharm. 1999, 42, 309.

[20] Dezutter, N. A. ; De Groot, T. J. ; Bormans, G. M. ; Verbruggen, A. M. ; Dom, R. J. Eur. J. Nucl. Med. Mol. Imaging 1999, 26, 1392.

[21] Vanderghinste, D. ; Eeckhoudt, M. V. ; Cleynhens, J. Technetium, Rhenium and Other Metals in Chemistry and Nuclear Medicine; Nicolini, M. ; Mazzi, U. , Eds. Italy, 2002; pp463-465.

[22] Tzanopoulou, S. ; Patsis, G. ; Sagnou, M. ; Pirmettis, I. ; Papadopoulos, M. ; pelecanou, M. Technetium, Rhenium and Other Metals in Chemistry and Nuclear Medicine; Mazzi, U. , Ed. Italy, 2006; pp103-104.

[23] Serdons, K. ; Landeloos, C. ; Terwinghe, C. ; Bormans, G. : Verbruggan, A. Technetium, Rhenium and Other Metals in Chemistry and Nuclear Medicine; Mazzi, U. , Ed. Italy, 2006: pp377-382.

[24] Serdons, K. ; Verduyckt, T. ; Cleynhens, J. ; Terwinghe, C. ; Mortelmans, L. ; Bormans, G. ; Verbruggen, A. Bioorg. Med. Chem. lett. 2007, 17, 6086.

[25] Zhuang, Z. P. ; Kung, M. P. ; Hou, C. ; Kung, H. F. J. Label. Compd. Radiopharm. 2005, 48, S251.

[26] Zhuang, Z. P. ; Kung, M. P. ; Hou, C. ; Ploessl, K. ; Kung, H. F. Nucl. Med. Biol. 2005, 32, 171.

[27] Chen, X. J. Theochem-J. Mol. Struct. 2006, 763, 83.

[28] For further information, see Supporting Information.

[29] Styren, S. D. ; Hamilton, R. L. ; Styren, G. C. ; Klunk, W. E. J. Histochem. Cytochem. 2000, 48, 1223.

[30] Hansch, C. ; Leo, A. J. In Substituent Constant for Correlation Analysis in Chemistry and Biology; Wiley: New York, 1979.

Copolymer-based Hepatocyte Asialoglycoprotein Receptor Targeting Agent for SPECT*

Abstract Poly(vinylbenzyl-O-β-D-galactopyranosyl-D-gluconamide) (PVLA) can be specifically internalized by hepatocytes via the asialoglycoprotein receptor. In this study, we synthesized and characterized galactose-carrying copolymers with hydrazinonicotinamide chains as bifunctional groups to radiolabel PVLA with ^{99m}Tc for SPECT targeting specific hepatocytes. Methods: Poly(N-p-vinylbenzyl-[O-β-D-galactopyranosyl-(1 → 4)-D-gluconamide]-co-N-p-vinylbenzyl-6-[2-(4-dimethylamino)benzaldehydehydrazono]nicotinate) (P(VLA-co-VNI)) was first prepared via copolymerization of N-p-vinylbenzyl-O-β-D-galactopyranosyl-D-gluconamide with 5% (mol) of N-p-vinylbenzyl-(4-dimethylaminobenzaldehyde hydrazono) nicotinamide. The copolymer was labeled with ^{99m}Tc using tricine as a coligand. Then ^{99m}Tc[P(VLA-co-VNI)] (tricine)$_2$ was evaluated by in vivo metabolic stability and biodistribution in normal mice. SPECT was performed in normal New Zealand White rabbits and rabbits with liver cancer. Results: ^{99m}Tc[P(VLA-co-VNI)] (tricine)$_2$ was prepared in high labeling yield(>95%) and radiochemical purity (>99%), with good stability. The results of biodistribution in mice demonstrated that the liver uptake was 125.33 ± 10.99 percentage injected dose per gram at 10min after injection and could be blocked significantly by preinjecting free neogalactosylalbumin or P(VLA-co-VNI). SPECT images with high quality were obtained at 15min, 30min, 60min, and 120min after injection of the radiotracer. Significant radioactivity defect was observed in the liver cancer model. Conclusion: The bifunctional coupling agent hydrazinonicotinamide was introduced to PVLA via copolymerization and labeled with ^{99m}Tc. The promising biologic properties of ^{99m}Tc[P(VLA-co-VNI)] (tricine)$_2$ afford potential applications for the assessment of hepatocyte function in the future.

Key words asialoglycoprotein receptor, P(VLA-co-VNI), ^{99m}Tc, animal imaging, radiopharmaceuticals

Asialoglycoprotein receptors(ASGP-Rs) are well known to exist on the mammalian liver[1], situated on the surface of hepatocyte membrane, and participate in the hepatic metabolism of serum proteins. Quantitative imaging of ASGP-Rs could estimate the function of the liver, a unique way to noninvasively diagnose disease. The ASGP-R imaging agent can assess the anatomy and function of the liver and help the early diagnosis of hepatic diseases and accurate evaluation of functional status[1,2]. Two other types of radiopharmaceutical are available for liver imaging[3,4] : one type, for hepatobiliary scintigraphy, is labeled lipophilic compounds(^{99m}Tc-dimethyliminodiacetic acid, ^{99m}Tc-diisopropyliminodiacetic acid, and others), which reflect the excretory function of

---

* Copartner: Yang Wenjiang, Mou Tiantian, Shao Guoqiang, Wang Feng, Zhang Xianzhong. Reprinted from *The Journal of Nuclear Medicine*, 2011, 52(6): 978-985.

hepatocytes and the abnormalities of the hepatobiliary system. Recently, de Graaf et al[5]. summarized the use of these ^{99m}Tc-labeled iminodiacetic acid derivates for the assessment of hepatic function in liver surgery and transplantation and compared them with ASGP-R-based imaging agents. Uptake of labeled lipophilic compounds can be influenced by hepatic blood flow, hypoalbuminemia, and high concentrations of bilirubin[5]. The other type of radiopharmaceutical available for liver imaging is labeled colloids (^{99m}Tc-sulfur colloid and ^{99m}Tc-albumin colloid) for colloid scintigraphy, which reflect mainly phagocytic function of the Kupffer cells and evaluate hepatic function indirectly. However, the ASGP-R-based imaging agents can be transferred to the hepatocyte via receptor-mediated endocytosis, which is a different principle from the other two types of liver imaging agents. These tracers are affected only by hepatocyte function and cannot be used to diagnose the biliary system but permit a noninvasive way to evaluate hepatic function and hepatic functional reserve directly and quantitatively. ASGP-Rs can recognize galactose or N-acetylgalactosamine residues of desialylated glycoproteins; materials having these ligands have been studied for targeting hepatocytes directly. In the 1980s, neogalactosylalbumin had been developed by attaching the galactosyl unit to human serum albumin(HSA) and then labeling it with ^{99m}Tc for imaging of the liver[6]. Later, to simplify the labeling procedure, diethylenetriaminepentaacetic acid(DTPA)galactosyl human serum albumin(GSA)was obtained and developed as an instant kit[7]. ^{99m}Tc-GSA has been used clinically as an ASGP-R-binding radiopharmaceutical in Japan since 1992. The derivates of glycoprotein radiolabeled with ^{99m}Tc[8-11], ^{125}I/^{131}I[12], and ^{111}In[13] have been reported for SPECT applications. ^{68}Ga-deferoxamine-neogalactosylalbumin[14] and ^{18}F-fluorobenzoate galactosyl-neoglycoalbumin[15] were introduced as PET agents. These agents are similar in that each uses a protein as the backbone for galactose or lactose. Recently, several researchers have reported the use of other polymer backbones, such as dextran and chitosan, instead of proteins to design new probes. Galactosylated dextran or chitosan were able to bind to hepatocytes because of the galactose residue positioned on the exterior of the polymer[16]. Radiotracers using dextran or chitosan as the backbone, such as ^{99m}Tc-labeled Cy5.5-DTPA-galactosyl-dextran[17], ^{99m}Tc-galactosyl-methylated chitosan[18], ^{99m}Tc-hydrazinonicotinamide-galactosylated chitosan[4], and ^{18}F-fluorobenzoate-galactosylated chitosan[19], have been developed.

The galactose-carrying polystyrene derivative poly[N-p-vinylbenzyl-(O-β-D-galactopyranosyl-(1→4)-D-gluconamide)](PVLA) has an amphiphilic structural unit composed of hydrophilic oligosaccharide side chains covalently bound to a hydrophobic polystyrene backbone[20]. PVLA can be specifically internalized by hepatocytes via the interaction between the ASGP-R and galactose residue of PVLA. PVLA radiolabeled with ^{125}I was reported by Mitsuaki et al[21]. The ^{125}I-PVLA was distributed rapidly and mainly to the liver within 15 min of administration, especially to parenchymal liver cells. In this study, we synthesized and characterized copolymer poly(N-p-vinylbenzyl-[O-β-D-galactopyranosyl-(1→4)-D-gluconamide]-co-N-p-vinylbenzyl-6-[2-(4-dimethylamino)benz aldehydehydrazono]nicotinate)P(VLA-co-VNI) and incorporated hydrazinonicotinamide-functionalized groups into PVLA for radiolabeling with ^{99m}Tc.

1 Materials and Methods

1.1 Materials

All chemicals obtained commercially were used without further purification. Tricine and 4-vinylbenzyl chloride were purchased from J&K Chemical Ltd. Instant thin-layer chromatography (ITLC) silica gel strips were purchased from Pall Life Sciences. The LabGEN 7 homogenizer was purchased from Cole-Parmer. Radio-high-pressure liquid chromatography (radio-HPLC) experiments were performed on a system with an SCL-10Avp HPLC pump (Shimadzu Corp.) and a liquid scintillation analyzer (Packard BioScience Co.). The reversed-phase Kromasil C-4 column (4.6mm × 250mm, 5μm particle size, 30nm (300Å); Eka Chemicals) was eluted at a flow rate of 1mL/min according to the procedure described in the Radiochemical Analysis section. The absorbance was monitored at 220nm. The HiTrap Desalting column (filled with Sephadex G25) was purchased from GE Healthcare, eluted with phosphate buffer (0.05mol/L; pH 7.5). Animal experiments were performed in Kunming mice (female; average weight, ~20g) and New Zealand White rabbits (female; average body weight, ~2.5kg), obtained from the Animal Center of Peking University. All biodistribution studies were performed under a protocol approved by the Beijing Administration Office of Laboratory Animal.

1.2 Synthesis of galactose-carrying monomer: VLA

The p-vinylbenzylamine was prepared from 4-vinylbenzyl chloride according to the Gabriel synthesis[22]. VLA was synthesized according to the reported method of Kobayashi et al[23]. Briefly, lactobionolactone was dissolved in refluxing methanol, and p-vinylbenzylamine solution in methanol was added. The mixed solution was refluxed for 120min. The white crystal was filtered and purified by recrystallization from methanol.

1.3 Synthesis of hydrazinonicotinamide-carrying monomer: VNI

Succinimidyl 6-[2-(4-dimethylamino)benzaldehydehydrazono]nicotinate was prepared from 6-hydrazinonicotinic acid according to the procedure of Harris et al[24]. p-vinylbenzylamine (2.55mmol) was added to a solution of succinimidyl 6-[2-(4-dimethylamino)benzaldehydrazono]nicotinate (2.65mmol) in 20mL of dry dimethyl formamide. The resulting mixture was stirred at room temperature for 24h. After the reaction, the solvent was removed under reduced pressure, and the residue was suspended in methanol (3mL). The solid was collected by filtration, washed with diethyl ether, and dried under a vacuum. The yield was 0.84g (82%). ^{1}H nuclear magnetic resonance (NMR) (400MHz, dimethylsulfoxide-d_6) δ: 3.16 (s, 6H, N(CH_3)$_2$), 4.44 (s, 2H, N—CH_2), 5.22 (d, 1H, H—C=C), 5.79 (d, 1H, H—C=C), 6.73 (m, 3H, H—C=C and N—phenyl), 7.18 (d, 1H, pyridinyl), 7.29 (d, 2H, phenyl), 7.43 (d, 2H, phenyl), 7.50 (d, 2H, phenyl), 7.97 (s, 1H, pyridinyl), 8.07 (d, 1H, pyridinyl), 8.64 (s, 1H, N=CH), 8.85 (s, 1H, NH—CO), 10.98 (s, 1H, NH—N=C); infrared spectroscopy (IR) (KBr, cm^{-1}) ν: 3454

(O—H), 1608(C═O).

1.4 Synthesis of galactose-carrying copolymer with hydrazinonicotinamide group

P(VLA-co-VNI) was obtained by a method similar to that used for PVLA[23]. Briefly, the mixture of VLA and VNI(5% mol) was dissolved in dimethylsulfoxide containing azobisisobutyronitrile as an initiator and stirred for 14h at 60℃ to allow copolymerization. The product was precipitated in methanol. The copolymer was dialyzed against water for 24h after being dissolved in water (molecular weight cutoff, 8-10 kDa), and the resulting solution was lyophilized.

1.5 Preparation of ^{99m}Tc-labeled P(VLA-co-VNI)

To a 10mL vial, 0.2mL of the P(VLA-co-VNI) solution (0.2mg in phosphate buffer (0.05mol/L), pH 6.0, about 4.25×10^{-9} mol), 0.5 mL of tricine solution (60mg/mL in phosphate buffer (0.05mol/L), pH 6.0), 10μL of $SnCl_2 \cdot 2H_2O$ solution (2mg/mL in HCl (0.1mol/L)), and 0.5 mL of freshly eluted $^{99m}TcO_4^-$ from a commercial generator (~37-370 MBq) were added in turn. The vial was sealed and heated for 20min at 100℃ to obtain the resulting complex $^{99m}Tc[P(VLA-co-VNI)](tricine)_2$.

After reaction, the crude product was passed through a 0.22μm Millipore filter and loaded onto a HiTrap desalting column and eluted with phosphate buffer (0.05mol/L), pH 7.5. After purification, the radiochemical purity (RCP) was evaluated by ITLC chromatography and radio-HPLC.

1.6 Radiochemical analysis

The chromatography analyses were performed on ITLC silica gel strips with acid-citrate-dextrose buffer (citrate (0.068mol/L), glucose (0.074mol/L), pH 5.0) as a mobile phase. ^{99m}Tc-labeled P(VLA-co-VNI) remained at the point of spotting ($R_f = 0$-0.1), whereas hydrolyzed ^{99m}Tc and other radioactive impurities moved with the solvent front ($R_f = 0.8$-1.0). Radio-HPLC was performed on a Shimadzu system with a reversed-phase Kromasil C-4 column (4.6mm × 250mm, 5μm). The column was eluted with 0.1% trifluoroacetic acid in water (solvent A) and 0.1% trifluoroacetic acid in acetonitrile (solvent B) at a flow rate of 1mL/min (gradient: 0-5min: 95% A, 5.01-30min: 70%-30% A).

After purification, $^{99m}Tc[P(VLA-co-VNI)](tricine)_2$ was incubated in saline at room temperature for 4h. The RCP was evaluated by ITLC and radio-HPLC. Plasma stability was determined by $^{99m}Tc[P(VLA-co-VNI)](tricine)_2$ diluted 20-fold with freshly prepared murine plasma, and the solutions were incubated at 37℃ for 4h. The RCP was evaluated by ITLC chromatography every hour. At the end of 4h, the plasma was passed through a Sep-Pak C_{18} cartridge (Waters) and washed with 0.5mL of water and eluted with 0.5mL of acetonitrile containing 0.1% trifluoroacetic acid. The combined aqueous and organic solutions were passed through a 0.22μm Millipore filter and evaluated by radio-HPLC.

1.7 Metabolic stability

Mice were intravenously injected with 7.4 MBq of $^{99m}Tc[P(VLA-co-VNI)](tricine)_2$ (about

4×10^{-9} mol of polymer/kg of body weight). The animals were sacrificed at 10 and 120min after tracer injection. Blood, liver, and urine were collected. Blood samples were immediately centrifuged for 5 min at 13200r/min. After removal of the supernatants, the pellets were washed with 250μL of phosphate-buffered saline(pH 7.5). Supernatants of both centrifugation steps were combined and passed through a Sep-Pak C_{18} cartridge. After washing with saline, the liver samples were homogenized in 4mL of phosphate-buffered saline(pH 7.5) using a polytron homogenizer(LabGEN 7) at full speed for 5 min. The resulting homogenate was centrifuged for 60min at 14000r/min. Supernatants were passed through a Sep-Pak C_{18} cartridge. The urine samples were directly diluted with 0.5mL of phosphate-buffered saline and then passed through a Sep-Pak C_{18} cartridge. All the cartridges were washed with 0.5mL of water and eluted with 0.5mL of acetonitrile containing 0.1% trifluoroacetic acid. The combined aqueous and organic solutions were passed through a 0.22μm Millipore filter and then injected onto radio-HPLC using a flow rate of 1mL/min and a gradient as described.

1.8 Biodistribution

The biodistribution study of $^{99m}Tc[P(VLA\text{-}co\text{-}VNI)](tricine)_2$ was performed in normal mice. About 0.185 MBq(in 100μL of solution) of $^{99m}Tc[P(VLA\text{-}co\text{-}VNI)](tricine)_2$ ($\sim 1 \times 10^{-9}$ mol of polymer/kg of body weight) was injected through the tail vein. At selected times(10min, 30min, and 120min), mice($n = 5$ at each time point) were sacrificed, and the tissues and organs of interest were collected, wet-weighed, and counted in a γ-counter. The percentage injected dose per gram(%ID/g) for each sample was calculated by comparing its activity with an appropriate standard of ID.

To further confirm that the $^{99m}Tc[P(VLA\text{-}co\text{-}VNI)](tricine)_2$ had specific receptor binding, a blocking study was performed by conducting the biodistribution experiment in the presence of free neogalactosylalbumin(200μg, $\sim 1.4 \times 10^{-7}$ mol/kg of body weight) and free cold P(VLA-co-VNI) (200μg, $\sim 2.1 \times 10^{-7}$ mol/kg of body weight) as a blocking agent. Five minutes after the first injection of blocking agent, about 0.185MBq (in 100μL of solution) of $^{99m}Tc[P(VLA\text{-}co\text{-}VNI)]$ $(tricine)_2$ was intravenously injected($\sim 1 \times 10^{-9}$ mol of polymer/kg of body weight). Mice were sacrificed at 10min, 30min, and 120min after injection($n = 5$). Results were expressed as the %ID/g. Averages and SD were calculated.

1.9 SPECT

The SPECT/CT(Millennium VG; GE Healthcare) scanner was equipped with a pinhole collimator to acquire planar images. The acquiring parameters were as follows: energy peak of 140keV, window width of 20%, matrix of 256×256, zoom of 1.0, and 10min/frame. SPECT/CT fusion images were acquired using parallel-hole collimators. New Zealand White rabbits were placed near the center of the field of view of the SPECT scanner, at which the highest image resolution and sensitivity are available. The rabbits were deeply anesthetized by intraperitoneal injection of sodium pentobarbital; anesthesia was supplemented as needed. About 60 MBq of $^{99m}Tc[P(VLA\text{-}co\text{-}VNI)](tricine)_2$ ($\sim 1.7 \times 10^{-9}$ mol of polymer/kg of body weight) in 1mL of phosphate buffer(0.05mol/L; pH 7.5) was injected in the marginal ear vein. Computer acquisition of the γ-camera data was initiated during administration of the radiopharmaceutical. Time-activity curves for the blood, kidney, and liver were generated by selecting regions of interest around the heart, kidney, and liver. Static images at 15min, 30min, 60min, and 120min after the tracer injection were acquired. The scan time was 2min. For comparison, a blocking study was performed using preinjected free neogalactosylalbumin(25mg, ~ 1.4×10^{-7} mol/kg of body weight) as a blocking agent 5min before intravenous injection of the radiotracer.

The liver cancer model was prepared by induced VX2 carcinoma in the liver of rabbits. Tumor size was checked by CT. The rabbits with 2cm to 3cm tumors were included in the experiment. SPECT was performed in the same manner as for the normal group. The SPECT/CT study was performed 30min after injection. After imaging studies, pathologic examination was performed to confirm the liver tumor.

1.10 Statistical analysis

Values are presented as mean ± SD. An unpaired 2-tailed t test was used to determine statistical significance using GraphPad InStat software(GraphPad Software). A P value of less than 0.05 was considered statistically significant.

2 Results

2.1 Chemistry

The monomers of both VLA and VNI were prepared with good yield and characterized by ^{1}H NMR and IR. VLA with VNI were copolymerized with azobisisobutyronitrile as an initiator at 60°C for 14h, as shown in Fig. 1. The pale yellow powdery polymers were isolated by freeze-drying from the aqueous solution. The P(VLA-co-VNI) was characterized by ^{1}H NMR and elemental analyses. The copolymer composition was estimated from the carbon-to-nitrogen (C/N) mass ratio, which was

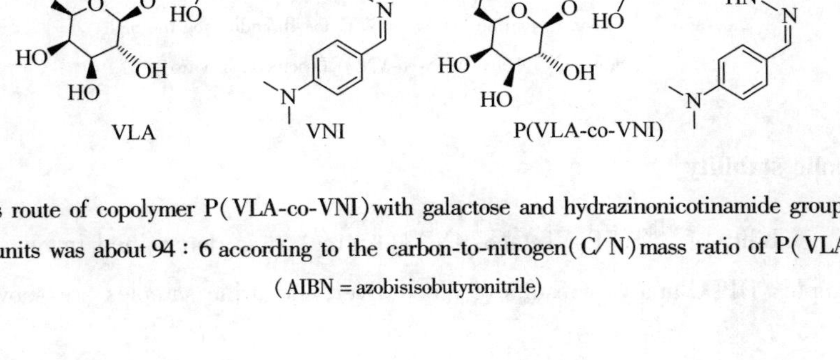

Fig. 1 Synthesis route of copolymer P(VLA-co-VNI) with galactose and hydrazinonicotinamide group(ratio of VLA units to VNI units was about 94 : 6 according to the carbon-to-nitrogen(C/N) mass ratio of P(VLA-co-VNI))
(AIBN = azobisisobutyronitrile)

14.25. The ratio of VLA units to VNI units in P(VLA-co-VNI) was about 94∶6 ($n∶m$). The theoretic molecular weight of P(VLA-co-VNI) was about 47kDa, calculated from the ratio of azobisisobutyronitrile initiator to monomer, and has narrow size distribution (Supplemental Fig. 1; supplemental materials are available online only at http://jnm.snmjournals.org). The number of galactose and hydrazinonicotinamide units per polymer was about 94 and 6, respectively.

2.2 Radiochemistry

The labeling conditions of ^{99m}Tc[P(VLA-co-VNI)](tricine)$_2$ were investigated, and optimized conditions were obtained under pH 6.0 for 20min at 100℃. The labeling yield was greater than 95%. After rapid purification with a HiTrap desalting column, the RCP of ^{99m}Tc[P(VLA-co-VNI)](tricine)$_2$ was more than 99%, as determined by ITLC and radio-HPLC. The HPLC analysis results of in vitro stability are shown in Fig. 2. The retention time of ^{99m}Tc[P(VLA-co-VNI)](tricine)$_2$ was 11.9min in our gradient. In vitro stability, the result measured by ITLC and HPLC, showed that ^{99m}Tc[P(VLA-co-VNI)](tricine)$_2$ could be stable over 4h in saline at room temperature. Stability in murine plasma was also high; ^{99m}Tc[P(VLA-co-VNI)](tricine)$_2$ was still intact after 4h at 37℃. Furthermore, the dilutional stability was also investigated using a 1,000-fold dilution with freshly prepared murine plasma to more closely reproduce the situation in humans. After a 4h incubation period at 37℃, the RCP was still more than 92%, as determined by ITLC.

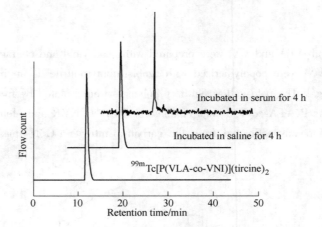

Fig. 2 High-performance liquid chromatograms of ^{99m}Tc[P(VLA-co-VNI)](tricine)$_2$
(Retention time was 11.9min both after storage in saline at room temperature for 4h and
after incubation in murine plasma at 37℃ for 4h, indicating the high
stability of ^{99m}Tc[P(VLA-co-VNI)](tricine)$_2$ in vitro)

2.3 Metabolic stability

The metabolic stability of ^{99m}Tc[P(VLA-co-VNI)](tricine)$_2$ was determined in mouse blood, liver, and urine samples. HPLC analysis results of blood, liver, and urine samples are shown in Fig. 3. At

10min after the injection of tracer, the small amount of ^{99m}Tc[P(VLA-co-VNI)](tricine)$_2$ that remained in the blood was still stable by HPLC analysis. The homogenate supernatant of liver samples at 10min and 120min after injection showed a major radioactivity peak at 11.9 min on the high-performance liquid chromatogram, meaning most of the ^{99m}Tc[P(VLA-co-VNI)](tricine)$_2$ was intact and not metabolized in the liver within 2h. The HPLC retention time of the urinary metabolite was 18.8min, meaning no intact ^{99m}Tc[P(VLA-co-VNI)](tricine)$_2$ was excreted in the urine.

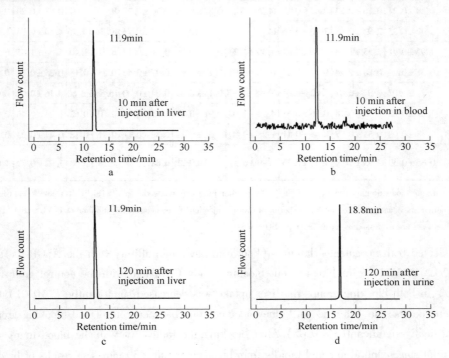

Fig. 3 HPLC profiles of metabolic stability of ^{99m}Tc[P(VLA-co-VNI)](tricine)$_2$. Liver was collected at 10min and 120min after injection of ^{99m}Tc[P(VLA-co-VNI)](tricine)$_2$, blood was collected at 10min after injection, and urine was collected at 120min after injection

a, c—^{99m}Tc[P(VLA-co-VNI)](tricine)$_2$ was intact in liver at 10min and 120min after injection, with retention time of 11.9min;

b, d—Retention times of blood and urine samples were 11.9min and 18.8min, respectively

2.4 Biodistribution

To evaluate tissue distribution characteristics of ^{99m}Tc[P(VLA-co-VNI)](tricine)$_2$, the biodistribution studies were performed using normal Kunming mice. The results are shown in Table 1. As described, ^{99m}Tc[P(VLA-co-VNI)](tricine)$_2$ had high liver accumulation, with good retention. The initial liver uptake was (125.32 ± 10.99)% ID/g at 10min after injection. At 30 and 120min after injection, the liver uptake was (122.64 ± 9.06)% ID/g and (114.74 ± 9.93)% ID/g, respectively. The blood and kidneys showed relatively low activity. Even at 30min, the ratio of liver to blood uptake and liver to kidney uptake could reach 1742 and 42—a ratio obviously higher than that of ^{99m}Tc-GSA(31.7 and 5.7, respectively, at 30min, in our experiments).

Table 1 Biodistribution of ^{99m}Tc[P(VLA-co-VNI)](tricine)$_2$ in normal mice

Tissue	Control/min			Block with P(VLA-co-VNI)/min			Block with neogalactosylalbumin/min		
	10	30	120	10	30	120	10	30	120
Heart	0.71 ±0.13	0.71 ±0.18	0.62 ±0.17	6.69 ±2.11	0.62 ±0.31	0.65 ±0.19	3.19 ±0.59	1.11 ±0.41	1.10 ±0.25
Liver	125.33 ± 10.99	122.64 ± 9.06	114.74 ± 9.93	34.12 ±3.24	67.81 ±5.28	63.33 ±5.21	58.05 ±5.27	75.30 ±4.99	56.45 ±8.53
Lung	0.88 ±0.24	0.76 ±0.24	0.42 ±0.18	14.79 ±5.22	0.87 ±0.19	0.70 ±0.27	9.19 ±1.04	0.83 ±0.20	0.96 ±0.36
Kidney	2.42 ±0.35	2.92 ±0.82	2.40 ±0.66	14.70 ±3.48	5.33 ±0.57	4.50 ±0.73	10.28 ±1.15	4.50 ±0.74	5.08 ±0.62
Spleen	1.92 ±0.91	2.53 ±1.13	1.35 ±0.71	4.49 ±1.23	0.61 ±0.08	0.54 ±0.12	6.71 ±1.40	6.56 ±1.55	6.89 ±2.47
Stomach	0.51 ±0.24	0.88 ±0.24	0.52 ±0.23	0.30 ±0.08	0.22 ±0.06	0.77 ±0.46	0.23 ±0.13	0.25 ±0.14	0.32 ±0.12
Blood	0.23 ±0.07	0.07 ±0.01	0.02 ±0.01	30.48 ±7.18	0.15 ±0.03	0.04 ±0.00	16.55 ±2.60	0.16 ±0.07	0.03 ±0.00
Bone	1.02 ±0.28	0.98 ±0.18	0.86 ±0.31	4.74 ±1.00	0.91 ±0.13	1.02 ±0.41	3.50 ±0.62	1.73 ±0.46	2.37 ±0.74
Muscle	0.38 ±0.08	0.51 ±0.27	0.43 ±0.11	1.28 ±0.50	0.54 ±0.17	0.39 ±0.11	0.81 ±0.18	0.47 ±0.23	0.55 ±0.30
Small intestine	0.85 ±0.30	2.93 ±2.49	1.46 ±0.35	1.79 ±0.37	1.93 ±0.21	0.88 ±0.28	1.38 ±0.08	1.92 ±0.72	1.03 ±0.27

Note: Data are biodistribution 10min, 30min, and 120min after intravenous injection of ^{99m}Tc[P(VLA-co-VNI)](tricine)$_2$ into normal mice, and results of blocking experiment with preinjection of free neogalactosylalbumin or P(VLA-co-VNI) as inhibitor. Data are expressed as mean %ID/g ± SD(n =5).

We selected both neogalactosylalbumin(known to have high affinity with the ASGP-Rs) and free cold P(VLA-co-VNI) as inhibitors in our blocking study. Compared with the control group (without blocking), in both blocking groups the liver uptake was decreased significantly (P <0.01) at all selected time points. The difference in kidney accumulation between control and blocking groups was also statistically significant (P <0.05). After blocking, the radioactivity in the blood increased only in the initial stages and eliminated rapidly from bloodstream after 30 min. The results of biodistribution showed that the ^{99m}Tc[P(VLA-co-VNI)](tricine)$_2$ has high affinity with the ASGP-R, and its uptake in the liver is receptor-mediated.

2.5 SPECT

We also evaluated ^{99m}Tc[P(VLA-co-VNI)](tricine)$_2$ in vivo with SPECT. For the dynamic SPECT study, the regions of interest were drawn on the heart, liver, and kidneys, and the time-activity curves were calculated for these regions. An analysis of the time-activity curves (Fig. 4) shows that the liver uptake in the rabbit peaked within 2.5-5min after injection, whereas the radioactivity in the blood was decreased rapidly. The blocking study results showed that the uptake rate in the liver was significantly reduced and the clearance of radioactivity in the blood was significantly slower than that in the control group. The results also showed that a portion of radioactivity accumulated in and was metabolized via the kidneys after inhibition. Representative planar images are displayed in Fig. 5. High liver activity accumulation was observed at 15min after injection. There was no significant uptake in the kidneys, spleen, or other organs in the abdomen. A small portion of radioactivity was eliminated from the liver at later time points, and the outline of the liver was still clear. The ab-

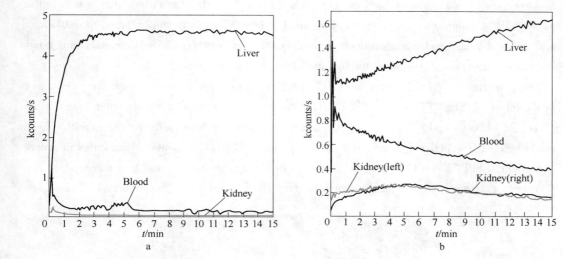

Fig. 4 Time-activity curves obtained from 15-min dynamic planar acquisition

a—In normal rabbit, liver uptake peaked rapidly within 2.5-5min, whereas radioactivity in blood decreased rapidly;
b—In blocking study, uptake rate in liver was significantly reduced by preinjected neogalactosylalbumin. Uptake in blood and kidneys was relatively high and decreased continuously over time

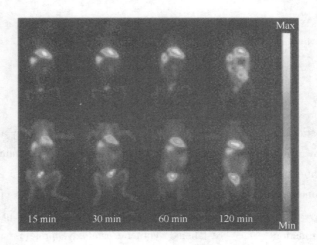

Fig. 5 Planar images of rabbit

(The upper row is control group; lower row is blocking group with free neogalactosylalbumin as blocking agent(10mg/kg of rabbit body weight). Times are times after injection)

dominal radioactivity was increased slightly, indicating that the radioactivity was eliminated from the liver by hepatobiliary excretion as the major route. In contrast, after inhibition liver accumulation was significantly lower, and the outline of the liver was vague at 15min (Fig. 5). After blocking, the cardiac level of activity was high, because of the nonspecific activity accumulated in the blood, and reduced gradually at later time points. The counts in the region of the liver accounted for 81% of the total radioactivity counts in the whole body at 30min in the control group, whereas after inhibition the counts in the region of the liver were reduced to 41% of the total radioactivity counts in the whole body at 30min after injection. We also found a high radioactivity concentration in the kidney and urinary bladder, and a gradual decrease in the uptake in the kidneys over time, indicating rapid clearance of the unbounded tracer through the kidneys after inhibition.

The potential of ^{99m}Tc[P(VLA-co-VNI)](tricine)$_2$ was also evaluated with the liver cancer model using SPECT/CT. The SPECT/CT study was performed at 30min after injection of ^{99m}Tc[P(VLA-co-VNI)](tricine)$_2$. As shown in Fig. 6, the tumor site in the rabbit's liver could be clearly visualized, despite the use of a clinical SPECT/CT scanner. Minor errors in superimposition of SPECT and CT data may be due to involuntary respiratory and bowel movement. The tumor was confirmed by pathologic examination after the imaging study(Supplemental Fig. 2).

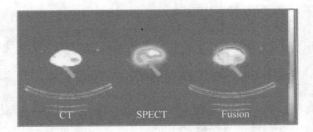

Fig. 6 SPECT/CT for detection and localization of liver cancer
(Liver is visualized as transaxial sections at 30min after injection merged with CT and SPECT images. Arrows indicate site of tumor)

3 Discussion

Measurement of hepatic functional reserve could be useful before hepatectomy and for the evaluation of posttransplantation residual hepatic functional reserve in donors[25]. The number of ASGP-Rs on the hepatocytes of patients with liver disease is reduced and thus considered a good indicator for the evaluation of liver function[26]. Noninvasive quantification of the liver uptake of radiolabeled asialoglycoproteins can provide a unique way to estimate liver function, especially in terms of spatial distribution of hepatic function[27-29]. ^{99m}Tc-labeled synthetic asialoglycoproteins such as neogalactosylalbumin and GSA were developed as ASGP-R imaging agents. ^{99m}Tc-GSA is used clinically and is at this time commercially available only in Japan. Those ASGP-R imaging agents using albumin as the backbone have encountered problems. There is a risk of viral breakthrough for the blood products of HSA, and therefore these products are closely regulated during good manufacturing practice and importation. A newly designed probe with nonalbumin backbone could avoid this problem. In addition, because of the natural properties of albumin those tracers using albumin as the backbone usually have high stability in the bloodstream and a long circulation time, which is unfavorable for cold area imaging in liver. Especially in patients with severe liver diseases, the liver images could be rather vague. Moreover, other problems exist, such as sensitivity to changes of temperature and pH and the introduction of toxic byproducts during synthesis. Recently, radiotracers using chitosan or dextran as the backbone were developed[4,17-19]. Those tracers showed specific and rapid targeting of hepatocytes. However, the tracers using chitosan as the backbone show high renal uptake, limiting their clinical applications for hepatic imaging. To overcome these problems, we developed a new AS-GP-R agent using PVLA as the backbone by incorporating hydrazinonicotinamide-functionalized groups for radiolabeling with ^{99m}Tc. Hydrazinonicotinamide is an attractive bifunctional coupling agent for preparing ^{99m}Tc-labeled antibodies, peptides, and small biomolecules[30]. We synthesized P(VLA-co-VNI) through polymerization of 2 monomers, VLA and VNI, with good yield and easy purification. This approach is feasible for introducing bifunctional coupling agents for the polymer radiopharmaceutical. At the same time, through copolymerization of a variety of different monomers, we could introduce affinity ligands, bifunctional groups, or organic dye, affording a potential method to prepare pharmaceuticals for targeted molecular imaging or even multifunction imaging.

^{99m}Tc[P(VLA-co-VNI)](tricine)$_2$ showed promising properties in preliminary biologic experiments and characteristics different from ^{99m}Tc-GSA. There are at least 2 processes after ^{99m}Tc-GSA enters the hepatocyte. First, the endocytosed receptor and ligand dissociate in lysosomes. The receptor recycles to the cell surface. Then, the backbone of ^{99m}Tc-GSA is degraded to yield DTPA-coupled lysine(^{99m}Tc-DTPA-lysine) in lysosome and is slowly released from the cell[31,32]. Similar to results with GSA, in our blocking experiments, the results of blockage were comparable using neogalactosylalbumin or P(VLA-co-VNI) as an inhibitor. After blocking, the radioactivity in the liver was increased gradually at the later time after injection, likely reflecting dissociation of the receptor and ligand(both neogalactosylalbumin and P(VLA-co-VNI)) and leading to the recycling of ASGP-Rs to the surface membrane. Furthermore, we investigated the metabolic stability of ^{99m}Tc[P(VLA-co-VNI)](tricine)$_2$ in the liver. No or little radiotracer was degraded in the liver for at least 2h. On the basis of the results of our blocking experiment and metabolic studies, we hypothesized that ^{99m}Tc[P(VLA-co-VNI)](tricine)$_2$ would be rapidly taken up by hepatocytes via ASGP-R-mediated endocytosis. Then the receptor-ligand complex was dissociated in lysosomes. The ASGP-Rs circulated to the surface of the cell membrane, whereas the intact ligand remained in the hepatocytes metabolized slowly. Because the protein backbone could be degraded in lysosomes, the biologic properties display a marked difference when adopting different bifunctional groups in a protein-based ASGP-R targeting agent[15]. In contrast, the bifunctional groups in this copolymer-based ASGP-R targeting agent may have only a minor impact on biodistribution because of the high stability of the copolymer backbone in the liver.

Furthermore, the in vivo SPECT evaluation of ^{99m}Tc[P(VLA-co-VNI)](tricine)$_2$ was observed in rabbits with high liver accumulation and good retention. The high radioactivity concentration in the urinary bladder after blocking indicated rapid clearance of the unbounded tracer through the kidney. The fact that ^{99m}Tc[P(VLA-co-VNI)](tricine)$_2$ had a short circulation time in the bloodstream could reduce the level of radioactivity in the hepatic blood pool and other vascularized organs. Thus, SPECT images with ambiguity caused by high radioactivity in blood, especially for the patients with severe liver disease, could be avoided when using ^{99m}Tc[P(VLA-co-VNI)](tricine)$_2$. The clear SPECT images of the liver were obtained both in the blocking group at 30 min after injection and in the rabbits with liver cancer. The novel tracer $^{99m}Tc[P(VLA-co-VNI)](tricine)_2$ shows good biologic properties that may provide clear images and detailed information. Moreover, $^{99m}Tc[P(VLA-co-VNI)](tricine)_2$ can be easy to formulate using a kit, making it possible for clinical applications in the future.

4 Conclusion

A receptor-specific ligand for the ASGP-R was developed via copolymerization. The bifunctional coupling agent hydrazinonicotinamide was introduced to a galactose-carrying polystyrene derivative. The copolymer P(VLA-co-VNI) was labeled with ^{99m}Tc, with high labeling yield. Biodistribution and imaging studies demonstrated hepatic targeting and rapid blood clearance for this tracer. $^{99m}Tc[P(VLA-co-VNI)](tricine)_2$ accumulated mainly in the liver. Clear images can be obtained using $^{99m}Tc[P(VLA-co-VNI)](tricine)_2$ because of the high liver-to-background ratio of the copolymer. The promising biologic properties of $^{99m}Tc[P(VLA-co-VNI)](tricine)_2$ afford potential applications for assessment of hepatocyte function in the future. In addition, the synthesis process could enable efficient modification of copolymers with other ligands and bifunctional coupling agents for nuclear medicine imaging of different targets.

Disclosure Statement

The costs of publication of this article were defrayed in part by the payment of page charges. Therefore, and solely to indicate this fact, this article is hereby marked "advertisement" in accordance with 18 USC section 1734.

Acknowledgements

We thank Dr. Yong He for his kind donation of dextran standards and Dr. Wei Fang, Wenyan Guo, Huihui Jing, and Zuoquan Zhao for their generous help. The project was sponsored by the Fundamental Research Funds for the Central Universities and supported partially by the National Natural Science Foundation of China (20871020), Beijing Natural Science Foundation (2092018), and the Scientific Research Foundation for the Returned Overseas Chinese Scholars, State Education Ministry.

References

[1] Kokudo N, Vera D R, Makuuchi M. Clinical application of TcGSA. Nucl Med Biol. 2003;30:845-849.
[2] Stadalnik R C, Vera D R. The evolution of ^{99m}Tc-NGA as a clinically useful receptor-binding radiopharmaceutical. Nucl Med Biol. 2001;28:499-503.
[3] Saha G B. Fundamentals of Nuclear Pharmacy. 4th ed. New York, NY: Springer New York; 1998.
[4] Kim E M, Jeong H J, Kim S L, et al. Asialoglycoprotein-receptor-targeted hepatocyte imaging using Tc-99m galactosylated chitosan. Nucl Med Biol. 2006;33:529-534.
[5] de Graaf W, Bennink R J, Vetelainen R, van Gulik T M. Nuclear imaging techniques for the assessment of hepatic function in liver surgery and transplantation. J Nucl Med. 2010;51:742-752.
[6] Vera D R, Stadalnik R C, Krohn K A. Technetium-99m galactosyl-neoglycoalbumin: preparation and pre-

clinical studies. J Nucl Med. 1985;26:1157-1167.

[7] Kubota Y, Kojima M, Hazama H, et al. A new liver function test using the asialoglycoprotein-receptor system on the liver cell membrane: I. Evaluation of liver imaging using Tc-99m-neoglycoprotein [in Japanese]. Kaka Igaku. 1986;23:899-905.

[8] Ono M, Arano Y, Uehara T, et al. Intracellular metabolic fate of radioactivity after injection of technetium-99m-labeled hydrazino nicotinamide derivatized proteins. Bioconjug Chem. 1999;10:386-394.

[9] Yang W, Mou T, Zhang X, Wang X. Synthesis and biological evaluation of ^{99m}Tc-DMP-NGA as a novel hepatic asialoglycoprotein receptor imaging agent. Appl Radiat Isot. 2010;68:105-109.

[10] Chaumet-Riffaud P, Martinez-Duncker I, Marty A L, et al. Synthesis and Application of lactosylated, ^{99m}Tc chelating albumin for measurement of liver function. Bioconjug Chem. 2010;21:589-596.

[11] Jeong J M, Hong M K, Lee J, et al. Tc-99m-neolactosylated human serum albumin for imaging the hepatic asialoglycoprotein receptor. Bioconjug Chem. 2004;15:850-855.

[12] Wakisaka K, Arano Y, Uezono T, et al. A novel radioiodination reagent for protein radiopharmaceuticals with L-lysine as a plasma-stable metabolizable linkage to liberate m-iodohippuric acid after lysosomal proteolysis. J Med Chem. 1997;40:2643-2652.

[13] Arano Y, Mukai T, Akizawa H, et al. Radiolabeled metabolites of proteins play a critical role in radioactivity elimination from the liver. Nucl Med Biol. 1995;22:555-564.

[14] Vera D R. Gallium-labeled deferoxamine-galactosyl-neoglycoalbumin: a radiopharmaceutical for regional measurement of hepatic receptor biochemistry. J Nucl Med. 1992;33:1160-1166.

[15] Yang W, Mou T, Peng C, et al. Fluorine-18 labeled galactosyl-neoglycoalbumin for imaging the hepatic asialoglycoprotein receptor. Bioorg Med Chem. 2009;17:7510-7516.

[16] Park I K, Yang J, Jeong H J, et al. Galactosylated chitosan as a synthetic extracellular matrix for hepatocytes attachment. Biomaterials. 2003;24:2331-2337.

[17] Vera D R, Hall D J, Hoh C K, Gallant P, McIntosh L M, Mattrey R F. Cy5. 5-DTPA-galactosyl-dextran: a fluorescent probe for in vivo measurement of receptor biochemistry. Nucl Med Biol. 2005;32:687-693.

[18] Kim E M, Jeong H J, Park I K, Cho C S, Kim C G, Bom H S. Hepatocyte-targeted nuclear imaging using Tc-99m-galactosylated chitosan: conjugation, targeting, and biodistribution. J Nucl Med. 2005;46:141-145.

[19] Yang W, Mou T, Guo W, et al. Fluorine-18 labeled galactosylated chitosan for asialoglycoprotein-receptor-mediated hepatocyte imaging. Bioorg Med Chem Lett. 2010;20:4840-4844.

[20] Hiratsuka T, Goto M, Kondo Y, Cho C S, Akaike T R. Copolymers for hepatocyte-specific targeting carrying galactose and hydrophobic alkyl groups. Macromol Biosci. 2008;8:231-238.

[21] Mitsuaki G, Hirohumi Y, Chia-Wun C, et al. Lactose-carrying polystyrene as a drug carrier: investigation of body distributions to parenchymal liver cells using ^{125}I-labelled lactose-carrying polystyrene. J Control Release. 1994;28:223-233.

[22] Kobayashi K, Sumitomo H, Ina Y A. Carbohydrate-containing synthetic polymer obtained from N-p-vinyl-benzyl-D-gluconamide. Polym J. 1983;15:667-671.

[23] Kobayashi K, Sumitomo H, Ina Y. Synthesis and functions of polystyrene derivatives having pendant oligosaccharides. Polym J. 1985;17:567-575.

[24] Harris T D, Sworin M, Williams N, et al. Synthesis of stable hydrazones of a hydrazinonicotinyl-modified peptide for the preparation of ^{99m}Tc-labeled radiopharmaceuticals. Bioconjug Chem. 1999;10:808-814.

[25] Clavien P A, Petrowsky H, DeOliveira M L, Graf R. Strategies for safer liver surgery and partial liver transplantation. N Engl J Med. 2007;356:1545-1559.

[26] Stockert R J. The asialoglycoprotein receptor: relationships between structure, function, and expression. Physiol Rev. 1995;75:591-609.

[27] Hwang E H, Taki J, Shuke N, et al. Preoperative assessment of residual hepatic functional reserve using ^{99m}Tc-DTPA-galactosyl-human serum albumin dynamic SPECT. J Nucl Med. 1999;40:1644-1651.

[28] Sugai Y, Komatani A, Hosoya T, Yamaguchi K. Response to percutaneous transhepatic portal embolization: new proposed parameters by ^{99m}Tc-GSA SPECT and their usefulness in prognostic estimation after hepatectomy. J Nucl Med. 2000;41:421-425.

[29] Shuke N, Okizaki A, Kino S, et al. Functional mapping of regional liver asialoglycoprotein receptor amount from single blood sample and SPECT. J Nucl Med. 2003;44:475-482.

[30] Liu S. 6-hydrazinonicotinamide derivatives as bifunctional coupling agents for ^{99m}Tc-labeling of small biomolecules. In: Topics in Current Chemistry, Contrast Agents Ⅲ. Heidelberg, Germany: Springer Berlin; 2005:117-153.

[31] Duncan J R, Welch M J. Intracellular metabolism of indium-111-DTPA-labeled receptor targeted proteins. J Nucl Med. 1993;34:1728-1738.

[32] Arano Y, Mukai T, Uezono T, et al. A biological method to evaluate bifunctional chelating agents to label antibodies with metallic radionuclides. J Nucl Med. 1994;35:890-898.

^{99m}Tc-and Re-labeled 6-dialkylamino-2-naphthylethylidene Derivatives as Imaging Probes for β-amyloid Plaques[*]

Abstract Based on the conjugate strategy, two neutral ^{99m}Tc labeled 2-(1-(6-(dialkylamino)naphthalen-2-yl)ethylidene)malononitrile (DDNP) and 1-(6-(dialkylamino)naphthalen-2-yl)ethanone (ENE)derivatives, and their corresponding rhenium complexes were synthesized. In vitro fluorescent staining indicated that the corresponding rhenium derivatives selectively stained the β-amyloid(Aβ) plaques in the brain sections of AD model mice with low background. Compared with FDDNP and FENE, the affinities of the corresponding rhenium derivatives to Aβ aggregates decreased about 10-14-fold. In vivo biodistribution experiments in normal mice showed that ^{99m}Tc-MAMA-ENE displayed medium initial brain uptake (0.65% ID/g at 2min) with a reasonable washout from the brain (0.19% ID/g at 2h) while ^{99m}Tc-MAMA-DDNP showed a low brain uptake (0.28% ID/g at 2min). Further optimize these ^{99m}Tc-labeled tracers in order to improve their binding affinities to Aβ plaques and diffusion through the blood brain barrier may generate useful imaging agents for SPECT.

Key words alzheimer's disease, β-amyloid, technium-99m, imaging agent

Senile plaques(SPs) composed of misfolded β-amyloid(Aβ) peptides and neurofibrillary tangles (NFTs) made of hyperphosphorylated tau aggregates are two of the pathological hallmarks of AD[1,2]. Development of imaging probes targeting amyloid plaques or neurofibrillary tangles for positron emission tomography(PET) and single photon emission computed tomography(SPECT) will be crucial for early diagnosis of AD[3]. Based on the structure of Thioflavin-T and Congo red, several types of Aβ imaging agents have been synthesized and evaluated in the past decade. Some of them have so far reached clinical stage, such as [^{18}F]FDDNP[4-6], [^{11}C]PIB[7,8], [^{11}C]SB-13[9,10], [^{11}C]BF-227[11], [^{18}F]AV-45[12,13], and [^{123}I]IMPY[14-16]. Although the clinical evaluations of these agents are encouraging, ^{11}C and ^{18}F have their distinct features, such as short half-life and high cost of nuclear production from cyclotron, which decreased the potential use of these agents in clinical. Among the medical radio-nuclides, ^{99m}Tc ($T_{1/2}$ = 6h, 140keV) is the most widely used radioisotope for in vivo diagnostic imaging in SPECT with some favorable factors: it can be easily produced from a ^{99}Mo/^{99m}Tc generator, the medium gamma-ray energy emitted by ^{99m}Tc(140keV) is suitable for gamma camera detection and the half-life is compatible with the biological localization and residence time required for imaging[17]. Thus, the successful developing of ^{99m}Tc labeled imaging agents for Aβ plaques

---

[*] Copartner: Cui Mengchao, Tang Ruikun, Li Zijing, Ren Huiying. Reprinted from *Bioorganic & Medicinal Chemistry Letters*, 2011, 21(3):1064-1068.

would provide a simple and widespread method for clinical diagnosis of AD using SPECT. However, in order to introduce the transition metal ^{99m}Tc to an organic molecule, a chelating structure is obligatory. Two broad strategies (conjugate approach or integrated approach) have been used in the design of ^{99m}Tc-labeled Aβ imaging agents in order not to reduce the affinity and selectivity or decrease the brain uptake[18]. Previous studies of ^{99m}Tc-labeled Congo Red and Chrysamine G derivatives based on the conjugate strategy have been reported[19,20]. But these ^{99m}Tc complexes are both too large and charged molecules and therefore, cannot penetrate the blood-brain barrier (BBB). Attempts to prepare small, neutral and more lipophilic ligands such as derivatives of benzothiazole aniline (BTA)[21-23], stilbene[24], flavone[25], chaclone[26], and biphenyl[27] labeled with ^{99m}Tc have been reported (Fig. 1). A biphenyl derivative based on the integrated strategy, showed high initial brain uptakes, with a brain uptake up to 1.18% dose/organ at 2min, but it failed to bind to the Aβ plaques in the postmortem human brain tissue of patients with confirmed AD[27]. Through the conjugate strategy, our group reported that ^{99m}Tc-labeled BTA derivatives using monoamide-monoamine (MAMA) as the chelator, displayed high brain uptake in normal mice (1.34% ID/g at 2min)[28].

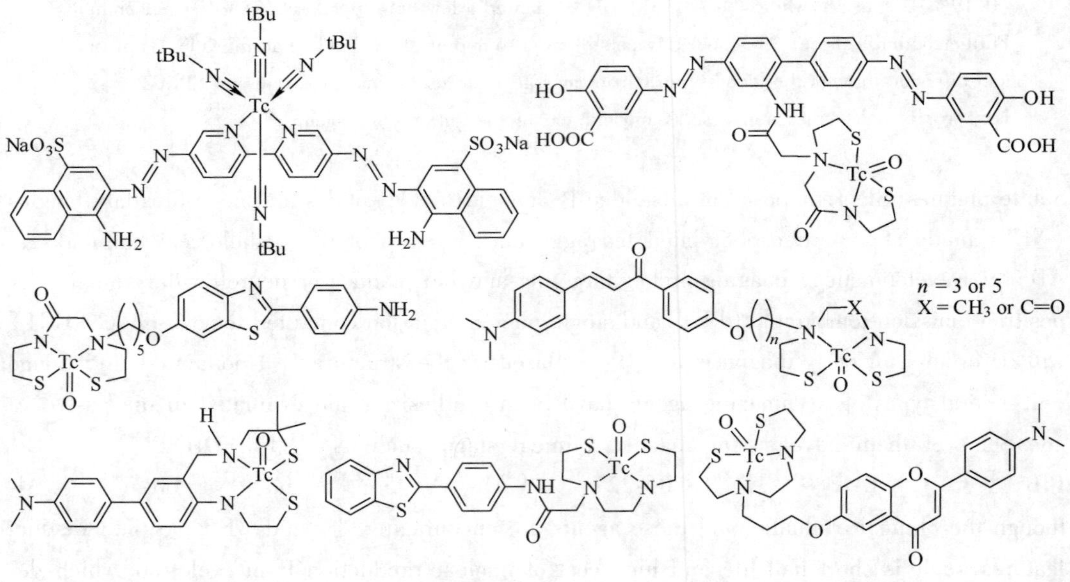

Fig. 1 Structures of ^{99m}Tc-labeled amyloid imaging probes

[^{18}F]FDDNP and its analog [^{18}F]FENE are two small molecules with compact naphthalene core and high affinity for both Aβ plaques and neurofibrillary tangles[4]. Previous reports indicated that they bind to Aβ aggregates on a distinct binding site[29]. The aim of this study was to synthesize neutral ^{99m}Tc/rhenium labeled derivatives of FDDNP and FENE, and to evaluate their biological characteristics.

The synthesis of the ^{99m}Tc- and Re-labeled 6-dialkylamino-2-naphthylethylidene derivatives is outlined in Scheme 1. Compounds 3 and 5 were obtained according to the literature previously reported[30]. Then bromination of the hydroxyl group was performed in CH_2Cl_2 with NBS, giving 6 and 7 with the yields of 79.3% and 76.7%, respectively. Then the compounds 6 and 7 were conjugated to MAMA (thiol groups were protected with p-methoxy benzyl to prevent oxidation) in CH_3CN with N,N-diisopropylethylamine (DIEA) as base to generate compounds 9 (yield 32.3%) and 10 (yield 25.3%). After deprotection of the thiol groups using trifluoro acetic acid (TFA) and CH_3SO_3H at 0℃, precursors 11 and 12 were obtained with the yields of 42.0% and 57.0%, respectively. Fluorinated compounds 4 (FENE) and 8 (FDDNP) were also obtained from 3 and 5 reacted with diethylamino sulfur trifluoride (DAST).

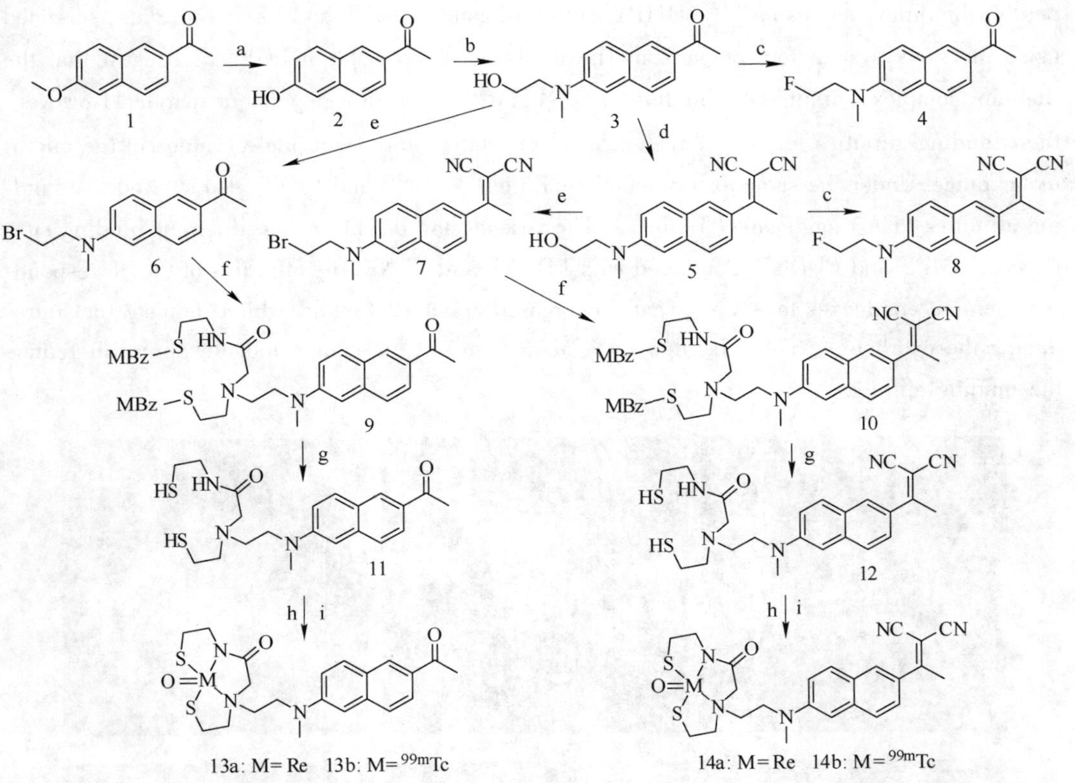

Scheme 1 Reagents and conditions

a—37% HCl, reflux; b—$CH_3NHCH_2CH_2OH$, $Na_2S_2O_5$, H_2O, 140℃; c—DAST, CH_2Cl_2, -78℃ to 0℃;
d—$CH_2(CN)_2$, pyridine, 110℃; e—CH_2Cl_2, NBS, PPh_3, 0-25℃; f—acetonitrile, MAMA-MBz
(p-methoxy benzyl), DIEA, reflux; g—Anisole, CH_3SO_3H, TFA, 0-25℃;
h—$ReOCl_3(PPh_3)_2$, CH_3OH, CH_2Cl_2, 90℃; i—^{99m}Tc-GH, 100℃

Rhenium complexes generally display similar physical and biodistributional properties to those of technetium complexes and have been normally adopted as non-radioactive surrogates for ^{99m}Tc complexes for structure identification. In this study, the corresponding rhenium complexes 13a (yield 23.2%) and 14a (yield 20.4%) were obtained from 11 and 12 by reacting with $ReOCl_3(PPh_3)_2$ in CH_2Cl_2.

^{99m}Tc-labeled complexes 13b and 14b were prepared by ligand exchange reaction employing the precursor ^{99m}Tc-glucoheptonate (GH), after heating in boiling water bath for 15min. The labeling yields detected by radio-HPLC were higher than 95%. After purification by HPLC, the radio-chemical purities of 13b and 14b were both greater than 98%. They were stable in saline more than 6h at room temperature. The radio-chemical identities of the ^{99m}Tc labeled tracers were verified by co-elution with the corresponding rhenium complexes on HPLC profiles(See in Supplementary data).

The affinity of rhenium complexes 13a and 14a for $A\beta_{1-42}$ aggregates was determined by competition binding assay using [^{125}I]IMPY as radio-ligand, FDDNP and FENE were also screened using the same system for comparison. The inhibition curves shown in Fig. 2 suggest that the rhenium complexes inhibit the binding of [^{125}I]IMPY in a dose-dependent manner. However, their binding affinities for $A\beta_{1-42}$ aggregates were relative lower with the K_i values in the micromolar range. Under the same experimental conditions FDDNP and FENE also showed low binding affinities to Aβ aggregates(Table 1). The reason may be due to the different binding sites between IMPY and FDDNP. Compared with FDDNP and FENE, the affinities of the corresponding rhenium complexes to Aβ aggregates decreased about 10-14-fold, which indicate that introducing the oxorhenium core chelated by a MAMA moiety through conjugate approach reduce the binding affinity.

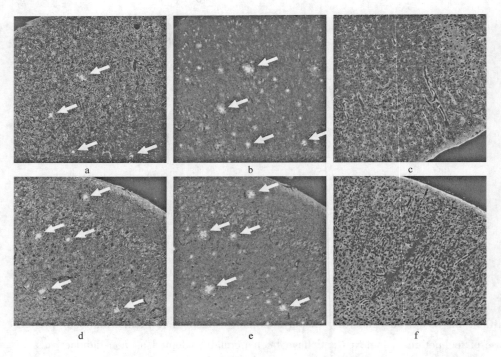

Fig. 2   Fluorescent staining of 13a(a) and 14a(d) on the brain slices(5μm) of a AD transgenic model(APP/PS1) and wilf control(c and f), plaques were also confirmed by the staining of the adjacent sections with Thioflavin-S(b and e)

Table 1 Inhibition constants (K_i, μM) for the binding to $A\beta_{1-42}$ aggregates versus [^{125}I] IMPY and HPLC profiles

Compound	$K_i^{①}$/μM	Retention time/min
4(FENE)	0.39 ± 0.05	n.d.
13a	3.65 ± 0.49	25.83
13b	n.d.	26.15
8(FDDNP)	0.33 ± 0.09	n.d.
14a	4.64 ± 0.77	24.92
14b	n.d.	25.35

① Measured in triplicate with results given as the mean ± SD.

The specific nature of rhenium complexes 13a and 14a to Aβ plaques was investigated by neuropathological staining with the brain sections of a 12-month-old AD transgenic model (Tg-C57), which encoded a double mutant form of APP and PS1. Many Aβ plaques were clearly stained by these compounds with low background (Fig. 2a and d). The similar pattern of Aβ plaques was also stained with Thioflavin-S using the adjacent brain sections (Fig. 2b and e). No plaques were found in control sections (Fig. 2c and f). The results indicated that these derivatives show specific binding to Aβ plaques. Because of the similar characteristics of technetium and rhenium, ^{99m}Tc-labeled complexes could also bind to Aβ plaques.

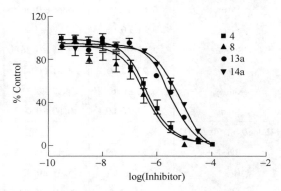

Fig. 3 Inhibition curves of 4, 8, 13a and 14a for the binding to aggregates of $A\beta_{1-42}$ versus [^{125}I] IMPY

It is believed that small molecular size (<600Da) and moderate lipophilicity (LogP values in the range of 1-3) are two of the major factors for BBB penetration[29,31]. The partition coefficients of the radiotracers were determined using a described procedure[32], LogD values of 13b and 14b were 1.70 ± 0.02 and 1.89 ± 0.04, respectively, which are in a good range for BBB penetration. As their molecular weight does not exceed 600 Da, 13b and 14b are expected to pass through the BBB and show better brain uptake.

In vivo biodistribution experiment was carried out on ICR normal mice (weight 18-22g). A saline solution (100μL) containing 7μCi purified radiotracer was injected directly into the tail vein. The mice were sacrificed at various time points after intravenous administration. The biodistribution data are shown in Table 2. Complex 13b displayed a medium initial brain uptake (0.65% ID/g at 2min pi) and a reasonable washout of the radioactivity from the brain (0.19% ID/g at 2h pi) while 14b showed a low brain uptake (0.28% ID/g at 2min). Although the lipophilicity of 13b and 14b was in a moderate range and their molecular weight did not exceed the threshold of 600 Da, they cannot cross the BBB to a sufficient degree, since the brain uptake was also dependent on other factors, such as hydrogen bonding, percentage of intact tracer in vivo, etc. Compared with that of the ^{99m}Tc-labeled BTA reported by our group[28], the initial brain uptake of these ^{99m}Tc-labeled DDNP derivatives appears insufficient for in vivo imaging. Therefore, further optimizations are needed to improve the pharmacokinetics of these ^{99m}Tc-labeled DDNP derivatives in vivo.

Table 2 Biodistribution experiments of 13b and 14b in normal mice①

Tissue	Time after injection/min				
	2	10	30	60	120
13b log$D$ = 1.89 ± 0.04					
Blood	5.81 ± 0.95	2.55 ± 0.81	2.07 ± 0.19	1.64 ± 0.27	1.22 ± 0.38
Heart	6.88 ± 1.03	3.84 ± 0.56	2.85 ± 0.25	2.26 ± 0.52	1.35 ± 0.4
Liver	9.31 ± 0.96	12.49 ± 1.85	13.85 ± 1.25	12.79 ± 2.97	8.63 ± 2.09
Spleen	2.78 ± 0.41	2.98 ± 0.36	2.51 ± 0.3	2.13 ± 0.21	1.21 ± 0.23
Lung	8.27 ± 2.03	4.31 ± 0.48	3.17 ± 0.13	2.66 ± 0.4	1.50 ± 0.34
Kidney	6.56 ± 0.63	4.67 ± 0.71	4.12 ± 0.62	3.66 ± 0.52	2.24 ± 0.53
Brain	0.65 ± 0.09	0.60 ± 0.04	0.44 ± 0.02	0.36 ± 0.06	0.19 ± 0.03
14b log$D$ = 1.70 ± 0.02					
Blood	6.35 ± 0.59	2.43 ± 0.39	1.71 ± 0.31	1.14 ± 0.13	0.73 ± 0.08
Heart	10.77 ± 1.1	5.47 ± 0.43	2.94 ± 0.46	1.86 ± 0.29	1.25 ± 0.3
Liver	18.31 ± 3.23	21.12 ± 3.83	16.95 ± 1.89	15.48 ± 3.63	13.9 ± 3.61
Spleen	3.57 ± 0.44	3.21 ± 0.35	2.34 ± 0.29	1.50 ± 0.23	1.08 ± 0.23
Lung	10.59 ± 1.74	4.67 ± 0.85	3.24 ± 0.43	2.29 ± 0.31	1.47 ± 0.21
Kidney	11.01 ± 1.88	8.56 ± 0.84	5.93 ± 0.87	4.18 ± 0.47	3.14 ± 0.74
Brain	0.28 ± 0.03	0.21 ± 0.02	0.17 ± 0.03	0.12 ± 0.02	0.10 ± 0.04

① All data are expressed as the mean percentage (n = 4) of the injected dose per gram of wet tissue (%ID/g) ± the standard deviation of the mean.

In conclusion, two ^{99m}Tc-labeled DDNP derivatives and their corresponding rhenium complexes were successfully synthesized through the conjugate approach. In experiments in vitro, the rhenium complexes 13a and 14a showed lower affinities for Aβ aggregates than FDDNP. However, both of them can bind to Aβ plaques in the brain sections of AD transgenic mouse. Due to the similar chemical and physical properties between rhenium and technetium, the ^{99m}Tc-labeled tracers 13b and 14b are expected to retain the binding affinity to Aβ plaques. Despite the fact that 13b and 14b displayed moderate lipophilicity and their molecular weights were less than 600 Da, in vivo biodistribution studies of 13b exhibited a medium initial brain uptake while 14b showed lower brain uptake. These results imply that these ^{99m}Tc-labeled DDNP derivatives probably require further refinement in order to improve their diffusion through the BBB and provide some useful information for the development of ^{99m}Tc-labeled probes for β-amyloid imaging.

Acknowledgements

We thank Dr. Lianfeng Zhang (Institute of Laboratory Animal Science, Chinese Academy of Medical Sciences and Comparative Medicine Center of Peking Union Medical College) for kindly providing the Paraffin-embedded brain sections of Tg-C57 mouse and Dr. Xiaoyan Zhang (College of Life Science, Beijing Normal University) for assistance in the in vitro neuropathological staining. This work was funded by NSFC(20871021).

Supplementary data

Supplementary data (procedure for the preparation of ^{99m}Tc-and Re-labeled DDNP derivatives, in vitro binding assay, fluorescent staining and biodistribution experiments) associated with this article can be found in the online version, at doi: 10.1016/j.bmcl.2010.11.096.

References

[1] Selkoe, D. J. JAMA 2000, 283, 1615.
[2] Hardy, J. A. ; Selkoe, D. J. Science 2002, 297, 353.
[3] Mathis, C. A. ; Wang, Y. ; Klunk, W. E. Curr. Pharm. Des. 2004, 10, 1469.
[4] Agdeppa, E. D. ; Kepe, V. ; Liu, J. ; Flores-Torres, S. ; Satyamurthy, N. ; Petric, A. ; Cole, G. M. ; Small, G. W. ; Huang, S. C. ; Barrio, J. R. J. Neurosci. 2001, 21, RC189.
[5] Shoghi-Jadid, K. ; Small, G. W. ; Agdeppa, E. D. ; Kepe, V. ; Ercoli, L. M. ; Siddarth, P. ; Read, S. ; Satyamurthy, N. ; Petric, A. ; Huang, S. C. ; Barrio, J. R. Am. J. Geriatr. Psychiatry 2002, 10, 24.
[6] Small, G. W. ; Kepe, V. ; Ercoli, L. M. ; Siddarth, P. ; Bookheimer, S. Y. ; Miller, K. J. ; Lavretsky, H. ; Burggren, A. C. ; Cole, G. M. ; Vinters, H. V. ; Thompson, P. M. ; Huang, S. C. ; Satyamurthy, N. ; Phelps, M. E. ; Barrio, J. R. N. Eng. J. Med. 2006, 355, 2652.
[7] Mathis, C. A. ; Wang, Y. ; Holt, D. P. ; Huang, G. F. ; Debnath, M. L. ; Klunk, W. E. J. Med. Chem. 2003, 46, 2740.
[8] Klunk, W. E. ; Engler, H. ; Nordberg, A. ; Wang, Y. ; Blomqvist, G. ; Holt, D. P. ; Bergstrom, M. ; Savitcheva, I. ; Huang, G. F. ; Estrada, S. ; Ausen, B. ; Debnath, M. L. ; Barletta, J. ; Price, J. C. ; Sandell, J. ; Lopresti, B. J. ; Wall, A. ; Koivisto, P. ; Antoni, G. ; Mathis, C. A. ; Langstrom, B. Ann. Neurol. 2004, 55, 306.
[9] Verhoeff, N. P. ; Wilson, A. A. ; Takeshita, S. ; Trop, L. ; Hussey, D. ; Singh, K. ; Kung, H. F. ; Kung, M. P. ; Houle, S. Am. J. Geriatr. Psychiatry 2004, 12, 584.
[10] Ono, M. ; Wilson, A. ; Nobrega, J. ; Westaway, D. ; Verhoeff, P. ; Zhuang, Z. P. ; Kung, M. P. ; Kung, H. F. Nucl. Med. Biol. 2003, 30, 565.
[11] Kudo, Y. ; Okamura, N. ; Furumoto, S. ; Tashiro, M. ; Furukawa, K. ; Maruyama, M. ; Itoh, M. ; Iwata, R. ; Yanai, K. ; Arai, H. J. Nucl. Med. 2007, 48, 553.
[12] Choi, S. R. ; Golding, G. ; Zhuang, Z. P. ; Zhang, W. ; Lim, N. ; Hefti, F. ; Benedum, T. E. ; Kilbourn, M. R. ; Skovronsky, D. ; Kung, H. F. J. Nucl. Med. 2009, 50, 1887.
[13] Kung, H. F. ; Choi, S. R. ; Qu, W. C. ; Zhang, W. ; Skovronsky, D. J. Med. Chem. 2010, 53, 933.
[14] Kung, M. P. ; Hou, C. ; Zhuang, Z. P. ; Zhang, B. ; Skovronsky, D. ; Trojanowski, J. Q. ; Lee, V. M. ; Kung, H. F. Brain Res. 2002, 956, 202.
[15] Zhuang, Z. P. ; Kung, M. P. ; Wilson, A. ; Lee, C. W. ; Plossl, K. ; Hou, C. ; Holtzman, D. M. ; Kung,

H. F. J. Med. Chem. 2003, 46, 237.

[16] Newberg, A. B. ; Wintering, N. A. ; Plossl, K. ; Hochold, J. ; Stabin, M. G. ; Watson, M. ; Skovronsky, D. ; Clark, C. M. ; Kung, M. P. ; Kung, H. F. J. Nucl. Med. 2006, 47, 748.

[17] Jurisson, S. S. ; Lydon, J. D. Chem. Rev. 1999, 99, 2205.

[18] Hom, R. K. ; Katzenellenbogen, J. A. Nucl. Med. Biol. 1997, 24, 485.

[19] Zhen, W. ; Han, H. ; Anguiano, M. ; Lemere, C. A. ; Cho, C. G. ; Lansbury, P. T. J. Med. Chem. 1999, 42, 2805.

[20] Dezutter, N. A. D. G. ; de Groot, T. J. ; Bormans, G. M. ; Verbruggen, A. M. ; Dom, R. J. Eur. J. Nucl. Med. Mol. Imaging 1999, 26, 1392.

[21] Serdons, K. ; Verduyckt, T. ; Cleynhens, J. ; Terwinghe, C. ; Mortelmans, L. ; Bormans, G. ; Verbruggen, A. Bioorg. Med. Chem. Lett. 2007, 17, 6086.

[22] Serdons, K. ; Verduyckt, T. ; Cleynhens, J. ; Terwinghe, C. ; Bormans, G. ; Verbruggen, A. J. Labelled Compd. Radiopharm. 2008, 51, 357.

[23] Lin, K. S. ; Debnath, M. L. ; Mathis, C. A. ; Klunk, W. E. Bioorg. Mde. Chem. Lett. 2009, 19, 2258.

[24] Zhuang, Z. P. ; Kung, M. P. ; Hou, C. ; Kung, H. F. J. Labelled Compd. Radiopharm. 2005, 48, S251.

[25] Ono, M. ; Ikeoka, R. ; Watanabe, H. ; Kimura, H. ; Fuchigami, T. ; Haratake, M. ; Saji, H. ; Nakayama, M. Bioorg. Med. Chem. Lett. 2010, 20, 5743.

[26] Ono, M. ; Ikeoka, R. ; Watanabe, H. ; Kimura, H. ; Fuchigami, T. ; Haratake, M. ; Saji, H. ; Nakayama, M. ACS Chem. Neurosci. 2010, 1, 598.

[27] Zhuang, Z. P. ; Kung, M. P. ; Hou, C. ; Ploessl, K. ; Kung, H. F. Nucl. Med. Biol. 2005, 32, 171.

[28] Chen, X. J. ; Yu, P. R. ; Zhang, L. F. ; Liu, B. L. Bioorg. Med. Chem. Lett. 2008, 18, 1442.

[29] Cai, L. S. ; Innis, R. B. ; Pike, V. W. Curr. Med. Chem. 2007, 14, 19.

[30] Liu, J. ; Kepe, V. ; Zabjek, A. ; Petric, A. ; Padgett, H. C. ; Satyamurthy, N. ; Barrio, J. R. Mol. Imaging Biol. 2007, 9, 6.

[31] Furumoto, S. ; Okamura, N. ; Iwata, R. ; Yanai, K. ; Arai, H. ; Kudo, Y. Curr. Top. Med. Chem. 2007, 7, 1773.

[32] Wu, C. Y. ; Wei, J. J. ; Gao, K. Q. ; Wang, Y. M. Bioorg. Med. Chem. 2007, 15, 2789.

Synthesis and Biological Evaluation of Novel Technetium-99m Labeled Phenylbenzoxazole Derivatives as Potential Imaging Probes for β-amyloid Plaques in Brain*

Abstract Two uncharged ^{99m}Tc-labeled phenylbenzoxazole derivatives were biologically evaluated as potential imaging probes for β-amyloid plaques. The ^{99m}Tc and corresponding rhenium complexes were synthesized by coupling monoamine-monoamide dithiol(MAMA) and bis(aminoethanethiol)(BAT) chelating ligand via a pentyloxy spacer to phenylbenzoxazole. The fluorescent rhenium complexes 6 and 9 selectively stained the β-amyloid plaques on the sections of transgenic mouse, and showed high affinity for Aβ$_{(1-42)}$ aggregates (K_i = 11.1nM and 14.3nM, respectively). Autoradiography in vitro indicated that [^{99m}Tc]6 clearly labeled β-amyloid plaques on the sections of transgenic mouse. Biodistribution experiments in normal mice revealed that [^{99m}Tc]6 displayed moderate initial brain uptake(0.81% ID/g at 2min), and quickly washed out from the brain(0.25% ID/g at 60min). The preliminary results indicate that the properties of [^{99m}Tc]6 are promising, although additional refinements are needed to improve the ability to cross the blood-brain barrier.

Key words alzheimer's disease, β-amyloid plaque, technetium-99m, phenylbenzoxazole, imaging probe

Alzheimer's disease(AD) is the leading cause of neurodegenerative disorder of the brain, accounting for most dementia cases after the age of 60. Histopathologically, the formation of extracellular β-amyloid(Aβ) plaques is one of the pathological hallmarks of this disease[1]. In clinic, it's very difficult to differentiate AD from other dementia cases. Currently, diagnosis of AD relies on the results from neuropsychological tests. However, this method is often complicated and unreliable, where the degree of accuracy ranges from 50% to 90%. A definite diagnosis of AD can only be confirmed by histopathological observation of Aβ plaques in the cerebral cortex of postmortem brain tissue. At present, the exact etiology of AD is not completely understood, the most widely accepted theory regarding to this disease is the amyloid cascade hypothesis[2-4]. Therefore, Aβ represents an important molecular target for AD, and the detection of deposited Aβ plaques with non-invasive techniques such as positron emission tomography(PET) or single photon emission computed tomography(SPECT) could achieve the differential diagnosis of AD in its pre-symptomatic stages[5,6].

Over the past decade, based on the scaffolds of thioflavin-T(ThT) and Congo Red(CR), (commonly used dyes for detection of Aβ plaques), several specific radiotracers were synthesized and evaluated as in vivo imaging probes for Aβ plaques with PET and SPECT. Some PET tracers such as [^{11}C]PIB[7,8], [^{18}F]FDDNP[9-11], [^{11}C]SB-13[12], [^{18}F]BAY94-9172[13],

---

* Copartner: Wang Xuedan, Cui Mengchao, Yu Pingrong, Li Zijing, Yang Yanping, Jia Hongmei. Reprinted from *Bioorganic & Medicinal Chemistry Letters*, 2012, 22(13):4327-4331.

[^{18}F]GE-067[14] and [^{18}F]AV-45[15,16] have been evaluated in clinical trials.

However, the development of SPECT tracers was lagging far behind, to the best of our knowledge, a radioiodinated ligand [^{123}I]IMPY, is the only SPECT tracer tested in humans. Some undesirable properties including high lipophilicity, in vivo instability together with insufficient target-to-background ratio may account for the failure of [^{123}I]IMPY in clinical trials[17-19]. Furthermore, some other radioiodinated ThT analogs for SPECT such as TZDM[20], IBOX[21], phenylbenzofuran derivatives[22] and phenylindole derivatives[23] have shown high affinity for Aβ aggregates in vitro and high initial uptake, but their potential use was limited by the slow washout rate from the brain and the lack of routine availability of ^{123}I radioisotope.

At present, the most widely used radionuclide in diagnostic nuclear medicine for SPECT imaging is technetium-99m (^{99m}Tc), mainly due to its optimal nuclear properties ($T_{1/2}$ = 6.01h, 141keV), easy availability (through commercial ^{99}Mo/^{99m}Tc generators), and low cost. Accordingly, the development of ^{99m}Tc-labeled imaging probes targeting Aβ plaques will provide simple, convenient, and widespread method for the diagnosis of AD. In the past few years, several ^{99m}Tc-labeled imaging probes have been developed (Fig. 1). Such as the initially reported ^{99m}Tc-labeled Congo Red[24,25] and Chrysamine G derivatives[26,27], which displayed high affinity for Aβ aggregates in vitro. However, due to the large molecular weight and ionized character at physiological pH, they cannot cross the blood-brain barrier (BBB). Attempts to prepare ^{99m}Tc-labeled nonpolar, neutral, small in size and more lipophilic ligands have been reported. Kung and his co-

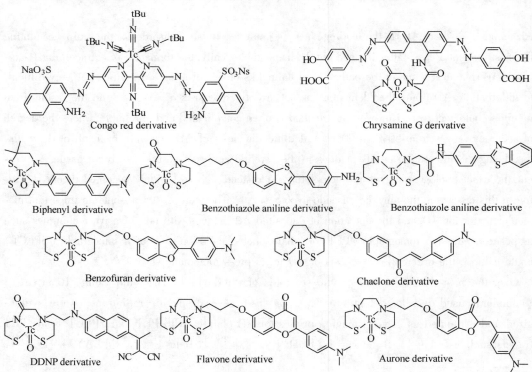

Fig. 1 Chemical structures of ^{99m}Tc-labeled Aβ imaging probes reported previously

workers reported the ^{99m}Tc-labeled N_2S_2-biphenyl derivatives, which showed high initial brain uptake(1.18% dose/organ at 2min), but it failed to bind to the Aβ plaques in the postmortem human brain tissue of patients with confirmed AD[28]. Some other groups reported a series of benzothiazole aniline(BTA)[29-31], phenylbenzofuran[32], DDNP[33], chaclone[34], flavone and aurone[35] derivatives conjugated with bis(aminoethanethiol)(BAT) or monoamine-monoamide dithiol(MAMA) ligands. Although it was demonstrated that these ^{99m}Tc-labeled derivatives bind to Aβ plaques in vitro, they still cannot cross BBB to a sufficient degree for the in vivo imaging.

Recently, we have synthesized and evaluated a series of ^{18}F-labeled phenylbenzoxazole derivatives as novel Aβ imaging probes. They showed high affinity for Aβ aggregates and favorable pharmacokinetics in vivo, which suggests that this class of tracers has potential applications for PET imaging[36]. Building on the previous results, we attempt to design novel ^{99m}Tc-labeled phenylbenzoxazole derivatives for SPECT imaging. In the present study, we report the synthesis of two phenylbenzoxazole derivatives conjugated to MAMA and BAT via a pentyloxy spacer as well as their complexation to Re and ^{99m}Tc. We also report the biological evaluation of these two probes as novel Aβ imaging probes.

The synthetic route of the phenylbenzoxazole derivatives is outlined in Scheme 1. The key step in the formation of the phenylbenzoxazole backbone was achieved by reacting 2-amino-4-methoxyphenol with 4-(dimethylamino) benzoic acid in PPA, compound **1** was obtained in a yield of 39.6%. Subsequent demethylation of the methoxyl group using BBr_3 in CH_2Cl_2 afforded **2** in a yield of 89.5%. The thiol-protected chelation ligands(PMB-MAMA and N-Boc-Tr-BAT) were synthesized according to the methods reported previously[37]. After introduction of the pentyloxy linker into **2** by reacting with 1,5-dibromo pentane, **3** were conjugated to PMB-MAMA or N-Boc-Tr-BAT to generate the compounds **4** and **7**, respectively. Compounds **5** and **8**(the precursors for ^{99m}Tc labeling) were obtained by deprotection of the thiol groups in **4** and **7**. The rhenium complexes(**6** and **9**) were prepared by the reaction of **4** and **7** with $(PPh_3)_2ReOCl_3$ in CH_2Cl_2. The corresponding ^{99m}Tc complexes [^{99m}Tc]**6** and [^{99m}Tc]**9** were prepared by a ligand exchange reaction employing the precursor ^{99m}Tc-glucoheptonate(GH) in boiling water, and the desired products were formed as major radioactive products. After purification by reversed-phase high-performance liquid chromatography(HPLC), the final products were obtained with radiochemical purity higher than 95%. The identity of the complexes were established by comparative HPLC using the corresponding Re complexes as reference. The retention time of [^{99m}Tc]**6** and [^{99m}Tc]**9** on HPLC(radioactivity) were 7.92min and 9.97min, respectively. The retention times of the corresponding rhenium complexes(**6** and **9**) on HPLC(UV detection) were 7.46min and 9.31min, respectively(Fig. S1 in the supplementary data).

The specific binding of the rhenium complexes **6** and **9** to Aβ plaques were evaluated by neuropathological staining with the brain sections from a transgenic mouse(C57BL6, APPswe/PSEN1, 12 months old), an animal model for AD. As shown in Fig. 2a and c, both Re complexes could clearly stain Aβ plaques on the brain sections with low background, and the similar stain pattern of Aβ plaques was consistent with that stained with thioflavin-S using the adjacent

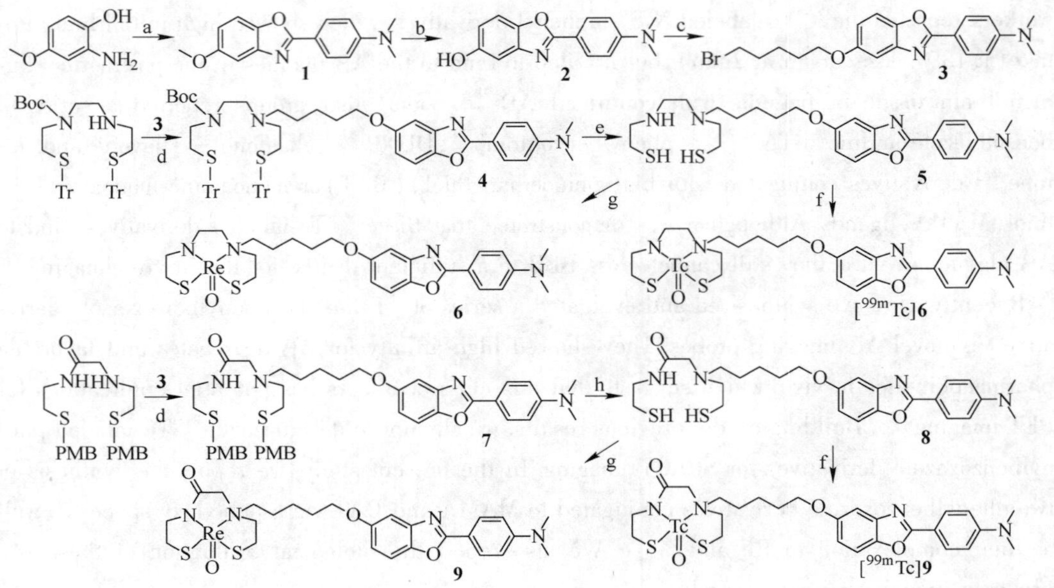

Scheme 1 Synthesis of ^{99m}Tc-labeled phenyl benzoxazole and the corresponding rhenium derivatives. Reagents and conditions

a—4-(dimethylamino)benzoic acid, PPA, 140 ℃; b—BBr$_3$, CH$_2$Cl$_2$; c—K$_2$CO$_3$, CH$_3$CN, 90 ℃; d—K$_2$CO$_3$, CH$_3$CN, 90 ℃; e—TFA, Et$_3$SiH; f—^{99m}Tc-GH, 100 ℃; g—ReOCl$_3$(PPh$_3$)$_2$, CH$_2$Cl$_2$, CH$_3$OH; h—Anisole, CH$_3$SO$_3$H, TFA

sections (Fig. 2b and d). The results indicated that these two Re complexes could bind specifically to Aβ plaques. Furthermore, the quantitative binding affinities of rhenium complexes **6** and **9** for Aβ$_{1-42}$ aggregates were determined by competition binding assay using [^{125}I]IMPY as radioligand, while IMPY was also screened using the same system for comparison. As shown in Table 1, complexes **6** and **9** displayed high affinity to Aβ$_{1-42}$ aggregates (K_i = 11.1 nM and 14.3 nM, respectively), which is comparable to the value of IMPY (K_i = 10.1 nM). Compared with the ^{99m}Tc-labeled phenylbenzothiazole[38] and phenylbenzofuran[32] derivatives reported previously, these phenylbenzoxazole derivatives also maintain high binding affinity to Aβ aggregates.

Table 1 Inhibition constants for the binding of [^{125}I]IMPY to Aβ$_{1-42}$ aggregates[①]

Compounds	K_i/nM
6	11.1 ± 3.5
9	14.3 ± 2.2
IMPY	10.1 ± 1.9

① Measured in triplicate with results given as the mean ± SD.

To further characterize the binding to Aβ plaques, in vitro autoradiography was performed by incubating [^{99m}Tc]**6** with sections of AD mouse brain. As shown in Fig. 3a and c, [^{99m}Tc]**6** displayed excellent labeling of Aβ plaques in the cortex region and a low background level in white matter in the AD mouse brain. Furthermore, the hot spots of radioactivity were perfectly accorded with the results of in vitro fluorescence staining in the same sections with thioflavin-S (Fig. 3b and d, red arrows). The results confirm the high binding affinity of [^{99m}Tc]**6** to Aβ$_{1-42}$ aggregates.

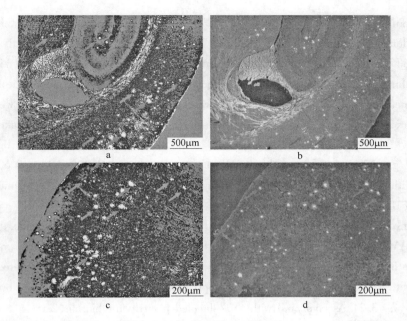

Fig. 2 Fluorescence staining of the rhenium complexes **6** and **9** on the sections from a transgenic model mouse (C57BL6, APPswe/PSEN1, 12 months old, 10μm thick) (a, c) and the presence and distribution of Aβ plaques on the adjacent sections were confirmed with thioflavin-S (b, d)

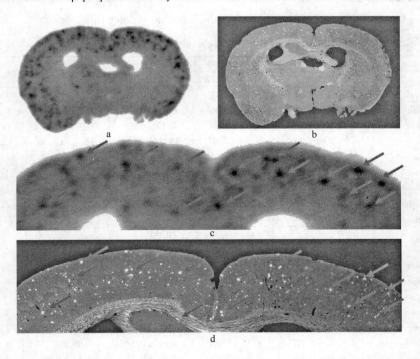

Fig. 3 Autoradiography of [^{99m}Tc]**6** in vitro on the section from a transgenic model mouse (C57BL6, APPswe/PSEN1, 12 months old, 10μm thick) (a), the presence and distribution of plaques in the same section were confirmed with thioflavin-S (b), a magnification view of the Aβ plaques labeled with [^{99m}Tc]**6** (c) and confirmed with thioflavin-S (d)

The lipophilicity (lgD) of $[^{99m}Tc]6$ and $[^{99m}Tc]9$ measured under experimental conditions showed moderate partition coefficients ($lgD = 2.73 \pm 0.02$ and 2.93 ± 0.09, respectively), which are in a good range for BBB penetration. High initial brain uptake and fast washout kinetics from the normal brain are two important properties for a promising Aβ imaging probe. The biodistribution experiments of $[^{99m}Tc]6$ and $[^{99m}Tc]9$ were performed in normal mice (5 weeks, male). Biodistribution study in mice is often used as a method to measure the brain uptake. As shown in Table 2, $[^{99m}Tc]6$ with a BAT chelation ligand displayed moderate initial brain uptake (0.81% ID/g at 2min), and quickly washout from the brain (0.25% ID/g at 60min). But $[^{99m}Tc]9$ with a MAMA chelation ligand displayed lower initial brain uptake at 2min postinjection (0.43% ID/g), and was washed out from the brain relatively slowly (0.30% ID/g at 60min). Since the only difference between $[^{99m}Tc]6$ and $[^{99m}Tc]9$ is the chelation ligands, the carbonyl group in MAMA moiety may form hydrogen bond with water and other biomoleculars in vivo which decrease the initial brain uptake of $[^{99m}Tc]9$. The ratio $brain_{2min}/brain_{60min}$ is considered an important index with which to select probes with appropriate kinetics in vivo. The $brain_{2min}/brain_{60min}$ ratio of $[^{99m}Tc]6$ was 3.24. As compared with ^{99m}Tc-labeled phenylbenzothiazole (2.06)[30] and phenylbenzofuran (2.39)[32], the ^{99m}Tc-labeled phenylbenzoxazole derivative 6 had superior $brain_{2min}/brain_{60min}$ ratio. Furthermore, the excretion of the ^{99m}Tc-labeled ligands was observed predominantly by the hepatobiliary and excretory systems. Liver uptake dropped gradually with time, while the accumulation of radioactivity was observed within the intestine at later time periods.

Table 2 Biodistribution of $[^{99m}Tc]6$ and $[^{99m}Tc]9$ in normal mice[①]

Organ	2min	10min	30min	60min	120min
$[^{99m}Tc]6$ ($lgD = 2.73 \pm 0.02$)					
Blood	3.68 ± 0.67	1.75 ± 0.17	1.29 ± 0.09	1.45 ± 0.45	1.14 ± 0.20
Brain	0.81 ± 0.01	0.49 ± 0.06	0.40 ± 0.03	0.25 ± 0.05	0.20 ± 0.02
Heart	10.09 ± 2.15	4.40 ± 0.84	2.13 ± 0.30	1.44 ± 0.32	0.98 ± 0.15
Liver	23.23 ± 1.89	21.34 ± 5.61	18.34 ± 2.54	15.29 ± 3.36	13.89 ± 3.37
Spleen	5.43 ± 0.55	3.02 ± 0.51	2.34 ± 0.20	1.65 ± 0.24	1.06 ± 0.22
Lung	9.38 ± 1.92	3.63 ± 0.98	3.04 ± 0.49	2.32 ± 0.60	1.57 ± 0.37
Kidney	9.98 ± 2.09	6.08 ± 0.82	3.42 ± 0.46	2.99 ± 0.48	2.24 ± 0.42
Stomach[②]	1.13 ± 0.20	1.68 ± 0.27	3.34 ± 0.95	3.30 ± 0.51	2.28 ± 0.35
Intestine[②]	2.86 ± 0.74	11.34 ± 2.45	21.99 ± 3.16	27.19 ± 2.64	21.43 ± 3.94
$[^{99m}Tc]9$ ($lgD = 2.93 \pm 0.09$)					
Blood	5.50 ± 0.30	3.27 ± 0.12	2.04 ± 0.66	3.13 ± 0.30	2.34 ± 0.16
Brain	0.43 ± 0.06	0.30 ± 0.01	0.23 ± 0.01	0.30 ± 0.04	0.28 ± 0.05
Heart	7.94 ± 0.44	5.12 ± 0.11	2.41 ± 0.50	2.76 ± 0.05	1.87 ± 0.08
Liver	21.08 ± 1.85	19.88 ± 0.67	9.01 ± 1.32	22.75 ± 2.99	14.54 ± 0.56
Spleen	5.66 ± 0.22	3.49 ± 0.31	2.01 ± 0.11	2.97 ± 0.26	1.82 ± 0.21
Lung	12.83 ± 0.98	5.28 ± 0.41	3.45 ± 0.04	3.20 ± 0.30	2.56 ± 0.17
Kidney	7.56 ± 0.32	6.15 ± 0.35	3.68 ± 0.14	5.29 ± 0.26	3.26 ± 0.27
Stomach[②]	1.44 ± 0.16	1.17 ± 0.09	2.05 ± 0.45	1.65 ± 0.06	4.38 ± 0.39
Intestine[②]	2.66 ± 0.29	4.59 ± 3.05	11.01 ± 0.80	12.07 ± 5.58	13.77 ± 0.49

① All data are expressed as the mean percentage ($n = 5$) of the injected dose per gram of wet tissue (%ID/g) ± the standard deviation of the mean.

② Expressed as % injected dose per organ.

In conclusion, two novel uncharged ^{99m}Tc-labeled phenylbenzoxazole derivatives and their corresponding rhenium complexes were successfully synthesized and biologically evaluated as potential imaging probes for β-amyloid plaques. In vitro competition binding assay indicated that the rhenium complexes displayed high affinity for Aβ$_{(1-42)}$ aggregates, and the high binding affinity was further confirmed by in vitro autoradiography, [^{99m}Tc]6 clearly labeled Aβ plaques in sections of brain tissue from AD mouse brain. In addition, [^{99m}Tc]6 exhibited moderate initial penetration(0.81% ID/g) of and fast clearance from the brain, with a brain$_{2min}$/brain$_{60min}$ ratio of 3.24. Taken together, the present results suggest that the ^{99m}Tc-labeled phenylbenzoxazole derivatives 6 may be a useful Aβ imaging probe for SPECT. But additional chemical modifications are required to further improve the diffusion through the BBB.

Acknowledgement

This study was funded by NSFC(20871021).

Supplementary data

Supplementary data(procedure for the preparation of ^{99m}Tc/Re labeled phenylbenzoxazole derivatives, in vitro binding assay, fluorescent staining and biodistribution experiments) associated with this article can be found in the online version, at http://dx.doi.org/10.1016/j.bmcl.2012.05.010.

References

[1] Selkoe,D. J. Physiol. Rev. 2001,81,741.
[2] Hardy,J. A. ;Higgins,G. A. Science 1992,256,184.
[3] Selkoe,D. J. JAMA J. Am. Med. Assoc. 2000,283,1615.
[4] Hardy,J. A. ;Selkoe,D. J. Science. 2002,297,353.
[5] Mathis,C. A. ;Wang,Y. ;Klunk,W. E. Curr. Pharm. Des. 2004,10,1469.
[6] Cai,L. ;Innis,R. B. ;Pike,V. W. Curr. Med. Chem. 2007,14,19.
[7] Mathis,C. A. ; Wang, Y. M. ; Holt, D. P. ; Huang, G. F. ; Debnath, M. L. ; Klunk, W. E. J. Med. Chem. 2003,46,2740.
[8] Klunk, W. E. ; Engler, H. ; Nordberg, A. ; Wang, Y. M. ; Blomqvist, G. ; Holt, D. P. ; Bergstrom, M. ; Savitcheva,I. ;Huang,G. F. ;Estrada,S. ;Ausen,B. ;Debnath,M. L. ;Barletta,J. ;Price,J. C. ;Sandell,J. ; Lopresti,B. J. ;Wall,A. ;Koivisto,P. ;Antoni,G. ;Mathis,C. A. ;Langstrom,B. Ann. Neurol. 2004,55,306.
[9] Agdeppa,E. D. ; Kepe,V. ; Flores-Torres,S. ; Liu,J. ; Satyamurthy,N. ; Petric,A. ; Small,G. W. ; Huang, S. C. ;Barrio,J. R. J. Nucl. Med. 2001,42,64.
[10] Agdeppa,E. D. ;Kepe,V. ;Shoghi-Jadid,K. ;Satyamurthy,N. ;Small,G. W. ;Petric,A. ;Vinters,H. V. ; Huang,S. C. ;Barrio,J. R. J. Nucl. Med. 2001,42,65.
[11] Shoghi-Jadid, K. ; Small, G. W. ; Agdeppa, E. D. ; Kepe, V. ; Ercoli, L. M. ; Siddarth, P. ; Read, S. ; Satyamurthy,N. ;Petric,A. ;Huang,S. C. ;Barrio,J. R. Am. J. Geriatr. Psychiatry 2002,10,24.
[12] Verhoeff,N. P. L. G. ;Wilson,A. A. ;Takeshita,S. ;Trop,L. ;Hussey,D. ;Singh,K. ;Kung,H. F. ;Kung, M. P. ;Houle,S. Am. J. Geriatr. Psychiatry 2004,12,584.
[13] Rowe,C. C. ; Ackerman, U. ; Browne, W. ; Mulligan, R. ; Pike, K. L. ; O' Keefe, G. ; Tochon-Danguy, H. ; Chan,G. ;Berlangieri,S. U. ;Jones,G. ;Dickinson-Rowe,K. L. ;Kung,H. P. ;Zhang,W. ;Kung,M. P. ;

Skovronsky, D. ; Dyrks, T. ; Holl, G. ; Krause, S. ; Friebe, M. ; Lehman, L. ; Lindemann, S. ; Dinkelborg, L. M. ; Masters, C. L. ; Villemagne, V. L. Lancet Neurol. 2008, 7, 129.

[14] Koole, M. ; Lewis, D. M. ; Buckley, C. ; Nelissen, N. ; Vandenbulcke, M. ; Brooks, D. J. ; Vandenberghe, R. ; Van Laere, K. J. Nucl. Med. 2009, 50, 818.

[15] Choi, S. R. ; Golding, G. ; Zhuang, Z. ; Zhang, W. ; Lim, N. ; Hefti, F. ; Benedum, T. E. ; Kilbourn, M. R. ; Skovronsky, D. ; Kung, H. F. J. Nucl. Med. 2009, 50, 1887.

[16] Kung, H. F. ; Choi, S. R. ; Qu, W. C. ; Zhang, W. ; Skovronsky, D. J. Med. Chem. 2010, 53, 933.

[17] Kung, M. P. ; Hou, C. ; Zhuang, Z. P. ; Zhang, B. ; Skovronsky, D. ; Trojanowski, J. Q. ; Lee, V. M. ; Kung, H. F. Brain Res. 2002, 956, 202.

[18] Zhuang, Z. P. ; Kung, M. P. ; Wilson, A. ; Lee, C. W. ; Plossl, K. ; Hou, C. ; Holtzman, D. M. ; Kung, H. F. J. Med. Chem. 2003, 46, 237.

[19] Newberg, A. B. ; Wintering, N. A. ; Plossl, K. ; Hochold, J. ; Stabin, M. G. ; Watson, M. ; Skovronsky, D. ; Clark, C. M. ; Kung, M. P. ; Kung, H. F. J. Nucl. Med. 2006, 47, 748.

[20] Zhuang, Z. P. ; Kung, M. P. ; Hou, C. ; Skovronsky, D. M. ; Gur, T. L. ; Plossl, K. ; Trojanowski, J. Q. ; Lee, V. M. Y. ; Kung, H. F. J. Med. Chem. 2001, 44, 1905.

[21] Zhuang, Z. P. ; Kung, M. P. ; Hou, C. ; Plossl, K. ; Skovronsky, D. ; Gur, T. L. ; Trojanowski, J. Q. ; Lee, V. M. Y. ; Kung, H. F. Nucl. Med. Biol. 2001, 28, 887.

[22] Ono, M. ; Kung, M. P. ; Hou, C. ; Kung, H. F. Nucl. Med. Biol. 2002, 29, 633.

[23] Watanabe, H. ; Ono, M. ; Haratake, M. ; Kobashi, N. ; Saji, H. ; Nakayama, M. Bioorg. Med. Chem. 2010, 18, 4740.

[24] Han, H. ; Cho, C. G. ; Lansbury, P. T. J. Am. Chem. Soc. 1996, 118, 4506.

[25] Zhen, W. ; Han, H. ; Anguiano, M. ; Lemere, C. A. ; Cho, C. G. ; Lansbury, P. T. J. Med. Chem. 1999, 42, 2805.

[26] Dezutter, N. A. ; Dom, R. J. ; de Groot, T. J. ; Bormans, G. M. ; Verbruggen, A. M. Eur. J. Nucl. Med. Mol. Imaging. 1999, 26, 1392.

[27] Dezutter, N. A. ; de Groot, T. J. ; Busson, R. H. ; Janssen, G. A. ; Verbruggen, A. M. J. Labelled Compd. Radiopharm. 1999, 42, 309.

[28] Zhuang, Z. P. ; Kung, M. P. ; Hou, C. ; Ploessl, K. ; Kung, H. F. Nucl. Med. Biol. 2005, 32, 171.

[29] Serdons, K. ; Verduyckt, T. ; Cleynhens, J. ; Terwinghe, C. ; Mortelmans, L. ; Bormans, G. ; Verbruggen, A. Bioorg. Med. Chem. Lett. 2007, 17, 6086.

[30] Chen, X. J. ; Yu, P. R. ; Zhang, L. F. ; Liu, B. Bioorg. Med. Chem. Lett. 2008, 18, 1442.

[31] Serdons, K. ; Verduyckt, T. ; Cleynhens, J. ; Bormans, G. ; Verbruggen, A. J. Labelled Compd. Radiopharm. 2008, 51, 357.

[32] Ono, M. ; Fuchi, Y. ; Fuchigami, T. ; Kobashi, N. ; Kimura, H. ; Haratake, M. ; Saji, H. ; Nakayama, M. Acs Med. Chem. Lett. 2010, 1, 443.

[33] Cui, M. C. ; Tang, R. K. ; Li, Z. J. ; Ren, H. Y. ; Liu, B. L. Bioorg. Med. Chem. Lett. 2011, 21, 1064.

[34] Ono, M. ; Ikeoka, R. ; Watanbe, H. ; Kimura, H. ; Fuchigami, T. ; Haratake, M. ; Saji, H. ; Nakayama, M. Acs Chem. Neurosci. 2010, 1, 598.

[35] Ono, M. ; Ikeoka, R. ; Watanabe, H. ; Kimura, H. ; Fuchigami, T. ; Haratake, M. ; Saji, H. ; Nakayama, M. Bioorg. Med. Chem. Lett. 2010, 20, 5743.

[36] Cui, M. C. ; Ono, M. ; Kimura, H. ; Ueda, M. ; Nakamoto, Y. ; Togashi, K. ; Okamoto, Y. ; Ihara, M. ; Takahashi, R. ; Liu, B. L. ; Saji, H. J. Med. Chem. Submitted for publication.

[37] Oya, S. ; Plossl, K. ; Kung, M. P. ; Stevenson, D. A. ; Kung, H. F. Nucl. Med. Biol. 1998, 25, 135.

[38] Lin, K. S. ; Debnath, M. L. ; Mathis, C. A. ; Klunk, W. E. Bioorg. Med. Chem. Lett. 2009, 19, 2258.

^{99m}Tc-labeled Dibenzylideneacetone Derivatives as Potential SPECT Probes for *in vivo* Imaging of β-amyloid Plaque[*]

Abstract Four ^{99m}Tc-labeled dibenzylideneacetone derivatives and corresponding rhenium complexes were successfully synthesized and biologically evaluated as potential imaging probes for Aβ plaques using SPECT. All rhenium complexes (5a-d) showed affinity for Aβ$_{(1-42)}$ aggregates (K_i = 13.6-120.9nM), and selectively stained the Aβ plaques on brain sections of transgenic mice. Biodistribution in normal mice revealed that [^{99m}Tc]5a-d exhibited moderate initial uptake ((0.31-0.49)%ID/g at 2min) and reasonable brain washout at 60min post-injection. Although additional optimizations are still needed to facilitate it's penetration through BBB, the present results indicate that [^{99m}Tc]5a may be a potential SPECT probe for imaging Aβ plaques in Alzheimer's brains.

Key words alzheimer's disease, β-amyloid plaque, dibenzylideneacetone, technetium-99m, SPECT imaging

1 Introduction

Alzheimer's disease (AD) is a neurodegenerative disorder characterized by the devastating loss of memory and cognitive abilities, and it will afflict about 63 million people by 2030, and 114 million by 2050 worldwide[1-3]. The deposition of extracellular β-amyloid (Aβ) plaques and intracellular neurofibrillary tangles (NFTs) containing highly phosphorylated tau protein in the brain have been regarded as the dominant factors driving AD pathogenesis[4]. Currently, the diagnosis of AD primarily depends on clinical evaluation, patient history, structural and functional imaging of the brain by computed tomography (CT) and magnetic resonance imaging (MRI)[5,6]. However, these methods are usually insufficient and unreliable in the definite diagnosis of AD, especially for patients in early stage[7]. The final diagnosis of AD can be accomplished only by postmortem examination of brain tissues. As the accumulation of Aβ plaques and NFTs in AD brains occurs 20-40 years before the onset of clinical symptoms[8], detecting their changes *in vivo* may impressively facilitate early diagnosis and monitoring the progression of AD. Therefore, there is an urgent need for *in vivo* imaging of Aβ plaques with the assistance of noninvasive techniques such as positron emission tomography (PET) or single photon emission computed tomography (SPECT).

Over recent years, a big variety of PET probes based on Congo Red (CR) and Thioflavin-T

---

[*] Copartner: Yang Yanping, Cui Mengchao, Jin Bing, Wang Xuedan, Li Zijing, Yu Pingrong, Jia Jianhua, Fu Hualong, Jia Hongmei. Reprinted from *European Journal of Medicinal Chemistry*, 2013, 64: 90-98.

(ThT) have been synthesized and evaluated for imaging Aβ plaques. Among them, [^{11}C]-4-*N*-methylamino-4′-hydroxystilbene([^{11}C]SB-13)[9,10], [^{11}C]-2-(4-(methylamino)phenyl)-6-hydroxybenzothiazol([^{11}C]PIB)[11,12], [^{18}F]-4-(*N*-methylamino)-4′-(2-(2-(2-fluoroethoxy)ethoxy)ethoxy)-stilbene([^{18}F]BAY94-9172)[13-15], [^{18}F]-2-(3-fluoro-4-methyaminophenyl)benzothiazol-6-ol([^{18}F]GE-067)[16] and [^{18}F]-(*E*)-4-(2-(6-(2-(2-fluoroethoxy)ethoxy)ethoxy)pyridin-3-yl)vinyl)-*N*-methylaniline([^{18}F]AV-45)[17-19] have been evaluated clinically and demonstrated potential utilization. The imaging of Aβ plaques *in vivo* with PET has achieved remarkable progress so far, while the development of SPECT tracers lags far behind. When compared with PET, SPECT have several certain advantages including lower cost, widespread availability, and the use of isotopes with longer half-lives. Hence extraordinary efforts should be spent on the innovation of SPECT probes for Aβ imaging.

Among the radioisotopes used in diagnostic imaging nuclear medicine by SPECT, ^{99m}Tc($T_{1/2}$ = 6.01h, γ(89%):141keV) is the most widely used and can even be regarded as a star radioisotope due to the following aspects: it is readily produced by a commercial ^{99}Mo/^{99m}Tc generator, besides, it's moderate gamma-ray energy and physical half-life extremely fulfill the SPECT imaging requirements by providing good biological localization and appropriate residence time. Thus, new ^{99m}Tc-labeled Aβ imaging probes will provide simple, convenient, economical and widespread SPECT-based imaging methods for the early diagnosis of AD. In the past decade, a great deal of research has been conducted in this field. Initially reported ^{99m}Tc-labeled CR[20,21] and Chrysamine G derivatives[22,23], which displayed high affinity for Aβ aggregates *in vitro*, failed to cross the blood-brain barrier(BBB) for their high molecular weight and charged character. Then further attempts were focused on smaller, neutral, and more lipophilic ^{99m}Tc-labeled ligands. Derivatives of biphenyl[24], benzothiazole aniline[25-27], chalcone[28], flavone, aurone[29], curcumin[30], benzofuran[31,32] and benzoxazole[33] conjugated with monoamine-monoamide dithiol(MAMA) and bis(aminoethanethiol)(BAT) chelating ligands(Fig. 1) have been synthesized and evaluated as potential Aβ imaging probes, unfortunately, none of them showed specific binding to Aβ plaques *in vivo* due to either low affinity or low uptake in the brain.

Recently, based on the structure of curcumin, an extensive series of dibenzylideneacetone derivatives developed as a novel molecular scaffold for Aβ imaging were reported[34]. Structure-activity relationships revealed that the para position on the phenyl ring of dibenzylideneacetone showed extraordinarily high tolerance for steric bulk substitutions. Increasing the size of the substituent has negligible impact on the affinity for Aβ aggregates, and even when the ligand was substituted by a bulky trityloxy group also exhibited high affinity(K_i = (2.8 ± 0.4)nM). This encouraging finding inspires us that the introduction of a large ^{99m}Tc chelating structure probably won't bring about a considerable change in the binding properties.

In the present study, we designed four novel ^{99m}Tc-labeled dibenzylideneacetone derivatives (Fig. 2) conjugated to MAMA and BAT via a propoxy or pentyloxy spacer for Aβ plaque imaging. Hereon, we report the synthesis and preliminary biological evaluation of these novel ^{99m}Tc-labeled dibenzylideneacetone derivatives as potential SPECT probes for imaging Aβ plaques in AD.

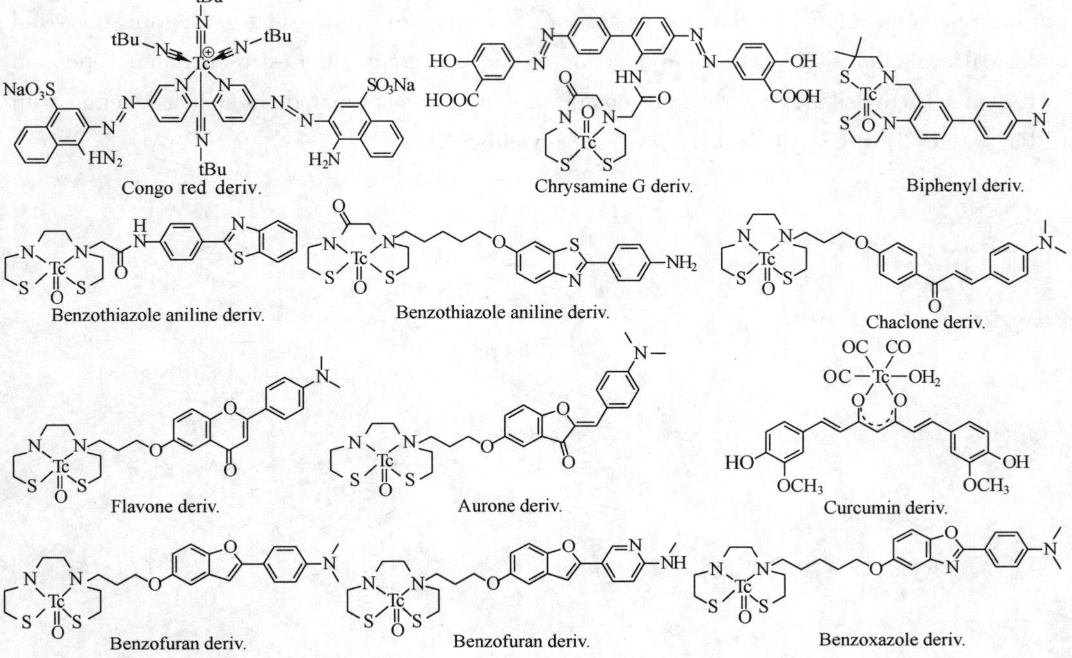

Fig. 1　Chemical structure of ^{99m}Tc-labeled Aβ imaging probes reported previously

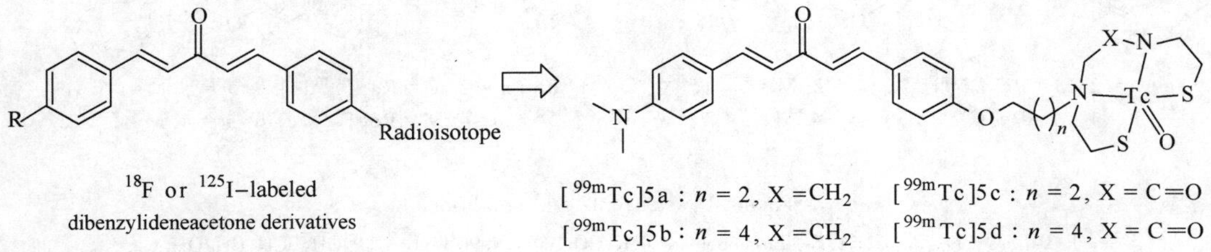

Fig. 2　Chemical structure of ^{99m}Tc-labeled dibenzylideneacetone derivatives reported in the present study

2　Results and Discussion

2.1　Chemistry

The synthesis of ^{99m}Tc/Re dibenzylideneacetone derivatives is outlined in Scheme 1. The key step in the formation of the dibenzylideneacetone skeleton was accomplished by base catalyzed Claisen condensation reaction starting from (E)-4-(4-(dimethylamino)phenyl)but-3-en-2-one and substituted benzaldehyde(2a or 2b). Claisen condensation afforded compound 3a and 3b in yields of 54.6% and 48.9%, respectively. Then, a BAT or MAMA chelating group was conjugated with bromo-aliphatic compounds 3a or 3b by reacting them with Tr-Boc-BAT or Tr-MA-MA, which achieved 4a-d in yields of 36.0%, 27.1%, 50.8% and 54.1%, respectively. The structural and biochemical properties of ^{99m}Tc correlate well with those of Re, thus Re complexes have been usually adopted as surrogates for ^{99m}Tc complexes for structure identifications and biological evaluations[35,36]. After deprotection of the thiol groups in 4a-d by trifluoroacetic acid (TFA) and triethylsilane, the rhenium complexes 5a-d were prepared through a reaction with $(PPh_3)_2ReOCl_3$ in CH_2Cl_2 in 15.5%-37.0% yields.

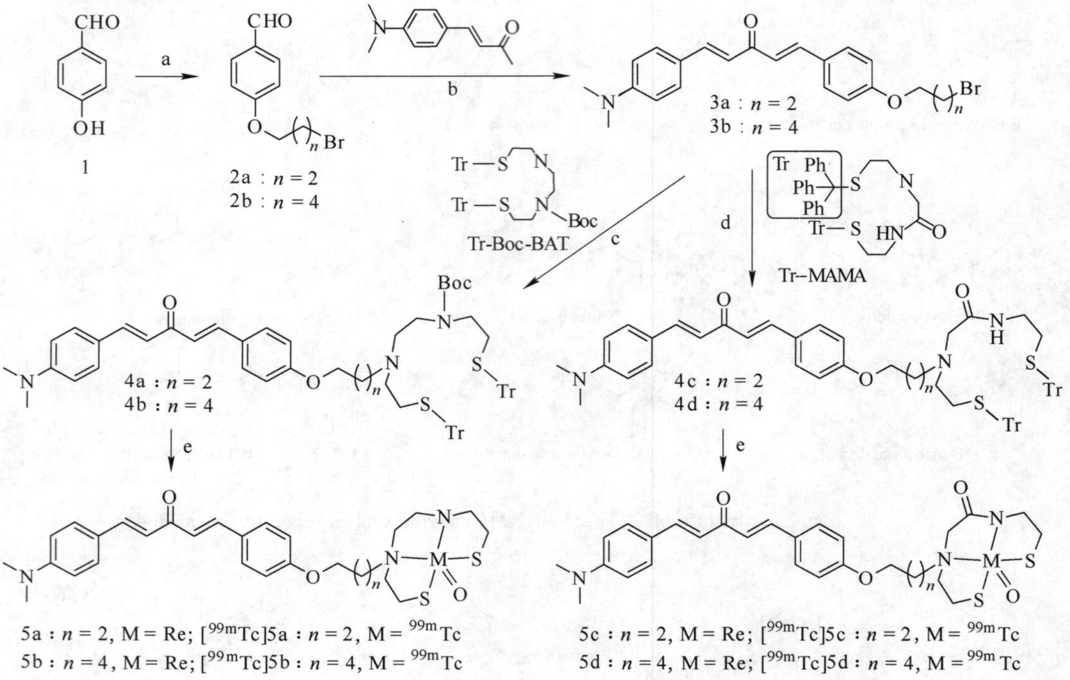

Scheme 1 Reagents and conditions

a—1,3-dibromopropane(or 1,5-dibromopentane), K_2CO_3, MeCN, 90℃; b—NaOCH$_3$, CH$_3$CH$_2$OH, r. t.;

c—Tr-Boc-BAT, K_2CO_3, MeCN, 90℃; d—Tr-MAMA, K_2CO_3, MeCN, 90℃;

e—(1)TFA, Et$_3$SiH; (2)^{99m}Tc-GH, 100℃; or(PPh$_3$)ReOCl$_3$, CH$_2$Cl$_2$, CH$_3$OH, 90℃

2.2 *In vitro* binding studies using Aβ$_{(1-42)}$ aggregates

The binding affinities of rhenium complexes 5a-d for Aβ$_{(1-42)}$ aggregates were evaluated by competition binding assay using [^{125}I]6-iodo-2-(4'-dimethylamino)-phenyl-imidazo[1,2-a]pyridine, [^{125}I]IMPY, as the competing radio-ligand according to conventional methods[34]. IMPY was also tested using the same system for comparison. As shown in Table 1, compounds with the MAMA cheating ligand, 5c and 5d, exhibited moderate binding affinity to Aβ$_{(1-42)}$ aggregates (K_i = 120.9 and 59.1nM, respectively), while compounds with the BAT cheating ligand, 5a and 5b, displayed high affinity(K_i = 24.7 and 13.6nM, respectively) which was comparable to the value of IMPY($K_i = 11.5$ nM), indicating that they hold sufficient affinity for $A\beta_{(1-42)}$ aggregates *in vitro*. This result demonstrates that the introduction of the bulky BAT chelating group didn't interfere with the binding to $A\beta_{(1-42)}$ aggregates observably. When considering the impact on affinity bring by the length of spacers, the result indicated that the affinity of compounds 5b and 5d, with longer pentyloxy spacer, was considerably higher than that of compounds with propoxy spacer. As expected, when the spacer getting longer, the interference of the large ^{99m}Tc chelating structure on the binding property of targeting molecule(dibenzylideneacetone) is smaller, and thus the affinity was maintained preferably.

Table 1 Inhibition constants for the binding of [^{125}I] IMPY to $A\beta_{(1-42)}$ aggregates

Compound	K_i/nM[①]
5a	24.7 ± 6.1
5b	13.6 ± 7.8
5c	120.9 ± 4.3
5d	59.1 ± 24.0
IMPY	11.5 ± 2.5

① Measured in triplicate with values given as the mean ± SD.

2.3 *In vitro* fluorescent staining

In vitro fluorescent staining of Aβ plaques on sections of brain tissue from a transgenic(Tg) model mice(C57BL6, APPswe/PSEN1, 11 months old) and an age-matched control mice were conducted to confirm the specific and high binding affinity of compounds 5a-d to Aβ plaques(Fig. 3). Fluorescent staining images revealed that these four rhenium complexes all displayed excellent labeling of Aβ plaques on the sections of Tg mice with low background (Fig. 3a, d, g and j), while there was no notable staining in the age-matched control mice (Fig. 3c, f, i and l). The distribution of Aβ plaques was consistent with the results of fluorescent staining with Thioflavin-S on the adjacent sections(Fig. 3b, e, h and k). The results incontestably manifested that these rhenium complexes could bind specifically to Aβ plaques with low background.

2.4 Radiolabeling

The novel radio-labeled ligands [^{99m}Tc]5a-d were prepared by a ligand exchange reaction employing the precursor 4a-d and ^{99m}Tc-glucoheptonate(^{99m}Tc-GH) in boiling water with radiochemical yields of 31.8%, 43.8%, 47.9% and 37.7%, respectively. After purification by high-performance liquid chromatography(HPLC), the radiochemical purity of these radiotracers was higher than 95%. The radiochemical identities of the ^{99m}Tc-labeled tracers were affirmed by comparison of the retention times on HPLC colum with that of the corresponding Re complexes

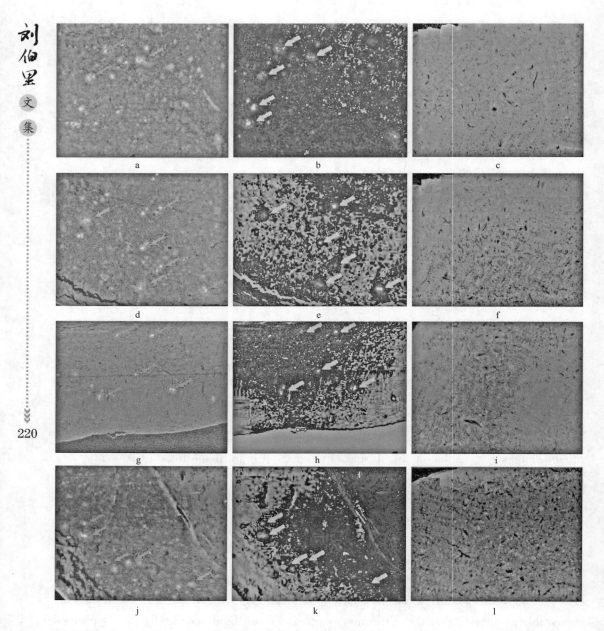

Fig. 3 *In vitro* fluorescent staining of Aβ plaques by compounds 5a-d on brain sections of a Tg model mouse (C57BL6, APPswe/PSEN1, 11 months old, male) with a DAPI filter (a, d, g and j), The presence and distribution of Aβ plaques on the adjacent sections were confirmed with Thioflavin-S with an AF filter (b, e, h and k) and fluorescence staining of compounds 5a-d on brain sections of a normal mouse was also carried out as control (c, f, i and l)

(Table 2, see supporting information). The retention times of [^{99m}Tc] 5a-d on HPLC column were 10.45min, 10.88min, 15.18min and 15.36min, and that of the corresponding rhenium complexes 5a-d were 8.97min, 9.29min 13.26min and 13.43min, respectively. The approximation of retention times between ^{99m}Tc-labeled tracers and corresponding rhenium complexes sufficiently suggested that the desired ^{99m}Tc-labeled dibenzylideneacetone derivatives were successfully generated.

2.5 Partition coefficient determination

The lg$D$ values of [^{99m}Tc]5a-d were 3.42 ± 0.02, 3.17 ± 0.04, 3.53 ± 0.01, and 3.57 ± 0.17, respectively (Table 2), which are all in an optimal range (1-3.5) for BBB penetration, indicating that the four tracers have appropriate lipophilicity suitable for brain imaging.

Table 2 HPLC retention times of ^{99m}Tc-labeled dibenzylideneacetone derivatives and corresponding Re complexes and lg$D$ values of [^{99m}Tc]5a-d

^{99m}Tc compound	Retention time/min	Re compound	Retention time/min	lg$D$ of ^{99m}Tc compound [3]
[^{99m}Tc]5a[1]	10.45	5a	8.97	3.36 ± 0.08
[^{99m}Tc]5b[1]	10.88	5b	9.29	3.17 ± 0.10
[^{99m}Tc]5c[2]	15.18	5c	13.26	3.52 ± 0.10
[^{99m}Tc]5d[2]	15.36	5d	13.43	3.57 ± 0.15

[1] Reversed-phase HPLC using a binary gradient system (acetonitrile/water: 80%/20%) at a 1.0 mL/min flow rate.
[2] Reversed-phase HPLC using a binary gradient system (acetonitrile/water: 70%/30%) at a 1.0 mL/min flow rate.
[3] Measured in triplicate with values expressed as the mean ± SD.

2.6 *In vivo* biodistribution in normal mice

Biodistribution experiments in normal ICR mice (5 weeks, male) were carried out to evaluate the pharmacokinetics of ^{99m}Tc-labeled dibenzylideneacetone tracers *in vivo* (Table 3). A biodistribution study provides crucial information on penetration of the BBB and radioactivity pharmacokinetics *in vivo*. [^{99m}Tc]5a-d all showed the highest initial uptake at 2min post-injection, 0.49% ID/g, 0.47% ID/g, 0.48% ID/g and 0.31% ID/g, respectively. Although the values were comparable or slightly lower than that of ^{99m}Tc-labeled tracers reported previously ((0.2-1.80)% ID/g)[24-26,28-32], [^{99m}Tc]5a and [^{99m}Tc]5b can also expect excellent labeling of Aβ plaques *in vivo* due to their high affinities. The brain$_{2min}$/brain$_{60min}$ ratio of [^{99m}Tc]5a-c were 6.13, 3.92 and 5.33, respectively, which were superior than that of ^{99m}Tc-labeled phenylbenzothiazole (2.06)[26], phenylbenzofuran (2.39)[31], pyridylbenzofuran (2.28)[32], phenylbenzoxazole (3.24)[33] and so on. This data undoubtedly indicated that all the four tracers possessed good clearance property from the normal brain. The accumulation of radioactivity was observed predominantly in liver and kidney at the early stage, and then dropped gradually with time, while continuous gastrointestinal accumulation of the radiotracers resulted in a high intestine uptake at later time periods.

Table 3 Biodistribution of radioactivity after injection of [^{99m}Tc]5a-d in normal ICR mice[①]

Organ	2min	10min	30min	60min
[^{99m}Tc]5a(lg$D$ = 3.36 ± 0.08)				
Blood	6.36 ± 1.19	1.61 ± 0.18	0.71 ± 0.09	0.61 ± 0.07
Brain	0.49 ± 0.08	0.23 ± 0.04	0.12 ± 0.03	0.08 ± 0.01
Heart	10.23 ± 1.20	2.43 ± 0.20	0.86 ± 0.15	0.53 ± 0.06
Liver	26.69 ± 4.08	26.22 ± 0.91	17.57 ± 1.42	12.58 ± 3.90
Spleen	3.03 ± 0.61	1.67 ± 0.37	0.95 ± 0.23	0.55 ± 0.14
Lung	17.66 ± 3.05	6.05 ± 1.82	1.44 ± 0.56	0.89 ± 0.14
Kidney	21.45 ± 2.66	13.2 ± 2.81	6.96 ± 1.76	5.97 ± 1.16
Pancreas	3.78 ± 0.66	2.74 ± 0.61	1.09 ± 0.37	0.52 ± 0.11
Stomach[②]	1.10 ± 0.18	1.64 ± 0.54	2.55 ± 0.78	1.53 ± 0.13
Intestine[②]	5.77 ± 0.72	12.02 ± 1.41	25.04 ± 4.70	54.01 ± 6.54
[^{99m}Tc]5b(lg$D$ = 3.17 ± 0.10)				
Blood	7.21 ± 1.38	1.25 ± 0.08	0.80 ± 0.05	0.52 ± 0.05
Brain	0.47 ± 0.11	0.27 ± 0.03	0.19 ± 0.03	0.12 ± 0.02
Heart	29.79 ± 5.78	11.72 ± 1.75	10.12 ± 3.29	6.63 ± 1.32
Liver	77.67 ± 18.54	56.22 ± 9.89	52.99 ± 4.78	43.84 ± 6.36
Spleen	16.64 ± 2.62	9.08 ± 1.25	5.68 ± 1.53	4.06 ± 0.71
Lung	11.87 ± 2.56	3.84 ± 1.09	2.42 ± 0.57	1.75 ± 0.16
Kidney	12.33 ± 3.20	7.93 ± 1.24	5.02 ± 0.27	3.41 ± 0.48
Pancreas	4.21 ± 0.76	4.80 ± 0.82	2.74 ± 0.87	1.44 ± 0.21
Stomach[②]	1.56 ± 0.75	3.72 ± 0.79	5.09 ± 2.15	1.57 ± 0.60
Intestine[②]	5.97 ± 1.22	25.67 ± 6.08	54.31 ± 6.14	60.14 ± 16.90
[^{99m}Tc]5c(lg$D$ = 3.52 ± 0.10)				
Blood	4.87 ± 0.76	1.30 ± 0.12	0.60 ± 0.05	0.42 ± 0.02
Brain	0.48 ± 0.06	0.19 ± 0.02	0.10 ± 0.02	0.09 ± 0.01
Heart	14.24 ± 1.21	3.89 ± 0.18	1.33 ± 0.28	0.90 ± 0.12
Liver	23.67 ± 2.24	21.38 ± 3.39	18.86 ± 0.71	16.61 ± 3.98
Spleen	4.40 ± 0.30	2.19 ± 0.26	0.94 ± 0.09	0.52 ± 0.11
Lung	24.52 ± 3.50	9.24 ± 2.20	3.06 ± 0.43	1.15 ± 0.31
Kidney	16.51 ± 0.66	9.03 ± 1.66	4.91 ± 0.76	3.86 ± 0.46
Pancreas	4.28 ± 0.30	3.46 ± 0.52	2.11 ± 0.34	1.24 ± 0.30
Stomach[②]	1.14 ± 0.09	1.40 ± 0.12	1.12 ± 0.05	1.68 ± 0.51
Intestine[②]	5.73 ± 0.62	12.09 ± 0.70	30.47 ± 6.84	39.15 ± 7.12
[^{99m}Tc]5d(lg$D$ = 3.57 ± 0.15)				
Blood	5.10 ± 0.37	2.32 ± 0.54	1.29 ± 0.15	1.21 ± 0.24
Brain	0.31 ± 0.06	0.22 ± 0.02	0.11 ± 0.04	0.15 ± 0.02

Continued 3

Organ	2min	10min	30min	60min
[^{99m}Tc]5d($\lg D = 3.57 \pm 0.15$)				
Heart	14.56 ± 0.67	4.17 ± 0.33	1.75 ± 0.41	1.14 ± 0.19
Liver	42.12 ± 9.68	43.73 ± 4.71	29.40 ± 4.31	25.97 ± 5.95
Spleen	5.97 ± 1.76	2.63 ± 0.24	1.67 ± 0.70	1.35 ± 0.49
Lung	12.03 ± 2.92	4.20 ± 0.49	1.64 ± 0.28	1.24 ± 0.22
Kidney	16.85 ± 1.16	12.07 ± 2.72	6.45 ± 1.41	4.85 ± 0.94
Pancreas	4.90 ± 0.69	5.50 ± 1.25	3.44 ± 0.91	1.93 ± 0.47
Stomach[②]	1.34 ± 0.13	2.05 ± 0.39	2.82 ± 0.72	6.42 ± 1.10
Intestine[②]	5.20 ± 0.58	21.46 ± 2.27	35.03 ± 5.70	54.66 ± 11.20

① Expressed as % injected dose per gram. Each value represents the mean ± SD for 4-5 mice.
② Expressed as % injected dose per organ.

3 Conclusion

In conclusion, four ^{99m}Tc-labeled dibenzylideneacetone derivatives and their corresponding rhenium complexes were successfully synthesized and biologically evaluated as SPECT imaging probes for Aβ plaques. *In vitro* binding studies indicated that rhenium complexes 5a-d displayed moderate to high affinity for Aβ$_{(1-42)}$ aggregates varied from 13.6 to 120.9nM. The high binding affinity was further confirmed by fluorescent staining, and all the dibenzylideneacetone derivatives specifically labeled Aβ plaques on the brain sections of Tg mice. Biodistribution in normal mice revealed that [^{99m}Tc]5a exhibited moderate initial uptake (0.49% ID/g at 2min) and fast washout from the brain, with a brain$_{2min}$/brain$_{60min}$ ratio of 6.13. All in all, the preliminary results suggest that the [^{99m}Tc]5a may be a potential SPECT probe for imaging Aβ plaques in Alzheimer's brains. But additional chemical refinements are earnestly required to further improve the penetration through BBB.

4 Experimental

4.1 General information

All reagents used for chemical synthesis were commercial products and were used without further purification. Na^{99m}TcO$_4$ was obtained from a commercial ^{99}Mo/^{99m}Tc generator, Beijing Atomic High-tech Co. All ^{1}H NMR spectra were acquired at 400MHz on Bruker Avance III NMR spectrometers in CDCl$_3$ solutions at room temperature with trimethylsilyl (TMS) as an internal standard. Chemical shifts were reported as δ values relative to the internal TMS, and Coupling constants were reported in hertz. The multiplicity is defined by s(singlet), d(doublet), t(triplet), and m(multiplet). Mass spectra were acquired with a micrOTOF-Q II instrument. Reactions were monitored by TLC (Silica gel 60 F$_{254}$ aluminum sheets, Merck) and compounds were visualized by illumination with a short wavelength Ultra-Violet lamp (λ = 254nm or

365nm). Column chromatography purification were performed on silica gel (54-74μm) from Qingdao Haiyang Chemical Co., Ltd. HPLC was performed on a Shimadzu SCL-20 AVP(which is equipped with a Bioscan Flow Count 3200 NaI/PMT γ-radiation scintillation detector and a SPD-20A UV detector, λ = 254nm) and a Venusil MP C18 reverse phase column(Agela Technologies, 5μm, ID = 4.6mm, length = 250mm) eluted with a binary gradient system at 1.0mL/min flow rate. Mobile phase A was water, while mobile phase B was acetonitrile. Fluorescent observation was performed by the Axio Observer Z1 inverted fluorescence microscope(Zeiss, Germany) equipped with a DAPI filter set(excitation, 405nm) and AF488 filter set(excitation, 495nm). Normal ICR mice (five weeks, male) were used for biodistribution experiments. Transgenic mice(C57BL6, APPswe/PSEN1, male, 11 months old), used as an Alzheimer's model, were purchased from the Institute of Laboratory Animal Science, Chinese Academy of Medical Sciences. All protocols requiring the use of mice were approved by the animal care committee of Beijing Normal University. The purity of all key compounds was proven to be more than 95% by HPLC.

4.2 4-(3-bromopropoxy)benzaldehyde(2a)

To a solution of 1(3.67g, 30mmol) in CH_3CN(50mL) was added 1,3-dibromopropane(9.10g, 45mmol) and K_2CO_3(8.29g, 60mmol). The mixture was stirred at 90℃ for 12h, after cooling to room temperature, solvent was removed by vacuum, and the residue was extracted with $CHCl_3$. The organic layer was dried over $MgSO_4$ and solvent was removed. The crude product was purified by silica gel chromatography (petroleum ether/AcOEt = 8/1, v/v) to give 2a (4.41g, 60.4%) as a colorless oil. ^{1}H NMR(400MHz, $CDCl_3$) δ 9.89(s, 1H, CHO), 7.84(d, J = 8.7Hz, 2H, ArH), 7.01(d, J = 8.7Hz, 2H, ArH), 4.20(t, J = 5.8Hz, 2H, CH_2), 3.61(t, J = 6.3Hz, 2H, CH_2), 2.39-2.33(m, 2H, CH_2). HRMS(ESI): m/z calcd for $C_{10}H_{12}BrO_2$ 245.0000; found 245.0091 $[M + H]^+$.

4.3 4-(5-bromopentyloxy)benzaldehyde(2b)

The procedure described above for the preparation of 2a was employed to give 2b(6.61g, 81.2%) as a colorless oil. ^{1}H NMR(400MHz, $CDCl_3$) δ 9.88(s, 1H, CHO), 7.83(d, J = 8.8Hz, 2H, ArH), 6.99(d, J = 8.7Hz, 2H, ArH), 4.06(t, J = 6.3Hz, 2H, CH_2), 3.45(t, J = 6.7Hz, 2H, CH_2), 2.00-1.91(m, 2H, CH_2), 1.90-1.82(m, 2H, CH_2), 1.70-1.60(m, 2H, CH_2). HRMS(ESI): m/z calcd for $C_{12}H_{16}BrO_2$ 271.0334; found 271.0438 $[M + H]^+$.

4.4 (1E,4E)-1-(4-(3-bromopropoxy)phenyl)-5-(4-(dimethylamino)phenyl)penta-1,4-dien-3-one(3a)

A mixture of 2a(3.02g, 12.4mmol), (E)-4-(4-(dimethylamino)phenyl)but-3-en-2-one (2.35g, 12.4mmol) and sodium methoxide(0.95g, 25.0mmol) in ethanol(50mL) was stirred at room temperature overnight. The precipitate was collected by filtration and then recrystallized from ethanol to give 3a(2.81g, 54.6%) as a yellow solid. ^{1}H NMR(400MHz, $CDCl_3$) δ 7.70

(d, J = 15.7Hz, 1H, *trans*—CH =CH—), 7.67 (d, J = 15.8Hz, 1H, *trans*—CH =CH—), 7.56(d, J = 8.7Hz, 2H, ArH), 7.52 (d, J = 8.8Hz, 2H, ArH), 6.97 (d, J = 15.8Hz, 1H, *trans*—CH =CH—), 6.93(d, J = 8.7Hz, 2H, ArH), 6.87 (d, J = 15.7Hz, 1H, *trans*—CH =CH—), 6.69 (d, J = 8.8Hz, 2H, ArH), 4.15 (t, J = 5.8Hz, 2H, CH_2), 3.61 (t, J = 6.4Hz, 2H, CH_2), 3.04(s, 6H, N(CH_3)$_2$), 2.37-2.31 (m, J = 6.1Hz, 2H, CH_2). HRMS(ESI): m/z calcd for $C_{22}H_{25}BrNO_2$ 414.1069; found 414.0955 [M + H]$^+$.

4.5 (1E,4E)-1-(4-(5-bromopentyloxy) phenyl)-5-(4-(dimethylamino) phenyl) penta-1,4-dien-3-one(3b)

The procedure described above for the preparation of 3a was employed to give 3b (2.22g, 48.9%) as a yellow solid. ^{1}H NMR(400MHz, CDCl$_3$) δ 7.70 (d, J = 15.7Hz, 1H, *trans*—CH =CH—), 7.67 (d, J = 15.8Hz, 1H, *trans*—CH =CH—), 7.55 (d, J = 8.7Hz, 2H, ArH), 7.51(d, J = 8.8Hz, 2H, ArH), 6.96 (d, J = 15.8Hz, 1H, *trans*—CH =CH—), 6.90 (d, J = 8.7Hz, 2H, ArH), 6.87 (d, J = 15.8Hz, 1H, *trans*—CH =CH—), 6.69 (d, J = 8.8Hz, 2H, ArH), 4.01 (t, J = 6.3Hz, 2H, CH_2), 3.44 (t, J = 6.7Hz, 2H, CH_2), 3.04 (s, 6H, N(CH_3)$_2$), 1.99-1.90 (m, 2H, CH_2), 1.88-1.79 (m, 2H, CH_2), 1.64 (m, 2H, CH_2). HRMS(ESI): m/z calcd for $C_{24}H_{29}BrNO_2$ 444.1361; found 444.1333 [M + H]$^+$.

4.6 Tert-butyl 2-((3-(4-((1E,4E)-5-(4-(dimethylamino) phenyl)-3-oxopenta-1,4-dienyl) phenoxy) propyl)(2-(tritylthio) ethyl) amino) ethyl (2-(tritylthio) ethyl) carbamate(4a)

A mixture of 3a (294.2mg, 0.71mmol), tert-butyl 2-(tritylthio) ethyl (2-(2-(tritylthio) ethylamino) ethyl) carbamate (539.2mg, 0.71mmol) and K_2CO_3 (193.6mg, 1.42mmol) in acetonitrile(50mL) was stirred at 90℃ for 48h. The solvent was removed by vacuum, and water was added. The mixture was extracted with CHCl$_3$. The organic layers were combined and dried over anhydrous Mg_2SO_4 and evaporated dry. The crude product was purified by silica gel chromatography(petroleum ether/AcOEt = 3/1, v/v) to give 4a (272.5mg, 36.0% yield) as yellow solid foams. ^{1}H NMR(400MHz, CDCl$_3$) δ 7.70 (d, J = 15.6Hz, 1H, *trans*—CH =CH—), 7.66 (d, J = 15.6Hz, 1H, *trans*—CH =CH—), 7.52 (d, J = 8.8Hz, 2H, ArH), 7.48 (d, J = 8.6Hz, 2H, ArH), 7.42-7.36 (m, 13H, ArH), 7.24-7.17 (m, 17H, ArH), 6.95 (d, J = 15.8Hz, 1H, *trans*—CH =CH—), 6.87 (d, J = 15.6Hz, 1H, *trans*—CH =CH—), 6.84 (d, J = 8.3Hz, 2H, ArH), 6.70 (d, J = 8.8Hz, 2H, ArH), 3.93 (t, J = 6.2Hz, 2H, CH_2), 3.04 (s, 6H, N(CH_3)$_2$), 2.99-2.89 (m, 4H, CH_2), 2.38-2.21 (m, 8H, CH_2), 1.71 (s, 2H, CH_2), 1.36 (s, 11H, CH_2 and tert-Bu). ^{13}C NMR (101MHz, CDCl$_3$) δ 188.84, 160.89, 155.01, 151.94, 145.01, 144.83, 143.73, 141.96, 130.24, 129.95, 129.64, 129.59, 127.91, 127.86, 126.70, 126.60, 123.63, 122.70, 121.11, 114.88, 111.90, 79.55, 77.38, 77.06, 76.74, 66.65, 53.40, 50.16, 47.21, 40.14, 30.05, 28.44, 27.22. HRMS(ESI): m/z calcd for $C_{71}H_{76}N_3O_4S_2$ 1098.5277; found 1098.5263 [M + H]$^+$.

4.7 Tert-butyl 2-((5-(4-((1E,4E)-5-(4-(dimethylamino)phenyl)-3-oxopenta-1,4-dienyl)phenoxy)pentyl)(2-(tritylthio)ethyl)amino)ethyl(2-(tritylthio)ethyl)carbamate(4b)

The procedure described above for the preparation of 4a was employed to give 4b(211.0mg, 27.1%) as yellow solid foams. ^{1}H NMR(400MHz, CDCl$_3$)δ 7.70(d, J = 15.8Hz, 1H, trans—CH=CH—), 7.68(d, J = 15.8Hz, 1H, trans—CH=CH—), 7.54(d, J = 7.7Hz, 2H, ArH), 7.52(d, J = 8.4Hz, 2H, ArH), 7.41-7.39(m, 13H, ArH), 7.24-7.15(m, 17H, ArH), 6.96(d, J = 15.8Hz, 1H, trans—CH=CH—), 6.89(d, J = 8.6Hz, 2H), 6.87(d, J = 15.9Hz, 1H, trans—CH=CH—), 6.69(d, J = 8.8Hz, 2H, ArH), 3.95(t, J = 6.4Hz, 2H, CH$_2$), 3.04(s, 6H, N(CH$_3$)$_2$), 2.98-2.79(m, 4H, CH$_2$), 2.39-2.17(m, 10H, CH$_2$), 1.77-1.68(m, 2H, CH$_2$), 1.36(s, 13H, CH$_2$ and tert-Bu). ^{13}C NMR(101MHz, CDCl$_3$)δ 188.81, 160.93, 155.01, 151.95, 145.04, 144.86, 143.73, 141.89, 130.23, 129.96, 129.67, 127.91, 127.85, 126.69, 126.61, 123.69, 122.72, 121.11, 114.89, 111.91, 79.47, 77.39, 77.07, 76.75, 68.04, 66.68, 53.38, 47.09, 40.13, 29.94, 29.07, 28.46, 27.12, 23.72. HRMS(ESI):m/z calcd for C$_{73}$H$_{80}$N$_3$O$_4$S$_2$ 1126.5590;found 1126.5538 [M+H]$^+$.

4.8 2-((3-(4-((1E,4E)-5-(4-(dimethylamino)phenyl)-3-oxopenta-1,4-dienyl)phenoxy)propyl)(2-(tritylthio)ethyl)amino)-N-(2-(tritylthio)ethyl)acetamide(4c)

A mixture of 3b(414.3mg, 1.0mmol), N-(2-(tritylthio)ethyl)-2-(2-(tritylthio)ethylamino)acetamide(622.8mg, 1.0mmol) and K$_2$CO$_3$(276.4mg, 2.0mmol) in acetonitrile(50mL) was stirred at 90℃ for 48h. The solvent was removed by vacuum, and water was added. The mixture was extracted with CHCl$_3$. The organic layers were combined and dried over Na$_2$SO$_4$ and evaporated dry. The crude product was purified by silica gel chromatography(petroleum ether/AcOEt = 3/1, v/v) to give 4c(502.3mg, 50.8%) as yellow solid foams. ^{1}H NMR(400MHz, CDCl$_3$) δ 7.71(d, J = 15.8Hz, 1H, trans—CH=CH—), 7.66(d, J = 15.8Hz, 1H, trans—CH=CH—), 7.52(d, J = 8.8Hz, 2H, ArH), 7.45(d, J = 8.5Hz, 2H, ArH), 7.38-7.33(m, 13H, ArH), 7.24-7.18(m, 17H, ArH), 6.94(d, J = 15.8Hz, 1H, trans—CH=CH—), 6.88(d, J = 15.7Hz, 1H, trans—CH=CH—), 6.82(d, J = 8.7Hz, 2H, ArH), 6.69(d, J = 8.9Hz, 2H, ArH), 3.95(t, J = 5.9Hz, 2H, CH$_2$), 3.04(s, 6H, N(CH$_3$)$_2$), 3.03-2.98(m, 2H, CH$_2$), 2.90(s, 2H, CH$_2$), 2.50(t, J = 6.9Hz, 2H, CH$_2$), 2.40(t, J = 6.5Hz, 2H, CH$_2$), 2.35(t, J = 6.3Hz, 2H, CH$_2$), 2.28(t, J = 6.5Hz, 2H, CH$_2$), 1.84-1.77(m, 2H, CH$_2$). ^{13}C NMR(101MHz, CDCl$_3$)δ 188.80, 170.96, 160.60, 151.96, 144.74, 143.81, 141.84, 130.26, 129.96, 129.56, 129.55, 127.94, 127.89, 126.74, 123.77, 122.66, 121.08, 114.85, 111.91, 77.41, 77.10, 76.78, 66.88, 66.78, 65.71, 58.47, 54.10, 51.19, 40.13, 38.06, 32.11, 30.09, 26.97. HRMS(ESI):m/z calcd for C$_{66}$H$_{66}$N$_3$O$_3$S$_2$ 1012.4546;found 1012.4502 [M+H]$^+$.

4.9 2-((5-(4-((1E,4E)-5-(4-(dimethylamino)phenyl)-3-oxopenta-1,4-dienyl)phenoxy)pentyl)(2-(tritylthio)ethyl)amino)-N-(2-(tritylthio)ethyl)acetamide(4d)

The procedure described above for the preparation of 4c was employed to give 4d(549.6mg,

54.1%) as yellow solid foams. ^{1}H NMR(400MHz, CDCl$_3$)δ 7.70(d, J = 15.7Hz, 1H, trans—CH=CH—), 7.67 (d, J = 15.8Hz, 1H, trans—CH=CH—), 7.52 (d, J = 8.7Hz, 2H, ArH), 7.52(d, J = 8.9Hz, 2H, ArH), 6.96(d, J = 15.8Hz, 1H, trans—CH=CH—), 6.87 (d, J = 15.6Hz, 1H, trans—CH=CH—), 6.85 (d, J = 8.6Hz, 2H, ArH), 6.69 (d, J = 8.8Hz, 2H, ArH), 3.89 (t, J = 6.3Hz, 2H, CH$_2$), 3.04 (s, 6H, N(CH$_3$)$_2$), 3.01 (m, 2H, CH$_2$), 2.85 (s, 2H, CH$_2$), 2.44-2.34 (m, 4H, CH$_2$), 2.33-2.22 (m, 4H, CH$_2$), 1.74-1.66 (m, 2H, CH$_2$), 1.38(s, 4H, CH$_2$). ^{13}C NMR(101MHz, CDCl$_3$)δ 188.82, 171.19, 160.82, 151.95, 144.77, 144.74, 143.79, 141.87, 130.25, 129.97, 129.56, 127.93, 127.77, 126.72, 123.71, 122.67, 121.07, 114.86, 111.90, 77.40, 77.08, 76.76, 67.86, 66.81, 66.74, 58.33, 54.72, 53.95, 40.14, 37.99, 32.07, 30.06, 29.01, 26.97, 23.79. HRMS(ESI): m/z calcd for C$_{68}$H$_{70}$N$_3$O$_3$S$_2$ 1040.4859; found 1040.4801 [M + H]$^+$.

4.10 (1E, 4E)-1-(4-(dimethylamino)phenyl)-5-(4-(3-((2-mercaptoethyl)(2-(2-mercaptoethylamino)ethyl)amino)propoxy)phenyl)penta-1,4-dien-3-one-rhenium(V)oxide(5a)

A solution of 4a(157.8mg, 0.15mmol) in TFA(2mL) was stirred at room temperature for 5min. Triethylsilane(60μL) was added and the mixture was stirred at 0℃ for 10 min, then the solvent was removed in vacuum. The residue was redissolved in a mixed solvent (10mL, CH$_2$Cl$_2$/CH$_3$OH = 10/1, v/v), (Ph$_3$P)$_2$ReOCl$_3$ (122.3mg, 0.15mmol) and sodium acetate in methanol(1M, 2mL) were added. The reaction mixture was heated to reflux for 4h. After cooling to room temperature, the solvent was evaporated. Then water was added and the resulting mixture was extracted by CH$_2$Cl$_2$(3 × 15mL), the combined organic layer was dried over anhydrous MgSO$_4$. After the solvent was removed, the residue was purified by silica gel chromatography (CH$_2$Cl$_2$/CH$_3$OH = 400/1, v/v) to give 5a(28.9mg, 27.6%) as an orange solid. m.p. 122.8-123.4℃. ^{1}H NMR(400MHz, CDCl$_3$)δ 7.71(d, J = 15.3Hz, 1H, trans—CH=CH—), 7.67 (d, J = 15.4Hz, 1H, trans—CH=CH—), 7.57 (d, J = 8.6Hz, 2H, ArH), 7.52 (d, J = 8.7Hz, 2H, ArH), 6.98(d, J = 15.8Hz, 1H, trans—CH=CH—), 6.90(d, J = 8.9Hz, 2H, ArH), 6.87(d, J = 16.6Hz, 1H, trans—CH=CH—), 6.69(d, J = 8.7Hz, 2H, ArH), 4.32-4.21(m, 1H, CH$_2$), 4.16(dd, J = 10.4, 6.2Hz, 2H, CH$_2$), 4.10(s, 2H, CH$_2$), 3.97-3.86 (m, 1H, CH$_2$), 3.85-3.74 (m, 2H, CH$_2$), 3.45-3.35 (m, 2H, CH$_2$), 3.32-3.24 (m, 1H, CH$_2$), 3.04(s, 6H, N(CH$_3$)$_2$), 2.83-2.74 (m, 1H, CH$_2$), 2.35-2.31 (m, 2H, CH$_2$), 1.79 (td, J = 11.9, 4.6Hz, 2H, CH$_2$), 0.90-0.78 (m, 2H, CH$_2$). ^{13}C NMR(101MHz, CDCl$_3$)δ 188.79, 159.79, 151.98, 143.93, 141.43, 130.29, 130.02, 128.57, 124.18, 122.66, 121.03, 114.80, 111.92, 77.31, 77.00, 76.68, 70.89, 65.48, 64.95, 62.67, 60.62, 59.14, 49.40, 40.11, 29.65, 23.41. HRMS (ESI): m/z calcd for C$_{28}$H$_{37}$N$_3$O$_3$ReS$_2$ 714.1834; found 714.2107 [M + H]$^+$.

4.11 (1E, 4E)-1-(4-(dimethylamino)phenyl)-5-(4-(5-((2-mercaptoethyl)(2-(2-mercaptoethylamino)ethyl)amino)pentyloxy)phenyl)penta-1,4-dien-3-one-rhenium(V)oxide(5b)

The procedure described above for the preparation of 5a was employed to give 5b(30.1mg,

29.6%) as an orange solid. m. p. 154.5-155.3℃. ^{1}H NMR(400MHz, CDCl$_3$) δ 7.70(d, J = 15.7Hz, 1H, trans—CH =CH—), 7.67(d, J = 15.8Hz, 1H, trans—CH =CH—), 7.56(d, J = 8.6Hz, 2H, ArH), 7.51(d, J = 8.8Hz, 2H, ArH), 6.97(d, J = 15.8Hz, 1H, trans—CH =CH—), 6.90(d, J = 8.7Hz, 2H, ArH), 6.87(d, J = 15.8Hz, 1H, trans—CH =CH—), 6.69(d, J = 8.8Hz, 2H, ArH), 4.18-4.09(m, 2H, CH$_2$), 4.02(t, J = 6.1Hz, 2H, CH$_2$), 3.89-3.82(m, 1H, CH$_2$), 3.77(dd, J = 11.2, 5.2Hz, 1H, CH$_2$), 3.61-3.51(m, 1H, CH$_2$), 3.41-3.31(m, 3H, CH$_2$), 3.29-3.18(m, 1H, CH$_2$), 3.04(s, 6H, N(CH$_3$)$_2$), 3.00-2.93(m, 2H, CH$_2$), 2.73(dd, J = 13.4, 3.1Hz, 1H, CH$_2$), 1.92-1.82(m, 4H, CH$_2$), 1.76-1.65(m, 2H, CH$_2$), 1.63-1.51(m, 2H, CH$_2$). ^{13}C NMR(101MHz, CDCl$_3$) δ 188.78, 160.58, 151.96, 143.85, 141.67, 130.24, 129.97, 128.02, 123.86, 122.64, 121.01, 114.84, 111.89, 77.33, 77.01, 76.69, 70.92, 67.53, 65.48, 63.60, 62.75, 58.89, 49.14, 40.12, 40.00, 28.83, 23.76, 22.83. HRMS(ESI): m/z calcd for C$_{30}$H$_{41}$N$_3$O$_3$ReS$_2$ 742.2147; found 742.2329 [M + H]$^+$.

4.12 2-((3-(4-((1E,4E)-5-(4-(dimethylamino)phenyl)-3-oxopenta-1,4-dienyl)phenoxy)propyl)(2-mercaptoethyl)amino)-N-(2-mercaptoethyl)acetamide-rhenium(Ⅴ)oxide(5c)

The procedure described above for the preparation of 5a was employed to give 5c(57.4mg, 15.5%) as an orange solid. m. p. 116.3-117.5℃. ^{1}H NMR(400MHz, CDCl$_3$) δ 7.71(d, J = 16.2Hz, 1H, trans—CH =CH—), 7.67(d, J = 16.6Hz, 1H, trans—CH =CH—), 7.57(d, J = 8.7Hz, 2H, ArH), 7.52(d, J = 8.8Hz, 2H, ArH), 6.98(d, J = 15.8Hz, 1H, trans—CH =CH—), 6.91(d, J = 8.7Hz, 2H, ArH), 6.87(d, J = 15.8Hz, 1H, trans—CH =CH—), 6.70(d, J = 8.7Hz, 2H, ArH), 4.73(d, J = 16.4Hz, 1H, CH$_2$), 4.60(dd, J = 11.7, 5.5Hz, 1H, CH$_2$), 4.15(d, J = 16.9Hz, 2H, CH$_2$), 4.12-4.07(m, 2H, CH$_2$), 3.85-3.75(m, 1H, CH$_2$), 3.48-3.40(m, 1H, CH$_2$), 3.32-3.24(m, 2H, CH$_2$), 3.23-3.19(m, 1H, CH$_2$), 3.05(s, 6H, N(CH$_3$)$_2$), 2.93-2.89(m, 1H, CH$_2$), 2.50-2.41(m, 1H, CH$_2$), 2.37-2.24(m, 3H, CH$_2$). ^{13}C NMR(101MHz, CDCl$_3$) δ 188.67, 186.86, 159.57, 155.59, 151.94, 143.95, 141.30, 130.24, 129.99, 128.53, 125.00, 124.30, 121.00, 114.77, 111.92, 77.31, 76.99, 76.68, 66.83, 64.54, 60.77, 59.88, 48.07, 40.15, 39.03, 32.20, 29.67, 26.40, 24.13, 23.41. HRMS(ESI): m/z calcd for C$_{28}$H$_{35}$N$_3$O$_4$ReS$_2$ 728.1626; found 728.1600 [M + H]$^+$.

4.13 2-((5-(4-((1E,4E)-5-(4-(dimethylamino)phenyl)-3-oxopenta-1,4-dienyl)phenoxy)pentyl)(2-mercaptoethyl)amino)-N-(2-mercaptoethyl)acetamide-rhenium(Ⅴ)oxide(5d)

The procedure described above for the preparation of 5a was employed to give 5d(56.0mg, 37.0%) as an orange solid. m. p. 121.7-122.3℃. ^{1}H NMR(400MHz, CDCl$_3$) δ 7.70(d, J = 15.6Hz, 1H, trans—CH =CH—), 7.67(d, J = 15.7Hz, 1H, trans—CH =CH—), 7.56(d, J = 8.5Hz, 2H, ArH), 7.52(d, J = 8.7Hz, 2H, ArH), 6.98(d, J = 15.9Hz, 1H, trans—CH =CH—), 6.90(d, J = 8.8Hz, 2H, ArH), 6.87(d, J = 15.8Hz, 1H, trans—CH =CH—), 6.69

(d, J = 8.6Hz, 2H, ArH), 4.66(d, J = 16.4Hz, 1H, CH_2), 4.58(dd, J = 11.5, 5.3Hz, 1H, CH_2), 4.12(d, J = 16.2Hz, 1H, CH_2), 4.08-3.93(m, 3H, CH_2), 3.59-3.52(m, 1H, CH_2), 3.41-3.33(m, 1H, CH_2), 3.27-3.17(m, 3H, CH_2), 3.04(s, 6H, $N(CH_3)_2$), 2.87(dd, J = 13.5, 4.1Hz, 1H, CH_2), 1.93-1.85(m, 4H, CH_2), 1.68-1.56(m, 4H, CH_2). ^{13}C NMR (101MHz, $CDCl_3$) δ 188.77, 186.94, 160.45, 151.94, 143.85, 141.61, 130.23, 129.98, 128.17, 123.94, 122.73, 121.06, 114.80, 111.93, 77.30, 76.98, 76.67, 67.40, 67.00, 64.21, 63.68, 59.83, 47.91, 40.14, 38.95, 29.67, 28.71, 23.73, 23.55. HRMS(ESI): m/z calcd for $C_{30}H_{39}N_3O_4ReS_2$ 756.1939; found 756.1945 $[M+H]^+$.

4.14 *In vitro* fluorescent staining using Tg mouse brain sections

The *trans*genic model mice(C57BL6, APPswe/PSEN1, 11 months old) and age-matched control mice(C57BL6, 11months old) were used for the studies. Paraffin-embedded brain sections (10μm thick) were deparaffinized with 3 ×10min washes in xylene, 2 ×5min washes in 100% ethanol, 5min wash in 80% ethanol/H_2O, 5 min wash in 60% ethanol/H_2O, and 10min wash in running tap water and then incubated in PBS(0.2M, pH = 7.4) for 30 min. The brain sections were incubated with a 10% ethanol solution(1.0μM) of 5a, 5b, 5c and 5d for 10min, respectively. The locations of plaques were confirmed by staining with Thioflavin-S(0.125%) in adjacent sections. Finally, the sections were washed with 40% ethanol for 10min. Fluorescent observation was performed by the Axio Observer Z1 inverted fluorescence microscope(Zeiss, Germany) equipped with DAPI(excitation, 405nm) and AF(excitation, 495nm) filter sets.

4.15 Binding assay *in vitro* using Aβ aggregates

Peptides Aβ$_{(1-42)}$ were purchased from Osaka Peptide Institute(Osaka, Japan). Aggregation was carried out by gently dissolving the peptide [0.56mg/mL for Aβ$_{(1-42)}$] in a buffer solution (pH = 7.4) containing 10mM sodium phosphate and 1 mM EDTA. The solution was incubated at 37℃ for 42h with gentle and constant shaking. Inhibition experiments were carried out in 12 ×75mm borosilicate glass tubes according to procedures described previously with some modification. Briefly, 100μL of Aβ$_{(1-42)}$ aggregates(28nM in the final assay mixture) was added into a mixture containing 100μL of radioligand([^{125}I]IMPY, 100,000cpm/100μL), 100μL of inhibitors(5a, 5b, 5c or 5d, 10^{-4} to $10^{-8.5}$ M in ethanol), and 700μL of PBS(0.2M, pH = 7.4) containing 0.5% BSA in a final volume of 1 mL. Non-specific binding was defined in the presence of 1μM IMPY. The mixture was incubated at room-temperature for 3 h, and then the bound and free radioactive fractions were separated by vacuum filtration through borosilicate glass fiberfilters(Whatman GF/B) using a Mp-48T cell harvester(Brandel, Gaithersburg, MD). The radioactivity of filters containing the bound ^{125}I-ligand was measured in an automatic γ-counter (WALLAC/Wizard1470, USA) with 70% efficiency. Under the assay conditions, the specifically bound fraction accounted for about 10% of total radioactivity. The half-maximal inhibitory concentration(IC_{50}) was determined from displacement curves of three independent experiments using GraphPad Prism 5.0, and the inhibition constant(K_i) was calculated using the Cheng-Prusoff inhibition constant equation: $K_i = IC_{50}/(1 + [L]/K_d)$, where $[L]$ represents the concentration of $[^{125}I]$IMPY used in the assay and K_d is the dissociation constant of $[^{125}I]$IMPY.

4.16 ^{99m}Tc labeling reaction and analysis by HPLC

Na^{99m}TcO$_4$ solution (74MBq, 200μL) was added to a GH kit and reacted at room temperature for 10min to give a ^{99m}Tc-GH solution. To a solution of precursor (4a-d, 0.5mg) in TFA (200μL) was added triethylsilane (2μL), and the mixture was left at room temperature for 10min, then the solvents were removed under a stream of nitrogen gas. The residue was redissolved in ethanol (200μL), and then mixed with the ^{99m}Tc-GH solution (200μL). The reaction mixture was heated to 100℃ for 10min. After cooling to room temperature, the product was extracted with CHCl$_3$ (2 × 1mL) and then the solvents were removed under a stream of nitrogen gas. The residue was redissolved in 100μL acetonitrile and purified by HPLC on a Venusil MP C18 reverse phase column (5μm, 4.6mm × 250mm) with a binary gradient system (acetonitrile/water: 70%/30%) at 1.0mL/min flow rate to give $[^{99m}Tc]$5a, $[^{99m}Tc]$5b, $[^{99m}Tc]$5c and $[^{99m}Tc]$5d, respectively. The identity of $[^{99m}Tc]$5a-d was verified by a comparison of the retention time with that of the nonradioactive rhenium complexes from the HPLC profiles. The four ^{99m}Tc-complexes were proven to show >95% radiochemical purity by HPLC.

4.17 Partition coefficient determination

The lg$D$ values of $[^{99m}Tc]$5a, $[^{99m}Tc]$5b, $[^{99m}Tc]$5c and $[^{99m}Tc]$5d were determined by measuring the distribution of the radiotracer between 1-octanol and PBS buffer at pH = 7.4. The two phases were pre-saturated with each other. 1-Octanol (3.6mL) and PBS (3.0mL) were pipetted into a 10mL plastic centrifuge tube containing 740kBq of ^{99m}Tc-labeled tracers. The mixture was vortexed for 5min, followed by centrifugation for 5min (3500r/min, Anke TDL80-2B, China). 50μL of the 1-octanol layer and 500μL of the buffer layer were deposed in two test tubes for counting. The amount of radioactivity in each tube was measured with an automatic γ-counter (Wallac 1470 Wizard, USA). The measurement was carried out in triplicate and repeated three times. The distribution coefficient was determined by calculating the ratio of cpm/mL of 1-octanol phase versus that of PBS phase and expressed as lg$D$. Then 2.0mL of the remaining 1-octanol layer was transferred into a new centrifuge tube, and new 1-octanol (1.6mL) and PBS (3.0mL) were pipetted into it. The vortexing, centrifuging, and counting protocols were repeated until consistent distribution coefficient values were obtained.

4.18 Biodistribution in normal mice

A saline solution containing the HPLC-purified ^{99m}Tc-labled tracer (100μL, 10% ethanol, 185kBq) was injected intravenously into ICR mice (five weeks, male) via the tail vein. The mice ($n = 5$ for each time point) were sacrificed by decapitation exactly at 2min, 10min, 30min and 60min post-injection. Samples of blood and organs of interest were removed, weighed and radioactivity was counted with an automatic γ-counter (Wallac 1470 Wizard, USA). The results were expressed in terms of the percentage of the injected dose per gram (%ID/g) of blood or organs.

Acknowledgements

The authors present special thanks to Dr. Jin Liu (College of Life Science, Beijing Normal University) for assistance in the *in vitro* neuropathological staining. This work was funded by National Natural Science Foundation of China (No: 21201019), Research Fund for the Doctoral Program of Higher Education of China (No: 20120003120013) and Fundamental Research Flunds for the Central Universities (No: 2012LYB19).

Appendix A. Supplementary data

Supplementary data related to this srticle can be found at http://dx.doi.org/10.1016/j.ejmech.2013.03.057.

References

[1] D. J. Selkoe, The origins of Alzheimer disease: A is for amyloid, JAMA 283 (2000) 1615-1617.

[2] D. J. Selkoe, Alzheimer's disease: genes, proteins, and therapy, Physiol. Rev. 81 (2001) 741-766.

[3] A. Wimo, B. Winblad, H. Aguero-Torres, E. von Strauss, The magnitude of dementia occurrence in the world, Alzheimer Dis. Assoc. Disord. 17 (2003) 63-67.

[4] J. Hardy, D. J. Selkoe, The amyloid hypothesis of Alzheimer's disease: progress and problems on the road to therapeutics, science 297 (2002) 353-356.

[5] J. L. Cummings, Alzheimer's disease, N. Engl. J. Med. 351 (2004) 56-67.

[6] B. C. P. Lee, M. Mintun, R. L. Buckner, J. C. Morris, Imaging of Alzheimer's disease, J. Neuroimaging 13 (2003) 199-214.

[7] V. L. Villemagne, C. C. Rowe, S. Macfarlane, K. E. Novakovic, C. L. Masters, Imaginem oblivionis: the prospects of neuroimaging for early detection of Alzheimer's disease, J. Clin. Neurosci. 12 (2005) 221-230.

[8] H. Braak, E. Braak, Neuropathological stageing of Alzheimer-related changes, Acta Neuropathol. 82 (1991) 239-259.

[9] M. Ono, A. Wilson, J. Nobrega, D. Westaway, P. Verhoeff, Z. P. Zhuang, M. P. Kung, H. F. Kung, ^{11}C-labeled stilbene derivatives as Aβ-aggregate-specific PET imaging agents for Alzheimer's disease, Nucl. Med. Biol. 30 (2003) 565-571.

[10] N. P. L. G. Verhoeff, A. A. Wilson, S. Takeshita, L. Trop, D. Hussey, K. Singh, H. F. Kung, M. P. Kung, S. Houle, In-Vivo imaging of Alzheimer disease [beta]-amyloid with [^{11}C] SB-13 PET, Am. J. Geriatr. Psychiatry 12 (2004) 584-595.

[11] W. E. Klunk, H. Engler, A. Nordberg, Y. Wang, G. Blomqvist, D. P. Holt, M. Bergström, I. Savitcheva, G. F. Huang, S. Estrada, B. Ausén, M. L. Debnath, J. Barletta, J. C. Price, J. Sandell, B. J. Lopresti, A. Wall, P. Koivisto, G. Antoni, C. A. Mathis, B. Långström, Imaging brain amyloid in Alzheimer's disease with Pittsburgh compound-B, Ann. Neurol. 55 (2004) 306-319.

[12] C. A. Mathis, Y. Wang, D. P. Holt, G. F. Huang, M. L. Debnath, W. E. Klunk, Synthesis and evaluation of ^{11}C-labeled 6-substituted 2-arylbenzothiazoles as amyloid imaging agents, J. Med. Chem. 46 (2003) 2740-2754.

[13] C. C. Rowe, U. Ackerman, W. Browne, R. Mulligan, K. L. Pike, G. O'Keefe, H. Tochon-Danguy, G. Chan,

S. U. Berlangieri, G. Jones, K. L. Dickinson-Rowe, H. P. Kung, W. Zhang, M. P. Kung, D. Skovronsky, T. Dyrks, G. Holl, S. Krause, M. Friebe, L. Lehman, S. Lindemann, L. M. Dinkelborg, C. L. Masters, V. L. Villemagne, Imaging of amyloid β in Alzheimer's disease with ^{18}F-BAY94-9172, a novel PET tracer: proof of mechanism, Lancet Neurol. 7(2008)129-135.

[14] V. L. Villemagne, K. Ong, R. S. Mulligan, G. Holl, S. Pejoska, G. Jones, G. O'Keefe, U. Ackerman, H. Tochon-Danguy, J. G. Chan, C. B. Reininger, L. Fels, B. Putz, B. Rohde, C. L. Masters, C. C. Rowe, Amyloid imaging with ^{18}F-florbetaben in Alzheimer disease and other dementias, J. Nucl. Med. 52(2011)1210-1217.

[15] M. T. Fodero-Tavoletti, D. Brockschnieder, V. L. Villemagne, L. Martin, A. R. Connor, A. Thiele, M. Berndt, C. A. McLean, S. Krause, C. C. Rowe, C. L. Masters, L. Dinkelborg, T. Dyrks, R. Cappai, In vitro characterization of [^{18}F]-florbetaben, an Aβ imaging radiotracer, Nucl. Med. Biol. 39(2012)1042-1048.

[16] M. Koole, D. M. Lewis, C. Buckley, N. Nelissen, M. Vandenbulcke, D. J. Brooks, R. Vandenberghe, K. Van Laere, Whole-body biodistribution and radiation dosimetry of ^{18}F-GE067: a radioligand for in vivo brain amyloid imaging, J. Nucl. Med. 50(2009)818-822.

[17] K. J. Lin, W. C. Hsu, I. T. Hsiao, S. P. Wey, L. W. Jin, D. Skovronsky, Y. Y. Wai, H. P. Chang, C. W. Lo, C. H. Yao, T. C. Yen, M. P. Kung, Whole-body biodistribution and brain PET imaging with [^{18}F]AV-45, a novel amyloid imaging agent-a pilot study, Nucl. Med. Biol. 37(2010)497-508.

[18] S. R. Choi, G. Golding, Z. Zhuang, W. Zhang, N. Lim, F. Hefti, T. E. Benedum, M. R. Kilbourn, D. Skovronsky, H. F. Kung, Preclinical Properties of ^{18}F-AV-45: a PET agent for Aβ plaques in the brain, J. Nucl. Med. 50(2009)1887-1894.

[19] D. F. Wong, P. B. Rosenberg, Y. Zhou, A. Kumar, V. Raymont, H. T. Ravert, R. F. Dannals, A. Nandi, J. R. Brašić, W. Ye, J. Hilton, C. Lyketsos, H. F. Kung, A. D. Joshi, D. M. Skovronsky, M. J. Pontecorvo, Vivo imaging of amyloid deposition in Alzheimer disease using the radioligand ^{18}F-AV-45 (Florbetapir F 18), J. Nucl. Med. 51(2010)913-920.

[20] H. Han, C. G. Cho, P. T. Lansbury, Technetium complexes for the quantitation of brain amyloid, J. Am. Chem. Soc. 118(1996)4506-4507.

[21] W. Zhen, H. Han, M. Anguiano, C. A. Lemere, C. G. Cho, P. T. Lansbury, Synthesis and amyloid binding properties of rhenium complexes: preliminary progress toward a reagent for SPECT imaging of Alzheimer's disease brain, J. Med. Chem. 42(1999)2805-2815.

[22] N. A. Dezutter, R. J. Dom, T. J. de Groot, G. M. Bormans, A. M. Verbruggen, ^{99m}Tc-MAMA-chrysamine G, a probe for beta-amyloid protein of Alzheimer's disease, Eur. J. Nucl. Med. 26(1999)1392-1399.

[23] N. A. Dezutter, T. J. de Groot, R. H. Busson, G. A. Janssen, A. M. Verbruggen, Preparation of ^{99m}Tc-N$_2$S$_2$ conjugates of chrysamine G, potential probes for the beta-amyloid protein of Alzheimer's disease, J. Labelled Comp. Radiopharm. 42(1999)309-324.

[24] Z. P. Zhuang, M. P. Kung, C. Hou, K. Ploessl, H. F. Kung, Biphenyls labeled with technetium 99m for imaging β-amyloid plaques in the brain, Nucl. Med. Biol. 32(2005)171-184.

[25] K. Serdons, T. Verduyckt, J. Cleynhens, C. Terwinghe, L. Mortelmans, G. Bormans, A. Verbruggen, Synthesis and evaluation of a ^{99m}Tc-BAT-phenylbenzothiazole conjugate as a potential in vivo tracer for visualization of amyloid β, Bioorg. Med. Chem. Lett. 17(2007)6086-6090.

[26] X. Chen, P. Yu, L. Zhang, B. Liu, Synthesis and biological evaluation of ^{99m}Tc, Re-monoamine-monoamide conjugated to 2-(4-aminophenyl)benzothiazole as potential probes for β-amyloid plaques in the brain, Bioorg. Med. Chem. Lett. 18(2008)1442-1445.

[27] K. Serdons, T. Verduyckt, J. Cleynhens, G. Bormans, A. Verbruggen, Development of ^{99m}Tc-thioflavin-T de-

rivatives for detection of systemic amyloidosis, J. Labelled Comp. Radiopharm. 51(2008)357-367.

[28] M. Ono, R. Ikeoka, H. Watanabe, H. Kimura, T. Fuchigami, M. Haratake, H. Saji, M. Nakayama, Synthesis and evaluation of novel chalcone derivatives with ^{99m}Tc/Re complexes as potential probes for detection of β-amyloid plaques, ACS Chem. Neurosci. 1(2010)598-607.

[29] M. Ono, R. Ikeoka, H. Watanabe, H. Kimura, T. Fuchigami, M. Haratake, H. Saji, M. Nakayama, ^{99m}Tc/Re complexes based on flavone and aurone as SPECT probes for imaging cerebral β-amyloid plaques, Bioorg. Med. Chem. Lett. 20(2010)5743-5748.

[30] M. Sagnou, D. Benaki, C. Triantis, T. Tsotakos, V. Psycharis, C. P. Raptopoulou, I. Pirmettis, M. Papadopoulos, M. Pelecanou, Curcumin as the OO bidentate ligand in "2 + 1" complexes with the $[M(CO)_3]^+$ (M = Re, ^{99m}Tc) tricarbonyl core for radiodiagnostic applications, Inorg. Chem. 50(2011)1295-1303.

[31] M. Ono, Y. Fuchi, T. Fuchigami, N. Kobashi, H. Kimura, M. Haratake, H. Saji, M. Nakayama, Novel benzofurans with ^{99m}Tc complexes as probes for imaging cerebral β-amyloid plaques, ACS Med. Chem. Lett. 1(2010)443-447.

[32] Y. Cheng, M. Ono, H. Kimura, M. Ueda, H. Saji, Technetium-99m labeled pyridyl benzofuran derivatives as single photon emission computed tomography imaging probes for β-amyloid plaques in Alzheimer's brains, J. Med. Chem. 55(2012)2279-2286.

[33] X. Wang, M. Cui, P. Yu, Z. Li, Y. Yang, H. Jia, B. Liu, Synthesis and biological evaluation of novel technetium-99m labeled phenylbenzoxazole derivatives as potential imaging probes for β-amyloid plaques in brain, Bioorg. Med. Chem. Lett. 22(2012)4327-4331.

[34] M. Cui, M. Ono, H. Kimura, B. Liu, H. Saji, Synthesis and structure-affinity relationships of novel dibenzylideneacetone derivatives as probes for β-amyloid plaques, J. Med. Chem. 54(2011)2225-2240.

[35] S. S. Jurisson, J. D. Lydon, Potential technetium small molecule radiopharmaceuticals, Chem. Rev. 99(1999)2205-2218.

[36] G. Tamagnan, R. M. Baldwin, N. S. Kula, R. J. Baldessarini, R. B. Innis, Cyclopentadienyltricarbonylrheniumbenzazepines: synthesis and binding affinity, Bioorg. Med. Chem. Lett. 10(2000)1113-1115.

Novel Cyclopentadienyl Tricarbonyl Complexes of ^{99m}Tc Mimicking Chalcone as Potential Single-Photon Emission Computed Tomography Imaging Probes for β-amyloid Plaques in Brain[*]

Abstract Rhenium and technetium-99m cyclopentadienyl tricarbonyl complexes mimicking the chalcone structure were prepared. These complexes were proved to have affinity to β-amyloid(Aβ) in fluorescent staining on brain sections of Alzheimer's Disease(AD) patient and binding assay using Aβ$_{1-42}$ aggregates, with K_i values ranging from 899 to 108nM as the extension of conjugated π system. In vitro autoradiograpy on sections of transgenic mouse brain confirmed the affinity of [^{99m}Tc]5 (K_i =108nM). In biodistribution, all compounds showed good initial uptakes into the brain and fast blood clearance, while the decreasing of initial brain uptakes correspond to increasing of conjugation length, from (4.10 ± 0.38)% ID/g([^{99m}Tc]3) to (1.11 ± 0.34)% ID/g([^{99m}Tc]5). These small technetium-99m complexes(<500Da) designed by an integrated approach provide encouraging evidence that development of a promising ^{99m}Tc-labeled agent for imaging Aβ plaques in the brain may be feasible.

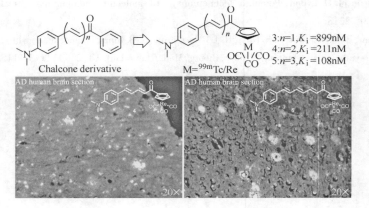

1 Introduction

Alzheimer's disease(AD) is a progressive neurodegenerative disorder pathologically characterized by deposition of misfolded β-amyloid(Aβ) peptides as senile plaques in the brain. Because the deposition of Aβ plaques is an early event in the development of AD, a validated biomarker of Aβ deposition in the brain might prove useful to identify and follow individuals at risk for AD

---

[*] Copartner: Li Zijing, Cui Mengchao, Dai Jiapei, Wang Xuedan, Yu Pingrong, Yang Yanping, Jia Jianhua, Fu Hualong, Masahiro Ono, Jia Hongmei, Hideo Saji. Reprinted from *J. Med. Chem.*, 2013, 56(2):471-482.

and to assist in the evaluation of new antiamyloid therapies currently under development[1-3].

A number of groups have reported radiolabeled Aβ imaging agents for positron emission tomography(PET) and single photon emission computed tomography(SPECT) in clinical trials such as [^{18}F]FDDNP([^{18}F]-2-(1-(6-[(2-fluoroethyl)(methyl)amino]-2-naphthyl)ethylidene)malononitrile)[4,5], [^{11}C]PIB([^{11}C]-2-(4′-methylaminophenyl)-6-oxybenzothiazole)[6,7], [^{11}C]SB-13([^{11}C]-4-N-methylamino-4′-hydroxystilbene)[8,9], [^{11}C]BF-227([^{11}C]-(2-[fluoro]ethoxy)benzoxazole)[10], [^{18}F]BAY94-9172([^{18}F]-4-(N-methylamino)-4′-(2-(2-(2-fluoroethoxy)ethoxy)ethoxy)-stilbene)[11], [^{18}F]AV-45([^{18}F]-(E)-4-(2-(6-(2-(2-(2-fluoroethoxy)ethoxy)ethoxy)pyridin-3-yl)vinyl)-N-methylaniline)[12,13], and [^{123}I]IMPY([^{123}I]-6-iodo-2-(4′-dimethylamino-)phenylimidazo[1,2]pyridine)[14-16]. Recent reports using these amyloid imaging agents have indicated that detecting Aβ plaques in the living human brain by PET or SPECT may lead to differentiation between AD patients and healthy human. However, the signal-to-noise ratio for plaque labeling of [^{123}I]IMPY, which is the only SPECT Aβ imaging agents in preclinical trail, was not robust in AD and healthy controls, while it was believed that the in vivo instability and fast metabolism of [^{123}I]IMPY may ultimately lead to a decreased signal. After that, there is no report of any Aβ imaging candidate for SPECT moving into clinical trial.

However, the radionuclide technetium-99m is among the most widely used isotopes in diagnostic nuclear medicine. Readily produced by a ^{99}Mo/^{99m}Tc generator, essentially 24h/day, ^{99m}Tc emits medium γ-ray energy suitable for detection, and its physical half-life is compatible with the biological localization and residence time required for imaging. New ^{99m}Tc-labeled imaging agents for Aβ plaques will provide simple, convenient and widespread SPECT-based imaging methods for detecting and eventually quantifying Aβ plaques in living brain tissue[17-19], Whereas in the past ^{99m}Tc-complexes were preferentially applied as perfusion agents. A challenge now lies in combining a ^{99m}Tc complex with a targeting molecule such as a small central nervous system(CNS) receptor-binding molecule.

Kung et al. reported that the dopamine transporter imaging agent [^{99m}Tc]TRODAT-1[20] is useful to detect the loss of dopamine transportes in Parkinson's disease. This is the first example of a ^{99m}Tc imaging agent that can penetrate the blood-brain barrier(BBB) via a simple diffusion mechanism and localize at receptor binding sites in the CNS. On the basis of this success, efforts were made to search for comparable ^{99m}Tc imaging agents that target binding sites on Aβ plaques in the brain of AD patients. Several ^{99m}Tc-labeled imaging probes have been developed (Fig. 1). Two of them are based on an integrated approach(the Congo Red derivative[21] and the biphenyl derivative[22]). The integrated approach involves the replacement of part of a known high-affinity receptor ligand with the requisite "unnatural" ^{99m}Tc chelate in such a way that there are minimal changes in size, conformation, and receptor binding affinity[23]. The others are based on a bifunctional approach(the chrysamine G derivative[24,25], the benzothiazole aniline derivative[26-28], the chalcone derivative[29], the flavone and aurone derivative[30], the benzoxazole derivative[31], the DDNP derivative[32], etc.). The bifunctional approach leads to volume expansion. This expansion twists the planar shape of the binding agent, which is very important for a molecular agent to fit into the planar gap on the Aβ plaques and also increases the molecular weight greatly. Conversely, the integrated approach strategy seems wiser but more difficult to practice. Unfortunately, no clinical study of both kinds has been reported because the low initial brain uptake is not sufficient for SPECT imaging, even if they display high affinities to Aβ.

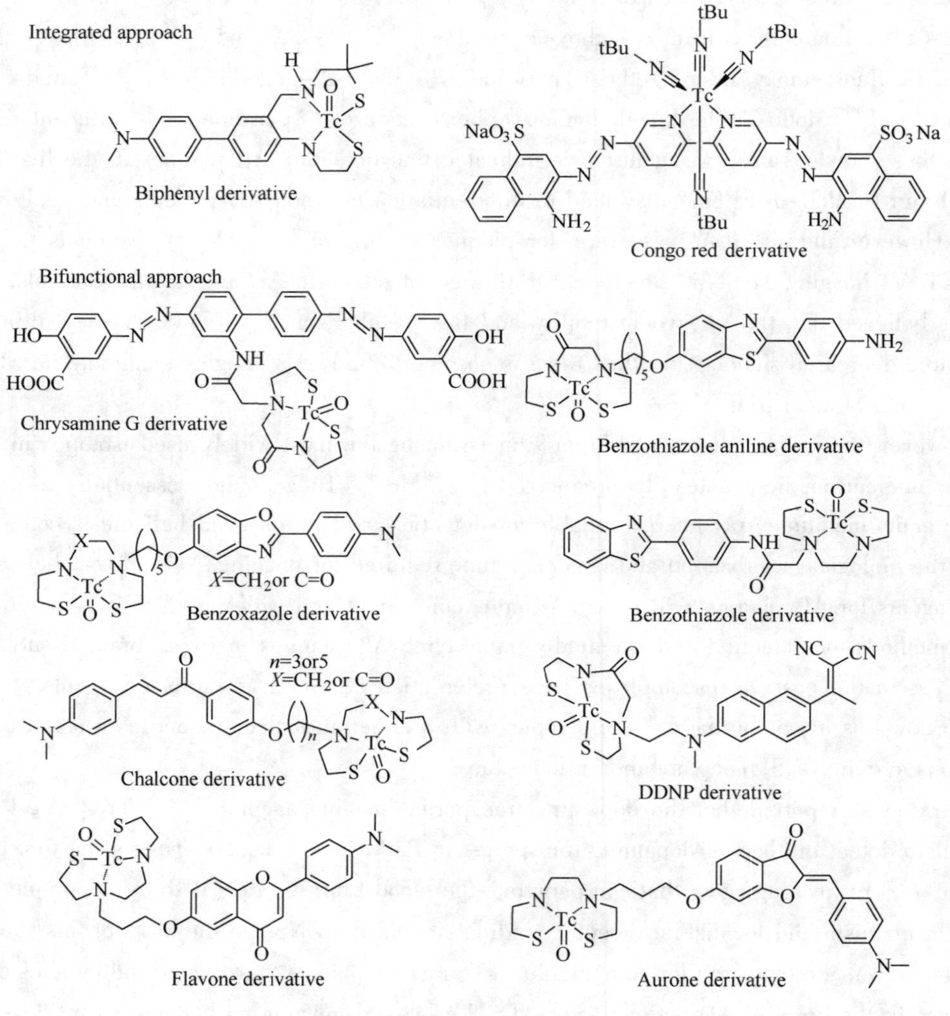

Fig. 1 Chemical structures of reported ^{99m}Tc labeled Aβ imaging probes

On the basis of the discovery so far, the best strategy to design Aβ imaging agents is to find an small, lipophilic, ^{99m}Tc-chelating core to substitute or mimic one part of the structure of a binding agent by integrated approach in order to maintain the planar shape of the ligand, minimize the molecular weight and maintain the ability to penetrate the BBB.

As early as 1992, Wenzel reported a double ligand transfer (DLT) reaction[33], which ultimately led to the formation of [Cp^{99m}Tc(CO)$_3$, Cp = cyclopentadienyl] starting from ^{99m}TcO$_4^-$.

Then the method of preparing [$Cp^{99m}Tc(CO)_3$] in water has been developed using sensitive organometallic ligands[34]. Apart from this feasibility consideration, the cyclopentadienyl tricarbonyl ligand offers very attractive properties as a ligand for radiopharmaceutical purposes. Its inherent advantages are the small size, the low molecular weight and the high stability of half-sandwich configuration and the minimized steric interference with the receptor binding moiety of a labeled biomolecule[35]. The "piano stool" organometallic core [$CpM(CO)_3$, M = Re(I), Tc(I)] is a neutral 18 electron species, with high stability resulting from the low-spin d^6 electron configuration, further stabilized by the cyclopentadienyl and tricarbonyl ligands. Although many highly stable chelate systems have been developed for ^{99m}Tc, the small size of the [$CpM(CO)_3$] core is advantageous for maintaining biological activity, particularly when labeling small molecules. The fact that [$CpM(CO)_3$] (M = Re and Tc) can be coupled to biomolecules by classical organometallic methods without affecting the bioactivity has been demonstrated by several groups[34]. Furthermore, [$Cp^{99m}Tc(CO)_3$] complexes are highly lipophilic, which makes them particularly promising as ^{99m}Tc-labeled, BBB-crossing molecules. So, taking the configuration and aromatic properties of [$Cp^{99m}Tc(CO)_3$] core into consideration, it is an excellent choice to substitute or mimic a benzene ring of a ligand by an integrated approach.

Ono et al. reported in 2007 that chalcone derivatives[36], whose backbone structure is considered to be a promising scaffold, showed excellent characteristics as new amyloid imaging agents, such as high binding affinity to Aβ aggregates, high uptake into the brain, and rapid clearance from the brain, besides it can easily be formed by one-pot condensation reaction.

After ^{125}I-, ^{11}C-, and ^{18}F-labeled chalcone derivatives were prepared and chalcone scaffold were proved promising candidate as AD imaging agent[36-38], additionally, in 2010, the same group reported synthesis of four chalcone derivatives with monoamine-monoamide dithiol (MAMA) and bis-amino-bis-thiol (BAT) selected as chelation ligands by bifunctional approach. For the first time, ^{99m}Tc/Re complexes as chalcone derivatives have been proposed as probes for the detection of Aβ plaques in the brain[29]. In their study, MAMA and BAT were selected as chelation ligands to form an electrically neutral complex with ^{99m}Tc, but these two big chelating groups increase the molecular weight, and also decrease the affinities and initial brain uptakes as well.

In the present study, we designed and synthesized novel chalcone-mimic complexes by introducing [$Cp^{99m}Tc(CO)_3$] core to substitute a benzene ring (Fig. 2) by an integrated approach, in order to acquire good brain uptake while keeping excellent affinities to Aβ plaques. Furthermore, the properties of this series of ^{99m}Tc/Re complexes were studied as the extension of conjugated π system.

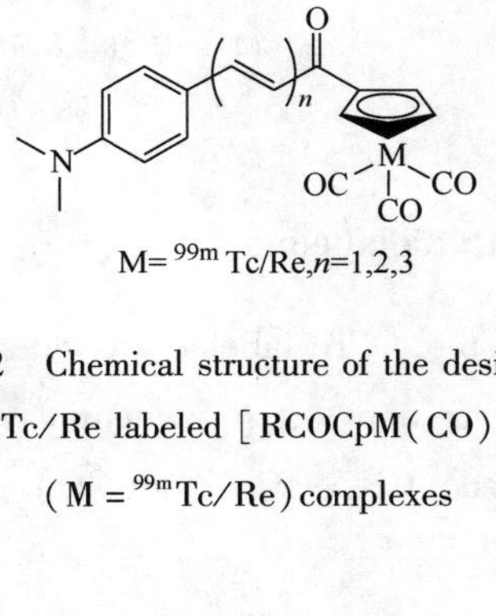

Fig. 2 Chemical structure of the designed ^{99m}Tc/Re labeled [$RCOCpM(CO)_3$] (M = ^{99m}Tc/Re) complexes

2 Results and Discussion

2.1 Chemistry

For characterization of the ^{99m}Tc complexes, we have prepared the corresponding rhenium complexes (Scheme 1). To collect the three corresponding aromatic aldehydes, the aromatic aldehyde 1 was prepared by Wittig reaction from (E)-3-(4-(dimethylamino)phenyl)acrylaldehyde, which is commercially available. (Cyclopentadienyl)tricarbonylrhenium was at first acetylated by acetyl chloride in ice bath to obtain (acetylcyclopentadienyl)tricarbonylrhenium(2) at 95% yield. This complex can react with three aromatic aldehydes respectively through base-catalyzed Claisen condensation to obtain final rhenium complexes (3, 4 and 5) of different π conjugation length at yields above 90%. By the way, two ferrocene complexes (6 and 7) were synthesized through the same method as precursors for [^{99m}Tc]4 and [^{99m}Tc]5. All these complexes were fully characterized by spectroscopic methods. Complexes 3 and 4 could be recrystallized by slow evaporation of an ethanol-CH_2Cl_2 solution to afford X-ray quality crystals. The structures were elucidated and their ORTEPs are given in Fig. 3 and Fig. 4, with relevant crystallographic data in Table 1.

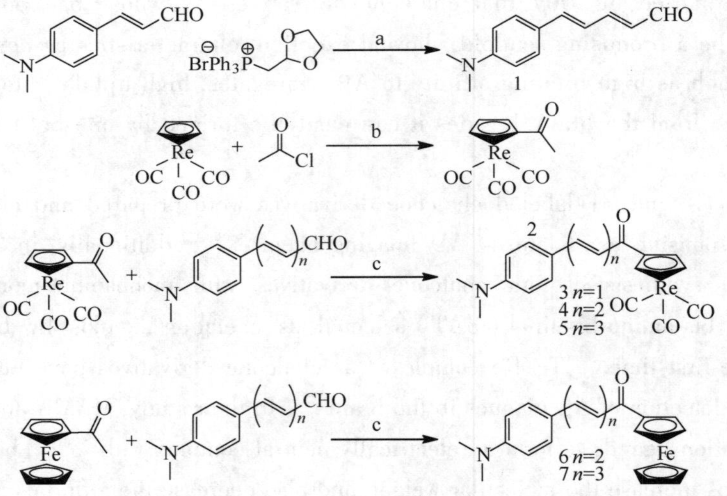

Scheme 1   Reagents and conditions

a—(1) NaH, 18-crown-6, dry THF, r. t. , (2) concentrated HCl, K_2CO_3 aq. , r. t. ;

b—$COCl_2$, $AlCl_3$, CH_2Cl_2; c—NaOH, EtOH, r. t

2.2 Radiolabeling

To get the ^{99m}Tc-labeled cyclopentadienyl tricarbonyl complexes [^{99m}Tc]3, [^{99m}Tc]4 and [^{99m}Tc]5, a two-step sequential reaction (Scheme 2) has to be applied. In the first step, we prepared [$CH_3COCp^{99m}Tc(CO)_3$] ([^{99m}Tc]2) under 150℃ heat for 20min through the DLT

Scheme 2 Reagents and conditions

a—Mn(CO)$_5$Br, ^{99m}TcO$_4^-$, H$_2$O, DMF, 150 ℃, 20min; b—NaOH, EtOH, r.t

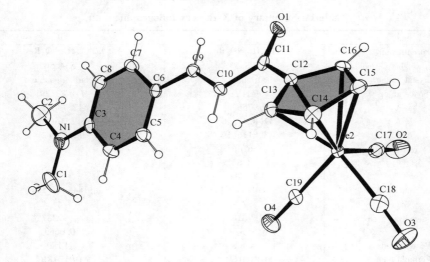

Fig. 3 Molecular structure of 3
(Thermal ellipsoids drawn at the 30% probability level)

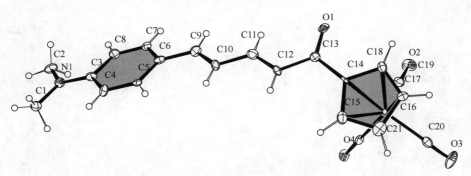

Fig. 4 Molecular structure of 4
(Thermal ellipsoids drawn at the 30% probability level)

method with an average radiochemical yield of 50% (no decay corrected). Then [^{99m}Tc]2 went through a base-catalyzed Claisen condensation with three corresponding aldehydes to yield [^{99m}Tc]3, [^{99m}Tc]4, and [^{99m}Tc]5. The final ^{99m}Tc-labeled products were purified by radio high performance liquid chromatography (HPLC), and the identity of [^{99m}Tc]3, [^{99m}Tc]4 and [^{99m}Tc]5 was verified by a comparison of the retention time with that of the nonradioactive rhenium compounds (Fig. 5-Fig. 7). In the second step, the radiochemical yields of [^{99m}Tc]3-5 were also about 50% (no decay corrected), to achieve a 25% total yield with radiochemical purity of >95% after HPLC purification. We took this two-step strategy because the direct DLT labeling method under 150℃ would destroy the ferrocene precursor (6 and 7) and give no products. Furthermore, this radiochemical synthesis by Claisen condensation strategy was proved brilliant in the following experiments because the reaction was not only fast enough for ^{99m}Tc labeling but also ready to be applied to label many different aldehyde or large molecules which contain aldehyde groups with [Cp^{99m}Tc(CO)$_3$] core at acceptable yields.

Table 1 Summary of X-ray crystallographic data

Item	3	4
Formula sum	$C_{19}H_{16}NO_4Re$	$C_{21}H_{18}NO_4Re$
Formula weight/g·mol^{-1}	508.53	534.56
Crystal system	triclinic	orthorhombic
Space group	$P\bar{1}(2)$	$Pca\,21(29)$
a/Å	5.9756(8)	12.730(2)
b/Å	11.8247(16)	6.3516(10)
c/Å	13.6116(19)	12.730(2)
α/(°)	66.46	
β/(°)	78.62	
γ/(°)	80.76	
Cell volume/Å^3	860.82(82)	1845.45(52)
Z	2	4
Calcd density/g·cm^{-3}	1.96182	1.92387
RAll	0.0379	0.0669
Pearson code	aP82	oP180
Formula type	NOP4Q16R19	NOP4Q18R21
Wyckoff sequence	i41	a45

Note: 1Å = 0.1nm.

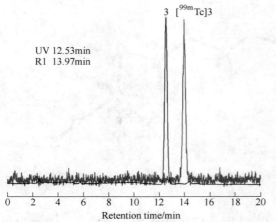

Fig. 5 HPLC profiles of 3 and [^{99m}Tc]3
(HPLC conditions: Venusil MP C18 column (Agela Technologies, 10mm × 250mm), CH$_3$CN/H$_2$O = 70/30, 4mL/min, UV, 254nm)

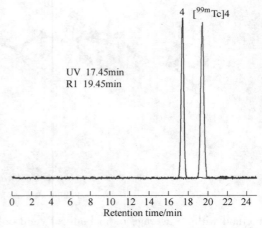

Fig. 6 HPLC profiles of 4 and [^{99m}Tc]4
(HPLC conditions: Venusil MP C18 column (Agela Technologies, 10mm × 250mm), CH$_3$CN/H$_2$O = 70/30, 4mL/min, UV, 254nm)

2.3 Biological evaluation

In vitro fluorescent staining of Aβ plaques in sections of brain tissue from AD patients and Tg model mice(C57BL6, APP-swe/PSEN1, 11 months old) were carried out to evaluate the binding affinity of complexes 3, 4 and 5 to Aβ plaques. As shown in Fig. 8a, d, specific staining of plaques was observed in the brain section of Tg mice for complex 4. The presence and distribution of Aβ plaques was consistent with the results of staining using thioflavin-S(a common dye for staining of Aβ plaques) on the adjacent section (Fig. 8b, e). Furthermore, intense labeling of plaques were observed in the brain section of an AD patient(Fig. 8g). In contrast, no

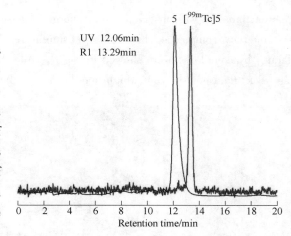

Fig. 7 HPLC profiles of 5 and [^{99m}Tc]5
(HPLC conditions: Venusil MP C18 column(Agela Technologies, 10mm × 250mm), CH_3CN/H_2O = 80/20, 4mL/min, UV, 254nm)

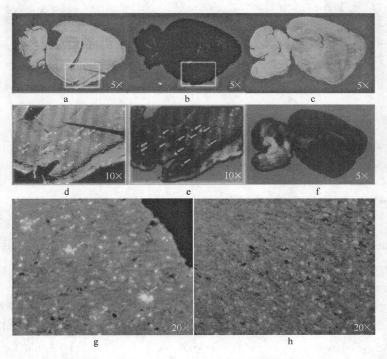

Fig. 8 In vitro fluorescent staining of Aβ plaques by complex 4
a, d—Complex 4 on brain section of a Tg model mouse(C57BL6, APPswe/PSEN1, 11 months old, male); b, e—The presence and distribution of plaques on the sections were confirmed by fluorescence staining using thioflavin-S on the adjacent section; c—Complex 4 on brain section of a normal mouse as control; f—Thioflavin-S on the adjacent brain section of the normal mouse; g—Complex 4 on brain section of an AD patient(91 years old, male); h—Complex 4 on brain section of a normal person(69 years old, female) as control

apparent labeling was observed in both normal mouse and normal adult brain sections (Fig. 8c, h) stained by complex 4 (Fig. 8f). The similar results of in vitro fluorescent staining of Aβ plaques by complex 5 were showed in Fig. 9, while the fluorescent signal stained by complex 3 was weak (data not shown), which may be due to the low affinity.

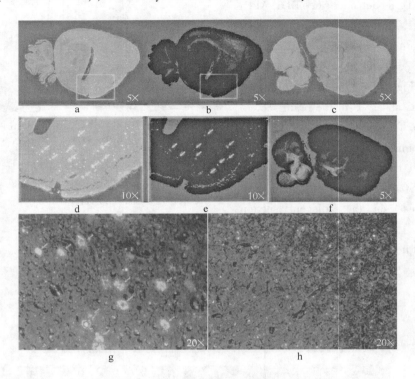

Fig. 9　In vitro fluorescent staining of Aβ plaques by complex 5

a, d—Complex 5 on brain section of a Tg model mouse (C57BL6, APPswe/PSEN1, 11 months old, male); b, e—The presence and distribution of plaques on the sections were confirmed by fluorescence staining using thioflavin-S on the adjacent section; c—Complex 5 on brain section of a normal mouse as control; f—Thioflavin-S on the adjacent brain section of the normal mouse; g—Complex 5 on brain section of an AD patient (91 years old, male); h—Complex 5 on brain section of a normal person (69 years old, female) as control

To quantitatively evaluate the binding affinities of these chalcone-mimic complexes to $A\beta_{1-42}$ aggregates, in vitro inhibition assay was carried out in solutions with $[^{125}I]$IMPY as the competing radioligand according to conventional methods. The three rhenium complexes (3, 4 and 5) inhibited the binding of $[^{125}I]$IMPY in a dose-dependent manner (Fig. 10). With the result shown in Table 2, rhenium complexes of different conjugation lengths showed moderate binding affinities to $A\beta_{1-42}$ aggregates ($K_i = (899 \pm 78)$ nM for 3, $K_i = (211 \pm 19)$ nM for 4, and $K_i = (108 \pm 16)$ nM for 5), which are not satisfactory but sufficiently high for Aβ aggregates comparable to the value determined under the same assay system for IMPY ($K_i = (10.5 \pm 1.0)$ nM) and chalcone derivatives (with K_i ranged from 2.9 to >10000 nM as reported[36]). The crystal structures elucidated in Fig. 3 and Fig. 4 implied that complex 4 is less distorted than complex 3, because the dihedral angle between the benzene plane (light-blue) and the Cp plane (light-red) is 43.69° for complex 3, while the same dihedral angle of complex 4 is 31.34°. So, we would like to blame the decreasing of affinity for the three "CO" stools of the $[Cp^{99m}Tc(CO)_3]$ core which distort the planar and flake-like configuration of chalcone. Because the K_i values are decreasing as the extension of conjugated π system, we can infer that the extension weakens the influence of the three "CO" stools of the $[Cp^{99m}Tc(CO)_3]$ core on the planar configuration. The affinity of the two ferrocene complexes whose structures have no stools still keep high affinity ($K_i = (3.36 \pm 0.30)$ nM for 6, $K_i = (5.08 \pm 1.74)$ nM for 7) also confirm our inference.

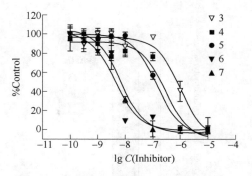

Fig. 10 Inhibition curves for the binding of $[^{125}I]$IMPY to Aβ$_{1-42}$ aggregates

Table 2 Inhibition constants (K_i, nM) for binding to aggregates of Aβ$_{1-42}$ versus $[^{125}I]$IMPY[①]

Compd	K_i/nM	Compd	K_i/nM
3	899 ± 78	6	3.36 ± 0.30
4	211 ± 19	7	5.08 ± 1.74
5	108 ± 16	IMPY	11.5 ± 2.5

① Measured in triplicate with results given as the mean ± SD.

In binding assays using the aggregated Aβ$_{1-42}$ peptides in solution, we also confirmed that the Aβ$_{1-42}$ aggregate-bound radioactivities (%) were varied differently in the four ^{99m}Tc-labeled complexes. In terms of Aβ$_{1-42}$ aggregate-bound radioactivity, the derivatives rank in the following order: $[^{99m}Tc]5(10.15\%) > [^{99m}Tc]4(3.87\%) > [^{99m}Tc]3(0.68\%) > [^{99m}Tc]2(0.32\%)$ (Fig. 11). Furthermore, the bound radioactivities indicate that $[^{99m}Tc]5$ and $[^{99m}Tc]4$ occupied the specific binding sites of Aβ$_{1-42}$ aggregates, while $[^{99m}Tc]3$ and $[^{99m}Tc]2$ showed no remarkable binding to Aβ$_{1-42}$ aggregates. This result suggests that the length of conjugated π system played an important role in the binding of Aβ$_{1-42}$ aggregates, which are also consistent with the binding affinities of these chalcone-mimic complexes in the inhibition assay. The results of blocking assay using excess of rhenium complexes 2-5 showed that the specific binding of $[^{99m}Tc]5$ to Aβ$_{1-42}$ aggregates was blocked about 80% by an excess of 5 (0.5μM); the binding of $[^{99m}Tc]4$ was blocked nearly a half by an excess of 4 at 1.0μM, while excess of complexes 3 and 2 at 1.0μM could not significantly block the binding radioactivity owing to their lower affinities. These results confirmed that complex 5 displayed specific and high binding to Aβ$_{1-42}$ aggregates.

In vitro autoradiography studies of $[^{99m}Tc]5$ were performed with sections from Tg mice (C57BL6, APPswe/PSEN1, 11 months old) and an age-matched control mice. As shown in Fig. 12, $[^{99m}Tc]5$ displayed good labeling of Aβ plaques in the cortical regions of Tg mice, and the control case was clearly void of any notable Aβ labeling. The distribution of Aβ plaques was

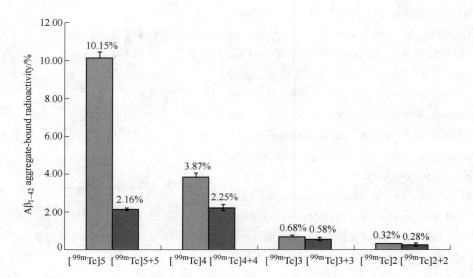

Fig. 11 Binding and blocking assay of [^{99m}Tc]2, [^{99m}Tc]3, [^{99m}Tc]4, and [^{99m}Tc]5 with Aβ$_{1-42}$ aggregates
(Values are the mean ± standard error of the mean for tent experiments. Black columns represent the Aβ$_{1-42}$ aggregate-bound radioactivities(%) of [^{99m}Tc]2-5. Gray columns represent the Aβ$_{1-42}$ aggregate-bound radioactivities(%) of [^{99m}Tc]2-5 blocked by complexes 2-4 (1.0 μM) and 5 (0.5 μM), respectively)

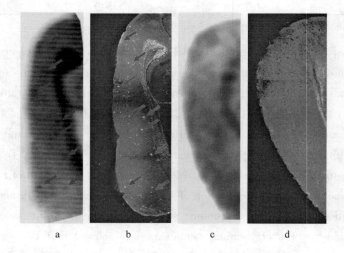

Fig. 12 In vitro autoradiography of [^{99m}Tc]5 on a Tg model mouse (C57BL6, APPswe/PSEN1, 11 months old, male) (a), the presence and distribution of plaques in the section A were confirmed by fluorescence staining using thioflavin-S on the same section with a filter set for GFP(b), in vitro autoradiography of [^{99m}Tc]5 on a brain section of a normal mouse (C57BL6, 11 months old, male) as control (c) and fluorescence staining using thioflavin-S on the section C with a filter set for GFP on a brain section of a normal mouse as control (d)

consistent with the results of fluorescent staining with thioflavin-S. Although the binding affinity of [^{99m}Tc]5 to Aβ aggregates was not potent (K_i = (108 ± 16) nM), [^{99m}Tc]5 was still able to label the plaques in sections of Tg mice.

The lg$D$ values(2.89 ± 0.09 for [^{99m}Tc]3, 3.61 ± 0.04 for [^{99m}Tc]4, and 3.45 ± 0.09 for [^{99m}Tc]5, respectively) shown in Table 3 indicate that complexes [^{99m}Tc]3, [^{99m}Tc]4, and [^{99m}Tc]5 have moderate lipophilicity suitable for brain imaging.

Table 3 lg$D$ value of compound [^{99m}Tc]3-5[①]

Compd	lg$D$
[^{99m}Tc]3	2.89 ± 0.09
[^{99m}Tc]4	3.61 ± 0.04
[^{99m}Tc]5	3.45 ± 0.09

① Measured in triplicate with results given as the mean ± SD.

Biodistribution experiments in normal male ICR mice (5 weeks, male) were carried out to evaluate the ability of these ^{99m}Tc cyclopentadienyl tricarbonyl complexes of different conjugation lengths ([^{99m}Tc]3, [^{99m}Tc]4, and [^{18}F]5) to penetrate the BBB and properties of clearance from the brain. High initial brain uptake and high brain$_{2min}$/brain$_{60min}$ ratio in normal mouse brain are considered to be important as pharmacokinetic indexes for selecting appropriate Aβ imaging tracers. As shown in Table 4-Table 6, [^{99m}Tc]3 with a short conjugated π system displayed a very high initial brain uptake ($n = 1$, (4.10 ± 0.38)% ID/g at 2min) than that of [^{99m}Tc]4 and [^{99m}Tc]5 with a longer conjugated π system ($n = 2$, (2.30 ± 0.27)% ID/g at 2min for [^{99m}Tc]4; $n = 3$, (1.11 ± 0.34)% ID/g at 2min for [^{99m}Tc]5). What impressed us was that [^{99m}Tc]3 exhibits such high initial brain uptake barely not seen before for a ^{99m}Tc-labeled receptor binding agent and that the decreasing of brain uptake is as sharp as the π system extension, where we could not give a good explanation by now. Compared with the ^{99m}Tc-labeled chalcone derivatives (0.22% ID/g, 0.78% ID/g, 0.62% ID/g, 1.48% ID/g at 2min) by bifunctional approach reported previously[29], the initial brain uptakes of [^{99m}Tc]3 and [^{99m}Tc]4 are apparently superior. Meanwhile, the brain$_{2min}$/brain$_{60min}$ ratio of 8.20, 4.18 and 2.18 for [^{99m}Tc]3, [^{99m}Tc]4, and [^{99m}Tc]5 were from good to acceptable. Furthermore, we can also conclude that [^{99m}Tc]3, [^{99m}Tc]4, and [^{99m}Tc]5 are metabolized by liver and small intestine, because the liver showed high uptakes with very slow washout and the small intestine uptakes kept increasing with time.

The biodistribution with permeability-glycoprotein 1 (PgP) blocked by cyclosporin A (an immunosuppressant drug known to block PgP activity) further describes their brain penetration abilities. The blood and brain uptakes at 2min were measured, and the results were shown in Table 7. After the PgP were blocked by cyclosporin A, the brain uptakes of [^{99m}Tc]3-5 increased obviously. This result may reveal [^{99m}Tc]3-5 to be substrates for the rodent PgP transporter.

Table 4 Biodistribution of [^{99m}Tc]3 in male ICR mice①

Organ	Time after injection				
	2min	10min	30min	60min	120min
Blood	2.02 ± 0.17	0.82 ± 0.05	0.51 ± 0.07	0.51 ± 0.06	0.44 ± 0.08
Brain	4.10 ± 0.38	2.27 ± 0.51	0.69 ± 0.09	0.50 ± 0.08	0.37 ± 0.08
Heart	11.47 ± 1.65	1.87 ± 0.36	0.98 ± 0.22	0.84 ± 0.16	0.55 ± 0.15
Liver	18.14 ± 1.77	24.75 ± 3.38	22.43 ± 4.39	25.96 ± 2.06	25.24 ± 5.17
Spleen	3.53 ± 0.52	1.93 ± 0.30	0.81 ± 0.14	0.64 ± 0.14	0.51 ± 0.15
Lung	7.39 ± 1.06	7.32 ± 0.71	5.52 ± 0.49	5.38 ± 0.38	4.79 ± 0.43
Kidney	13.36 ± 0.99	5.38 ± 0.66	3.87 ± 0.79	3.17 ± 0.16	2.56 ± 0.40
Stomach②	1.69 ± 0.16	3.06 ± 0.90	2.18 ± 0.39	1.38 ± 0.55	0.75 ± 0.16
Small intestine②	6.81 ± 1.57	17.98 ± 1.58	37.71 ± 7.94	34.75 ± 2.26	26.22 ± 4.20

① Expressed as % injected dose per gram. Average for 5 mice ± standard deviation.
② Expressed as % injected dose per organ.

Table 5 Biodistribution of [^{99m}Tc]4 in male ICR mice①

Organ	Time after injection				
	2min	10min	30min	60min	120min
Blood	3.80 ± 0.71	1.25 ± 0.17	0.98 ± 0.49	0.64 ± 0.15	0.55 ± 0.07
Brain	2.30 ± 0.27	1.85 ± 0.25	0.93 ± 0.09	0.55 ± 0.08	0.49 ± 0.11
Heart	12.94 ± 2.16	3.58 ± 0.51	1.92 ± 0.26	1.36 ± 0.13	0.92 ± 0.12
Liver	32.64 ± 4.34	28.81 ± 4.57	34.32 ± 5.35	35.93 ± 4.25	29.59 ± 5.35
Spleen	4.50 ± 0.91	4.34 ± 0.92	2.84 ± 0.90	1.25 ± 0.23	1.00 ± 0.16
Lung	9.08 ± 2.20	6.18 ± 1.59	4.03 ± 0.83	3.29 ± 1.45	2.74 ± 0.38
Kidney	15.59 ± 1.89	6.56 ± 1.01	5.56 ± 0.94	4.31 ± 0.83	3.82 ± 0.68
Stomach②	1.11 ± 0.24	1.28 ± 0.33	1.86 ± 0.60	1.90 ± 0.22	1.38 ± 1.63
Small intestine②	5.27 ± 0.46	15.12 ± 5.32	24.42 ± 4.62	34.63 ± 4.72	29.94 ± 5.47

① Expressed as % injected dose per gram. Average for 5 mice ± standard deviation.
② Expressed as % injected dose per organ.

Table 6 Biodistribution of [^{99m}Tc]5 in male ICR mice①

Organ	Time after injection				
	2min	10min	30min	60min	120min
Blood	13.53 ± 1.37	1.01 ± 0.11	0.70 ± 0.08	0.82 ± 0.06	0.98 ± 0.22
Brain	1.11 ± 0.34	0.40 ± 0.05	0.38 ± 0.05	0.51 ± 0.08	0.64 ± 0.11
Heart	11.48 ± 1.82	3.50 ± 0.23	2.61 ± 0.28	1.70 ± 0.14	1.73 ± 0.24
Liver	52.40 ± 3.64	67.08 ± 3.54	57.20 ± 3.09	40.49 ± 6.00	42.91 ± 3.43
Spleen	13.14 ± 2.57	21.59 ± 2.91	20.95 ± 2.72	18.63 ± 4.75	11.65 ± 4.25
Lung	31.98 ± 4.58	13.89 ± 1.39	8.66 ± 0.99	9.69 ± 0.52	9.18 ± 1.87
Kidney	7.11 ± 0.80	2.83 ± 0.33	2.87 ± 0.73	2.98 ± 0.36	4.12 ± 0.56
Stomach②	0.63 ± 0.10	0.50 ± 0.06	0.74 ± 0.08	0.93 ± 0.12	1.15 ± 0.13
Small intestine②	2.30 ± 0.45	3.84 ± 0.48	9.85 ± 0.72	16.64 ± 2.29	14.93 ± 2.70

① Expressed as % injected dose per gram. Average for 5 mice ± standard deviation.
② Expressed as % injected dose per organ.

Table 7 Biodistribution of [^{99m}Tc]3-5 at 2min with/without PgP blocked by cyclosporin A in male ICR mice[①]

Organ	[^{99m}Tc]3	[^{99m}Tc]3[②]	[^{99m}Tc]4	[^{99m}Tc]4[②]	[^{99m}Tc]5	[^{99m}Tc]5[②]
Blood	2.02 ± 0.17	4.20 ± 0.37	3.80 ± 0.71	4.24 ± 0.19	13.53 ± 1.37	13.07 ± 1.01
Brain	4.10 ± 0.38	6.34 ± 0.81	2.30 ± 0.27	3.68 ± 0.07	1.11 ± 0.34	1.64 ± 0.17

① Expressed as % injected dose per gram. Average for 5 mice ± standard deviation.
② Biodistribution of [^{99m}Tc]3-5 at 2min with PgP blocked by Cyclosporin A in male ICR mice.

3 Conclusions

Synthesis of organometallic complexes mimicking chalcone structure with ^{99m}Tc cyclopentadienyl tricarbonyl core using a two-step sequential reaction is described in this article. The first step is to prepare the intermediate ([CH$_3$COCp^{99m}Tc(CO)$_3$]) through the DLT method, and the second step is base-catalyzed Claisen condensation with appropriate aldehydes. We want to emphasize that this pathway can probably be applied to label other aldehydes with a [Cp^{99m}Tc(CO)$_3$] core, especially biomacromolecule, by avoiding the damage of 150℃ heat to the aldehyde in one-step pathway. In vitro fluorescent staining pictures of Aβ plaques on brain sections of patients diagnosed with AD and Tg mouse were clear for complexes 4 and 5. The binding assay using Aβ$_{1-42}$ aggregates indicated that the K_i value ranges from 899 to 108nM as the extension of conjugated π system, among which complex 5 has the highest affinity. In vitro autoradiography on section of transgenic mouse brain also confirmed the affinity of [^{99m}Tc]5 (K_i = 108nM). In biodistribution, [^{99m}Tc]3 ((4.10 ± 0.38)% ID/g at 2min, brain$_{2min}$/brain$_{60min}$ ratio: 8.20) and [^{99m}Tc]4 ((2.30 ± 0.27)% ID/g at 2min, brain$_{2min}$/brain$_{60min}$ ratio: 4.18) showed excellent initial uptakes and fast clearance in the brain, while [^{99m}Tc]5 ((1.11 ± 0.34)% ID/g at 2min, brain$_{2min}$/brain$_{60min}$ ratio: 1.73) was also good as an ^{99m}Tc-labeled ligand for Aβ imaging. Meanwhile, [^{99m}Tc]3-5 are probably substrates for the rodent PgP transporter. Therefore, the pretreating BBB abilities of these complexes are more remarkable with the PgP blocked. These findings suggest that additional effort should be made to explore why these kinds of complexes can penetrate the BBB more efficiently than other ^{99m}Tc-labeled ligands, which may lead to some new suggestions about how to design ^{99m}Tc-labeled CNS probes. In conclusion, these small technetium-99m complexes (< 500 Da) designed by an integrated approach mimicking chalcone provide encouraging evidence that development of a ^{99m}Tc-labeled agent for imaging Aβ plaques in the brain may be feasible.

4 Experimental Section

4.1 General information

All reagents used in the synthesis were commercial products and were used without further purification unless otherwise indicated. The ^{99m}Tc-pertechnetate was eluted from a commercial

^{99}Mo/^{99m}Tc generator which was obtained from Beijing Atomic High-tech Co. The ^{1}H NMR spectra were obtained at 400MHz on a Bruker spectrometer in CDCl$_3$ at room temperature with TMS as an internal standard. Chemical shifts were reported as δ values with respect to residual solvents. The ^{13}C NMR spectra were obtained at 100MHz on Bruker spectrometer in CDCl$_3$ at room temperature. Chemical shifts were reported as δ values with respect to residual solvents. The multiplicity is defined by s(singlet), d(doublet), t(triplet), m(multiplet). Mass spectrometry was acquired under the Surveyor MSQ Plus(ESI)(Waltham, MA, USA) instrument. X-ray crystallography data were collected on a Bruker Smart APEX Ⅱ diffractometer (Bruker Co., Germany). Reactions were monitored by TLC(TLC Silica gel 60 F$_{254}$, Merck). Radiochemical purity was determined by HPLC performed on a Shimadzu SCL-20 AVP equipped with a Bioscan Flow Count 3200 NaI/PMT γ-radiation scintillation detector. Separations were achieved on a Venusil MP C18 column(Agela Technologies, 10μm, 10mm × 250mm) eluted with a binary gradient system at a 4.0mL/min flow rate. Mobile phase A was water, while mobile phase B was acetonitrile. Fluorescent observation was performed by Axio Oberver Z1(Zeiss, Germany) equipped with DAPI(excitation, 405nm) and GFP filter sets(excitation, 505nm). The purity of the synthesized key compounds was determined using analytical HPLC and was found to be more than 95%. Normal ICR mice(five weeks, male) were used for biodistribution experiments. All protocols requiring the use of mice were approved by the animal care committee of Beijing Normal University. Post-mortem brain tissues from an autopsy-confirmed case of AD(91-year-old, male, 35μm, prefrontal cortex) and a control subject(69-year-old, female, 35μm, prefrontal cortex) were kindly gifted from Dr. Jiapei Dai, which were obtained from the Netherlands Brain Bank(NBB) by autopsy. Transgenic mice brain tissues (C57BL6, APPswe/PSEN1, 11 months old, male) were purchased from Institute of Laboratory Animal Sciences, Chinese Academy of Medical Sciences.

4.2 Chemistry

(2E,4E)-5-(4-(Dimethylamino)phenyl)penta-2,4-dienal(1). To a stirring solution of(E)-3-(4-(dimethylamino)phenyl)acrylaldehyde(526mg, 3.0mmol) in dry THF, (1,3-dioxolan-2-yl)methyltriphenylphosphonium(1.55g, 3.6mmol) was added to form a suspension at room temperature, followed by 18-crown-6(79mg, 0.3mmol) and NaH(60% in paraffin wax, 200mg, 5.0mmol) in a ice bath. The reaction was then stirred under room temperature for 2h until quenched by adding 2.0mL of concentrated hydrochloric acid. Thirty minutes later, the solution was neutralized with saturated K$_2$CO$_3$ aqueous solution and extracted with ethyl acetate(3 × 50mL). The organic layer was dried over Na$_2$SO$_4$. After the solvent was removed, the residue was purified by silica gel chromatography(ethyl acetate/petroleum ether = 6∶1, v/v) to afford the final products(yield 55%). ^{1}H NMR(CDCl$_3$, 400MHz): δ 3.05(s,6H), 6.15-6.20(m, 1H), 6.51-6.57(m, 1H), 6.68-6.96(m, 3H), 7.22-7.46(m, 3H), 9.55-9.60(m, 1H). MS: m/z calcd for [C$_{13}$H$_{15}$NO + H]$^+$ 202.1; found 202.4.

(Acetylcyclopentadienyl)tricarbonylrhenium(2). To a stirring solution of(cyclopentadienyl) tricarbonylrhenium(200mg, 0.6mmol) in dry CH_2Cl_2 in ice bath, acetyl chloride (94mg, 1.2mmol) was added dropwise. The reaction mixture was stirred for 30min at 0℃ and 30min at room temperature. After adding 50mL of water, the mixture was extracted with ethyl acetate (3×30mL). The organic layer was dried over Na_2SO_4. The solvent was removed to afford the white solid products(yield 95%). ^{1}H NMR(CDCl$_3$,400MHz): δ 2.34(s,3H),5.40(t, J = 2.3Hz,2H), 5.98(t, J =2.3Hz,2H). MS: m/z calcd for $[C_{10}H_7O_4^{187}Re+H]^+$ 379.0;found 379.1.

1-[(2E)-1-Oxo-3-(4-dimethylaminophenyl)-2-propenyl]-(cyclopentadienyl) tricarbonylrhenium (3). Complex 2 (50mg, 0.15mmol), 4-(dimethylamino) benzaldehyde (30mg, 0.2mmol), and solid NaOH(40mg, 1.0mmol) were dissolved in anhydrous ethanol(5mL). The mixture was stirred at room temperature for 6h until stopping by adding 30mL of water. After extraction with ethyl acetate, the solvent was removed under reduced pressure. The residue was purified by silica gel chromatography(ethyl acetate/petroleum ether = 1∶1, v/v) to afford a yellow solid. The solid was crystallized from a mixture of ethanol and dichloromethane as light-yellow crystals(yield 91%). mp180.2-182.4℃. ^{1}H NMR(CDCl$_3$,400MHz): δ 3.05(s,6H), 5.41(t, J = 2.3Hz,2H),6.09(t, J = 2.3Hz,2H),6.68(d, J = 8.9Hz,2H),6.73(d, J = 15.3Hz,2H),7.50(d, J = 8.8Hz,1H),7.78(d, J = 15.3Hz,1H). ^{13}C NMR(CDCl$_3$, 100MHz): δ 40.1, 84.6, 87.8, 98.8, 111.8, 114.9, 122.0, 130.7, 145.6, 152.3, 183.8, 192.3. HRMS: m/z calcd for $[C_{19}H_{16}NO_4^{187}Re+H]^+$ 510.0715;found 510.0717.

1-[(2E,4E)-1-Oxo-5-(4-dimethylaminophenyl)-2,4-pentadienyl]-(cyclopentadienyl) tricarbonylrhenium(4). Complex 4 was prepared following the procedure used for 3. The residue was purified by silica gel chromatography(ethyl acetate/petroleum ether = 1∶1, v/v) to afford an orange solid. The solid was crystallized from a mixture of ethanol and dichloromethane as light-orange crystals(yield 87%);mp217.6-218.4℃. ^{1}H NMR(CDCl$_3$,400MHz): δ 3.02(s, 6H),5.40(t, J = 2.3Hz,2H),6.04(t, J = 2.3Hz,2H),6.39(d, J = 14.6Hz,1H),6.67(d, J = 8.8Hz,2H),6.77(dd, J_1 = 15.3Hz, J_2 = 11.3Hz,1H),6.97(d, J = 15.4Hz,1H),7.39 (d, J = 8.8Hz,2H),7.61(dd, J_1 = 14.6Hz, J_2 = 11.3Hz,1H). ^{13}C NMR(CDCl$_3$,100MHz): δ 40.2, 85.0, 87.7, 98.5, 111.9, 120.6, 121.8, 123.7, 129.1, 144.3, 146.2, 151.3, 183.8, 192.2. HRMS: m/z calcd for $[C_{21}H_{18}NO_4^{187}Re+H]^+$ 536.0872;found 536.0859.

1-[(2E,4E,6E)-1-Oxo-7-(4-dimethylaminophenyl)-2,4,6-heptatrienyl)]-(cyclopentadienyl)tricarbonylrhenium(5). Complex 5 was prepared following the procedure used for 3. The residue was purified by silica gel chromatography(ethyl acetate/petroleum ether = 1∶1, v/v) to afford a red solid(yield 85%);mp205.2-207.2℃. ^{1}H NMR(CDCl$_3$,400MHz): δ 3.00(s, 6H),5.40(s,2H),6.04(s,2H),6.32-6.44(m,2H),6.67(d, J = 8.8Hz,2H),6.72-6.74 (m,2H),6.82-6.88(m,1H),7.34-7.40(m,2H),7.53(dd, J_1 = 14.5Hz, J_2 = 11.8Hz, 1H). ^{13}C NMR(CDCl$_3$,100MHz): δ 40.2, 85.1, 87.7, 98.3, 112.1, 121.5, 123.9, 124.7, 127.9, 128.8, 138.9, 145.1, 145.5, 150.8, 183.8, 192.1. HRMS: m/z calcd for $[C_{23}H_{20}NO_4^{187}Re+H]^+$ 562.1028;found 562.1030.

1-[(2E,4E)-1-Oxo-5-(4-dimethylaminophenyl)-2,4-pentadienyl]-ferrocene(6). Acetylferrocene (35mg, 0.15mmol), (E)-3-(4-(dimethylamino) phenyl) acrylaldehyde (35mg,

0.20mmol), and solid NaOH(40mg, 1.0mmol) were dissolved in anhydrous ethanol(5mL). The mixture was stirred at room temperature for 6h. After filtration, the solvent was removed under reduced pressure. The residue was purified by silica gel chromatography(ethyl acetate/petroleum ether = 1 : 1, v/v) to afford a dark-red solid (yield 90%). ^{1}H NMR (CDCl$_3$, 400MHz): δ 3.01(s,6H), 4.20(s,5H), 4.53(s,2H), 4.86(s,2H), 6.53-6.98(m,5H), 7.32-7.79(m,3H). MS: m/z calcd for $[C_{23}H_{23}FeNO_4 + H]^+$ 386.1; found 386.5.

1-[(2E, 4E, 6E)-1-Oxo-7-(4-dimethylaminophenyl)-2,4,6-heptatrienyl)]-ferrocene (7). Complex 7 was prepared following the procedure used for 6. The residue was purified by silica gel chromatography(ethyl acetate/petroleum ether = 1 : 1, v/v) to afford a brown solid(yield 88%). ^{1}H NMR(CDCl$_3$, 400MHz): δ 3.01(s,6H), 4.20(s,5H), 4.54(s,2H), 4.85(s, 2H), 6.53-6.50(m,1H), 6.56(d, J = 14.8Hz,1H), 6.64-6.86(m,4H), 6.79-6.86(m, 1H), 7.33-7.37(m,2H), 7.47-7.54(m,1H). MS: m/z calcd for $[C_{25}H_{25}FeNO_4 + H]^+$ 412.1; found 412.6.

4.3 X-ray crystallography

Single-crystal X-ray diffraction measurements were carried out on a Bruker Smart APEXII CCD diffractometer at 150(2)K using graphite monochromated Mo Kα radiation(λ = 0.071070nm). An empirical absorption correction was applied using the SADABS program[39]. All structures were solved by direct methods and refined by full-matrix least-squares on F^2 using the SHELXL-97 program package[40]. All of the hydrogen atoms were geometrically fixed using the riding model.

4.4 In vitro fluorescent staining of Aβ plaques in transgenic mouse brain sections and human brain materials

Paraffin-embedded brain sections of Tg mouse(C57BL6, APPswe/PSEN1, 11 months old, male, 6μm) were used for the fluorescent staining. The brain sections were deparaffinized with 2 × 20min washes in xylene, 2 × 5min washes in 100% ethanol, 5min washes in 90% ethanol/H$_2$O, 5min wash in 80% ethanol/H$_2$O, 5min wash in 60% ethanol/H$_2$O, and running tap water for 10min and then incubated in PBS(0.2M, pH = 7.4) for 30min. The sections of human prefrontal cortex(35μm) were obtained from the Netherlands Brain Bank(NBB) by autopsy, and the sections were stored at 0-4℃ in 50% glycerol diluted with 0.05M TBS before use.

The brain sections were incubated with 10% ethanol solution(1.0μM) of 3, 4 and 5 for 10min. The localization of plaques was confirmed by staining with thioflavin-S(0.125%) on the adjacent sections. Finally, the sections were washed with 50% ethanol and PBS(0.2M, pH = 7.4) for 10min. Fluorescent observation was performed by Axio Oberver Z1 (Zeiss, Germany) equipped with DAPI(excitation, 405nm) and GFP filter sets(excitation, 505nm).

4.5 Binding assay in vitro using Aβ aggregates

The trifluoroacetic acid salt forms of peptides Aβ$_{1-42}$ were purchased from AnaSpec. Aggregation of peptides was carried out by gently dissolving the peptide [0.25mg/mL for $A\beta_{1-42}$] in a buffer solution(pH = 7.4) containing 10mM potassium dihydrogen phosphate and 1mM EDTA. The solutions were incubated at 37℃ for 42h with gentle and constant shaking. Inhibition experiments were carried out in 12mm × 75mm borosilicate glass tubes according to procedures described previously with some modification[17]. Briefly, 100μL of aggregated Aβ fibrils(28nM in the final assay mixture) was added to a mixture containing 100μL of radioligand([^{125}I]IMPY, 100000cpm/100μL), 100μL of inhibitors(complex 3, 4 or 5, 10^{-4}M to 10^{-10}M in ethanol), and 700μL of PBS(0.2M, pH = 7.4) in a final volume of 1.0mL. Nonspecific binding was defined in the presence of 1μM IMPY. The mixture was incubated for 2h at 37℃, and then the bound and free radioactive fractions were separated by vacuum filtration through borosilicate glass fiber filters(Whatman GF/B) using a Mp-48T cell harvester(Brandel, Gaithersburg, MD). The radioactivity from filters containing the bound [^{125}I]-ligand was measured in a γ-counter(WALLAC/Wizard1470, USA) with 70% efficiency. Under the assay conditions, the specifically bound fraction accounted for about 10% of total radioactivity. The inhibitory concentration(IC_{50}) was determined using Graph Pad Prism 4.0, and the inhibition constant(K_i) was calculated using the Cheng-Prusoff inhibition constant equation: $K_i = IC_{50}/(1 + [L]/K_d)$[41].

4.6　Preparation of [^{99m}Tc]3, [^{99m}Tc]4, and [^{99m}Tc]5

To an orange solution of 1.0mg of acetylferrocene and 1.0mg of $Mn(CO)_5Br$ in 1mL of DMF, 1.0 mL of $^{99m}TcO_4^-$ aqueous solution(10mCi) were added. The reaction mixture was kept at 150℃ for 20min in a sealed vial. After extraction with CH_2Cl_2 and water, inorganic salts were separated from the product while the intermediate [^{99m}Tc]2 was saved in CH_2Cl_2. After removing the CH_2Cl_2 under nitrogen gas, the residue was dissolved again in 1.0mL of absolute ethanol. The appropriate aldehyde(3.0mg) was added into the reaction, as well as 0.5mg NaOH to catalyze the Claisen Condensation. The reaction was kept under room temperature for 30min. After extraction by CH_2Cl_2, the solvent was evaporated under nitrogen gas and the residue was dissolved in CH_3CN and purified by radio-HPLC under conditions as following: Venusil MP C18 column(Agela Technologies, 10mm × 250mm), CH_3CN/H_2O = 70/30 for [^{99m}Tc]3, [^{99m}Tc]4, CH_3CN/H_2O = 80/20 for [^{99m}Tc]5, 4mL/min, UV = 254nm.

4.7　Binding assay using Aβ aggregates with [^{99m}Tc]2-5

The binding assay was performed by mixing 100μL of [^{99m}Tc]2-5 (100000cpm/100μL), 100μL of $A\beta_{1-42}$ aggregates(7.27μg/mL), and 800μL of 10% ethanol in 12mm × 75mm borosilicate glass tubes. The blocking assay was performed by conducting the binding assay in the presence of excess of rhenium complexes 2-5 (1μM for 2-4, 0.5μM for 5) as blocking agents. After incubation for 2h at room temperature, the mixture was filtered through GF/B filters (Whatman GF/B) using a Mp-48T cell harvester(Brandel, Gaithersburg, MD). Filters containing the bound ^{99m}Tc-labeled form were examined in an automatic γ-counter (Wallac 1470 Wizard, USA).

4.8 Autoradiography in vitro using brain sections of human and transgenic model mouse

Paraffin-embedded brain sections of Tg and control mice were deparaffinized with 2 × 20min washes in xylene, 2 × 5min washes in 100% ethanol, a 5min wash in 90% ethanol/H$_2$O, a 5min wash in 80% ethanol/H$_2$O, a 5min wash in 60% ethanol/H$_2$O, and a 10min wash in running tap water and then incubated in PBS(0.2M, pH = 7.4) for 30min. The sections were incubated with [^{99m}Tc]5 (370KBq/100μL) for 1h at room temperature. They were then washed in 40% EtOH before being rinsed with water for 1min. After drying, the sections were exposed to a phosphorus plate (Perkin-Elmer, USA) for 4h. In vivo autoradiographic images were obtained using a phosphor imaging system (Cyclone, Packard). After autoradiographic examination, the same mouse brain sections were stained by thioflavin-S to confirm the presence of Aβ plaques. After drying, the fluorescent observation was performed by Axio Oberver Z1 (Zeiss, Germany) equipped with DAPI (excitation, 405nm) and GFP filter sets (excitation, 505nm).

4.9 Biodistribution studies

A saline solution containing the HPLC-purified ^{99m}Tc-labled tracer (100μL, 10% ethanol, 5μCi) was injected via tail vein of ICR mice (five weeks, male). The mice were sacrificed exactly at 2min, 10min, 30min, 60min, and 120min. Samples of blood and organs of interest were removed, weighed and counted in an automatic γ-counter (Wallac 1470 Wizard, USA). The results were expressed in terms of the percentage of the injected dose per gram (%ID/g) of blood or organs.

The biodistribution with PgP blocked was conducted using male ICR mice (5 weeks, male, $n = 5$) pretreated with Cyclosporin A (manufactured by Nanjing Duly biotech Co., Ltd. USP grade). Each mouse was injected via tail vein with 50mg/kg (100μL solution consisting of 10% EtOH, 15% saline and 75% propylene glycol), 1h prior to administration of [^{99m}Tc]3-5[42]. The blood and brain uptakes at 2min were measured the same way narrated before.

4.10 Partition coefficient determination

The determination of partition coefficients of ^{99m}Tc-labeled complexes were performed according to the procedure previously reported[19]. A solution of ^{99m}Tc-labeled complexes (1.5MBq) was added to a premixed suspensions containing 3.0g n-octanol and 3.0g PBS (0.05M, pH = 7.4) in a test tube. The test tube was vortexed for 3min at room temperature, followed by centrifugation for 5min at 3000 rpm. Two samples from the n-octanol (50μL) and water (500μL) layers were measured. The partition coefficient was expressed as the logarithm of the ratio of the count per gram from n-octanol versus PBS. Samples from the n-octanol layer were repartitioned until consistent partition coefficient values were obtained. The measurement was done in triplicate and repeated three times.

Associated Content

ⓈSupporting information

Purities of key target compounds; HPLC profiles of 3, [^{99m}Tc]3, 4, [^{99m}Tc]4, 5, [^{99m}Tc]5; ^{1}H NMR spectra, ^{13}C NMR spectra and HRMS data of rhenium complexes; crystal parameters for complex 3 and 4. This material is available free of charge via the Internet at http://pubs.acs.org.

Author Information

Corresponding Author

*Phone: +86-10-58808891. Fax: +86-10-58808891. E-mail: cmc@bnu.edu.cn(M.C.); liuboli@bnu.edu.cn(B.L.).

Notes

The authors declare no competing financial interest.

Acknowledgements

We present special thanks to Dr. Jin Liu(College of Life Science, Beijing Normal University) for assistance in the in vitro neuropathological staining, and Dr. Xuebing Deng(College of Chemistry, Beijing Normal University) for assistance in the X-ray diffraction. This work was supported by National Natural Science Foundation of China(grants 21201019, 20871021, and 31070961) and Research Fund for the Doctoral Program of Higher Education of China(20120003120013).

Abbreviations Used

Aβ, β-amyloid; AD, Alzheimer's Disease; PET, positron emission tomography; SPECT, single photon emission computed tomography; CNS, central nervous system; BBB, blood-brain barrier; DLT, double ligand transfer; Cp, cyclopentadienyl; MAMA, monoamine-monoamide dithiol; BAT, bis-amino-bis-thiol; HPLC, high performance liquid chromatography; NBB, Netherlands Brain Bank; PgP, permeability-glycoprotein 1.

References

[1] Selkoe, D. J. Imaging Alzheimer's amyloid. Nature Biotechnol. 2000, 18, 823-824.

[2] Mathis, C. A.; Wang, Y.; Klunk, W. E. Imaging beta-amyloid plaques and neurofibrillary tangles in the aging human brain. Curr. Pharm. Des. 2004, 10, 1469-1492.

[3] Nordberg, A. PET imaging of amyloid in Alzheimer's disease. Lancet Neurol. 2004, 3, 519-527.

[4] Agdeppa, E. D.; Kepe, V.; Liu, J.; Flores-Torres, S.; Satyamurthy, N.; Petric, A.; Cole, G. M.; Small, G. W.; Huang, S. C.; Barrio, J. R. Binding characteristics of radiofluorinated 6-dialkylamino-2-naphthylethylidene derivatives as positron emission tomography imaging probes for beta-amyloid plaques in Alzheimer's disease. J Neurosci. 2001, 21, RC189.

[5] Shoghi-Jadid, K.; Small, G. W.; Agdeppa, E. D.; Kepe, V.; Ercoli, L. M.; Siddarth, P.; Read, S.; Satyamur-

thy, N. ; Petric, A. ; Huang, S. C. ; Barrio, J. R. Localization of neurofibrillary tangles and beta-amyloid plaques in the brains of living patients with Alzheimer disease. Am. J. Geriatr. Psychiatry 2002,10,24-35.

[6] Mathis, C. A. ; Wang, Y. ; Holt, D. P. ; Huang, G. F. ; Debnath, M. L. ; Klunk, W. E. Synthesis and evaluation of 11C-labeled 6-substituted 2-arylbenzothiazoles as amyloid imaging agents. J. Med. Chem. 2003,46, 2740-2754.

[7] Klunk, W. E. ; Engler, H. ; Nordberg, A. ; Wang, Y. ; Blomqvist, G. ; Holt, D. P. ; Bergstrom, M. ; Savitcheva, I. ; Huang, G. F. ; Estrada, S. ; Ausen, B. ; Debnath, M. L. ; Barletta, J. ; Price, J. C. ; Sandell, J. ; Lopresti, B. J. ; Wall, A. ; Koivisto, P. ; Antoni, G. ; Mathis, C. A. ; Langstrom, B. Imaging brain amyloid in Alzheimer's disease with Pittsburgh Compound-B. Ann. Neurol. 2004, 55,306-319.

[8] Ono, M. ; Wilson, A. ; Nobrega, J. ; Westaway, D. ; Verhoeff, P. ; Zhuang, Z. P. ; Kung, M. P. ; Kung, H. F. 11C-labeled stilbene derivatives as Abeta-aggregate-specific PET imaging agents for Alzheimer's disease. Nucl. Med. Biol. 2003, 30,565-571.

[9] Verhoeff, N. P. ; Wilson, A. A. ; Takeshita, S. ; Trop, L. ; Hussey, D. ; Singh, K. ; Kung, H. F. ; Kung, M. P. ; Houle, S. In vivo imaging of Alzheimer disease beta-amyloid with [11C] SB-13 PET. Am. J. Geriatr. Psychiatry 2004,12,584-595.

[10] Okamura, N. ; Shiga, Y. ; Furumoto, S. ; Tashiro, M. ; Tsuboi, Y. ; Furukawa, K. ; Yanai, K. ; Iwata, R. ; Arai, H. ; Kudo, Y. ; Itoyama, Y. ; Doh-ura, K. In vivo detection of prion amyloid plaques using [(11)C] BF-227 PET. Eur. J. Nucl. Med. Mol. Imaging 2010,37,934-941.

[11] Rowe, C. C. ; Ackerman, U. ; Browne, W. ; Mulligan, R. ; Pike, K. L. ; O'Keefe, G. ; Tochon-Danguy, H. ; Chan, G. ; Berlangieri, S. U. ; Jones, G. ; Dickinson-Rowe, K. L. ; Kung, H. P. ; Zhang, W. ; Kung, M. P. ; Skovronsky, D. ; Dyrks, T. ; Holl, G. ; Krause, S. ; Friebe, M. ; Lehman, L. ; Lindemann, S. ; Dinkelborg, L. M. ; Masters, C. L. ; Villemagne, V. L. Imaging of amyloid beta in Alzheimer's disease with 18F-BAY94-9172, a novel PET tracer:proof of mechanism. Lancet Neurol. 2008, 7,129-135.

[12] Choi, S. R. ; Golding, G. ; Zhuang, Z. ; Zhang, W. ; Lim, N. ; Hefti, F. ; Benedum, T. E. ; Kilbourn, M. R. ; Skovronsky, D. ; Kung, H. F. Preclinical properties of 18F-AV-45: a PET agent for Abeta plaques in the brain. J. Nucl. Med. 2009, 50,1887-1894.

[13] Kung, H. F. ; Choi, S. R. ; Qu, W. ; Zhang, W. ; Skovronsky, D. 18F stilbenes and styrylpyridines for PET imaging of A beta plaques in Alzheimer's disease: a miniperspective. J. Med. Chem. 2010,53,933-941.

[14] Kung, M. P. ; Hou, C. ; Zhuang, Z. P. ; Zhang, B. ; Skovronsky, D. ; Trojanowski, J. Q. ; Lee, V. M. ; Kung, H. F. IMPY: an improved thioflavin-T derivative for in vivo labeling of beta-amyloid plaques. Brain Res. 2002, 956,202-210.

[15] Zhuang, Z. P. ; Kung, M. P. ; Wilson, A. ; Lee, C. W. ; Plossl, K. ; Hou, C. ; Holtzman, D. M. ; Kung, H. F. Structure-activity relationship of imidazo[1,2-a]pyridines as ligands for detecting beta-amyloid plaques in the brain. J. Med. Chem. 2003,46,237-243.

[16] Newberg, A. B. ; Wintering, N. A. ; Plossl, K. ; Hochold, J. ; Stabin, M. G. ; Watson, M. ; Skovronsky, D. ; Clark, C. M. ; Kung, M. P. ; Kung, H. F. Safety, biodistribution, and dosimetry of 123I-IMPY: a novel amyloid plaque-imaging agent for the diagnosis of Alzheimer's disease. J. Nucl. Med. 2006, 47,748-754.

[17] Qu, W. ; Kung, M. P. ; Hou, C. ; Benedum, T. E. ; Kung, H. F. Novel styrylpyridines as probes for SPECT imaging of amyloid plaques. J. Med. Chem. 2007,50,2157-2165.

[18] Watanabe, H. ; Ono, M. ; Haratake, M. ; Kobashi, N. ; Saji, H. ; Nakayama, M. Synthesis and characterization of novel phenylindoles as potential probes for imaging of beta-amyloid plaques in the brain. Bioorg. Med. Chem. 2010,18,4740-4746.

[19] Cui, M. ; Ono, M. ; Kimura, H. ; Kawashima, H. ; Liu, B. L. ; Saji, H. Radioiodinated benzimidazole deriva-

tives as single photon emission computed tomography probes for imaging of beta-amyloid plaques in Alzheimer's disease. Nucl. Med. Biol. 2011, 38, 313-320.

[20] Kung, H. F.; Kim, H. J.; Kung, M. P.; Meegalla, S. K.; Plossl, K.; Lee, H. K. Imaging of dopamine transporters in humans with technetium-99m TRODAT-1. Eur. J. Nucl. Med. 1996, 23, 1527-1530.

[21] Han, H.; Cho, C. G.; Lansbury, P. T. Technetium complexes for the quantitation of brain amyloid. J. Am. Chem. Soc. 1996, 118, 4506-4507.

[22] Zhuang, Z. P.; Kung, M. P.; Hou, C.; Ploessl, K.; Kung, H. F. Biphenyls labeled with technetium 99m for imaging beta-amyloid plaques in the brain. Nucl. Med. Biol. 2005, 32, 171-184.

[23] Liu, S.; Edwards, D. S. 99mTc-Labeled small peptides as diagnostic radiopharmaceuticals. Chem. Rev. 1999, 99, 2235-2268.

[24] Dezutter, N. A.; Dom, R. J.; de Groot, T. J.; Bormans, G. M.; Verbruggen, A. M. 99mTc-MAMA-chrysamine G, a probe for beta-amyloid protein of Alzheimer's disease. Eur. J. Nucl. Med. 1999, 26, 1392-1399.

[25] Dezutter, N. A.; Sciot, R. M.; de Groot, T. J.; Bormans, G. M.; Verbruggen, A. M. In vitro affinity of 99Tcm-labelled N2S2 conjugates of chrysamine G for amyloid deposits of systemic amyloidosis. Nucl. Med. Commun. 2001, 22, 553-558.

[26] Serdons, K.; Verduyckt, T.; Cleynhens, J.; Terwinghe, C.; Mortelmans, L.; Bormans, G.; Verbruggen, A. Synthesis and evaluation of a 99mTc-BAT-phenylbenzothiazole conjugate as a potential in vivo tracer for visualization of amyloid beta. Bioorg. Med. Chem. Lett. 2007, 17, 6086-6090.

[27] Chen, X. J.; Yu, P. R.; Zhang, L. F.; Liu, B. Synthesis and biological evaluation of Tc-99m, Re-monoamine-monoamide conjugated to 2-(4-aminophenyl) benzothiazole as potential probes for beta-amyloid plaques in the brain. Bioorg. Med. Chem. Lett. 2008, 18, 1442-1445.

[28] Lin, K. S.; Debnath, M. L.; Mathis, C. A.; Klunk, W. E. Synthesis and beta-amyloid binding properties of rhenium 2-phenylbenzothiazoles. Bioorg. Med. Chem. Lett. 2009, 19, 2258-2262.

[29] Ono, M.; Ikeoka, R.; Watanabe, H.; Kimura, H.; Fuchigami, T.; Haratake, M.; Saji, H.; Nakayama, M. Synthesis and Evaluation of Novel Chalcone Derivatives with (99m) Tc/Re Complexes as Potential Probes for Detection of beta-Amyloid Plaques. ACS. Chem. Neurosci. 2010, 1, 598-607.

[30] Ono, M.; Ikeoka, R.; Watanabe, H.; Kimura, H.; Fuchigami, T.; Haratake, M.; Saji, H.; Nakayama, M. Tc-99m/Re complexes based on flavone and aurone as SPECT probes for imaging cerebral beta-amyloid plaques. Bioorg. Med. Chem. Lett. 2010, 20, 5743-5748.

[31] Wang, X.; Cui, M.; Yu, P.; Li, Z.; Yang, Y.; Jia, H.; Liu, B. Synthesis and biological evaluation of novel technetium-99m labeled phenylbenzoxazole derivatives as potential imaging probes for beta-amyloid plaques in brain. Bioorg. Med. Chem. Lett. 2012, 22, 4327-4331.

[32] Cui, M.; Tang, R.; Li, Z.; Ren, H.; Liu, B. 99mTc-and Re-labeled 6-dialkylamino-2-naphthylethylidene derivatives as imaging probes for beta-amyloid plaques. Bioorg. Med. Chem. Lett. 2011, 21, 1064-1068.

[33] Wenzel, M. Tc-99m labelling of Cymantrene analogues with different substituents. a new approach to Tc-99m radiodiagnostics. J. Labelled Compd. Radiopharm. 1992, 31, 641-650.

[34] Bernard, J.; Ortner, K.; Spingler, B.; Pietzsch, H. J.; Alberto, R. Aqueous synthesis of derivatized cyclopentadienyl complexes of technetium and rhenium directed toward radiopharmaceutical application. Inorg. Chem. 2003, 42, 1014-1022.

[35] N'Dongo, H. W.; Raposinho, P. D.; Fernandes, C.; Santos, I.; Can, D.; Schmutz, P.; Spingler, B.; Alberto, R. Preparation and biological evaluation of cyclopentadienyl-based 99mTc-complexes [(Cp-R) 99mTc(CO)$_3$] mimicking benzamides for malignant melanoma targeting. Nucl. Med. Biol. 2010, 37, 255-264.

[36] Ono, M.; Hori, M.; Haratake, M.; Tomiyama, T.; Mori, H.; Nakayama, M. Structure-activity relationship of chalcones and related derivatives as ligands for detecting of beta-amyloid plaques in the brain. Bioorg. Med. Chem. 2007, 15, 6388-6396.

[37] Ono, M.; Haratake, M.; Mori, H.; Nakayama, M. Novel chalcones as probes for in vivo imaging of beta-amyloid plaques in Alzheimer's brains. Bioorg. Med. Chem. 2007, 15, 6802-6809.

[38] Ono, M.; Watanabe, R.; Kawashima, H.; Cheng, Y.; Kimura, H.; Watanabe, H.; Haratake, M.; Saji, H.; Nakayama, M. Fluoro-pegylated chalcones as positron emission tomography probes for in vivo imaging of beta-amyloid plaques in Alzheimer's disease. J. Med. Chem. 2009, 52, 6394-6401.

[39] Sheldrick, G. M. SADABS, Program for Empirical Absorption Correction of Area Detector Data; University of Gottingen: Gottingen, Germany, 1996.

[40] Sheldrick, G. M. SHELXL-97 Program for the Refinement of Crystal Structure from Diffraction Data; University of Gottingen: Gottingen, Germany, 1997.

[41] Cheng, Y.; Prusoff, W. Relationship between the inhibition constant (K_i) and the concentration of inhibitor which causes 50% inhibition (I_{50}) of an enzymatic reaction. Biochem. Pharmacol. 1973, 22, 3099-3108.

[42] Shao, X.; Carpenter, G. M.; Desmond, T. J.; Sherman, P.; Quesada, C. A.; Fawaz, M.; Brooks, A. F.; Kilbourn, M. R.; Albin, R. L.; Frey, K. A.; Scott, P. J. H. Evaluation of [11C] N-Methyl Lansoprazole as a Radiopharmaceutical for PET Imaging of Tau Neurofibrillary Tangles. ACS Med. Chem. Lett. 2012, 3, 936-941.

Labelling of 6-Iodocholestercl in Melt*

Key words labelled compound, 6-iodocholestercl, melt method, iodine-125, isotope exchange, radiopharmaceutical

1 Introduction

It is well known that isotopic exchange of iodine in melt is a good approach to iodine labelling when the exchange in solution can not conveniently be achieved[1-4]. The advantage of this method is very efficient and rapid labelling technique, and the ease of preparation for labelling without addition or removal of solvent. However, it essentially required that the compound of interest is to be stable at the molten state. Another approach has been reported for iodine labelling of compounds melt with decomposition by using an unreactive solvent with a low melting point such as acetamide[2]. Recently a melt method has been also adapted to non-isotopic exchange such as the I-for-Br and the F-for-Br exchange[5,6].

The radioiodine analogue of 6-iodocholesterol(CL-6-I), synthesized by Liu Bo-Li, et al.,[7] has been used clinically as a new diagnostic agent for adrenal gland in China. This labelled compound was prepared by exchange of unlabelled compound with radioactive iodine using n-butanol as a solvent. However, longer reaction time was necessary(90% exchange after 3 hours at refluxing temperature). Therefore, the need was felt for a rapid method for labelling with high and reproducible yields, especially when ^{123}I should be used as label. Considering the notable thermal stability of CL-6-I, we tried to label CL-6-I with Na^{125}I by means of melt method.

2 Experimental

6-Iodocholesterol(CL-6-I) was synthesized starting from cholesterol as described in a previous work[7]. Labelling was performed in a open pyrex tube. Aqueous solution(pH = 11) of no carrier added Na^{125}I obtained from CIS(France) was used. The Na^{125}I solution was carefully transferred to CL-6-I in the bottom of the tube with a micropipette. The tube was then placed in a heated oil bath maintained at 160℃ ± 1℃. After cooling of the reaction mixture, a few milliters of ether and water were added and shaken to dissolve the solid. Activity of aqueous and organic phases were determined in a well type automatic scintillation counter and the labelling yield was calculated. Both organic and aqueous phases were also analyzed by thin-layer chromatography(TLC)

---

* Copartner: Jin Yutai, Minoru MAEDA, Masaharu KOJIMA. Reprinted from *Radioisotopes*, 1981, 30(5):275-277.

(silica gel 60F$_{254}$, 0.5mm, E. Merck). TLC showed a single radioactive spot coincident with the spot corresponding to cold CL-6-I(R_f 0.42).

3 Results and Discussion

The results of the time dependence of labelling yield from 0.5min to 30min at a temperature of 160℃ is shown in Fig. 1, indicating a very fast reaction. It can be seen that with the first 3min. the yield of ^{125}I-6-iodocholesterol(CL-6-^{125}I) reaches a saturation value of about 80%. It is supposed that the exchange reaction is a homogeneous reaction. Accordingly, the exchange rate should be expressed by the following equation:

$$-\ln(1-F) = \frac{[A]+[B]}{[A][B]}Rt$$

where, [A] and [B] indicate the concentration of CL-6-I and NaI respectively; R is the overall rate of exchange; F is the exchange fraction at time t. Since [A], [B] and R are all constant for a given experimental conditions, it might be expressed that a semilogarithmic plot of $(1-F)$ versus time t would be approximately a straight line. As shown in Fig. 2, the passing of the straight line through the origin confirms that the reaction in melt approximately followed the exponential exchange law, the half-time for the exchange being about 36s.

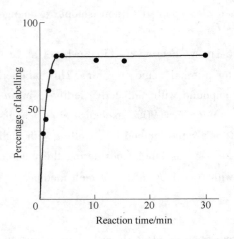

Fig. 1 Time dependence of iodine exchange of 6-iodocholesterol(CL-6-I) in melt
(Reaction conditions: 20mg CL-6-I, 10μg NaI-carrier, 160℃)

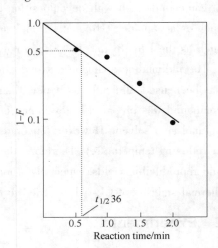

Fig. 2 A semilogarithmic plot of 1-F versus reaction time for iodine exchange of 6-iodocholesterol(CL-6-I) in melt
(Reaction conditions: 20mg CL-6-I, 10μg NaI-carrier, 160℃)

The dependence of the labelling yields of CL-6-^{125}I in melt on the addition of NaI-carrier is illustrated in Fig. 3. The labelling yield decreases by adding carrier and addition of more than about 100μg NaI-carrier leads to a considerable decrease of yield.

The influence of the amount of CL-6-I on labelling yield without addition of NaI-carrier was also investigated. In view of time dependence, 5min. was chosen as reaction time. In this case

CL-6-^{125}I was practically isolated after reaction and the respective radiochemical yield was calculated. Changing the amount of CL-6-I from 1mg to 20mg did not affect the labelling yield at all and in all runs the radiochemical yield is kept constant at about 97%.

From the above results, there are some advantages of molten method in the iodine exchange of CL-6-I with Na^{125}I. The rate of exchange by melt method is considerably greater than that of exchange using *n*-butanol as solvent; the half-time for the exchange in the former is only 36s. in the presence of 10μg of NaI-carrier while that in the latter is one hour. Further, the radiochemical purity of CL-6-^{125}I obtained by the present method, after extraction

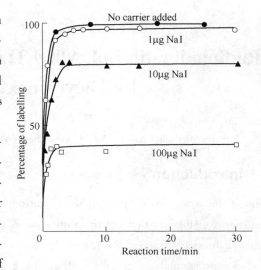

Fig. 3   Carrier dependence of iodine exchange of 6-iodocholesterol(CL-6-I) in melt
(Reaction conditions: 20mg CL-6-I, 160℃)

with a mixture of ether and water, proved sufficient for direct medical application. The optimum reaction time in melt is 5min. and longer heating time more than 60min. results in a only slight decomposition. In conclusion, we can say that the iodine exchange of CL-6-I in melt is an ideal method of labelling CL-6-I for routine production runs because of its rapidness and a high yield.

References

[1] H. Elias:Proc. 9 th Japan Conf. on Radioisotopes,p. 538,Tokyo(1969).
[2] H. Elias,Ch. Arnold and G. Kloss:Int. J. Appl. Radiat. Isotopes,24,463(1973).
[3] R. Ghandadian,S. L. Waters,M. L. Thakur and G. D. Ckrisholm:ibid. ,26,343(1975).
[4] G. D. Jr. Robinson,and A. W. Lee:J. Nucl. Med. ,16,17(1975).
[5] M. L. Thakur,B. M. Chauser and R. F. Hudson:Int. J. Appl. Radiat. Isotopes,26,319(1975).
[6] G. Stöcklin,E. J. Knust and M. Schuller:J. Labelled Compds. Radiopharm. ,17,353(1980).
[7] Liu Boli,Jin Yutai,Liu Zhenghao,Sun Zhaoxiang,Lu Gongxu and Li Taihau:J. Nuclear and Radiochemistry(China),2,107(1980).

Radioiodination of Alkyl Halides with Na^{125}I Catalyzed by Dicyclohexyl-18-crown-6[*]

Key words   alkyl halides, Na^{125}I, dicyclohexyl-18-crown-6, radioiodination

1   Introduction

One of the most important methods for introducing radioactive halogen into a molecule is the halogen exchange carried out in solution or in molten organic compounds[1]. However, in some cases of practical interest the rate of the halogen exchange in solution is slow. In recent years phase-transfer catalysts or crown ethers have become increasingly important tools in organic chemistry. The application of these catalysts to radiosynthetic works, especially in ^{13}F-and ^{131}I labellings, has also been reported[2-5]. However, no or little attention has been paid to the use of crown ethers in the halogen exchanges in radiosynthetic work. It is the purpose of this paper to report a method or labelling of halogenated hydrocarbons with radio-iodine in the presence of crown ether which plays an important role in the exchange at relatively low temperatures. Dicyclohexyl-18-crown-6 was chosen as a crown ether because of its low melting point, and simple alkyl halides were used as substrates for optimizing the reaction conditions.

2   Experimental

Commercial alkyl halides used were reagent grade and used without further purincation. The isotopic and non-isotopic iodine exchange were performed as follows. A homogeneous mixture of alkyl halides in neat or organic solvents and dicyciohexyl-18-crown-6(DCC) was placed in an open glass tube containing dry no carrier added Na^{125}I(0.37-0.74MBq)(10-20μCi)(CIS, France). When the experiments were carried out at a definite temperature, the glass tube was placed in an oil bath maintained at a given temperature. The exchange was also performed with stirring under the conditions usually employed in phase-transfer process between immiscible aqueous(2mL) and organic phases(2mL). For radioactivity measurement at given intervals of time, samples from the organic liquid were withdrawn and analyzed by thin-layer chromatography(silica gel 60F$_{254}$, 0.5mm, Merck) using benzene-ethyl acetate (9 : 1). The radioactivity of each aliquot was measured with a well type automatic scintillation counter. In a biphase system activity of aqueous and organic phases were both determined.

3   Results and Discussion

A nucleophilic ^{125}I-for-Br exchange reaction was conducted by mixing 1-bromooctane in neat

---

[*]   Copartner: Jin Yutai, Minoru MAEDA, Masaharu KOJIMA. Reprinted from *Radiosotopes*, 1981, 30(7) :391-393.

with dry Na^{125}I in the presence of dicyciohexyl-18-crown-6 (DCC). The reaction mixture was homogeneous at all times. The time dependence of labelling yield from 1min to 120min at 18℃ is shown in Fig. 1 together with the results obtained in the absence of DCC for comparison. It can be seen that the exchange reaction proceeded rapidly even at 18℃ and the labelling yield reached a saturation value of about 80% within 30min. On the other hand, no exchange reaction took place in the absence of DCC under the identical conditions covering the same period of time. Examples illustrating the utility of DCC as a catalytic medium in the iodine exchange are provided in Table 1. However, DCC exhibited no catalytic activity in the exchange between 1-chlorooctane and Na^{125}I under the same conditions.

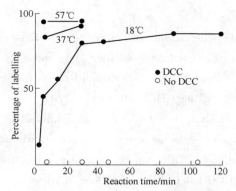

Fig. 1　Time and temperature dependence of the I-tor-Br exchange of 1-bromooctane
(Reaction conditions: 10μL 1-bromooctane, 16-25mg DCC, no carrier added Na^{125}I 0.37-0.74MBq(10-20μCi))

Table 1　Iodine labelling of alkyl iodides in neat with Na^{125}I in the presence of dicyclohexyl-18-crown-6[①]

Alkyl iodides	Labelling yield/%
2-Iodobutane	78
1-Iodohexane	96
1-Iododecane	85
Ethyl iodide	70

① Alkyl iodide 10μL dicyclohexyl-18-crown-6(DCC) 16mg, 40℃, 30min, no carrier added Na^{125}I 0.37-0.74MBq(10-20μCi).

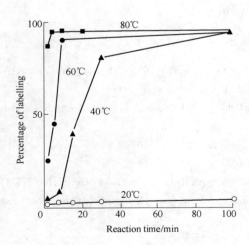

Fig. 2　Time and temperature dependence of iodine exchange of 1-iododecane
(Reaction conditions: 10μL 1-iododecane, 16mg DCC, no carrier added Na^{125}I 0.37-0.74MBq(10-20μCi))

The effect of temperature on the exchange yield was studied. The exchange yield for the run with 1-bromooctane increased from 45% at 18℃ to 95% at 57℃ in 5min as shown in Fig. 1. The similar result was also obtained in the iodine exchange of 1-iododecane as shown in Fig. 2, indicating that at elevated temperatures the exchange was completed within less than 10min. The influence of DCC concentration added to the reaction system on the yield was also examined. In view of the time dependence, 30min at 18℃ was chosen as reaction time. However, the change of the amount of DCC from 2mg to 45mg did not affect the labelling yield at all. This demonstrates that DCC can be used as an excellent medium in the halogen exchange reaction.

For the purpose of comparison, we have carried out the exchange between 1-bromooctane and Na^{125}I at 18℃ as liquid-liquid phase-transfer reaction using $CHCl_3$-H_2O or CCl_4-H_2O and DCC as catalyst. In a biphase system, only 20%-30% exchange was achieved after 30min without the addition of DCC. On the other hand, the presence of DCC led to the remarkable increase of activity (60%-70%) in the organic phase, but no significant increase of labelling yield was noticed. Thus, DCC was proved to be totally ineffective at relatively low temperatures in a biphase system. When the same reaction was performed in benzene solution containing DCC, the addition of small amount of DCC resulted in the great increase of the labelling yield as shown in Fig. 3, though its catalytic activity was lower than that without solvent. With further increase of the amount of DCC, however, the exchange yield decreased.

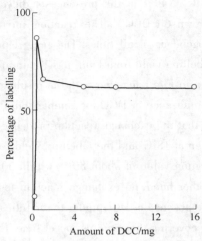

Fig. 3 Influence of the amount of DCC on the labelling yield in the I-for-Br exchange of 1-bromooctane in benzene
(Reaction conditions: 10μL 1-bromooctane, 50μL benzene, 30min, 20℃, no carrier added Na^{125}I 0.37-0.74MBq(10-20μCi))

In conclusion, the present results indicate that DCC exhibits good catalytic action towards the halogen exchange such as the I-for-I and the I-for-Br exchange when used as a catalytic medium, presumably due to not only the increased iodide reactivity but also the enhanced degree of ion dissociation. This method would be effectively used in the radioiodine labelling of long chain halogenated hydrocarbons soluble in crown ether.

References

[1] H. Elias, Ch. Arnold and G. Kloss: Int. J. Appl. Radiat. Isotopes, 24, 463(1973); H. Elias and F. Lotterhos: Chem. Ber., 109, 1580(1976); G. Stöcklin: Int. J. Appl. Radiat. Isotopes, 28, 131(1977).

[2] G. D. Jr. Robinson: J. Nucl. Med., 16, 561(1975); J. P. Dekleijin, J. W. Seetz, J. F. Zawierko and B. van Zanten: Int. J. Appl. Radiat. Isotopes, 28, 591(1977).

[3] M. Maeda, H. Shimoirisa, H. Komatsu and M. Kojima: ibid., 30, 255, 713(1979).

[4] L. A. Spitznagle and C. A. Marine: Steroids, 30, 2112(1977).

[5] T. Irie, K. Fukushi, T. Ido, T. Nozaki and Y. Kashida: 2nd Int. Symp. Radio-pharm. Chem., Oxford, July(1978), Abstract No. 8; T. Irie, K. Fukushi, T. Ido and T. Nozaki: 3rd Int. Symp. Radiopharm. Chem., St. Louis, June(1980), Abstract No. 16.5.

卤代烷烃与 ^{125}I, ^{77}Br 快速交换的新方法[*]

1 引言

迄今为止,把放射性卤素特别是它的短寿命核素引入有机化合物的主要方法是同位素交换法。但在实际工作中对于某些体系,由于交换速度太小而常常不便使用,近年来发展了若干快速交换法,如衰变诱导交换[1~3]、酶促交换[4,5]、熔融交换[6,7]和其他方法[8,9]。这些方法在某些情况下使用,确较一般同位素交换法更为有效。

众所周知,冠醚可以作为相转移剂将碱金属卤化物从水相转入有机相,从而在有机相内与卤代烷烃进行同位素交换,但与此同时,也伴随有一定量的溶剂化水分子进入有机相,这在一定程度上影响交换反应的速度[10~13]。本文首次提出以冠醚为溶剂,将固体碱金属卤化物和被标记物直接溶于冠醚溶剂,实现了非溶剂化和"裸露"卤离子的快速同位素交换,由于这种冠醚效应,交换过程甚至在低温下也能快速进行。实验表明,这一新方法对长链卤代烷烃、长链脂肪酸以及有关化合物的快速标记,具有实际意义。特别是对短寿命卤素^{18}F、^{34m}Cl、$^{75-77}$Br、^{123}I 的快速交换,将有明显的应用价值。

2 实验部分

2.1 方法

以二环己基 18-冠醚-6(简称 DCC)为溶剂,溴代正辛烷或其他卤代烷烃为溶质(Dicyc-18-CR-6 由 Aldrich Chemical Company Inc. 生产,1-Bromooctane 由 Wako Pure Chemical industries 生产),在不同温度下与^{125}I 和^{77}Br 进行交换,先将无载体 Na^{125}I 或 Na^{77}Br 水溶液在适当条件下脱水,然后加入 10μL CH$_3$(CH$_2$)$_6$CH$_2$Br 及不同量 DCC,在混合器上进行交换。在不同交换时间取样,用薄层层析法进行分析(硅胶板为 60F$_{254}$,0.5mm),以苯和乙酸乙酯 9∶1 为展开剂,从层析板上碘(溴)离子和^{125}I(^{77}Br)-正辛烷的放射性峰值计算标记百分率。放射性强度用自动井形闪烁计数器测定。

2.2 结果及讨论

2.2.1 标记百分率与交换速度的关系

溴代辛烷与^{125}I$^-$进行卤原子交换。实验表明,当体系内不加入 DCC 时,几乎不发生交换反应,当以 DCC 为溶剂时,由于固体 Na^{125}I 和溴代辛烷均能直接溶解在冠醚中,反应在 18℃ 的低温条件下也能快速进行,结果表明,反应时间为 30min 时标记百分率已达饱和值(80% 标记)。实验结果如图 1 所示。

---

[*] 本文合作者:金昱泰、前田稔、小嶋正治。原发表于《核化学与放射化学》,1981, 3(4): 242~246。

此外，在 DCC 存在条件下，对其他碘代烷烃也进行了同位素交换，结果如表 1 所示。

溴代辛烷与 $^{125}I^-$ 在水相中进行交换（18℃），反应同样不能发生。当用 $CHCl_3$ 或 CCl_4 为有机相，以 DCC 为相转移剂时，则有机相的表观放射性较不加 DCC 时显著增加，但从有机相的层析分析表明，溴代辛烷的标记百分率实际并没有增加。实验还表明，在冠醚存在条件下，氯代辛烷与 $^{125}I^-$ 之间也不发生交换反应。正如预期的那样，这一方面是因为 C—Cl 键较 C—I 为强，同时也由于 I^- 的体积远较 Br^-、Cl^-、F^- 为大，其"裸露"离子效应不如 Br^-、Cl^-、F^- 明显。

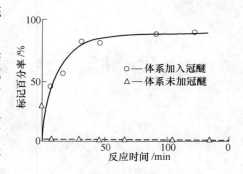

图 1 标记百分率与反应时间的关系
（反应条件：溴代辛烷 10μL, DCC 25mg,
反应温度 18℃，无载体 $Na^{125}I$)

表 1 各种卤代烷烃与 $^{125}I^-$ 的同位素交换
（反应条件：碘代烷烃各 10μL, DCC 16mg, 反应时间 30min, 反应温度 40℃）

化 合 物	标记率/%
碘乙烷	70①
2-碘丁烷	78
1-碘己烷	96
1-碘癸烷	85
碘苯	≈0

① 薄层层析过程中，碘乙烷有部分挥发。

2.2.2 标记百分率与反应温度的关系

图 2 表示，当反应温度升高至 57℃ 时，溴代辛烷的标记百分率在 5min 内就达到 95%。这显示了在 DCC 存在条件下，反应温度对反应速度的显著影响。当然 DCC 在较高温度下由于黏度的降低也有利于 $^{125}I^-$ 的扩散，这也是促使反应速度加快的原因之一。

碘癸烷在不同温度下的交换反应也得到了类似的规律。实验结果如图 3 所示。

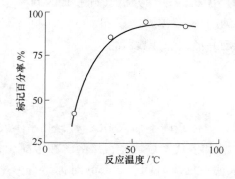

图 2 温度对溴代辛烷和 $^{125}I^-$ 交换反应的影响
（反应条件：溴代辛烷 10μL, DCC 16mg,
反应时间 5min）

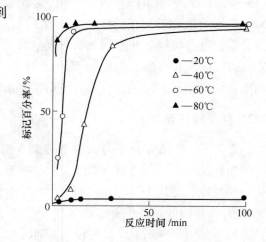

图 3 温度对碘癸烷和 $^{125}I^-$ 交换反应的影响
（反应条件：碘癸烷 10μL, DCC 16mg）

2.2.3 标记百分率与冠醚（DCC）量的关系

实验表明，以冠醚为溶剂时，当 DCC 的加入量超过 2mg 时，并不能导致标记百分率的增加。与此相反，正如图 4 所示，若降低 DCC 的含量，标记百分率反而有所提高。还进一步比较了以苯或乙腈为溶剂，当加入不同量 DCC 时对标记百分率的影响。结果如图 5 和图 6 所示。数据表明，在非极性溶剂苯中，当加入 0.5mg DCC 时，标记百分率急剧上升，但随着 DCC 量的进一步增加，标记百分率逐渐下降至一定值。极性溶剂乙腈的情况与此类似，但随着 DCC 量的增加，标记百分率一开始就逐渐降低，这可能是由于乙腈本身能溶解一定量的碱金属卤化物，因此在纯乙腈溶剂中 $^{125}I^-$ 离子能比较自由地与溴代辛烷发生交换，当引入 DCC 后 $^{125}I^-$ 离子的交换反而受到一定的阻碍。但从图 1 和图 6 的数据不难看出，在以纯 DCC 为溶剂时，溴代辛烷的标记百分率仍大于纯乙腈的标记百分率。

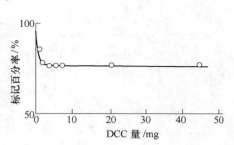

图 4　加入不同冠醚量（DCC）对标记百分率的影响

（反应条件：溴代辛烷 10μL，反应时间 30min，反应温度 18℃）

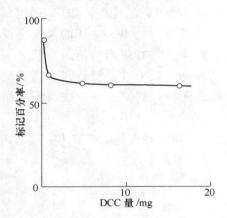

图 5　苯溶液中不同量的 DCC 对标记百分率的影响

（反应条件：溴代辛烷 10μL，苯 50μL，反应时间 30min，反应温度 20℃）

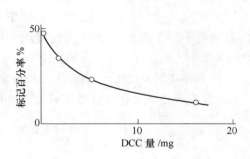

图 6　乙腈溶液中不同量的 DCC 对标记百分率的影响

（反应条件：溴代辛烷 10μL，乙腈 50μL，反应时间 30min，反应温度 20℃）

2.2.4　KI 载体量对标记百分率的影响

在无载体 $Na^{125}I$ 水溶液中，加入不同量的 KI 载体，然后脱水进行交换实验，由于 DCC 对 K^+ 离子有很强的配合作用，因此载体量在 100μg 时，标记百分率仍为 90% 以上，实验结果如图 7 所示。

2.2.5　溴代辛烷与 $^{77}Br^-$ 的同位素交换反应

以冠醚 DCC 为溶剂，在室温下（22℃）进行溴代辛烷与 $^{77}Br^-$ 之间的同位素交换，其规律与 Br、$^{125}I^-$ 卤原子交换，I、$^{125}I^-$ 同位素交换相似，结果如图 8 所示。

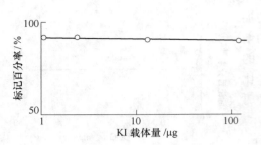

图 7　KI 载体量对标记百分率的影响
（反应条件：溴代辛烷 10μL, DCC 16mg,
反应时间 5min, 反应温度 70℃）

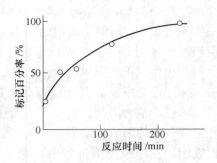

图 8　标记百分率与反应时间的关系
（反应条件：溴代辛烷 10μL, DCC 16mg,
无载体 Na^{77}Br, 反应温度 22℃）

3　小结

从上述实验结果可以看出：用冠醚作为溶剂（有时作为固体相转移剂）是标记卤代烷烃的一种新方法。这种方法使卤原子同位素交换反应能够在室温下快速进行，而这些反应用通常的交换方法是难于实现的。它具有一些明显的优点，如反应温和、快速、交换百分率高等。可以预期，在短寿命卤素 ^{18}F、^{34m}Cl、$^{75-77}$Br 和 ^{123}I 的快速交换中以及对需要在低温下进行生物活性物质的标记工作中，具有应用价值。目前这一方法的使用还受到某些标记物在冠醚中是否溶解的限制。如果进一步研究不同类型的冠醚对不同类型标记物的同位素或非同位素交换，则将会促使这一方法得到广泛的应用。

作者对日本国理化学研究所 Nozaki 教授和东北大学加速器同位素中心 Ido 教授在 ^{77}Br 的制备和交换工作中所给予的大力协作，表示衷心感谢。

参 考 文 献

[1]　M. J. Welch, J. Am. Chem. Soc., 92, 408(1970).

[2]　R. M. Lambrecht et al., J. Nucl. Med., 13, 266(1972).

[3]　M. Elgarhy et al., Radiochem. Radioanal. Lett., 18, 318(1974).

[4]　H. Elias, Proc. 9th Japan Conf. on Radioisotopes, Tokyo, 538(1968).

[5]　R. M. Lambrecht, J. Nucl. Med., 15, 863(1974).

[6]　G. J. Stocklin, J. Labelled Compounds and Radiopharmaceuticals, 17(3), 353(1980).

[7]　J. Nunez, J. Labelled Compd., 1, 128(1965).

[8]　St. Wong, Int. J. Appl. Radiat. Isot. 27, 19(1976).

[9]　G. Stocklin, Int. J. Appl. Radiat. Isot. 28, 131(1977).

[10]　L. J. Arsenault, J. Radioanal. Chem., 57, 543(1980).

[11]　V. D. Eckehard, Angew. Chem., 16, 493(1977).

[12]　D. Landini, J. Chem. Soc. Chem. Commun., 950(1975).

[13]　D. Landini, J. Chem. Soc. Chem. Commun., 879(1974).

A Novel Method of Labelling Halogenated Hydrocabon with ^{125}I and ^{77}Br

Liu Boli, Jin Yutai

(Chemistry Department, Division of Radiochemistry, Beijing Normal University, Beijing, 100875, China)

MINORU MAEDA, MASAHARU KOJIMA

(Faculty of Pharmaceutical Sciences Kyushu University, Fukuoka 812, Japan)

Abstract Here is reported a novel method of labelling halogenated hydrocarbon with ^{125}I and ^{77}Br in the presence of crown ether Dicyclohexyl 18-crown-6(DCC), which is used as a solvent to dissolve the substrate and alkali halide. The unsolvated and "naked radioactive anion" has more reactive ability than the anion in the conventional condition.

The isotopic exchange between n-octylbromide and ^{125}I iodide or others was performed in the presence of DCC. The time dependence of labelling of n-octylbromide with ^{125}I and ^{77}Br, temperature dependence on the labelling percentage, DCC amounts dependence on the labelling percentage, and carrier amount KI on the percentage of labelling were studied. Our results show that there is a big crown ether effect in the halogen atom exchange or isotopic exchange between the halogsnated hydrocarbon and radioactive halide in relatively low temperature. It may be expected that this method could be used in the labelling of halogen atom exchange process, particularly in the labelling of long chain fatty acid with ^{18}F.

Radiobromine Labeled Cholesterol Analogs Synthesis and Tissue Distribution of Bromine-82 Labeled 6-Bromocholesterol[*]

Abstract Bromine-82 labeled 6-bromocholesterol (CL-6-Br-82) was prepared by the reaction of 6-chloromercurycholesterol with bromine-82 labeled bromine in chloroform. Tissue distribution of CL-6-Br-82 in rats was determined. The adrenal uptake of CL-6-Br-82 reached a maximum of 136% dose/g at 5 days after injection. The adrenal-to-liver ratio increased from 57 at 3 days to 141 at 5 days. The substitution of radiobromine for radioiodine in 6-iodo-cholesterol (CL-6-I) resulted in an agent which demonstrates less affinity for the adrenal gland than CL-6-I itself.

Key words adrenal imaging agent, ^{82}Br-6-bromocholesterol, tissue distribution

1 Introduction

The demand for a scintigraphic agent for adrenal visualization has led to the development of various radiolabeled analogs of cholesterol[1,2]. The first successful images of human adrenals were obtained with ^{131}I-19-iodocholesterol (CL-19-^{131}I) synthesized by Counsell et al[3]. Thereafter, a homoallylic isomer of this compound, ^{131}I-6β-iodomethyl-19-norcholest-5(10)-en-3β-ol (NCL-6-^{131}I) has shown to provide better human adrenal images than CL-19-^{131}I because of its more rapid uptake and greater target-to-background ratio[4,5], and is the current agent of choice for adrenal imaging. On the other hand, ^{131}I-6-iodocholesterol (CL-6-^{131}I), synthesized by Wang and Liu et al[6,7], has been used clinically as a new diagnostic agent for adrenal gland in China due to its notable stability in-vivo and in-vitro, though demonstrating lower adrenal-liver and adrenal-kidney radioactivity ratios than NCL-6-^{131}I[8]. However, the limitations of diagnostic imaging of the adrenal gland with these radioiodinated cholesterol analogs are the time required and the high radiation dose to patients.

The potential use of radioactive isotope of bromine as an alternative to radioiodine in nuclear medicine has been pointed out[9]. ^{75}Br ($T_{1/2} = 101$ min), ^{76}Br ($T_{1/2} = 15.9$h) and ^{77}Br ($T_{1/2} = 56$h) are the most suitable nuclides in-vivo as radiopharmaceutical labels. It is of interest to use a bromine labeled cholesterol analog which would give a radiopharmaceutical with more favorable characteristics. As a part of study of analog synthesis of cholesterol and subsequent structure-activity relationship evaluation, we planned to synthesize ^{82}Br labeled 6-bromocholesterol (CL-6-^{82}Br) and to evaluate its ability to selectively localize in adrenals. The rationale for se-

---

[*] Copartner: Jin Yutai, Pan Zhongyun, Minoru Maeda, Masaharu Kojima. Reprinted from *Journal of Labelled Compounds and Radiopharmaceuticals*, 1982, XIX(9):1089-1096.

lecting ^{82}Br($T_{1/2}$ =35.4h) in the present research was that ^{82}Br is readily available and a suitable model for ^{75}Br, ^{76}Br or ^{77}Br.

2 Results and Discussion

The nonlabeled compound, 6-bromocholesterol, was prepared by the reaction of 6-chloromercurycholesterol with Br_2. Initial attempts to obtain ^{82}Br-6-bromocholesterol(CL-6-^{82}Br) by the Br-for-Br exchange using Na^{82}Br in various solvents or in a melt resulted in markedly resistant incorporation of the label. The preparation of CL-6-^{82}Br by the direct bromination of 6-chloromercurycholesterol with ^{82}Br$_2$ in chloroform was attempted. The results of the time dependence of labeling yield is shown in Fig. 2, indicating that the yield of CL-6-^{82}Br reaches a theoretical maximum value of about 46% in 70min. Using 18mCi of ^{82}Br$_2$ we have synthesized 8.1mCi of CL-6-^{82}Br with more than 95% radiochemical purity and a specific activity of 0.87mCi/mg.

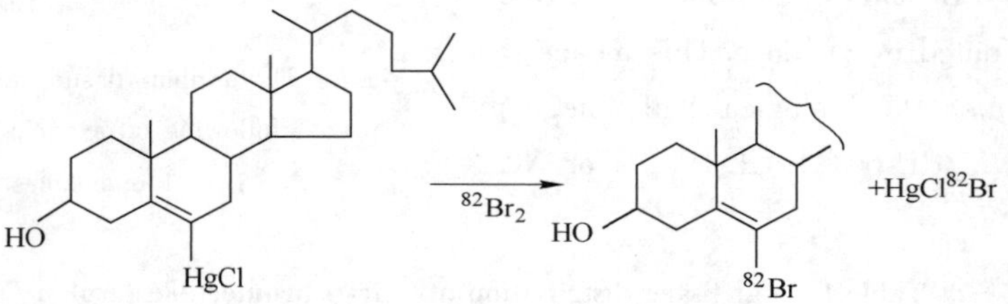

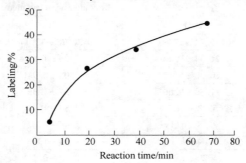

Fig. 1 Reaction scheme for the synthesis of ^{82}Br-6-bromocholesterol(CL-6-^{82}Br)

Fig. 2 Time dependence of the reaction of 6-chloromercury-cholesterol with ^{82}Br$_2$

The distribution of radioactivity in tissues of male rats was determined at time intervals from 1 day to 5 days following the administration of CL-6-^{82}Br and the results are summarized in Table 1. The distribution profile of radioactivity from CL-6-^{82}Br showed many similarities with the data previously reported for the other radiolabeled cholesterol analogs. This result clearly indicates a considerable selective localization of radioactivity in the adrenals. Apart from the adrenals, the other tissue showing high level of radioactivity was the spleen. However, the high splenic uptake considerably decreased at 3 days after injection. The adrenal uptake of CL-6-^{82}Br reached a maximum of 136% dose/g at 5 days after injection. It must be mentioned that the adrenal concentration of radioactivity is less than that achieved with CL-6-^{131}I[8] or NCL-6-^{131}I[4] over the 5-day period, although it is at higher level than that observed with CL-19-

^{131}I[4]. The relative concentration of radioactivity in adrenals, compared with nearby organs, is very important for adrenal imaging. The adrenal to non-target ratios for several selected tissues are presented in Fig. 3. The adrenal-to-liver ratio increased from 57 at 3 days to 141 at 5 days, which are comparable to those of CL-6-^{131}I, but these values are considerably lower than those observed with CL-19-^{131}I and NCL-6-^{131}I. The value is probably not high enough to allow clean imaging of adrenals by conventional techniques. Measurements of the whole-body activity following the injection of CL-6-^{82}Br shows that 50% of the radioactivity is eliminated in 2.5 days. This means that CL-6-^{82}Br has a shorter biological half-time, compared with that of CL-6-^{131}I, CL-19-^{131}I or NCL-6-^{131}I.

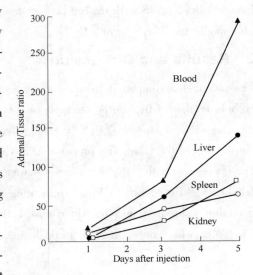

Fig. 3 The adrenal/tissue ratios of selected tissues following intravenous injection of ^{82}Br-6-bromocholesterol

Table 1 Rat tissue distribution of ^{82}Br-6-bromocholesterol (CL-6-^{82}Br) at various time intervals①

Tissue	Days after administration		
	1 day	3 days	5 days
Adrenal	115.74 ± 22.02	130.78 ± 72.14	136.28 ± 41.10
Liver	21.44 ± 9.26	2.26 ± 1.02	0.96 ± 0.18
Kidney	10.20 ± 6.12	3.08 ± 0.94	2.24 ± 0.62
Lung	21.30 ± 9.66	5.34 ± 1.86	3.28 ± 0.86
Spleen	55.72 ± 21.74	4.84 ± 1.98	1.72 ± 0.32
Testicle	2.38 ± 0.38	1.44 ± 0.32	1.42 ± 0.32
Blood	6.84 ± 1.44	1.64 ± 1.16	0.46 ± 0.08
Thyroid	4.20 ± 1.06	0.92 ± 0.12	0.86 ± 0.68

① Values represent mean % administered dose per gram of tissue for 3 rats at 1 day and 3 days, for 5 rats at 5 days, with SD of mean.

The present results indicate that the substitution of radiobromine for radioiodine in the CL-6-I results in an agent which demonstrates less affinity for the adrenal gland than CL-6-I itself. By the use of CL-6-Br labeled with the positron emitter such as ^{76}Br and positron emission tomography, however, it may be possible to visualize tomographically adrenal masses.

3 Experimental

3.1 6-bromocholesterol (CL-6-Br)

6-Chloromercurycholesterol (CL-6-HgCl) prepared according to the method described by Mertz

(10) was recrystallized from acetic acid. To a solution of CL-6-HgCl(300mg) was added dropwise a solution of Br_2 (200mg) in chloroform (5mL). The mixture was stirred at 25℃ for 2h. After filtration, the filtrate was washed with 0.5% Na_2SO_3, extracted with ether, and dried (Na_2SO_4). The residue, after removal of the solvent, was chromatographed on silica gel (Silica ARCC-4, 100 mesh, Mallinckrodt) eluting with chloroform to give 6-bromocholesterol (CL-6-Br) (199mg) as colorless needles, m. p. 141-142℃, after recrystallization from ethanol; IR ν_{max} (Nujol) 3400cm^{-1}, 3240cm^{-1} and 1650cm^{-1}. Anal. Calcd. for $C_{27}H_{45}OBr$: C, 69.66; H, 9.74. Found: C, 69.58; H, 9.73.

3.2 ^{82}Br-6-bromocholesterol(CL-6-^{82}Br)

An aqueous solution of ^{82}Br-NaBr (20mCi, 5mCi/mg) was added to a separatory funnel with chloroform (4mL) and an aqueous solution of $KBrO_3$ (0.8mL) (3.45mg/mL) was then added to the mixture. Ten drops of conc. H_2SO_4 was further added to the mixture. The $^{82}Br_2$ (18mCi) in chloroform solution was thus obtained. A solution of $^{82}Br_2$ in chloroform was added dropwise to a solution of CL-6-HgCl (13mg) in chloroform (1mL) with stirring at room temperature. The mixture was allowed to stand for 2h at room temperature. The filtered solution was streaked on silica gel glass plates (Silica gel 60F 254, 0.5mm layer, E. Merck) and developed in benzene-ethylacetate (9:1). The separated CL-6-^{82}Br was scraped off and eluted with chloroform to give CL-6-^{82}Br (8.1mCi) with a specific activity of 0.87mCi/mg. It showed a single radioactive peak coincident with unlabeled CL-6-Br on thin-layer chromatography.

3.3 Tissue distribution

CL-6-^{82}Br was dissolved in ethanol (0.4mL), and Tween 80 (0.5mL) and sufficient saline were added to give a 10% ethanol solution having a radioactive concentration of 100μCi/mL. Eleven male Wistar rats weighing 120-150g received through the tail vein a dose of 20μCi/animal of CL-6-^{82}Br. Three or five rats were killed at 1 day, 3 and 5 days after the injection. Major organs were excised, weighed and placed in small counting vials. Samples of tissues were counted in a gamma well counter, after which corrections were made for radioactive decay and counting efficiency. The concentration in each tissue was expressed as percentage of injected dose per gram.

References

[1] Beierwaltes W. H., Wieland D. M., and Swanson D. Radiopharmaceuticals: Structure-Activity Relationships, ed., by Spencer R. P., Grune and Stratton, Inc., New York, 1981, p395.

[2] Counsell R. E. and Ice R. D. Drug Design, vol. 6, ed., by Ariens E. J., Chap. 4, Academic Press, New York, 1975.

[3] Counsell R. E., Ranade V. V., Blair R. J., Beierwaltes W. H., and Weinhold P. A. Steroids 16: 317 (1970).

[4] Kojima M., Maeda M., Ogawa H., Nitta K., and Ito T. J. Nucl. Med. 16: 666 (1975); idem., Chem. Pharm. Bull. (Tokyo) 24: 2322 (1976).

[5] Sarkar S. D., Beierwaltes W. H., Ice R. D., Basmadjian G. P., Hetzel K. R., Kennedy W. P., and Mason M. M. J. Nucl. Med. 16:1038(1975).

[6] Wang Shizen, Liu Chengbin, and Chen Zhong. Radioisotope and Radiation Application Exhibition Information Pamphlet(1972).

[7] Liu Boli, Jin Yutai, Liu Zhenghao, Sun Zhaoxiang, Lu Gongxu, and Li Taihau. J. Nucl. and Radiochem. (China)2:107(1980).

[8] Jixiao M. and Ruisen Z. Chinese Med. J. 92:237(1979); Pan Zhongyun, Li Guanzhong, Liang Shitian, Liu Zhongmin, Wang Yongchao, Xia Zhenmin, Zhou Qian, and Liu Xiujie. J. Chinese Nucl. Med. 1:85(1981).

[9] Stöcklin G. Int. J. appl. Radiat. Isotopes 28:131(1977).

[10] Merz W. Z. Physiol. Chem. 154:225(1926).

无载体 ^{18}F 的制备及快速标记*

摘 要 研究了以苯并-12-冠-4、15-冠-5、18-冠-6、二苯并-24-冠-8 的锂配合物为靶子，在反应堆内照射，通过 $^6Li(n,t)^4He$，$^{16}O(t,n)^{18}F$ 二次核反应制备无载体 ^{18}F，并以靶子物冠醚为溶剂直接进行了 ^{18}F 的快速标记。由于 $^{18}F^-$ 的"裸露"离子效应，加速了 $^{18}F^-$ 与 $CH_2Br(CH_2)_{14}CH_3$ 化合物中 Br^- 之间的交换，标记率 >90%，产额 (80±5)%。本方法把 ^{18}F 的制备、分离和标记统一起来，操作时间较短，适合于短半衰期 ^{18}F 快速标记的需要。

1 引言

由于 ^{18}F 在核医学应用中具有较理想的核性质，如半衰期短（$T_{1/2}=(109.72\pm0.06)$ min）、β^+ 辐射以及某些氟标记物所具有的良好的生物活性等，因此有关它的制备及标记受到各国广泛的注意。目前用于放射性药物的 ^{18}F 大部分由加速器生产[1~3]，但利用反应堆二次核反应 $^6Li(n,t)^4He$，$^{16}O(t,n)^{18}F$ 来制备 ^{18}F 的工作也日益受到重视[4,5]。一般可以选用 Li_2CO_3、Li_2O 或 $LiOH\cdot H_2O$ 为靶子，照射后先进行分离纯化，然后再将无载体 ^{18}F 在适当条件下标记。通常无载体 ^{18}F 的标记产额低，费时长。鉴于以冠醚为溶剂，可以实现卤代烷烃与卤原子间的快速交换[6,7]，因此本文探索了以锂冠醚配合物为靶子、利用锂被冠醚配合后受周围氧原子包围的特点，实现 $^6Li(n,t)^4He$，$^{16}O(t,n)^{18}F$ 二次核反应。生成的无载体 ^{18}F 由于仍处在冠醚介质中，从而形成具有高反应活性的"裸露" $^{18}F^-$ 离子，这时若把标记物注入照射的石英管中，就能直接进行 ^{18}F 的同位素交换。上述过程由于把 ^{18}F 的制备、分离和标记统一起来，预期可以达到快速高效的目标。本工作以锂的苯并-12-冠-4、15-冠-5、18-冠-6 和二苯并-24-冠-8 配合物为靶子，选用 $CH_2Br(CH_2)_{14}CH_3$ 为标记物，作为模式反应来研究，结果表明，达到了预期的效果。

2 实验方法

2.1 靶子的制备

以克式量相当的 Li_2CO_3 和苯并-12-冠-4、15-冠-5、18-冠-6 以及二苯并-24-冠-8 分别溶解于水相（Li_2CO_3）和 $CHCl_3$ 相（冠醚）内，充分混合搅拌，将水及 $CHCl_3$ 缓慢蒸干。然后将上述样品分别密封在石英照射管内，置重水反应堆照射 20min，中子通量为 8.3×10^{13} 中子/$(cm^2\cdot s)$。每一样品均伴有相同质量的 Li_2CO_3 对照样品并进行照射。所配锂冠醚配合物的靶子成分如表1所示。照射后的靶子冷却 2h，然后用带有 Ge(Li) 探头的多道能谱仪测定 0.511MeV γ 射线强度，计算 ^{18}F 的相对产额。

---

* 本文合作者：韩俊、刁国平、董桂芝。原发表于《核化学与放射化学》，1982, 4(2)：100~103。

表1 锂冠醚配合物的靶子成分

靶子物	Li_2CO_3 量/mg	冠醚量/mg
锂-苯并-12-冠-4	8	54
锂-15-冠-5	8	48
锂-苯并-15-冠-5	8	64
锂-18-冠-6	8	58
锂-二苯并-24-冠-8①	8	78
Li_2CO_3	8	0

① 未完全配合。

2.2 ^{18}F 的交换反应

由于 $^6Li(n,t)^4He$ 核反应的截面较大（936 靶，0.025eV），在无冷却系统的条件下，靶子物温度上升较快，为了避免靶子物黏结在石英管壁，可在交换前先用少量 $CHCl_3$ 溶解，挥发干后再注入标记物 $CH_2Br(CH_2)_{14}CH_3$ 100μL。在混合器上进行振荡交换，不同时间取样，用 SiO_2 薄层层析法进行分析，以苯和乙酸乙酯 9∶1 为展开剂，从层析板上的 ^{18}F 离子和 $CH_2^{18}F(CH_2)_{14}CH_3$ 的放射性峰值计算标记百分率，放射性强度用闪烁计数器测定。

3 结果及讨论

3.1 应用不同锂冠醚配合物为靶子时相对产额❶的比较

锂被不同冠醚配合后，Li^+ 离子被不同数目的氧原子所包围，因此在不同的锂冠醚配合物靶子内，t 与 ^{16}O 核反应发生的几率就不同，而且将随着氧原子数目的增加而有所增大。但另一方面冠醚的空穴直径也将随氧原子数目的增加而加大，这样，锂离子与氧原子之间的距离也随之加大，这又使 t 与 ^{16}O 核反应发生的几率减小。因此在相对产额与冠醚氧原子数的曲线上呈现一个最大值。实验结果如图1所示。

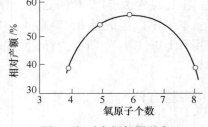

图1 相对产额与冠醚内氧原子个数的关系

3.2 ^{18}F 与 $CH_2Br(CH_2)_{14}CH_3$ 标记物中 Br 的交换

在进行交换反应前，先取少量靶子物进行薄层层析，观察有无明显的冠醚辐解标记产物，实验结果如图2所示。

锂冠醚配合物靶子经辐照后，在 $CHCl_3$ 中的溶解性能较辐照前不同，其中苯并-12-冠-4 锂配合物靶子的溶解性能较好，其余几种均较差，因此 $^{18}F^-$ 与 $CH_2Br(CH_2)_{14}CH_3$ 之间的交换反应选择在苯并-12-冠-4 体系中进行，结果如图3所示。

---

❶ 与相同量的纯 Li_2CO_3 产额之比。

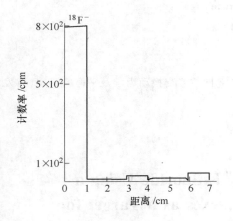

图2 锂冠醚配合物经反应堆辐照后的薄层层析图

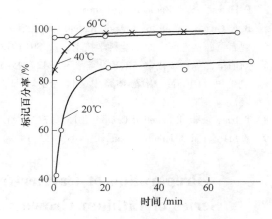

图3 不同温度下标记百分率与反应时间的关系

图3表明以苯并-12-冠-4锂配合物为靶子，它一方面作为制备无载体^{18}F的靶子物，同时它又作为进行^{18}F标记的溶剂体系，其中又包含着"裸露"氟离子，因此即使在常温下（20℃），Br^-与$^{18}F^-$间的交换也能快速进行。标记率≥90%（60℃），产额为(80±5)%，克服了通常在无载体^{18}F加热蒸干时，由容器表面吸附所引起的大量损失[8,9]。文献中曾报道了各种材料表面如玻璃、石英、铂、锆、聚氟乙烯、石墨等对无载体^{18}F因吸附而引起的损失，但均未获满意解决的结果[4]。^{18}F标记产物的薄层层析结果见图4。

在用二次核反应制备^{18}F的过程中，一般产生相同强度的^{3}H，标记物在上述体系中进行标记时发现有^{3}H的沾污。如有必要可对标记物进行去除^{3}H的处理。

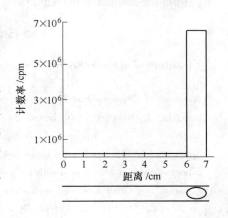

图4 标记产品$CH_2^{18}F(CH_2)_{14}CH_3$的薄层层析图

本方法还存在以下的一些限制：由二次核反应产生的^{18}F，一般处于$^{18}F^-$状态，因此只对亲核的交换反应适用。此外还受到标记物是否在冠醚中溶解的限制。由于目前的堆照条件，靶子物在照射时没有冷却装置，因此苯并-12-冠-4尚有一定程度的分解（颜色变深）。

作者对姜延林同志提供部分冠醚和张蕴辉、丁负吾同志协助进行γ能谱的测定工作表示衷心感谢。

参 考 文 献

[1] R. M. Lambrecht, A. P. Wolf. Radiopharmaceuticals and Labelled Compounds Vol. 1, International Atomic Energy Agency, Vienna(1973).

[2] T. J. Ruth, A. P. Worf. Radiochem. Acta, 26, 21(1979).

[3] F. Helus, et al. Radiochem. Radioanal. Lett., 38, 395(1979).
[4] B. E. Gnade, et al. Int. J. Appl. Radiat. Isot. 32, 91(1981).
[5] 魏启慧，贺先运，原子能科学技术，1，89(1978).
[6] 刘伯里，等．核化学与放射化学，3，242(1981).
[7] 刘伯里，国玉智．冠醚体系中卤代烷烃与^{125}I、^{82}Br、^{18}F同位素交换动力学及平衡的研究(待发表).
[8] T. Irie, et al. J. Labelled Compd., 16, 17(1979).
[9] T. Ido, et al. J. Labelled Compd., 16, 153(1979).

Preparation of Carrier-free ^{18}F-fluoride Using a Series of Lithium Crown Complexes as a Target for Rapid Labelling of Halogenated Hydrocarbons

Liu Boli, Han Jun

(Beijing Normal University, Division of Radiochemistry, Beijing, 100875, China)

Diao Guoping, Dong Guizhi

(Institute of Atomic Energy, Academia Sinica, P. O. Box 275, Beijing, 10875, China)

Abstract A new method is described for the preparation of reactor-produced carrier—free ^{18}F using Li_2CO_3 complexed with benzo-12-crown-4, 15-crown-5, 18-crown-6, dibenzo-24-crown-8 as a target for subsequent synthesis of labelled organic compounds. The relative yields of these targets are approximately 38% for benzo-12-crown-4, 53% for 15-crown-5, 55% for 18-crown-6 and 37% for dibenzo-24-crown-8 compared with the pure Li_2CO_3 target for the same quantity of lithium. After irradiation, this target was used as a solvent for synthesis of halogenated compounds. The labelling yields of $CH_2^{18}F(CH_2)_{14}CH_3$ typically reach 80 ± 5% within 40min at 60℃. The combination of preparation and labelling in one step makes this method very attractive for labelling of halogenated hydrocarbons with carrier-free ^{18}F.

Halogen Exchanges Using Crown Ethers: Synthesis and Preliminary Biodistribution of 6-[^{211}At]-astatomethyl-19-norcholest-5(10)-en-3β-ol[*]

Abstract Halogen exchange labeling of 6β-iodomethyl-19-norcholest-5(10)-en-3β-ol(NCL-6-I) with ^{82}Br, ^{131}I and ^{123}I was carried out using benzo-12-crown-4 or 18-crown-6 as a catalyst. The reaction proceeded rapidly and efficiently within 10min at 70℃. The reaction conditions have been adapted to the preparation of 6-[^{211}At] astatomethyl-19-norcholest-5(10) en-3β-ol (NCL-6-^{211}At), an agent that may be useful as an adrenal therapeutic drug. Preliminary tissue distribution study of the astatine compound in mice showed localization of radioactivity in the adrenal glands.

1 Introduction

There is currently an interest in the use of compounds labeled with ^{211}At for therapeutic applications such as immunosuppression and tumor therapy due to the favorable decay properties of this nuclide ($t_{1/2}$ = 7.2h, 100% α-decay, average energy = 6.8MeV)[1-7]. Radioiodinated 6β-iodomethyl-19-norcholest-5(10)-en-3β-ol(NCL-6-I) is currently used as a scintiscanning agent for the adrenal gland[8]. In earlier publications we and others have described the preparation and tissue uptake characteristics of structurally modified cholesterol and 19-norcholesterol analogs labeled with radioiodine, radiobromine or radloselenium[9-13]. In view of the high adrenal specificity associated with the 6-substituted 19-norcholesterol structure, the replacement of radioiodine with ^{211}At in NCL-6-I might result in the development of a new drug with therapeutic applications.

There have been several reports of the preparation of organic astatine compounds, where radioiodination techniques have been applied to labeling with ^{211}At[2-6,14-16]. Among these techniques are halogen exchange reactions and these represent one of the more attractive methods for introducing ^{211}At into a molecule. It is preferable to employ a rapid exchange method giving a high radiochemical yield, in view of the short half-life of ^{211}At. Recently efforts to improve the radiochemical yields and decrease reaction times for halogen exchange procedures have led to the development of a variety of catalysts[17-22]. In the present work, we have found that the halogen exchange labeling of NCL-6-I with ^{82}Br, ^{131}I and ^{123}I using a crown ether as a medium enables much shorter reaction times to be used than previously reported for the exchange method in an organic solvent. The preparation of 6-[^{211}At] astatomethyl-19-norcholest-5(10)-en-3β-ol

---

[*] Copartner: Jin Yutai, Liu Zhenghao, Luo Cheng, Masaharu Kojima, Minoru Maeda. Reprinted from *Int. J. Appl. Radiat. Isot.*, 1985, 36(7):561-563.

(NCL-6-^{211}At) using this labeling procedure and its preliminary biodistribution in mice are described in this paper.

2 Experimental

Astatine-211 was prepared by the ^{209}Bi(α, 2n)$^{[211]}$At reaction, using the 1.2m cyclotron at the Institute of Atomic energy, Beijing. Bismuth electrodeposited onto supporting copper foils was irradiated with 28MeV α-particles. The astatine-211 produced was extracted from the irradiated target by dry distillation at 560-700℃ into 0.1M NaOH containing 10mmol Na$_2$SO$_3$[2]. Routinely 75-80μCi of activity was prepared in alkaline solution per μAh of cyclotron irradiation. The ^{211}At activity was monitored by counting the Po x rays using a 2 in. NaI well-type detector. Na^{131}I, Na^{123}I and Na^{82}Br were provided by the Institute of Atomic Energy, Beijing. Benzo-12-crown-4 (m.p. 44-45℃) and 18-crown-6 (m.p. 37-39℃) were purchased from Shanghai First Chemical Reagent Factory, China.

2.1 Labeling technique

An aqueous solution of radioactive sodium halide (20-30μCi Na^{131}I, 15-20μCi Na^{123}I, 50μCi Na^{82}Br, 200μCi Na^{211}At) was placed in a small test tube and evaporated to dryness by gentle heating in an oil bath. The crown ether (30mg) was added to the test tube and stirred. NCL-6-I (1mg) was then added, and the exchange reaction allowed to proceed at a constant temperature (70℃) with samples being taken at various time intervals. The product was separated by thin-layer chromatography (silica gel; solvent system, benzene-ethyl acetate [9 : 1 v/v]). The separated radioactive species was scraped off and extracted with ether. In all cases a single radioactive peak was observed, and in the iodine and bromine cases the peak was coincident with authentic NCL-6-I and 6-bromomethyl-19-norcholest-5(10)-en-3β-ol (NCL-6-Br)[9,23]. The radioactive species had a R_f value of 0.50-0.55 in all cases.

2.2 Tissue distribution

NCL-6-^{211}At was dissolved in ethanol (0.2mL) and Tween 80 (0.05mL), and saline was added to give a final ethanol concentration of 5% with the radioactive concentration being 30μCi/mL. The solution was administered to mice (20-22g) by injection through the tail vein, each mouse receiving a dose of 2-10μCi of NCL-6-^{211}At. These mice received no thyroid blocking agents. The mice were sacrificed at 3h, 6h, 12h and 24h after injection, four at each time interval. The major organs were excised, weighed and placed in small counting vials and counted in a γ well counter. As the thyroid was difficult to remove, it was taken out with trachea. After correction for radioactive decay, the concentration in each tissue was expressed as a percentage of injected dose per gram. Radioactivity in the thyroid tissue was expressed as a percentage of the injected dose.

3 Results and Discussion

We have described the use of dicyclohexyl-18-crown-6 both as a solvent and catalyst for I-for-

I or I-for-Br exchange of alkyl halides[24]. This observation prompted us to label 6-halogeno-methyl-19-norcholesterol by halogen exchange using a crown eher. Halogen exchange labeling of NCL-6-I with ^{82}Br, ^{131}I, ^{123}I and ^{211}At was carried out using the crown ethers benzo-12-crown-4 and/or 18-crown-6, these crowns being chosen because of their low melting points and easy availability. In all cases the reaction mixtures appeared homogeneous. The time dependence of the isolated radiochemical yield from 1 to 10min at 70℃ is shown in Fig. 1. It can be seen that the exchange reactions proceeded rapidly and the yields reached a value of about 80% within 10min and the order of the rate of exchange being ^{211}At > ^{131}I, ^{123}I > ^{82}Br. The two crown ethers used as catalytic media produced almost the same exchange efficiencies.

Table 1 Mice tissue distribution of NCL-6-^{211}At at various time intervals[①]

Tissue	Hours after administration			
	3	6	12	24
Adrenal	88.92 ± 32.81	136.87 ± 33.09	139.95 ± 44.13	181.89 ± 28.33
Liver	23.96 ± 2.36	16.27 ± 1.22	9.99 ± 0.89	3.11 ± 0.36
Kidney	7.64 ± 1.15	7.21 ± 0.77	6.48 ± 0.37	4.30 ± 0.36
Lung	21.78 ± 3.65	16.72 ± 0.98	12.98 ± 0.95	6.01 ± 0.80
Spleen	23.97 ± 5.29	18.46 ± 1.61	17.48 ± 1.88	7.90 ± 1.66
Stomach	23.58 ± 6.88	27.47 ± 6.24	8.62 ± 1.38	6.12 ± 0.32
Heart	6.42 ± 1.32	5.46 ± 0.66	5.17 ± 0.34	3.60 ± 0.41
Blood	8.03 ± 0.75	5.98 ± 0.55	4.65 ± 0.82	2.60 ± 0.38
Thyroid[②]	1.8	3.2	4.8	1.4

① Number of mice: $n = 4$. Values represent mean % administered dose per gram of tissue with standard deviations of mean.

② Percent of injected dose. Number of mice: $n = 1$.

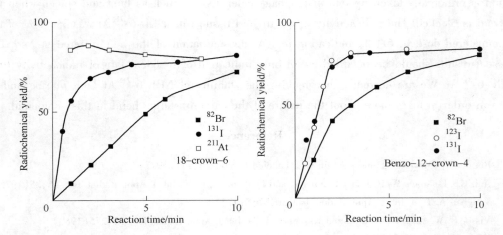

Fig. 1 Time dependence of the halogen exchanges of 6β-iodomethyl-19-norcholest-5(10)-en-3β-ol(NCL-6-I) in crown ethers

As has previously been reported, the preparation of NCL-6-^{131}I and NCL-6-^{123}I by the halogen exchange in organis solvents such as acetone and acetonitrile required long reaction times

(4h and 80min) to obtain a satisfactory yield of the labeled steroids[25,26]. Although the previous synthesis of NCL-6-^{82}Br involved a nucleophilic process with p-toluenesulfonyl as the leaving group, a long reaction time (6h, 72%) was also needed[9]. The present halogen exchange method to produce NCL-6-^{131}I, NCL-6-^{123}I and NCL-6-^{82}Br has a much shorter reaction time, probably due not only to the increased halide reactivity but also the enhanced degree of ionic dissociation. In particular, with 18-crown-6, heating for only 2min enabled a rapid and efficient preparation of NCL-6-^{211}At to be achieved. Using 200μCi of Na^{211}At, 160μCi of NCL-6-^{211}At was obtained with more than 95% radiochemical purity after separation of the reaction mixture by thin layer chromatography (TLC). However, the isolated NCL-6-^{211}At was contaminated by NCL-6-I because complete TLC separation of the two compounds having almost the same R_f value was hard to achieve.

The preliminary tissue distribution of NCL-6-^{211}At in mice was determined at time intervals from 3 to 24h following the injection with the resuits being shown in Table. 1. The distribution of NCL-6-^{211}At resembles that of NCL-6-^{131}I or NCL-6-^{82}Br. The concentration of radioactivity in the adrenal gland greatly exceeded that in other organs; the average concentration in the adrenal gland ranged from 136 to 181%/g over an initial 24h period. When organic astatine compounds are used for medical applications, it is important that the carbon-astatine bond is not labile *in vivo*. It is known that many organic astatinated compounds are susceptible to both *in vivo* and *in vitro* dehalogenation, although several compounds such as 6-[^{211}At] astatocholesterol[6] and 6-[^{211}At] astato-2-methyl-1,4-naphthoquinol bis(diphosphate salt)[7] have been reported to be stable. Free astatine was shown to accumulate in the thyroid gland, like free iodine, by Hamilton[27] and Visser et al[6]. A recent report by Brown et al[7]. has demonstrated that the ^{211}At anion is principally taken up into macrophage-laden tissues such as lung and spleen when the thyroid is blocked. The radioactivity in the thyroid tissue with NCL-6-^{211}At was 3%-5% of the administered dose at 6-12h, indicating *in vivo* deastatination of the administered steroid. The present results should be regarded as an only indication of biolgical fate of radioactivity from NCL-6-^{211}At. We are currently investigating the stability of NCL-6-^{211}At both *in vivo* and *in vitro* in order to learn more about the nature of the carbon-astatine bond in this compound.

References

[1] Rossler K., Meyer G. J. and stöcklin G. J. Labeled Compd. 13, 271(1977).

[2] Aaij C., Tschroots W. R. J. M., Lindner L. and Feltkamp T. E. W. Int. J. Appl. Radiat. Isot. 26, 25(1975).

[3] Vanghan A. T. M. Int. J. Appl. Radiat. Isot. 30, 576(1979).

[4] Visser G. W. M., Diemer E. L. and Kaspersen F. M. Int. J. Appl. Radiat. Isot. 31, 275(1980).

[5] Visser G. W. M., Diemer E. L. and Kaspersen F. M. Int. J. Appl. Radiat. Isot. 30, 749(1979).

[6] Visser G. W. M., Diemer E. L., Vos C. M. and Kaspersen F. M. Int. J. Appl. Radiat. Isot. 32, 913(1981).

[7] Brown I. J. Labeled Compd. Radiopharm. 19, 1389(1982); Brown I. Int. J. Appl. Radiat. Isot. 33, 75(1982); Brown I., Carpenter R. N. and Mitchell J. S. Int. J. Appl. Radiat. Isot. 35, 843(1984).

[8] Gross M. D., Swanson D. P., Wieland D. M. and Beierwaltes W. H. Studies of Cellular Function Using Radiotracers chap. 8. (Ed. Billinghurst M. W.) (CRC press, Boca Raton, Florida 1982).

[9] Kojima M., Maeda M., Komatsu H., Shimoirisa H., Ogawa H., Nitta K. and Ito T. Steroids 29,443(1977).

[10] Ito T., Ogawa H., Maeda M, and Kojima M. Yakugaku Zasshi,103,644(1983).

[11] Ito T., Ogawa H., Maeda M. and Kojima M. Steroids 41,131(1983)(and references cited therein).

[12] Riley A. L. M. J. Labeled Compd. Radiopharm. 16,14(1979).

[13] Beierwaltes W. H., Wieland D. M., Yu T., Swanson D. P. and Mosley S. T. Semin Nucl. Med. 8,1(1978).

[14] Vaughan A. T. M. and Fremlin J. H. Int. J. Appl. Radiat. Isot. 28,595(1977).

[15] Vaughan A. T. M. Int. J. Nucl. Med. Biol. 5,229(1978).

[16] Shiue C. Y., Meyer G. J., Ruth T. J. and Wolf A. P. J. Labeled Compd. Radiopharm. 18,1039(1981).

[17] Maeda M., Shimoirisa H., Komatsu H. and Kojima M. Int. J. Appl. Radiat. Isot. 30,255(1979).

[18] Hawkins L. A., Elliot A. T., Dyke L. J. and Barker F. J. Labeled Compd. Radiopharm. 18,126(1981).

[19] Tarle M., Padovan R. and Spaventi S. J. Labeled Compd. Radiopharm. 15,7(1978).

[20] Maeda M., Shimoirisa H., Komatsu H. and Kojima M. Int. J. Appl. Radiat. Isot. 30,713(1979).

[21] Shiue C. Y. and Wolf A. P. J. Labeled Compd. Radiopharm. 20,1363(1982).

[22] Mangner T. J., Wu J. L. and Wieland D. M. J. Org. Chem. 47,1484(1982).

[23] Maeda M., Kojima M., Ogawa H., Nitta K. and Ito T. Steroids 26,241(1975).

[24] Liu B. L., Jin Y. T., Maeda M. and Kojima M. Radioisotopes 30,391(1981).

[25] Nitta K., Ogawa H., Ito T., Maeda M. and Kojima M. Chem. pharm. Bull. 24,2322(1976).

[26] Ido T., Irie T., Suzuki K., Fukushi K., Rikitake T., Tateno Y., Iwata R., Kashida Y., Kojima M. and Maeda M. Jpn. J. Nucl. Med. 15,840(1978).

[27] Hamilton J. G., Asling C. W., Garrison W. M. and Scott K. G. Univ. Calif. Publ. 2,283(1953).

复相体系同位素交换的研究
热液熔融快速标记法*

摘 要 本文初步探讨了热液熔融法在胆固醇、马尿酸、玫瑰红和脂肪酸四种放射性药物标记中的应用，分析了它们在热液熔融状态下，复相同位素交换反应的动力学规律。实验结果表明该法具有速度快、产额高、纯度好、宜于大批量生产的优点，在一定程度上具有普遍应用的意义。

关键词 热液熔融法 复相同位素交换 ^{131}I-马尿酸 ^{131}I-玫瑰红 ^{131}I-6-胆固醇 ^{131}I-脂肪酸

1 引言

随着短寿命核素标记的需要，当前世界各国正在不断地发展各种快速、简易的标记方法[1~4]，以利于放射性药物的质量控制和生产，也利于广大核医学家的临床使用。近年来在熔融标记法的基础上，正在探索一种新的快速交换法——热液熔融法（hydrothermal melt method）[5]，但迄今为止未见详细报道。本文初步研究了热液熔融法在胆固醇、玫瑰红、脂肪酸和马尿酸四种放射性药物标记中的应用，首次探讨了在热液熔融状态下，复相同位素交换反应的动力学规律。实验结果表明，该法具有快速、安全、适宜于大量生产的特点，在一定程度上具有普遍应用的意义。

2 实验部分

2.1 实验方法

（1）将2mg 碘-6-胆固醇（CL-6-I）、15μL 无水乙醇和2μL Na^{131}I 水溶液放入10mL 安瓿中密封，在恒温油浴（150℃）中加热30min，反应毕用乙醚和水溶解，氯仿（或乙醚）萃取，用薄层层析法分别测定有机相及水相的 CL-6-^{131}I 和 ^{131}I$^-$ 的含量，并计算标记率和产额。所用薄层板为硅胶 G 板，展开剂为苯∶乙酸乙酯 = 9∶1（V/V）。

（2）将0.2mg 玫瑰红、28μL 95%乙醇和7μL Na^{131}I 密封入10mL 安瓿中，在恒温油浴（100℃）中加热15min，反应完毕，用纸层析分别测定^{131}I$^-$和标记玫瑰红的含量，并计算标记率和产额。层析纸为新华1号纸，展开剂为正戊醇∶浓氨水∶水 = 0.5∶2∶100（V/V）。

（3）将0.5mg ω-溴代壬酸、15μL 无水乙醇和2μL Na^{131}I 密封入10mL 安瓿中，在恒温油浴（90℃）中加热20min，反应完毕，用硅胶 G 薄层层析板分别测定^{131}I$^-$和ω-碘代壬酸的含量，并计算标记率和产额，展开剂为苯∶乙酸乙酯∶甲醇 = 9∶1∶0.5（V/V）。

---

* 本文合作者：金昱泰、李太华。原发表于《核化学与放射化学》，1985，7(1)：29~34。

（4）将 1mg 邻碘马尿酸、13μL 0.2M HCl 和 13μL Na^{131}I 密封入 10mL 安瓿中，在恒温油浴（154℃）中加热 20min，反应完毕，用硅胶 G 薄层层析板分别测定 ^{131}I$^-$、O-^{131}I-HA 和 O-^{131}I-BA 的含量，并计算标记率和产额，展开剂为三氯甲烷：冰醋酸 = 9：2（V/V）。

2.2 结果及讨论

2.2.1 交换体系中不同水量对 CL-6-I 标记率的影响

在热液熔融法中，反应温度一般低于标记物的熔点，但高于常压下溶剂（反应介质）的沸点，反应瓶在 150℃ 油浴中加热，其中 2μL 水和 15μL 乙醇均为过热的蒸气相，^{131}I$^-$（Na^{131}I）和 CL-6-I 在过热蒸气的搅动下不断地混合，进行交换。图 1 表明交换体系中含水量太少或过多都明显影响胆固醇的标记率。

进一步的实验表明，单纯乙醇和单纯水的效率都不如在一定比例时的乙醇溶液好，其实验条件和结果示于图 2 和表 1。

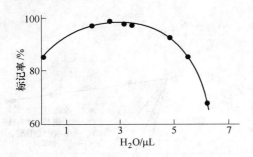

图 1　交换体系中不同水量对
CL-6-I 标记率的影响
（反应条件：2mg CL-6-I，5μL 无水乙醇，
温度：150℃，反应时间：30min）

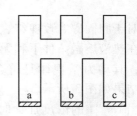

图 2　密封三通反应管
（温度 150℃；反应时间 30min）
a—固体胆固醇；固体 Na^{131}I；
b—乙醇；c—水

表 1　水蒸气、乙醇蒸气对 6-碘胆固醇标记的影响

反应条件	a	a + b	a + c	a + b + c
	2mg CL-6-I Na^{131}I	2mg CL-6-I Na^{131}I 15μL C$_2$H$_5$OH	2mg CL-6-I Na^{131}I 15μL H$_2$O	2mg CL-6-I Na^{131}I 15μL H$_2$O 15μL C$_2$H$_5$OH
标记率/%	61	86	83	97

表 1 数据表明，在热液熔融法中，必须同时在过热水蒸气和乙醇蒸气作用下进行同位素交换才比较完全，这可能是因为乙醇蒸气仅对胆固醇固体表面湿润，而水蒸气只对 Na^{131}I 固体表面湿润，因此只有在乙醇蒸气和水蒸气的混合作用下，才能对两相同位素交换产生有利的影响。

2.2.2 热液熔融交换的反应动力学

对于固-液、固-气非均相交换反应，一般不服从均相同位素交换的指数定律，同位

素的交换速度主要取决于放射性核素在两相内的扩散速度和相界面的交换速度，体系的总交换速度由最慢的阶段所决定。当两相中的扩散速度远大于界面的交换速度时，则交换速度是过程的限制阶段，反之亦然。在热液熔融条件下，胆固醇和 $Na^{131}I$ 均处于固相状态，但在湿热水蒸气和乙醇蒸气作用下，胆固醇的表面处于一种连续平衡的但又是瞬时的热液状态，这时无载体 $Na^{131}I$ 的量由于极微而处于热液层中。由于可交换的无载体 $^{131}I^-$ 仅占胆固醇表面稳定碘的很小一部分，因此交换过程可以忽略 $^{131}I^-$ 在胆固醇固相中的扩散过程。在这种特定条件下，体系总的交换速度仅由界面的交换速度所决定，可以遵从均相交换的指数定律：

$$-\ln(1-F) = \frac{A+B}{AB}Rt \tag{1}$$

对于复相交换体系，式（1）中的浓度 A、B 可以换成与它成正比的固相表面和溶液单位体积（cm^3）中的可交换原子数 $N_固$ 和 $N_液$，而交换总速度 R 可以换成在单位时间内可交换的原子数 N'，式（1）变换为：

$$-\ln(1-F) = \frac{N_固 + N_液}{N_固 N_液}N't \tag{2}$$

对于胆固醇、玫瑰红、脂肪酸和马尿酸体系，在热液熔融条件下和无载体 $^{131}I^-$ 进行交换反应，其动力学规律是符合方程式（2）的。实验结果如图 3 所示。

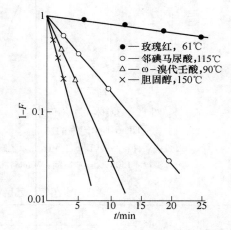

图 3　$1-F$ 与 t 的关系

当 $Na^{131}I$ 的浓度加大时（加入载体），在热液熔融条件下，两相界面的交换速度仍远大于在固相中的扩散速度，这时体系的交换速度取决于 $^{131}I^-$ 在固相中的扩散速度，$^{131}I^-$ 在固相中的分布随其所在坐标位置（$\vec{r}$）和时间（t）而变化，整个固相的放射性比度 $S_固$ 随时间的变化率应为[6]：

$$\frac{\partial S_固}{\partial t} = D\nabla S_固 \tag{3}$$

式中，$S_固 = N_固^*/N_固$；D 为 $^{131}I^-$ 在固相中的扩散系数；∇ 为 Laplace 算符；t 为扩散时间。

若 $t=0$ 时，所有 $^{131}I^-$ 都处于热液层中（或另一相中），其比度为 S_0，则：

$$1-F = 1 - \frac{A_{固(t)}}{A_{固(\infty)}}$$

其中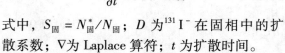

式中，$A_{固(t)}$ 为在时间 t 时固相的活度；$V_固$ 为固相体积；$A_{固(\infty)} = \lim_{t \to \infty} A_固$，即固相达到平衡时的活度。

由 $^{131}I^-$ 的物料平衡可得：

$$A_0 = A_{固(\infty)} + A_{液(\infty)} = V_{液} N_{液} S_0 = V_{固} N_{固} S_{固(\infty)} + V_{液} N_{液} S_{液(\infty)}$$

$$S_{液(\infty)} = S_{固(\infty)} = \frac{S_0}{1+C}$$

式中，$C = \dfrac{V_{固} N_{固}}{V_{液} N_{液}}$，即交换原子数之比。

当用 2mg CL-6-I 和 100μg Na^{131}I（NaI）进行热液熔融（150℃）交换时，其 $S_{液}$ 和 $S_{固}$ 随反应时间的变化如图4所示。

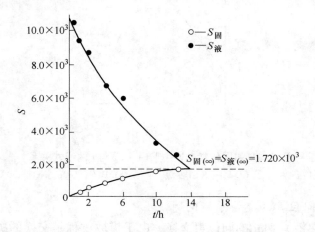

图4　胆固醇 $S_{固}$ 与 $S_{液}$（Na^{131}I）随时间的变化

(反应条件：CL-6-I，2mg；Na^{131}I，4μL（加100μg NaI 载体）；无水乙醇，20μL；
反应温度：150℃。虚线为计算值；实线为实验值)

方程式（3）只对那些具有简单形状的固相（如薄片、球形、圆柱形）才能有近似解，其形式为：

$$1 - F = \frac{2(v+\eta)}{v(v+\eta)+\alpha^2} \exp(-\alpha^2 \tau) \tag{4}$$

式中，F 为交换度；$\tau = Dt/r^2$，r 为圆柱形、球形半径，对于薄片为 $2r$；$v = C$（薄片），$2C$（圆柱形），$3C$（球形）；η 对于薄片、圆柱形和球形其值分别为 1、2、3；α 是由 C、v 决定的数值。

式（4）可简化为：

$$\lg(1-F) = a + bt \tag{5}$$

其中

$$a = \lg \frac{2(v+\eta)}{v(v+\eta)+\alpha^2}；\quad b = -\frac{\alpha^2 D}{r^2} \times 0.4343$$

设上述胆固醇的形状为薄片，则 α 可由下式求出：

$$\alpha \mathrm{ctan}\alpha + C = 0 \tag{6}$$

对于 2mg CL-6-I 和 100μg Na^{131}I（NaI）的交换体系，$C = 5.9$，$v = C$，$\alpha = 2.71$，$\eta = 1$。

实验结果如图5所示。

由直线部分外推所得截距为 0.16，由式（5）计算所得的 a 值为 0.54，经过多次近似的计算值与实验值比较，结果尚较一致。

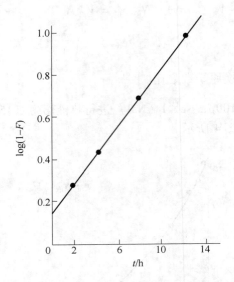

图 5 复相同位素交换中 $\lg(1-F)$ 与 t 的线性关系
（反应条件：2mg CL-6-I；20μL 无水乙醇；4μL H₂O；100μg NaI；150℃）

2.2.3 复相同位素的交换平衡问题

对于固-液（热液）两相的同位素交换，由于扩散过程是交换的决定步骤，因此体系要真正达到热力学平衡需时较长。众所周知，同位素交换达到平衡时存在下列关系：

$$^*N_1/N_1 = {}^*N_2/N_2$$

式中，$^*N_1$、$^*N_2$ 表示胆固醇和 NaI 中的放射性碘原子数；N_1、N_2 表示胆固醇和 NaI 中参与交换的碘原子总数。

达到平衡时的最大标记率为：

$$标记率 = \frac{N_1}{N_1 + N_2} \times 100\% \tag{7}$$

对于一定量的胆固醇，改变不同的 NaI 载体浓度，可由式（7）计算相应同位素交换最大标记率。图 6 实验结果表明，对于复相同位素交换反应，体系是可以达到热力学平衡的，只是需时较长。由此可见，用热液熔融法标记放射性药物，其比度只是由载体的含量所决定。

2.2.4 温度和 pH 对标记率的影响

热液熔融法的反应温度，一般低于标记物的熔点，如碘代胆固醇的熔点为 162℃，而实际采用的是 150℃，图 7 实验结果表明，当反应温度高于 100℃ 时，标记率明显增加，这可能是由于超过水的沸点后，产生的过热水蒸气和乙醇蒸气有利于胆固醇和 NaI 的均匀混合，使产额增加。

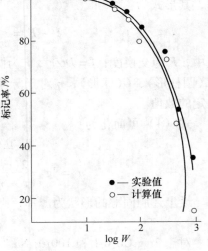

图 6 NaI 载体含量（W，μg）对标记率的影响
（反应条件：2mg CL-6-I；4μL Na¹³¹I（无载体）；反应温度：150℃；反应时间：15h）

图8的实验结果还表明,对于胆固醇体系,溶液起始的pH值不同,并不显著影响标记率,而对于像马尿酸、玫瑰红等体系,可能由于碘的解离活性受不同pH值的影响较大,因而标记率对pH值的依赖关系十分明显。

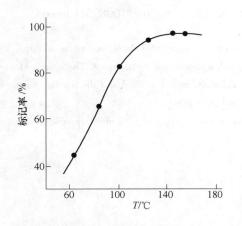

图7 热液熔融法的反应温度对标记率的影响
（反应条件：2mg CL-6-I；2μL Na^{131}I；15μL 无水乙醇；反应时间：30min）

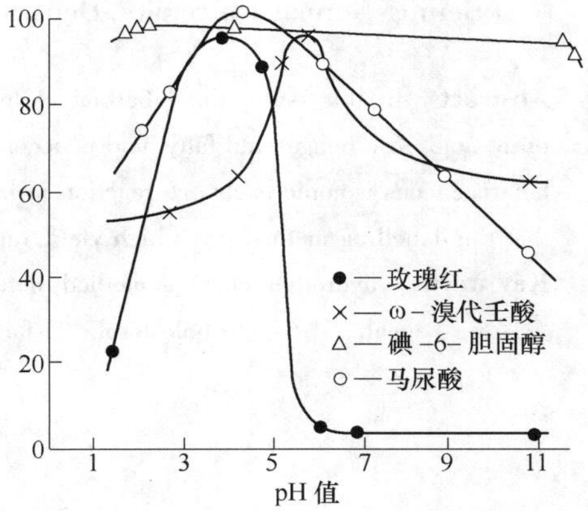

图8 pH值对标记率的影响

3 结论

（1）实验结果表明,当所用的Na^{131}I为无载体时,热液熔融的交换过程服从均相同位素交换的指数定律。加入NaI载体后,热液熔融交换过程基本遵从复相交换的动力学方程。

（2）由于热液熔融标记法没有像熔融法中Na^{131}I的脱水过程,因此在生产上避免了^{131}I$^-$在烘干时的溅出、污染和损失。

（3）热液熔融标记体系多数采用水和乙醇作为溶剂,标记率很高,因此标记后的产品一般不需要作纯化处理,宜于放射性"药箱"的应用。

（4）热液熔融标记法克服了熔融法中载体对标记率有明显影响的缺点[7],同时交换反应又较熔融法温和（比熔点低10℃）。

（5）热液熔融法具有交换速度快、产额高、方法简单等特点,在一定程度上具有普遍应用的意义,但使用热液熔融法的限制条件是要求标记物有较好的热稳定性。

参 考 文 献

[1] Elias H. Proc. 9th Japan Conf. on Radioisotopes, Tokyo, 538(1968).
[2] Stocklin G. Int. J. Appl. Radiat. Isot., 28, 131(1977).
[3] 刘伯里,等. 核化学与放射化学, 3, 242(1981).
[4] 国铳智,等. 核化学与放射化学, 5, 1(1983).
[5] Sinn H, et al. Int. J. Appl. Radiat. Isot., 28, 809(1977).
[6] Несмеянов, Ан. Н., Радиохимия, Издательство《Химия》,1978, crp. 35.
[7] Liu Boli, et al. Radioisotopes, 30, 275(1981).

A Study on Heterogeneous Isotopic Exchange Reaction

Liu Boli, Jin Yutai, Li Taihua

(Beijing Normal University, Department of Chemistry, Beijing, 100875, China)

Abstract In this paper the labelling of four radiopharmaceuticals, 6-iodocholesterol, o-iodohippuric acid, rose bengal and fatty acid is accomplished with hydrothermal melt method. The kinetic of heterogeneous isotopic exchange reaction in hydrothermal melt state is studied and the results show that this labelling method gives high yield, rapid rate and good purity.

Key words hydrothermal melt method, heterogeneous isotopic exchange, ^{131}I-oiodohippuric acid, ^{131}I-rose bengal, ^{131}I-6-iodocholesterol, ^{131}I-fatty acid

A Fast, Kit-type Radioiodination Procedure for ^{127}I-for-^{123}I Exchange[*]

Abstract A simple, fast, reliable, and efficient method for the radioiodination of HIPDM was developed and fully tested for its potential practical use in clinical applications. The method is based upon a heterogeneous isotopic exchange of ^{127}I-for-^{123}I at submelting temperatures and was easily adapted into a kit-form. In this manner, the uses of [^{123}I] HIPDM for the study of brain function can be optimized and expanded. Furthermore, the kit-form method is expected to facilitate the use of HIPDM labeled with high-purity ^{123}I.

1 Introduction

Following its proposal by Kung et al[1]. in 1983, the promises of [^{123}I] HIPDM (N, N, N′-trimethyl-N′-[2-hydroxy-3-methyl-5-⟨^{123}I⟩-iodobenzyl]-1, 3-propane-diamine) as a regional brain-perfusion agent for clinical use, is yet to materialize. As of this date, its availability has been less than adequate. This is in part due to the need to label HIPDM with scarcely-available high-purity ^{123}I in order to optimize imaging resolution and reduce radiation dose to patients[2]. In addition, the availability of HIPDM as a precursor has been severely restricted due to the current exclusive interest in its commercialization as a single-or as a multiple-dose radiopharmaceutical product. Since its development, however, the limited use of [^{123}I] HIPDM has been sufficient to document its usefulness and desirability for the diagnosis of brain function in conjunction with single-photon-emission-computerized tomography (SPECT). The preliminary reports also seems to justify the need of a large and reliable [^{123}I] HIPDM supply. In addition to the work by Kung et al. [1], several other methods for the radioiodination of HIPDM have been reported[3,4]. All these methods are based upon the ^{127}I-for-^{123}I isotopic exchange. However, labeling efficiencies obtained with these methods are less than quantitative resulting in radiochemical impurities ranging from 2% to 8%. Because the use of [^{123}I] HIPDM with SPECT systems require injecting multi-millicurie doses (3-5mCi), several hundreds microcuries of radiochemical contaminants (mostly as iodide) would then be present. The labeled [^{123}I] HIPDM would need to be purified prior to its use in patients. The needs for purification clearly defeat some of the advantages and the simplicity derived from the use of isotopic exchange labeling in kit form. Purification invariably results in losses due to decay and/or to physical handling and manipulations. A reliable, single-step labeling method which results in ^{123}I-labeled HIPDM of

---

[*] Copartner: OMAR F. CARVACHO, MANUEL C. LAGUNAS-SOLAR, Yutai Jin, Zhaoxiang Sun. Reprinted from *Appl. Radiat. Isot.*, 1986, 37(8):883-888.

sufficient radiochemical purity (i. e. >99%) and stability to allow for its immediate use is, therefore, highly desirable. Most labeling methods which satisfy this criterion normally allow the preparation of cold kits which are used to synthesize and formulate radiopharmaceuticals for clinical use. Because there are no thermodynamic effects in isotopic exchange reactions, the completion of these reactions depends solely on kinetic factors. Liu et al[5]. in 1985, reported kinetic and mechanistic factors in the exchange labeling of HIPDM. Based upon this latter work, we formulated, tested, and validated a kit-form method for the synthesis of [^{123}I] HIPDM. Other methods were also tested as means of large-scale synthesis. The results of this work are reported here.

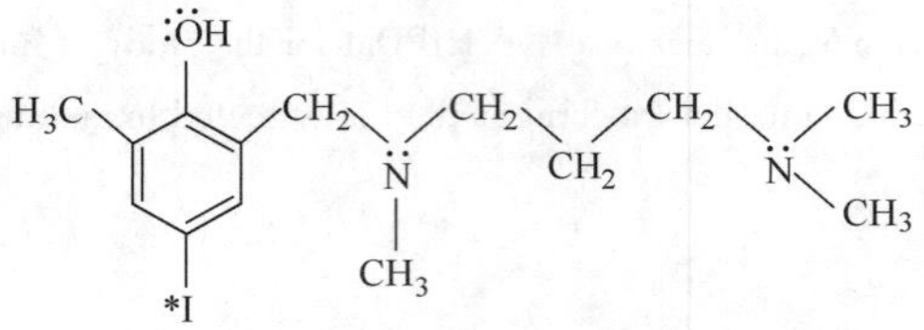

Fig. 1 The chemical structure of N, N, N'-trimethyl-N'-[2-hydroxy-3-methyl-5-iodobenzyl]-1,3-propanediamine (HIPDM)

2 Experimental Methods

2.1 Materials

HIPDM is not presently available commercially. The HIPDM (~3g) was obtained from the Department of Nuclear Medicine, State University of New York (SUNY), in Buffalo, New York, Mass., u. v., and i. r. spectrometry; HPLC; and NMR methods were used to analyze the chemical composition of HIPDM. These analyses were all in agreement with the structural formula for HIPDM as given in Fig. 1. Aqueous solutions of HIPDM (10-30mg/mL) adjusted, with 0.01 N NaOH, to pH 5-6 were shown to have excellent stability (>1y) based on periodic u. v. (optical density; peak maxima) and HPLC analyses. Chemical and identity assays were performed before and after labeling with radioiodine in order to evaluate any structural changes that might have occurred.

No-carrier-added [^{123}I] sodium iodide was produced in the 1.93-m isochronous cyclotron of the Crocker Nuclear Laboratory, University of California, Davis[6,7]. Iodine-123 was prepared indirectly from its 2.08-h ^{123}Xe parent produced via the ^{127}I(p,5n) reaction[6,7] and using a continuous-flow, generator production mode[8]. Iodine-123 in several different solvents (i. e. 0.1 N NaOH; ethanol; 0.1 N HCl), with and without sodium thiosulfate (as a reducing agent), was used in all the experiments. All the glassware (i. e. reaction vials) was rigorously cleaned with acid (HNO$_3$ or HCl), rinsed with triply-distilled, deionized H$_2$O, and dried in an oven at elevated temperatures (~250℃). All the seals for the reaction vials were Teflon-coated to avoid ^{123}I losses and to prevent the introduction of organic impurities. Isotonic saline(0.9% NaCl, sterile, pyrogen-free) was used to dissolve (or dilute) the [^{123}I] HIPDM. These solutions were sterilized by filtration, and sterility and apyrogenicity were tested using standard methods.

2.2 Labeling of HIPDM

The isotopic exchange reaction between HIPDM(1-10mg) and ^{123}I(1-1500mCi) was investigated at different conditions. All the experiments were carried out in open or sealed glass vials(3-15mL). Temperatures lower than the melting point of HIPDM(<180℃) and higher than the boiling point of the solvents were used. The temperature for these exchange experiments ranged between 105 and 175℃, although some experiments were carried out at room temperature(22-25℃) and at the melting point of HIPDM(~180℃). A large 200mL silicone-oil bath was used as a heat source, so as to prevent significant temperature changes when introducing the colder reaction vials. All the experiments were carried out inside a well-vented radioisotope glove box.

2.3 Radioanalytical assays

High-performance thin-layer chromatography (HPTLC) and thin-layer chromatography (Merck Silica Gel 60 F-254) or ascending paper chromatography (Whatman No. 1) were used to separate the different radiochemical forms obtained in the isotopic exchange labeling of HIPDM. In an 80:20:1 (v/v) solvent mixture of chloroform, ethanol, and water the R_f values for free iodide and labeled HIPDM were 0-0.5 and 0.9-0.95, respectively. Using an 80:20:0.1 (v/v) solvent mixture of chloroform, ethanol, and ammonia with both TLC methods, the separation was faster and more consistent with R_f values of 0.1(iodide), 0.2(iodate) and 0.8-0.9(HIPDM). A semi-remote radiochemical quality control system described by Harris et al[9]. was used to scan all the radiochromatograms, and an automatic ND-66 multi-scaling system [also described by Harris et al. [9]] was used for the identification and quantitation of the different radiochemicals. Typical results as given by this system are shown in Fig. 2.

3 Results and Discussion

The different procedures and factors investigated in this work were all directed towards the development of a rapid and efficient method for the exchange labeling of HIPDM. Therefore, time and easiness of the procedures were important parameters considered in the experimental design. For all the experiments, the different components used were prepared and made available as parts of a "cold" (i.e. non radioactive, long storage) kit. Different solutions of ^{123}I with radioactivities ranging from 1 to 1500mCi (radioactivity concentrations from 1×10^2 to 1×10^3 mCi/mL at the time of synthesis) were used in order to test the possible applications of these labeling methods to single-and/or multiple-dose batch preparations. Because of the scarcity of HIPDM, an effort was made to minimize its use, although because of its lack of pharmacological action (as compared with amphetamine derivatives) up to 10mg per injected dose is considered ac-

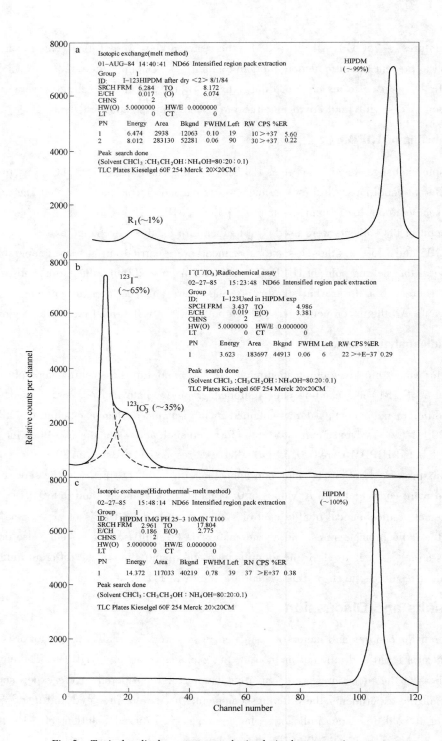

Fig. 2 Typical radiochromatograms obtained via the automatic scanner-
multi-scaling radiochemical control system

a—HIPDM obtained via the homogeneous/solid-phase exchange method. Note the presence of about 1% impurities;
b—Assay of radiochemical composition of ^{123}I solutions in 0.1N NaOH; c—HIPDM obtained via the HTM
method showing quantitative exchange. No impurities were detected using the automatic quality control system

ceptable for human use. The efficiencies for the isotopic-exchange labeling of HIPDM obtained under the different experimental conditions are given in Table 1 and Table 2. Table 1 has a summary of all the experiments which were started in a homogeneous single liquid phase, at different temperatures, pH values, and HIPDM concentrations, but finished in a solid-phase media for the completion of the reaction. These experiments were conducted in open reaction vessels. Table 2 contains a summary of the efficiencies in labeling HIPDM under heterogeneous kinetic conditions. Both the higher temperatures and the nature of the solvents helped to rapidly reach the conditions for solid-phase exchange in the melt. Other reactions were conducted in closed systems using a container with a volume large enough to allow for a complete evaporation of the solvents. In this manner, a solid and gas (vapor) phase were formed. In these experiments, both solid and a gas (vapor) phases containing the reactants were prevalent during most of the reaction time.

Table 1 Experimental conditions studied for the synthesis of [^{123}I]HIPDM by homogeneous and solid-phase exchange

HIPDM/Solvent/mg	^{123}I Solvent	pH	[^{123}I]HIPDM /%	Others/%	Samples	Time/min
Temperature range: 22-25℃							
10/H$_2$O	0.1N NaOH/S$_2$O$_3^{2-}$	5-6	50-80	20-50	12	5-15
10/H$_2$O	0.1N NaOH	5-6	50-80	20-50	12	5-15
Temperature range: 138-142℃							
5/H$_2$O	0.1N NaOH	5.5	>98	<2	4	2-5
5/H$_2$O	0.1N NaOH/S$_2$O$_3^{2-}$	5.5	>98	<2	4	2-5
10/H$_2$O	0.1N NaOH	5.5	>98	<2	5	2-5
10/H$_2$O	0.1N NaOH/S$_2$O$_3^{2-}$	5.5	>98	<2	5	2-5
10/H$_2$O	0.1N NCl	3.0	>99	<1	4	2-5
Temperature range: 148-155℃							
5/H$_2$O	0.1N NaOH	4-6	>95	<5	11	<2
0.5/H$_2$O	CH$_3$CH$_2$OH	10	<1	>99	3	<2
3.7/H$_2$O	0.1N NaOH	5-6	>97	<3	3	<2
Temperature range: 170-172℃							
5/H$_2$O	0.1N NaOH	6-7	>96	<4	5	<2
5/H$_2$O	0.1N NaOH	5.0	>97	<3	5	<2
5/H$_2$O	0.1N NaOH/S$_2$O$_3^{2-}$	6-7	>96	<3	5	<2
5/H$_2$O	0.1N NaOH/S$_2$O$_3^{2-}$	5.0	>98	<2	6	<2
10/H$_2$O	0.1N NaOH	5.0	>98	<2	4	<2
10/H$_2$O	0.1N NaOH/S$_2$O$_3^{2-}$	5.0	>98	<2	4	<2

Table 2 Experimental conditions for the synthesis of [^{123}I] HIPDM by heterogeneous isotopic exchange

HIPDM/Solvent/mg	^{123}I Solvent	pH	[^{123}I] HIPDM /%	Others/%	Samples	Time/min
Exchange in the melt (Temperature range: 178-182℃)						
10/H$_2$O	0.1N NaOH	8.0	~70	~30	2	<2
10/H$_2$O	0.1N NaOH	5.0	~65	~30	2	<2
Hydrothermal-melt method (Temperature range: 110-115℃)						
1/H$_2$O	0.1N NaOH	2-3	>99.9	<0.1	24	<2

In all the different methods investigated (see Table 1 and Table 2), the reaction time was less than 5min (except room temperature experiments, see Table 1). When best labeling yields were obtained, the reaction time was usually less than 2min. This clearly indicated that a fast, kit-type system for synthesizing [^{123}I] HIPDM was possible.

The exchange in solution (homogeneous kinetics) is appreciable even at room temperature (see Table 1) but a wide range (50%-80%) of labeling efficiencies were obtained at pH 5-6. This indicates that the aromatic ring is activated, most likely to nucleofilic attack. No effect was detected due to the presence of thiosulfate (used as a reducing agent). As expected, when the temperature of the reaction was increased to 138-142℃, labeling increased significantly (~98%) and was reliably achieved in a shorter time. The mass of HIPDM (5-10mg) did not have a marked effect, but lowering pH (from 5-6 to 3) increased the labeling to >99%. At higher temperatures (148-155℃) the effects of pH, carrier HIPDM, and ^{123}I solvents were also tested. Ethanol was found to greatly reduce the labeling efficiency, mostly because of pH effects. The "ethanol" (or pH) effect was also demonstrated when traces of ethanol were added to a reaction mixture that otherwise produced high labeling. The above results suggested that 5-10mg of HIPDM dissolved in H$_2$O and adjusted to pH 5-6 were the best conditions for HIPDM labeling. Because of this, these conditions were tested at higher temperatures, in an attempt to complete the exchange and achieve a preparation that would not require further purification. In the temperature range of 170-172℃, however, the results were similar (~98% labeling, <2min). It was noted, however, that a light-yellow tinted solution was obtained when the labeled product was redissolved. This indicated that the high temperature resulted in some decomposition of HIPDM.

3.1 Synthesis of [^{123}I] HIPDM via the homogeneous and solid-phase exchange method

Single-dose (5-30mCi) as well as multi-dose batches (up to 1.5 Ci) were then prepared at (140±2)℃, with ^{123}I in 0.1 NNaOH or 0.1 N HCl, and using 5-10mg of HIPDM in H$_2$O, and adjusted to pH 5.5. The labeling efficiency was found to be >98% ($n=30$). A typical radio-chromatogram of [^{123}I] HIPDM prepared with this method is shown in Fig. 2a. Radiochemical impurities commonly associated with no-carrier-added ^{123}I (i.e. iodate) had no effect on labeling yield or in [^{123}I] HIPDM radiochemical purity. Although the radioanalytical method used (see Section Ⅱ. c) clearly has the capability of detecting the presence of iodide/iodate in the ^{123}I solutions (see Fig. 2b), iodate was not found on any of the final [^{123}I] HIPDM preparations. Therefore, we concluded that this method reduces the ^{123}I oxidized forms during the exchange making unnecessary the addition of reducing agents (i. e. thiosulfate). Several analytical techniques (NMR, i. r., u. v., HPLC) showed no detectable chemical modifications to the HIPDM.

Because the above methods failed to provide a [^{123}I] HIPDM preparation of >99% radiochemical purity that would require no purification, other methods based on heterogeneous kinetics were attempted. As suggested by Seevers and Counsell in 1982[10], exchange in the melt is normally a better alternative than exchange in solution. However, when melting HIPDM its decomposition was clearly extensive and eliminated this possibility. In addition, a larger amount of HIPDM (10mg) was found to be necessary for reproducible results. The results in Table 2, at 178-182℃, clearly showed this effect, and suggested that labeling of HIPDM must be carried out at submelting temperatures. However, the above results also showed the importance of having the reactants in a solid-phase where the data also suggests that the exchange was more efficient. As a way to achieve these conditions, Liu et al[5]. suggested the use of a closed system but with enough volume to allow for a complete removal of solvents to a gas (vapor) phase and thus forming a reactive solid-phase to enhance the isotopic exchange. This required that for a specified reaction vessel, the volumes of both HIPDM and ^{123}I solutions (i. e. reactants total volume) would have a maximum. If H_2O is used as a solvent, a sealed vial sufficiently large to allow for the evaporation of H_2O, also allows the overheated vapor to enhance solubilities (or mixing) of both reactants (HIPDM) and ^{123}I. A series of experiments resulted in formulating the best conditions for labeling HIPDM via this procedure. These conditions are given in Table 2. As expected, reaction time, temperature, and pH, are factors affecting the labeling efficiency. These effects are shown in Fig. 3. Liu et al[5]. suggested this method to be known as "hydrothermal-melt" (HTM) method. It appears that the HTM method does in fact promote a rapid exchange in the interphase between the solid and gas (vapor) phase. However, if the interphase is the only cause of this labeling enhancement, the fact that only 1mg of HIPDM is needed for >99.9% labeling cannot be explained based upon our previous results that showed a marked carrier-HIPDM effect (see Table 1). As the amount of HIPDM was increased, it seemed to provide an adequately larger reacting (or mixing) surface. With only 1mg of HIPDM the methods described in Table 1 yielded low (<90%) and inconsistent results. It appears that solvation effects may also have a role in the lack of complete exchange. In the "open" methods (Table 1, and melt method in Table 2), the solvent was simply driven off the solid surfaces and may have in fact been responsible for some of the ^{123}I losses that were observed. In the HTM method and once the solvent is evaporated, the solid-phase surface should provide an active site for exchange, while the closed system assures that any of the "evaporated" ^{123}I (in droplets) will still be available for exchange when effective collisions take place in the interphase boundary. Nevertheless the results shown in Table 2 (n = 24) clearly indicated the efficiency of the HTM method.

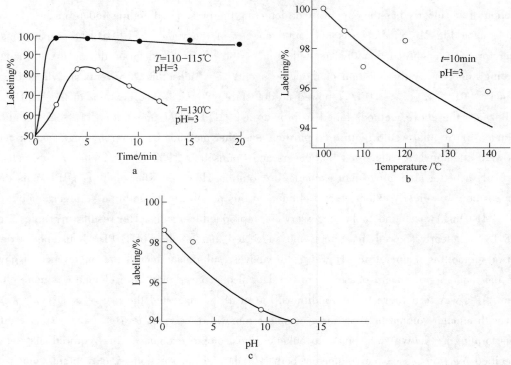

Fig. 3　Effects of reaction time(a), temperature (b), and pH on the labeling efficiency(%)(c) of HIPDM using the hydrothermal-melt(HTM) method

Based upon these results, a single-dose HIPDM labeling kit was prepared as follows:

(1) A solution of HIPDM in H_2O (10mg/mL) adjusted to pH 2-3 by addition of 0.1N HCl. This solution could be freeze-dried for long-storage purposes.

(2) A 3mL sterile, pyrogen-free glass ampoule as a reaction vessel, and

(3) A 5mL sterile, pyrogen-free vial with Teflon-coated seal discs as a final container to receive the final radiopharmaceutical preparation.

In addition, a silicone-oil bath or any other form of dry heat(metal blocks) capable to provide approximately 130℃ is adequate for this kit. Several other items(i.e. syringes and needles, membrane filters, etc.) and the ability to perform radiochemical and biological quality-control assays are also needed. However, those are common items and capabilities in any clinical facility.

3.2　Synthesis of [^{123}I]HIPDM with the hydrothermal-melt method

Only 100μL of the HIPDM (10mg/mL) solution are needed for the synthesis of a single dose. This solution is transferred to a sterile, pyrogen-tree 3mL glass ampoule. However, with certain limitations(see below) the same amount of HIPDM(i.e. 1mg) can be used for the synthesis of multiple doses. The required amount of ^{123}I in 0.1N NaOH(maximum volume 200μL) is added to the ampoule. The ampoule is sealed with a flame. The ampoule is then submerged into the silicone-oil bath (preheated to 110-115℃) for 2min. After this reaction time, the ampoule is removed, cooled down rapidly in H_2O, and opened. Two fractions of isotonic saline (0.5-1mL) are used to remove the [^{123}I]HIPDM, and both washes are transferred to a sterile/pyrogen-free vial via a sterilizing 0.22μm membrane filter. As-ceptically, the required volume for quality controls is then removed. The quality control assays would normally require approximately 1h. However, using HPTLC and an automatic radiochromatography system with multi-scaling capabilities, the radiochemical test can be completed in less than 45min which is also sufficient time for the performance of the pyro-genicity test using the LAL method. Sterility can be confirmed using standard assays.

The only potential practical limitation of the HTM method is due to the requirement of a low volume of the reaction mixture (i.e. 300μL into a 3mL vial). For the synthesis of a single-dose (≈5mCi at time of injection) batch, the ^{123}I solution would have to be prepared with >400mCi/mL (assuming a 40h of decay from end of processing). This can in fact be accomplished with some of the current techniques used in the production of high-purity ^{123}I[7,8]. For large, multi-Curie batches intended for the distribution of HIPDM as a finished radiopharmaceutical, the homogeneous/solid-phase exchange method seems more practical because of the limitations in the concentration (Ci/mL) of the ^{123}I solutions. These latter solutions can be obtained in concentrations of up to 5Ci/mL. Both methods (i.e. homogeneous/ solid-phase exchange and HTM) yielded [^{123}I]-HIPDM which does not decompose or deiodinate when tested in plasma (incubated at 35℃) and for radiochemical purity under room-temperature storage (>3 days). Therefore both preparations are expected to have good *in vivo* stability.

4 Conclusions

Two methods have been developed for the synthesis of radiopharmaceutical-quality ^{123}I-labeled HIPDM. The simplicity, reproducibility, and efficiency of the methods reported here are clearly adequate to increase the future availability of [^{123}I]HIPDM. Single-or multiple-dose [^{123}I]HIPDM radiopharmaceutical preparations in an easy-to-use kit form is therefore, forthcoming. These methods are also applicable to the synthesis of other activated aromatic iodine chemicals, and can be used in fast radioiodination procedures with other radioiodines such as ^{122}I(3.6min), which has recently been proposed as a new positron-emitting radionuclide for PET studies[10,11].

Acknowledgements

The authors wish to thank the staff of Crocker Nuclear Laboratory for the considerable support during these experiments. Special thanks to Dr Hank Kung, of the State University of New York, in Buffalo, for supplying the HIPDM used in this project, and for his considerable help in discussing and encouraging this work.

References

[1] Kung H. F., Tramposch K. M. and Blau M. J. Nucl. Med. 24,66(1983).

[2] Lagunas-Solar M. C. and Hines H. H. Proc. Int. Symposium on The Developing Role of Short-Lived Radionuclides in Nuclear Medicine Practice(Eds Paras P. and Thiessen J. W.). CONF-820523(DE82008258), pp. 423-444(1985).

[3] Mangner T. J., Tobes M. C., Wieland D. M. and Sisson J. C. J. Nucl. Med. 25, P123(1984)(Abstr.).

[4] Lagunas-Solar M. C., Harris L. J. and Carvacho O. F. Fifth Int. Symposium on Radiopharmaceutical Chemistry, 9-13 July 1984, Tokyo, Japan. p. 155(1984)(Abstr.).

[5] Liu B-L., Chang J., Sun J. S., Billings J., Blau M. and Kung H. F. (Abstract). J. Nucl. Med. 26, P123 (1985).

[6] Jungerman J. A. and Lagunas-Solar M. C. J. Radioanal. Chem. 65, 31(1981).

[7] Lagunas-Solar M. C. Proc. Int. Symposium on The Developing Role of Short-Lived Radionuclides in Nuclear Medicine Practice(Eds Paras P. and Thiessen J. W.). CONF 820523(DE82008258), pp. 203-221(1985).

[8] Lagunas-Solar M. C., Thibeau H. L., Goodart C. E., Little F. E., Navarro N. J. and Hartnett D. E. Proc. Int. Symposium on The Developing Role of Short-Lived Radionuclides in Nuclear Medicine Practice (Eds Paras P. and Thiessen J. W.). CONF 820523(DE82008258), pp. 190-202(1985).

[9] Harris L. J., Avila M. J., Shadoan D. J., Essert T. K. and Lagunas-Solar M. C. J. Nucl. Med. Technol. 12, 126(1984).

[10] Seevers R. H. and Counsell R. E. Chem. Rev. 82, 575(1982).

[11] Mathis C. M., Lagunas-Solar M. C., Sargent T., Yano Y., Vuletich A. and Harris L. J. Appl. Radiat. Isot. 37, 258(1986).

[12] Lagunas-Solar M. C., Carvacho O. F., Harris L. J. and Mathis C. M. Appl. Radiat. Isot. 37, 835(1986).

Cyclotron Production of High-purity ^{123}I

I. A Revision of Excitation Functions, Thin-target and Cumulative Yields for ^{127}I(p, xn) Reactions*

Abstract Excitation functions, thin-target and cumulative yields for the proton-induced reactions on ^{127}I targets were measured in the 67.5-to 5.3-MeV energy region. These results were used primarily to define the proton-energy ranges and target thicknesses to optimize radionuclide yields and purities for ^{123}I production from its 2.08-h ^{123}Xe parent. Other reactions producing radioxenons of interest in nuclear medicine (i. e. 36.406-d ^{127}Xe, 17.3-h ^{125}Xe, 20.1-h ^{122}Xe, and 38.85-min ^{121}Xe), were also measured. These results are compared to other previously reported values.

1 Introduction

Iodine-123 radiopharmaceuticals are currently receiving considerable attention due to an improved knowledge in ^{123}I-labeling/syntheses chemistry, and to the increased availability of (p, 5n)-made ^{123}I for use in radiopharmaceutical and clinical research. As a consequence, a series of ^{123}I-labeled radiopharmaceuticals for imaging and functional studies of the brain (IMP, HIPDM); myocardium (fatty acids, MIBG); tumor detection (antibodies); and a series of specific receptor-binding agents, are currently being evaluated. This potential has also prompted the private sector to invest in 70-MeV accelerator facilities with the potential to produce larger amounts of high-purity (p, 5n)-made ^{123}I. In addition, and as a response to increased needs by the research community, several national and university accelerator facilities have also made substantial commitments to develop increased production capabilities for high-purity ^{123}I. Several authors have reported excitation functions for the ^{127}I(p, xn) (x = 1-7) nuclear reactions[1-4]. Q-value and product-nuclide data for these proton-induced reactions are given in Table 1. With a few exceptions, most authors have given particular attention to the ^{127}I(p, 5n)^{123}Xe→^{123}I and the ^{127}I(p, 3n)^{125}Xe→^{125}I reactions, which are both of importance for the maximization of yields and radionuclidic purities for (p, 5n)-made ^{123}I. A summary of the reported cross-section values for these reactions is given in Table 2. When comparing these values, many discrepancies can be found, in particular, in the maximum values for the ^{127}I(p, 5n) reaction. In addition, integration of ^{123}Xe-and ^{125}Xe-yields in the region-of-interest for ^{123}I production (i. e. 67- and 46-MeV) resulted in significant differences. Therefore, we decided to reexamine this matter by executing a series of high-energy resolution experiments using a low-intensity 67.5 MeV proton

---

* Copartner: MANUEL C. LAGUNAS-SOLAR, OMAR F. CARVACHO, Yutai Jin, Zhaoxiang Sun. Reprinted from *Appl. Radiat. Isot.*, 1986, 37(8):823-833.

beam, and a stack of thin-, hermetically-sealed NaI targets. In addition, we minimized and, or corrected for many of the typical sources of error for this type of measurements. The results of these experiments and a discussion of their significance, are reported here.

Table 1 Product nuclides and Q values for proton-induced reactions on ^{127}I targets①

Reaction	Product nuclides	Q/MeV
(p, n)	^{127}Xe(36.406d) →^{127}I(stable)	−1.45
(p, 2n)	^{126}Xe(stable)	−9.21
(p, 3n)	^{125}Xe(17.3h) →^{125}I(60.25d) →^{125}Te(stable)	−19.33
(p, 4n)	^{124}Xe(stable)	−27.06
(p, 5n)	^{123}Xe(2.08h) →^{123}I(13.22h) →^{123}Te(stable)	−37.29
(p, 6n)	^{122}Xe(20.0h) →^{122}I(3.63min) →^{122}Te(stable)	−45.50
(p, 7n)	^{121}Xe(38.85min) →^{121}I(2.12h) →^{121}Te(16.78d) →^{121}Sb(stable)	−56.40
(p, 8n)	^{120}Xe(40min) →^{120}I(1.35h) →^{120}Te(stable)	−64.96

① For Q-value calculations, the atomic masses were taken from *Nuclear Data Tables*, 1971, 9: 4-5.

Table 2 Comparison of reported literature values for proton-induced reactions on ^{127}I targets

Reaction	Cross-section values (mbarn)(Uncertainty)	Proton energy/MeV	Reference
^{127}I(p,5n)^{123}Xe	490 ± 59(12%)	55-56(max.)	Wilkins et al. [1]
	377 ± 47(12.5%)	56-57(max.)	Paans et al. [2]
	290 ± 29(10.1%)	56-57(max.)	Syme et al. [3]
	440 ± 53(12%)	55-56(max.)	Hegedus [4]
^{127}I(p,3n)^{125}Xe	210 ± 25(10%)	50	Wilkins et al. [1]
	115 ± 27(12.5%)	50	Paans et al. [2]
	660 ± 63(9.6%)	32(max.)	Syme et al. [3]
	150 ± 14(9.6%)	50	Syme et al. [3]

2 Experimental Methods

2.1 Targetry and irradiations

When measuring excitation functions (i.e. cross sections as a function of particle energy) and radionuclide yields, there are many sources of errors that contribute to the overall uncertainties. These uncertainties have already been discussed in detail in earlier publications[5,6]. The measurement of excitation functions and yields from reactions producing radioxenons, adds another potential source of errors. Xenon isotopes are gases, and thus once formed in a target, physical losses may occur, thus decreasing the measured activities. However, because most of the formation of the radioxenons occurs inside the NaI crystal lattice, one could expect good retention of radioxenons if the lattice remains intact[7]. Low beam currents (i.e. nanoamperes) and thin (i.e. 0.045cm) NaI targets were then used in order to prevent any possible lattice alteration due to heating caused by beam-energy deposition. Despite this precaution, and because some radioxenon production also occurs at the surface of each NaI target, it was also necessary to physically contain each NaI target in a sealed enclosure, in order to avoid potential surface losses. The NaI targets were then prepared with anhydrous NaI (Mallinckrodt Chemical, St Louis, Mo.), pressed into a circular cavity (0.95-cm i.d., 0.045-cm thick) of an Al disk (2.54-cm o.d.), and sealed on both sides with 0.00254-cm thick Al foils using a fast-drying epoxy resin as an adhesive. The Al surface surrounding the NaI pellet provided an adequate area for sealing the targets. Because NaI is hygroscopic several targets were punctured to cause a pinhole and then subjected to prolonged heating to determine water-weight losses. These tests indicated an average of 4.5% loss in weight. This average weight-loss was used to correct the measured thickness (g/cm^2) of each target, from which the number of ^{127}I-target atoms was calculated. A stack of 35 NaI sealed targets were used and mounted in a special low-intensity-beam target system as is shown in Fig. 1. The target system is provided with an electrically-insulated Faraday cup to monitor the charge deposited by the beam; a magnet to attract secondary electrons produced by strayed beam that could otherwise produce charge-reading errors by neutralizing positively charged beams: and two thick collimators (0.64-cm thick Ta, 0.32-cm i.d.; and 2.54-cm thick C, 24-cm. i.d.). This arrangement was used to assure that all of the integrated beam interacted in a small area ($<0.08 cm^2$) of the NaI targets. In this manner, the location of the induced radioactivities on each target allowed for a nearly ideal point-source-counting geometry. However, the proton beam was swept electronically in order to minimize the beam-power density. Under these conditions, the beam power deposited per target was $<2.5 W/cm^2$. No tar-

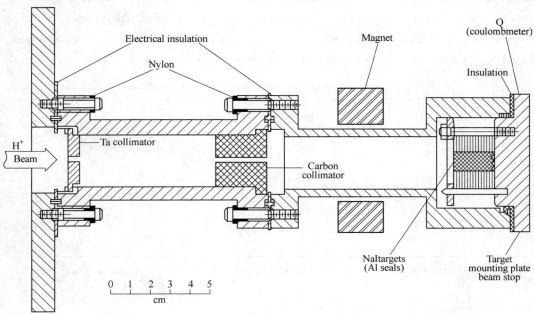

Fig. 1 Schematic of UC Davis's low-intensity-beam experimental target

get cooling was used. All the irradiations were conducted at the Crocker Nuclear Laboratory, University of California, Davis's 1.93-m(76-in.) isochronous cyclotron. The proton energy was measured as (67.5 ±0.5) MeV, using both cyclotron parameters and γ-flash techniques[8]. The proton energy at the entrance and exit surfaces of each NaI target was calculated using the range-energy data of Janni[9].

2.2 Radionuclide assays

The radioassay of each single NaI target was conducted with a high-purity Ge detector coupled to a Nuclear Data (ND 66) 8192-multichannel analyzer. This detector has been calibrated (for photon energies and point-source detector efficiencies) with National Bureau of Standards reference materials. This radioassay system has been described previously[6]. The γ rays and their probabilities per decay, and the half lives of the radioxenons used in this work are given in Table 3. These values were taken from Lederer and Shirley[10] for consistency, although the γ-ray abundance for the 133-keV γ ray from the 38.85-min ^{121}Xe, was taken from the work reported by Syme et al[3]. All the radioassays were conducted with the idea of maximizing the counting statistics for each of the photons of interest. Therefore, single NaI targets were radioassayed at different times after the end of bombardment (EOB), and for a sufficiently long time to obtain >10000 net counts in the region of interest. In addition, and in order to detect the different radioxenons as close to the threshold energy for their production, the NaI targets were radioassayed in ascending order of their calculated proton energies. With only one exception (see Section Ⅲ, Results and Discussion) all of the primary and secondary photons used in these radioassays were easily resolved with the Ge-detector system (resolution 2.5keV FWHM, 662keV). By using half-life determinations, all of the γ rays observed were identified with a particular radioxenon. Finally, in order to test any radioxenon losses post bombardment, several targets were assayed over a 4-day period. No losses (other than decay) were observed.

Table 3 Nuclear properties used in data evaluation

Radionuclide	Half life	Gamma rays (probability per decay)/keV
^{127}Xe	36.406d	202(68.1%)
^{125}Xe	17.3h	188(55%);243(28.7%)
^{123}Xe	2.08h	178(14.9%);330(8.6%)
^{122}Xe	20.1h	350(8%)
^{121}Xe	38.85min	133(11.4%)①

① The abundance of the 133-keV ^{121}Xe γ ray was taken from a measured value reported by Syme et al.[1].

2.3 Data analysis and uncertainties

The data from the radioassays was analyzed with the ND 66 peak-integration routine, using a linear-background subtraction. This less sophisticated method was compared with a PDP 11/44 computer-programmed routine for peak extraction and integration using gaussian; gaussian-plus-exponential-tail; and a least-squares routine to minimize errors in curve fitting. This comparison showed no significant differences and indicated that the simpler ND 66 method can be utilized when the photon spectra are sufficiently resolved, and the radioassays are per-formed on single photopeaks with adequate counting statistics (<1% errors). Most photopeaks used in data analysis complied with the above criteria. The integrated net areas were converted to counts-per-second(cps) and corrected for detector efficiency; probability per decay(i. e. abundance); decay losses; and detector dead time. A simple counting test using non-irradiated NaI targets as absorbers, indicated that a measurable fraction(2%-4%) of photons were attenuated by the Al-sealed NaI targets(i. e. self-absorption). Therefore, an experimentally measured correction was also used. This correction ranged from 2.4% at 330keV to 3.4% at 172keV. The NaI-target thicknesses(g/cm^2) were also corrected for the thicknesses occupied by the two 0.00254-cm thick Al seals on each NaI target. These seals represented 11.3% more NaI on each target, and thus the induced radioactivities would have been a factor of 1.113 greater. This later correction does not affect the cross-section values, but it represented a large source of errors in thick-target yield integrations. Cross sections (mbarn) and radionuclide yields ($\mu Ci/(\mu A \cdot h)$ or $mCi/(\mu A \cdot h)$) were calculated from this data by using the formulation reported by Lagunas-Solar et al[11]. The total uncertainties were obtained by summation of the squares of each uncertainty (expressed as percentages), and then taking the square-root of the sum. The uncertainties included in the energy calculations(based upon the one-standard-deviation criteria), were: incident beam energy($\pm 1\%$); range values(9)($\pm 1\%$); NaI target density($\pm 1\%$); and NaI target thickness(weighing errors)($\pm 1\%$). The total uncertainties in energy calculations ranged from $\pm 1\%$ (one-standard-deviation) at 67MeV to $\pm 29\%$ at 7.5MeV. For the cross-section and yield calculations, the following individual uncertainties were included: beam-current integration ($\pm 3\%$); calibration sources ($\pm 3\%$); detector efficiency ($\pm 5\%$); counting geometry ($\pm 1\%$); decay corrections ($\pm 3\%$); γ-ray abundances ($\pm 5\%$); and integration routines ($\pm 3\%$)(av.)($\pm 8\%$)(max.). The total uncertainty in the cross-section and radionuclide-yield values was calculated as $\pm 9\%$ (one-standard-deviation). Recoil losses and photon summing(source-to-detector distance >10cm) were assumed to be negligible.

3 Results and Discussion

Thirty-five NaI targets were used to cover the 67.5-5.3-MeV energy range. The excitation functions reported here have energy uncertainties ranging from ± 0.55MeV at 66.9MeV, to ± 2.2MeV at 7.5MeV. In the evaluation of the energy uncertainty, straggling effects were not included and therefore, the uncertainties may be underestimated particularly at the low end of the energy range. However, straggling have little or no effect on average energies on which all proton energy calculations were based[12]. Radio-nuclide yields for the production of ^{127}Xe (36.406d); ^{125}Xe(17.3h); ^{123}Xe(2.08h); ^{122}Xe(20.1h); and ^{121}Xe(38.85min) were calculated as a function of proton energies. The cross-section values for the reactions studied were obtained from the measured radio-nuclide yields and combined with the calculated proton energies on each NaI target to obtain the excitation functions. These results are particularly relevant to the assessment and fine-tuning of ^{123}I-production methods. In order for these results to be practical to use, single and cumulative values for NaI-target thicknesses and for radionuclide yields are also given. The excitation functions for the production of ^{127}Xe, ^{125}Xe, ^{123}Xe, ^{122}Xe and ^{121}Xe, over the 67.5-5.3-MeV energy range are shown in Fig. 2. All of the reaction's Q values are also indicated and agreed well with calculated values as given in Table 1. Thin-target and cumulative radionuclide yields are shown in Fig. 3. The results of this work are compared with several other measured values reported in the literature. However, because none of the referenced reports[1-4] included all the excitation functions reported here, the comparisons will be made based upon the individual xenon radionuclides.

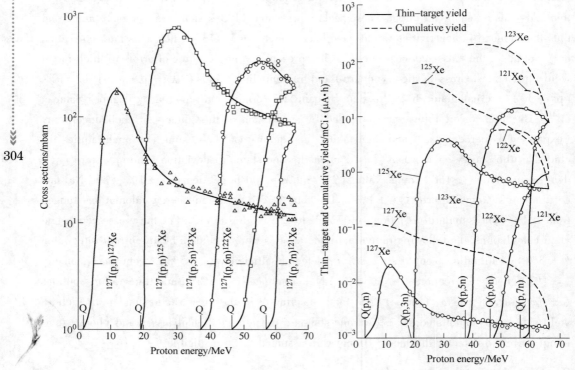

Fig. 2 Excitation functions for the production of ^{127}Xe, ^{125}Xe, ^{123}Xe, ^{122}Xe, and ^{121}Xe, over the 67.5-5.3-MeV proton energy range
(Calculated Q values(Q) for each reaction are also indicated. Experimental data points used in determining the eye-fitted curves are shown for each of the reactions studied. Below 20MeV, the ^{127}I(p,n) excitation function was found to be underestimated when compared to other reported values(see Ref. 13))

Fig. 3 Thin-target and cumulative yields (μCi/(μA·h),mCi/(μA·h)) for the production of ^{127}Xe, ^{125}Xe, ^{123}Xe, ^{122}Xe and ^{121}Xe, are given as a function of proton energies(MeV)
(Cumulative yields were calculated by summation of thin target yields in the direction of the proton beam transversing a thick NaI target)

3.1 Xenon-127

The excitation function for the ^{127}I(p, n)^{127}Xe($Q = -1.45$MeV) reaction was measured in the entire 67.5-5.3-MeV energy range. The radioassays were conducted using the 202-keV (68.1%) γ-ray. These measurements were started 3 days after EOB to allow for decay of other radionuclides produced. The results are given in Table 4. Other authors referenced in this work[1-4] did not report this reaction. Our measurements indicated that the maximum cross section is (178 ± 16) mbarn for the ^{127}I(p, n) reaction and occurs in the 11-12MeV range. This cross-section value is somewhat smaller than the value of (471 ± 40) mbarn at 9.92MeV, reported by Colle and Kishore[13] in 1974. This indicates that our cross-section values were underestimated due to the low-energy resolution achieved for this reaction at energies below 20MeV. Above 20-MeV, our results are in excellent agreement with the Colle and Kishore values. In the region-of-interest for ^{123}I production (i.e. 67-46-MeV), the ^{127}Xe yield is (32.5 ± 2.9) μCi/(μA · h). This is a relatively small yield and therefore, the formation of ^{127}Xe constitutes only a handling/storage problem easily circumvented by proper system design. No radio-contaminants are produced from ^{127}Xe because it decays to stable ^{127}I. This latter process, however, results in a decrease of ^{123}I specific activities, although this source of carrier ^{127}I should be considered as less significant when compared to other sources (i.e. reagents).

Table 4 Cross sections (mbarn) and yields (μCi/(μA · h)) for ^{127}I(p, n)^{127}Xe①

Proton energy/MeV		NaI target/g · cm^{-2}		Cross sections /mbarn	^{127}Xe yields /μCi · (μA · h)$^{-1}$	
In/Out	Av.	Single	Cumulative		Single	Cumulative
67.5-66.4	66.9	0.203	0.203	13	1.5	1.5
66.3-65.3	65.8	0.198	0.401	14	1.6	3.1
65.2-64.1	64.6	0.202	0.603	15	1.6	4.7
64.0-62.9	63.4	0.203	0.806	14	1.6	6.3
62.8-61.8	62.3	0.176	0.982	12	1.4	7.7
61.7-60.6	61.2	0.195	1.177	14	1.6	9.3
60.5-59.4	60.0	0.193	1.370	15	1.7	11.0
59.3-58.1	58.7	0.215	1.585	14	1.6	12.6
58.0-56.8	57.4	0.203	1.788	14	1.6	14.2
56.7-55.6	56.1	0.193	1.981	14	1.6	15.8
55.4-54.2	54.8	0.203	2.184	15	1.7	17.5
54.1-53.0	53.5	0.188	2.372	15	1.7	19.2
52.8-51.5	52.2	0.223	2.595	16	1.7	20.9
51.3-50.0	50.7	0.207	2.802	15	1.7	22.6
49.9-48.6	49.3	0.199	3.001	20	2.1	24.7
48.5-47.1	47.8	0.206	3.207	15	1.7	26.4
47.0-45.8	46.4	0.176	3.383	19	2.0	28.4
45.7-44.4	45.0	0.185	3.568	20	2.1	30.5

Continued 4

Proton energy/MeV		NaI target/g·cm^{-2}		Cross sections /mbarn	^{127}Xe yields /μCi·(μA·h)$^{-1}$	
In/Out	Av.	Single/Cumulative			Single/Cumulative	
44.2-42.8	43.5	0.210	3.778	19	2.1	32.6
42.6-41.2	41.9	0.198	3.976	17	1.9	34.5
41.0-39.6	40.3	0.193	4.169	20	2.2	36.7
39.4-37.8	38.6	0.206	4.375	21	2.2	38.9
37.7-36.0	36.8	0.206	4.581	21	2.3	41.2
35.9-34.2	35.0	0.204	4.785	23	2.5	43.7
34.0-32.6	33.3	0.169	4.954	18	2.1	45.8
32.4-30.7	31.5	0.188	5.142	23	2.5	48.3
30.5-28.8	29.7	0.181	5.323	25	2.7	51.0
28.6-26.8	27.7	0.187	5.510	27	2.7	53.7
26.6-24.6	25.6	0.195	5.705	28	2.8	56.5
24.3-22.2	23.3	0.195	5.900	35	3.9	60.4
21.9-19.6	20.8	0.196	6.096	40	4.3	62.7
19.3-16.8	18.1	0.193	6.289	63	6.8	69.5
16.5-13.7	15.1	0.195	6.484	116	12.6	82.1
13.3-10.1	11.7	0.191	6.675	178	19.3	101.4
9.6-5.3	7.5	0.199	6.874	61	6.6	108.0

① Proton energies (MeV) are given for the incident (In) and exit (Out) energies from single NaI targets. Average proton energies (Av.) at the geometric center of each NaI target, are also given. Cumulative NaI target thicknesses and yields were obtained by summation of single values.

3.2 Xenon-125

The production of ^{125}Xe via the ^{127}I(p, 3n) ($Q = -19.33$ MeV) reaction was studied in the 67.5-16.8-MeV energy range, and can be compared with other reported values as given in Table 2. The ^{125}Xe yield was measured using both the 188-keV (55%) and the 243-keV (28.7%) γ rays with similar results. Radioassays were started 1 day after EOB. These results are given in Table 5.

Table 5 Cross sections (mbarn) and yields (mCi/(μA·h)) for ^{127}I(p, 3n)^{125}Xe①

Proton energy/MeV		NaI target/g·cm^{-2}		Cross sections /mbarn	^{125}Xe yields /mCi·(μA·h)$^{-1}$	
In/Out	Av.	Single/Cumulative			Single/Cumulative	
67.5-66.4	66.9	0.203	0.203	89	0.48	0.48
66.3-65.3	65.8	0.198	0.401	86	0.47	0.95
65.2-64.1	64.6	0.202	0.603	93	0.50	1.45
64.0-62.9	63.4	0.203	0.806	90	0.49	1.94
62.8-61.8	62.3	0.176	0.982	99	0.47	2.41
61.7-60.6	61.2	0.195	1.177	110	0.57	2.98

Continued 5

Proton energy/MeV		NaI target/g·cm^{-2}	Cross sections /mbarn	^{125}Xe yields /mCi·(μA·h)$^{-1}$	
In/Out	Av.	Single/Cumulative		Single/Cumulative	
60.5-59.4	60.0	0.193	1.370	114	0.59	3.57
59.3-58.1	58.7	0.215	1.585	113	0.65	4.22
58.0-56.8	57.4	0.203	1.788	108	0.59	4.81
56.7-55.6	56.1	0.193	1.981	116	0.59	5.40
55.4-54.2	54.8	0.203	2.184	118	0.64	6.04
54.1-53.0	53.5	0.188	2.372	123	0.62	6.66
52.8-51.5	52.2	0.223	2.595	124	0.74	7.40
51.3-50.0	50.7	0.207	2.802	130	0.72	8.12
49.9-48.6	49.3	0.199	3.001	150	0.80	8.92
48.5-47.1	47.8	0.206	3.207	129	0.70	9.62
47.0-45.8	46.4	0.175	3.383	189	0.90	10.52
45.7-44.4	45.0	0.185	3.568	196	0.97	11.49
44.2-42.8	43.5	0.210	3.778	217	1.19	12.68
42.6-41.2	41.9	0.198	3.976	225	1.19	13.87
41.0-39.6	40.3	0.193	4.169	281	1.45	15.32
39.4-37.8	38.6	0.206	4.375	346	1.91	17.23
37.7-36.0	36.8	0.206	4.581	444	2.45	19.68
35.9-34.2	35.0	0.204	4.785	590	3.22	22.90
34.0-32.6	33.3	0.169	4.954	619	2.81	25.71
32.4-30.7	31.5	0.188	5.142	731	3.69	29.40
30.5-28.8	29.7	0.181	5.323	727	3.53	32.93
28.6-26.8	27.7	0.187	5.510	684	3.43	36.36
26.6-24.6	25.6	0.195	5.705	479	2.50	38.86
24.3-22.2	23.3	0.195	5.900	297	1.55	40.41
21.9-19.6	20.8	0.196	6.096	61	0.32	40.73
19.3-16.8	18.1	0.193	6.289	3	0.02	40.75
16.5-13.7	15.1	0.195	6.484	0	0.00	40.75

① See footnote on Table 4.

Xenon-125 is the source of the only significant radiocontaminant (60.25-d ^{125}I) present in high-purity(p,5n)-made ^{123}I. However, ^{125}I decays by electron capture(100%) emitting a low-energy 35.5-keV(6.7%) γ ray, characteristic Te X rays (3.77keV, 27.2keV, 27.5keV and 31.0keV) and some low-energy Auger and conversion electrons. For these reasons, the presence of ^{125}I does not influence the resolution of scintillation-camera images using(p,5n)-made ^{123}I radio-pharmaceuticals[14]. In spite of this, some questions exist on the potential radiobiological effects of ^{125}I decay at the cellular level, where a large fraction of the energy deposition from its emissions occurs. In addition, and due to the emergence of several ^{123}I radiopharmaceutical requiring several millicurie-doses (i.e. [^{123}I]IMP; [^{123}I]HIPDM; [^{123}I]MIBG; etc.), lowering the ^{125}I levels in ^{123}I radiopharmaceuticals would also reduce the potential of environmental/

storage problems that may result from the administration/use of 5-7-mCi doses containing 50-70-μCi(1%) of ^{125}I. For all of these reasons, the measurement of cross sections and yields of ^{125}Xe was given special attention.

Table 6 Cross sections(mbarn) and yields(mCi/(μA·h)) for ^{127}I(p, 5n)^{123}Xe[①]

Proton energy/MeV		NaI target/g·cm^{-2}		Cross sections /mbarn	^{125}Xe yields /mCi·(μA·h)$^{-1}$	
In/Out	Av.	Single	Cumulative		Single	Cumulative
67.5-66.4	66.9	0.203	0.203	220	9.10	9.10
66.3-65.3	65.8	0.198	0.401	238	9.62	18.72
65.2-64.1	64.6	0.202	0.603	253	10.49	29.21
64.0-62.9	63.4	0.203	0.806	246	10.18	39.39
62.8-61.8	62.3	0.175	0.982	346	12.45	51.84
61.7-60.6	61.2	0.195	1.177	309	12.32	64.16
60.5-59.4	60.0	0.193	1.370	338	13.31	77.47
59.3-58.1	58.7	0.215	1.585	359	15.80	93.27
58.0-56.8	57.4	0.203	1.788	349	14.48	107.75
56.7-55.6	56.1	0.193	1.981	380	14.94	122.69
55.4-54.2	54.8	0.203	2.184	348	14.39	137.08
54.1-53.0	53.5	0.188	2.372	341	13.10	150.18
52.8-51.5	52.2	0.223	2.595	315	14.35	164.53
51.3-50.0	50.7	0.207	2.802	258	10.93	175.46
49.9-48.6	49.3	0.199	3.001	251	10.23	185.69
48.5-47.1	47.8	0.206	3.207	138	5.80	191.49
47.0-45.8	46.4	0.176	3.383	137	4.94	196.43
45.7-44.4	45.0	0.185	3.568	74	2.80	199.23
44.2-42.8	43.5	0.210	3.778	28	1.18	200.41
42.6-41.2	41.9	0.198	3.976	7	0.28	200.69
41.0-39.6	40.3	0.193	4.169	2	0.09	200.78
39.4-37.8	38.6	0.206	4.375	0	0.00	200.78

① See footnote on Table 4.

The maximum of the excitation function for the ^{127}I(p, 3n)^{125}Xe reaction was found in the 31-32-MeV energy range, where a cross section equal to (731 ± 66) (9%) mbarn was measured. This cross-section value is 11% greater than the value of (660 ± 63) mbarn reported by Syme et al.[3], although it agrees within the stated error limits. In the energy range of interest for ^{123}I production(i.e. 67-46-MeV), the yield of ^{125}Xe was measured as (10.52 ± 0.95) mCi/μA·h(see Table 5), and differs considerably from some of the integrated values obtained from the cross-section data already reported. In the region-of-interest, the data of Wilkins et al.[1] results in a ^{125}Xe yield of (14.2 ± 2.1) mCi/(μA·h); Paans et al.[2] of (7.8 ± 0.98) mCi/(μA·h); while Syme et al.[3] reported a ^{125}Xe yield of (10.3 ± 0.99) mCi/(μA·h). Our values agreed quite well with those obtained from the data of Syme et al.[3], but are 26% lower than those reported by Wilkins et al.[1], and 35% larger than the calculated yield obtained from the Paans et al.[2] cross-section results.

3.3 Xenon-123

The production of ^{123}Xe via the ^{127}I(p, 5n) ($Q = -37.29$ MeV) reaction was studied in the 67.5-37.8-MeV energy range. The radionuclide-yield and cross-section values were obtained by assaying both the 178-keV (14.9%) and the 330-keV (8.6%) γ rays, rather than its most abundant 149-keV (49%) γ ray. The radioassays started approximately 30min after EOB. No significant differences were obtained when either the 178-keV or the 330-keV radioassays were utilized in the calculation of ^{123}Xe yields. However, larger values were obtained in some of the higher-energy (>56MeV) NaI targets indicating a discrepancy that required further investigation. We believe that this discrepancy was due to the presence of 20.1-h ^{122}Xe, which emits a 148.8-keV γ ray in 3.12% of its disintegrations. Despite the low abundance of this emission, its presence can affect the radioassay of the 149-keV (49%) ^{123}Xe γ ray in several ways. First, in the 46-67.5-MeV energy range (see Fig. 2), the ^{122}Xe yield increases and does not reach a maximum. Secondly, and due to a longer half life (20.1h), the ^{122}Xe/^{123}Xe activity ratio also increases with time. Because the NaI-targets's radioassay sequence (see above, Section Ⅲ b) was in an ascending order of the proton energy, these factors clearly combined and resulted in increasing radioassay errors of the 149-keV γ ray. Below the 56-MeV proton energy, all of the ^{123}Xe γ rays being assayed (i.e. 149keV, 178keV and 330keV) showed excellent agreement.

The maximum for the ^{127}I(p, 5n) reaction was measured as (380 ± 34) mbarn in the 56-57-MeV range. This value is in excellent agreement with the measurement of Paans et al.[2], but is 22.5% lower than Wilkins et al.[1] and 31% larger than the value reported by Syme et al[3]. (see Table 2). Our result is also within one-standard deviation with the value obtained by Hegedus[4]. In the region-of-interest for ^{123}I production (i.e. 67-46-MeV), the ^{123}Xe yield was measured as (196.4 ± 17.7) mCi/(μA·h). This result is 13.5% lower than the (227.1 ± 34.1) mCi/(μA·h) value reported by Wilkins et al.[1], and 5.1% higher than the Paans et al. value of (186.8 ± 23.4) mCi/(μA·h). All these values, however, agreed within the one-standard-deviation criteria, although the Syme et al. value of (143.8 ± 12.9) mCi/(μA·h), is 26.8% lower than our measurement.

3.4 Xenon-122

The ^{127}I(p, 6n)^{122}Xe ($Q = -45.50$ MeV) reaction was studied in the 67.5-45.8-MeV energy range, using the 350-keV (8%) γ ray and began 1 day after EOB (simultaneously with ^{125}Xe).

The results for this reaction are given in Table 7. The decay of ^{122}Xe results in a 3.6-min ^{122}I daughter, which in turns decays (β +77%; EC 23%) to stable ^{122}Te. The production of ^{122}Xe in the region-of-interest for ^{123}I production has not been reported. However, recently the ^{122}Xe-^{122}I parent-daughter system has been proposed as a source of ^{122}I radiopharmaceuticals for positron emission tomography (PET) studies[15-17]. Therefore, the production of ^{122}Xe, the techniques for handling, and rapid syntheses procedures for ^{122}I radiopharmaceuticals are being reported[18-20]. In terms of ^{123}I production, the presence of a rapidly decaying 3.6-min ^{122}I only causes temporary problems in the quantitation (γ ray spectrometry or ionization-chamber measurements) of ^{123}I, immediately after the end-of-processing. After a about 30-min decay time, however, this problem disappears due to the decay of ^{122}I.

The cross-section maximum for the ^{127}I(p,6n) reaction is expected to be at a higher energy (72-74-MeV), which is above the energy range covered in this work. In the 67-50-MeV energy range, however, a ^{122}Xe yield of (5.6 ± 0.50) mCi/(μA · h) was measured. This yield is more than adequate to support an initial R&D effort to develop and evaluate ^{122}I-labeled "PET" radiopharmaceuticals.

Table 7 Cross sections (mbarn) and yields (mCi/(μA · h)) for ^{127}I(p,6n)^{122}Xe①

Proton energy/MeV		NaI target/g · cm^{-2}		Cross sections /mbarn	^{125}Xe yields /mCi · (μA · h)$^{-1}$	
In/Out	Av.	Single	Cumulative		Single	Cumulative
67.5-66.4	66.9	0.203	0.203	197	0.92	0.92
66.3-65.3	65.8	0.198	0.401	181	0.83	1.75
65.2-64.1	64.6	0.202	0.603	176	0.82	2.57
64.0-62.9	63.4	0.203	0.806	164	0.77	3.34
62.8-61.8	62.3	0.176	0.982	155	0.64	3.98
61.7-60.6	61.2	0.195	1.177	107	0.49	4.47
60.5-59.4	60.0	0.193	1.370	81	0.36	4.83
59.3-58.1	58.7	0.215	1.585	66	0.33	5.16
58.0-56.8	57.4	0.203	1.788	42	0.20	5.36
56.7-55.6	56.1	0.193	1.981	23	0.10	5.46
55.4-54.2	54.8	0.203	2.184	14	0.06	5.52
54.1-53.0	53.5	0.188	2.372	9	0.04	5.56
52.8-51.5	52.2	0.223	2.595	4	0.01	5.57
51.3-50.0	50.7	0.207	2.802	2	0.01	5.58
49.9-48.6	49.3	0.199	3.001	0	0.00	5.58

① See footnote on Table 4.

3.5 Xenon-121

The reaction ^{127}I(p,7n)^{121}Xe (Q = −56.40MeV) was studied in the 67.5-55.6-MeV energy range by radioassaying the 133-keV (11.4%) γ ray. Radioassays began 30-min after EOB due to the relatively short (i.e. 38.85min) half life of ^{121}Xe. The results of these measurements are given in Table 8. Clearly, this excitation function is well below its maximum (expected above 90MeV). The ^{121}Xe yield was measured as (29.0 ± 2.6) mCi/(μA · h) in the 67.5-59.4-MeV energy range. Despite this large yield, the production of ^{121}Xe reaches steady state approximately 2-h after start of irradiation, at a level of about 90mCi/μA, and is the source of the trace-level (<0.001%) of ^{121}Te(16.78d) impurity found in (p,5n)-made ^{123}I[7].

Table 8 Cross sections (mbarn) and yields (mCi/(μA·h)) for ^{127}I(p,7n)^{121}Xe①

Proton energy/MeV		NaI target/g·cm^{-2}		Cross sections /mbarn	^{121}Xe yields /mCi·(μA·h)$^{-1}$	
In/Out	Av.	Single/Cumulative			Single/Cumulative	
67.5-66.4	66.9	0.203	0.203	78	8.28	8.28
66.3-65.3	65.8	0.198	0.401	61	6.02	14.30
65.2-64.1	64.6	0.202	0.603	53	5.79	20.09
64.0-62.9	63.4	0.203	0.806	44	4.31	24.40
62.8-61.8	62.3	0.176	0.982	26	2.78	27.18
61.7-60.6	61.2	0.195	1.177	18	1.86	29.04
60.5-59.4	60.0	0.193	1.370	0	0.00	29.04

① See footnote on Table 4.

3.6 Comparison with other reported cross-section values

As indicated earlier, one of the purposes of this work was to measure the different cross sections for the proton-induced reactions on ^{127}I targets, and in particular, for those reactions of importance for the production of high-purity ^{123}I. In the previous Sections [Ⅲ(a) to (e)], a comparison has been made based upon cross-section maxima and integrated yields for the different radioxenons measured. These comparisons can also be made in the 67-46-MeV energy region at which ^{123}I production can be maximized. Fig. 4 shows the comparison of the previously reported values with the ones reported here. As seen in Fig. 4, a revision of these values was necessary.

3.7 Comparison with experimental ^{123}I yields

The yield data presented here were also compared with experimental yields obtained at UC Davis from the ^{123}I production program. Routinely, for a 14-mm thick NaI target (65.5-45-MeV energy range), an 8-h bombardment followed by a 3.5-h ingrowth time, produces an average of (22 ± 2) mCi/(μA·

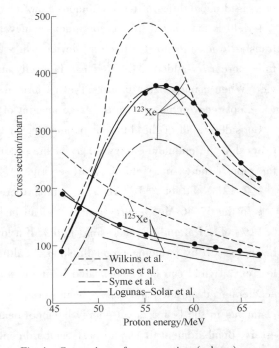

Fig. 4 Comparison of cross sections (mbarn) as a function of proton energies (MeV) for the production of ^{123}Xe and ^{125}Xe in the region-of-interest (67-46-MeV) for ^{123}I production

(The cross-section values reported in this work agree well with other reported values, although the best agreements corresponded to different reports)

h) of ^{123}I. Using the data reported here, and under the same production parameters, a ^{123}I yield of (20.9 ± 1.9) mCi/(μA·h) can be calculated. These calculated and experimental yields are in excellent agreement.

3.8 Optimization of ^{123}I yields and purities

Because of the indirect nature of this ^{123}I production method, many time-dependent parameters (i. e. time of bombardment; time of decay), as well as other physical parameters (i. e. beam energy; target thickness; solid or liquid targets; etc.) will ultimately affect the effective yields and purities obtained in large-scale production operations. For these reasons, the production rate of ^{123}I varies significantly at the different accelerator facilities presently involved in its production. Batch processes (i. e. irradiation followed by target-radiochemical processing), are clearly a less desirable production mode. This is due to the fact that the production of ^{123}I is strongly reduced by the decay of ^{123}Xe during bombardment. In batch production, short-duration runs are clearly favorable, with the consequent limitations in batch-production levels. By comparison, a continuous-flow method using a generator system[21], favors longer bombardment times with proportionally shorter ingrowth times for maximum ^{123}I yields. This allows a larger-scale production of ^{123}I. The increased capabilities of the continuous-flow/generator mode is mostly due to the fact that the ^{123}I yield is maximized by the continuous removal of the ^{123}Xe produced in the target, its rapid transfer to a collection (generator) device where most of the ^{123}Xe decays. After the ingrowth time, no-carrier-added (NCA) ^{123}I can be easily and efficiently recovered from this generator device. When using a batch process, the ^{123}I formed by ^{123}Xe decaying during bombardment is lost (i. e. not recoverable) due to the large amount of ^{127}I carrier in the target material.

Considering all of the different parameters for the production of medical radionuclides, it appears that the optimum energy range for the production of ^{123}I should not exceed 68 MeV due to the rapidly increasing yield of ^{121}Xe (see Fig. 3). The ^{121}Xe produced results in the formation of ^{121}I (2.12h) and of ^{121}Te (16.78d), thus increasing the amounts of radio-contaminants in ^{123}I. In the 67-46-MeV energy range, ^{123}I can be produced with $<0.001\%$ ^{121}Te and 0.07%-0.12% of ^{125}I at end-of-processing (EOP). If a lower ^{125}I content is desired, the exit energy can be increased (by reducing target thickness), although at the expense of lower ^{123}I yields. This loss of ^{123}I yield can be compensated by increasing the operational beam intensity (μA) of the accelerator, if an adequate excess beam-intensity capability is available. In addition, minor adjustments in the beam/target energy balance (beam input energy vs target cooling) may be necessary. Bombardment times and post-bombardment processing modes offers other options to minimize the ^{125}I content. In both cases, shorter bombardment and ingrowth times can help minimize ^{125}I content. If this latter option is chosen, a multi-generator operational capability is needed to allow for frequent collection and processing of several (but smaller) ^{123}I batches. The effect of these ^{123}I-production variables (i. e. bombardment and ingrowth times) on the ^{125}I content at EOP is shown in Fig. 5. If proven necessary by radiobiological and, or environmental concerns, the levels of ^{125}I in the (p, 5n)-made ^{123}I can be reduced by factors of 5-10 (to 0.007% at

EOP) by a combination of the above mentioned bombardment/handling variables. Recently, the reaction $^{124}Xe(p,2n)^{123}Cs(5.9min) \rightarrow ^{123}Xe(2.08h) \rightarrow ^{123}I(Q = -15.92MeV)$, using a 30-MeV proton beam on 99.7%-enriched ^{124}Xe gas targets, is being utilized to produce ^{123}I with <0.2% of total (mostly 16.78-d ^{121}Te) radiocontaminants at time-of-calibration[22]. The use of this reaction, however, requires the irradiation and processing of gas targets, the recirculation of expensive enriched ^{124}Xe gas, and the recovery of ^{123}I from the target vessel where approximately 80% of the ^{123}I yield is obtained (about 20% of the yield is obtained from undecayed ^{123}Xe). Its operation is based on batch processing, but the ^{123}I formed in the target during bombardment is recoverable due to the absence of ^{127}I carrier materials. The production of ^{123}I is estimated as 4-8-mCi/(μA·h) (depending on target gas pressure), and batches of up to 8Ci/run are expected in the near future if long, higher-pressure ^{124}Xe targets can be developed and operated reliably to withstand up to 100μA beams[22]. By contrast, the use of solid-or molten-NaI targets have been proved to operate successfully and reliably at elevated temperatures (500-700℃) and are expected to withstand up to 200μA of 68-MeV (exit at 46MeV) proton beams. Under these conditions, the production of ^{123}I can reach 4Ci/h, and batches of about 40Ci/10-h (0.05%-0.07% ^{125}I at EOP) runs are possible.

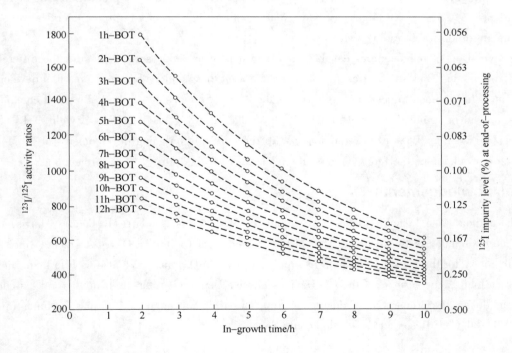

Fig. 5 Effects of bombardment times (h), and ingrowth times (h) on the ^{125}I content (%) at end-of-processing for $^{127}I(p,5n)$-made ^{123}I

3.9 Effect of other $^{121}I(p,xn)$ reactions

Other reactions forming radioxenons [i.e. $^{127}I(p,6n)^{122}Xe$, $^{127}I(p,7n)^{121}Xe$, and $^{127}I(p,n)$

^{127}Xe] do not affect the clinical applicability and imaging properties of ^{123}I. However, these reactions are important factors to consider when designing radioactivity handling, remote processing, and quality control procedures. With the exception of ^{127}Xe(yield = (32.5 ± 2.9) μCi/(μA·h);67-46-MeV range), both ^{122}Xe(yield = (5.6 ± 0.50) mCi/(μA·h)) and ^{121}Xe (yield = (29.0 ± 2.6) mCi/(μA·h)) are produced in sufficiently large amounts to be included in radiation-safety analyses, and system-design considerations.

4 Conclusions

The present work provides new and revised data on the use of high-energy proton beams on ^{127}I targets. This work confirms that the use of the ^{127}I(p,5n) reaction is an efficient method for the production of high purity ^{123}I. By a combination of the (p,5n) method, and possibly the use of ^{124}Xe(p,2n) reaction, the future availability of ^{123}I should not be a limiting factor in the development of ^{123}I radiopharmaceuticals. With an appropriate R&D effort, it appears that both of these reactions can be utilized, in tandem targets, to simultaneously produced high-purity ^{123}I. A degraded proton beam from the 68-46-MeV energy range for the ^{127}I(p,5n) method, into a second ^{124}Xe target, appears to have a good potential use for large-scale ^{123}I production. The highly-enriched (99.7%) ^{124}Xe target can be irradiated with a wider spectra of proton energies without inducing undesirable radioactivities.

Several optional procedures can be utilized to maximize yields and purities for the different ^{123}I-production modes. However, the implementation of these options requires a parallel development of fully-remote processing systems. For the ^{127}I(p,5n)-made ^{123}I, the ^{125}I level of contamination can be reduced to <0.007% (10 times) at EOP, while keeping the ^{123}I yield at 13-15-mCi/(μA·h) (30%-40% reduction). With the currently on-going developments of 70-MeV proton accelerators, the future large-scale availability of high-purity ^{123}I is forthcoming.

Acknowledgements

The authors wish to acknowledge the support of Mr Charles Goodart, Mr Lennox Harris, Mr Walter Kemmler, of the technical staff of the University of California, Davis's Crocker Nuclear Laboratory for the considerable help in support of thes R&D project. We also wish to thank Professor John A. Jungerman, of the UC Davis's Physics Department, for the many helpful discussions during the preparation of this manuscript. This work was supported, in part, by funds from the 83/84 CNL Research & Development Grants.

References

[1] Wilkins S. R., Shimose S. T., Hines H. H., Jungerman J. A., Hegedus F. and DeNardo G. L. Int. J. Appl. Radiat. Isot. 26,279(1975).

[2] Paans A. M. J., Vaalburg W., van Herk G. and Woldring M. G. Int. J. Appl. Radiat. Isot. 27,465(1976).

[3] Syme D. B., Wood E., Blair I. M., Kew S., Perry M. and Cooper P. Int. J. Appl. Radiat. Isot. 29,29(1978).

[4] Hegedus F. EIR, Swiss Federal Institute for Reactor Research, 5303 Wurenlingen, Switzerland. (Private

communication).

[5] Little F. E. and Lagunas-Solar M. C. Int. J. Appl. Radiat. Isot. 34,631(1983).

[6] Johnson P. C. ,Lagunas-Solar M. C. and Avila M. J. Int. J. Appl. Radiat. Isot. 35,371(1984).

[7] Jungerman J. A. and Lagunas-Solar M. C. J. Radioanal. Chem. 65,31(1981).

[8] Jungerman J. A. ,Romero J. L. and Uhlenkott R. Nucl. Jnstrum. Methods 204,41(1982).

[9] Janni J. F. Calculations of Energy Loss,Range,Path-length,Straggling,Multiple Scattering,and the Probabiiity of Inelastic Nuclear Collision for 0. 1-1000-MeV Protons. Air Force Weapons Laboratory Technical Report No. AFWL-TR-65-150(September 1966). See also:Energy Loss Distributions of Heavy Particles in Thick Absorbers. C. Tschalar. Rutherford Laboratory Report,RHEL/R 146(1967).

[10] Lederer C. M. and Shirley V. M. (Eds)Table of Isotopes. 7th edn(John Wiley,New York,1978).

[11] Lagunas-Solar M. C. , Jungerman J. A. , Peek N. F. and Theus R. M. Int. J. Appl. Radiat. Isot. 29, 159 (1978).

[12] Segre Nuclei and Particles(W. A. Bejamin,New York,1964).

[13] Colle R. and Kishore R. Phys. Rev. C9,2166(1974).

[14] Lagunas-Solar M. C. and Hines H. H. Effects of Radionuclidic Composition on Dosimetry and Scintillation-Imaging Characteristics of I-127(p,5n)-and Te-124(p,2n)-Made Iodine-123 for Nuclear Medicine Applications. Proc. Int. Symp. on the Developing Role of Short-Lived Radionuclides in Nuclear Medicine Practice. Washington D. C. ,3-5 May 1982. DOE Conf-820523,pp. 423-444(1985).

[15] Richards P. and Ku T. H. Int. J. Appl. Radiat. Isot. 30,250(1979).

[16] Mathis C. A. ,Lagunas-Solar M. C,Sargent T. III. , Yano Y. , Vuletich A. and Harris L. J. Int. J. Appl. Radiat. Isot. 37,258(1986).

[17] Lagunas-Solar M. C. ,Carvacho O. F. ,Harris L. J. and Mathis C. A. Appl. Radial. Isot. 37,835(1986).

[18] Mathis C. A. ,Shulgin A. T. and Sargent III T. J. Labeled Compd. Radiopharm. (Submitted).

[19] Mathis C. A. ,Sargent III T. and Shulgin A. T. J. Nucl. Med. In press.

[20] Mathis C. A. , Sargent III T. , Shulgin A. T. , Yano Y. , Budinger T. F. and Lagunas-Solar M. C. J. Nucl. Med. 26,P69(1985). Abstr.

[21] Lagunas-Solar M. C. ,Thibeau H. L. ,Goodart C. E. , Little F. E. ,Navarro N. J. and Hartnett D. E. "An Integrally Shielded, Transportable, Multi-Curie Xenon-123-> Iodine-123 Generator System". Proc. Int. Symp. on the Development Role of Short-Lived Radionuclides in Nuclear Medicine Practice. Washington D. C. May 3-5,1982. DOE Conf-820523,pp. 190-202(1985).

[22] Trevena I. AECL Radiochemical Co. Ontario,Canada. (Private Communication).

Comparison of [^{82}Br]4-Bromoantipyrine and [^{125}I]4-iodoantipyrine: the Kinetics of Exchange Reaction and Biodistribution in Rats[*]

Abstract Kinetics and mechanism of isotope exchange reaction between [^{82}Br]bromide anion and 4-bromoantipyrine(BrAP), and the iodine-bromine exchange reaction between [^{125}I]iodide anion and BrAP were studied. The preparation of [^{82}Br]BrAP followed by exponential exchange law, the kinetics of the exchange reaction is a second-order reaction with an activation energy of 23.3kcal/mol. The optimal exchange condition for halogen exchange between [^{125}I]iodide and BrAP was by a hydrothermal melt method at 110℃ and 5min reaction time. The partition coefficient at pH 7.0 for IAP and BrAP was 20.9 and 13.5, respectively. However, BrAP, which displayed the lower partition coefficient, showed higher brain uptake in rats than that for IAP(2.0% dose/organ vs 1.74% dose/organ), at 2min after an i.v. injection.

1 Introduction

Antipyrine(1,5-dimethyl-2-phenyl-3-pyrazolone) and its halogen derivatives(4-iodoantipyrine, IAP; 4-bromoantipyrine, BrAP; 4-fluoroantipyrine, FAP; see Scheme 1) are lipid soluble molecules which can cross cell membrane by a simple diffusion mechanism.

Iodine-123 IAP[1], [^{82}Br]BrAP[2], and [^{18}F]FAP[3,4] have been reported as useful tracers for measuring regional cerebral blood flow using autoradiographic or imaging techniques[5-7]. The radioactive iodine labeled IAP is usually prepared by an iodine-iodine isotope exchange reaction or by an iodine-bromine exchange reaction between radioactive iodide and 4-bromoantipyrine(BrAP). The labeling reactions have been evaluated by Robinson[8,9], Forrester[10], and Boothe[11]. In these reports, the radiochemical yield has been studied in different solvent systems, pHs, reaction times, or concentrations of IAP. However, the reaction kinetics and mechanism are still uncertain.

Preparation of [^{82}Br]BrAP has been reported by a bromine-bromine exchange reaction[2] using three different techniques: the melt method, the isotope exchange in acidic medium method, and the silica gel catalyzed method.

In this article, we describe the comparative study of the kinetics of isotope exchange reaction between [^{82}Br]bromide anion and BrAP, and the iodine-bromine exchange reaction between [^{125}I]iodide and BrAP. The optimum condition for preparing [^{125}I]IAP by a hydrothermal melt technique was determined. In addition, comparative studies on biodistribution of [^{125}I]IAP and

---

[*] Copartner: HANK F. KUNG, JEFFREY BILLINGS, MONTE BLAU. Reprinted from *Nucl. Med. Biol.*, 1987, 14(1):69-74.

[^{82}Br]IAP in rats and the partition coefficients are reported.

2 Materials and Methods

4-Bromoantipyridine (BrAP) purchased from Aldrich Chemical Company was used directly without further purification. The radiochemical yield was determined by TLC (Merck silica gel 60 F-254) using a solvent system of toluene-ethylacetate (1 : 1). In this TLC system, free bromide and free iodide have a R_f value equal to 0.0-0.1; [^{125}I]IAP and [^{82}Br]BrAP have a R_f value equal to 0.5-0.6. Samples of [^{125}I]IAP and [^{82}Br]BrAP for animal study were evaluated using HPLC: Varian 5000, Hamilton PRP-l column, solvent of acetonitrile: water (85 : 15), and the purity for all of them was greater than 98%. For detecting radioactivity in the HPLC system a flow cell was placed in a well of a gamma counter and the eluent counts were recorded by a multichannel analyzer (ND-100, Nuclear Data Systems) set in a multiscalar mode. Reaction temperature was controlled by a constant temperature bath (+0.1℃, Haake model 4000). The pH values were measured by a pH meter (Radio meter, pHM 64). Radioactivity was determined by an automatic gamma counter (Beckman 4000). Radioisotope ammonium [^{82}Br]bromide (>3Ci/mg), and sodium [^{125}I]iodide (>17Ci/mg) were purchased from New England Nuclear Inc.

2.1 Labeling kinetic studies

Ammonium [^{82}Br]bromide or sodium [^{125}I]iodide, ~100μCi of radioactivity in ~40μL of 0.1N sodium hydroxide solution, was placed in a 10mL round-bottom flask, to which a fixed amount of sodium bromide or sodium iodide carrier was added. The pH was adjusted by a 0.1N hydrogen chloride solution or by a 0.1 N sodium hydroxide solution. Only the initial pH of the reaction mixture was measured. The total volume of each exchange solution was 6 mL. The solution was placed in a Haake constant temperature bath, and after equilibration at a specific temperature, a fixed amount of BrAP was added. The exchange reaction was allowed to proceed with constant magnetic stirring and at different times samples (1μL each) were removed and analyzed.

2.2 Partition coefficients

The partition coefficient was measured by mixing [^{125}I]IAP or [^{82}Br]BrAP with 3g each of 1-octanol and buffer (0.1 M phosphate) in a test tube. This test tube was vortexed (3×1min) at room temperature and then centrifuged for 5min. Two weighted samples (0.5g each) from the 1-octanol and buffer layers were counted in a well counter. The partition coefficient was determined by calculating the ratio of cpm/g of octanol to that of buffer. Samples from the octanol layer were repartitioned until consistent partition coefficient values were obtained. Usually the measurement was repeated three times.

2.3 Animal distribution

Rats. Sprague-Dawley male rats (220-300g) under light ether anesthesia were injected intravenously with 0.2 mL of a saline solution containing both $[^{125}I]$IAP ($\approx 1\mu$Ci, 12.5μg) and $[^{82}Br]$BrAP(1-5μCi, 12.5μg). At different time periods after injection the rats were anesthetized by ether and killed by cardiectomy. Organs of interest were excised, weighed and counted with a two channel gamma counter. The % dose/organ was determined by comparison of tissue radioactivity to suitably diluted aliquots of the injected dose. Spill-over counts into each window were corrected by a computer program.

3 Results and Discussion

3.1 Isotope exchange reaction between $[^{82}Br]$bromide and BrAP

Due to the rapid isotope exchange reaction at the no-carrier added level, an appropriate amount of bromide carrier (sodium bromide) was added to slow down the reaction. The Br-Br isotope exchange was carried out under the following condition: pH = 2.0, total volume = 6.0mL, sodium bromide = 3.09mg (0.03mmol), BrAP = 15.6mg (0.06mmol) and the reaction temperature was at 99℃, 92℃ and 77℃, respectively.

The radiochemical yield (%) and reaction time curve is shown in Fig. 1. From the ratio of free bromide and bound bromine in BrAP, the radiochemical yield of equilibrium was calculated. The radiochemical yield (% labeling) was very close to the theoretical value (Table 1).

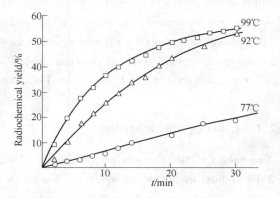

Fig. 1 Radiochemical yield(%) vs time for isotope exchange reaction between $[^{82}Br]$bromide and BrAP

Table 1 Effects of sodium bromide carrier on $[^{82}Br]$BrAP radiochemical yield(%)①

BrAP	NaBr	Theory②	Expt
5mg	0.103mg	94.9	94.7
5mg	0.515mg	78.9	76.7
5mg	3.09mg	38.2	34.4
5mg	6.18mg	23.7	22.3

① Reaction condition: pH = 2.4, at 100℃ for 90min, total volume 6mL.

② Theory = $\frac{[BrAP]}{[BrAP] + [NaBr]}$.

In an aqueous medium, the exchange reaction between $[^{82}Br]$bromide and BrAP is a simple homogeneous radioisotope exchange reaction and its kinetics will follow the exponential exchange law:

$$-\ln(1-F) = \frac{[A]+[B]}{[A][B]}Rt$$

where, F = fraction of exchange; $[A]$ = the concentration of sodium bromide; $[B]$ = the concentration of BrAP; R = isotope exchange rate; t = time.

The data in Fig. 1 is recalculated to F (fraction of exchange) and plotted using $-\ln(1-F)$ and t (min) as the axis. The result is shown in Fig. 2. It is clearly demonstrated that the exchange reaction between [^{82}Br] bromide and BrAP follows the exponential exchange law; the straight line passes through the origin. In the exchange reaction, bromide (Br^-) is unlikely to be oxidized to bromine (Br_2); therefore, it is possible that the exchange reaction is a nucleophilic reaction. By using the exponential exchange law, it is easy to determine that the order of the exchange reaction is second-order:

$R = k(A)(B)$, k = reaction rate constant

$$-\ln(1-F) = kat$$

where $a = [A] + [B] = [BrAP] + [NaBr]$

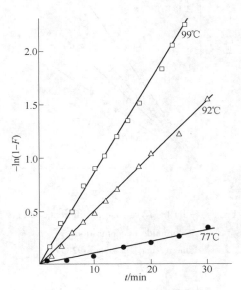

Fig. 2 $-\ln(1-F)$ vs time for isotope exchange reaction between [^{82}Br] bromide and BrAP (F = fraction of exchange)

The exchange reactions were carried out at 92℃ with a fixed concentration of BrAP and various amounts of sodium bromide (1.67×10^{-3} M to 1.67×10^{-2} M) in the reaction mixture. When the data were plotted using $-\ln(1-F)$ vs time (Fig. 3), straight lines were obtained. The average rate constant k was equal to $(3.16 \pm 0.3) M^{-1} \cdot min^{-1}$ (see Table 2). The slope is proportional to ka, this was proven by using various concentration ratios of $[A]$ and $[B]$ as shown in Fig. 3 and Table 2. At different temperature but the same $[A]$ and $[B]$ different reaction rate constant, k value, can be obtained. Based on the Arrhenius Equation:

$$k = A \cdot \exp\left(-\frac{E}{RT}\right)$$

where, A = frequency factor; E = activation energy; R = gas constant; T = temperature.

Plotting $\ln k$ vs $1/T$ (Fig. 4), the activation energy of this exchange reaction was calculated to be $E = 23.3$ kcal/mol.

3.2 Iodine-bromine exchange reaction between [^{125}I] iodide and BrAP

The exchange reaction between [^{125}I] iodide and BrAP for [^{125}I] AP preparation is different from that between iodide and bromoalkyl compound, such as the iodo fatty acids, because the bromine atom on BrAP is attached to a double bond. The exchange reaction is studied under the following condition: Total volume of the aqueous solution = 6mL, pH = 2.4, sodium iodide =

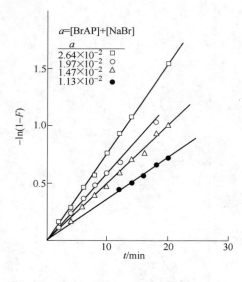
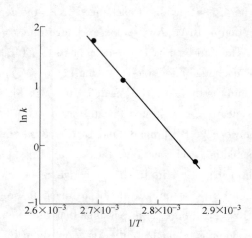

Fig. 3 $-\ln(1-F)$ vs time at different sodium bromide concentrations
(the reaction temperature: 92℃)

Fig. 4   Relationship between reaction rate constant k and reaction temperature ($1/T$)
(The activation energy (E) for isotope exchange reaction between [^{82}Br] bromide and BrAP is 23.3 kcal/mol)

4.5 mg, BrAP = 16.4 mg, reaction temperature = 80℃.

Table 2   The proportional relation between slope m and total concentration C①

BrAP	9.67×10^{-3} M	9.67×10^{-3} M	9.67×10^{-3} M	9.67×10^{-3} M
NaBr	1.67×10^{-3} M	4.998×10^{-3} M	1.002×10^{-2} M	1.67×10^{-2} M
$a =$ [BrAP] + [NaBr]	1.134×10^{-2} M	1.467×10^{-2} M	1.969×10^{-2} M	2.63×10^{-2} M
m	0.03725	0.0518	0.0517	0.0768
$k = \dfrac{m}{a}$	3.28	3.53	2.90	2.91

Avg. $k = (3.16 \pm 0.31)$ M^{-1} · min^{-1}

① Reaction temperature: 92℃.

The radiochemical yield (%) at different reaction time was obtained and treated as that for [^{82}Br] bromide and BrAP. The results are shown in Fig. 5. The exchange reaction between iodide and brominated compounds is a heterogeneous halogen exchange (Br-I) reaction rather than a simple Br-Br isotope exchange. The exponential exchange law can not be applied in this case, even though an apparent straight line was obtained in Fig. 3, when plotting $-\ln(1-F)$ vs time. This straight line usually leads to the incorrect conclusion that the exchange rection between halogens also follow the same principle for isotope exchange reactions. Recently, Feng et al.[12] have reported a general equation for the kinetic of halogen exchange reactions. The differences between these two exchange reactions and the kinetics from a theoretical point of view was elucidated (see Appendix).

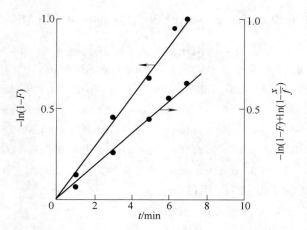

Fig. 5 　$-\ln(1-F)$ and $-\ln(1-F)+(1-x(t)/f)$ vs time for halogen exchange reaction between $[^{125}I]$iodide and BrAP

(F = fraction of exchange)

Using this kinetic model:

$$^{125}I^- + BrAP \underset{k_2}{\overset{k_1}{\rightleftharpoons}} {}^{125}IAP + Br^-$$

The rate constant for forward k_1 and backward k_2 reaction can be calculated (Table 3). $k_1 = 20.8 \text{mol}^{-1} \cdot \text{min}^{-1}$.

$$k_2 = 1.03 \text{mol}^{-1} \cdot \text{min}^{-1}$$

Table 3　Fraction of exchange vs time

Time(min)	% Labeling	Fraction of exchange(F)	$-\ln(1-F)$	$-\ln(1-F)+\ln\left[1-\dfrac{x(t)}{f}\right]$
2	0.32	0.33	0.40	0.23
4	0.49	0.51	0.71	0.47
6	0.63	0.66	1.08	0.76
8	0.72	0.86	1.39	1.00
10	0.84	0.86	2.21	1.66
720	0.96			
720	0.96			

$a = [I^-]_0 + [BrAP]_0 = 0.010M + 0.00479M = 1.479 \times 10^{-2} M.$

$b = [I^-]_0 \cdot [BrAP]_0 = 4.79 \times 10^{-5} M.$

$x_t = [IAP]_t = [Br^-]_t = 4.60 \times 10^{-3} M.$

$K_{eq} = 20.3.$

$f = 1.096 \times 10^{-2}.$

$g = 0.460 \times 10^{-2}.$

3.3 Hydrothermal melt exchange between [^{125}I]iodide and BrAP

Considering the fact that BrAP is relatively heat stable and the exchange with radioactive bromide or iodide is a nucleophilic reaction, it would be convenient to use a faster exchange reaction condition, namely the hydrothermal melt method, in this method, the reaction is carried out in a sealed vial at a temperature lower than the melting point of the compound to be labeled, but higher than the boiling point of the solvent. The exchange takes place in this closed reaction system. Usually a small amount of water(10μL) is employed as the solvent, but a mixed solvent system such as water-ethanol mixture can also be used.

Using 2mg of BrAP and 10μL of [^{125}I]sodium iodide sealed in an ampule, effects of the reaction time, temperature and pH were evaluated. In Fig. 6 the effects of pH (only the starting pH was measured) on the radiochemical yield is demonstrated. At low pH(1-3), the radiochemical yield was high. There was a sharp drop of the radiochemical yield when the initial pH was above 3. This observation is similar to that reported by Robinson et al.[9]

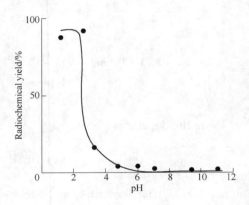

Fig. 6 Radiochemical yield(%) vs pH for halogen exchange reaction between [^{125}I]iodide and BrAP using hydrothermal melt method

The optimum pH for the exchange reaction was between 1 and 3. Under hydrothermal reaction conditions, the exchange was fast. At 5 min the reaction has reached equilibrium (Fig. 7). Therefore, for routine preparation of radioactive IAP, a reaction time of 10 min will be sufficient. Temperature has a notable effect on reaction rate (Fig. 8). When the reaction temperature is above the boiling point of the solvent (water), the radiochemical yield increases significantly. This implies that in this sealed vial, the overheated vapor of water or other solvent may enhance the exchange reaction. However, when the reaction temperature was too high, it may induce the deiodination reaction or decomposition, resulting in a decreased radiochemical yield. The ideal condition for preparation of radioactive IAP was at 110℃ for 5min.

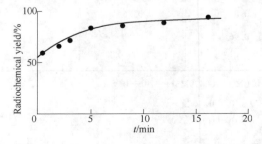

Fig. 7 Radiochemical yield(%) vs time for halogen exchange reaction between [^{125}I] iodide and BrAP using hydrothermal melt method

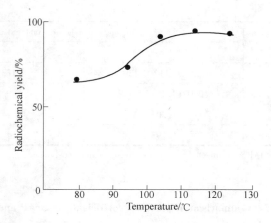

Fig. 8 Radiochemical yield(%) vs temperature for halogen exchange reaction between [^{125}I]iodide and BrAP using hydrothermal melt method

3.4 Biodistribution in rats

The biodistribution in rats was studied by simultaneous i. v. injection of [^{82}Br]BrAP and [^{125}I] IAP(Table 4). At 2min after injection, both BrAP and IAP showed high uptake in brain, 2.07% dose/organ and 1.74% dose/organ, respectively. At pH 7.0, partition coefficient of IAP and BrAP is 20.9 and 13.5, respectively(Table 5). In view of the fact that IAP is more lipid-soluble than BrAP, the partition coefficient may not be the only determining factor for the total brain uptake. The kinetics of efflux and influx of these freely diffusible tracers into brain tissue may be of more importance. High initial uptake was also observed in muscle, heart, liver, and skin. In general, the initial distribution pattern is similar to the blood supply for each organ.

Table 4 Comparison of biodistribution of [^{82}Br]bromo-antipyrine(BrAP) and [^{125}I] iodoantipyrine(IAP) in rats(dual isotope experiment)[1]

Organ	% dose/organ(average of 3 rats and range)2min	
	BrAP	IAP
Blood	18.60(17.44-20.25)	18.94(18.05-20.49)
Muscle	20.29(11.92-33.00)	18.29(10.90-29.69)
Heart	1.12(0.985-1.23)	1.09(0.964-1.21)
Lungs[2]	1.56(1.38-1.68)	1.61(1.45-1.71)
Spleen	0.632(0.629-0.637)	0.581(0.576-0.587)
Kidneys	2.08(1.98-2.26)	2.07(1.96-2.23)
Stomach	1.71(1.56-1.84)	1.59(1.44-1.73)
Small intestines	5.36(4.48-6.11)	4.19(3.69-4.77)
Liver	18.68(17.85-19.91)	18.00(17.28-18.99)
Skin	11.13(8.03-13.12)	11.76(8.34-13.02)

Organ	% dose/organ (average of 3 rats and range) 2min	
	BrAP	IAP
Thyroid	0.126(0.112-0.139)	0.133(0.122-0.149)
Brain	2.07(1.72-2.55)	1.74(1.44-2.13)
Brain/blood[2] ratio	1.004	0.8272
^{82}Br/^{125}I brain ratio	1.193	

[1] [^{82}Br]BrAP and [^{125}I]IAP were injected into femoral vein simultaneously.

[2] $\dfrac{\text{Brain dose/g}}{\text{Blood dose/g}}$.

Table 5 Comparison of partition coefficient for BrAP and IAP

(1-octanol: buffer)

Item	pH 7.0	pH 7.4
BrAP	13.5	13.5
IAP	20.9	21.0

In summary, the radiolabeling reaction of BrAP using the Br-Br exchange reaction follows the exponential exchange law. This reaction is a second order reaction with an activation energy of 23.3 kcal/mol. The preparation of [^{125}I]IAP, using the I-Br exchange reaction, follows a more complex halogen exchange kinetics with k_1 (forward reaction) = 20.8 mol^{-1} · min^{-1} and k_2 (backward reaction) = 1.03 mol^{-1} · min^{-1}. The biodistribution of [^{125}I]IAP and [^{82}Br]BrAP in rats showed high brain uptake at 2 min.

Acknowledgements

Authors thank Dr C-Y Shiue for his helpful discussion, Mrs Elongia Farrell for her technical help, and Ms Rebecca Hoffner for her secretarial assistance.

References

[1] Uszler J. M., Bennett L. R., Mean I. et al. Radiology 115,197(1975).

[2] Shiue C. Y. and Wolf A. P. J. Labelled Compd. Radiopharm. 20,1363(1983).

[3] Robbins P. J., Fortman D. J., Scholz K. L. et al. J. Nucl. Med. 19,1346(1978).

[4] Shie C. Y. and Wolf A. P. J. Labelled Compd. Radio-pharm. 18,1059(1981).

[5] Sakurada O., Kennedy C., Jehle J. et al. Am. J. Physiol. 234,H59(1978).

[6] Lear J. L., Jone S. C., Greenberg J. H. et al. Stroke 12,589(1981).

[7] Mies G., Niebuhr I. and Hossman K.-A. Stroke 12,581(1981).

[8] Robinson G. D. and Lee A. W. J. Nucl. Med. 17,1093(1976).

[9] Robinson G. D. and Lee A. W. Int. J. Appl. Radiat. Isot. 30,365(1979).

[10] Forrester D. W., Spence V. A., Bell I. et al. Eur. J. Nucl. Med. 5,145(1980).

[11] Boothe T. E., Campbell J. A., Djermonni B. et al. Int. J. Appl. Radiat. Isot. 32,153(1981).

[12] Feng Xizhang, Liu Boli and Go Yuzhi. Bull. Beijing Normal University 3,(1983).

Appendix

For the following reaction:

$$^{125}I^- + BrAP \underset{k_2}{\overset{k_1}{\rightleftharpoons}} {}^{125}IAP + Br^-$$

$$-\ln(1-F) + \ln\left(1 - \frac{x(t)}{f}\right) = (k_1 - k_2)(f-g)t$$

from this equation, it is evident that a plot of

$$-\ln(1-F) + \ln\left(1 - \frac{x(t)}{f}\right)$$

vs time t should be a straight line with slope of $(k_1 - k_2)(f-g)$ (Fig. 5).

$$m = (k_1 - k_2)(f - g)$$

Experimentally, from the labeling efficiency (%) at equilibrium, the K_{eq} can be calculated.

$$K_{eq} = \frac{x_\infty^2}{b - ax_\infty + x_\infty^2} \quad K_{eq} = \frac{k_1}{k_2}$$

From K_{eq}, the values for f and g can be obtained.

$$f = \frac{1}{2}\left[\frac{k_1 a}{k_1 - k_2} + \sqrt{\left(\frac{k_1 a}{k_1 - k_2}\right)^2 - \frac{4k_1 b}{k_1 - k_2}}\right]$$

$$= \frac{1}{2}\frac{K_{eq} a}{K_{eq} - 1} + \sqrt{\left(\frac{K_{eq} a}{K_{eq} - 1}\right)^2 - \frac{4K_{eq} b}{K_{eq} - 1}}$$

$$g = \frac{1}{2}\left[\frac{k_1 a}{k_1 - k_2} - \sqrt{\left(\frac{k_1 a}{k_1 - k_2}\right)^2 - \frac{4k_1 b}{k_1 - k_2}}\right]$$

$$= \frac{1}{2}\frac{K_{eq} a}{K_{eq} - 1} - \sqrt{\left(\frac{K_{eq} a}{K_{eq} - 1}\right)^2 - \frac{4K_{eq} b}{K_{eq} - 1}}$$

$$a = [I^-]_0 + [BrAP]_0$$

$$b = [I^-]_0 \cdot [BrAP]_0$$

$$x_t = [IAP]_t = [Br^-]_t; t: \text{reaction time 0 to } \infty \text{ (infinity)}.$$

Radioactive Iodine Exchange Reaction of HIPDM: Kinetics and Mechanism*

Abstract In conjunction with single photon emission computed tomography(SPECT), iodine-123 (^{123}I) labeled N, N, N′-trimethyl-[2-hydroxy-3-methyl-5-iodobenzyl]-1, 3-propanediamine (HIPDM) has been used clinically as a regional cerebral perfusion imaging agent. The [^{123}I] HIPDM can be prepared by a simple aqueous exchange reaction in a kit form. We synthesized unlabeled HIPDM by condensation of 2-hydroxy-3-methyl-5-iodobezaldehyde and N, N, N′-trimethyl-1, 3-propanediamine, followed by a sodium borohydride reduction reaction. The kinetics of the radioactive iodine exchange reaction for the preparation of [^{123}I] HIPDM is controlled by the pH, the temperature, and the presence of reductant(sodium bisulfite), and oxidant(sodium iodate). The reaction is a second order iodine-iodine exchange with an activation energy of 30.6 kcal/mol. The mechanism of this reaction probably involves the formation of an active I$^+$ or iodine free radical, which is sensitive to the presence of a reductant, such as sodium bisulfite.

The lipid-soluble diamine N, N, N′-trimethyl-[2-hy-droxy-3-methyl-5-iodobenzyl]-1, 3-propanediamine(HIPDM) crosses the blood-brain barrier with high first pass extraction ratio and displays a long brain retention time[1]. In conjunction with single photon emission computed tomography(SPECT), [^{123}I] HIPDM has been used clinically for studying regional cerebral perfusion[2-5] and is currently in the second phase of clinical study.

The major advantage of [^{123}I] HIPDM, as compared with the other iodinated brain imaging agent—a monoamine, [^{123}I] IMP(N-isopropyl-4-iodoamphetamine)[6-10]—is the ease of labeling with [^{123}I] sodium iodide by a simple aqueous exchange reaction in a kit form. The hydroxy group on the benzene ring of HIPDM activates the ring thereby rendering the iodine atom more readily exchangeable(Scheme 1). In spite of the simplicity of the exchange reaction, low labeling yields have sometimes been encountered when using supposedly pure Na^{123}I. Contamination of the NaI with metal ions from the tellurium-127(^{127}Te) target during the production of ^{123}I by

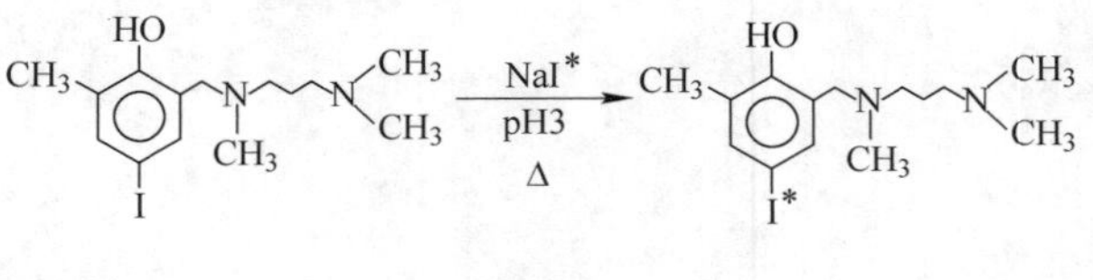

Scheme 1

---

* Copartner: J. Chang, J. S. Sun, J. Billings, A. Steves, R. Ackerhalt, M. Molnar, and H. F. Kung. Reprinted from *The Joumal of Nuclear Medicine*, 1987, 28(3): 360-365.

a(p,5n) reaction or from rinsing the xenon trap with aqueous sodium hydroxide during the collection of Na ^{123}I may affect the exchange reaction.

In this report, the chemical synthesis of cold HIPDM is illustrated. The effect of the low level metal ions, temperature, and pH were evaluated. The action of the reducing agent, sodium bisulfite (NaHSO$_3$), was also studied, as this reagent is sometimes used in the [^{123}I] sodium iodide solution as a preservative. The activation energy was also determined. The information presented in this paper may be useful in understanding the basic kinetics of the iodine-iodine exchange reaction, and may also be valuable for formulating convenient kits for ^{123}I-labeled radiopharmaceuticals.

1 Methods

The cold HIPDM was prepared by a method similar to that reported earlier for other iodinated phenolic diamine derivatives[11] (Scheme 2).

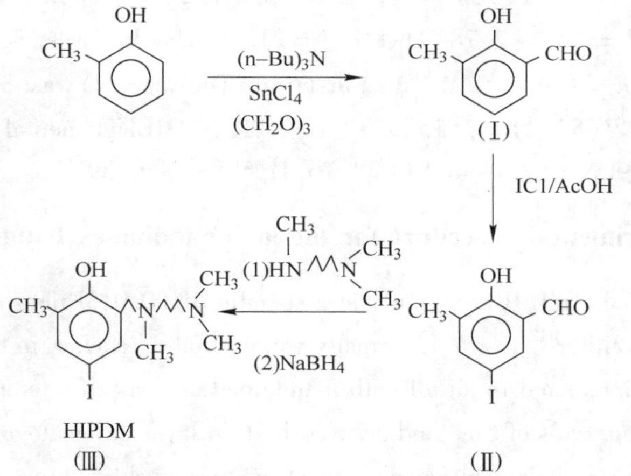

Scheme 2

1.1 Synthesis of 5-iodo-3-methyl-salicylaldehyde(Ⅱ)

A solution of the 3-methyl-salicylaldehyde, I, (15g, 110mmol), prepared as described previously[11], was added dropwise to a solution of ICl (24.3g, 150mmol) in 100mL of glacial acetic acid at 60℃. After the addition was completed, the dark mixture was heated for 2h at 75-80℃, then stirred at 55-60℃ for 18h, after which most of the solvents were evaporated. The residue was diluted with cold water(50mL) and filtered. The brown solid was sublimated at 55℃ (0.1-0.2torr) to give 15g(57mmol) of pure product(yield 52%)[11].

1.2 Synthesis of N, N, N′-trimethyl-N′-(2-hydroxy-3-methyl-5-iodobenzyl)-1, 3-propanediamine(Ⅲ)

A solution of the iodinated aldehyde, Ⅱ, (8.13g, 31mmol) and N, N, N′-trimethylpropane-1, 3-diamine(4.1g, 35mmol) in 100mL of benzene was refluxed for 30min. The solvent was evaporated under reduced pressure to give a yellow oil. The oil was dissolved in ethanol(75mL) and NaBH$_4$ (1.0g, 26mmol) was added by a spatula in small portions over 20min. The reaction mixture was stirred at room temperature for 18h. The resulting clear solution was concentrated under reduced pressure, and a solution of saturated sodium bicarbonate(100mL) and carbon tetrachloride(150mL) was added. The carbon tetrachloride layer was separated and the aqueous layer was re-extracted with carbon tetrachloride (2 × 50mL). The combined organic extracts were dried over anhydrous sodium sulfate and evaporated under reduced pressure to give a clear oil. The oil was redissolved in 60mL of absolute ethanol and the product was converted to the dihydrochloride salt by passing dry HCl gas through the solution at 0℃. The solution was treated with a small amount of hexane until it turned cloudy. After cooling to 4℃ for 18h, the white crystals were filtered, dried, and recrystallized from ethanol-hexane to give 7.85g(18mmol) of pure dihydrochloride salt (yield 58%). The following data were recorded: ^{1}H-NMR δ(DMSO-d$_6$)2.23(S,3H,CH$_3$);2.71(S,3H,CH$_3$)2.78(S,6H,2CH$_3$);3.23(M,6H);4.36(S,2H,CH$_2$)7.51(D,1H,J=2Hz);7.78(D,1H,J=2Hz). The UV was measured as (in 0.9% NaCl):λ_{max} 289nm(E = 1.9 × 10^{6}M^{-1}), min266nm. The MS(CI) was; 364(15.6,M^{+1});363(100);362(50.9)247(52.2);87(150). C$_{14}$H$_{23}$H$_2$IO · 2HCl elemental analysis showed: theory C: 38.64, H: 5.79, N: 6.44; found C: 38.76, H: 5.88, N: 6.39.

1.3 General experimental procedure for the iodine-iodine exchange reaction

To a solution of HIPDM(2mL, 0.25mg/mL) at a specific pH(0.01M phosphate buffer), one drop of radioactive iodide(either ^{125}I or ^{123}I, normally without iodide carrier, in 0.1N NaOH) was added. The solution was incubated in an oil bath maintained at a specific temperature. Samples were removed at different intervals of time and assayed by thin layer chromatography(TLC) using Merck TLC plates 60F-254 and a freshly made solvent mixture of chloroform：ethanol：concentrated ammonia (8：1.8：0.2 v/v). After developing, the TLC plates were cut into 0.5cm sections and counted in a gamma counter; R_f values for free iodide and labeled HIPDM were 0.01 and 0.5-0.8, respectively. Iodine-125 and ^{123}I gave the same results for the exchange reaction.

Most of the kinetic studies reported in this paper were carried out at a lower concentration (0.25mg of HIPDM/mL) than that of a clinical kit(2mg/mL). At the lower concentration, the reaction rate is slower, so the reaction kinetics can be measured more adequately.

1.4 Effects of temperature

The reaction solution was buffered at pH 3.0 and immersed in a constant temperature bath. The reaction solution was sampled at various time points and analyzed using the TLC method described above. The exchange reaction was studied at 40℃, 50℃, 60℃, 70℃, 80℃, 90℃, and 100℃.

1.5 Effects of sodium iodide concentration

The exchange reaction was carried out in the presence of increasing amounts of sodium iodide to evaluate if it is a second order reaction. The sodium iodide concentration varied from $0\mu g/mL$ (carrier-free) to $66\mu g/mL$.

1.6 Effects of pH

The exchange reaction was studied at various pHs ranging between 1 and 14. At lower pHs (pH 1 and 2) a citrate buffer was employed, whereas at higher pHs (pH 13 and 14) a solution of sodium hydroxide was used. Phosphate buffers were used for pHs between 3 and 12. Reaction temperature was held at 100℃ and samples were analyzed at 30min.

1.7 Effects of trace metal ions

The reaction was carried out in the presence of different metal ions at concentrations indicated in Table 1. The reaction temperature, time, and pH were 100℃, 30min, and 3.0, respectively.

Table 1　Effects of sodium iodide concentration

No.	HIPDM[①]			NaI[②]			Expected yield[③]	Yield found
	Volume/μL	Weight/μg	Final concentration /μM	Volume/μL	Weight/μg	Final concentration /μM		
A	100	100	2.30	0	0	0	100.0	97.2
B	100	100	2.09	10	3.4	0.22	90.9	86.1
C	100	100	1.91	20	6.8	0.38	83.4	78.1
D	100	100	1.84	25	8.5	0.45	80.1	72.9
E	100	100	1.53	50	17.0	0.75	66.9	58.9
F	100	100	1.15	100	34.0	1.13	50.3	44.9
G	50	50	0.77	100	34.0	1.51	33.6	30.4

① HIPDM stock 1.0mg/mL + 0.5μg/mL NaI(kit).
② NaI stock 0.340mg/mL.
③ The expected yield is ([HIPDM]/[HIPDM] + [I^-]) × 100%.

1.8 Effects of sodium bisulfite or potassium iodate

The exchange reaction was studied under the following condition: HIPDM (1mL, 2mg/mL in 0.01M phosphate buffer), pH 3, heated at 100℃ for 30min. Each reaction was carried out in the presence of reducing agent or oxidizing agent at a concentration specified in Table 2.

Table 2　Effect of trace metal ions on the HIPDM exchange reaction

Trace metal	Chemical formula	Quantity/mg·mL^{-1}	Labeling yield/%
None	—	—	98.3
Fe^{3+}	$Fe(NO_3)_3 \cdot 9H_2O$	1.47	95.2
Cu^{2+}	$CuSO_4 \cdot 5H_2O$	1.31	91.5
Al^{3+}	$Al_2(SO_4)_3$	1.63	96.8
Zn^{2+}	$Zn(Ac)_2 \cdot 7H_2O$	1.51	95.6
Mg^{2+}	$MgCl_2 \cdot 6H_2O$	1.55	96.7
Ni^{2+}	$Ni(NO_3)_2 \cdot 6H_2O$	1.48	97.6

2 Results and Discussion

2.1 Effects of temperature

There is a significant effect of temperature on the exchange reaction rate (Fig. 1). When the reaction temperature is lower than 90℃, the reaction did not reach 95% in <1h of heating. At 100℃, the reaction reached 95% in <30min. The fraction of exchange (F) was calculated based on the following equation:

$$F = \frac{[\text{HIPDM}^*]_t - [\text{HIPDM}^*]_0}{[\text{HIPDM}^*]_\infty - [\text{HIPDM}^*]_0} \quad (1)$$

$[\text{HIPDM}^*]_t$ = radioactive HIPDM at time t

$[\text{HIPDM}^*]_0$ = radioactive HIPDM at time 0

$[\text{HIPDM}^*]_\infty$ = radioactive HIPDM at time ∞

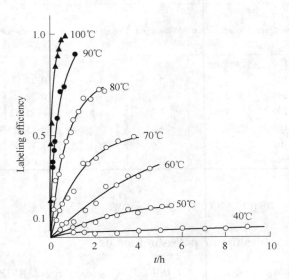

Fig. 1 Effects of temperature on labeling efficiency at various time points

The $[\text{HIPDM}^*]_\infty$ was determined at 24h after the initiation of the exchange reaction. At time zero, the amount of radiolabeled HIPDM was close to zero. Therefore:

$$F = \frac{[\text{HIPDM}^*]_t}{[\text{HIPDM}^*]_\infty}$$

When $\ln(1 - F)$ is plotted as a function of time, a straight line is obtained for all reaction temperatures studied (Fig. 2). This strongly suggests that the mechanism of this exchange reaction is a simple second order iodine-iodine isotope exchange reaction. The reaction rate appeared to be very dependent on the concentration of HIPDM and sodium iodide (see data presented below). At higher HIPDM comcentration (2mg/mL), a concentration used in the current kit formulation, the exchange reaction is usually completed within 5min at 100℃.

2.2 Effects of sodium iodide concentration

The concentration of sodium iodide clearly showed a significant effect on the labeling yield. It is apparent from the data presented in Table 1 that the reaction is a second order reaction. The expected yield is quite consistent with the calculated yield based on secondorder reaction kinetics.

2.3 Effects of pH

When the reaction pH is below 6, the HIPDM exchange reaction proceeds smoothly, as there is no significant difference in the labeling yield. When the pH is above 6, however, the exchange reaction shows a sharp drop in the labeling yield (Fig. 3). This is probably due to the formation of I^- anion at pHs higher than 6, which prevents the exchange reaction. The optimum pH for this exchange reaction is around pH 3. In this report, except as noted, all of the HIPDM exchange reactions are carried out in pH 3.0 buffered solutions. The huge difference in labeling efficiency between pH 7 and pH 8 may be related to the deprotonation of amines. A similar type of curve was reported for labeling ^{123}I iodoantipyrine, but the dramatic shift in labeling yield occurred at lower pH ($\approx$ pH 3-4)[12].

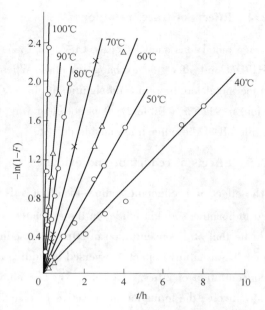

Fig. 2　Effects of temperature on labeling efficiency
(The same data points from Fig. 1 but presented as a plot of $\ln(1-F)$ vs. time (F: fraction exchanged). The straight lines for every temperature studied indicate that the reaction is a simple iodine-iodine exchange reaction)

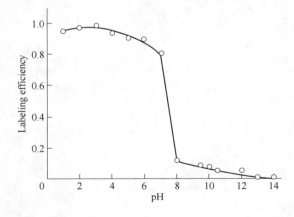

Fig. 3　Effects of pH on the labeling yield
(The labeling yield declines significantly when the reaction pH is above 6)

2.4 Effects of trace metal ions

Trace metal ions such as Fe^{3+}, Cu^{2+}, Al^{3+}, Zn^{2+}, Mg^{2+}, and Ni^{2+}, may form a complex with HIPDM and affect the exchange reaction. When the reaction is carried out in the presence of trace metal ion, there are no significant changes. The only exception is Cu^{2+}, which lowers the yield to 91.5% (Table 2). It is not clear why only Cu^{2+} has the ability to inhibit the reaction while all of the other trace metals do not.

2.5 Effects of reductant and oxidant

The effect of a reductant, sodium bisulfite($NaHSO_3$), a common preservative for no carrier added radioactive sodium iodide, on the exchange reaction was evaluated. The results in Table 2 indicate that at a concentration of 20μg/mL, sodium bisulfite effectively stops the exchange reaction. The inhibitive effect is reversed by adding an oxidant such as potassium iodate(KIO_3), at a concentration of 20μg/mL(Table 3). The above results suggest that the reaction mechanism may involve the formation of an active I^+ or iodine free radical. In the presence of a reductant, the I^+ or iodine free radical can no longer exist and inhibition of the exchange reaction occurs. However, when the reducing effect is neutralized by an oxidant, then the exchange reaction does take place again in high yield(see Table 3). The presence of I^+ as an active intermediate is an important factor to consider when designing a kit formulation for routine nuclear medicine use. Proper precaution should be taken to avoid the contamination of reductants, which will inhibit the exchange reaction.

Table 3 Effect of reductant and oxidant[1] on the HIPDM exchange reaction

$NaHSO_3$/μg	Labeling yield/%[2]	Reductant $NaHSO_3$/μg	Oxidant KIO_3/μg	Labeling yield/%[2]
0	98.1	0	20	97.9
1	95.5	20	0	0.6
5	97.5	20	20	96.5
10	97.4	20	40	96.5
20	0.5	20	100	96.6
30	0.3			
40	0.2			
50	0.1			
200	0.4			
500	0.6			
1000	0.0			

[1] Sodium bisulfite was the oxidant used.
[2] Average of two experiments.

2.6 Activation energy for the exchange reaction

The carrier-added exchange reactions, using 0.1mg of potassium iodide and 0.5mg/mL of HIPDM at pH 3, were studied at three different temperatures, 79℃, 89℃, and 94℃. The reaction rate at each temperature was measured and the second-order rate constants (k) were determined (Fig. 4). By measuring the slope for a plot of ln k versus $1/T$ and fitting the Arrhenius equation:

$$\ln k = \frac{E_a}{RT} + \ln A \qquad (2)$$

where, R is the gas constant, T is the absolute temperature, the activation energy E_a can be calculated. As a result, E_a was determined to be 30.6kcal/mol (Fig. 5).

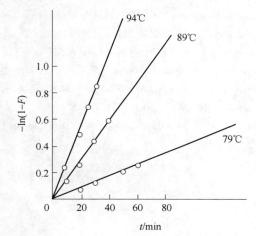
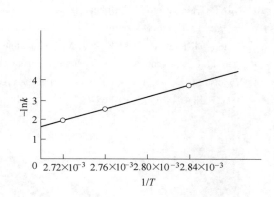

Fig. 4　Measurement of reaction rates (the slope of each line) at different temperatures under a carrier-added condition

Fig. 5　A plot of reaction rates vs. $1/T$ (T: absolute temperature, K. The activation energy E_a is determined by fitting the data with the Arrhenius equation. Slope = 1.541×10^4; $\Delta E_a = 30.6$kcal/mol)

The kinetics of the iodine-iodine exchange reaction of HIPDM have been evaluated. Based on the effects of pH and temperature and the presence of reductants and oxidants, the reaction probably is a second order iodine-iodine exchange reaction with an activation energy of 30.6kcal/mol. The mechanism of this reaction may involve the formation of an active I^+ or iodine free radical. The information may be of use in designating new radioiodine labeled radiopharmaceuticals in a kit form for regular nuclear medicine clinics.

References

[1] Kung H F, Tramposch K M, Blau M. A new brain perfusion imaging agent: [I-123]HIPDM: N,N,N'-trimethyl-N'-[2-hydroxy-3-methyl-5-iodobenzyl]1,3-propanediamine. J Nucl Med 1983;24:66-72.

[2] Fazio F, Lenzi G L, Gerundini P, et al. Tomographic assessment of regional cerebral perfusion using intravenous I-123HIPDM and a rotating gamma camera. J Comp Asst Tomogr 1984;8:911-921.

[3] Fazio F, Lenzi G L, Gerundini P, et al. Assessment of regional cerebral perfusion with SPECT and ^{123}I-HIPDM in patients with EC-IC bypass. Monogr Neu-rol Sci 1984;11:98-103.

[4] Drayer B, Jaszczak R, Freidman A, et al. In vivo quantitation of regional cerebral blood flow in glioma and cerebral infarction: Validation of the HIPDM-SPECT method. Am J Neurol Rad 1983;4:572-576.

[5] Lucignani G, Nehlig A, Blasberg R, et al. Metabolic and kinetic consideration in the use of $[^{125}I]$HIPDM for quantitative measurement of regional blood flow. J Cerebral Flow Metab 1985;5:86-96.

[6] Winchell H S, Baldwin R M, Lin T H. Development of I-123 labeled amines for brain studies: Localization of I-123 iodopheylalkylamines in rat brain. J Nucl Med 1980;21:940-946.

[7] Winchell H S, Horst W D, Braun L. N-isopropyl$[^{123}I]$-p-iodoamphetamine: single-pass brain uptake and washout. binding to brain synaptosomes, and localization in dog and monkey brain. J Nucl Med 1980;21:947-952.

[8] Kuhl D E, Barrio J R, Huang S C, et al. Quantifying local cerebral blood flow by N-isopropyl-p-$[^{123}I]$iodoamphetamine(IMP)tomography. J Nucl Med 1982;23:196-203.

[9] Lassen N A, Henriksen L, Holm S, et al. Cerebral blood flow tomography by SPECT(single photon emission tomography): Xenon-133 compared to isopropyl-amphetamine-iodine-123. Ann Radiol 1983;26:53.

[10] Hill T C, Magistretti P L, Holman B L, et al. Assessment of regional cerebral blood flow(rCBF)in stroke using SPECT and N-isopropyl-(I-123)-p-iodoamphetamine(IMP). Stroke 1984;15:40.

[11] Tramposch K M, Kung H F, Blau M. Radioiodine labeled N,N-dimethyl-N'-(2-hydroxy-3-alkyl-5-iodobenzyl)-1,3-propanediamines for brain perfusion imaging. J Med Chem 1983;28:121-125.

[12] Robinson G D, Lee A W. Reinvestigation of the prep-aration of ^{131}I-4-iodoantipyrine from ^{131}I-iodide. Int J Appl Radiat Isot 1979;30:365-367.

A Kit Formulation for Preparation of Iodine-123-IBZM: a New CNS D-2 Dopamine Receptor Imaging Agent[*]

Abstract A method for the preparation of iodine-123IBZM, a central nervous system D-2 imaging agent, is reported. By using a rapid filtration technique to remove the unreacted iodide, the preparation can be completed in less than 20min (overall yield > 60%). The product, with high purity (≥95%) and specific activity, is suitable for human use.

IBZM (S)-(−)-3-iodo-2-hydroxy-6-methoxy-N-[(1-ethyl-2-pyrrodinyl) methyl] benzamide belongs to a group of structurally related benzamides which display significant antidopaminergic activity[1]. The iodine-125-IBZM has been used as a probe for the study of D-2 receptors (K_d = 0.4nM in rat striatum tissue preparation)[2-4], particularly for the detergent-solu-bilized D-2 receptors[5]. The results from the in vivo and in vitro data all indicated that IBZM binds specifically to the central nervous system (CNS) dopamine D-2 receptor with high affinity and stereospecificity. The [^{123}I]IBZM is a highly selective CNS D-2 dopamine receptor ligand useful for SPECT imaging studies in normal subjects[6,7] and is potentially useful for patients with various neurologic disorders.

Preparation of the no-carrier-added (NCA) [^{123}I]IBZM has been achieved by an oxidative iodination of BZM (the uniodinated starting material) with sodium [^{123}I]iodide and peracetic acid as an oxidant[8]. The final product (NCA [^{123}I]IBZM) is obtained through several steps of organic extraction and high-performance liquid chromatography (HPLC) purification. The multistep synthesis and purification procedure requires at least 2h. Using this procedure, a limited Phase 1 clinical trial has been accomplished[6]. In order to optimize this procedure for the routine nuclear medicine clinic use, a faster and easier method for preparing [^{123}I]IBZM is preferred. In this paper, we report a simplified and expeditious procedure for the preparation of [^{123}I]IBZM based on a kit formulation for routine clinical use. For convenience, the products prepared by using HPLC separation and by using a kit separation are designated as [^{125}I] or [^{123}I]IBZM-NCA and IBZM-KIT, respectively.

1 Materials and Methods

1.1 Reagents

The uniodinated starting material, BZM(S)-(−)-2-hydroxy-6-methoxy-N-[(1-ethyl-2-pyrroli-

---

[*] Copartner: Mei-Ping Kung, Yun-Yun Yang, Jeffrey J. Billings, Hank F. Kung. Reprinted from *The Journal of Nuclear Medicine*, 1991, 32(2):339-342.

dinyl)methyl]benzamide and nonradioactive IBZM were prepared by the method described previously[2]. Sodium [^{125}I]iodide was obtained from NEN/Dupont in a NCA form (specific activity 17Ci/mg; 2200Ci/mmol). Sodium [^{123}I]iodide was obtained from Atomic Energy of Canada Ltd. (specific activity 2.4×10^5 Ci/mmol). Peracetic acid (32 wt% solution in dilute acetic acid) was obtained from Aldrich Chemicals, (St. Louis, MO) and diluted from stock solution with distilled water before use. The Accubond-C4 (butyl) columns (100mg/mL) were obtained from Bodman Chemicals, Aston, PA. The columns were prewashed with 1mL of absolute ethanol followed by 2mL of distilled water before use. Ethyl acetate and acetonitrile were of HPLC grade and purchased from J. T. Baker. All other chemicals were of reagent grade and purchased commercially.

1.2 Radiolabeling

NCA Preparation (IBZM-NCA). The [^{125}I] and [^{123}I] IBZM were prepared by the method reported previously[8]. Briefly, 50μg of the starting material, BZM, was reacted with radioactive iodide in the presence of ammonium acetate and peracetic acid as the oxidant. After quenching and neutralization, the reaction mixture was extracted with ethyl acetate. The final product, after purification with HPLC, was designated as IBZM-NCA.

Kit Formulation (IBZM-KIT). The method was developed as follows: a small amount of BZM (4μg) dissolved in 20μL absolute ethanol was mixed with 100μL of 0.5M ammonium acetate, pH 4.0, and 20μL of radioactive sodium ([^{123}I] or [^{125}I]) iodide. The oxidative iodination was initiated by the addition of 50μL of diluted peracetic acid (0.3 wt% from stock solution prepared freshly every day) and the mixture was allowed to proceed at room temperature for 5min. After quenching with 100μL of sodium bisulfite (300mg/mL), the reaction was neutralized by addition of 0.7mL of saturated sodium bicarbonate solution. To remove the unreacted radioactive iodide, the reaction mixture was passed through a prewashed accubond-C4 column and washed with distilled water (2mL). The desired product, radioactive IBZM, was eluted from the column with absolute ethanol (1mL) and its radiochemical purity was analyzed on HPLC (PRP-1 column, acetonitrile/5.0mM dimethylglutaric acid buffer pH 7.0, 90:10) as described previously[8]. The IBZM obtained through this method was designated as IBZM-KIT.

1.3 In vivo study

Male Sprague-Dawley rats (225-300g) which were allowed access to food and water ad libitum were used for all of the studies. The total brain uptake and the regional brain distribution of [^{125}I]IBZM in rats were obtained after an i.v. injection of 0.2mL saline containing 1-2μCi of [^{125}I]IBZM-NCA or [^{125}I]IBZM-KIT. The rats were sacrificed at various time points postinjection by cardiac excision under ether anesthesia. After dissecting, weighing and counting samples from different brain regions (cortex, striatum, hippocampus and cerebellum), % dose/g of samples was calculated by comparing the sample counts with the counts of the diluted initial dose. The ratio of uptake in each region was obtained by dividing % dose/g of each region with that of the cerebellum.

1.4 In vitro binding study

Rat striatal tissue homogenates were prepared as described previously[2]. The binding assays were performed by incubating 50μL of the striatal membrane preparation containing 40-60μg protein with increasing amounts of [^{125}I]IBZM-NCA or [^{125}I]IBZM-KIT (for saturation analysis) or appropriate amounts of labeled ligands and nonradioactive IBZM (for displacement study) in a total volume of 0.2mL of the assay buffer (50mM Tris buffer, pH 7.4, 120mM NaCl, 5mM KCl, 2mM $CaCl_2$, and 1mM $MgCl_2$). The incubation and the following procedure were carried out as described previously[2]. The nonspecific binding was determined in the presence of 10μM of spiperone and the data points were analyzed using the iterative, nonlinear least-square's curvefitting program, LIGAND[9].

1.5 Ultraviolet determination of BZM

The amount of BZM present in the final product, radiolabeled IBZM, was determined by measuring the absorbance at 308nm (the characteristic wavelength for BZM) using a spectrophotometer (Beckman DU-7). The measurement was done with an adequate sample for the blank (the blank also went through the same process as the samples except BZM was eliminated in the oxidative iodination). From the calibration curve generated with various amounts of BZM (2-16μg), the amount of BZM in the samples (radiolabeled IBZM) was thus calculated.

2 Results

By using a smaller amount of the starting material, BZM, 4μg in this method for preparation of IBZM-KIT as compared to the 50μg used in the preparation of IBZM-NCA, the labeling yield remained consistently high (70%-80%). This method eliminates the solvent extraction and the HPLC purification steps, and the simplified procedure can be completed in less than 20min. The radiochemical purity of the final product, IBZM-KIT, analyzed on HPLC, shown in Fig. 1, is high (95%-98%, n = 10) and the HPLC profiles are consistent with those using the previous meth-

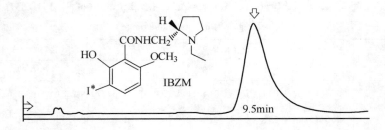

Fig. 1 HPLC of kit-formulated [^{123}I]IBZM (gamma detector) with reverse-phase column (PRP-1), solvent system of acetonitrile: dimethyl glutaric acid (5mM, pH 7.0)/ 90 : 10, flow rate = 1.0mL/min

od. The amount of uniodinated BZM present with the radioactive IBZM product was determined by UV measurement and calculated from the BZM calibration curve (Fig. 2). The value was found to be $(3.14 \pm 0.16)\mu g (n=5)$.

IBZM prepared with this new method (IBZM-KIT) was compared to the IBZM-NCA prepared by HPLC purification using the in vivo biodistribution and in vitro binding studies. Table 1 shows the brain uptake in rats with [^{125}I]IBZM prepared from the two methods. Similar values were obtained with IBZM-KIT and IBZM-NCA at 30min and 60min postinjection. The striatum/cerebellum ratios in rat brain were 2.86, 4.18 and 2.67, 4.61 for [^{125}I]IBZM-NCA and

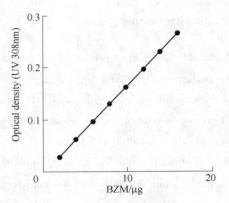

Fig. 2 Calibration curve for determination of amount of BZM in the final product
(The absorption at 308nm (characteristic wavelength for BZM) was chosen for the measurement)

[^{125}I]IBZM-KIT at 30min and 60min, respectively (Table 2). Furthermore, comparable K_d values (0.31nM, 0.24nM for NCA and KIT) were obtained using the in vitro binding assay in rat striatum membrane preparation (Table 3). The displacement studies with nonradioactive IBZM also gave similar K_i values (0.29nM for NCA and 0.27nM for KIT). Since the K_i values are almost identical to the K_d values, it is reasonable to assume that both preparations have very similar specific activity and are probably carrier-free (reaching the theoretical value-the specific activity of radioactive iodine, i.e., 2200Ci/mmol for ^{125}I and 240000Ci/mmol for ^{123}I, respectively).

Table 1 Brain uptake (% dose/organ) in rats after i. v. injection of [^{125}I]IBZM

(average of three rats and range)

Compound	30min	60min
[^{125}I]IBZM-NCA	1.84(1.70-2.00)	1.20(1.13-1.31)
[^{125}I]IBZM-KIT	1.73(1.52-2.12)	0.98(0.96-0.99)

Table 2 Regional brain ratio in rats after i. v. injection of [^{125}I]IBZM

(average of three rats and range)

Item	[^{125}I]IBZM-NCA		[^{125}I]IBZM-KIT	
	30min	60min	30min	60min
ST/CB	2.86(2.67-3.03)	4.18(4.11-4.20)	2.67(2.49-2.81)	4.61(4.20-5.01)
HP/CB	1.51(1.44-1.59)	1.74(1.72-1.76)	1.48(1.40-1.58)	1.79(1.75-1.83)
CX/CB	1.80(1.63-1.94)	1.76(1.72-1.83)	1.64(1.51-1.80)	1.70(1.66-1.73)

Note: ST = striatum; HP = hippocampus; CX = cortex; and CB = cerebellum.

Table 3 In vitro binding constants of [^{125}I]IBZM using rat striatal membrane preparation

Method	K_d(nM ± s.e.m.)①	K_i(nM ± s.e.m.)②
IBZM-NCA	0.31 ± 0.06	0.29 ± 0.05
IBZM-KIT	0.24 ± 0.05	0.27 ± 0.04

Note: Each value represents the mean ± s.e.m. of 3-5 determinations.

① K_d values were determined from Scatchard analysis of saturation isotherms.

② 0.20-0.30nM [^{125}I]IBZM-KIT or [^{125}I]IBZM-NCA was incubated in the presence of nonradioactive IBZM in 7-11 concentrations.

3 Discussion

In order to image D-2 dopamine receptors in the brain with iodinated ligands, it is essential to use selective D-2 ligands with high specific activity. In the basal ganglia of normal human brain, there are about 300pM (15pM in about 20mL of basal ganglia) of D-2 receptors[10-12]. One major objective of this paper is to demonstrate that it is possible to prepare radioactive labeled [^{123}I] or [^{125}I]IBZM with high specific activity using a kit formulation. Since the isotopic exchange labeling generally will not give a product with a high specific activity, direct oxidative iodination of BZM is the method of choice for preparing IBZM with high specific activity.

Using the previously reported procedure, [^{123}I] or [^{125}I]IBZM-NCA can readily be prepared. The procedure started with 50μg of uniodinated BZM. After the oxidative iodination step, the unreacted starting material and polar radioactive impurities were removed by extraction with organic solvents followed by a HPLC purification. The desired [^{123}I] or [^{125}I]IBZM-NCA could thus be obtained (yield ≈50%, >95% purity). The major disadvantages of this procedure are: (1) the time required for completing the steps is excessively long (≈2h) and (2) an adequate HPLC setup is needed for purification of the product. Both of these may prevent the widespread application of this agent in regular nuclear medicine clinics.

In order to eliminate the time-consuming steps, such as organic extraction and HPLC purification, for the preparation of NCA radioactive IBZM[8], the iodination conditions were optimized. The amount of starting material, buffer, and oxidant as well as the reaction time have been systematically investigated. The removal of unreacted radioactive iodide and polar side products can be easily achieved by passing the reaction mixture through a small Accubond-C4 column and the desired radioactive IBZM product can be recovered by eluting the column with ethanol. One major difference between IBZM-NCA and IBZM-KIT preparation is the presence of a small amount of uniodinated BZM in the latter preparation. Since the difference in binding affinity for D-2 receptors between IBZM and BZM is high (50 versus 1)[2], the small amount of uniodinated BZM together with the radioactive IBZM product will not affect the binding or imaging studies of the labeled IBZM. Further studies confirm this assumption. The results for the in vivo biodistribution and the in vitro binding study indicated that the IBZM-KIT, even in the presence of small amount of BZM, behaves similarly to the IBZM-NCA. Tests on sterility and pyrogenicity suggested that as expected, the kit preparation produced a sterile and pyrogen-free [^{123}I]IBZM, suitable for human use.

Recently, several new iodinated benzamides with higher affinity than IBZM were reported[13,14]. The newer benzamides are labeled by an iododestannylation reaction. Since the tri-n-butyltin compounds and the products, iodinated benzamides, show a large difference in their lipid-solubility, the type of kit formulation reported for IBZM preparation may also be applicable for these newer benzamides.

In conclusion, a method for the preparation of [^{123}I]IBZM is reported. By eliminating the time-consuming steps of organic extraction and HPLC purification, the kit preparation can be completed in less than 20min. The studies suggest that the IBZM prepared by the kit formulation is of high specific activity and is suitable for routine clinical use.

Acknowledgements

This work is supported by grants from the National Institute of Health (NS-23458 and MH 43880). The authors thank Catherine Cartwright for her assistance in preparing this manuscript.

References

[1] Hogberg T, Ramsby S, Ogren S O, Norinder U. New selective dopamine D-2 antagonists as antipsychotic agents. Acta Pharm Suec 1987;24:289-328.

[2] Kung H F, Kasliwal R, Pan S, Kung M P, Mach R H, Guo Y Z. Dopamine D-2 receptor imaging radiopharmaceuticals: synthesis, radiolabeling, and in vitro binding of (R)-(+)-and(S)-(−)-3-iodo-2-hydroxy-6-methoxy-N-[(1-ethyl-2-pyrrolidinyl)methyl]benzamide. J Med Chem 1988;31:1039-1043.

[3] Kung H F, Billings J J, Guo Y Z, Mach R H. Comparison of in vivo D-2 dopamine receptor binding of IBZM and NMSP in rat brain. Nucl Med Biol 1988;15:203-206.

[4] Brücke T, Tsai Y F, McClellan C, et al. In vitro-binding properties and autoradiographic imaging of 3-iodo-benzamide ([^{125}I]-IBZM): a new imaging ligand for D-2 dopamine receptors in SPECT. Life Sci 1989;42:2097-2104.

[5] Schonwetter B S, Luedtke R R, Kung M P, Billings J J, Kung H F, Molinoff P B. Characterization of membrane-bound and soluble dopamine D2 receptors in canine caudate using [^{125}I]IBZM. J Pharmacol Exp Ther 1989;250:100-116.

[6] Kung H F, Alavi A, Chang C W, et al. In vivo SPECT imaging of CNS D-2 dopamine receptors: initial studies with [^{123}I]IBZM in humans. J Nucl Med 1990;31:573-579.

[7] Brücke T, Podreka I, Angelberger P, et al. Dopamine D-2 receptor imaging with ^{123}I-labeled iodobenzamide in SPECT[Abstract]. J Nucl Med 1989;30:31.

[8] Kung M P, King H F. Peracetic acid as a superior oxidant for preparation of [^{123}I]IBZM: a potential dopamine D-2 receptor imaging agent. J Lab Compd Radiopharm 1989;27:691-700.

[9] Munson D J, Rodbard D. LIGAND: a versatile computerized approach for characterization of ligand binding systems. Anal Biochem 1980;107:220-239.

[10] Wong D F, Wagner H N, Dannals R J, et al. Effects of age on dopamine and serotonin receptors measured by positron tomography in the human brain. Science 1984;226:1393-1396.

[11] Wong D F, Wagner H N J, Tune L E, et al. Positron emission tomography reveals elevated D_2 dopamine receptors in drugnative schizophrenics. Science 1986;243:1558-1563.

[12] Farde L, Halldin C, Stone-Elander S, et al. PET analysis of human dopamine receptor subtypes using ^{11}C-SCH-23390 and ^{11}C-raclopride. Psychopharmacol 1987;92:278-284.

[13] de Paulis T, Janowsky A, Kessler R M, et al. (S)-N-[(1-Ethyl-2-pyrrolidinyl)methyl]-5-[^{125}I]iodo-2-methoxybenzamide hydochloride: a new selective radioligand radioligand for dopamine D-2 receptors. J Med Chem 1988;31:2027.

[14] Murphy R A, Kung H F, Kung M P, Billings J J. Synthesis and characterization of iodobenzamide analogs: potential D-2 dopamine receptor imaging agents. J Med Chem 1990;33:171-178.

新型肾上腺显影剂——6-(^{82}Br)溴甲基胆固醇的合成及其在动物组织内的分布[*]

关键词 肾上腺显影剂　6-(^{82}Br)溴甲基胆固醇　组织分布

1　引言

自从 1968 年 Nagai 等[1,2]首次将^{131}I 标记胆固醇作为肾上腺示踪剂以来，1970 年 Counsell 等[3]报道了 19-(^{131}I)碘胆固醇的合成和在动物肾上腺中有明显的浓集。1975 年 Kojima 等[4,5]制备了 6-(^{131}I)碘甲基去甲胆固醇，动物实验表明它具有更高的亲肾上腺组织特性。其后，王世真[6,7]、刘伯里等[8,9]研制了 6-(^{131}I)碘胆固醇，显示有较好的亲肾上腺功能和更为优良的热稳定性和辐射稳定性[10]。为了寻找更佳的肾上腺显影剂，将胆固醇的结构进行了一系列改造，合成了若干种胆固醇类肾上腺显影剂[5,11~16]。

在卤素的放射性核素中，缺中子核素^{77}Br 和^{123}I 是较适合体内研究和诊断用的两种核素[17]。溴标记的胆固醇类肾上腺显影剂 NCL-6-^{82}Br[15]、CL-6-^{82}Br[18]和 NCL-6-Et ^{82}Br[19]等均已被合成并研究了它们在动物体内的分布。为了进一步考察胆固醇类肾上腺显影剂的构效关系和适应开展^{77}Br 肾上腺显影剂诊断工作的需要，我们合成了一种新型肾上腺显影剂 6-(^{82}Br)溴甲基胆固醇，并研究了它在小白鼠体内的组织分布。选择^{82}Br($t_{1/2}$ = 35.4h)作为标记核素是由于它易处理和实验结果适用于^{77}Br[18]。

2　实验部分

2.1　试剂

胆固醇为 Merck 公司产品，其他试剂均为国产分析纯产品。^{82}Br 由中国原子能科学研究院提供，化学形式是溴化钾。为了除去大部分放射性钾（主要是^{42}K），先将溴化钾溶解于少量水中，然后经过氢型阳离子交换树脂，将流出液收集后用稀氢氧化钠溶液中和至中性，浓缩待用。

2.2　(^{82}Br)溴甲基胆固醇(6-(^{82}Br) Bromomethyl-cholest-5-en-3β-ol) 的合成

(^{82}Br) 溴甲基胆固醇的合成按图 1 所示进行。

（1）Cholestan-3β, 5α, 6β-triol (B) 的合成[18]。40.0g 胆固醇 (A) 在 400mL 88% 的甲酸中的悬浮液，用蒸汽浴加热到所有固状物消失而形成油状物，让其自然冷却到室温。将所有块状物捣碎后，加入 40mL 30% 的过氧化氢。上述混合物在高温下放置 5.5h，其间经常进行搅拌。然后加入 600mL 沸水，让悬浮液自然冷到室温。将形成

---

[*] 本文合作者：张小祥。原发表于《核化学与放射化学》，1991 (1)：49~53。

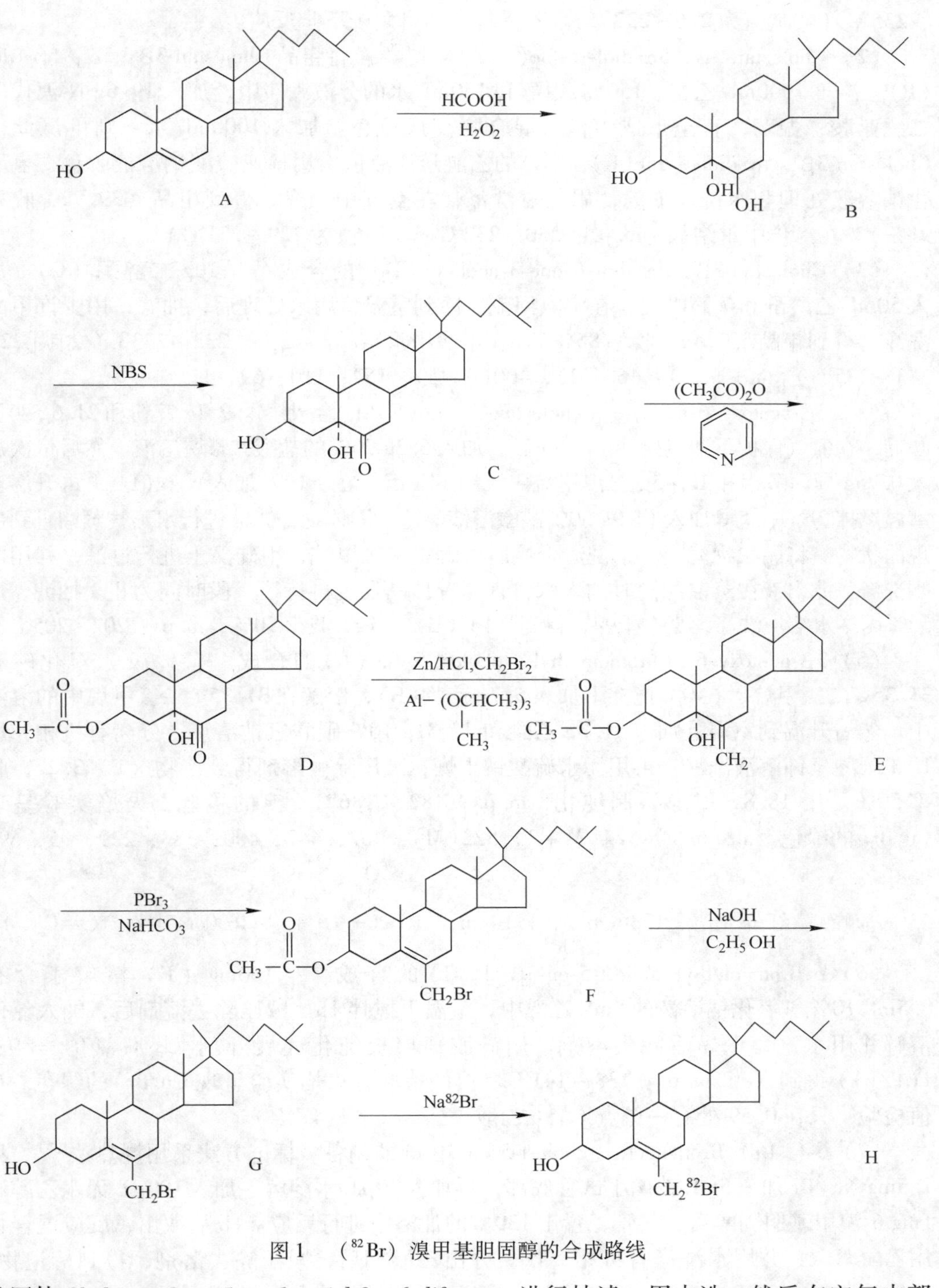

图1 (^{82}Br) 溴甲基胆固醇的合成路线

的固体 Cholestan-3β, 5α, 6β-triol-3, 6-diformate 进行抽滤, 用水洗, 然后在空气中部分干燥后, 溶解于1200mL 的甲醇中, 并加入 40mL 25% 的氢氧化钾溶液, 然后在蒸汽浴中加热 10min, 用盐酸酸化, 再用 400mL 水进行稀释。将得到的悬浮液冷却后, 进行抽滤, 得到粗品 Cholestan-3β, 5α, 6β-triol(B)。水洗后, 在空气中干燥, m.p.: 228

~235℃（文献值为 226~232℃）。不需提纯即可用于下步反应。

（2）Cholestan-3β，5α-diol-6-one（C）的合成[20]。将粗品 Cholestan-3β，5α，6β-triol（B）悬浮于 1000mL 乙醚、150mL 甲醇和 150mL 水的分液漏斗中，加入 19.6g N-溴代丁二酰亚胺，振荡，直至形成溶液，混合物变成橙色。加入 1000mL 水，则可沉淀出 Cholestan-3β，5α-diol-6-one（C）。（C）的乙醚悬浮液依次用稀亚硫酸钠溶液脱色，氢氧化钠溶液处理和水洗。抽滤，用乙醚洗涤，在空气中干燥，得到粗品（39.5g，收率 91%）。在氯仿中重结晶，m.p.：230~232℃（文献值为 229~231℃）。

（3）Cholestan-3β，5α-diol-6-one-3-acetate（D）的合成[21]。10g 二醇酮（C），加入 50mL 乙酸酐和 7 滴吡啶，在沸点回流。冷到室温，用水处理后，抽滤，用少许甲醇洗涤，得到粗品（9.4g，收率 85%）。在甲醇中重结晶。m.p.：232~233℃（文献[22] 233~235℃）。MS(m/e)：460(M)，400(基峰)，154，110，69，41。

（4）3β-acetoxy-6-methylene cholestan-5α-ol（E）的合成[23]。9.7g 锌粉和 21.6g 四氢呋喃，在氮气保护下回流搅拌 10 min 后，加入 5.2g 20% 的盐酸二噁烷溶液。然后依次加入 0.75g 异丙醇铝和 10.45g 二溴甲烷。大约 14h 后，在 -10℃ 加入 6.3g（D），再升温至室温搅拌 2h。在 5℃ 加入 14.9g 50% 醋酸溶液。滤去固体物，滤液进行水蒸气蒸馏直到出现晶状物。将晶状物滤出、水洗、干燥后溶解于二氯甲烷，用硅藻土进行过滤，并用丙酮洗涤。滤出液在蒸馏的同时加入水，直到有结晶发生再持续一段时间为止。抽滤，用 50% 丙酮水溶液洗涤，真空干燥，得到产物（E）5.7g，收率 90%，m.p.：202~205℃。

（5）3β-acetoxy-6-(bromomethyl)cholest-5-ene（F）的合成[23]。5.7g 三溴化磷和 31.35g 二氯甲烷，在 0℃ 搅拌下加入 5.7g 产物（E），溶解在 31.35g 二氯甲烷中的溶液中。然后升温到室温，加入 51.3g 乙醚和 47.31g 10% 碳酸氢钠溶液。分离有机相，用 10% 碳酸氢钾溶液洗涤，再用无水硫酸钠干燥，减压除去溶剂得到产物（F）3.2g，收率 50%。用 18.8g 甲醇研制纯化，m.p.：182~186℃。溴的检验：火焰实验呈现 Cu_2Br_2 的绿色。MS(m/e)：524(M+2)，522(M)，462，442，400，383，229，95，69，43（基峰）。红外光谱，1736cm^{-1}，1714cm^{-1}（—C(=O)—O—），1630cm^{-1}（C=C）。

（6）6-Bromomethyl-cholest-5-en-3β-ol（G）的合成[24]。100mg（F）溶解于含有 0.5mL 10% 氢氧化钠溶液的 5mL 乙醇中，室温下搅拌 1h。减压除去溶剂后，加入氯仿溶解并用水洗，干燥后除去溶剂。用硅胶薄层板纯化（展开剂，苯：氯仿 = 9：1(V/V)）得到（G）。m.p.：138~141℃。溴的检验：火焰实验呈现 Cu_2Br_2 的绿色。R_f 值检验：只在 0.59 处有一斑点（苯：乙酸乙酯 = 6：1(V/V)）。

（7）6-(^{82}Br) Bromomethyl-cholest-5-en-3β-ol 的制备。标记方法采用湿热法[25]。将 1.5mg Na^{82}Br 加入 5mL 玻璃小试管底部，再加入 20μL 水溶解。加入 100μL 无水乙醇和 1mg 6-溴甲基胆固醇后，封管，置于 130℃ 的油浴中进行反应。1h 后取出，反应混合物用乙醚萃取，并用水洗涤有机相 2 次。分出乙醚相后，减压除去溶剂，得到标记物，比度约为 18.13MBq/mg（490μCi/mg）。

2.3 动物实验

将标记物与 1.6% tween-80、10% C_2H_5OH 的生理盐水溶液配制成澄清的注射针剂。将9只小白鼠（每只重20~22g）分成3组，每组为3只。每只小白鼠从尾静脉注射约0.74MBq（20μCi）针剂。注射后的第一天、第三天和第五天分别解剖一组小白鼠。在这期间让小白鼠自由进食、进水。解剖所取脏器为：肾上腺、肝、肾、脾、肺和血，随后进行称重和测量。

3 结果和讨论

6-(^{82}Br)溴甲基胆固醇在小白鼠体内的组织分布列入表1。动物实验采集了3个时间的数据。结果表明6-(^{82}Br)溴甲基胆固醇对肾上腺组织有明显的亲和力。从注射后的第一天到第五天，肾上腺组织的摄取值依次增大，第五天时摄取达到最大值为92.08% dose/g（平均值）。除了肾上腺以外，6-(^{82}Br)溴甲基胆固醇在肺中也有较高的浓集，但是随着时间呈减小的趋势，从注射后第一天的11.42% dose/g 降到第五天的 6.12% dose/g。这与 6-(^{82}Br)溴乙基去甲胆固醇类似[19]。应指出的是 6-(^{82}Br)溴甲基胆固醇亲肾上腺组织的能力虽然强于 19-(^{131}I)碘胆固醇[5]，但是弱于6-(^{131}I)碘胆固醇[26]和6-(^{131}I)碘甲基去甲胆固醇[5]。

放射性在肾上腺及其邻近组织（肝、肾、脾等）中摄取值的相对比对于得到清晰的肾上腺显影是至关重要的。图2表示肾上腺摄取值与某些非靶器官摄取值之比随注射后时间的变化。肾上腺/肝比从注射后第一天的4.20增加到第五天的48.72。这个比值与其他几种常见的碘代胆固醇[5,26]或溴代胆固醇[15,18]相比都较低。从比值的变化趋势来看，类似于6-(^{82}Br)溴胆固醇[16]。

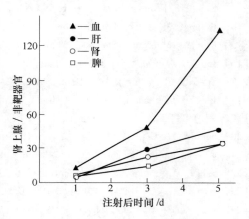

图2 肾上腺与非靶器官摄取值之比
随注射后时间的变化

表1 6-(^{82}Br)溴甲基胆固醇在小白鼠体内的组织分布 （% dose/g[①]）

脏器	注射后时间		
	一天	三天	五天
肾上腺	38.92 (31.05~48.20)	74.23 (66.40~80.95)	92.08 (79.24~105.01)
肝	9.26 (7.24~12.35)	2.52 (2.09~3.16)	1.89 (1.62~2.06)
肾	6.21 (4.36~7.29)	3.25 (2.65~3.96)	2.71 (2.23~3.11)
脾	5.72 (5.34~6.21)	4.72 (3.94~5.92)	2.71 (2.13~3.65)
肺	11.42 (7.90~14.01)	8.00 (6.91~9.02)	6.12 (5.13~6.92)
血	3.14 (2.96~3.30)	1.51 (1.21~1.95)	0.68 (0.47~0.89)

注：表中数据为3只小白鼠的平均值，括号中数据表示最小值和最大值区间。
① 为单位组织剂量百分数。

参 考 文 献

[1] Nagai, T. et al., J. Nucl. Med., 9, 576(1968).

[2] Nagai, T. et al., J. Nucl. Med., 11, 217(1970).
[3] Counsell, R. E. et al., Steroids, 16, 317(1970).
[4] Kojima, M. et al., J. Chem. Soc. Chem. Comm., 47(1975).
[5] Kojima, M. et al., J. Nucl. Med., 16, 666(1975).
[6] 王世真等，放射性同位素与射线应用展览会资料(1972).
[7] 王世真等，中国医学科学院学报，1, 29(1979).
[8] 刘伯里等，北京师范大学学报，2, 63(1979).
[9] 刘伯里等，核化学与放射化学，2, 107(1980).
[10] 刘伯里，刘正浩，核化学与放射化学，4, 48(1982).
[11] Ito, T. et al., Steroids, 41, 131(1983).
[12] Riley, A. L. M., J. Lab. Comp. Radiopharm., 16, 28(1979).
[13] Knap, F. F. and Ambrose, K. R., J Nucl. Med., 18, 600(1977); 21, 251(1980).
[14] Kojima, M. et al., Radioisotopes, 25, 222(1976).
[15] Kojima, M. et al., Steroids, 29, 443(1977).
[16] Liu, B. L. et al., Int. J. Appl. Radiat. Isot., 36, 561(1985).
[17] Stöcklin, G., Int. J. Appl. Radiat. Isot., 28, 131(1977).
[18] Liu, B. L. et. al., J. Lab. Comp. Radiopharm., 19, 1089(1982).
[19] 刘伯里等，核技术，12, 54(1989).
[20] Schultz, R. G., J. Org. Chem., 24, 1955(1959).
[21] Fieser, F. and Rajagopalan, S., J. Am. Chem. Soc., 17, 3938(1949).
[22] Salch, A. A. et. al., Z. Naturforsch, 35b, 102(1980).
[23] Nysted, L. N., U. S, P., 3960904; U. S. P. 4001220.
[24] Kobayashi, T. et al., Steroids, 39, 585(1982).
[25] 刘伯里等，核化学与放射化学，7, 29(1985).
[26] Jixiao, M. and Ruisen, Z., Chinese Med. J., 92, 237(1979).

New Cholesterol Analogue as Adrenal Imaging Agent——synthesis and Tissue distribution of Bromine-82 Labeled 6-Bromomethyl Cholesterol

Zhang Xiaoxiang   Liu Boli

(The Radio and Radiation Chemistry Division, Beijing Normal University, Beijing, 100875, China)

Abstract   6-Bromomethyl cholesterol(CL-6-MeBr) is synthesized from cholesterol by six step reactions. Bromine-82 labeled 6-bromomethyl cholesterol (CL-6-MeBr-82) is prepared by means of hydrothermal melting method. Tissue distribution of CL-6-MeBr-82 in mice is determined. The adrenal uptake of CL-6-MeBr-82 reaches a maximum of 92.08% dose/g at the 5th day after injection, and the adrenal to liver ratio increases from 4.2 at 1 day to 48.72 at five days. CL-6-MeBr-82 shows less affinity for the adrenal gland than CL-6-Br-82.

Key words   adrenal imaging agent, 6-(^{82}Br)bromomethyl cholesterol, tissue distribution

Novel Anilinophthalimide Derivatives as Potential Probes for β-amyloid Plaque in the Brain*

Abstract A group of novel 4,5-dianilinophthalimide derivatives has been synthesized in this study for potential use as β-amyloid (Aβ) plaque probes. Staining of hippocampus tissue sections from Alzheimer's disease (AD) brain with the representative compound 9 indicated selective labeling of it to Aβ plaques. The binding affinity of radioiodinated [^{125}I]9 for AD brain homogenates was 0.21nM (K_d), and of other derivatives ranged from 0.9 to 19.7nM, except for N-methyl-4,5-dianilinophthalimide (K_i > 1000nM). [^{125}I]9 possessed the optimal lipophilicity with Log P value of 2.16, and its in vivo biodistribution in normal mice exhibited excellent initial brain uptake (5.16% ID/g at 2min after injection) and a fast washout rate (0.56% ID/g at 60min). The encouraging results suggest that this novel derivative of [^{123}I]9 may have potential as an in vivo SPECT probe for detecting amyloid plaques in the brain.

Key words Alzheimer's disease, β-amyloid plaque, imaging probe, binding affinity, brain uptake

1 Introduction

One of the key pathological features in the Alzheimer's disease (AD) brain is the presence of abundant senile plaques (SP)[1]. The plaques are composed of β-sheet-containing fibrils formed by the aggregation of short β-amyloid (Aβ) peptides[2,3]. As Aβs are at the centre of pathogenesis of AD, non-invasive detection of them using imaging probes will be a powerful strategy for early diagnosis of dementia[4].

Aβ plaque imaging probes applicable for positron emission tomography (PET) or single photon emission computed tomography (SPECT) require two essential criteria including the high binding affinity to Aβ plaques and the adequate permeability of blood-brain barrier (BBB)[5]. The consideration of in vitro binding affinity as the first prerequisite to image plaques, the plaque-binding dyes of Congo red (CR) and Thioflavin T (ThT) were consequently designed to develop several types of core structures for Aβ plaque imaging probes[6-8]. Among them, [^{11}C]PIB (K_i = 4.3nM)[9], [^{11}C]SB-13 (K_i = 1.2nM)[10] and [^{123}I]IMPY (K_i = 15.0nM)[11] have been clinically utilized to in vivo detect SPs in the living AD brain[12-14] (Fig. 1). It is worth noting that all these three molecules share similarity in core structure with the presence of two phenyl rings (aromatic section) combined with a variety of linkers, such as a thiazole ring in [^{11}C]PIB and an ethylene group in [^{11}C]SB-13. It is evident that these two phenyl rings

---

* Copartner: Duan Xinhong, Qiao Jinping, Yang Yang, Cui Mengchao, Zhou Jiangning. Reprinted from *Bioorganic & Medicinal Chemistry*, 2010, 18: 1337-1343.

should be essential for a probe's affinity to Aβ plaques; on the other hand, this linker may serve to a great extent as an affinity-modifier based on its flexible spatial arrangement. For example, it contributes to the probe molecule as a planar form entering the Aβ plaque's binding pockets where binding will take place through the π-π interaction (intermolecular overlapping of p-orbitals) between the probe molecule and the aromatic side chains of Aβ such as Phe 19[8,15-17]. This structural information may provide the basis for a rational probe design effort centered on identifying the core structures, which could be utilized to make potential Aβ plaque probes.

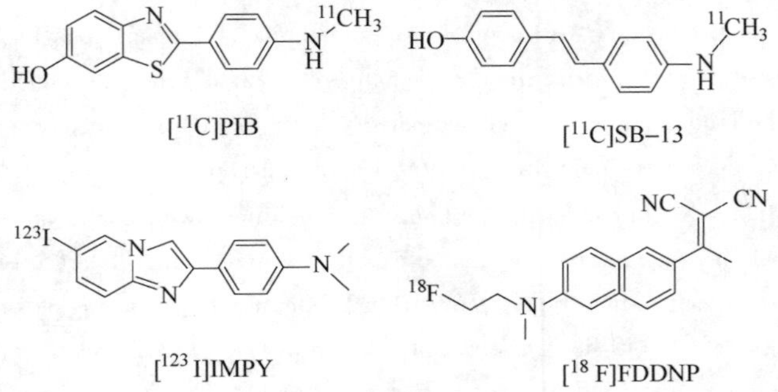

Fig. 1 Chemical structures of Aβ plaque imaging probes for clinical use

In fact, research has also been underway to target new core structures based on this rationale for Aβ plaque imaging probes. Several potent aromatic-based β-sheet inhibitors such as Naproxen[18,19], Flavone[20] and Curcumin[21], were currently reported as novel core structures of Aβ plaque probes with their derivatives exerting good Aβ affinities[22-25]. Naproxen[23], for example, based on which inhibits and binds plaque (K_i = 5.7nM), was designed through substitutional-group modifications to be a PET radioligand of [^{18}F]FDDNP (K_i = 2.7nM) for clinical use[26] (Fig. 1).

4,5-Dianilinophthalimide (DAPH), a three-phenyl-ring-bearing compound, is recently identified as a promising lead β-sheet inhibitor because of its strong reversal of β-sheet formation (IC_{50} ~ 15μM), as well as ability to destabilize preformed fibrils[27]. Thus, such anti-aggregation action was believed to involve specific Aβ plaque-binding[15,16]. This prompted us to apply it as a novel core structure for Aβ plaque imaging probes. As a matter of fact, potential probes designed for detecting Aβ plaques in vivo must also cross the intact BBB. It is generally acceptable that an Aβ plaque imaging molecule with sufficient brain uptake has to be relatively small (mol wt < 600), lipophilic (log$P$ = 2-3) and neutral[8,28]. However, DAPH was not expected to efficiently enter the brain because of its ionizable group of the acidic phthalimide NH. We thus hypothesized that its N-methylated analogue (MDAPH) should be better BBB-permeable (as shown in Scheme 1). To develop better and more versatile core structure suitable for providing Aβ plaque probes with high affinity and permeability of BBB, we proceeded to design a novel MAPH core by removal of an aniline group from MDAPH (Scheme 1). There are several reasons for such an optimized modification as Aβ-plaque probe's core structure. Firstly, this remaining backbone with two phenyl rings (aromatic section) retains MDAPH's essential structural elements which seem to be responsible for the plaque-binding affinity. Secondly, such core has only a NH group of linker, whose conformational flexibility may obtain the best fit of the probe molecules in the binding pocket so as to maximize the binding affinity. Thirdly, it contains decreased molecular weight, which may increase the extent of penetration of brain tissue. In addition, its derivatives can be prepared via a simplified coupling reaction scheme, which facilitates introduction of a wide variety of substituents including—CH_3, —Br, —I, —OH, —OCH_3, and the chelate used for further radiolabeling.

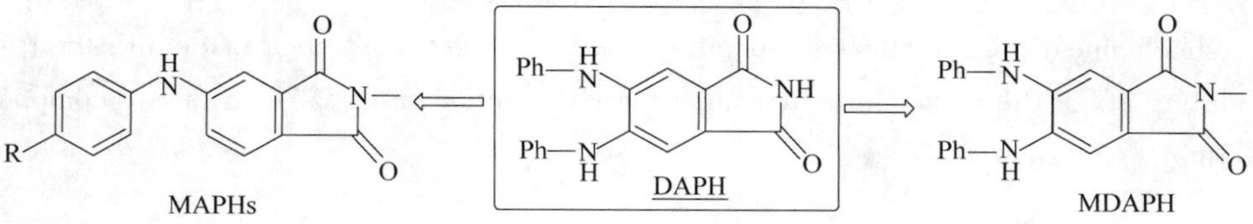

Scheme 1 Design considerations of DAPH derivatives

Because none of Aβ plaque imaging probes has been reported so far based on this type of core structure, we initially utilized the derivative of N-methyl-4-(4-iodoanilino) phthalimide (MAPH-I) as a representative by performing a tissue staining assay to test whether the compound is capable of binding specifically to Aβ plaques. Once it was confirmed to possess the plaque-binding potential, we then proceeded to prepare other derivatives and the radiolabeled tracer for further biological evaluation.

In comparison with PET, SPECT is a more widely accessible and cost-effective technique regardless of its limitations. Consequently, we have a strong desire to develop SPECT probes labeled with ^{123}I ($T_{1/2}$, 13h, 159keV) or ^{99m}Tc ($T_{1/2}$, 6h, 140keV). Reported herein are the synthesis and biological evaluations of a novel class of DAPH-based compounds and especially, the radioiodinated derivative as a prospective SPECT tracer for imaging amyloid plaques in the brain.

2 Results and Discussion

2.1 Chemistry

MDAPH was achieved by the methylation of DAPH(Scheme 2), and MAPHs(**6-10**) containing N-methyl-4-anilinophthalimide core was synthesized through a coupling reaction between **4** and the suitable commercially available p-substituted iodobenzenes(**5a-e**). Intermediate **4** was prepared in three steps as outlined in Scheme 2. Phthalimide 1 was first N-methylated to form N-methylphthalimide **2**, and this compound was then treated by nitration to yield N-methyl-4-nitrophthalimide 3. Finally, reduction of **3** with $SnCl_2$ afforded the compound of N-methyl-4-aminophthalimide **4**. The coupling was performed by reacting intermediate **4** with **5a-e** in DMF at 130℃ in the presence of Cu powder and K_2CO_3 (Ref. [29] with some modifications, Scheme

2). In this reaction, the desired compounds **6-10** were achieved in yields ranging from 7.8% to 27.8%. However, the coupling of p-OCH$_3$-substituted iodobenzenes with **4** yielded appreciable amounts of disubstituted compounds, the desired products were formed only in negligible amounts(7.8%). It is probably because this electron-donating group OCH$_3$ might increase the nucleofugacity of p-position iodo group, and the unexpected disubstituted products were greatly achieved. The free phenol derivative, **11**, was obtained from the corresponding **10** by a demethylation reaction using BBr$_3$ (Scheme 2). The tributyltin derivative **12** was prepared from the bromo precursor **8** using an exchange reaction catalyzed by Pd(0) (10.8% yield), and its radioiodination was successfully carried out by the standard iododestannylation reaction, using hydrogen peroxide as the oxidant (Scheme 3). The final HPLC-purified [^{125}I]**9** showed greater than 98% radiochemical purity with high specific activities ($\approx$ 2000Ci/mmol). Identification of the peak eluting at 7.30min was done by comparing its retention time (t_R) with that of compound **9** at 7.05min (Fig. 2).

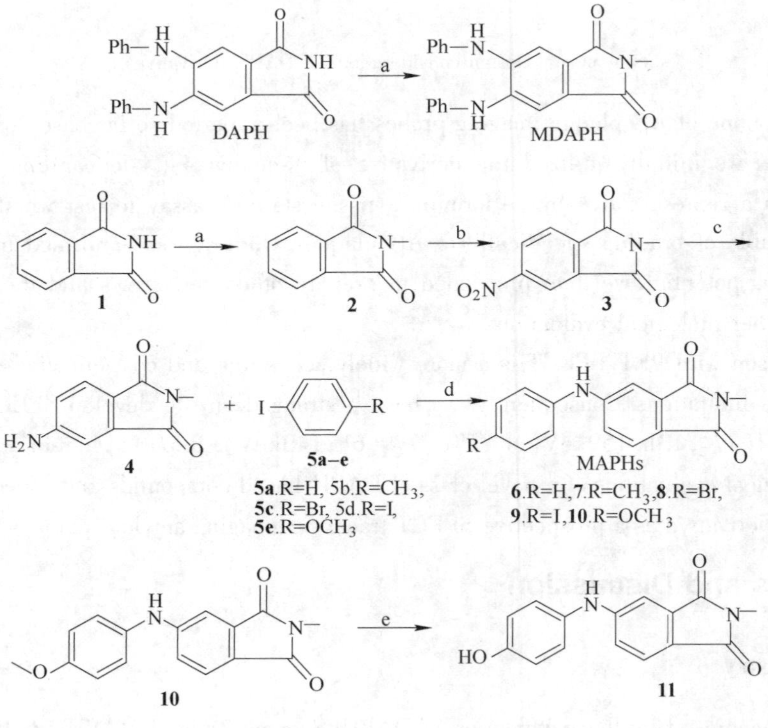

Scheme 2 Reagents and conditions
a—CH$_3$I, DMF, K$_2$CO$_3$, rt; b—H$_2$SO$_4$, HNO$_3$, 65℃; c—SnCl$_2$, HCl, H$_2$O, rt;
d—K$_2$CO$_3$, Cu, DMF, 130℃; e—BBr$_3$, CH$_2$Cl$_2$, -70℃

2.2 Staining on AD human brain sections

To assess whether these novel DAPH-based derivatives interacted with Aβ plaques in vitro, the tissue staining on AD human brain sections (Female, 72 years old, hippocampus) was firstly uti-

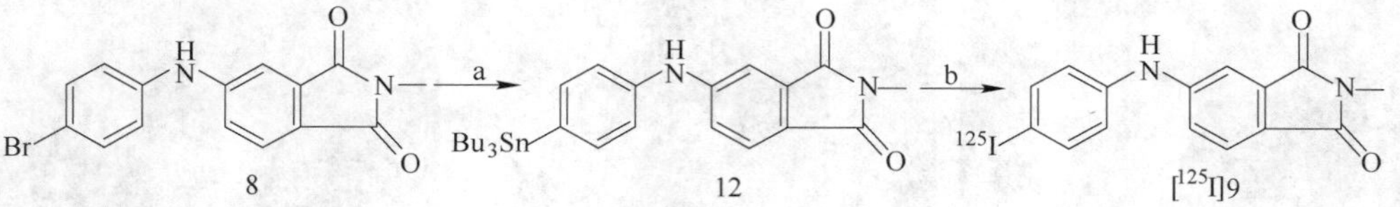

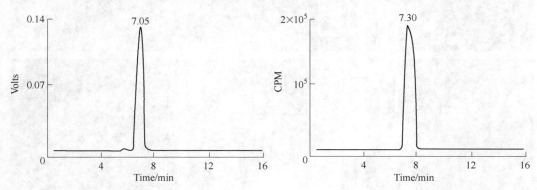

Scheme 3 Reagents and conditions

a—$(Bu_3Sn)_2$, $Pd(PPh_3)_4$, toluene, 110℃; b—1—$[^{125}I]NaI$, H_2O_2, HCl, rt; 2—NaOH

Fig. 2 HPLC profiles of compound 9(top) and $[^{125}I]9$(bottom)
(HPLC condition: Agilent TC-C18 column(analytical 4.6×150mm) CH_3OH/H_2O =8/2, 1mL/min, 360nm, t_R = (UV)7.05min, (γ)7.30min)

lized to test their specificity for Aβ plaques. As shown in Fig. 3, the pretreated brain section indeed showed lack of autofluorescence(Fig. 3, A1, A2 or A3) while a positive staining can be observed when stained with the representative compound **9**. The staining with **9** was localized in Aβ plaques in hippocampus tissue sections(Fig. 3, B3), and moreover, relatively lower background staining was observed. The identity of the plaques stained with **9** was confirmed by staining serial sections with Thioflavin S(ThS) to Aβ (Fig. 3, C1, C2 and C3). In a qualitative way, the compound demonstrated its specific affinity for Aβ plaques, even in the complex milieu of human. The choice of compounds **7** and **8** for additional studies showed nearly the same staining pattern of SPs as **9**(data not shown).

2.3 In vitro binding studies

Encouraged by above promising results, we then quantified the binding affinities of this class of compounds to Aβ plaques by an in vitro binding assay. The binding study of $[^{125}I]9$ to homogenates of cortex tissue from postmortem AD brain(Female, 91 years old, Superior Temporalis Gyrus) was carried out. This ligand displayed a saturable binding, and Scatchard analysis showed highly binding affinity to AD brain homogenates with a K_d of(0.21±0.07)nM(Fig. 4). Binding affinities(K_i) of MAPHs(compounds **6-11**), DAPH and MDAPH for AD brain homogenates in competition with $[^{125}I]9$ were also evaluated(Table 1). All the MAPHs demonstrated excellent affinity to Aβ plaques with K_i values ranging from 0.9 to 19.7nM. The derivative of N-

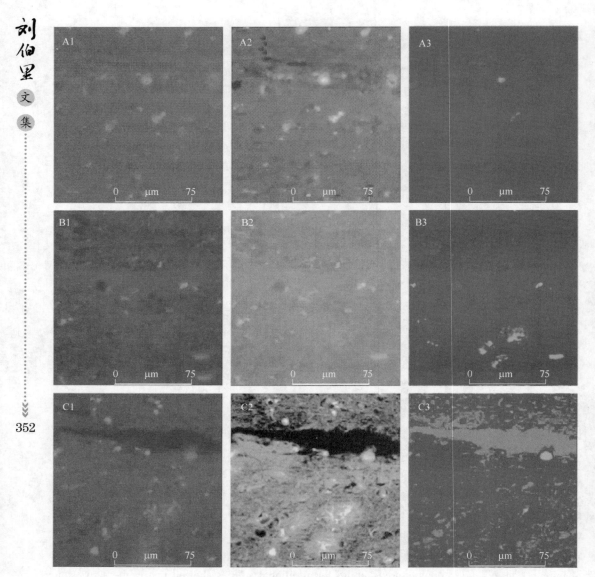

Fig. 3 Fluorescence micrographs of AD brain sections (6μm thick) blank (A1, A2 and A3), stained with **9** (B1, B2 and B3) and Thioflavin S (C1, C2 and C3)

(The images were obtained under similar incubatin conditions using a concentration of 50μM. The pretreated brain section shows lack of autofluorescence on any channel of RFP(A1), GFP(A2) and DAPI(A3) while a positive staining is observed when stained with compound **9** at the DAPI(B3) channel or ThS at the channel of RFP(C1), GFP(C2) and DAPI(C3). SPs labeled with compound **9** are identical to staining in a serial section with ThS. Bar = 75μm)

methyl-4-anilinophthalimide (**6**) without a substituent at 4-position on the aniline ring showed a desirable affinity (7.6nM). When introducing a higher hydrophobic substituent such as CH_3 (**7**, $K_i = 4.1$nM), Br(**8**, $K_i = 3.8$nM) or I group (**9**, $K_i = 0.9$nM), a comparable increase in binding affinity was observed. However, the presence of a lower hydrophobic group such as OCH_3 (**10**, $K_i = 16.3$nM) or OH (**11**, $K_i = 19.7$nM) slightly decreased the affinity. The relative binding affinity order for this class of derivatives was I > Br > CH$_3$ > H > OCH$_3$ > OH. In this study, it was shown that binding affinity of the derivatives to Aβ plaques was related to their substituent's hydrophobicity. More importantly, the good binding affinity of [^{125}I]MAPH-I towards Aβ plaques indicates that there is obviously a large degree of bulk tolerance in the 4-substituent moiety for in vitro binding to the Aβ plaques. As a result, this [^{125}I]9 could serve as a model compound for ^{123}I-labeled SPECT imaging studies.

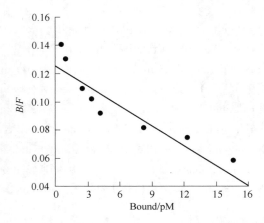

Fig. 4 Scatchard plot of the binding of [^{125}I]9 to the gray matter homogenates of AD brain
(This ligand displayed highly binding with a K_d of (0.21 ±0.07)nM)

Table 1 Inhibition constants(K_i) of DAPHs[①] and of PIB and IMPY

Compound	K_i/nM[①]	Compound	K_i/nM[①]
6	7.6 ±0.8	DAPH	>1000
7	4.1 ±0.4	MDAPH	>1000
8	3.8 ±0.9	CR	>1000
9	0.9 ±0.3	ThT	>1000
10	16.3 ±4.6	PIB	>1000
11	19.7 ±5.4	IMPY	>1000

① Measured using amyloid plaques in AD brain homogenates and [^{125}I]9 as the ligand; each value represents the mean ± SEM of three experiments.

Surprisingly, despite definite interaction with Aβ fibrils[27], DAPH (and its derivative MDAPH) showed in this study no specific binding signal for amyloid plaques (both K_i values > 1000nM). Like [^{11}C]PIB or [^{11}C]SB-13, [^{125}I]9 is a long, thin and flat molecule[8]. Comparatively, DAPH has a large asymmetric propeller-shape conformation[30]. Therefore, an possibility may be that DAPH did not match the tracer's binding pockets where a thin and flat conformation should be capable of entering to bind (through π-π interaction)[8,15-17], and therefore result in lack of binding competition. Indeed, recent studies demonstrated that DAPH interacts with the Aβ42 fibrils by selectively targeting their intermolecular contacts (the Phe stacking), while leaving particular intramolecular contacts intact[31]. That is to say, such a characteristic on Aβ fibrils seemed different from that of MAPHs on amyloid plaques.

In addition, the competition of CR and ThT to [^{125}I]9 binding on AD brain homogenates was examined with high K_i values (>1000nM), indicating no binding competition. This finding suggests that MAPHs did not share a same binding site with CR or ThT.

2.4 Biodistribution study

One of the key prerequisites for an in vivo Aβ plaque imaging probe is its ability to cross the intact BBB after an iv injection. Actually, a relatively high (optimal) lipophilicity should be critical for a probe's initial brain permeability in addition to its neutral and small properties[32]. The lipophilicity of [^{125}I]9 was 2.16 (measured by a partition between 1-octanol and pH 7.4 phosphate buffer), with a calculated lg P value (c lg P) of 3.73 (calculated using ACD software package version 8.0)[8], indicating a potentially BBB-permeable ability. A biodistribution study in normal mice after an iv injection showed that [^{125}I]9 exhibited indeed an excellent brain uptake of 5.16% ID/g at 2min and a fast washout rate of 0.59% ID/g at 60min (Table 2). Moreover, the blood levels of the compound were relatively low throughout the time course measured [(2.47-2.78)% ID/g].

Table 2 Biodistribution in normal mice after iv injection of [^{125}I]9 (% ID/g, avg of 4 mice ± SD) and its partition coefficient

Organ	2min	10min	30min	60min
	[^{125}I]9 (lg$P$ = 2.16 ± 0.03)			
Blood	2.47 ±0.30	2.74 ±0.39	2.69 ±0.06	2.78 ±0.38
Heart	7.40 ±1.34	3.41 ±0.14	2.83 ±0.31	2.24 ±0.14
Lung	6.77 ± 1.51	6.35 ± 1.04	3.47 ± 0.14	2.82 ± 0.43
Kidney	9.81 ±1.48	6.55 ±0.22	3.59 ±0.22	2.68 ±0.21
Spleen	5.64 ±0.57	6.87 ±0.37	3.70 ±0.36	2.92 ±0.53
Liver	11.98 ±1.28	14.38 ±0.27	5.84 ±0.63	3.37 ±0.24
Brain	5.16 ±1.36	3.41 ±0.15	1.16 ±0.14	0.59 ±0.05

A good initial brain uptake combined with a rapid washout in normal mouse brain (no Aβ plaques for extra binding of the probes) is a highly desirable property for Aβ plaque-targeting probes. At present, [^{123}I]IMPY is the most promising SPECT tracer for clinically detecting amyloid plaques because of its desirable properties: high binding affinity for amyloid plaques (K_i = 15nM) and fast kinetics of high initial brain uptake (2.88% ID/organ at 2min) and rapid washout (0.21% ID/organ at 60min) in normal brain. Similar to the radioiodinated IMPY, [^{125}I]9 also showed high in vitro binding affinity, desirable kinetic properties of high brain uptake and rapid washout in the non-Aβ plaque-containing areas. Therefore, [^{123}I]9 is most likely to be successful as a probe for Aβ plaques in the brain. A more detailed biodistribution study using transgenic mice containing an excess amount of Aβ aggregate deposition in the brain is currently under way.

3 Conclusion

In summary, MAPHs represent a novel class of probes for Aβ plaques. They bound efficiently to Aβ plaques and showed high in vitro binding affinity with K_i in the nM range. In particular, [^{125}I]**9**, because of specific in vitro labeling of amyloid plaques, and together with its excellent BBB permeability and fast washout, is a promising candidate probe for SPECT imaging in the brain.

4 Experimental

4.1 General information

Unless otherwise noted, starting materials were purchased from Sigma-Aldrich and Alfa Asia organics and were used without further purification. [^{125}I]NaI(2200Ci/mmol) was obtained from PerkinElmer Life and Analytical Sciences, Japan. The AD human brain homogenates and paraffin brain sections were obtained through Hefei National Laboratory for Physical Sciences at Microscale and School of Life Sciences, University of Science & Technology of China, in cooperation with the Netherlands Brain Bank(Coordinator, Dr. I. Huitinga). ^{1}H NMR spectra were obtained on Avance-500 Bruker(500MHz) spectrometer with TMS as an internal standard. Mass spectra were obtained on a GC 2000/TRACE^{TM}(EI-MS) and LCMS-2010(ESI-MS). Elemental analyses were obtained on a Elementar Vario EI.

4.1.1 N-Methylphthalimide(2)

Phthalimide **1**(147mg, 1mmol) was added to a mixture of DMF(15mL) and K_2CO_3(13.8mg), then a solution of CH_3I(213mg, 1.5mmol) in DMF was added dropwise to the resulting solution. The solvent was removed after the mixture was stirred for 5h at room temperature. The residue was purified by recrystallization with methanol to give **2** as a colorless needles(79mg, 49.2%), mp 132-133℃.

4.1.2 N-Methyl-4-nitrophthalimide(3)

To a solution of **2**(161mg, 1mmol) dissolved in H_2SO_4(0.4mL), 0.2mL HNO_3 was added dropwise over 5 min. After 1h, the reaction mixture was poured into crushed ice and then filtrated. Compound **3**(183mg, 88.9% yield) was obtained as a pale yellow needle after recrystallization with methanol, mp 175-176℃.

4.1.3 N-Methyl-4-aminophthalimide(4)

Compound 3(144mg, 0.7mmol) was added hydrochloric acid(2mL) and followed by $SnCl_2 \cdot 2H_2O$(400mg). The mixture was stirred at room temperature until the white solid was largely formed. The resulting residue was washed with water and purified by recrystallization with ethanol to afford compound **4**(90.2mg, 73.2%) as a yellow solid, mp 242-243℃.

4.1.4 General procedure for preparing DAPHs(6-10)

The suitable **4** (176mg, 1mmol), p-substituted-iodobenzene **5a-e** (1.1mmol), Cu powder (64mg, 1mmol) and K_2CO_3 (138mg, 1mmol) in 12mL DMF was heated to 130℃ for about 6h. The mixture was filtered and organic layer was concentrated under vacuum. Crude materials were purified by column chromatography on silica gel (petroleum ether/AcOEt, 80/20).

4.1.5 N-Methyl-4-anilinophthalimide(6)

Yield 13.6%, mp 171-172℃. ^{1}H NMR(500MHz, CDCl$_3$, δppm) 7.68(1H, d, J = 8.2Hz), 7.42 (1H, s), 7.39(2H, d, J = 7.8Hz), 7.22(2H, d, J = 7.7Hz), 7.17(1H, t, J = 7.4Hz), 7.13(1H, d, J = 8.2Hz), 6.27(1H, b, NH), 3.16(3H, s). EI-MS: m/z(%) = 252 [M$^+$]. Anal. Calcd for $C_{15}H_{12}N_2O_2$: C, 71.42; H, 4.79; N, 11.10. Found: C, 71.41; H, 4.86; N, 10.99.

4.1.6 N-Methyl-4-(4-methylanilino)phthalimide(7)

Yield 17.2%, mp 195-196℃. ^{1}H NMR(500MHz, CDCl$_3$, δppm) 7.65(1H, d, J = 8.2Hz), 7.31(1H, s), 7.21(2H, d, J = 8.0Hz), 7.12(2H, d, J = 7.6Hz), 7.06(1H, d, J = 7.6Hz), 3.15(3H, s), 2.37(3H, s). EI-MS: m/z(%) = 266[M$^+$]. Anal. Calcd for $C_{16}H_{14}N_2O_2$: C, 72.16; H, 5.30; N, 10.52. Found: C, 72.12; H, 5.43; N, 10.49.

4.1.7 N-Methyl-4-(4-bromoanilino)phthalimide(8)

Yield 27.8%, mp 204-205℃. ^{1}H NMR(500MHz, CDCl$_3$, δppm) 7.69(1H, d, J = 8.1Hz), 7.50(2H, d, J = 8.3Hz), 7.37(1H, s), 7.13(1H, d, J = 8.3Hz), 7.10(2H, d, J = 8.4Hz), 6.21(1H, b, NH), 3.17(3H, s). EI-MS: m/z(%) = 332 [MH$^+$]. Anal. Calcd for $C_{15}H_{12}BrN_2O_2$: C, 54.40; H, 3.35; N, 8.46. Found: C, 54.19; H, 3.26; N, 8.89.

4.1.8 N-Methyl-4-(4-iodoanilino)phthalimide(9)

Yield 25.5%, mp 212-213℃. ^{1}H NMR(500MHz, CDCl$_3$, δppm) 7.70(1H, d, J = 7.9Hz), 7.69(2H, d, J = 8.3Hz), 7.38(1H, s), 7.14(1H, d, J = 8.1Hz), 6.98(2H, d, J = 8.4Hz), 6.21(1H, b, NH), 3.17(3H, s). EI-MS: m/z(%) = 378 [MH$^+$]. Anal. Calcd for $C_{15}H_{11}IN_2O_2$: C, 47.64; H, 2.93; N, 7.41. Found: C, 47.70; H, 2.57; N, 6.93.

4.1.9 N-Methyl-4-(4-methoxyanilino)phthalimide(10)

Yield 7.8%, mp 173-174℃. ^{1}H NMR(500MHz, CDCl$_3$, δppm) 7.63(1H, d, J = 8.2Hz), 7.20 (1H, s), 7.17(2H, d, J = 8.6Hz), 6.95(3H, d, J = 8.6Hz), 6.07(1H, b, NH), 3.86(3H, s), 3.14(3H, s). EI-MS: m/z(%) = 282 [M$^+$]. Anal. Calcd. for $C_{16}H_{14}N_2O_3$: C, 68.07; H, 5.00; N, 9.92. Found: C, 68.09; H, 5.12; N 9.78.

4.1.10 N-Methyl-4-(4-hydroxyanilino)phthalimide(11)

To a solution of **10**(341mg, 1.21mmol) in CH_2Cl_2(15mL) cooled to −70℃, BBr_3(12mL, 1M solution in CH_2Cl_2) was added dropwise. The reaction mixture was then allowed to warm to room temperature and stirred for 12h. The precipitate was formed after neutralization with 1N NaOH, then filtered and washed with water. The compound **11** was obtained as a yellow solid by recrystallized with methanol (220mg, 68.1% yield), mp 224-225℃. ^{1}H NMR (500MHz, CDCl$_3$, δppm) 7.61(1H, d, J = 8.2Hz), 7.17(1H, d, J = 8.0Hz), 7.10(2H, d, J = 8.6Hz), 6.94(1H, d, J = 8.0Hz), 6.87(2H, d, J = 8.6Hz), 3.12(3H, s). EI-MS: m/z(%) = 268 [M$^+$]. Anal. Calcd for $C_{15}H_{12}N_2O_3$: C, 67.16; H, 4.51; N, 10.44. Found: C, 67.41; H, 4.81; N, 10.47.

4.1.11 N-Methyl-4,5-dianilinophthalimide(MDAPH)

The same reaction described above to prepare **2** was used, and 5.9mg of MDAPH was obtained in a 85.8% yield from DAPH (6.6mg, 0.02mmol), mp 192-193℃. ^{1}H NMR (500MHz, CDCl$_3$, δppm)7.64(2H,s),7.37(4H, t, J = 7.7Hz),7.07(6H,s),5.89(2H,b,NH),3.14 (3H,s). ESI-MS: $m/z(\%)$ = 344 [MH$^+$]. Anal. Calcd for C$_{21}$H$_{17}$N$_3$O$_2$: C, 73.45; H, 4.99; N, 12.24. Found: C, 73.42; H, 5.08; N, 11.99.

4.1.12 N-Methyl-4-(4-tributylstannylanilino)phthalimide(12)

A mixture of **8** (33.1mg, 0.1mmol), bis(tributyltin)(290mg, 0.5mmol), and Pd(Ph$_3$P)$_4$ (12.0mg, 0.01mmol) in toluene(5mL) was stirred at 110℃ overnight. After removing the solvent in vacuo, the crude products were purified by column chromatography(petroleum ether/ AcOEt, 80/20) to give **12** as an orange-colored solid with a yield of 10.8% (5.8mg). ^{1}H NMR (500MHz, CDCl$_3$, δppm)7.67(1H, d, J = 8.2Hz),7.48(2H, d, J = 8.2Hz),7.41(1H, s), 7.18(2H, d, J = 8.1Hz),7.15(1H, d, J = 8.2Hz),6.22(1H, b, NH),3.16(3H, s),1.58 (6H, m),1.37(6H, m),1.09(6H, m),0.93(9H, m). ESI-MS: $m/z(\%)$ = 543 [MH$^+$].

4.2 N-methyl-4-(4-[^{125}I]iodoanilino)phthalimide([^{125}I]9)

One-hundred microliters of H$_2$O$_2$(3%) were added to a mixture of **12** (100μg/100μL EtOH), [^{125}I]NaI(600μCi, specific activity 2200Ci/mmol), and 100μL of 1N HCl in a sealed vial. The reaction was allowed to proceed at room temperature for 10min and terminated by addition of 100 μL saturated NaHSO$_3$ solution. After neutralized with 1N NaOH, the mixture was loaded on a small preconditioned C-18 Sep-pak minicolumn and rinsed sequentially with H$_2$O and 30% EtOH solution. Then, rinsing of this column with 80% EtOH solution obtained the desired product of [^{125}I]9(380μCi). The radiochemical purity was checked by HPLC on a reversed-phase column(Agilent TC-C18 analytical column, 4.6mm × 150mm, CH$_3$OH/H$_2$O 8/2 flow rate 1mL/min). This no-carrier-added product was stored at −20℃ for in vitro binding assay and in vivo biodistribution study.

4.3 Tissue staining

Postmortem brain tissues from an autopsyconfirmed case of AD were used. 6μm-thick tissue was processed for staining according to previously described methods[33]. Firstly, paraffin sections were taken through two 10min washes in xylene, two 10min washes in 100% ethanol, a 5min sequential wash in 95%, 90%, 80% and 70% EtOH/H$_2$O, and a 5min wash under MQ water and then in PBS(0.01M, pH 7.4). Secondly, quenching of autofluorescence was performed by the following method: the sections were blanched in 0.05% potassium permanganate solution for 20min. They were then washed three times in PBS for 5min and treated with 0.2% potassium metabisulfite and 0.2% oxalic acid in PBS for 1min, followed by washing three times in PBS for 5min. Then, quenched tissues were immersed in the solution of 50 μM **9** or Thioflavin S. To ensure the full solubilization of the compound 40% EtOH in PBS was used. Finally, the sections were washed in 75% EtOH/H$_2$O for several seconds and PBS for 15min. Fluorescent sections were viewed using an Olympus IX 71 fluorescence microscope (Olympus, Tokyo) with the following filter sets: DAPI(U-MNUA2, Excitation Spectral Region: Ultraviolet, narrow Excitation Range, Dichromatic Mirror Cut-On Wavelength: 400nm); GFP(U-MWIBA2, Excitation Spectral Region: blue, wide Excitation Range, Dichromatic Mirror Cut-On Wavelength: 505nm); RFP(U-MWIG2, Excitation Spectral Region: green, wide Excitation Range, Dichromatic Mirror Cut-On Wavelength: 565nm).

4.4 Preparation of brain tissue homogenates

AD brain homogenates were prepared from dissected gray and white matters from pooled control and AD patients in phosphate buffered saline (PBS, pH 7.4) at the concentration of approximately 100 mg wet tissue/mL (motor-driven glass homogenizer with setting of 6 for 30s). The homogenates were aliquoted into 1-mL portions and stored at $-80°C$.

4.5 In vitro binding assays with AD brain homogenates

Binding studies were carried out in 12mm × 75mm borosilicate glass tubes according to the method described in Ref. [34] with some modification. For saturation study, 100μL AD brain homogenates were added to a mixture containing 0.01-2.0nM of [^{125}I]9 in PBS(pH 7.4; final volume of 1mL). The competition binding assays were performed by mixing 100μL AD brain homogenates, 100μL 0.04 nM [^{125}I]9, 100μL of inhibitor($3 \times 10^{-6} - 1 \times 10^{-10}$M) and 700μL PBS in a final volume of 1mL. Nonspecific binding was defined in the presence of 100μL 9 (3×10^{-6} mol/L in PBS) in the same assay tubes. The mixture was incubated at 37°C for 2h, the bound and the free radioactivity was separated by vacuum filtration through Whatman GF/B filters followed by 2-3mL washes of cold PBS for three times at room temperature. Filters containing the bound I-125 ligand were counted in a gamma counter (USTC Zonkia GC-1200) with 65% counting efficiency. Under the assay conditions, the specifically bound fraction was less than 15% of the total radioactivity. Values for the half-maximal inhibitory concentration (IC_{50}) were determined from displacement curves of three independent experiments using GraphPad Prism, and those for the inhibition constant (K_i) were calculated using the Cheng-Prusoff equation[35]: $K_i = IC_{50}/(1 + [L]/K_d)$, where [L] is the concentration of [^{125}I]9 used in the assay, and K_d is the dissociation constant of compound 9.

4.6 In vivo biodistribution in normal mice

Experiments were performed under the regulations of the ethics committee of Beijing Normal University. A saline solution (100μL) containing [^{125}I]9 (2μCi) was injected directly into the tail vein of Kunming mice (female, average weight 18-22g). The mice ($n = 4$ for each time point) were sacrificed at designated time points postinjection. The organs of interest were removed and weighed, and the radioactivity was counted with an automatic γ counter (PerkinElmer Life). The % ID/g of samples was calculated by comparing the sample counts with the count of the diluted initial dose. The experiment was duplicated for three times.

4.7 Partition coefficient determination

Partition coefficient was measured by mixing the [^{125}I]**9** with 3g each of 1-octanol and buffer (0.1M PBS, pH 7.4) in a test tube. The test tube was vortexed for 3min at room temperature, followed by centrifugation for 5min. Two weighed samples (0.5g each) from the 1-octanol and buffer layers were counted in a well counter. The partition coefficient was determined by calculating the ratio of cpm/g of the 1-octanol to that of buffer. Sample from the 1-octanol layer was repartitioned until consistent partitions of coefficient values were obtained. The measurement was done in triplicate and repeated three times.

Acknowledgements

This work was funded by NSFC(20871021). We thank Professor Duanzhi Yin of Shanghai Institute of Applied Physics and Ms. Xiaoyan Zhang of College of Life Science for providing the equipment of the Zeiss 510 META.

References

[1] Hardy. J. ;Selkoe,D. J. Science 2002,297,353.

[2] Tycko,R. Curr. Opin. Struct. Biol. 2004,14,96.

[3] Kang,J. ; Lemaire,H. G. ; Unterbeck,A. ; Salbaum,J. M. ; Masters,C. L. ; Grzeschik,K. H. ; Multhaup, G. ;Beyreuthe,K. ;Muller-Hill,B. Nature 1987,325,733.

[4] Nordberg,P. Lancet Neurol. 2004,3,519.

[5] Johannsen,B. ;Pietzsch,H. J. Eur. J. Nucl. Med. 2002,29,263.

[6] Mathis,C. A. ;Wang,Y. ;Klunk,W. E. Curr. Pharm. Des. 2004,10,1469.

[7] Lockhart,A. Drug Discovery Today 2006,11,1093.

[8] Cai,L. S. ;Innis,R. B. ;Pike,V. W. Curr. Med. Chem. 2007,14,19.

[9] Mathis,C. A. ; Wang,Y. ; Holt,D. P. ; Huang,G. F. ; Debnath,M. L. ; Klunk,W. E. J. Med. Chem. 2003, 46,2740.

[10] Zhang,W. ;Oya,S. ;Kung,M. P. ;Hou,C. ;Maier,D. L. ;Kung,H. F. J. Med. Chem. 2005,48,5980.

[11] Kung, M. P. ; Hou, C. ; Zhuang, Z. P. ; Zhang, B. ; Skovronsky, D. ; Trojanowski, J. Q. ; Lee, V. M. Y. ; Kung,H. F. Brain Res. 2002,956,202.

[12] Klunk, W. E. ; Engler, H. ; Nordberg, A. ; Wang, Y. ; Blomqvist, G. ; Holt, D. P. ; Bergström, M. ; Savitcheva,I. ; Huang,G. F. ; Estrada,S. ; Ausén,B. ; Debnath,M. L. ; Barletta,J. ; Price,J. C. ; Sandell, J. ;Lopresti,B. J. ;Wall,A. ;Koivisto,P. ;Antoni,G. ;Mathis,C. A. ;Lagström,B. Amm. Neurol. 2004, 55,306.

[13] Ono,M. ; Wilson,A. ; Nobrega,J. ; Westaway,D. ; Verhoeff,P. ; Zhuang,Z. P. ; Kung,M. P. ; Kung,H. F. Nucl. Med. Biol. 2003,30,565.

[14] Newberg, A. B. ; Wintering, N. A. ; Plossl, K. ; Hochold, J. ; Stabin, M. G. ; Watson, M. ; Skovronsky, D. ; Clark,C. M. ;Kung,M. P. ;Kung,H. F. J. Nucl. Med. 2006,47,748.

[15] Duan,X. H. ;Liu,B. L. J. Sci. China Ser. B Chem. 2008,51,801.

[16] Krebs,M. R. H. ;Bromley,E. H. C. ;Donald,A. M. J. Struct. Biol. 2005,149,30.

[17] Stains,C. I. ;Mondal,K. ;Ghosh,I. Chem Med Chem. 2007,2,1674.

[18] Thomas, T. ; Nadackal, T. G. ; Thomas, K. Neuroreport 2001, 12, 3263.
[19] Masters, C. L. ; Cappai, R. ; Barnham, K. J. ; Villemagne, V. L. J. Neurochem. 2006, 97, 1700.
[20] Ono, K. ; Yoshiike, Y. ; Takashima, A. ; Hasegawa, K. ; Naiki, H. ; Yamada, M. J. Neurochem. 2003, 87, 172.
[21] Yang, F. S. ; Lim, G. P. ; Begum, A. N. ; Ubeda, O. J. ; Simmons, M. R. ; Ambegaokar, S. S. ; Chen, P. ; Kayed, R. ; Glabe, C. G. ; Frantschy, S. A. ; Cole, G. M. J. Biol. Chem. 2005, 280, 5892.
[22] Shoghi-Jadid, K. ; Small, G. W. ; Agdeppa, E. D. ; Kepe, V. ; Ercoli, L. M. ; Siddarth, P. ; Read, S. ; Satyamurthy, N. ; Petric, A. ; Huang, S. C. ; Barrio, J. R. Am. J. Geriatr. Psychiatry 2002, 10, 24.
[23] Agdeppa, E. D. ; Kepe, V. ; Petric, A. ; Satyamurthy, N. ; Liu, J. ; Huang, S. C. ; Small, G. W. ; Cole, G. M. ; Barrio, J. R. Neuroscience 2003, 117, 723.
[24] Ono, M. ; Yoshida, N. ; Ishibashi, K. ; Haratake, M. ; Arano, Y. ; Mori, H. ; Nakayama, M. J. Med. Chem. 2005, 48, 7253.
[25] Ryu, E. K. ; Choe, Y. S. ; Lee, K. H. ; Choi, Y. ; Kim, B. T. J. Med. Chem. 2006, 49, 6111.
[26] Agdeppa, E. D. ; Kepe, V. ; Liu, J. ; Small, G. W. ; Huang, S. C. ; Petrič, A. ; Satyamurthy, N. ; Barrio, J. R. Mol. Imaging Biol. 2003, 5, 404.
[27] Blanchard, B. J. ; Chen, A. ; Rozeboom, L. M. ; Stafford, K. A. ; Weigele, P. ; Ingram, V. M. Proc. Natl. Acad. Sci. U. S. A. 2004, 101, 14326.
[28] Chandra, R. ; Oya, S. ; Kung, M. P. ; Hou, C. ; Jin, L. W. ; Kung, H. F. J. Med. Chem. 2007, 50, 2415.
[29] Marcoux, J. F. ; Wagaw, S. ; Buchwald, S. L. J. Org. Chem. 1997, 62, 1568.
[30] Trinks, U. ; Buchdunger, E. ; Furet, P. ; Kump, W. ; Mett, H. ; Meyer, T. ; Miiller, M. ; Regenass, U. ; Rihs, G. ; Lydon, N. ; Traxler, P. J. Med. Chem. 1994, 37, 1015.
[31] Wang, H. ; Duennwald, M. L. ; Roberts, B. E. ; Rozeboom, L. M. ; Zhang, Y. L. ; Steele, A. D. ; Krishnan, R. ; Su, L. J. ; Griffin, D. ; Mukhopadhyay, S. ; Hennessy, E. J. ; Weigele, P. ; Blanchard, B. J. ; King, J. ; Deniz, A. A. ; Buchwald, S. L. ; Ingram, V. M. ; Lindquist, S. ; Shorter, J. Proc. Natl. Acad. Sci. U. S. A. 2008, 105, 7159.
[32] Dishino, D. D. ; Welch, M. J. ; Kilbourn, M. R. ; Raichle, M. E. J. Nucl. Med. 1983, 24, 1030.
[33] Styren, S. D. ; Hamilton, R. L. ; Styren, G. C. ; Klunk, W. E. J. Histochem. Cytochem. 2000, 48, 1223.
[34] Klunk, W. E. ; Wang, Y. ; Huang, G. F. ; Debnath, M. L. ; Holt, D. P. ; Shao, L. ; Hamilton, R. L. ; Ikonomovic, M. ; Dekosky, S. T. ; Mathis, C. A. J. Neurosci. 2003, 23, 2086.
[35] Cheng, Y. ; Prusoff, W. H. Biochem. Pharmacol. 1973, 22, 3099.

Synthesis and Evaluation of Novel Benzothiazole Derivatives Based on the Bithiophene Structure as Potential Radiotracers for β-amyloid Plaques in Alzheimer's Disease[*]

Abstract In this study, six novel benzothiazole derivatives based on the bithiophene structure were developed as potential β-amyloid probes. In vitro binding studies using Aβ aggregates showed that all of them demonstrated high binding affinities with K_i values ranged from 0.11 to 4.64nM. In vitro fluorescent staining results showed that these compounds can intensely stained Aβ plaques within brain sections of APP/PS1 transgenic mice, animal model for AD. Two radioiodinated compounds [^{125}I]-2-(5'-iodo-2,2'-bithiophen-5-yl)-6-methoxybenzo[d]thiazole [^{125}I]**10** and [^{125}I]-2-(2,2'-bithiophen-5-yl)-6-iodobenzo[d]thiazole [^{125}I]**13** were successfully prepared through an iododestannylation reaction. Furthermore, in vitro autoradiography of the AD model mice brain sections showed that both [^{125}I]**10** and [^{125}I]**13** labeled the Aβ plaques specifically with low background. In vivo biodistribution studies in normal mice indicated that [^{125}I]**13** exhibited high brain uptake(3.42% ID/g at 2min) and rapid clearance from the brain(0.53% ID/g at 60min), while [^{125}I]**10** showed lower brain uptake(0.87% ID/g at 2min). In conclusion, these preliminary results of this study suggest that the novel radioiodinated benzothiazole derivative [^{125}I]**13** may be a candidate as an in vivo imaging agent for detecting β-amyloid plaques in the brain of AD patients.

Key words Alzheimer's disease, β-amyloid, imaging agent, binding assay, autoradiography

1 Introduction

Alzheimer's disease(AD) is an age-related, irreversible, progressive brain disease which slowly destroys the cognitive functions of the brain by causing memory loss, disorientation and language impairment, etc[1]. Several studies indicate that the accumulation of senile plaques(SPs) composed of misfolded β-amyloid(Aβ) peptides and neurofibrillary tangles(NFTs) made of hyperphosphorylated tau aggregates are two of the pathological hallmarks of AD which begins 10-20 years before any symptoms are obvious[2-4]. Development of imaging probes that target Aβ plaques or neurofibrillary tangles for positron emission tomography(PET) or single photon emission computed tomography(SPECT) will be important for early diagnosis of AD, and it will also provide significant information to evaluate the efficacy of therapies of AD[5-8].

In the past 10 years a large number of radiolabeled ligands targeting Aβ plaques have been reported. Some of them have so far reached the clinical stage, such as [^{18}F]-2-(1-(2-(N-(2-

---

[*] Copartner: Cui Mengchao, Li Zijing, Tang Ruikun. Reprinted from *Bioorganic & Medicinal Chemistry*, 2010, 18(7): 2777-2784.

fluoroethyl)-N-methylamino)-naphthalene-6-yl)ethylidene)malononitrile ([^{18}F]FDDNP)[9-11], 2-(4'-[^{11}C]methylaminophenyl)-6-hydroxybenzothiazole ([^{11}C]PIB)[12,13], 4-N-[^{11}C]methyl-amino-4'-hydroxystilbene ([^{11}C]SB-13)[14,15], [^{11}C]-2-(2-[2-dimethylaminothiazol-5-yl]ethenyl)-6-(2-[fluoro]ethoxy)benzoxazole ([^{11}C]BF-227)[16], [^{18}F]-4-(N-methylamino)-4'-(2-(2-(2-fluoroethoxy)ethoxy)ethoxy)-stilbene ([^{18}F]BAY94-9172)[17] for PET imaging (Fig. 1). However, SPECT scanners are widely equipped in hospitals. It is more practical for routine clinical diagnostic. Development of SPECT imaging probes for Aβ plaques will benefit a large number of AD patients. Several groups have reported a lot of radioiodinated ligands for Aβ plaques[18-20]. But most of these ligands had unfavorable in vivo pharmacokinetics which preventing their further development as potential in vivo imaging agents. Up to now, [^{123}I]-6-iodo-2-(4'-dimethylamino-)phenyl-imidazo[1,2]pyridine ([^{123}I]IMPY) is the first SPECT imaging probe tested in human and the results are encouraging[21-23].

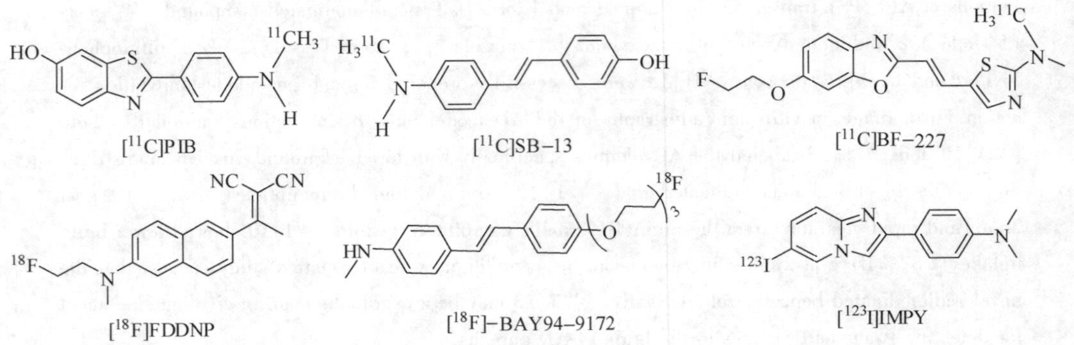

Fig. 1 Chemical structure of Aβ imaging probes clinically tested

As one of the known Aβ probes, (E)-2-((5'-(4-hydroxystyryl)-2,2'-bithiophen-5-yl)methylene)malononitrile (NIAD-4) is a near infrared fluorescent contrast agent for in vivo optical imaging Aβ plaques (Fig. 2). This compound contains a highly hydrophobic bithiophene structure and exhibits high binding affinity to Aβ aggregates ($K_i = 10nM$) in the same binding site as ThT. In addition, this agent

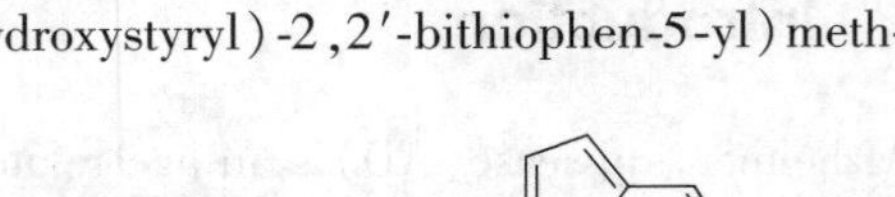

Fig. 2 Chemical structure of NIAD-4

readily crosses the blood-brain barrier (BBB) and specifically labels the Aβ plaques in the living brain of AD model mice[24].

Recently, Nilsson and his co-workers identified a class of luminescent conjugated polythiophenes (LCPs) as novel conformation sensitive optical probes for selective staining of protein aggregates[25-27]. Under physiological conditions, the pentameric thiophene LCPs showed a striking specificity for both Aβ plaques and NFTs. Two of them also crossed the BBB to a sufficient degree[28].

Thus, we conclude that the polythiophene moiety may play an important role in maintaining the specific binding ability to Aβ plaques. This prompted us to apply it as a core structure for Aβ imaging agents. Herein we report the synthesis and the initial biological characterizations of the novel benzothiazole derivatives containing the bithiophene structure as probes for Aβ plaques in the brain. To our knowledge, this is the first time bithiophene structures have been introduced to radiotracers for β-amyloid plaques in AD.

2 Results and Discussion

2.1 Chemistry

The benzothiazole derivatives (**6-10**) were prepared by a condensation reaction with the chemical yields ranged from 42.5% to 66.3% as depicted in Scheme 1. The 5-substituted-o-aminothiophenols (**1,2**) and 5′-substituted-2,2′-bithiophene-5-carbaldehyde (**3-5**) were prepared according to the reported procedures[29,30]. The tributyltin precursors (**11,12**) were prepared from the corresponding bromo compounds (**7,9**) under a bromo to tributyltin exchange reaction catalyzed by Pd(PPh$_3$)$_4$ with yields of 14.7%, 21.7% respectively. The tributyltin derivatives can be readily used as starting materials for preparing [^{125}I]**10** and [^{125}I]**13**. Moreover, the tributyltin precursor **11** was also served as an intermediate in iododestannylation reaction to produce the corresponding iodinated compound **13** with the yield of 30.2%.

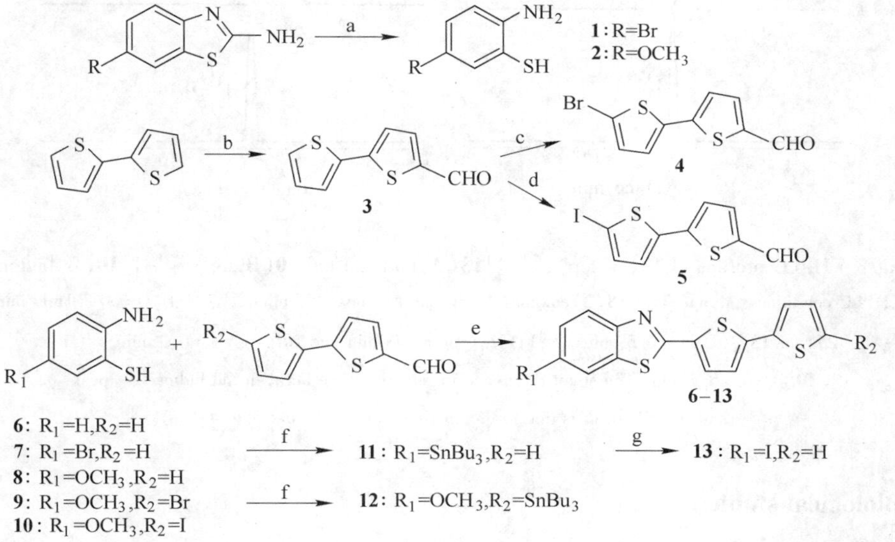

Scheme 1 Reagents and conditions
a—KOH, H$_2$O, reflux; b—POCl$_3$, DMF, 60℃; c—AcOH, Br$_2$; d—HgO, I$_2$, toluene;
e—DMSO, 160℃; f—(Bu$_3$Sn)$_2$, (Ph$_3$P)$_4$Pd, toluene, reflux; g—I$_2$, CHCl$_3$

2.2 Radiolabeling

The two radioiodinated ligands [^{125}I]**10** and [^{125}I]**13** were prepared from the corresponding tributyltin precursors through an iododestannylation reaction using hydrogen peroxide as oxidant with the radiochemical yields of 53.7% and 66.3%, respectively (Scheme 2). After purification by HPLC, the radio-chemical purities of [^{125}I]**10** and [^{125}I]**13** were both greater than 98%. The specific activity of the no-carrier-added preparation was comparable to that of Na^{125}I, 2200Ci/mmol. Finally, the radiochemical identities of the radioiodinated ligands were verified by co-injection with nonradioactive compounds by HPLC profiles (Fig. 3).

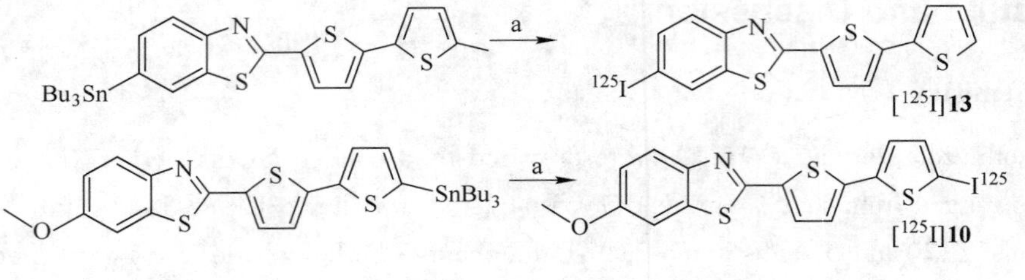

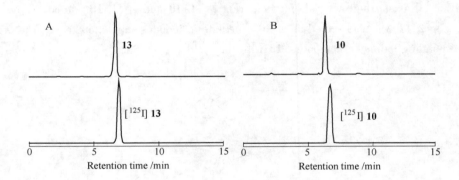

Scheme 2 Reagents
a—[^{125}I]NaI, HCl(1M), H$_2$O$_2$(3%)

Fig. 3 HPLC profiles of **13**(A, top), [^{125}I]**13**(A, bottom) and **10**(B, top), [^{125}I]**10**(B, bottom) (HPLC conditions: Agilent TC-C18(2) column (Analytical 4.6mm × 150mm) CH$_3$CN/H$_2$O = 9/1, 1mL/min, 254nm, **13**t_R(UV) = 6.59min, [^{125}I]**13**t_R(γ) = 7.00min and **10**t_R(UV) = 6.38min, [^{125}I]**10**t_R(γ) = 6.77min. The slight difference in retention time between the radioactive peak and the UV peak is due to the configuration of the detector system)

2.3 Biological studies

2.3.1 Binding assays using the aggregated Aβ(1-40) or Aβ(1-42) peptide in solution

[^{125}I]-4-(6-Iodobenzo[*d*]thiazol-2-yl)-*N*,*N*-dimethylaniline ([^{125}I]TZDM) was reported to be a selective ligand for in vitro binding of Aβ aggregates with high affinities[18]. So the binding affinities of these benzothiazole derivatives were evaluated by inhibition assays against [^{125}I]TZDM using Aβ(1-40) and Aβ(1-42) aggregates as described in the literatures[18,19,31].

The saturation binding of [^{125}I]TZDM to Aβ(1-40) and Aβ(1-42) showed almost the same K_d values as previously reported [K_d = 0.13nM for Aβ(1-42), K_d = 0.09nM for Aβ(1-40), see in Supplementary data]. The K_i values shown in Table 1 suggest that these new compounds share the same binding site as ThT and have excellent binding affinities for both Aβ(1-40) and Aβ(1-42) aggregates. The K_i values of compounds **6, 7, 8, 9, 10** and **13** were 4.64nM, 0.60nM, 2.09nM, 0.21nM, 0.11nM and 0.25nM for Aβ(1-40) aggregates and 3.94nM, 0.57nM, 0.60nM, 1.28nM, 0.61nM and 0.31nM for Aβ(1-42) aggregates, respectively. After analyzing the structures and the corresponding K_i values, we primarily concluded that functional groups may exert moderate effects on the binding affinity. Halogenations at the 5 position(Br or I) on the benzothiazole moiety enhanced the binding affinity(compounds **7** and **13**), while replacing the 5-halogen with a methoxy group also improved the binding affinity(compound **8**). Compounds **9** and **10** with the halogen at 2 position of the bithiophene moiety also displayed high binding affinity. Compared with [^{125}I]TZDM [0.9nM for Aβ(1-40), 2.2nM for Aβ(1-42)][18], [^{11}C]PIB [0.77nM for Aβ(1-42)][32] and [^{123}I]IMPY [15nM for Aβ(1-40)][21], the K_i values of compounds **10** and **13** were superior. In addition, there is no significant difference in the binding affinities between Aβ(1-40) and Aβ(1-42) aggregates.

Table 1 Inhibition constants(K_i) of compounds on ligand binding to aggregates of Aβ(1-40) and Aβ(1-42) versus [^{125}I]TZDM

Compounds	K_i/nM	
	Aβ(1-42)	Aβ(1-40)
6	3.94 ± 0.06	4.64 ± 0.10
7	0.57 ± 0.08	0.60 ± 0.06
8	0.60 ± 0.06	2.09 ± 0.09
9	1.28 ± 0.08	0.21 ± 0.01
10	0.61 ± 0.09	0.11 ± 0.01
13	0.31 ± 0.05	0.25 ± 0.07
TZDM[①]	2.2 ± 0.4	0.9 ± 0.2
PIB[②]	0.77	—
IMPY[③]	—	15.0 ± 5.0

①-③ Data from Ref. [18], Ref. [30], Ref. [21].

2.3.2 In vitro fluorescent staining of Aβ plaques in transgenic mouse brain sections
Since the in vitro binding assays demonstrated high binding affinities of these compounds for Aβ(1-40) and Aβ(1-42) aggregates, the affinities of compounds **6, 7, 8, 9, 10** and **13** for Aβ plaques were investigated by neuropathological staining with the brain sections of a 12-month-old transgenic model mouse, Tg-C57, which encoded a double mutant form of APP and PS1. Many Aβ plaques were clearly stained by these compounds with low background(Fig. 4a, c, e, g, i and k). The similar pattern of Aβ plaques was consistent with that stained with thioflavin-S using the adjacent brain section(Fig. 4b, d, f, h, j and l). Such results indicate that these derivatives displayed specific binding affinity for Aβ plaques as supported by the binding assay

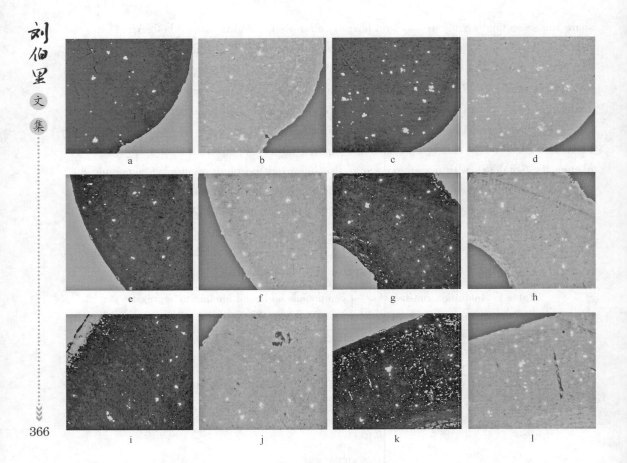

Fig. 4 Fluorescence staining of compounds **6, 7, 8, 9, 10** and **13** (a, c, e, g, i and k) on the sections from a Tg-C57 (APP/PS1) mouse (25 μm thick) with the filter set of DAPI, and the labeled plaques were confirmed by staining of the adjacent sections by thioflavin S (b, d, f, h, j and l) with the filter set of GFP

(Table 1), and may be applied as radiotracers to in vivo amyloid imaging.

2.3.3 In vitro labeling of brain sections from transgenic mouse by autoradiography

To further characterize the specific nature of Aβ plaques binding, in vitro autoradiography was performed by incubating [^{125}I]**10** and [^{125}I]**13**, with the APP/PS1 transgenic mice brain sections in the absence or presence of thioflavin-T (10mM). Both [^{125}I]**10** and [^{125}I]**13** showed excellent binding to Aβ plaques in the brain sections with a lot of spots of plaque signals in the cortex region and minimal background signals were observed (Fig. 5a and b). The same sections were also stained with thioflavin-S and the localizations of Aβ plaques were accordance with the results of autoradiography (Fig. 5c and d red arrow). In order to confirm that [^{125}I]**10** and [^{125}I]**13** had specific binding to Aβ plaques, blocking study was performed in the presence of thioflavin-T (10mM) as a blocking agent. The results showed that there was no notable Aβ plaques observed (Fig. 5 e and f). In conclusion, [^{125}I]**10** and [^{125}I]**13** were specific for Aβ plaques, which were consistent with the observation that compounds **10** and **13**

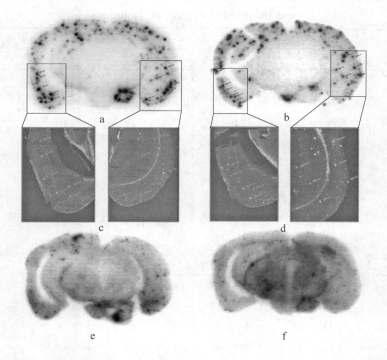

Fig. 5 In vitro labeling of brain sections from Tg-C57(APP/PS1)mice(6μm thick)by autoradiography. [^{125}I]**10** and [^{125}I]**13** strongly labeled the Aβ plaques in the cortex of the brain(a, b), the same sections were also stained with thioflavin-S(c ,d) and the localizations of Aβ plaques were accordance with the results of autoradiography(red arrows) ,and the brain sections were pretreated with thioflavin-T(10mM) (e, f)

do bind to Aβ(1-40)and Aβ(1-42)aggregates with high affinity.

2.3.4 In vivo biodistribution in normal mice

The ability to cross the intact BBB is one of the key qualifications for brain imaging agents. The log$D$ values of [^{125}I]**10** and [^{125}I]**13** were 3.27, 3.65 respectively(measured by a partition between 1-octanol and pH 7.4 phosphate buffer) which indicate a better BBB penetration. To evaluate brain uptake of the two radioiodinated ligands, [^{125}I]**10** and [^{125}I]**13** were injected intravenously into normal mice for biodistribution experiments. As shown in Table 2, [^{125}I]**13** displayed rapid brain entry at early time intervals. The initial brain uptake was (3.42 ±0.45)% ID/g at 2 min post-injection, which is favorable for potential clinical imaging studies. The brain radioactivity concentration decreased rapidly to (1.08 ±0.07)% ID/g at 30min and (0.53 ±0.08)% ID/g at 60min for there is no Aβ plaques in normal mice. These results indicate that [^{125}I]**13** possess favorable pharmacokinetics in brain. However, the initial brain uptake of [^{125}I]**10** was (0.87 ±0.16)% ID/g at 2 min and displayed a slow washout rate from the brain with (0.77 ±0.16)% ID/g at 60min, which is unfavorable for an Aβ imaging agent.

Table 2 Biodistribution in normal mice after iv injection of $[^{125}I]13$, $[^{125}I]10$ and the lipophilicity ($\lg D$) of the ligands[①]

Organ	2min	30min	60min	120min	240min
$[^{125}I]13$ ($\lg D = 3.65 \pm 0.04$)					
Blood	9.02 ± 2.33	2.12 ± 0.30	1.96 ± 0.16	1.76 ± 0.20	1.36 ± 0.29
Brain	3.42 ± 0.45	1.08 ± 0.07	0.53 ± 0.08	0.51 ± 0.09	0.50 ± 0.11
Heart	4.39 ± 0.72	2.49 ± 0.77	2.65 ± 0.26	1.50 ± 0.21	1.31 ± 0.27
Liver	53.64 ± 5.24	39.81 ± 3.51	29.12 ± 2.24	22.47 ± 4.10	13.82 ± 1.43
Spleen	12.75 ± 4.82	6.78 ± 0.89	8.61 ± 2.02	3.50 ± 0.99	2.07 ± 0.57
Lung	17.51 ± 2.44	8.04 ± 0.35	8.01 ± 2.22	5.76 ± 1.35	3.72 ± 0.62
Kidney	4.00 ± 0.29	7.22 ± 0.39	8.52 ± 1.52	7.33 ± 1.59	5.76 ± 1.09
Stomach[②]	0.39 ± 0.09	2.47 ± 0.74	4.05 ± 1.11	4.39 ± 1.25	3.24 ± 1.15
Muscle	0.65 ± 0.14	1.01 ± 0.02	1.47 ± 0.21	1.23 ± 0.27	0.87 ± 0.15
$[^{125}I]10$ ($\lg D = 3.27 \pm 0.01$)					
Blood	23.34 ± 2.79	6.55 ± 0.34	6.58 ± 0.54	6.62 ± 1.19	5.05 ± 0.73
Brain	0.87 ± 0.16	0.64 ± 0.09	0.77 ± 0.16	0.68 ± 0.11	0.53 ± 0.08
Heart	16.13 ± 2.86	5.94 ± 0.46	4.74 ± 0.83	3.48 ± 1.35	2.50 ± 0.69
Liver	35.25 ± 1.78	6.98 ± 0.51	5.76 ± 0.56	4.33 ± 0.83	3.23 ± 0.37
Spleen	19.04 ± 3.73	7.11 ± 0.83	6.24 ± 0.71	4.92 ± 0.80	4.49 ± 0.91
Lung	43.6 ± 2.51	10.28 ± 1.13	8.64 ± 0.75	7.7 ± 1.02	6.12 ± 1.12
Kidney	14.3 ± 1.09	6.49 ± 0.61	6.28 ± 0.67	5.67 ± 1.15	4.53 ± 0.58
Stomach[②]	1.31 ± 0.22	15.84 ± 2.80	20.24 ± 1.57	33.95 ± 5.73	19.37 ± 3.31
Muscle	2.73 ± 0.44	3.42 ± 0.18	3.56 ± 0.30	2.46 ± 0.60	1.95 ± 0.38

① Expressed as % injected dose per gram. Average of 5 mice ± standard deviation.
② Expressed as % injected dose per organ.

$[^{123}I]$IMPY is the most promising SPECT imaging probe which display high binding affinity for Aβ plaques ($K_i = 15$nM), excellent brain uptake (2.88% ID/organ at 2min) and fast washout (0.21% ID/organ at 60min). In comparison with $[^{123}I]$IMPY, $[^{125}I]13$ displayed superior binding affinity and also illustrated suitable in vivo pharmacokinetic properties for Aβ imaging.

Another noteworthy feature of the radioiodinated imaging agent is the stability in vivo. As shown in Table 2, for both of the radioiodinated ligands $[^{125}I]10$ and $[^{125}I]13$, accumulations of radioactivity in stomach were low, (1.31 ± 0.22)% ID/organ and (0.39 ± 0.09)% ID/organ at 2min, however $[^{125}I]10$ exhibited a significant increase to (20.24 ± 1.57)% ID/organ at 1h and (33.95 ± 5.73)% ID/organ at 2h. This observation demonstrated that $[^{125}I]10$ is not stable in vivo, which may result from deiodination caused by oxidation of the sulfur atom in a metabolically active organ[33]. To further illustrate this observation, we performed the stability study of $[^{125}I]10$ in plasma and liver homogenate from mouse. The results indicate that $[^{125}I]10$ is very stable in plasma, for two hours it does not show any decomposition (Fig. 6e). But in liver homogenate, $[^{125}I]$**10** quickly deiodinated. After two hours incubation, about 71% of $[^{125}I]$**10** transformed to free iodine (Fig. 6j). From several published articles, the uptake of free iodine in the glandular stomach was described for mice, rats, hamsters and cats[34-36]. So we can conclude that $[^{125}I]$**10** first undergoes a deiodination process in liver then accumulated in stomach.

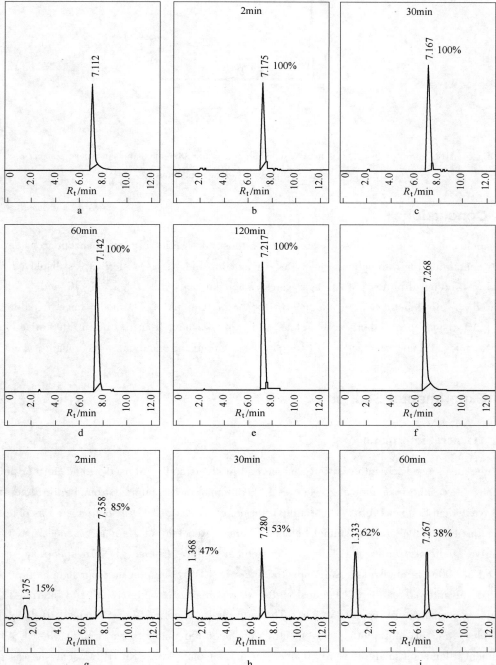

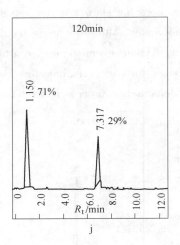

Fig. 6　HPLC profiles of [^{125}I]**10** in 10% ethanol(a,f), and after incubation in mouse plasma(b-e) and liver homogenate(g-j) at 37 ℃ for 2min, 30min, 1h and 2h

3　Conclusions

In conclusion, we successfully designed and synthesized a series of novel benzothiazole derivatives containing the bithiophene moiety as Aβ imaging agents. These derivatives displayed high in vitro binding affinities for Aβ aggregates. When labeled with ^{125}I, [^{125}I]**10** and [^{125}I]**13** showed specific labeling of Aβ plaques in the brain sections of AD model mice. In addition, [^{125}I]**13** displayed excellent brain uptake and rapid washout from the brain after injection in normal mice. Taken together, [^{125}I]**13** may be a promising candidate Aβ imaging agent for SPECT.

4　Experimental Section

4.1　General information

All chemicals used in synthesis were commercial products and were used without further purification unless otherwise indicated. Na^{125}I(2200Ci/mmol) was obtained from PerkinElmer Life and Analytical Sciences, USA. The double transgenic(APP/PS1) AD model mouse was obtained from Institute of Laboratory Animal Science, Chinese Academy of Medical Sciences and Comparative Medicine Center of Peking Union Medical College, China. ^{1}H-NMR spectra were obtained at 400MHz on Bruker NMR spectrometers in CDCl$_3$ solution at room temperature with TMS as an internal standard. Chemical shifts are reported as δ values relative to internal TMS. Coupling constants are reported in hertz. The multiplicity is defined by s(singlet), d(doublet), t(triplet), and m(multiplet). Mass spectra were acquired under SurveyorMSQ Plus(ESI) instrument. Elemental analyses were obtained on a Elementar Vario EI. HPLC analysis was performed on a Shimadzu SCL-10 AVP equipped with a Packard 500 TR series flow scintillation analyzer, conditions: Agilent TC-C18 reverse-phase analytical column (5μm, ID = 4.6mm, length = 150mm), 90∶10 CH_3CN/H_2O, 1.0mL/min, UV 254nm. Separations were achieved on a Alltech HPLC pump Model 626 equipped with LINEAR UVIS-201 and BIOSCAN flow-counter, conditions: Alltech C18 reverse-phase column (5μm, ID = 4.6mm, length = 250mm) eluted with acetonitrile at a 2.0mL/min flow rate.

4.2 Synthesis of 2-amino-5-bromobenzenethiol(1)

2-Amino-5-bromobenzenethiol was synthesized according to the literature[29]. Briefly 2-amino-5-bromobenzothiazole(9.16g, 40mmol) was suspended in 50% KOH (80g KOH dissolved in 80mL water). The suspension was heated to reflux for 24 h, and was cooled to room temperature. After that, the residue was filtered and the filtrate was neutralized with acetic acid(30% in water, 100mL) to give the crude product which was used without further purification(66.7% yield).

4.3 Synthesis of 2-amino-5-methoxybenzenethiol(2)

The same reaction as described above for preparing **1** was employed, and a yellow solid **2** was obtained(51.3% yield).

4.4 Synthesis of 2,2′-bithiophene-5-carbaldehyde(3)

According to the literature[30], $POCl_3$ (2.30g 15.0mmol) was added to DMF (1.10g 15.0mmol) at 0℃ and the mixture was stirred for 15min at 0℃. After that 2,2′-bithiophene (2.00g 12.0mmol) dissolved in 1,2-dichloroethane (40mL) were added dropwise with stirring. The reaction mixture was then heated 2h at 60℃. The organic layer was washed with saturated $NaHCO_3$ aqueous solution, and was dried with anhydrous Na_2SO_4. After the solvent was removed, the residue was purified by column chromatography(petroleum ether/AcOEt, 8∶1) to give 1.47g of **3** as a yellow solid(63.1% yield). MS(ESI): m/z calcd for $C_9H_6OS_2$ 193.99; found 194.9($M + H^+$).

4.5 Synthesis of 5′-bromo-2,2′-bithiophene-5-carbaldehyde(4)

To a stirring solution of 2,2′-bithiophene-5-carboxaldehyde (1.94g, 10.00mmol) in $CHCl_3$ (20mL) at 0℃, was added bromine(1.60g, 10.00mmol) dissolved in $CHCl_3$(20mL). The mixture was stirred at room temperature for 2h. The organic layer was washed with saturated $NaHCO_3$ aqueous solution, and was dried with anhydrous Na_2SO_4. Evaporation of the solvent under reduced pressure gave 4 as a yellow solid(86.3% yield). MS(ESI): m/z calcd for $C_9H_5BrOS_2$ 271.90; found 272.9($M + H^+$).

4.6 Synthesis of 5′-iodo-2,2′-bithiophene-5-carbaldehyde(5)

To a solution of 2,2′-bithiophene-5-carboxaldehyde(1.94g, 10.00mmol) in toluene(20mL) at

0 ℃, was added small portions of mercury(Ⅱ) oxide(2. 23g, 10. 29mmol, yellow powder) and iodine(2. 87g, 11. 32mmol). The mixture was stirred at room temperature for 8h, and the orange precipitate was filtered off and washed with ethyl acetate. The filtrate and washings were combined and washed with aqueous sodium thiosulfate and dried over magnesium sulfate. Solvent was removed by rotary evaporation and the solid residue was recrystallized from ethanol to give 2. 37 g of **5**(72% yield). MS(ESI) :m/z calcd for $C_9H_5IOS_2$ 319. 88 ;found 320. 8($M+H^+$).

4. 7 Synthesis of 2-(2 ,2′-bithiophen-5-yl) benzo[d]thiazole(6)

A mixture of 2-aminobenzenethiol (125mg, 1mmol) and 2, 2′-bithiophene-5-carbaldehyde (194mg, 1mmol) in DMSO was heated at 160 ℃ for 25min. Ice water(10mL) was added after the mixture was cooled. The solid was filtered and recrystallized in ethanol to give 168mg of **6** (56. 2% yield). ^{1}H NMR(400MHz, CDCl$_3$) δ8. 05(d, J = 8. 1Hz, 1H) ,7. 86(d, J = 8. 0Hz, 1H) ,7. 61(d, J = 3. 7Hz, 1H) ,7. 50(t, J = 7. 1Hz, 1H) ,7. 39(t, J = 7. 1Hz, 1H) ,7. 31(d, J = 4. 4Hz, 2H) ,7. 21(d, J = 3. 9Hz, 1H) ,7. 08(t, J = 4. 4Hz, 1H). MS(ESI) :m/z calcd for $C_{15}H_9NS_3$ 298. 99 ;found 300. 3($M+H^+$). Anal. Calcd: C, 60. 17; H, 3. 03; N, 4. 68. Found: C, 59. 63; H, 3. 41; N, 4. 72. M_p165. 3-167. 0 ℃.

4. 8 Synthesis of 2-(2 ,2′-bithiophen-5-yl) -6-bromobenzo[d]thiazole(7)

The same reaction as described above for preparing **6** was employed, a yellow solid **7** was obtained(66. 3% yield). ^{1}H NMR(400MHz, CDCl$_3$) δ7. 98(d, J = 1. 7Hz, 1H) ,7. 85(d, J = 8. 7Hz, 1H) ,7. 57(dd, J = 8. 7, 1. 8Hz, 1H) ,7. 54(d, J = 3. 9Hz, 1H) ,7. 31(d, J = 5. 4Hz, 2H) ,7. 19(d, J = 3. 9Hz, 1H) ,7. 07(t, J = 4. 2Hz, 1H). MS(ESI) :m/z calcd for $C_{15}H_8BrNS_3$ 376. 90 ;found 378. 2($M+H^+$). Anal. Calcd: C, 47. 62; H, 2. 13; N, 3. 70. Found: C, 47. 56; H, 2. 38; N, 3. 61. M_p205. 6-207. 0 ℃.

4. 9 Synthesis of 2-(2 ,2′-bithiophen-5-yl) -6-iodobenzo[d]thiazole(13)

To a solution of **11**(59mg, 0. 1mmol) in CHCl$_3$ (10mL) was added a solution of iodine(127mg dissolved in 10mL CHCl$_3$) dropwise at room temperature. The resulting mixture was stirred at room temperature for 30min and quenched by addition of 2mL saturated NaHSO$_3$ solution. The organic phase was separated, dried over MgSO$_4$, filtered, and concentrated to give the crude product which was recrystallized in AcOEt to give 13mg of **13** (30. 2% yield). ^{1}H NMR (400MHz, CDCl$_3$) δ8. 18(s ,1H) ,7. 77(s ,2H) ,7. 64(d, J = 3. 9Hz, 2H) ,7. 31-7. 32(m, 2H) ,7. 20(d, J = 3. 9Hz, 1H) ,7. 07(dd, J = 5. 0, 3. 9Hz, 1H). MS(ESI) :m/z calcd for $C_{15}H_8INS_3$ 424. 89; found 425. 9($M+H^+$). Anal. Calcd: C, 42. 36; H, 1. 90; N, 3. 29; Found: C, 42. 31; H, 1. 78; N, 3. 33. M_p216. 0-217. 2 ℃.

4. 10 Synthesis of 2-(2 ,2′-bithiophen-5-yl) -6-methoxybenzo[d]thiazole(8)

The same reaction as described above for preparing **6** was employed, and a yellow solid **8** was obtained(42. 5% yield). ^{1}H NMR(400MHz, CDCl$_3$) δ7. 95(d, J = 8. 9Hz, 1H) ,7. 58(d, J =

3.9Hz,1H),7.31(dd, J = 6.2, 3.0Hz,3H),7.20(d, J = 3.9Hz,1H),7.10(dd, J = 9.0, 2.5Hz,1H),7.07(d, J = 4.6Hz,1H),3.90(s,3H). MS(ESI): m/z calcd for $C_{16}H_{11}NOS_3$ 329.00; found 330.3(M + H$^+$). Anal. Calcd: C, 58.33; H, 3.37; N, 4.25; Found: C, 58.21; H, 3.69; N, 4.21. M_p 166.6-167.2 ℃.

4.11 Synthesis of 2-(5′-bromo-2,2′-bithiophen-5-yl)-6-methoxybenzo[d]thiazole(9)

The same reaction as described above for preparing **6** was employed, and a brown solid **9** was obtained(52.1% yield). ^{1}H NMR(400MHz, CDCl$_3$) δ7.94(d, J = 8.8Hz,1H),7.55(s,1H),7.30(s,1H),7.09(d, J = 10.5Hz,2H),7.02(s,2H),3.89(s,3H). MS(ESI): m/z calcd for $C_{16}H_{10}BrNOS_3$ 406.91; found 408.3(M + H$^+$). Anal. Calcd: C, 47.06; H, 2.47; N, 3.43; Found: C, 47.16; H, 2.39; N, 3.61. M_p 208.0-209.6 ℃.

4.12 Synthesis of 2-(5′-iodo-2,2′-bithiophen-5-yl)-6-methoxybenzo[d]thiazole(10)

The same reaction as described above for preparing **6** was employed, and a yellow solid **10** was obtained(47.2% yield). ^{1}H NMR(400MHz, CDCl$_3$) δ7.94(d, J = 9.0Hz,1H),7.56(d, J = 3.3Hz,1H),7.31(s,1H),7.20(d, J = 3.0Hz,1H),7.12(d, J = 3.0Hz,1H),7.09(d, J = 8.8Hz,1H),6.95(s,1H),3.90(s,3H). MS(ESI): m/z calcd for $C_{16}H_{10}INOS_3$ 454.90 found 456.3(M + H$^+$). Anal. Calcd: C, 42.20; H, 2.21; N, 3.08; Found: C, 42.27; H, 2.46; N 2.91. M_p 262.1-263.2 ℃.

4.13 Synthesis of 2-(2,2′-bithiophen-5-yl)-6-(tributylstannyl)-benzo[d]thiazole(11)

A mixture of **7** (189mg, 0.5mmol), bis(tributyltin)(580mg, 1.0mmol), and Pd(Ph$_3$P)$_4$ (12.0mg, 0.01mmol) in toluene was stirred at 110℃ for 12h. The solvent was removed, and the residue was purified by column chromatography(petroleum ether/AcOEt, 10 : 1) to give 86.5mg of **11** as a yellow solid(14.7% yield). ^{1}H NMR(400MHz, CDCl$_3$) δ7.90(d, J = 7.9Hz,1H),7.85(s,1H),7.48(d, J = 3.9Hz,1H),7.46(d, J = 4.7Hz,1H),7.21(d, J = 4.4Hz,1H),7.11(d, J = 3.9Hz,1H),7.00-6.98(m, 1H),1.53-1.45(m, 6H),1.33-1.23(m, 6H),1.07-1.04(m, 6H),0.83(t, J = 8.4Hz,9H). MS(ESI): m/z calcd for $C_{27}H_{35}NS_3Sn$ 589.10; found 590.0(M + H$^+$).

4.14 Synthesis of 6-methoxy-2-(5′-(tributylstannyl)-2,2′-bithiophen-5-yl)benzo[d]thiazole(12)

The same reaction as described above for preparing **11** was employed, and a yellow solid **12** was obtained from **9**(21.7% yield). ^{1}H NMR(400MHz, CDCl$_3$) δ7.89(d, J = 8.9Hz,1H),7.47(d, J = 3.9Hz,1H),7.39(d, J = 3.4Hz,1H),7.31(d, J = 2.5Hz,1H),7.17(d, J = 3.9Hz,1H),7.10(d, J = 3.4Hz,1H),7.07(dd, J = 9.0, 2.6Hz,1H),3.89(s,3H),1.64-1.54(m, 6H),1.41-1.31(m,6H),1.17-1.08(m,6H),0.91(t, J = 7.3Hz,9H). MS(ESI): m/z calcd for $C_{28}H_{37}NOS_3Sn$ 619.11; found 620.2(M + H$^+$).

4.15 Preparation of radioiodinated ligands

The radioiodinated compounds [^{125}I]10 and [^{125}I]13 were prepared from the corresponding tributyltin derivatives by an iododestannylation according to the procedure described previously with some modifications. Briefly, 100μL of H_2O_2(3%) was added to a mixture of a tributyltin derivative(0.1 mg/100μL in ethanol), 240μCi sodium [^{125}I]iodide(specific activity 2200Ci/mmol), and 100μL of 1M HCl in a sealed vial. The reaction was allowed to proceed at room temperature for 15min and then quenched by addition of 50μL saturated $NaHSO_3$ solution. The reaction mixture, after neutralization with 1M NaOH was purified by HPLC on a Alltech C18 reverse-phase column(5μ, ID = 4.6mm, length = 250mm) with the solvent of acetonitrile(100%) at a flow rate of 2.0mL/min. The desired fractions containing the product were evaporated to dryness and redissolved in 100% ethanol. The final product was stored at −20℃ for in autoradiography and biodistribution studies.

4.16 Binding assays using the aggregated Aβ(1-40) or Aβ(1-42) peptide in solution

The trifluoroacetic acid salt forms of peptides Aβ(1-40) and Aβ(1-42) were purchased from GL biochem(Shanghai, China). Aggregation of peptides was carried out by gently dissolving the peptide [0.25mg/mL for Aβ(1-40) and Aβ(1-42)] in a buffer solution(pH 7.4) containing 10mM sodium phosphate and 1mM EDTA. The solutions were incubated at 37℃ for 42h with gentle and constant shaking. Inhibition studies were carried out in 12 ×75mm borosilicate glass tubes according to the procedure described previously with some modifications[18,31]. Fifty microliters of aggregated Aβ fibrils(28nM in the final assay mixture) were added to the mixture containing 100μL of radioligands([^{125}I]TZDM) with appropriate concentration, 10μL of inhibitors(10^{-5}-10^{-10} M in ethanol) and 840μL PBS(0.2M, pH 7.4) in a final volume of 1mL. Nonspecific binding was defined in the presence of 100nM TZDM. The mixture was incubated for 2h at 37℃, then the bound and free radioactivities were separated by vacuum filtration through borosilicate glass fiber filters(0.1μm, JiuDing, China) using a ZT-II cell harvester (Shaoxing, China) followed by 3 ×4mL washes of cold PBS(0.2M, pH 7.4) containing 10% ethanol at room temperature. Filters containing the bound ^{125}I ligand were counted in a γ-counter(WALLAC/Wizard 1470, USA) with 71% counting efficiency. Under the assay conditions, the percent of the specifically bound fraction was about 20% of the total radioactivity. The half maximal inhibitory concentration(IC_{50}) values were determined using GraphPad Prism 4.0, and those for the inhibition constant(K_i) were calculated using the Cheng-Prusoff equation: K_i = $IC_{50}/(1 + [L]/K_d)$[37].

4.17 In vitro fluorescent staining of Aβ plaques in transgenic mouse brain sections

The transgenic model mice, Tg-C57, which encodes a double mutant form of APP and PS1 was used for the studies. Mice were sacrificed by cervical dislocation and brains were rapidly removed and frozen. Coronal 25μm sections were cut in a Leica-CM1900 cryostat from frozen

(-25℃), unfixed brains, and thaw mounted onto Poly-L-lysine coated microscope slides (Ultra-Lab, China). The brain sections were incubated with 10% ethanol solution (100nM) of **6**, **7**, **8**, **9**, **10** and **13** for 10min. The localization of plaques was confirmed by staining with thioflavin-S (1μm) on the adjacent sections. Finally, the sections were washed with 50% ethanol and PBS (0.2M, pH 7.4) for 10min. Fluorescent observation was performed by the LSM 510 META (Zeiss, Germany) equipped with DAPI (excitation, 405nm) and GFP filter sets (excitation, 505nm).

4.18 In vitro labeling of brain sections from transgenic mouse by autoradiography

Paraffin-embedded brain sections of Tg-C57 mouse (6μm) which were from Chinese Academy of Medical Sciences were used for the autoradiography. The brain sections were deparaffinized with 2 × 20min washes in xylene; 2 × 5min washes in 100% ethanol; 5min washes in 90% ethanol/H_2O; 5min washes in 80% ethanol/H_2O; 5min wash in 60% ethanol/H_2O; running tap water for 10min, and then incubated in PBS (0.2M, pH 7.4) for 30min. The sections were incubated with [^{125}I]**10** and [^{125}I]**13** in the absence or in the presence of thioflavin-T (10mM) for 30min at room temperature. The sections were then washed with saturated Li_2CO_3 in 40% ethanol for 3min and washed with 40% ethanol for 3min, followed by rinsing with water for 30s. After drying, the ^{125}I-labeled sections were exposed to phosphorus film for 2h then scanned with the phosphor imaging system (Cyclone, Packard) at the resolution of 600 dpi. The presence and localization of plaques were confirmed by the fluorescent staining with thioflavin-S (1μm) on the same sections using Stereo Discovery V12, excitation: 450-490nm, emission: 500-550nm, (Zeiss, Germany).

4.19 In vivo biodistribution in normal mice

In vivo biodistribution Studies were performed in female Kunming normal mice (average weight about 20g) and was approved by the animal care committee of Beijing Normal University. A saline solution (100μL) containing [^{125}I]**10** or [^{125}I]**13** (1μCi) was injected directly into the tail vein of mice. The mice were sacrificed at various time points post-injection. The organs of interest were removed and weighed, and the radioactivity was counted with an automatic γ-counter (WALLAC/Wizard 1470, USA). The percentage dose per gram of wet tissue was calculated by a comparison of the tissue counts to suitably diluted aliquots of the injected material.

4.20 Partition coefficient determination

The determination of partition coefficient of [^{125}I]**10** and [^{125}I]**13** were performed according to the procedure previously reported with some modifications[38]. Five μCi of [^{125}I]**10** or [^{125}I]**13** was added to the premixed suspensions containing 3 × of n-octanol and 3g of PBS (0.05M, pH 7.4) in a test tube. The test tube was vortexed for 3min at room temperature, followed by centrifugation for 5min at 3500 rpm. Two weighted samples from the n-octanol (50μL) and buffer (800μL) layers were counted. The partition coefficient was expressed as the logarithm of the ratio of the counts per gram from *n*-octanol versus that of PBS. Samples from *n*-octanol layer were repartitioned until consistent partitions of coefficient values were obtained. The measurement was done in triplicate and repeated three times.

4.21 Stability study

The stability in mouse plasma and liver homogenate was determined by incubating 10μCi purified [^{125}I]**10** in the solution of 100μL mouse plasma or mouse liver homogenate at 37℃ for 2min, 30min, 60min and 120min. Proteins were precipitated by adding 200μL acetonitrile, after centrifugation at 5000 rpm for 15min at −4℃. The radiochemical purity was analyzed by HPLC.

Acknowledgements

The authors thank Dr. Lianfeng Zhang(Institute of Laboratory Animal Science, Chinese Academy of Medical Sciences and Comparative Medicine Center of Peking Union Medical College) for kindly provided the Paraffin-embedded brain sections of AD model mice, Dr Hongmei Jia (College of Chemistry, Beijing Normal University) for many thoughtful discussions and preparation of the manuscript and Dr. Xiaoyan Zhang(College of Life Science, Beijing Normal University) for assistance in the in vitro neuropathological staining. The authors also thank Dr. Yan Cheng(Department of Patho-Functional Bioanalysis, Graduate School of Pharmaceutical Sciences, Kyoto University) for her help with preparation of the manuscript. This work was funded by NSFC (20871021).

Supplementary data

Supplementary data associated with this article can be found, in the online version, at doi: 10.1016/j.bmc.2010.02.002.

References

[1] Hardy,J. A. ;Higgins,G. A. Science. 1992,256,184.
[2] Selkoe,D. J. JAMA-J. Am. Med. Assoc. 2000,283,1615.
[3] Hardy,J. A. ;Selkoe,D. J. Science. 2002,297,353.
[4] Selkoe,D. J. Physiol. Rev. 2001,81,741.
[5] Mathis,C. A. ;Wang,Y. ;Klunk,W. E. Curr. Pharm. Des. 2004,10,1469.
[6] Nordberg,A. Curr. Opin. Neurol. 2007,20,398.
[7] Furumoto,S. ;Okamura,N. ;Iwata,R. ;Yanai,K. ;Arai,H. ;Kudo,Y. Curr. Top. Med. Chem. 2007,7, 1773.
[8] Cai,L. S. ;Innis,R. B. ;Pike,V. W. Curr. Med. Chem. 2007,14,19.
[9] Small,G. W. ;Kepe,V. ;Ercoli,L. M. ;Siddarth,P. ;Bookheimer,S. Y. ;Miller,K. J. ;Lavretsky,H. ;Burggren,A. C. ;Cole,G. M. ;Vinters,H. V. ;Thompson,P. M. ;Huang,S. C. ;Satyamurthy,N. ;Phelps,M. E. ; Barrio,J. R. N. Engl. J. Med. 2006,355,2652.
[10] Shoghi-Jadid,K. ;Small,G. W. ;Agdeppa,E. D. ;Kepe,V. ;Ercoli,L. M. ;Siddarth,P. ;Read,S. ;Saty-

amurthy, N. ; Petric, A. ; Huang, S. C. ; Barrio, J. R. Am. J. Geriatr. Psychiat. 2002, 10, 24.

[11] Agdeppa, E. D. ; Kepe, V. ; Liu, J. ; Flores-Torres, S. ; Satyamurthy, N. ; Petric, A. ; Cole, G. M. ; Small, G. W. ; Huang, S. C. ; Barrio, J. R. J. Neurosci. 2001, 21, RC189.

[12] Klunk, W. E. ; Engler, H. ; Nordberg, A. ; Wang, Y. ; Blomqvist, G. ; Holt, D. P. ; Bergstrom, M. ; Savitcheva, I. ; Huang, G. F. ; Estrada, S. ; Ausen, B. ; Debnath, M. L. ; Barletta, J. ; Price, J. C. ; Sandell, J. ; Lopresti, B. J. ; Wall, A. ; Koivisto, P. ; Antoni, G. ; Mathis, C. A. ; Langstrom, B. Ann. Neurol. 2004, 55, 306.

[13] Mathis, C. A. ; Wang, Y. ; Holt, D. P. ; Huang, G. F. ; Debnath, M. L. ; Klunk, W. E. J. Med. Chem. 2003, 46, 2740.

[14] Verhoeff, N. P. ; Wilson, A. A. ; Takeshita, S. ; Trop, L. ; Hussey, D. ; Singh, K. ; Kung, H. F. ; Kung, M. P. ; Houle, S. Am. J. Geriatr. Psychiatry 2004, 12, 584.

[15] Ono, M. ; Wilson, A. ; Nobrega, J. ; Westaway, D. ; Verhoeff, P. ; Zhuang, Z. P. ; Kung, M. P. ; Kung, H. F. Nucl. Med. Biol. 2003, 30, 565.

[16] Kudo, Y. ; Okamura, N. ; Furumoto, S. ; Tashiro, M. ; Furukawa, K. ; Maruyama, M. ; Itoh, M. ; Iwata, R. ; Yanai, K. ; Arai, H. J. Nucl. Med. 2007, 48, 553.

[17] Rowe, C. C. ; Ackerman, U. ; Browne, W. ; Mulligan, R. ; Pike, K. L. ; O'Keefe, G. ; Tochon-Danguy, H. ; Chan, G. ; Berlangieri, S. U. ; Jones, G. ; Dickinson-Rowe, K. L. ; Kung, H. F. ; Zhang, W. ; Kung, M. P. ; Skovronsky, D. ; Dyrks, T. ; Holl, G. ; Krause, S. ; Friebe, M. ; Lehman, L. ; Lindemann, S. ; Dinkelborg, L. M. ; Masters, C. L. ; Villemagne, V. L. Lancet Neurol. 2008, 7, 129.

[18] Zhuang, Z. P. ; Kung, M. P. ; Hou, C. ; Skovronsky, D. M. ; Gur, T. L. ; Plossl, K. ; Trojanowski, J. Q. ; Lee, V. M. Y. ; Kung, H. F. J. Med. Chem. 2001, 44, 1905.

[19] Zhuang, Z. P. ; Kung, M. P. ; Hou, C. ; Plossl, K. ; Skovronsky, D. ; Gur, T. L. ; Trojanowski, J. Q. ; Lee, V. M. Y. ; Kung, H. F. Nucl. Med. Biol. 2001, 28, 887.

[20] Lee, C. W. ; Kung, M. P. ; Hou, C. ; Kung, H. F. Nucl Med Biol. 2003, 30, 573.

[21] Kung, M. P. ; Hou, C. ; Zhuang, Z. P. ; Zhang, B. ; Skovronsky, D. ; Trojanowski, J. Q. ; Lee, V. M. ; Kung, H. F. Brain Res. 2002, 956, 202.

[22] Zhuang, Z. P. ; Kung, M. P. ; Wilson, A. ; Lee, C. W. ; Plossl, K. ; Hou, C. ; Holtzman, D. M. ; Kung, H. F. J. Med. Chem. 2003, 46, 237.

[23] Newberg, A. B. ; Wintering, N. A. ; Plossl, K. ; Hochold, J. ; Stabin, M. G. ; Watson, M. ; Skovronsky, D. ; Clark, C. M. ; Kung, M. P. ; Kung, H. F. J. Nucl. Med. 2006, 47, 748.

[24] Nesterov, E. E. ; Skoch, J. ; Hyman, B. T. ; Klunk, W. E. ; Bacskai, B. J. ; Swager, T. M. Angew. Chem. 2005, 117, 5588.

[25] Herland, A. ; Nilsson, K. P. R. ; Olsson, J. D. M. ; Hammarstrom, P. ; Konradsson, P. ; Inganas, O. J. Am. Chem. Soc. 2005, 127, 2317.

[26] Åslund, A. ; Herland, A. ; Hammarstrom, P. ; Nilsson, K. P. R. ; Jonsson, B. -H. ; Inganas, O. ; Konradsson, P. Bioconjugate Chem. 2007, 18, 1860.

[27] Nilsson, K. P. R. ; Åslund, A. ; Berg, I. ; Nystrom, S. ; Konradsson, P. ; Herland, A. ; Inganas, O. ; Stabo-Eeg, F. ; Lindgren, M. ; Westermark, G. T. ; Lannfelt, L. ; Nilsson, L. N. G. ; Hammarstrom, P. ACS Chem. Biol 2007, 2, 553.

[28] Åslund, A. ; Sigurdson, C. J. ; Klingstedt, T. ; Grathwohl, S. ; Bolmont, T. ; Dickstein, D. L. ; Glimsdal, E. ; Prokop, S. ; Lindgren, M. ; Konradsson, P. ; Holtzman, D. M. ; Hof, P. R. ; Heppner, F. L. ; Gandy, S. ; Jucker, M. ; Aguzzi, A. ; Hammarstrom, P. ; Nilsson, K. P. R. ACS Chem. Biol 2009, 4, 673.

[29] Zheng, M. Q. ; Yin, D. Z. ; Qiao, J. P. ; Zhang, L. ; Wang, Y. X. J. Fluorine Chem. 2008, 129, 210.

[30] Raposoa, M. M. ; Kirschb, G. Tetrahedron 2003, 59, 4891.
[31] Klunk, W. E. ; Debnath, M. L. ; Pettegrew, J. W. Biol. Psychiat. 1994, 35, 627.
[32] Chun, Y. S. ; Lim, S. J. ; Oh, S. J. ; Moon, D. H. ; Kim, D. J. ; Cho, C. G. ; Yoo, K. H. Bull. Korean Chem. Soc. 2008, 29, 1765.
[33] Goodman, M. M. ; Kirsch, G. ; Knapp, F. F. J. Heterocycl. Chem. 1984, 21, 1579.
[34] Brown, J. Endocrinology 1956, 58, 68.
[35] Brown, G. K. Physiol. Rev. 1961, 41, 189.
[36] Leder, O. Histochemistry 1982, 74, 585.
[37] Cheng, Y. ; Prusoff, W. Biochem. Pharmacol. 1973, 1973, 3099.
[38] Wu, C. Y. ; Wei, J. J. ; Gao, K. Q. ; Wang, Y. M. Bioorg. Med. Chem. 2007, 15, 2789.

Novel Imaging Agents for β-amyloid Plaque Based on the *N*-benzoylindole Core[*]

Abstract We report the synthesis and evaluation of a series of *N*-benzoylindole derivatives as novel potential imaging agents for β-amyloid plaques. In vitro binding studies using Aβ$_{1-40}$ aggregates versus [^{125}I]TZDM showed that all these derivatives demonstrated high binding affinities (K_i values ranged from 8.4 to 121.6 nM). Moreover, two radioiodinated compounds [^{125}I]**7** and [^{125}I]**14** were prepared. Autoradiography for [^{125}I]**14** displayed intense and specific labeling of Aβ plaques in the brain sections of AD model mice (C57, APP/PS1) with low background. In vivo biodistribution in normal mice exhibited sufficient initial brain uptake for imaging (2.19% and 2.00% ID/g at 2min postinjection for [^{125}I]**7** and [^{125}I]**14**, respectively). Although additional modifications are necessary to improve brain uptake and clearance from the brain, the *N*-benzoylindole may be served as a backbone structure to develop novel β-amyloid imaging probes.

Key words Alzheimer's disease, β-amyloid, binding affinity, imaging agent

Alzheimer's disease (AD), the most common form of dementia in elderly people, is characterized by cognitive impairment and memory loss. Currently, a definite confirmation of AD is attained only by post-mortem histopathology of the brain. It is generally accepted that the accumulation of senile plaques (SPs) and neurofibrillary tangles (NFTs) are two pivotal clinical pathological features of AD, and the formation of amyloid plaques is prior to the onset of clinical symptoms[1,2]. Therefore, noninvasive detection of Aβ plaques in vivo by PET (positron emission tomography) or SPECT (single photon emission computed tomography) would be very useful for early diagnosis of AD[3,4].

Currently, a lot of radiolabeled ligands derived from Congo Red (CR) or Thioflavin T (ThT) have been developed as Aβ plaques imaging probes. Several PET tracers, such as 2-(4'-[^{11}C]methylaminophenyl)-6-hydroxybenzothiazole ([^{11}C]PIB)[5,6], 4-*N*-[^{11}C]methylamino-4'-hydroxystilbene ([^{11}C]SB-13)[7,8], [^{18}F]-4-(*N*-methylamino)-4'-(2-(2-(2-fluoroethoxy)ethoxy)ethoxy)-stilbene ([^{18}F]BAY94-9172)[9], [^{18}F]-2-(3-fluoro-4-(methylamino)phenyl)benzo[d]thiazol-6-ol ([^{18}F]GE-067)[10], [^{18}F]-(*E*)-4-(2-(6-(2-(2-(2-fluoroethoxy)ethoxy)ethoxy)pyridin-3-yl)vinyl)-*N*-methylaniline ([^{18}F]AV-45)[11,12] have been evaluated in human studies. However, [^{123}I]-6-iodo-2-(4'-dimethylamino-)phenyl-imidazo[1,2]pyridine ([^{123}I]IMPY)[13-15], the only SPECT tracer tested in human studies, has failed because of its high lipophilicity, in vivo instability and insufficient target-to-background ratio (Fig. 1). In

---

[*] Copartner: Yang Yang, Duan Xinhong, Deng Junyuan, Jin Bing, Jia Hongmei. Reprinted from *Bioorganic & Medicinal Chemistry Letters*, 2011, 21: 5594-5597.

comparison with PET, SPECT is a more widely accessible and cost-effective technique in terms of routine diagnostic use. Therefore, the development of more useful Aβ plaques imaging agents for SPECT has been a critical issue. During the past decade, many SPECT tracers with new backbone structures have been reported as Aβ probes, such as radio-iodinated aza-diphenylacetylenes[16], styrylpyridines[17], flavone derivatives[18,19], aurone derivatives[20,21], chalcones derivatives[22,23], (E)-5-styryl-2,2'-bithiophen derivatives[24], as well as ^{99m}Tc labeled chalcone[25], flavone and aurone[26].

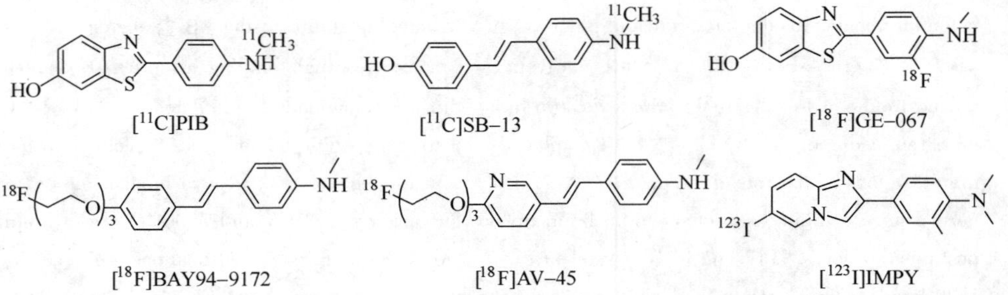

Fig. 1 Chemical structure of Aβ imaging probes for clinical study

In the search for novel Aβ imaging agents with improved properties, many small-molecule Aβ inhibitors, including nonsteroidal anti-inflammatory drugs (NSAIDs), have been developed with necessary modifications in order to improve the affinities to Aβ plaques and enhance permeability through the BBB[27]. Agdeppa et al. have reported that naproxen and ibuprofen shared the same binding sites of [^{18}F]FDDNP on Aβ fibrils. The K_i values of (S)-naproxen, (R)-ibuprofen, and (S)-ibuprofen were (5.70 ± 1.31) nM, (44.4 ± 17.4) μM and (11.3 ± 5.20) μM, respectively[28]. Indomethacin has been known to inhibit Aβ fibril formation from Aβ at required concentrations, and the anti-aggregation effect of it may due to its interaction with Aβ[29]. In our present study, we selected indomethacin as the lead compound aiming to develop radiotracers with a new core structure to image Aβ plaques. In order to simplify the synthesis process and enhance permeability through the BBB, methyl and CH_2COOH groups were removed from the chemical structure. Furthermore, "C=C" was introduced to improve the affinities to Aβ plaques (Scheme 1). A series of the resulted N-benzoylindole-based compounds were synthesized and their binding affinities for Aβ$_{1-40}$ aggregates were measured. Moreover, two radio-iodinated derivatives were prepared and evaluated as potential SPECT tracers for imaging amyloid plaques in the brain.

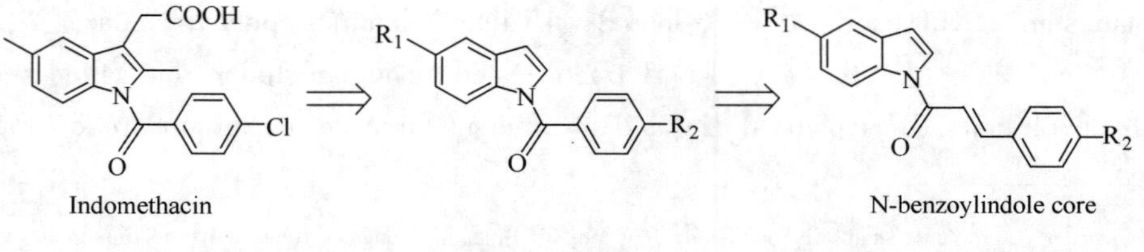

Scheme 1 Design considerations of N-benzoylindole-based derivatives

The synthetic route of *N*-benzoylindole derivatives (**3-17**) is outlined in Scheme 2. The key step was the Schotten-Baumann reaction between the suitable 5-substituted indoles and **2**. The tributyltin derivatives (**18**,**19**) were prepared from the bromo-precursors (**6**,**13**) using an exchange reaction catalyzed by Pd(0) with the yields of 15.8% and 22.6%, respectively. [^{125}I]**7** and [^{125}I]**14** were prepared via the iododestannylation reaction using hydrogen peroxide as the oxidant (Scheme 3). The reaction was quenched with saturated $NaHSO_3$ solution. The resulting mixture was purified by HPLC using a reverse-phase column and mobile phase consisting of acetonitrile with a flow rate of 1 mL/min. The radio-ligands were co-injected and co-eluted with the corresponding nonradioactive **7** and **14**. The radiochemical yields of [^{125}I]**7** and [^{125}I]**14** were 57%-72% and 50%-71%, respectively. The radiochemical purities of [^{125}I]**7** and [^{125}I]**14** were both greater than 98%. The specific activity of the no-carrier-added [^{125}I]NaI was >2200Ci/mmol at time of delivery. The specific activities of the products were not determined. The log *D* values of [^{125}I]**7** and [^{125}I]**14** were 2.32±0.02 and 2.53±0.04 (measured by a partition between *n*-octanol and pH 7.4 PBS buffer), respectively, which indicate a potential BBB permeability.

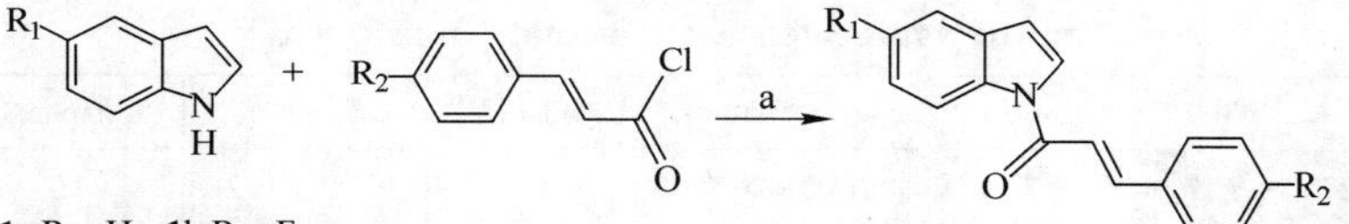

1a.R_1=H, **1b.**R_1=F
1c.R_1=Cl, **1d.**R_1=Br
1e.R_1=I, **1f.**R_1=CH_3
1g.R_1=OCH_3, **1h.**R_1=OH

2a.R_2=OCH_3
2b.R_2=H

3.R_1=H, R_2=OCH_3 **4.**R_1=F, R_2=OCH_3; **5.**R_1=Cl, R_2=OCH_3;
6.R_1=Br, R_2=OCH_3 —b→ **18.**R_1=$SnBu_3$, R_2=OCH_3;
7.R_1=I, R_2=OCH_3; **8.**R_1=CH_3, R_2=OCH_3 **9.**R_1=OCH_3, R_2=OCH_3;
10.R_1=H, R_2=H; **11.**R_1=F, R_2=H; **12.**R_1=Cl, R_2=H;
13.R_1=Br, R_2=H —b→ **19.**R_1=$SnBu_3$, R_2=H;
14.R_1=I, R_2=H; **15.**R_1=CH_3, R_2=H; **16.**R_1=OCH_3, R_2=H;
17.R_1=OH, R_2=H

Scheme 2 Reagents and conditions
a—Et_3N, CH_2Cl_2, reflux; b—$(Bu_3Sn)_2$, $(Ph_3P)_4Pd$, toluene, reflux

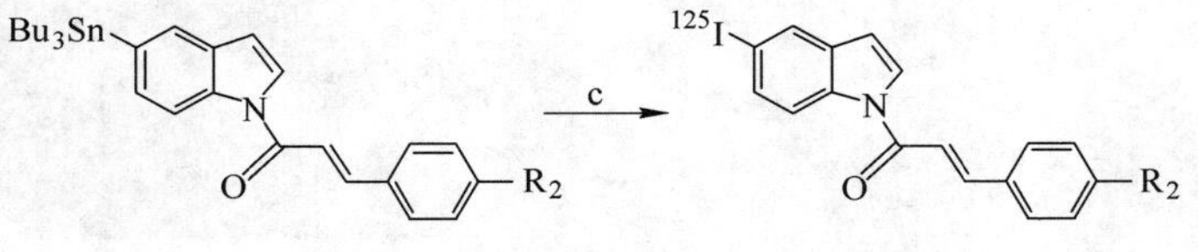

[^{125}I]**7**.R_2=OCH_3
[^{125}I]**14**.R_2=H

Scheme 3 Reagents and conditions
c—[^{125}I]NaI, H_2O_2, HCl, rt

The affinities of these N-benzoylindole derivatives (3-17) for $A\beta_{1-40}$ aggregates were determined by competition binding assay using $[^{125}I]$TZDM as the radiolabeled standard[30]. As shown in Table 1, the binding affinities of these N-benzoylindole derivatives for $A\beta_{1-40}$ aggregates varied from 8.4 to 121.6nM, suggesting that all these compounds share the same binding site with ThT. (E)-1-(1H-Indol-1-yl)-3-phenylprop-2-en-1-one (10) without any substituents showed good affinity (K_i = 15.9nM). The introduction of methoxy group into the R_2 group in the phenyl-ring effected the affinities slightly (3, K_i = 27.3nM). The introduction of halogen, such as F, Cl, Br, into the R_1 group in the indole-ring resulted in slight difference in affinities (K_i = 32.4nM, 18.1nM, 25.4nM, 32.7nM, 36.4nM and 18.9nM for compounds **4, 5, 6, 11, 12** and **13**, respectively), while modification with methoxy or methyl group at the above position resulted to a significant reduction in binding affinities (K_i = 103.3nM, 60.8nM, 127.7nM and 68.2nM for compounds **8, 9, 15** and **16**, respectively). Since derivatives **7** and **14** with iodine-substituent on the indole-ring displayed the highest binding affinities (K_i = 10.2 and 8.4nM, respectively), we prepared $[^{125}I]$**7** and $[^{125}I]$**14** for further biological studies.

Table 1 K_i values of N-benzoylindole derivatives on $[^{125}I]$TZDM binding to the aggregated $A\beta_{1-40}$ peptide in solution

Compound	$K_i^{①}$/nM	Compound	$K_i^{①}$/nM	Compound	$K_i^{①}$/nM	Compound	$K_i^{①}$/nM
3	27.3 ± 1.3	7	10.2 ± 1.7	11	32.7 ± 1.2	15	127.7 ± 1.5
4	32.4 ± 1.8	8	103.3 ± 1.7	12	36.4 ± 1.6	16	68.2 ± 1.3
5	18.1 ± 2.0	9	60.8 ± 1.2	13	18.9 ± 1.2	17	21.6 ± 4.6
6	25.4 ± 1.6	10	15.9 ± 1.6	14	8.4 ± 1.4		

① Measured in triplicate with results given as the mean ± SD.

The radioiodinated probes $[^{125}I]$**14** was investigated by in vitro autoradiography in the brain sections of a transgenic model mouse (C57, APP/PS1, 12 months). As shown in Fig. 2, autoradiographic images of $[^{125}I]$**14** showed excellent labeling of Aβ plaques in the cortex region of the brain sections, and no remarkable accumulation of radioactivity were observed in white mat-

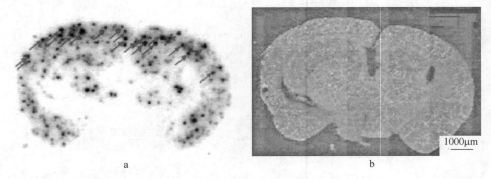

Fig. 2 Autoradiography of $[^{125}I]$**14** in vitro in Tg model mouse (C57-APP/PS1, 12 months old, male) brain sections (a), and the presence and distribution of plaques in the sections were confirmed with thioflavin-S (b) (red arrows)

ter. The same sections were also stained with thioflavin-S and the localizations of Aβ plaques were in accord with the results of autoradiography. These results demonstrated that $[^{125}I]$**14** was specific for Aβ plaques, which was consistent with high bind affinity of compound **14** to Aβ$_{1-40}$ aggregates.

Biodistribution experiments were performed in normal mice in order to evaluate the pharmacokinetic properties of these derivatives. $[^{125}I]$**7** and $[^{125}I]$**14** were injected intravenously into normal mice for biodistribution studies. As shown in Table 2, the initial brain uptake for $[^{125}I]$**7** and $[^{125}I]$**14** were 2.19 and 2.00% ID/g at 2 min postinjection, respectively, indicating a level sufficient for brain imaging probe. But the washout of these probes from the brain in normal mice appears to be relatively slow (1.60% and 1.57% ID/g at 30 min postinjection, respectively), suggesting the high nonspecific binding of these probes. In addition, the initial blood uptake of the two probes was higher (13.91% ID/g for $[^{125}I]$**7** and 14.13% ID/g for $[^{125}I]$**14**), which may be due to the relatively high lipophilicity. It has been suggested that the lipophilicity of Aβ imaging agents may play an important role in uptake and washout. Additional structural modifications, such as introducing a hydrophilic group to decrease the lipophilicity, are needed to achieve the fast washout rate from brain and blood for the *N*-benzoylindole-based compounds.

Table 2 Biodistribution in normal mice after iv injection of $[^{125}I]$**7** and $[^{125}I]$**14**
(%ID/g, avg of 5 mice ± SD) **and its partition coefficient** (*D*)

Organ	2 min	15 min	30 min	60 min	120 min	240 min
$[^{125}I]$**7** ($\lg D = 2.32 \pm 0.02$)						
Blood	13.91 ± 1.86	5.82 ± 0.31	4.90 ± 0.44	3.79 ± 0.71	3.12 ± 0.48	2.78 ± 0.35
Brain	2.19 ± 0.16	2.29 ± 0.13	1.60 ± 0.14	1.04 ± 0.17	0.42 ± 0.08	0.21 ± 0.04
Heart	12.64 ± 0.78	3.52 ± 0.26	2.45 ± 0.21	1.53 ± 0.19	1.21 ± 0.45	1.17 ± 0.07
Liver	24.76 ± 3.16	18.05 ± 1.12	14.05 ± 1.09	12.30 ± 1.32	9.30 ± 1.12	7.65 ± 1.12
Spleen	6.00 ± 1.74	8.04 ± 0.45	7.12 ± 2.39	7.16 ± 3.55	6.18 ± 2.39	4.45 ± 1.22
Lung	21.76 ± 4.31	11.13 ± 0.60	9.05 ± 1.44	6.39 ± 1.19	4.31 ± 0.71	2.95 ± 0.57
Kidney	12.01 ± 0.28	9.31 ± 0.38	8.21 ± 0.93	5.71 ± 0.36	4.21 ± 0.66	4.05 ± 0.72
Stomach[①]	0.71 ± 0.18	4.06 ± 0.09	4.92 ± 1.06	5.00 ± 1.27	5.31 ± 1.12	4.14 ± 1.23
Muscle	2.73 ± 0.37	2.62 ± 0.38	1.63 ± 0.32	1.30 ± 0.20	0.65 ± 0.15	0.92 ± 0.18
$[^{125}I]$**14** ($\lg D = 2.53 \pm 0.04$)						
Blood	14.13 ± 0.44	3.91 ± 0.48	3.75 ± 0.29	2.88 ± 0.92	1.61 ± 0.24	1.07 ± 0.14
Brain	2.00 ± 0.17	1.98 ± 0.13	1.57 ± 0.22	0.77 ± 0.12	0.33 ± 0.04	0.17 ± 0.03
Heart	10.72 ± 0.48	2.90 ± 0.42	2.07 ± 0.11	1.53 ± 0.23	0.99 ± 0.12	0.61 ± 0.16
Liver	15.89 ± 1.38	7.73 ± 0.83	6.48 ± 0.52	4.09 ± 0.85	4.86 ± 1.84	2.23 ± 0.48
Spleen	4.08 ± 0.60	2.44 ± 0.62	1.71 ± 0.23	1.85 ± 1.31	0.63 ± 0.54	0.59 ± 0.12
Lung	14.50 ± 3.50	6.73 ± 0.89	4.88 ± 0.78	2.66 ± 1.00	2.10 ± 0.22	1.27 ± 0.48
Kidney	11.22 ± 1.08	9.68 ± 1.03	10.16 ± 1.12	6.14 ± 0.30	3.54 ± 0.33	2.22 ± 0.72
Stomach[①]	1.09 ± 0.53	1.52 ± 0.21	1.72 ± 0.42	2.08 ± 0.68	2.29 ± 0.67	1.30 ± 0.38
Muscle	3.97 ± 0.32	2.47 ± 0.55	2.06 ± 0.53	1.06 ± 0.16	0.84 ± 0.53	0.19 ± 0.16

① Expressed as %ID/organ.

In summary, *N*-benzoylindole-based compounds have been synthesized and evaluated as novel imaging probes for Aβ plaques. They showed high binding affinities with K_i values in the nM range in vitro. Autoradiography for [^{125}I]**14** indicated that it stained Aβ plaques in Tg2576 mouse brain clearly. *N*-benzoylindole-based derivatives penetrated the brain was encouraging. Although additional modifications are necessary to improve the in vivo properties of these derivatives, *N*-benzoylindole may be served as a new backbone structure to develop β-amyloid imaging probes.

Acknowledgments

This work was funded by NSFC(20871021) and the Fundamental Research Funds for the Central Universities. The authors thank Dr. Mengchao Cui (College of Chemistry, Beijing Normal University) for his kindness to provide the Paraffin-embedded brain sections of AD model mice. The authors also thank Mr. Jin Liu (College of Life Science, Beijing Normal University) for providing the equipment of the Zeiss Oberver Z1.

Supplementary data

Supplementary data (procedure for the preparation of new *N*-benzoylindole-based derivatives, in vitro binding studies, in vitro autoradiography using Tg2576 mouse brain sections and biodistribution studies in normal mice) associated with this article can be found, in the online version, at doi: 10. 1016/j. bmcl. 2011. 06. 077.

References

[1] Selkoe,D. J. J. Am. Med. Assoc. 2000,283,1615.
[2] Hardy,J. ;Selkoe,D. J. Science 2002,297,353.
[3] Nordberg,A. Lancet Neurol. 2004,3,519.
[4] Cai,L. S. ;Innis,R. B. ;Pike,V. W. Curr. Med. Chem. 2007,34,19.
[5] Mathis, C. A. ; Wang, Y. M. ; Holt, D. P. ; Huang, G. F. ; Debnath, M. L. ; Klunk, W. E. J. Med. Chem. 2003,46,2740.
[6] Klunk, W. E. ; Engler, H. ; Nordberg, A. ; Wang, Y. M. ; Blomqvist, G. ; Holt, D. P. ; Bergström, M. ; Savitcheva,I. ;Huang,G. F. ;Estrada,S. ;Ausén,B. ;Debnath,M. L. ;Barletta,J. ;Price,J. C. ;Sandell, J. ;Lopresti,B. J. ;Wall,A. ;Koivisto,P. ;Antoni,G. ;Mathis,C. A. ;Langström,B. Ann. Neurol. 2004,55, 306.
[7] Ono,M. ;Wilson,A. ;Nobrega,J. ;Westaway,D. ;Verhoeff,P. ;Zhuang,Z. P. ;Kung,M. P. ;Kung,H. F. Nucl. Med. Biol. 2003,30,565.
[8] Verhoeff,N. P. ;Wilson,A. A. ;Takeshita,S. ;Trop,L. ;Hussey,D. ;Singh,K. ;Kung,H. F. ;Kung,M. P. ; Houle,S. Am. J. Geriat. Psychiat. 2004,12,584.
[9] Rowe, C. C. ; Ackerman, U. ; Browne, W. ; Mulligan, R. ; Pike, K. L. ; O'Keefe, G. ; Tochon-Danguy, H. ; Chan, G. ; Berlangieri, S. U. ; Jones, G. ; Dickinson-Rowe, K. L. ; Kung, H. F. ; Zhang, W. ; Kung, M. P. ; Skovronsky, D. ; Dyrks, T. ; Holl, G. ; Krause, S. ; Friebe, M. ; Lehman, L. ; Lindemann, S. ; Dinkelborg, L. M. ;Masters,C. L. ;Villemagne,V. L. Lancet Neurol. 2008,7,129.
[10] Koole, M. ; Lewis, D. M. ; Buckley, C. ; Nelissen, N. ; Vandenbulcke, M. ; Brooks, D. J. ; Vandenberghe,

R. ; Laere, K. V. J. Nucl. Med. 2009, 50, 818.
[11] Choi, S. R. ; Golding, G. ; Zhuang, Z. P. ; Zhang, W. ; Lim, N. ; Hefti, F. ; Benedum, T. E. ; Kilbourn, M. R. ; Skovronsky, D. ; Kung, H. F. J. Nucl. Med. 2009, 50, 1887.
[12] Kung, H. F. ; Choi, S. R. ; Qu, W. C. ; Zhang, W. ; Skovronsky, D. J. Med. Chem. 2010, 53, 933.
[13] Kung, M. P. ; Hou, C. ; Zhuang, Z. P. ; Zhang, B. ; Skovronsky, D. ; Trojanowski, J. Q. ; Lee, V. M. ; Kung, H. F. Brain Res. 2002, 956, 202.
[14] Zhuang, Z. P. ; Kung, M. P. ; Wilson, A. ; Lee, C. W. ; Plössl, K. ; Hou, C. ; Holtzman, D. M. ; Kung, H. F. J. Med. Chem. 2003, 46, 237.
[15] Newberg, A. B. ; Wintering, N. A. ; Plössl, K. ; Hochold, J. ; Stabin, M. G. ; Watson, M. ; Skovronsky, D. ; Clark, C. M. ; Kung, M. P. ; Kung, H. F. J. Nucl. Med. 2006, 47, 748.
[16] Qu, W. C. ; Kung, M. P. ; Hou, C. ; Jin, L. W. ; Kung, H. F. Bioorg. Med. Chem. Lett. 2007, 17, 3581.
[17] Qu, W. C. ; Kung, M. P. ; Hou, C. ; Benedum, T. E. ; Kung, H. F. J. Med. Chem. 2007, 50, 2157.
[18] Ono, K. ; Yoshiike, Y. ; Takashima, A. ; Hasegawa, K. ; Naiki, H. ; Yamada, M. J. Neurochem. 2003, 87, 172.
[19] Ono, M. ; Yoshida, N. ; Ishibashi, K. ; Haratake, M. ; Arano, Y. ; Mori, H. ; Nakayama, M. J. Med. Chem. 2005, 48, 7253.
[20] Ono, M. ; Maya, Y. ; Haratake, M. ; Ito, K. ; Mori, H. ; Nakayama, M. Biochem. Biophys. Res. Commun. 2007, 361, 116.
[21] Maya, Y. ; Ono, M. ; Watanabe, H. ; Haratake, M. ; Saji, H. ; Nakayama, M. Bioconjugate. Chem. 2009, 20, 95.
[22] Ono, M. ; Haratake, M. ; Mori, H. ; Nakayama, M. Bioorg. Med. Chem. 2007, 15, 6802.
[23] Ono, M. ; Hori, M. ; Haratake, M. ; Tomiyama, T. ; Mori, H. ; Nakayama, M. Bioorg. Med. Chem. 2007, 15, 6388.
[24] Cui, M. C. ; Li, Z. J. ; Tang, R. K. ; Jia, H. M. ; Liu, B. L. Euro. J. Med. Chem. 2011, 46, 2908.
[25] Ono, M. ; Ikeoka, R. ; Watanabe, H. ; Kimura, H. ; Fuchigami, T. ; Haratake, M. ; Saji, H. ; Nakayama, M. ACS Chem. Neurosci. 2010, 1, 598.
[26] Ono, M. ; Ikeoka, R. ; Watanabe, H. ; Kimura, H. ; Fuchigami, T. ; Haratake, M. ; Saji, H. ; Nakayama, M. Bioorg. Med. Chem. Lett. 2010, 20, 5743.
[27] Duan, X. H. ; Liu, B. L. Science in China Series B: Chemistry, 2008, 51, 801.
[28] Agdeppa, E. D. ; Kepe, V. ; Petric, A. ; Satyamurthy, N. ; Liu, J. ; Huang, S. C. ; Small, G. W. ; Cole, G. M. ; Barro, J. R. Neuroscience, 2003, 117, 723.
[29] Hirohata, M. ; Ono, K. ; Naiki, H. ; Yamada, M. Neuropharmacology 2005, 49, 1088.
[30] Zhuang, Z. P. ; Kung, M. P. ; Hou, C. ; Skovronsky, D. M. ; Gur, T. L. ; Plössl, K. ; Trojanowski, J. Q. ; Lee, V. M. ; Kung, H. F. J. Med. Chem. 2001, 44, 1905.

Novel (E)-5-styryl-2,2'-bithiophene Derivatives as Ligands for β-amyloid Plaques[*]

Abstract In continuation of our investigation on the bithiophene structure as potential β-amyloid probes, a series of (E)-5-styryl-2,2'-bithiophene(SBTP) derivatives was designed and synthesized. *In vitro* binding showed that all of them displayed high binding affinities to Aβ$_{1-42}$ aggregates (K_i = 0.10-41.05nM). Moreover, two radio-iodinated probes, [^{125}I]-(E)-5-(4-iodostyryl)-2,2'-bithiophene ([^{125}I]**8**) and [^{125}I]-(E)-5-iodo-5'-(4-methoxystyryl)-2,2'-bithiophene ([^{125}I] **31**) were prepared. Both of them displayed specific labeling of Aβ plaques in the brain sections of AD model mice with low background. *In vivo* biodistribution in normal mice indicated that [^{125}I]**8** exhibited high initial brain uptake (2.11% ID/g at 2min) and rapid clearance (0.41% ID/g at 30min). These preliminary results suggest that SBTP derivatives may be served as novel β-amyloid imaging probes.

Key words Alzheimer's disease, β-amyloid plaques, imaging agent, bithiophene, binding assay

1 Introduction

Alzheimer's disease (AD) is an irreversible neurodegenerative disorder characterized by progressive decline in cognitive function of brain and behavior. Histopathologically, β-amyloid plaques (Aβ) and neurofibrillary tangles (NFTs) are found in the brain of patients suffering from AD[1]. Although the etiology of AD is not completely understood, the most widely accepted theory concerning the etiology of AD is the amyloid cascade hypothesis[2,3]. The clinical diagnosis of this disease is made through the neurological observations and history of patients, which is often difficult and unreliable. To facilitate the early diagnosis of AD, functional imaging techniques such as positron emission tomography (PET) and single photon emission computed tomography (SPECT) have been employed. For this purpose, a radionuclide-labeled probe that specifically binds to Aβ plaques in the brain may greatly facilitate the diagnosis of AD[4,5].

During the past decade, several attempts have been made to develop specific radiotracers for *in vivo* imaging Aβ plaques with PET and SPECT. Most of the reported probes are based on the core structure of Congo Red (CR), Thioflavin T (ThT) and DDNP. Several PET tracers such as 2-(4'-[^{11}C]methylaminophenyl)-6-hydroxybenzothiazole ([^{11}C] PIB)[6,7], [^{18}F]-2-(1-(2-(N-(2-fluoroethyl)-N-methylamino)-naphthalene-6-yl) ethylidene) malononitrile ([^{18}F] FDDNP)[8-10], 4-N-[^{11}C]methylamino-4'-hydroxystilbene ([^{11}C] SB-13)[11,12], [^{18}F]-4-(N-methylamino)-4'-(2-(2-(2-fluoroethoxy)ethoxy)ethoxy)-stilbene ([^{18}F] BAY94-9172)[13] and [^{18}F]-(E)-4-(2-(6-(2-(2-(2-fluoroethoxy)ethoxy)ethoxy)pyridin-3-yl)vinyl)-N-meth-

---

[*] Copartner: Cui Mengchao, Li Zijing, Tang Ruikun, Jia Hongmei. Reprinted from *European Journal of Medicinal chemistry*, 2011, 46(7):2908-2916.

ylaniline([^{18}F]AV-45)[14,15] have been evaluated in human studies. Some of the candidates have already moved into Phase Ⅱ, Phase Ⅲ and even finalization stages(such as Florbetapir, [^{18}F]AV-45, is waiting for the approval from FDA). On the other hand, to the best of our knowledge, the only SPECT tracer tested in human studies is [^{123}I]-6-iodo-2-(4′-dimethylamino-)phenyl-imidazo[1,2]pyridine([^{123}I]IMPY)[16-18]. However, some undesirable properties including high lipophilicity, *in vivo* instability together with insufficient target-to-background ratio may account for the failure of [^{123}I]IMPY in clinical trials. After that, there is no report of any Aβ imaging candidate for SPECT moving into clinical trial. For the routine clinical applications, imaging with SPECT has some advantages over PET such as more widespread availability, no need for an on-site cyclotron and lower cost. From this point of view, there is an urgent need for developing Aβ imaging agents for SPECT. In the past few years, lots of the research related to Aβ imaging agents for SPECT have been reported. For example, Qu et al. reported radio-iodinated aza-diphenylacetyllenes[19] and styrylpyridines[20] as SPECT imaging agents for Aβ plaque detection. Ono et al. designed and synthesized radio-iodinated aurone derivatives[21,22], chalcones and their related derivatives[23,24], as well as ^{99m}Tc labeled flavonoids including chalcone[25], flavone and aurone[26] as probes for SPECT imaging of Aβ plaques.

In a search on novel Aβ binding probes, Nesterov et al. first proposed, synthesized and evaluated bithiophene type molecules as fluorescent imaging markers for Aβ plaques[27]. A bithiophene molecule, NIAD-4 had been screened out as a simple fluorescent marker for optical imaging of amyloid plaques in brain(Fig. 1). Later, Nilsson et al. identified that a class of luminescent conjugated polythiophenes(LCPs) with a striking specificity for Aβ plaques as shown in Fig. 1[28-30]. In our previous work, a series of benzothiazole derivatives based on the bithiophene structure(TZBP) was evaluated for the imaging of Aβ plaques. These derivatives displayed excellent binding affinity to Aβ aggregates[31]. Inspired by these successful results we decided to perform extensive research on the ligands with bithiophene structure, which may play an important role in maintaining the specific binding ability to Aβ plaques. Herein, we report a series of bithiophene derivatives combined with a vinylbenzene structure [(*E*)-5-styryl-2,2′-bithiophene(SBTP)] as novel Aβ imaging agents.

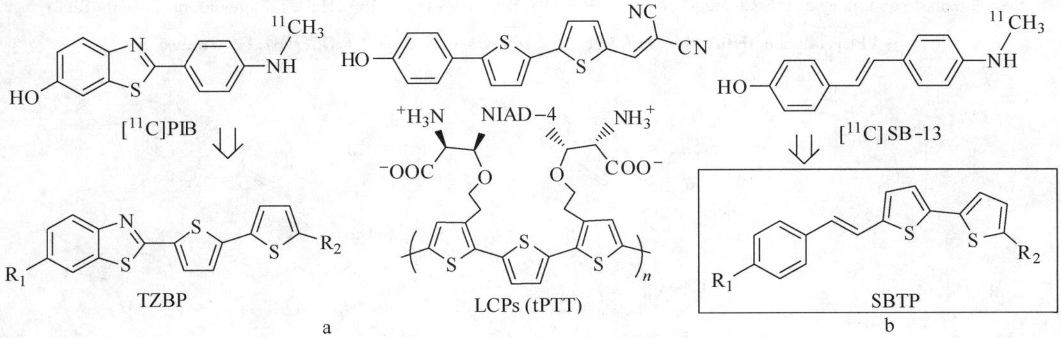

Fig. 1 Structures of [^{11}C]PIB, [^{11}C]SB-13, LCPs(tPTT) and TZBP(a) and structure of the designed SBTP derivatives(b)

2 Results and Discussion

2.1 Chemistry

The synthetic route of the SBTP derivatives is outlined in Scheme 1. The key step was the base-catalyzed Wittig reaction between substituted triphenyl phosphonium ylide and 5′-substituted-2, 2′-bithiophene-5-carbaldehydes. The substituted triphenyl phosphonium ylides reacted with 5′-substituted-2, 2′-bithiophene-5-carbaldehydes in the presence of CH_3ONa under reflux in THF to form the target SBTP derivatives. The chemical yields of compounds **4-33** were ranged from 14% to 90%. Reduction of the nitro group of **9** provided the amino-substituted compound **10**. The synthetic route of the tributyltin precursors (**34**, **35**) is shown in Scheme 2. The yields of **34** and **35** from the corresponding bromo compounds (**7**, **23**) under a bromo to tributyltin exchange reaction catalyzed by $Pd(PPh_3)_4$ were 21.7% and 39.3%, respectively. The tributyltin derivatives can be readily used as starting materials to prepare the radio-iodinated ligands.

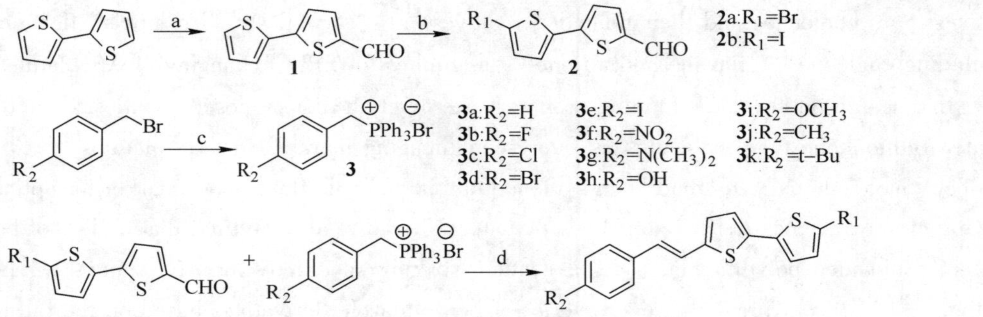

2a: R_1=Br			
2b: R_1=I			

3a: R_2=H	3e: R_2=I	3i: R_2=OCH$_3$
3b: R_2=F	3f: R_2=NO$_2$	3j: R_2=CH$_3$
3c: R_2=Cl	3g: R_2=N(CH$_3$)$_2$	3k: R_2=t-Bu
3d: R_2=Br	3h: R_2=OH	

4: R_1=H R_2=H	12: R_1=H R_2=OCH$_3$	20: R_1=Br R_2=NO$_2$	28: R_1=I R_2=Cl
5: R_1=H R_2=F	13: R_1=H R_2=CH$_3$	21: R_1=Br R_2=N(CH$_3$)$_2$	29: R_1=I R_2=Br
6: R_1=H R_2=Cl	14: R_1=H R_2=t-Bu	22: R_1=Br R_2=OH	30: R_1=I R_2=I
7: R_1=H R_2=Br	15: R_1=Br R_2=H	23: R_1=Br R_2=OCH$_3$	31: R_1=I R_2=OCH$_3$
8: R_1=H R_2=I	16: R_1=Br R_2=F	24: R_1=Br R_2=CH$_3$	32: R_1=I R_2=CH$_3$
9: R_1=H R_2=NO$_2$	17: R_1=Br R_2=Cl	25: R_1=Br R_2=t-Bu	33: R_1=I R_2=t-Bu
10: R_1=H R_2=NH$_2$	18: R_1=Br R_2=Br	26: R_1=I R_2=H	
11: R_1=H R_2=N(CH$_3$)$_2$	19: R_1=Br R_2=I	27: R_1=I R_2=F	

Scheme 1 Synthesis of the SBTP derivatives (**4-33**)

Reagents and conditions: a—$POCl_3$, DMF, 60 ℃; b—AcOH, Br_2, room temp (**2a**), HgO, I_2, toluene, room temp (**2b**); c—PPh_3, toluene, reflux; d—THF, CH_3ONa, reflux; e—$SnCl_2·2H_2O$, EtOH, HCl, reflux

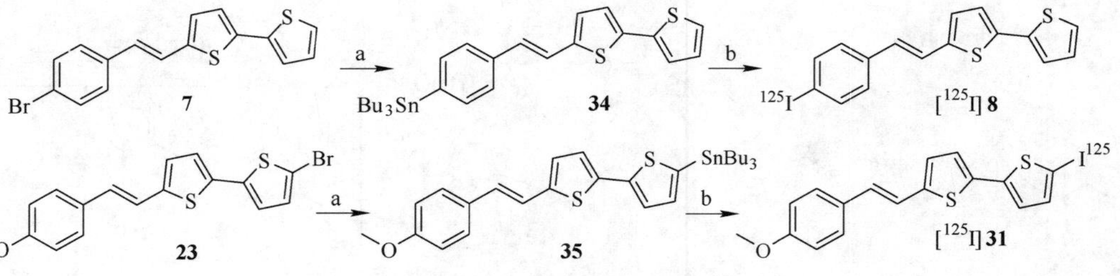

Scheme 2 Radiolabeling of ligands **8** and **31**

Reagents and conditions: a—$(Bu_3Sn)_2$, $(PPh_3)_4Pd$, toluene; b—[^{125}I]NaI, HCl(1M), H_2O_2(3%)

2.2 Binding studies using Aβ aggregates in solution

The affinity of these SBTP derivatives(**4-33**) for $A\beta_{1-42}$ aggregates was determined by competition binding assay using [^{125}I]TZDM as radio-ligand. The K_i values shown in Table 1 suggest that all of these new compounds inhibit the binding of [^{125}I]TZDM in a dose-dependent manner with a high binding affinity for $A\beta_{1-42}$ aggregates($K_i < 10$nM) except for compound **10** (K_i = 41.05nM). The affinity was increased by bromination or iodination of the 2 position of bithiophene moiety(e. g. , 4-5, 15-16, 26-27). Bromanation or iodination of the *para* position of phenyl ring also increased the binding affinity. The tertiary *N, N*-dimethylamino analogs of **11** and **21** were found to have a higher affinity than the primary amino analog **10**, which is in accord with the previously limited data on tertiary and primary amino analogs[5,23]. Methylation of the hydroxy group improved the binding affinity [e. g. , OH(**22**) versus OCH_3(**23**)]. The ligands with both electron-withdrawing substituents [e. g. , NO_2(**9**,**20**)] and electron-donating substituents [e. g. , OCH_3(**12**,**23**), $N(CH_3)_2$(**11**,**21**)] showed high affinities to Aβ aggregates, especially compound **20** with a nitro group displayed the highest affinity(K_i = 0.10nM).

Table 1 Inhibition constants(K_i, nM) for binding to aggregates of $A\beta_{1-42}$ versus [^{125}I]TZDM①

Compound	R_1	R_2	K_i/nM	Compound	R_1	R_2	K_i/nM
4	H	H	4.78 ± 1.01	19	Br	I	0.51 ± 0.11
5	H	F	4.06 ± 0.32	20	Br	NO_2	0.10 ± 0.02
6	H	Cl	0.74 ± 0.12	21	Br	$N(CH_3)_2$	0.36 ± 0.08
7	H	Br	0.99 ± 0.24	22	Br	OH	3.60 ± 0.79
8	H	I	0.92 ± 0.08	23	Br	OCH_3	0.72 ± 0.31
9	H	NO_2	0.56 ± 0.13	24	Br	CH_3	0.45 ± 0.22
10	H	NH_2	41.05 ± 3.65	25	Br	*t*-Bu	0.95 ± 0.17
11	H	$N(CH_3)_2$	0.73 ± 0.27	26	I	H	0.24 ± 0.09
12	H	OCH_3	1.48 ± 0.19	27	I	F	0.55 ± 0.19
13	H	CH_3	1.15 ± 0.34	28	I	Cl	1.10 ± 0.07
14	H	*t*-Bu	0.73 ± 0.05	29	I	Br	0.19 ± 0.02
15	Br	H	1.32 ± 0.06	30	I	I	0.63 ± 0.28
16	Br	F	0.81 ± 0.16	31	I	OCH_3	0.30 ± 0.05
17	Br	Cl	1.02 ± 0.34	32	I	CH_3	0.80 ± 0.03
18	Br	Br	0.22 ± 0.03	33	I	*t*-Bu	1.58 ± 0.24

① Measured in triplicate with results given as the mean ± SD.

In addition, the high affinities of compounds **14, 25** and **33** containing a bulk *tert*-butyl group indicated high tolerance for steric bulk on the *para* position of phenyl ring. A long polyethylene glycol chain, which is used to improve the pharmacokinetic properties of ^{18}F labeled radiotracers[32], may be introduced at this position to design new SBTP PET probes. More importantly, it is possible to design ^{99m}Tc labeled SBTP SPECT probes for Aβ imaging.

2.3 Radiochemistry

Due to the high binding affinities observed for iodinated ligands 8 (K_i = 0.92nM) and 31 (K_i = 0.30nM), these two ligands were chosen for radiolabeling and further biological evaluations. The novel radio-iodinated ligands [^{125}I]8 and [^{125}I]31 were prepared via an iododestannylation reaction using hydrogen peroxide as an oxidant. The reaction was quenched with saturated NaHSO$_3$. The resulting mixture was purified by HPLC. The radioactive product was co-injected and co-eluted with the corresponding nonradioactive compound. The radio-iodinated products were obtained in 39.8%-46.3% radiochemical yields with radiochemical purities of >98% after purification by HPLC. It is anticipated that the specific activity of the no-carrier-added preparation was comparable to that of [^{125}I]NaI(2200Ci/mmol). Under the experimental conditions, the log D values of [^{125}I]8 and [^{125}I]31 were 3.31 ± 0.10 and 3.07 ± 0.08, respectively (measured by a partition between 1-octanol and pH 7.4 phosphate buffer), which is desirable for blood-brain barrier (BBB) penetration.

2.4 Neuropathological staining on brain sections of double transgenic model mice

To confirm the specific binding of these SBTP derivatives to Aβ plaques, *in vitro* neuropathological staining was performed on the brain sections of a transgenic model mouse (C57, APP/PS1, 12 months). As shown in Fig. 2, compounds **11, 14, 20** and **21** distinctively stained Aβ plaques on the brain sections with low background (Fig. 2a-d). The similar pattern of Aβ plaques was consistent with that stained with thioflavin-S using the adjacent sections (Fig. 2e-h). The results

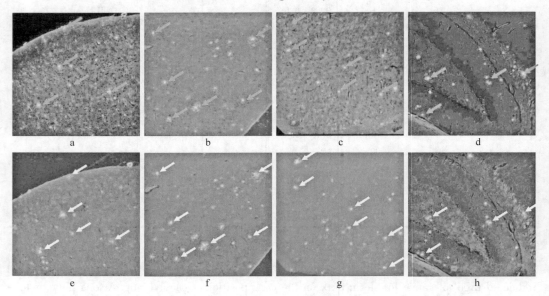

Fig. 2 Fluorescence staining of compounds **11, 14, 20** and **21** (a, b, c and d) on the sections from a transgenic model mouse (C57, APP/PS1, 10μm thick) with the filter set of GFP, and the plaques were confirmed by staining of the adjacent sections using thioflavin-S (e, f, g and h) with the filter set of GFP

from neuropathological staining indicate that the binding of these SBTP derivatives is specific for Aβ plaques.

2.5 Autoradiography *in vitro* using brain sections of double transgenic model mice

Next, the radio-iodinated probes [^{125}I]**8** and [^{125}I]**31** were investigated for their binding to Aβ plaques by *in vitro* autoradiography in the brain sections of a transgenic model mouse (C57, APP/PS1, 12 months). As shown in Fig. 3 autoradiographic images of [^{125}I]**8** and [^{125}I]**31** showed excellent labeling of Aβ plaques in the cortex region of the brain sections, and no remarkable accumulation of radioactivity was observed in the white matter (Fig. 3a and c). The same sections were also stained with thioflavin-S and the localizations of Aβ plaques were in accord with the results of autoradiography (Fig. 3b and d, red arrows). These results demonstrate that [^{125}I]**8** and [^{125}I]**31** were specific for Aβ plaques, which was consistent with high bind affinity of compounds **8** and **31** to Aβ$_{1-42}$ aggregates.

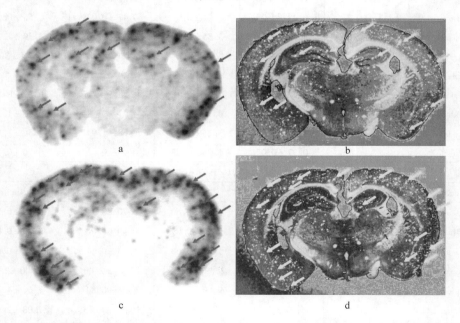

Fig. 3 In vitro labeling of brain sections from transgenic model mouse
(C57, APP/PS1, 10 μm thick) by autoradiography

([^{125}I]**8** and [^{125}I]**31** labeled the Aβ plaques in the cortex of the brain (a, c), while the white matter was clearly void of any notable labeling. The same sections were also stained with thioflavin-S (b, d) and the localizations of Aβ plaques were in accordance with the results of autoradiography (red arrows). For interpretation of the references to colour in this figure legend, the reader is referred to the web version of this article)

2.6 Biodistribution experiments with normal mice *in vivo*

To evaluate pharmacokinetic properties of the SBTP derivatives, *in vivo* biodistribution experiments were performed in normal mice with two radio-iodinated probes, [^{125}I]**8** and [^{125}I]**31**. As shown in Table 2, probe [^{125}I]8 with a radio-iodine atom on the phenyl ring displayed good

BBB penetration with the initial brain uptake (2.11 ± 0.23) % ID/g at 2min post-injection. Since there are no Aβ plaques in normal mice, the brain radioactivity concentration decreased rapidly with (0.41 ± 0.06) % ID/g at 30min. However, the radio-iodinated probe [^{125}I]**31** with a radio-iodine atom on the thiophene ring displayed lower initial brain uptake $((0.50 \pm 0.06)$ % ID/g at 2min$)$ and a relatively slow washout rate from the brain with (0.39 ± 0.09) % ID/g at 30min, suggesting the high non-specific binding of this probe. In addition, compared with the brain uptake, the initial blood uptake of the two probes was higher (12.1% ID/g for [^{125}I]**8** and 11.2% ID/g for [^{125}I]**31**), which may due to the high lipophilicity ($\log D = 3.31$ and 3.07, respectively).

Table 2 Biodistribution of [^{125}I]8 and [^{125}I]31 in normal mice[①]

Organ	2min	15min	30min	60min	120min
[^{125}I]**8** ($\lg D = 3.31 \pm 0.10$)					
Blood	12.10 ± 2.75	1.38 ± 0.22	1.16 ± 0.17	1.29 ± 0.13	1.76 ± 0.56
Brain	2.11 ± 0.23	0.42 ± 0.02	0.41 ± 0.06	0.47 ± 0.03	0.55 ± 0.06
Heart	5.53 ± 2.23	3.37 ± 0.39	3.35 ± 0.35	2.09 ± 0.33	3.39 ± 1.39
Liver	47.12 ± 8.11	40.18 ± 8.13	42.17 ± 7.14	47.70 ± 5.22	45.74 ± 3.42
Spleen	11.46 ± 1.81	12.61 ± 1.81	14.82 ± 3.66	14.58 ± 2.76	21.98 ± 5.32
Lung	20.89 ± 0.97	18.99 ± 0.97	15.69 ± 6.02	12.99 ± 4.75	6.29 ± 0.56
Kidney	4.69 ± 0.92	4.37 ± 0.92	4.56 ± 1.16	4.91 ± 0.81	6.35 ± 1.27
Stomach[②]	0.61 ± 0.16	1.48 ± 0.46	3.25 ± 1.26	4.28 ± 1.94	3.36 ± 1.56
Muscle	0.85 ± 0.07	0.97 ± 0.07	1.25 ± 0.45	0.93 ± 0.16	0.76 ± 0.12
[^{125}I]**31** ($\lg D = 3.07 \pm 0.08$)					
Organ	2min	15min	30min	60min	120min
Blood	11.21 ± 0.66	7.99 ± 1.68	6.18 ± 0.60	6.70 ± 0.91	6.85 ± 0.57
Brain	0.50 ± 0.06	0.44 ± 0.08	0.39 ± 0.09	0.50 ± 0.06	0.59 ± 0.07
Heart	5.08 ± 0.77	6.25 ± 1.70	5.55 ± 0.68	5.33 ± 0.61	4.18 ± 0.52
Liver	45.73 ± 3.78	46.86 ± 4.61	26.91 ± 3.27	15.74 ± 2.06	11.00 ± 2.41
Spleen	9.14 ± 2.11	11.20 ± 3.29	9.40 ± 1.02	6.38 ± 0.84	7.21 ± 4.43
Lung	26.17 ± 3.11	16.37 ± 3.42	12.95 ± 1.73	10.45 ± 0.75	7.08 ± 2.90
Kidney	4.82 ± 0.45	6.61 ± 1.43	5.92 ± 0.86	6.42 ± 0.98	6.33 ± 0.59
Stomach[②]	0.39 ± 0.18	2.60 ± 0.51	5.11 ± 2.17	8.90 ± 3.10	20.22 ± 5.59
Muscle	1.61 ± 0.65	2.23 ± 0.86	2.12 ± 0.39	3.95 ± 1.13	2.83 ± 0.69

①Expressed as % injected dose per gram. Average of 5 mice ± standard deviation.
②Expressed as % injected dose per organ.

Previously, we reported an iodinated benzothiazole derivative based on the bithiophene structure, [^{125}I]-2-(2,2'-bithiophen-5-yl)-6-iodobenzo[d]thiazole ([^{125}I]TZBP-PI) as a potential Aβ imaging agent. [^{125}I]TZBP-PI showed high initial brain uptake and rapid washout from the normal mouse brain[31]. Compared to [^{125}I]TZBP-PI, [^{125}I]**8** also possesses favorable *in vivo* pharmacokinetics in normal mice. It is worth mentioning that the introduction of a bithiophene moiety and iodine atom into one molecule will largely increase the lipophilicity of the molecule, which is detrimental to the *in vivo* biological activities of this Aβ imaging probe. Thus, further refinement in order to reduce the lipophilicity of this probe is needed.

3 Conclusion

A new series of novel SBTP derivatives, was successfully designed and synthesized as ligands for Aβ plaques. They displayed high *in vitro* binding affinities for Aβ aggregates. The *para* position of the phenyl ring was found to have high tolerance for steric bulk substituents. Two radiolabeled probes [^{125}I]**8** and [^{125}I]**31** showed specific labeling of Aβ plaques in the brain sections of AD model mice. Furthermore, [^{125}I]**8** displayed favorable *in vivo* pharmacokinetics in normal mice. The obtained results suggest that the novel SBTP derivatives may serve as potential candidates for diagnosis of AD.

4 Experimental

4.1 General information

All chemicals used in synthesis were commercial products and were used without further purification unless otherwise indicated. [^{125}I]NaI(2200Ci/mmol) was obtained from PerkinElmer Life and Analytical Sciences, USA. The double transgenic (APP/PS1) mouse model was obtained from Institute of Laboratory Animal Science, Chinese Academy of Medical Sciences and Comparative Medicine Center of Peking Union Medical College, China. ^{1}H-NMR spectra were obtained at 400MHz on Bruker NMR spectrometers in CDCl$_3$ solution at room temperature with TMS as an internal standard. Chemical shifts are reported as δ values relative to internal TMS. Coupling constants are reported in Hertz. The multiplicity is defined by s(singlet), d(doublet), t(triplet), and m(multiplet). Mass spectra were acquired under Survey or MSQ Plus (ESI)instrument. HPLC analysis was performed on a Shimadzu SCL-10 AVP equipped with a Packard 500 TR series flow scintillation analyzer, conditions: Agilent TC-C18 reverse-phase analytical column(5μm, ID=4.6mm, length=150mm), 90/10 CH$_3$CN/H$_2$O, 1.0mL/min, UV 254nm. Separations were achieved on a Alltech HPLC pump Model 626 equipped with LINEAR UVIS-201 and BIOSCAN flow-counter, conditions: Alltech C18 reverse-phase column(5μm, ID=4.6mm, length=250mm) eluted with acetonitrile at a 2.0mL/min flow rate. All key compounds were proven by analytical HPLC analysis to show ≥95% purity(Supporting information).

4.2 2,2′-bithiophene-5-carbaldehyde (1)

Compound 1 was synthesized according to the literature with some modification[33], POCl$_3$ (2.30g 15.0mmol)was added to DMF(1.10g 15.0mmol) at 0℃ and the mixture was stirred for 15min at 0℃. Then 2,2′-bithiophene(2.00g 12.0mmol) dissolved in 1,2-dichloroethane (40mL) was added drop-wise with stirring. The reaction mixture was then heated for 2h at

60 ℃. The organic layer was washed with saturated NaHCO$_3$ aqueous solution, and dried with anhydrous Na$_2$SO$_4$. After the solvent was removed, the residue was purified by column chromatography (petroleum ether/AcOEt, 8/1) to give 1.47g of 3 as a yellow solid (yield 63.1%).

4.3 5′-bromo-2,2′-bithiophene-5-carbaldehyde (2a)

To a stirring solution of 2,2′-bithiophene-5-carbaldehyde (1.94g, 10.00mmol) in CHCl$_3$ (20mL) at 0 ℃, was added bromine (1.60g, 10.00mmol) dissolved in CHCl$_3$ (20mL). The mixture was stirred at room temperature for 2h. The organic layer was washed with saturated NaHCO$_3$ aqueous solution, and dried with anhydrous Na$_2$SO$_4$. Evaporation of the solvent under reduced pressure gave **4** as a yellow solid (yield 86.3%). ^{1}H NMR (400MHz, CDCl$_3$) δ 9.79 (s, 1H), 7.59 (d, J = 4.0Hz, 1H), 7.11 (d, J = 3.9Hz, 1H), 7.03 (d, J = 3.8Hz, 1H), 6.96 (d, J = 3.8Hz, 1H). MS (ESI) : m/z calcd for C$_9$H$_5$BrOS$_2$ 271.90; found 272.9 [M + H]$^+$.

4.4 5′-iodo-2,2′-bithiophene-5-carbaldehyde (2b)

To a solution of 2,2′-bithiophene-5-carbaldehyde (1.94g, 10.00mmol) in toluene (20mL) at 0 ℃, was added small portions of mercury (Ⅱ) oxide (2.23g, 10.29mmol, yellow powder) and iodine (2.87g, 11.32mmol). The mixture was stirred at room temperature for 8h. The orange precipitate was filtered off and washed with ethyl acetate. The filtrate and washings were combined and washed with aqueous sodium thiosulfate and dried over magnesium sulfate. Solvent was removed by rotary evaporation and the solid residue was recrystallized from ethanol (2.37g, 72%). ^{1}H NMR (400MHz, CDCl$_3$) δ 9.87 (s, 1H), 7.67 (d, J = 3.8Hz, 1H), 7.23 (d, J = 3.7Hz, 1H), 7.19 (d, J = 3.8Hz, 1H), 7.02 (d, J = 3.6Hz, 1H). MS (ESI) : m/z calcd for C$_9$H$_5$IOS$_2$ 319.88; found 320.8 [M + H]$^+$.

4.5 General procedure for the synthesis of 4-substituted-benzyltriphenyl phosphonium salts (3a-3k)

Triphenylphosphine (5mmol) was added to a solution of 4-substituted-benzyl halide (5mmol) in toluene (50mL). The mixture was heated to reflux for 6h and then cooled to room temperature. The precipitate was collected, recrystallized from ethanol to give the products.

(1) Benzyltriphenylphosphonium bromide (**3a**). White solid, yield 92%. ^{1}H NMR (400MHz, CDCl$_3$) δ 7.86-7.47 (m, 15H), 7.20 (ddd, J = 8.6, 4.0, 1.9Hz, 1H), 7.15-7.03 (m, 4H), 5.35 (d, J = 14.4Hz, 2H).

(2) (4-fluorobenzyl) triphenylphosphonium bromide (**3b**). White solid, yield 89%. ^{1}H NMR (400MHz, CDCl$_3$) δ 7.84-7.52 (m, 15H), 7.22-7.07 (m, 2H), 6.78 (t, J = 8.6Hz, 2H), 5.53 (d, J = 14.3Hz, 2H).

(3) (4-chlorobenzyl) triphenylphosphonium chloride (**3c**). White solid, yield 84%. ^{1}H NMR (400MHz, CDCl$_3$) δ 7.91-7.48 (m, 15H), 7.13 (d, J = 6.7Hz, 2H), 7.04 (d, J = 8.2Hz, 2H), 5.72 (d, J = 14.9Hz, 2H).

(4) (4-bromobenzyl) triphenylphosphonium bromide (**3d**). White solid, yield 94%. ^{1}H NMR

(400MHz, CDCl$_3$) δ 7.88-7.43 (m, 15H), 7.17 (d, J = 7.8Hz, 2H), 7.06 (dd, J = 8.5, 2.6Hz, 2H), 5.57 (d, J = 14.8Hz, 2H).

(5) (4-iodobenzyl) triphenylphosphonium bromide (**3e**). White solid, yield 82%. ^{1}H NMR (400MHz, CDCl$_3$) δ 7.74-7.53 (m, 15H), 7.35 (d, J = 7.6Hz, 2H), 6.86 (dd, J = 8.4, 2.5Hz, 2H), 5.48 (d, J = 14.8Hz, 2H).

(6) (4-nitrobenzyl) triphenylphosphonium chloride (**3f**). Brown solid, yield 81%. ^{1}H NMR (400MHz, CDCl$_3$) δ 7.92-7.79 (m, 9H), 7.75 (dd, J = 8.2, 6.7Hz, 2H), 7.60 (td, J = 7.8, 3.5Hz, 6H), 7.46 (dd, J = 8.7, 2.4Hz, 2H), 6.09 (d, J = 15.9Hz, 2H).

(7) (4-(dimethylamino)benzyl) triphenylphosphonium bromide (**3g**). White solid, yield 27%. ^{1}H NMR (400MHz, CDCl$_3$) δ 7.92-7.49 (m, 15H), 6.90 (d, J = 6.9Hz, 2H), 6.54 (s, 2H), 5.14 (d, J = 13.2Hz, 2H), 2.89 (s, 6H).

(8) (4-hydroxybenzyl) triphenylphosphonium bromide (**3h**). White solid, yield 18%. ^{1}H NMR (400MHz, DMSO) δ 7.96-7.60 (m, 15H), 6.74 (d, J = 6.6Hz, 2H), 6.58 (d, J = 8.2Hz, 2H), 5.00 (d, J = 14.8Hz, 2H).

(9) (4-methoxybenzyl) triphenylphosphonium bromide (**3i**). White solid, yield 88%. ^{1}H NMR (400MHz, CDCl$_3$) δ 7.86-7.56 (m, 15H), 7.02 (dd, J = 8.8, 2.6Hz, 2H), 6.65 (d, J = 8.3Hz, 2H), 5.33 (d, J = 13.7Hz, 2H), 3.72 (s, 3H).

(10) (4-methylbenzyl) triphenylphosphonium chloride (**3j**). White solid, yield 86%. ^{1}H NMR (400MHz, CDCl$_3$) δ 7.88-7.48 (m, 15H), 6.94 (d, J = 8.1Hz, 2H), 6.90 (d, J = 7.9Hz, 2H), 5.37 (d, J = 14.2Hz, 2H), 2.23 (s, 3H).

(11) (4-tert-butylbenzyl) triphenylphosphonium bromide (**3k**). White solid, yield 91%. ^{1}H NMR (400MHz, CDCl$_3$) δ 7.91-7.44 (m, 15H), 7.14 (d, J = 7.9Hz, 2H), 7.01 (d, J = 8.1Hz, 2H), 5.33 (d, J = 14.0Hz, 2H), 1.23 (s, 9H).

4.6 General procedure for the synthesis of SBTPs (4-9, 11-33)

To a stirring solution of 5'-substituted-2,2'-bithiophene-5-carbaldehyde (1.00mmol) and 4-substituted-benzyl halide (1.00mmol) in THF (50mL), was added CH$_3$ONa (1.00mmol). The mixture was then refluxed for 2-6h. After cooling, water (20mL) was added to the reaction mixture. The resulting mixture was extracted by dichloromethane (3 times) and the organic layer was washed with brine and dried over anhydrous MgSO$_4$. The organic solvent was removed in vacuum and the residue was purified by column chromatography.

(1) (*E*)-5-styryl-2,2'-bithiophene (**4**). Yellow solid, yield 85%. ^{1}H NMR (400MHz, CDCl$_3$) δ 7.40 (d, J = 7.7Hz, 2H), 7.28 (t, J = 7.5Hz, 2H), 7.20 (d, J = 5.4Hz, 1H), 7.15 (d, J = 5.3Hz, 1H), 7.12 (d, J = 4.4Hz, 1H), 7.11 (d, J = 16.0Hz, 1H), 7.01 (d, J = 3.7Hz, 1H), 6.96 (t, J = 4.3Hz, 1H), 6.90 (d, J = 3.6Hz, 1H), 6.82 (d, J = 16.0Hz, 1H). MS(EI): m/z calcd for C$_{16}$H$_{12}$S$_2$ 268.04; found 268.25. m.p.: 106.4-107.3℃.

(2) (*E*)-5-(4-fluorostyryl)-2,2'-bithiophene (**5**). Yellow solid, yield 87%. ^{1}H NMR (400MHz, CDCl$_3$) δ 7.36 (dd, J = 7.1, 6.1Hz, 2H), 7.15 (d, J = 5.0Hz, 1H), 7.12 (d, J = 3.3Hz, 1H), 7.01 (d, J = 16.2Hz, 1H), 7.00 (d, J = 3.9Hz, 2H), 6.97 (s, 1H), 6.95 (d, J = 3.8Hz,

1H),6.88(d, J = 3.5Hz,1H),6.78(d, J = 15.8Hz,1H). MS(EI): m/z calcd for $C_{16}H_{11}FS_2$ 286.03; found 285.9. m.p.: 115.5-128.6℃.

(3) (E)-5-(4-chlorostyryl)-2,2′-bithiophene (**6**). Yellow solid, yield 81%. ^{1}H NMR (400 MHz, CDCl$_3$) δ 7.32(d, J = 8.3Hz,2H),7.24(d, J = 8.2Hz,2H),7.16(d, J = 5.1Hz,1H), 7.12(d, J = 3.5Hz,1H),7.08(d, J = 16.0Hz,1H),7.01(d, J = 3.6Hz,1H),6.96(t, J = 4.3Hz,1H),6.90(d, J = 3.6Hz,1H),6.76(d, J = 16.0Hz,1H). MS(EI): m/z calcd for $C_{16}H_{11}ClS_2$ 302.00; found 302.17. m.p.: 137.5-139.8℃.

(4) (E)-5-(4-bromostyryl)-2,2′-bithiophene (**7**). Yellow solid, yield 83%. ^{1}H NMR (400MHz, CDCl$_3$) δ 7.40(d, J = 8.5Hz,2H),7.26(d, J = 8.5Hz,2H),7.16(dd, J = 5.1, 1.1Hz,1H),7.13(dd, J = 3.6, 1.1Hz,1H),7.10(d, J = 16.1Hz,1H),7.01(d, J = 3.7Hz, 1H),6.96(dd, J = 5.1, 3.6Hz,1H),6.91(d, J = 3.7Hz,1H),6.74(d, J = 16.1Hz,1H). MS (EI): m/z calcd for $C_{16}H_{11}BrS_2$ 345.95; found 346.14. m.p.: 151.3-152.2℃.

(5) (E)-5-(4-iodostyryl)-2,2′-bithiophene (**8**). Yellow solid, yield 60%. ^{1}H NMR (400MHz, CDCl$_3$) δ 7.67(d, J = 8.4Hz,2H),7.24(dd, J = 5.1, 1.1Hz,1H),7.21(d, J = 8.6Hz,2H),7.20(d, J = 3.6Hz,1H),7.18(d, J = 16.2Hz,1H),7.08(d, J = 3.7Hz,1H), 7.04(dd, J = 5.1, 3.6Hz,1H),6.98(d, J = 3.8Hz,1H),6.79(d, J = 16.1Hz,1H). MS (EI): m/z calcd for $C_{16}H_{11}IS_2$ 393.93; found 394.12. m.p.: 160.7-161.3℃.

(6) (E)-5-(4-nitrostyryl)-2,2′-bithiophene (**9**). Dark red solid, yield 51%. ^{1}H NMR (400MHz, CDCl$_3$) δ 8.22(d, J = 8.5Hz,2H),7.58(d, J = 8.5Hz,2H),7.35(d, J = 16.0Hz,1H),7.27(d, J = 4.5Hz,1H),7.24(d, J = 3.5Hz,1H),7.12(d, J = 3.6Hz,1H), 7.08(d, J = 3.8Hz,1H),7.06(t, J = 4.3Hz,1H),6.90(d, J = 16.1Hz,1H). MS(ESI): m/z calcd for $C_{16}H_{11}NO_2S_2$ 313.02; found 313.9(M + H$^+$). m.p.: 169.2-170.9℃.

(7) (E)-4-(2-(2,2′-bithiophen-5-yl)vinyl)aniline (**10**). Yellow solid, yield 27%. ^{1}H NMR (400MHz, CDCl$_3$) δ 7.51(d, J = 8.2Hz,2H),7.44(d, J = 4.8Hz,1H),7.29(d, J = 16.1Hz, 1H),7.25(d, J = 2.1Hz,1H),7.27(d, J = 3.2Hz,1H),7.10(d, J = 8.1Hz,2H),7.07(d, J = 3.4Hz,1H),7.03(t, J = 4.1Hz,1H),6.85(d, J = 16.1Hz,1H). MS(EI): m/z calcd for $C_{16}H_{13}NS_2$ 283.05; found 283.25. m.p.: 158.6-159.2℃.

(8) (E)-4-(2-(2,2′-bithiophen-5-yl)vinyl)-N,N-dimethylaniline (**11**). Orange solid, yield 33%. ^{1}H NMR (400MHz, CDCl$_3$) δ 7.29(d, J = 8.4Hz,2H),7.12(d, J = 5.1Hz,1H),7.09 (d, J = 3.4Hz,1H),6.98(d, J = 3.6Hz,1H),6.95(d, J = 3.9Hz,1H),6.91(d, J = 15.9Hz,1H),6.80(d, J = 3.6Hz,1H),6.76(d, J = 16.0Hz,1H),6.63(d, J = 8.2Hz,2H), 2.92(s,6H). MS(ESI): m/z calcd for $C_{18}H_{17}NS_2$ 311.08; found 311.9[M + H]$^+$.

(9) (E)-5-(4-methoxylstyryl)-2,2′-bithiophene (**12**). Yellow solid, yield 86%. ^{1}H NMR (400MHz, CDCl$_3$) δ 7.41(d, J = 8.7Hz,2H),7.22(dd, J = 5.1, 1.0Hz,1H),7.18(dd, J = 3.6, 1.0Hz,1H),7.07(d, J = 3.6Hz,1H),7.04(d, J = 3.5Hz,1H),7.02(d, J = 3.7Hz, 1H),6.93(d, J = 3.8Hz,1H),6.90(d, J = 8.8Hz,2H),6.86(d, J = 16.1Hz,1H),3.84(s, 3H). MS(EI): m/z calcd for $C_{17}H_{14}OS_2$ 298.05; found 298.21. m.p.: 143.8℃.

(10) (E)-5-(4-methylstyryl)-2,2′-bithiophene (**13**). Yellow solid, yield 78%. ^{1}H NMR (400MHz, CDCl$_3$) δ 7.36(d, J = 7.6Hz,2H),7.21(d, J = 5.0Hz,1H),7.18(d, J = 6.8Hz,

1H),7.16(d, J = 7.8Hz,2H),7.13(d, J = 16.0Hz,1H),7.07(d, J = 3.5Hz,1H),7.03(d, J = 4.9Hz,1H),6.94(d, J = 3.4Hz,1H),6.87(d, J = 16.0Hz,1H),2.35(s,3H). MS(EI): m/z calcd for $C_{17}H_{14}S_2$ 282.05;found 282.23. m.p.:119.1-119.8℃.

(11) (E)-5-(4-tert-butylstyryl)-2,2′-bithiophene (**14**). Yellow solid, yield 86%. ^{1}H NMR (400MHz, CDCl$_3$) δ 7.40(d, J = 8.6Hz,2H),7.37(d, J = 8.6Hz,2H),7.21(d, J = 5.1Hz, 1H),7.18(d, J = 3.6Hz,1H),7.13(d, J = 16.0Hz,1H),7.07(d, J = 3.7Hz,1H),7.02 (dd, J = 5.1, 3.7Hz,1H),6.94(d, J = 3.7Hz,1H),6.88(d, J = 16.1Hz,1H),1.33(s, 9H). MS(EI): m/z calcd for $C_{20}H_{20}S_2$ 324.10;found 324.31. m.p:112.0-112.6℃.

(12) (E)-5-bromo-5′-styryl-2,2′-bithiophene (**15**). Yellow solid, yield 83%. ^{1}H NMR (400MHz, CDCl$_3$) δ 7.46(d, J = 7.4Hz,2H),7.35(t, J = 7.2Hz,2H),7.26(d, J = 13.3Hz, 1H),7.16(d, J = 16.1Hz,1H),6.99(d, J = 11.4Hz,2H),6.94(d, J = 10.3Hz,2H),6.89 (d, J = 16.9Hz,1H). MS(EI): m/z calcd for $C_{16}H_{11}BrS_2$ 345.95;found 345.90. m.p.:138.2-139.9℃.

(13) (E)-5-bromo-5′-(4-fluorostyryl)-2,2′-bithiophene (**16**). Yellow solid, yield 81%. ^{1}H NMR(400MHz, CDCl$_3$) δ 7.43(t, J = 6.3Hz,2H),7.09-6.97(m,5H),6.93(d, J = 4.0Hz, 2H),6.84(d, J = 16.0Hz,1H). MS(EI): m/z calcd for $C_{16}H_{10}BrFS_2$ 363.94;found 364.13. m.p.:159.0-159.9℃.

(14) (E)-5-bromo-5′-(4-chlorostyryl)-2,2′-bithiophene(**17**). Yellow solid, yield 60%. ^{1}H NMR(400MHz, CDCl$_3$) δ 7.38(d, J = 7.9Hz,2H),7.31(d, J = 7.9Hz,2H),7.13(d, J = 16.2Hz,1H),6.99(d, J = 10.9Hz,2H),6.94(d, J = 11.1Hz,2H),6.83(d, J = 16.0Hz, 1H). MS(EI): m/z calcd for $C_{16}H_{10}BrClS_2$ 379.91;found 380.04. m.p.:187.3-188.8℃.

(15) (E)-5-bromo-5′-(4-bromostyryl)-2,2′-bithiophene (**18**). Yellow solid, yield 86%. ^{1}H NMR(400MHz, CDCl$_3$) δ 7.47(d, J = 7.5Hz,2H),7.32(d, J = 8.0Hz,2H),7.15(d, J = 16.0Hz,1H),6.99(d, J = 10.5Hz,2H),6.95(d, J = 12.3Hz,2H),6.81(d, J = 15.9Hz,1H). MS(EI): m/z calcd for $C_{16}H_{10}Br_2S_2$ 423.86;found 424.06. m.p.:199.9-201.6℃.

(16) (E)-5-bromo-5′-(4-iodostyryl)-2,2′-bithiophene (**19**). Yellow solid, yield 79%. ^{1}H NMR(400MHz, CDCl$_3$) δ 7.67(d, J = 6.7Hz,2H),7.19(d, J = 8.0Hz,2H),7.14(s,1H), 7.05-6.91(m,4H),6.79(d, J = 15.7Hz,1H). MS(EI): m/z calcd for $C_{16}H_{10}BrIS_2$ 471.85; found 472.06. m.p.:212.3-214.1℃.

(17) (E)-5-bromo-5′-(4-nitrostyryl)-2,2′-bithiophene(**20**). Dark red solid, yield 45%. ^{1}H NMR(400MHz, CDCl$_3$) δ 8.21(d, J = 8.0Hz,2H),7.57(d, J = 8.2Hz,2H),7.32(d, J = 16.0Hz,1H),7.05(s,2H),6.98(d, J = 11.8Hz,2H),6.90(d, J = 16.1Hz,1H). MS(EI): m/z calcd for $C_{16}H_{10}BrNO_2S_2$ 390.93;found 391.12. m.p.:202.2-203.7℃.

(18) (E)-4-(2-(5′-bromo-2,2′-bithiophen-5-yl)vinyl)-N,N-dimethylaniline (**21**). Yellow solid, yield 31%. ^{1}H NMR(400MHz, CDCl$_3$) δ 7.38(d, J = 6.5Hz,2H),7.00-6.96(m, 3H),6.90(d, J = 3.4Hz,1H),6.84(d, J = 16.0Hz,2H),6.72(s,2H),3.01(s,6H). MS (EI): m/z calcd for $C_{18}H_{16}BrNS_2$ 388.99;found 389.18. m.p.:207.1-208.9℃.

(19) (E)-4-(2-(5′-bromo-2,2′-bithiophen-5-yl)vinyl)phenol (**22**). Yellow solid, yield 14%. ^{1}H NMR(400MHz, CDCl$_3$) δ 7.36(d, J = 8.5Hz,2H),7.02(d, J = 16.5Hz,1H),6.99

(d, J = 4.8Hz, 1H), 6.97(d, J = 3.9Hz, 1H), 6.91(d, J = 2.7Hz, 1H), 6.90(d, J = 3.4Hz, 1H), 6.83(d, J = 16.3Hz, 1H), 6.81(d, J = 8.5Hz, 2H). MS(EI): m/z calcd for $C_{16}H_{11}BrOS_2$ 361.94; found 362.18. m.p.: 152.6℃.

(20) (E)-5-bromo-5'-(4-methoxystyryl)-2,2'-bithiophene (**23**). Yellow solid, yield 86%. ^{1}H NMR(400MHz, CDCl$_3$) δ 7.33(d, J = 8.3Hz, 2H), 6.96(d, J = 16.0Hz, 1H), 6.92(d, J = 3.6Hz, 1H), 6.90(d, J = 3.8Hz, 1H), 6.83(s, 3H), 6.81(s, 1H), 6.78(d, J = 16.2Hz, 1H), 3.76(s, 3H). MS(EI): m/z calcd for $C_{17}H_{13}BrOS_2$ 375.96; found 376.13. m.p.: 178.3-180.2℃.

(21) (E)-5-bromo-5'-(4-methylstyryl)-2,2'-bithiophene(**24**). Yellow solid, yield 79%. ^{1}H NMR(400MHz, CDCl$_3$) δ 7.36(d, J = 7.6Hz, 2H), 7.16(d, J = 7.7Hz, 2H), 7.11(d, J = 16.1Hz, 1H), 6.98(d, J = 9.6Hz, 2H), 6.92(s, 2H), 6.87(d, J = 16.0Hz, 1H), 2.35(s, 3H). MS(EI): m/z calcd for $C_{17}H_{13}BrS_2$ 359.96; found 360.20. m.p.: 188.1-188.9℃.

(22) (E)-5-bromo-5'-(4-tert-butylstyryl)-2,2'-bithiophene (**25**). Yellow solid, yield 85%. ^{1}H NMR(400MHz, CDCl$_3$) δ 7.40(d, J = 8.7Hz, 2H), 7.37(d, J = 8.7Hz, 2H), 7.12(d, J = 16.0Hz, 1H), 7.00(d, J = 3.7Hz, 1H), 6.97(d, J = 3.9Hz, 1H), 6.93(d, J = 3.8Hz, 1H), 6.91(d, J = 3.9Hz, 1H), 6.88(d, J = 16.1Hz, 1H), 1.33(s, 9H). MS(EI): m/z calcd for $C_{20}H_{19}BrS_2$ 402.01; found 402.22. m.p.: 138.1℃.

(23) (E)-5-iodo-5'-styryl-2,2'-bithiophene(**26**). Yellow solid, yield 88%. H NMR(400MHz, CDCl$_3$) δ 7.46(d, J = 7.7Hz, 2H), 7.35(t, J = 7.5Hz, 2H), 7.25(d, J = 16.0Hz, 1H), 7.17(d, J = 5.8Hz, 1H), 7.15(d, J = 6.4Hz, 1H), 7.01(d, J = 3.7Hz, 1H), 6.95(d, J = 3.7Hz, 1H), 6.89(d, J = 16.0Hz, 1H), 6.85(d, J = 3.7Hz, 1H). MS(EI): m/z calcd for $C_{16}H_{11}IS_2$ 393.93; found 394.14. m.p.: 160.3℃.

(24) (E)-5-(4-fluorostyryl)-5'-iodo-2,2'-bithiophene (**27**). Yellow solid, yield 86%. ^{1}H NMR(400MHz, CDCl$_3$) δ 7.43(d, J = 5.6Hz, 1H), 7.41(d, J = 5.6Hz, 1H), 7.16(d, J = 3.8Hz, 1H), 7.07(d, J = 16.2Hz, 1H), 7.05(d, J = 8.5Hz, 2H), 7.01(d, J = 4.3Hz, 1H), 6.94(d, J = 3.7Hz, 1H), 6.85(d, J = 3.8Hz, 1H), 6.84(d, J = 16.0Hz, 1H). MS(EI): m/z calcd for $C_{16}H_{10}FIS_2$ 411.93; found 412.11. m.p.: 169.8℃.

(25) (E)-5-(4-chlorostyryl)-5'-iodo-2,2'-bithiophene (**28**). Yellow solid, yield 90%. ^{1}H NMR(400MHz, CDCl$_3$) δ 7.38(d, J = 7.8Hz, 2H), 7.31(d, J = 8.3Hz, 2H), 7.16(d, J = 3.7Hz, 1H), 7.13(d, J = 16.0Hz, 1H), 7.01(d, J = 3.7Hz, 1H), 6.95(d, J = 3.6Hz, 1H), 6.85(d, J = 3.7Hz, 1H), 6.83(d, J = 16.2Hz, 1H). MS(EI): m/z calcd for $C_{16}H_{10}ClIS_2$ 427.90; found 428.08. m.p.: 187.6℃.

(26) (E)-5-(4-bromostyryl)-5'-iodo-2,2'-bithiophene (**29**). Yellow solid, yield 89%. ^{1}H NMR(400MHz, CDCl$_3$) δ 7.46(d, J = 8.3Hz, 2H), 7.32(d, J = 8.2Hz, 2H), 7.16(d, J = 2.4Hz, 1H), 7.14(d, J = 14.8Hz, 1H), 7.01(d, J = 3.7Hz, 1H), 6.96(d, J = 3.8Hz, 1H), 6.85(d, J = 3.8Hz, 1H), 6.81(d, J = 16.0Hz, 1H). MS(EI): m/z calcd for $C_{16}H_{10}BrIS_2$ 471.85; found 472.10. m.p.: 179.7℃.

(27) (E)-5-iodo-5'-(4-iodostyryl)-2,2'-bithiophene (**30**). Yellow solid, yield 77%. ^{1}H NMR(400MHz, DMSO-d^6) δ 7.71(d, J = 8.2Hz, 2H), 7.45(d, J = 16.2Hz, 1H), 7.38(d,

$J = 8.3$Hz,2H),7.32(d, $J = 3.7$Hz,1H),7.24(d, $J = 3.6$Hz,1H),7.17(d, $J = 3.6$Hz, 1H),7.05(d, $J = 3.6$Hz,1H),6.90(d, $J = 16.2$Hz,1H). MS(EI): m/z calcd for $C_{16}H_{10}I_2S_2$ 519.83; found 519.97.

(28) (E)-5-iodo-5'-(4-methoxystyryl)-2,2'-bithiophene (**31**). Yellow solid, yield 85%. ^{1}H NMR(400MHz, CDCl$_3$) δ 7.40(d, $J = 8.6$Hz,2H),7.16(d, $J = 3.8$Hz,2H),7.03(d, $J = 16.1$Hz,1H),7.00(d, $J = 3.7$Hz,1H),6.90(d, $J = 2.6$Hz,1H),6.89(d, $J = 8.4$Hz,1H), 6.85(d, $J = 15.8$Hz,1H),6.84(d, $J = 3.9$Hz,1H),3.83(s,3H). MS(EI): m/z calcd for $C_{17}H_{13}IOS_2$ 423.95; found 424.11. m.p.: 199.5-201.6℃.

(29) (E)-5-iodo-5'-(4-methylstyryl)-2,2'-bithiophene (**32**). Yellow solid, yield 82%. ^{1}H NMR(400MHz, CDCl$_3$) δ 7.36(d, $J = 7.8$Hz,2H),7.16-7.14(m,3H),7.11(d, $J = 16.0$Hz, 1H),7.00(d, $J = 3.8$Hz,1H),6.92(d, $J = 3.7$Hz,1H),6.87(d, $J = 16.5$Hz,1H),6.84(d, $J = 3.9$Hz,1H),2.35(s,3H). MS(EI): m/z calcd for $C_{17}H_{13}IS_2$ 407.95; found 408.14. m.p.: 237.6-238.6℃.

(30) (E)-5-(4-tert-butylstyryl)-5'-iodo-2,2'-bithiophene (**33**). Yellow solid, yield 88%. ^{1}H NMR(400MHz, CDCl$_3$) δ 7.40(d, $J = 8.7$Hz,2H),7.37(d, $J = 8.7$Hz,2H),7.16 (d, $J = 3.8$Hz,1H),7.12(d, $J = 16.0$Hz,1H),7.00(d, $J = 3.7$Hz,1H),6.93(d, $J = 3.8$Hz,1H),6.88(d, $J = 16.1$Hz,1H),6.84(d, $J = 3.8$Hz,1H),1.33(s,9H). MS(EI): m/z calcd for $C_{20}H_{19}IS_2$ 450.00; found 450.23. m.p.: 152.4-154.0℃.

4.7 (E)-(4-(2-([2,2'-bithiophen]-5-yl)vinyl)phenyl)tributylstannane (34)

A mixture of **7**(123mg, 0.5mmol), (Bu$_3$Sn)$_2$(580mg, 1.0mmol) and (Ph$_3$P)$_4$Pd(57.8mg, 0.05mmol) in toluene(10mL) was stirred under reflux for 10h. The solvent was removed, and the residue was purified by silica gel chromatography to give 60.5mg of **34**. Yellow oil, yield 21.7%. ^{1}H NMR(400MHz, CDCl$_3$) δ 7.46(d, $J = 8.0$Hz,2H),7.42(d, $J = 8.1$Hz,2H), 7.22(dd, $J = 3.6, 1.2$Hz,1H),7.19(dd, $J = 3.6, 1.1$Hz,1H),7.19(d, $J = 16.0$Hz,1H), 7.08(d, $J = 3.7$Hz,1H),7.03(dd, $J = 5.1, 3.6$Hz,1H),6.96(d, $J = 3.8$Hz,1H),6.88(d, $J = 16.0$Hz,1H),1.84-0.57(m,27H). MS(EI): m/z calcd for $C_{20}H_{19}IS_2$ 558.14; found 558.45.

4.8 (E)-tributyl(5'-(4-methoxystyryl)-[2,2'-bithiophen]-5-yl)stannane (35)

The reaction described for **34** was used, 115.5mg of **35** was obtained as a yellow oil, yield 39.3%. ^{1}H NMR(400MHz, CDCl$_3$) δ 7.39(d, $J = 8.8$Hz,2H),7.29(d, $J = 3.4$Hz,1H), 7.06(d, $J = 3.7$Hz,1H),7.05(d, $J = 3.9$Hz,1H),7.04(d, $J = 16.4$Hz,1H),6.91(d, $J = 3.9$Hz,1H),6.88(d, $J = 8.8$Hz,2H),6.84(d, $J = 16.0$Hz,1H),3.83(s,3H),1.74-0.78 (m,27H). MS(EI): m/z calcd for $C_{20}H_{19}IS_2$ 588.15; found 588.38.

4.9 Iododestannylation reaction

The radio-iodinated ligands **8** and **31** were prepared from the corresponding tributyltin derivatives by iododestannylation according to a procedure described previously with some modifications[34].

Briefly, 50μL of H_2O_2 (3%) was added to a mixture of a tributyltin derivative 100μL (1mg/mL in EtOH), [^{125}I] NaI (200μCi, specific activity 2200Ci/mmol), and 100μL of 1M HCl in a sealed vial. The reaction was allowed to proceed at room temperature for 15min and terminated by addition of 50μL saturated $NaHSO_3$. After neutralization with sodium bicarbonate, the reaction mixture was extracted with ethyl acetate (3 × 1mL). The combined extracts were evaporated to dryness. The residues were dissolved in 100μL of EtOH and purified by HPLC using a Alltech C18 reversed phase column (4.6mm × 250mm, 5μm) with the solvent of acetonitrile (100%) at a flow rate of 2.0mL/min. Finally, the radiochemical identity of the radio-labeled tracers were verified by co-injection and co-ecution with nonradioactive compound by HPLC (Agilent TC-C18, 4.6 × 150mm, 5μm, CH_3CN/H_2O = 9/1 at the flow rate of 1.0mL/min). The no-carrier-added products were stored at −20℃ up to 6 weeks for biodistribution experiments, binding assay, and autoradiography studies.

4.10 Binding assay *in vitro* using Aβ aggregates

The trifluoro acetic acid salt forms of peptide $Aβ_{1-42}$ were purchased from GL biochem (Shanghai, China). Aggregation of peptides was carried out by gently dissolving the peptide (0.25mg/mL) in a buffer solution (pH = 7.4) containing 10mM sodium phosphate and 1mM EDTA. The solutions were incubated at 37℃ for 42h with gentle and constant shaking. Inhibition studies were carried out in 12mm × 75mm borosilicate glass tubes according to the procedure described previously with some modifications[31,34,35]. 50μL aggregated $Aβ_{1-42}$ fibrils (28nM in the final assay mixture) were added to the mixture containing 100μL of radio-ligands ([^{125}I] TZDM) with appropriate concentration, 10μL of inhibitors (10^{-5}-10^{-10} M in EtOH) and 840μL PBS (0.2M, pH = 7.4) in a final volume of 1mL. Non-specific binding was defined in the presence of 100nM TZDM. The mixture was incubated at 37℃ for 2h, then the bound and free radioactivities were separated by vacuum filtration through borosilicate glass fibre filters (0.1μm, JiuDing, China) using a ZT-II cell harvester (Shaoxing, China) followed by 3 × 4mL washes of cold PBS (0.2M, pH = 7.4) containing 10% ethanol at room temperature. Filters containing the bound ^{125}I ligand were counted in a γ-counter (WALLAC/Wizard 1470, USA) with 70% counting efficiency. The half maximal inhibitory concentration (IC_{50}) values were calculated using GraphPad Prism 4.0, and those for the inhibition constant (K_i) were calculated using the Cheng-Prusoff equation: $K_i = IC_{50}/(1 + [L]/K_d)$ [36].

4.11 *In vitro* fluorescent staining using brain sections of transgenic model mouse

Paraffin-embedded brain sections of Tg-C57 mouse (10μm) were used for the autoradiography. The brain sections were deparaffinized with 2 × 20min washes in xylene, 2 × 5min washes in 100% ethanol, 5min washes in 90% ethanol/H_2O, 5min washes in 80% ethanol/H_2O, 5min wash in 60% ethanol/H_2O, running tap water for 10min, and then incubated in PBS (0.2M, pH = 7.4) for 30min. The brain sections were incubated with 20% ethanol solution (1μM) of compounds for 10min. The localization of plaques was confirmed by staining with thioflavin-S(1μM) on the adjacent sections. Finally, the sections were washed with 50% ethanol and PBS(0.2M, pH = 7.4) for 10min. Fluorescent observation was performed by the LSM 510 META(Zeiss, Germany) equipped with a LP 505 filter set(excitation, 405nm; long-pass filter, 505nm).

4.12 Autoradiography *in vitro* using brain sections of transgenic model mouse

The brain sections mentioned above were incubated with $[^{125}I]$**8** or $[^{125}I]$**31**(5μCi/100μL) for 30min at room temperature. The sections were then washed with saturated Li_2CO_3 in 40% EtOH for 3min and washed with 40% EtOH for 3min, followed by rinsing with water for 30s. After drying, the ^{125}I-labeled sections were exposed to phosphorus film for 2h then scanned with the phosphor imaging system(Cyclone, Packard) at the resolution of 600 dpi. The presence and localization of plaques was confirmed by the fluorescent staining with thioflavin-S(1μM) on the same sections using Stereo Discovery V12, excitation: 450-490nm, emission: 500-550nm(Zeiss, Germany).

4.13 Biodistribution experiments with normal mice

In vivo biodistribution studies were performed in female KunMing normal mice(average weight about 20g) and approved by the national laws related to the care and experiments on laboratory animal. A saline solution(100μL) containing $[^{125}I]$**8** or $[^{125}I]$**31** (1μCi) was injected directly into the tail vein of mice. The mice were sacrificed at various time points post-injection. The organs of interest were removed and weighed, and the radioactivity was counted with an automatic γ-counter(WALLAC/Wizard 1470, USA). The percentage dose per gram of wet tissue was calculated by a comparison of the tissue counts to suitably diluted aliquots of the injected material.

4.14 Partition coefficient determination

The partition coefficient of $[^{125}I]$**8** and $[^{125}I]$**31** was determined according to the procedure previously reported with some modifications[37]. Five μCi of tracer was added to the premixed suspensions containing 3g of *n*-octanol and 3g of PBS(0.05M, pH = 7.4) in a test tube. The test tube was vortexed for 3min at room temperature, followed by centrifugation for 5min at 3500rpm. Two weighted samples from the *n*-octanol(100μL) and buffer(500μL) layers were counted. The partition coefficient was expressed as the logarithm of the ratio of the counts per gram from *n*-octanol versus that of PBS. Samples from *n*-octanol layer were repartitioned until consistent partitions of coefficient values were obtained. The measurement was done in triplicate and repeated three times.

Acknowledgment

The authors thank Dr. Xiaoyan Zhang(College of Life Science, Beijing Normal University) for assistance in the *in vitro* neuropathological staining. This work was funded by NSFC (20871021) and the Funddmental Research Funds for the Cental Universities.

Appendix. Supporting Information

Supporting information related to this article can be found online at doi: 10. 1016/j. ejmech. 2011. 04. 015.

References

[1] D. J. Selkoe, Alzheimer's disease: genes, proteins, and therapy, Physiol. Rev. 81(2001)741-766.

[2] J. A. Hardy, G. A. Higgins, Alzheimer's disease: the amyloid cascade hypothesis, Science 256(1992) 184-185.

[3] J. A. Hardy, D. J. Selkoe, The amyloid hypothesis of Alzheimer's disease: progress and problems on the road to therapeutics, Science 297(2002)353-356.

[4] C. A. Mathis, Y. Wang, W. E. Klunk, Imaging β-amyloid plaques and neurofibrillary tangles in the aging human brain, Curr. Pharm. Des. 10(2004)1469-1492.

[5] L. S. Cai, R. B. Innis, V. W. Pike, Radioligand development for PET imaging of β-amyloid (Aβ)-current status. , Curr. Med. Chem. 14(2007)19-52.

[6] C. A. Mathis, Y. Wang, D. P. Holt, G. F. Huang, M. L. Debnath, W. E. Klunk, Synthesis and evaluation of ^{11}C-labeled 6-substituted 2-arylbenzothiazoles as amyloid imaging agents, J. Med. Chem. 46(2003)2740-2754.

[7] W. E. Klunk, H. Engler, A. Nordberg, Y. Wang, G. Blomqvist, D. P. Holt, M. Bergstrom, I. Savitcheva, G. F. Huang, S. Estrada, B. Ausen, M. L. Debnath, J. Barletta, J. C. Price, J. Sandell, B. J. Lopresti, A. Wall, P. Koivisto, G. Antoni, C. A. Mathis, B. Langstrom, Imaging brain amyloid in Alzheimer's disease with Pittsburgh Compound-B, Ann. Neurol. 55(2004)306-319.

[8] E. D. Agdeppa, V. Kepe, J. Liu, S. Flores-Torres, N. Satyamurthy, A. Petric, G. M. Cole, G. W. Small, S. C. Huang, J. R. Barrio, Binding characteristics of radiofluorinated 6-dialkylamino-2-naphthylethylidene derivatives as positron emission tomography imaging probes for β-amyloid plaques in Alzheimer's disease, J. Neurosci. 21(2001)RC189.

[9] K. Shoghi-Jadid, G. W. Small, E. D. Agdeppa, V. Kepe, L. M. Ercoli, P. Siddarth, S. Read, N. Satyamurthy, A. Petric, S. C. Huang, J. R. Barrio, Localization of neurofibrillary tangles and β-amyloid plaques in the brains of living patients with Alzheimer disease, Am. J. Geriatr. Psychiatry 10(2002)24-35.

[10] G. W. Small, V. Kepe, L. M. Ercoli, P. Siddarth, S. Y. Bookheimer, K. J. Miller, H. Lavretsky, A. C. Burggren, G. M. Cole, H. V. Vinters, P. M. Thompson, S. C. Huang, N. Satyamurthy, M. E. Phelps, J. R. Barrio, PET of brain amyloid and tau in mild cognitive impairment, N. Engl. J. Med. 355(2006)2652-2663.

[11] M. Ono, A. Wilson, J. Nobrega, D. Westaway, P. Verhoeff, Z. P. Zhuang, M. P. Kung, H. F. Kung, ^{11}C-Labeled stilbene derivatives as abeta-aggregate-specific PET imaging agents for Alzheimer's disease, Nucl. Med. Biol. 30(2003)565-571.

[12] N. P. Verhoeff, A. A. Wilson, S. Takeshita, L. Trop, D. Hussey, K. Singh, H. F. Kung, M. P. Kung, S. Houle, In vivo imaging of Alzheimer disease β-amyloid with [^{11}C]SB-13 PET, Am. J. Geriatr. Psychiatry 12(2004)584-595.

[13] C. C. Rowe, U. Ackerman, W. Browne, R. Mulligan, K. L. Pike, G. O'Keefe, H. Tochon-Danguy, G. Chan, S. U. Berlangieri, G. Jones, K. L. Dickinson-Rowe, H. F. Kung, W. Zhang, M. P. Kung, D. Skovronsky, T. Dyrks, G. Holl, S. Krause, M. Friebe, R. Lehman, S. Lindemann, L. M. Dinkelborg, C. L. Masters, V. L. Villemagne, Imaging of amyloid β in Alzheimer's disease with 18F-BAY94-9172, a novel PET trac-

er: proof of mechanism, Lancet Neurol. 7(2008) 129-135.

[14] S. R. Choi, G. Golding, Z. P. Zhuang, W. Zhang, N. Lim, F. Hefti, T. E. Benedum, M. R. Kilbourn, D. Skovronsky, H. F. Kung, Preclinical properties of 18F-AV-45: a PET Agent for Aβ plaques in the Brain, J. Nucl. Med. 50(2009) 1887-1894.

[15] H. F. Kung, S. R. Choi, W. C. Qu, W. Zhang, D. Skovronsky, 18F stilbenes and styrylpyridines for PET imaging of Aβ plaques in Alzheimer's disease: a miniperspective, J. Med. Chem. 53(2010) 933-941.

[16] M. P. Kung, C. Hou, Z. P. Zhuang, B. Zhang, D. Skovronsky, J. Q. Trojanowski, V. M. Lee, H. F. Kung, IMPY: an improved thiofiavin-T derivative for in vivo labeling of beta-amyloid plaques, Brain Res. 956 (2002) 202-210.

[17] Z. P. Zhuang, M. P. Kung, A. Wilson, C. W. Lee, K. Plossl, C. Hou, D. M. Holtzman, H. F. Kung, Structure-activity relationship of imidazo[1,2-a]pyridines as ligands for detecting β-amyloid plaques in the brain, J. Med. Chem. 46(2003) 237-243.

[18] A. B. Newberg, N. A. Wintering, K. Plossl, J. Hochold, M. G. Stabin, M. Watson, D. Skovronsky, C. M. Clark, M. P. Kung, H. F. Kung, Safety, biodistribution, and dosimetry of ^{123}I-IMPY: a novel amyloid plaque-imaging agent for the diagnosis of Alzheimer's disease, J. Nucl. Med, 47(2006) 748-754.

[19] W. C. Qu, M. P. Kung, C. Hou, L. W. Jinc, H. F. Kung, Radioiodinated aza-diphenylacetylenes as potential SPECT imaging agents for b-amyloid plaque detection, Bioorg. Med. Chem. Lett. 17(2007) 3581-3584.

[20] W. C. Qu, M. P. Kung, C. Hou, T. E. Benedum, H. F. Kung, Novel styrylpyridines as probes for SPECT imaging of amyloid plaques, J. Med. Chem. 50(2007) 2157-2165.

[21] M. Ono, Y. Maya, M. Haratake, K. Ito, H. Mori, M. Nakayama, Aurones serve as probes of β-amyloid plaques in Alzheimer's disease, Biochem. Biophys. Res. Commun. 361(2007) 116-121.

[22] Y. Maya, M. Ono, H. Watanabe, M. Haratake, H. Saji, M. Nakayama, Novel radioiodinated aurones as probes for SPECT imaging of β-amyloid plaques in the brain, Bioconjug. Chem. 20(2009) 95-101.

[23] M. Ono, M. Haratake, H. Mori, M. Nakayama, Novel chalcones as probes for in vivo imaging of β-amyloid plaques in Alzheimer's brains, Bioorg. Med. Chem. 15(2007) 6802-6809.

[24] M. Ono, M. Hori, M. Haratake, T. Tomiyama, H. Mori, M. Nakayama, Structure-activity relationship of chalcones and related derivatives as ligands for detecting of β-amyloid plaques in the brain, Bioorg. Med. Chem. 15(2007) 6388-6396.

[25] M. Ono, R. Ikeoka, H. Watanabe, H. Kimura, T. Fuchigami, M. Haratake, H. Saji, M. Nakayama, Synthesis and evaluation of novel chalcone derivatives with 99mTc/Re complexes as potential probes for detection of β-amyloid plaques, ACS Chem. Neurosci. 1(2010) 598-607.

[26] M. Ono, R. Ikeoka, H. Watanabe, H. Kimura, T. Fuchigami, M. Haratake, H. Saji, M. Nakayama, 99mTc/Re complexes based on flavone and aurone as SPECT probes for imaging cerebral β-amyloid plaques, Bioorg. Med. Chem. Lett. 20(2010) 5743-5748.

[27] E. E. Nesterov, J. Skoch, B. T. Hyman, W. E. Klunk, B. J. Bacskai, T. M. Swager, In vivo optical imaging of amyloid aggregates in Brain: design of fluorescent markers, Angew. Chem. Int. Ed. Engl. 117(2005) 5588-5592.

[28] A. Åslund, A. Herland, P. Hammarstrom, K. P. R. Nilsson, B. H. Jonsson, O. Inganas, P. Konradsson, Studies of luminescent conjugated polythiophene derivatives: enhanced spectral discrimination of protein conformational states, Bioconjug. Chem. 18(2007) 1860-1868.

[29] K. P. R. Nilsson, A. Åslund, I. Berg, S. Nystrom, P. Konradsson, A. Herland, O. Inganas, F. Stabo-Eeg, M. Lindgren, G. T. Westermark, L. Lannfelt, L. N. G. Nilsson, P. Hammarstrom, Imaging distinct conformational states of amyloid-β fibrils in Alzheimer's disease using novel luminescent probes, ACS Chem. Biol.

2(2007)553-560.

[30] A. Åslund, C. J. Sigurdson, T. Klingstedt, S. Grathwohl, T. Bolmont, D. L. Dickstein, E. Glimsdal, S. Prokop, M. Lindgren, P. Konradsson, D. M. Holtzman, P. R. Hof, F. L. Heppner, S. Gandy, M. Jucker, A. Aguzzi, P. Hammarstrom, K. P. R. Nilsson, Novel pentameric thiophene derivatives for *in vitro* and *in vivo* optical imaging of a plethora of protein aggregates in cerebral amyloidoses, ACS Chem. Biol. 4 (2009)673-684.

[31] M. Cui, Z. Li, R. Tang, B. Liu, Synthesis and evaluation of novel benzothiazole derivatives based on the bithiophene structure as potential radiotracers for β-amyloid plaques in Alzheimer's disease, Bioorg. Med. Chem. 18(2010)2777-2784.

[32] K. A. Stephenson, R. Chandra, Z. P. Zhuang, C. Hou, S. Oya, M. P. Kung, H. F. Kung, Fluoro-pegylated (FPEG)imaging agents targeting Aβ aggregates, bioconjug. Chem. 18(2007)238-246.

[33] M. M. Raposoa, G. Kirschb, Formylation, dicyanovinylation and tricyanovinylation of 5-alkoxy-and 5-amino-substituted 2,2'-bithiophenes, Tetrahedron 59(2003)4891-4899.

[34] Z. P. Zhuang, M. P. Kung, C. Hou, D. M. Skovronsky, T. L. Gur, K. Plossl, J. Q. Trojanowski, V. M. Y. Lee, H. F. Kung, Radioiodinated styrylbenzenes and thioflavins as probes for amyloid aggregates, J. Med. Chem. 44(2001)1905-1914.

[35] M. P. Kung, D. M. Skovronsky, C. Hou, Z. P. Zhuang, T. L. Gur, B. Zhang, J. Q. Trojanowski, V. M. Y. Lee, H. F. Kung, Detection of amyloid plaques by radioligands for Aβ 40 and Aβ 42, J. Mol. Neurosci. 20 (2003)15-23.

[36] Y. Cheng, W. Prusoff, Relationship between the inhibition constant(K_i) and the concentration of inhibitor which causes 50 percent inhibition(I50) of an enzymatic reaction, Biochem. Pharmacol. 22(1973)3099-3108.

[37] C. Y. Wu, J. J. Wei, K. Q. Gao, Y. M. Wang, Dibenzothiazoles as novel amyloid-imaging agents, Bioorg. Med. Chem. 15(2007)2789-2796.

Synthesis and Evaluation of Novel ^{18}F Labeled 2-pyridinylbenzoxazole and 2-pyridinylbenzothiazole Derivatives as Ligands for Positron Emission Tomography(PET) Imaging of β-amyloid Plaques[*]

Abstract A series of fluoro-pegylated(FPEG) 2-pyridinylbenzoxazole and 2-pyridinylbenzothiazole derivatives were synthesized and evaluated as novel β-amyloid(Aβ) imaging probes for PET. They displayed binding affinities for $Aβ_{1-42}$ aggregates that varied from 2.7 to 101.6nM. Seven ligands with high affinity were selected for ^{18}F labeling. In vitro autoradiography results confirmed the high affinity of these radiotracers. In vivo biodistribution experiments in normal mice indicated that the radiotracers with a short FPEG chain($n = 1$) displayed high initial uptake into and rapid washout from the brain. One of the 2-pyridinylbenzoxazole derivatives, [^{18}F]-5-(5-(2-fluoroethoxy)benzo[d]oxazol-2-yl)-N-methylpyridin-2-amine([^{18}F]32) ($K_i = (8.0 ± 3.2)$nM) displayed a brain$_{2min}$/brain$_{60min}$ ratio of 4.66, which is highly desirable for Aβ imaging agents. Target specific binding of [^{18}F]32 to Aβ plaques was validated by ex vivo autoradiographic experiment with transgenic model mouse. Overall, [^{18}F] 32 is a promising Aβ imaging agent for PET and merits further evaluation in human subjects.

Ⓢ Supporting Information

1 Introduction

Alzheimer's disease (AD) is the most common form of dementia accounting for between 50% and over 70% of all cases among elderly people and is becoming an extensive health problem with the ever-increasing aging population. Clinically, AD is a lethal, progressive neurodegenerative disorder that leads to a decline in memory and many cognitive deficits, such as irreversible memory loss, impaired judgment, and problems with language. At present there are no treat-

---

[*] Copartner: Cui Mengchao, Wang Xuedan, Yu Pingrong, Zhang Jinming, Li Zijing, Zhang Xiaojun, Yang Yanping, Masahiro Ono, Jia Hongmei, Hideo Saji. Reprinted from *Journal of Medicinal Chemistry*, 2012, 55(21):9283-9296.

ments available to reverse or halt the progression of the disease. Although some drugs may provide temporary relief from symptoms, the benefit is limited. The best hope for progress might be to detect the disease at an early stage and initiate treatment before the brain damage is widespread. However, clinical diagnosis of AD is very difficult, and only possible or probable AD can be routinely diagnosed, while a definite diagnosis can only be made on biopsy and post-mortem brain tissue to confirm the existence of β-amyloid(Aβ)plaques and neurofibrillary tangles(NFTs). Although the etiology of AD has not been definitively established, the amyloid cascade hypothesis is the most prevailing theory to explain the pathogenesis of AD, which posits that the deposition of the Aβ peptide in the brain is a crucial step that ultimately leads to AD[1-5]. Therefore, the development of novel imaging agents specifically targeting Aβ plaques may lead to early diagnosis of AD and monitoring of the effectiveness of novel therapies for this devastating disease[6-8].

In past decades, a number of radiolabeled Aβ imaging agents based on highly conjugated Aβ specific dyes, such as Congo Red(CR) and thioflavin T(Th-T), have been developed and reported for single photon emission computed tomography(SPECT) and positron emission tomography(PET)[9,10], some of which have been tested in humans(Fig. 1). Among dozens of radio-iodinated ligands, [^{123}I]-6-iodo-2-(4'-dimethylamino)phenylimidazo[1,2] pyridine([^{123}I]**1**, IMPY), a unique Th-T derivative with a [1,2,a] imidazopyridine ring, was the only SPECT tracer tested in humans. However, the signal-to-noise ratio for plaque labeling of [^{123}I]**1** in AD and healthy controls was not robust, while it was believed that the in vivo instability and fast metabolism of [^{123}I]**1** may ultimately lead to a decreasing of signal[11-13]. After that, there is no report of Aβ imaging candidate for SPECT moving into clinical trial. In contrast, the development of radiotracers for PET imaging was more successful. [^{11}C]-2-(4-(Methylamino)phenyl)-6-hydroxybenzothiazol([^{11}C]**2**, PIB), to date, is the most suitable and extensively studied PET radioligand for amyloid imaging and can distinguish AD from control cases clearly[14-16]. Recently, Andersson et al. reported a close analogue of [^{11}C]**2**, [^{11}C]-2-[6-(methylamino)pyridin-3-yl]-1,3-benzothiazol-6-ol([^{11}C]**3**, AZD2184), by replacing the phenyl group with a

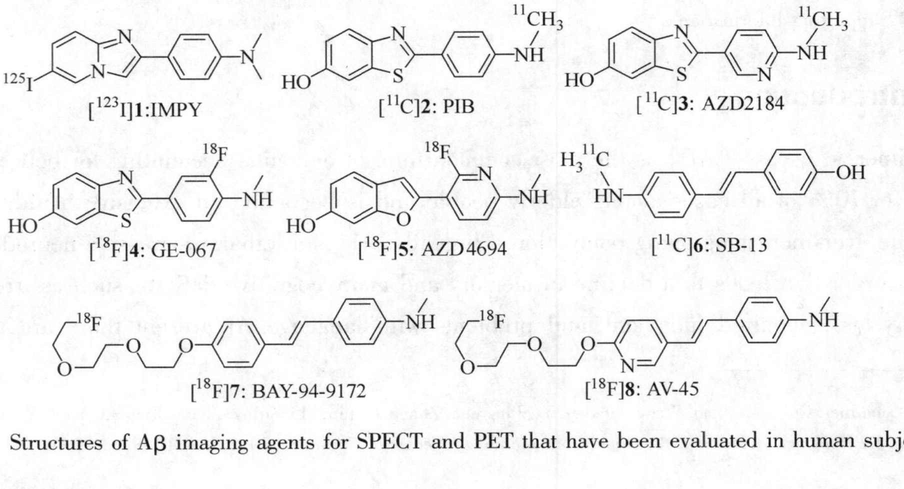

Fig. 1 Structures of Aβ imaging agents for SPECT and PET that have been evaluated in human subjects

pyridyl group, which displayed a decreased nonspecific binding in white matter[17,18]. Additionally, the stilbene scaffold, which could be considered as a more abbreviated form of CR, has also been selected for developing Aβ imaging agents. Kung et al. reported the first ^{11}C-labeled stilbene derivative, [^{11}C]-4-N-methylamino-4′-hydroxystilbene([^{11}C]6, SB-13). Initial PET imaging in vivo with[^{11}C]6 demonstrated potential usefulness in detecting Aβ plaques in the human brain[19,20]. However, the short half-life of ^{11}C($T_{1/2}$ = 20min) of the ^{11}C-labeled tracers limits their usefulness for widespread clinical application. Thus, great efforts have been directed to developing imaging agents labeled with ^{18}F($T_{1/2}$ = 110min). Two ^{18}F-labeled PIB analogues, [^{18}F]-2-(3-fluoro-4-methyaminophenyl)-benzothiazole-6-ol([^{18}F]4, GE-067)[21] and [^{18}F]-2-(2-fluoro-6-(methylamino)pyridin-3-yl)benzofuran-6-ol([^{18}F]5, AZD4694)[22], are currently under phase Ⅱ clinical testing in Europe. Meanwhile, two ^{18}F-labeled stilbene derivatives with a short length of polyethylene glycol(PEG) units, [^{18}F]-4-(N-methylamino)-4′-(2-(2-(2-fluoroethoxy)ethoxy)ethoxy)stilbene([^{18}F]7, BAY94-9172)[23] and [^{18}F]-(E)-4-(2-(6-(2-(2-(2-fluoroethoxy)ethoxy)ethoxy)pyridin-3-yl)vinyl)-N-methylaniline([^{18}F]8, AV-45)[24-26], are under commercial development. More recently, [^{18}F]8, brand-named Amyvid by Eli Lilly, has been approved by the U. S. Food and Drug Administration(FDA).

Despite the fact that great success has been achieved on ^{18}F-labeled tracers for amyloid imaging, it is clear that all ^{18}F-labeled tracers show high levels of nonspecific white matter retention than[^{11}C]2. In AD patients, the average cortical binding is similar to or less than the uptake in white matter and may limit the sensitivity of PET imaging[27]. The mechanism of white matter retention seems to be owing to nonspecific binding that to some degree is governed by the lipophilicity of the ligand[22]. Thus, further modifications in order to decrease lipophilicity are needed. Two approaches were used to decrease the lipophilicity of ^{18}F-labeled ligand: (1)introducing a short length of PEG(n = 2-12) and capping the end of the ethylene glycol chain with a fluorine atom[fluoropolyethylene glycol(FPEG)], which will provide a flexible tool to adjust lipophilicity and to maintain a relatively small size[28]; (2)displacing one of the benzene rings of ligand with one pyridine ring. Previous work indicates that the in vivo pharmacokinetics of the fluoro-pegylated pyridylbenzofuran derivatives([^{18}F]11, [^{18}F]12)[29,30] were greatly improved compared with the phenylbenzofuran derivatives([^{18}F]9, [^{18}F]10)[31]. Tracer[^{18}F]12 displayed the best pharmacokinetics of radioactivity in the brain, with the brain$_{2min}$/brain$_{60min}$ ratio being 2.34. Kung et al. also reported a series of fluoro-pegylated phenylbenzothiazole derivatives ([^{18}F]15-17) with improved properties compared with the original radioiodinated ligand[28] (Fig. 2).

In an attempt to further explore more promising ^{18}F-labeled tracers, we extended our fluoro-pegylation approach from benzofuran to benzoxazole. The strategy of FPEG was successfully applied to the benzoxazole scaffold, and it has been demonstrated that the fluoro-pegylated phenylbenzoxazole derivatives([^{18}F]13, [^{18}F]14) showed more promising in vivo kinetics than the corresponding radioiodinated ligand and the benzofuran derivatives[32]. Tracer[^{18}F]13 with a monomethylamino group exhibited high affinity for Aβ aggregates(K_i = 9.3nM) and high initial

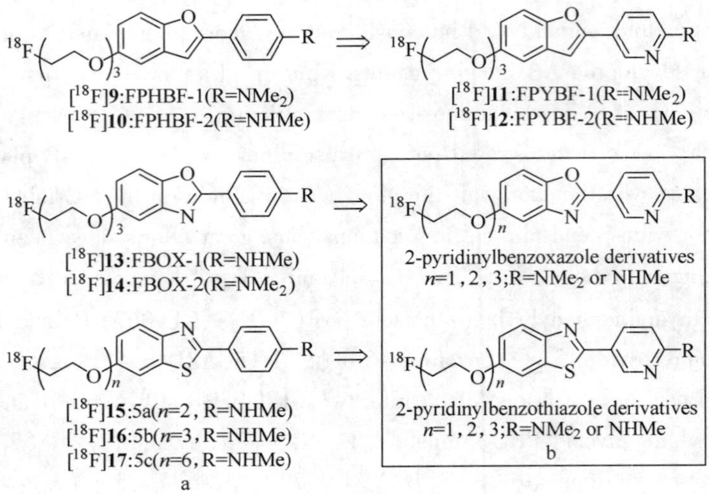

Fig. 2 Structures of radiolabeled benzofuran, bezoxazole, and benzothiazole derivatives (a) and structures of the designed ^{18}F-labeled 2-pyridinylbenzoxazole and 2-pyridinylbenzothiazole derivatives (b)

uptake into the brain and displayed excellent binding to Aβ plaques in ex vivo autoradiographic experiments with Tg2576 mice. What's more, small-animal PET studies demonstrated significant differences between Tg2576 and wild-type mice in the clearance profiles after the administration of [^{18}F]13. Although the preliminary results indicate that [^{18}F]13 is a promising PET tracer for imaging Aβ plaques, the brain$_{2min}$/brain$_{60min}$ ratio of [^{18}F]13(2.67) was still lower than that of [^{18}F]8(3.90). To further improve the pharmacokinetic profile of the fluoro-pegylated phenylbenzothiazole and phenylbenzoxazole derivatives, we decided to use the strategy of bioisosteric replacement by displacing the phenyl group with a pyridyl group to reduce nonspecific binding in white matter and enhance the signal-to-noise ratio. We report herein the synthesis and the initial biological evaluation of fluoro-pegylated pyridylbenzothiazole and pyridylbenzoxazole derivatives as novel PET imaging probes targeting Aβ plaques in the brain.

2 Results and Discussion

2.1 Chemistry

The synthesis of fluoro-pegylated pyridylbenzothiazole and pyridylbenzoxazole derivatives is outlined in Scheme 1. The intermediates **20** and **21** were obtained by ready fluorination of the oligoethylene glycol ditosylate ester (**18**, **19**) using anhydrous tetra-*n*-butylammonium fluoride in THF. The desired 2-aminopyridylbenzothiazole and 2-aminopyridylbenzoxazole core structures (**23** and **35**) were obtained by using a previously reported one-step reaction between 5-(trifluoromethyl)pyridin-2-amine and 2-amino-4-methoxyphenol (**22**) or 2-amino-5-methoxybenzenethiol (**34**) in aqueous NaOH solution with excellent yields (89% and 50%, respectively)[33].

Conversion of **23** and **35** to the monomethylamino derivatives **24** and **36** was achieved by the monomethylation of the amino group by using a method previously reported in yields of 98% and 98%, respectively[34]. The dimethylamino derivatives **25** and **37** were prepared by an efficient dimethylation method with paraformaldehyde, sodium cyanoborohydride, and acetic acid (89% and 91%, respectively). The *O*-methyl groups of **24, 25, 36** and **37** were removed by reacting with BBr_3 in CH_2Cl_2 to give **26, 27, 38** and **39** in yields of 97%, 89%, 88%, and 68%, respectively. Then the hydroxy groups were coupled with 1-bromo-2-fluoroethane to give the fluorinated derivatives (**32, 33, 44**, and **45**) with one ethoxy unit as the PEG linkage. The corresponding fluorinated derivatives with two or three ethoxy units (**28-31** and **40-43**) were prepared by the hydroxyl compounds with K_2CO_3 and the intermediate **20** or **21** in DMF (yield, 40%-75%).

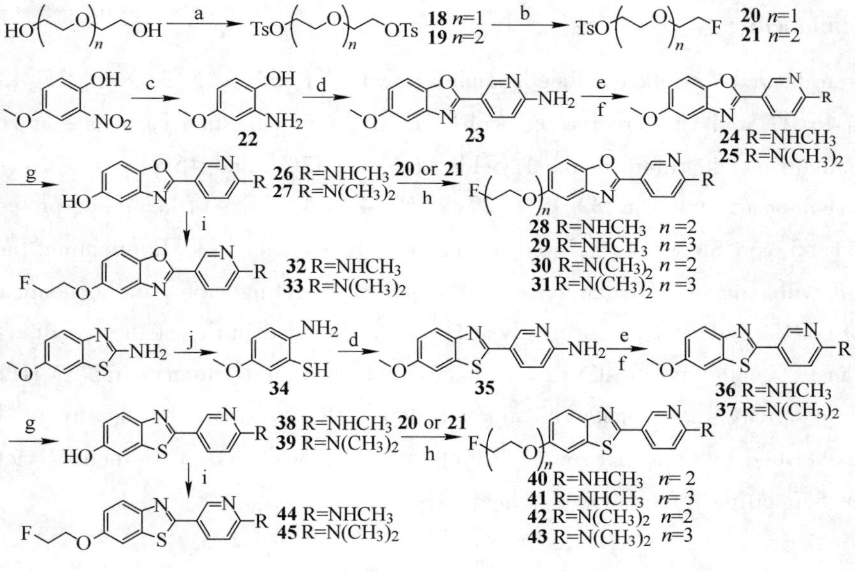

Scheme 1 Reagents and conditions

a—TsCl, KOH, CH_2Cl_2, 0℃ to rt; b—TBAF(1M in THF), THF, reflux; c—Pd/C, MeOH, rt; d—2-amino-5-(trifluoromethyl)pyridine, NaOH(1M), H_2O, reflux; e—(1) NaOMe, MeOH, $(CH_2O)_n$, reflux; (2) $NaBH_4$, rt; f—$NaBH_3CN$, $(CH_2O)_n$, CH_3COOH, rt; g—BBr_3(1M in CH_2Cl_2), CH_2Cl_2, −78℃ to rt; h—K_2CO_3, DMF, 110℃; i—1-bromo-2-fluoroethane, K_2CO_3, DMF, 110℃; j—KOH, H_2O, reflux

The synthesis of tosylate precursors is outlined in Scheme 2. The free hydroxy groups present in **26, 27, 38**, and **39** were coupled with ethane-1,2-diyl bis(4-methylbenzenesulfonate) or **19** in acetone catalyzed by 18-crown-6 to give **46-49** and **52-54** (yield, 28%-49%). For the monomethylated precursors, the methylamino groups of intermediates **48, 49, 53**, and **54** were protected by a butyloxycarbonyl(BOC) group to give the tosylate precursors **50, 51, 55**, and **56** in yields of 67%, 23%, 80%, and 63%, respectively. Compared with the synthesis routes for the precursors of benzofuran and benzoxazole derivatives, the procedures reported here are much simpler, while some protection and deprotection steps were not needed.

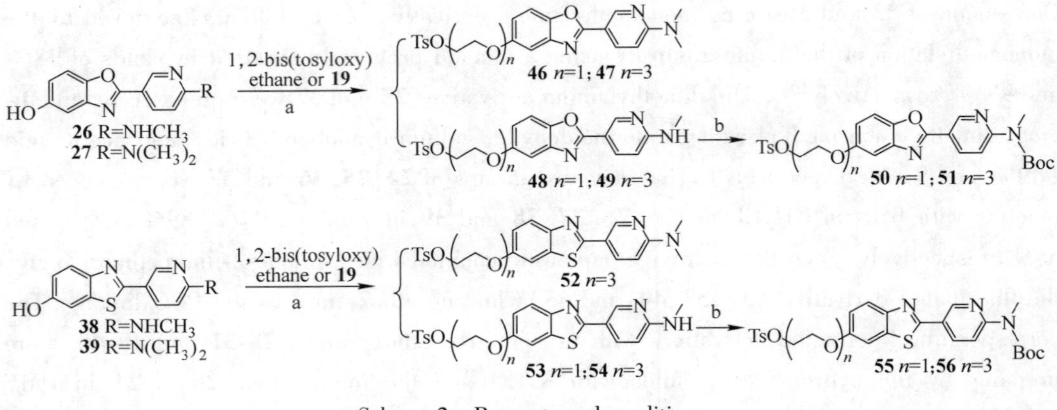

Scheme 2 Reagents and conditions
a—K_2CO_3, 18-crown-6, acetone, reflux; b—$(Boc)_2O$, THF, reflux

2.2 Radiolabeling

To make the desired ^{18}F-labeled dimethylamino tracers [^{18}F]33, [^{18}F]31, and [^{18}F]43, tosylate precursors 46, 47, and 52 were reacted with [^{18}F]fluoride/potassium carbonate and Kryptofix 222 in acetonitrile with radiochemical yields of 53%, 57%, and 46%, respectively. For the monomethylamino tracers [^{18}F]32, [^{18}F]29, [^{18}F]44, and [^{18}F]41, the N-Boc-protected tosylates 50, 51, 55, and 56 were employed as the precursors. After [^{18}F]fluorination, the mixture was treated with aqueous HCl to remove the N-Boc-protecting group (radiochemical yields, 41%, 21%, 33%, and 27%, respectively). The radiochemical purity of these radiotracers was greater than 98% after purification by high performance liquid chromatography (HPLC), and their specific activity was estimated as approximately 200GBq/μmol. The identity of ^{18}F-labeled tracers was verified by a comparison of the retention time with that of the nonradioactive compound (see Supporting Information) (Scheme 3).

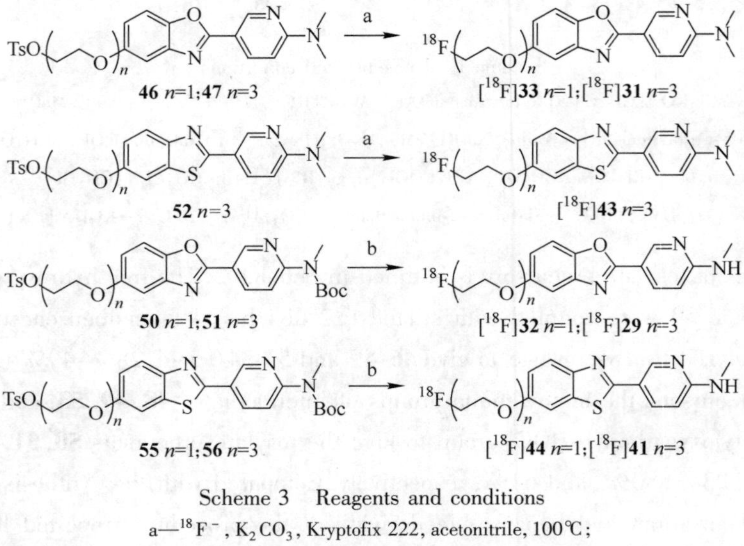

Scheme 3 Reagents and conditions
a—$^{18}F^-$, K_2CO_3, Kryptofix 222, acetonitrile, 100℃;
b—(1) $^{18}F^-$, K_2CO_3, Kryptofix 222, acetonitrile, 100℃; (2) HCl(1M), 100℃

2.3 In vitro binding studies

The affinities of these fluoro-pegylated pyridylbenzothiazole and pyridylbenzoxazole derivatives (**28-33** and **40-45**) for Aβ$_{1-42}$ aggregates were examined with competition binding assays using [^{125}I]**1** as the competing radioligand. IMPY and PIB were also screened using the same system for comparison. As shown in Table 1, all ligands with a dimethylamino group (**30**, **31**, **33**, **42**, **43**, and **45**) displayed high binding affinity to Aβ$_{1-42}$ aggregates (K_i < 10.0 nM). Similar to the findings reported previously[30], the tertiary N, N-dimethylamino analogues were found to have higher affinities than their secondary N-methylamino analogues, especially for the pyridylbenzoxazole derivatives (e.g., **28** vs **30**, **29** vs **31**, **32** vs **33**). The length of the FPEG chain did not bring about an appreciable progress in the binding affinity to amyloid aggregates for the pyridylbenzothiazole derivatives (**40-45**). However, the length of the FPEG chain is very important for the secondary N-methylamino analogues of pyridylbenzoxazole. The binding affinities were reduced dramatically with the addition of the FPEG chain [e.g., **28** (K_i = (101.6 ± 15.3) nM) and **29** (K_i = (76.9 ± 15.5) nM)], while **32** retained high binding affinity for the n = 1 FPEG conjugates (K_i = (8.0 ± 3.2) nM). On the basis of the high binding affinity to Aβ$_{1-42}$ aggregates, ligands **29**, **31**, **32**, **33**, **41**, **43**, and **44** were selected for ^{18}F labeling.

Table 1 Inhibition constants of fluorinated 2-pyridinylbenzoxazole and 2-pyridinylbenzothiazole derivatives for the binding of [^{125}I] IMPY to Aβ$_{1-42}$ aggregates

Compd	K_i ± SEM/nM[①]	Compd	K_i ± SEM/nM[①]
28	101.6 ± 15.3	41	9.3 ± 0.1
29	76.9 ± 15.5	42	4.6 ± 0.6
30	7.3 ± 0.8	43	5.8 ± 1.8
31	9.9 ± 0.5	44	10.1 ± 2.4
32	8.0 ± 3.2	45	2.7 ± 0.6
33	2.7 ± 0.7	IMPY	10.5 ± 1.0[②]
40	29.7 ± 7.1	PIB	9.0 ± 1.3[②]

① The K_i values were determined in three independent experiments (n = 3) unless otherwise noted.
② Data from Ref. [30].

2.4 In vitro autoradiography studies

In vitro autoradiographic studies of the ^{18}F-labeled tracers were first performed with sections of Tg mice (C57BL6, APPswe/PSEN1, 12 months old) and an age-matched control mice. As shown in Fig. 3 and Fig. 4, all of the ^{18}F-labeled tracers displayed excellent labeling of Aβ plaques in the hippocampus and cortical regions of the Tg mice and lack of any notable Aβ labeling in control mice. The distribution of Aβ plaques was consistent with the results of fluorescent staining with thioflavin S. A previous report suggested that the configuration/folding of Aβ plaques in Tg mice might be different from the tertiary/quaternary structure of Aβ plaques in AD brains[35]. Thus, the binding information for Aβ plaques in human AD brain is important. Tracer [^{18}F]**29**, [^{18}F]**32**, [^{18}F]**41**, and [^{18}F]**44** with a monomethylamino group were selected for autoradiographic studies in sections of human brain tissue. As shown in Fig. 5, a highly dense labeling of plaques was observed in AD brain sections. In contrast, no apparent labeling was observed in normal adult brain sections. Although the binding affinity of $[^{18}F]$**29** to Aβ aggregates was not potent ($K_i = (76.9 \pm 15.5)$ nM), $[^{18}F]$**29** was still able to label the plaques in sections of Tg mice and AD.

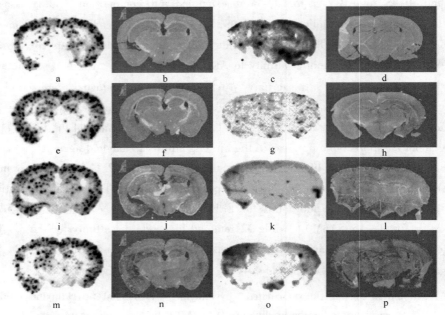

Fig. 3 In vitro autoradiography of ^{18}F-labeled 2-pyridinylbenzoxazole derivatives ($[^{18}F]$**29**, $[^{18}F]$**31**, $[^{18}F]$**32**, and $[^{18}F]$**33**) on a Tg model mouse (C57BL6, APPswe/PSEN1, 12 months old, male) (a, e, i, m) and normal control mouse (C57BL6, 12 months old, male) (c, g, k, o). The presence and distribution of plaques in the sections were confirmed by fluorescence staining using thioflavin S on the same sections with a filter set for GFP (b, d, f, h, j, l, n, p)

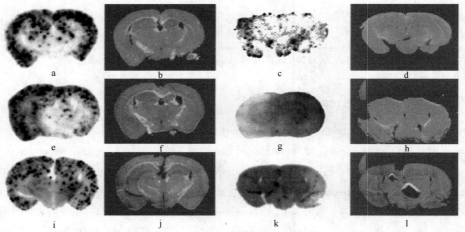

Fig. 4 In vitro autoradiography of ^{18}F-labeled 2-pyridinylbenzothiazole derivatives ($[^{18}F]$**41**, $[^{18}F]$**43**, and $[^{18}F]$**44**) on a Tg model mouse (C57BL6, APPswe/PSEN1, 12 months old, male) (a, e, i) and normal control mouse (C57BL6, 12 months old, male) (c, g, k). The presence and distribution of plaques in the sections were confirmed by fluorescence staining using thioflavin S on the same sections with a filter set for GFP (b, d, f, h, j, l)

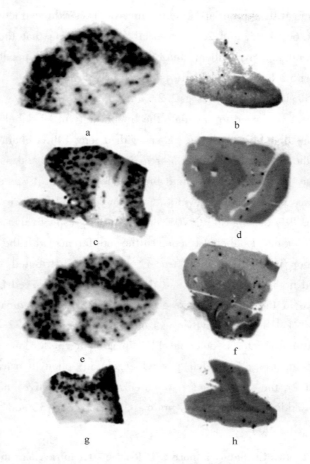

Fig. 5　In vitro autoradiography of [^{18}F]**29**(a), [^{18}F]**32**(c),
[^{18}F]**41**(e), and [^{18}F]**44**(g) on human brain sections

a, c, e, g—Postmortem AD cases; b, d, f, h—Normal control

2.5　In vivo biodistribution studies

Biodistribution study of the radiolabeled tracers in normal mice is often used as a test to measure the initial brain uptake [ability to penetrate intact blood-brain barrier (BBB)] and washout kinetics from the normal brain. The time-activity data on the brain permeation as well as the other organ distribution of the ^{18}F-labeled tracers acquired in normal ICR mice after intravenous administration are summarized in Table 2-Table 4. [^{18}F]**32**, [^{18}F]**33**, and [^{18}F]**44** with a short FPEG chain ($n = 1$) displayed high uptakes into the brain (7.23% ID/g, 7.27% ID/g, and 7.87% ID/g, respectively) at 2min postinjection and fast washout kinetics from the healthy brain (1.55% ID/g, 1.47% ID/g, and 2.83% ID/g at 60min, respectively). These values were comparable to those reported for the ^{18}F-labeled pyridylbenzofuran ([^{18}F]**10**, [^{18}F]**12**)[30] and phenylbenzoxazole ([^{18}F]**13**)[32] derivatives as well as [^{18}F]**7** and [^{18}F]**8**, which are under commercial development. The ratio brain$_{2min}$/brain$_{60min}$ is considered as an important index with which to select tracers with appropriate kinetics in vivo. These three tracers showed brain$_{2min}$/brain$_{60min}$ ratios of 4.66, 4.95, and 2.78, respectively, indicating that the pyridylbenzoxazole scaffold is more potent than pyridylbenzothiazole. However, the BBB penetration ability of the ^{18}F-labeled tracers with a long FPEG chain ($n = 3$) ([^{18}F]**29**, [^{18}F]**31**, [^{18}F]**41**, and [^{18}F]**43**) was decreased (4.05% ID/g, 3.79% ID/g, 2.23% ID/g, and 3.26% ID/g at 2min, respectively). The FPEG strategy was used to modulate the lipophilicity of ^{18}F-labeled tracers. As reflected in lg$D$ values, the lipophilicity of the tracers with a long FPEG chain decreased slightly, which may explain the lower brain uptake. Moreover, the flexiblility and polar properties of the long FPEG chain may also interfere and limit brain penetration[28]. It was also observed that in vivo defluorination was likely occurring with [^{18}F]**31** and [^{18}F]**44** because the bone uptakes increased with time (6.39% dose/g and 6.70% dose/g at 1h, respectively). However, they were minimal compared to other ^{18}F-labeled tracers, so the interference with the imaging is expected to be relatively minor. These ^{18}F-labeled tracers were also distributed to several other organs. The high initial uptakes in the liver and the kidneys were followed by a moderate washout, while the uptake in the intestines showed an accumulation of radioactivity over time. In summary, two of the ^{18}F-labeled pyridylbenzoxazole derivatives ([^{18}F]**32** and [^{18}F]**33**) meet the preliminary requirements for a potential PET imaging agent for in vivo detection of Aβ plaque. Compared with the ^{18}F-labeled pyridylbenzofuran, phenylbenzoxazole, and phenylbenzothiazol derivatives, the ^{18}F-labeled pyridylbenzoxazole derivatives had greatly improved pharmacokinetics. In addition, the brain$_{2min}$/brain$_{60min}$ ratio of [^{18}F]**32** and [^{18}F]**33** was superior than that of [^{18}F]**8** (3.90).

Table 2 Biodistribution in normal ICR mice after intravenous injection of the ^{18}F-labeled 2-pyridinylbenzoxazole tracers[①]

Organ	2min	10min	30min	60min
[^{18}F]**29** (lg$D$ = 2.86 ± 0.12)				
Blood	5.07 ± 0.41	3.04 ± 0.29	3.16 ± 0.26	2.74 ± 0.24
Brain	4.05 ± 0.36	2.05 ± 0.27	2.06 ± 0.29	1.78 ± 0.21
Heart	5.50 ± 0.70	3.35 ± 0.37	3.16 ± 0.28	2.50 ± 0.27
Liver	14.22 ± 2.65	9.32 ± 0.79	6.47 ± 0.72	4.21 ± 0.66
Spleen	4.13 ± 0.32	2.64 ± 0.35	2.18 ± 0.11	1.79 ± 0.16
Lung	5.30 ± 0.31	3.15 ± 0.46	2.81 ± 0.24	2.34 ± 0.19
Kidney	8.51 ± 0.97	4.79 ± 0.57	3.73 ± 0.20	2.54 ± 0.33
Stomach[②]	5.08 ± 0.98	12.32 ± 0.85	12.93 ± 1.25	12.02 ± 1.82
Intestine[②]	10.81 ± 0.86	22.90 ± 1.90	28.65 ± 2.66	25.16 ± 0.97
Bone	1.75 ± 0.25	1.92 ± 0.65	2.46 ± 0.33	3.93 ± 0.48

Continued 2

Organ	2min	10min	30min	60min
[^{18}F]**31** ($\lg D = 3.26 \pm 0.04$)				
Blood	5.47 ± 0.56	3.05 ± 0.25	2.74 ± 0.14	2.87 ± 0.12
Brain	3.79 ± 0.41	1.97 ± 0.14	1.85 ± 0.06	1.82 ± 0.07
Heart	5.24 ± 0.63	3.51 ± 0.11	2.97 ± 0.18	2.73 ± 0.13
Liver	13.25 ± 3.13	5.41 ± 0.67	7.24 ± 2.05	3.46 ± 1.52
Spleen	3.00 ± 0.63	2.69 ± 0.65	2.41 ± 0.10	2.34 ± 0.26
Lung	3.03 ± 0.63	2.34 ± 0.66	2.64 ± 0.29	1.87 ± 0.66
Kidney	8.27 ± 2.03	5.14 ± 0.30	4.20 ± 0.42	3.21 ± 0.46
Stomach[②]	2.66 ± 0.60	5.10 ± 0.80	9.26 ± 2.63	6.45 ± 1.79
Intestine[②]	8.13 ± 0.79	14.45 ± 2.65	28.15 ± 3.11	28.54 ± 4.18
Bone	1.79 ± 0.36	2.70 ± 0.47	4.71 ± 0.34	6.39 ± 0.57
[^{18}F]**32** ($\lg D = 3.52 \pm 0.13$)				
Blood	3.36 ± 0.15	2.55 ± 0.15	2.55 ± 0.17	3.20 ± 0.48
Brain	7.23 ± 0.04	2.24 ± 0.12	1.69 ± 0.17	1.55 ± 0.17
Heart	4.66 ± 0.34	2.34 ± 0.17	1.84 ± 0.19	2.05 ± 0.21
Liver	10.48 ± 1.65	8.50 ± 0.72	3.95 ± 0.66	2.64 ± 0.86
Spleen	3.32 ± 0.50	2.13 ± 0.24	1.63 ± 0.28	1.64 ± 0.13
Lung	4.14 ± 0.24	2.80 ± 0.48	2.29 ± 0.09	2.57 ± 0.40
Kidney	7.45 ± 0.35	3.52 ± 0.52	2.17 ± 0.22	3.04 ± 0.99
Stomach[②]	3.24 ± 0.62	5.59 ± 1.45	3.11 ± 0.94	3.89 ± 1.58
Intestine[②]	9.87 ± 0.97	19.61 ± 2.42	33.66 ± 5.26	39.05 ± 6.70
Bone	1.18 ± 0.21	1.11 ± 0.36	2.34 ± 0.88	2.37 ± 0.86
[^{18}F]**33** ($\lg D = 3.53 \pm 0.04$)				
Blood	3.57 ± 0.39	2.05 ± 0.10	1.89 ± 0.17	2.06 ± 0.26
Brain	7.27 ± 0.19	3.41 ± 0.36	1.76 ± 0.16	1.47 ± 0.25
Heart	5.95 ± 0.95	2.44 ± 0.18	1.83 ± 0.24	1.93 ± 0.21
Liver	15.70 ± 2.71	10.72 ± 1.44	4.39 ± 0.80	3.19 ± 0.61
Spleen	3.22 ± 0.52	2.02 ± 0.25	1.60 ± 0.14	1.32 ± 0.26
Lung	5.53 ± 0.50	2.58 ± 0.31	1.85 ± 0.46	1.79 ± 0.38
Kidney	8.48 ± 0.77	3.82 ± 0.37	2.44 ± 0.59	1.82 ± 0.34
Stomach[②]	2.44 ± 0.76	2.55 ± 0.82	2.54 ± 0.48	2.79 ± 0.57
Intestine[②]	8.41 ± 1.83	21.45 ± 6.22	38.88 ± 2.94	37.02 ± 6.37
Bone	2.08 ± 0.73	1.27 ± 0.15	1.56 ± 0.53	1.88 ± 0.39

① Expressed as % injected dose per gram unless otherwise indicated. Data are the average for five mice ± standard deviation.
② Expressed as % injected dose per organ.

Table 3 Biodistribution in normal ICR mice after intravenous injection of the ^{18}F-labeled 2-pyridinylbenzothiazole tracers[1]

Organ	2min	10min	30min	60min
[^{18}F]**41** ($\lg D = 2.85 \pm 0.07$)				
Blood	6.92 ± 0.60	4.25 ± 0.32	3.49 ± 0.11	2.94 ± 0.19
Brain	2.23 ± 0.33	1.63 ± 0.15	1.31 ± 0.17	1.26 ± 0.15
Heart	3.53 ± 0.36	2.16 ± 0.16	1.96 ± 0.13	1.85 ± 0.13
Liver	15.22 ± 2.34	9.38 ± 0.90	8.16 ± 0.46	4.12 ± 0.65
Spleen	2.68 ± 0.35	2.05 ± 0.24	1.74 ± 0.07	1.49 ± 0.15
Lung	5.53 ± 0.26	3.36 ± 0.50	3.11 ± 0.06	1.93 ± 0.43
Kidney	6.02 ± 0.98	4.04 ± 0.39	3.12 ± 0.25	2.66 ± 0.23
Stomach[2]	2.96 ± 0.62	9.44 ± 1.92	11.46 ± 3.60	8.78 ± 1.46
Intestine[2]	5.69 ± 0.98	10.06 ± 0.62	21.68 ± 2.74	18.60 ± 4.73
Bone	1.19 ± 0.42	0.86 ± 0.11	1.33 ± 0.32	1.42 ± 0.17
[^{18}F]**43** ($\lg D = 3.08 \pm 0.01$)				
Blood	9.79 ± 0.81	5.44 ± 0.16	4.96 ± 0.87	3.79 ± 0.37
Brain	3.26 ± 0.21	2.64 ± 0.19	1.98 ± 0.20	1.84 ± 0.22
Heart	4.50 ± 0.67	3.30 ± 0.17	2.62 ± 0.11	2.02 ± 0.98
Liver	12.90 ± 0.62	11.60 ± 0.71	6.79 ± 0.80	4.75 ± 0.39
Spleen	3.14 ± 0.32	2.97 ± 0.23	2.59 ± 0.55	1.47 ± 0.77
Lung	6.69 ± 0.53	4.59 ± 0.34	3.28 ± 0.95	2.33 ± 0.94
Kidney	6.82 ± 0.61	5.45 ± 0.32	3.88 ± 0.89	3.01 ± 0.61
Stomach[2]	2.55 ± 0.54	8.98 ± 0.14	14.13 ± 0.61	11.34 ± 0.98
Intestine[2]	4.68 ± 0.23	10.4 ± 1.27	19.57 ± 1.68	26.66 ± 2.90
Bone	1.86 ± 0.27	0.95 ± 0.48	1.64 ± 0.28	1.58 ± 0.28
[^{18}F]**44** ($\lg D = 3.59 \pm 0.02$)				
Blood	4.31 ± 0.34	3.94 ± 0.46	4.45 ± 0.28	4.45 ± 0.51
Brain	7.87 ± 0.48	4.54 ± 0.37	3.38 ± 0.25	2.83 ± 0.15
Heart	5.59 ± 0.69	3.11 ± 0.81	3.33 ± 0.43	3.70 ± 0.43
Liver	10.40 ± 0.83	5.94 ± 0.97	4.28 ± 0.47	3.06 ± 0.55
Spleen	4.52 ± 0.46	3.29 ± 0.75	2.87 ± 0.23	2.86 ± 0.51
Lung	5.85 ± 0.68	3.02 ± 0.60	3.57 ± 0.21	2.72 ± 0.59
Kidney	9.57 ± 0.68	4.42 ± 0.53	3.62 ± 0.19	3.45 ± 0.88
Stomach[2]	3.61 ± 0.45	5.61 ± 0.81	4.63 ± 0.69	3.67 ± 1.11
Intestine[2]	12.06 ± 0.35	11.15 ± 1.30	25.23 ± 3.23	19.00 ± 4.14
Bone	2.04 ± 0.33	2.43 ± 0.60	3.77 ± 0.97	6.70 ± 1.21

[1] Expressed as % injected dose per gram unless otherwise indicated. Data are the average for five mice ± standard deviation.
[2] Expressed as % injected dose per organ.

Table 4 Comparison of inhibition constants (K_i, nM) and brain kinetics between ^{18}F-labeled tracers in this work and other tracers

Compd	K_i/nM	2min①	60min①	brain$_{2min}$/brain$_{60min}$
[^{18}F]7②	6.7 ± 0.3	8.14 ± 2.03	2.60 ± 0.22	3.13
[^{18}F]8③	2.87 ± 0.17	7.33 ± 1.54	1.88 ± 0.14	3.90
[^{18}F]11④	2.0 ± 0.5	2.88 ± 0.46	2.80 ± 0.06	1.03
[^{18}F]12④	3.9 ± 0.2	8.18 ± 0.59	3.87 ± 0.42	2.11
[^{18}F]13④	1.0 ± 0.2	5.16 ± 0.30	2.44 ± 0.36	2.11
[^{18}F]14④	2.4 ± 0.1	7.38 ± 0.84	3.15 ± 0.10	2.34
[^{18}F]16⑤	9.5 ± 1.3	8.12 ± 0.51	3.04 ± 0.16	2.67
[^{18}F]17⑤	3.9 ± 0.2	5.29 ± 0.19	2.12 ± 0.08	2.50
[^{18}F]19⑥	2.2 ± 0.5	10.27 ± 1.30	3.94 ± 0.04	2.61
[^{18}F]20⑥	3.8 ± 0.5	5.53 ± 0.56	2.18 ± 0.09	2.54
[^{18}F]21⑥	4.7 ± 0.9	2.57 ± 0.12	1.80 ± 0.25	1.43
[^{18}F]29	76.9 ± 15.5	4.05 ± 0.36	1.78 ± 0.21	2.28
[^{18}F]31	9.9 ± 0.5	3.79 ± 0.41	1.82 ± 0.07	2.08
[^{18}F]32	8.0 ± 3.2	7.23 ± 0.04	1.55 ± 0.17	4.66
[^{18}F]33	2.7 ± 0.7	7.27 ± 0.19	1.47 ± 0.25	4.95
[^{18}F]41	9.3 ± 0.1	2.23 ± 0.33	1.26 ± 0.15	1.77
[^{18}F]43	5.8 ± 1.8	3.26 ± 0.21	1.84 ± 0.22	1.77
[^{18}F]44	10.1 ± 2.4	7.87 ± 0.48	2.83 ± 0.15	2.78

① Expressed as % injected dose per gram. ② Data from Ref. [40]. ③ Data from Ref. [25]. ④ Data from Ref. [30]. ⑤ Data from Ref. [32]. ⑥ Data from Ref. [28].

2.6 Ex vivo autoradiography studies

A previous report suggested that a rapid metabolism in vivo of the dimethylamino group in the tracer may occur[36], and most of the radiotracers for Aβ plaques in clinical trials have a monomethylamino group. For the same reason, [^{18}F]32 was selected for further ex vivo autoradiography studies in Tg mice (C57BL6, APPswe/PSEN1, 10 months old) and wild-type mice (C57BL6, 10 months old). As shown in Fig. 6a and Fig. 6c, the Aβ plaques in the cortical regions of Tg mice were clearly labeled by [^{18}F]32, while wild-type mouse brain showed no such labeling (Fig. 6e). The labeling of the plaques was confirmed by costaining with thioflavin S on the same sections (Fig. 6b and Fig. 6d).

2.7 In vitro stability studies

In vitro stability study of [^{18}F]32 in saline solution indicated that [^{18}F]32 solutions were stable during storage for at least 6h at room temperature. Moreover, [^{18}F]32 was found to have high in vitro stability in mice plasma; more than 95% of the activity was identified as the intact tracer after 60min of incubating with mice plasma at 37℃ (see Supporting Information).

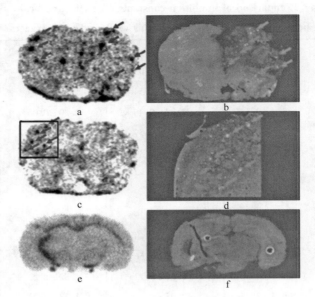

Fig. 6 Autoradiography of [^{18}F]**32** ex vivo using Tg model mouse
(C57BL6, APPswe/PSEN1, 10 months old, male) (a, c) and
wild-type controls (C57BL6, 10 months old) (e)
(Plaques were also confirmed by the staining of the same sections with thioflavin S)

3 Conclusions

A series of fluoro-pegylated (FPEG) 2-pyridinylbenzoxazole and 2-pyridinylbenzothiazole derivatives had been successfully prepared and evaluated as PET imaging tracers for Aβ plaques. In binding studies, they displayed binding affinities for Aβ aggregates that varied from 2.7 to 101.6 nM. Seven ligands with high affinity (**29**, **31**, **32**, **33**, **41**, **43**, and **44**) were selected for ^{18}F labeling. In vitro autoradiography with sections of postmortem AD brain and Tg mouse brain confirmed the affinity of these tracers. In biodistribution experiments using normal mice, the radiotracers with a short FPEG chain ($n = 1$) displayed high initial uptake into and rapid washout from the brain in normal mice. However, the BBB penetration ability of the ^{18}F-labeled tracers with a long FPEG chain ($n = 3$) was not robust. Compared with the ^{18}F-labeled pyridylbenzofuran, phenylbenzoxazole, and phenylbenzothiazol derivatives, the ^{18}F-labeled pyridylbenzoxazole derivatives had greatly improved pharmacokinetics. On the basis of the above results, one of the 2-pyridinylbenzoxazole derivative, [^{18}F]**32** ($K_i = (8.0 \pm 3.2)$ nM), displayed a brain$_{2\text{min}}$/brain$_{60\text{min}}$ ratio of 4.66, which is highly desirable for Aβ imaging agents. Target specific binding of [^{18}F]**32** to Aβ plaques was validated by ex vivo autoradiographic experiments with Tg mouse. Furthermore, [^{18}F]**32** displayed high in vitro stability in saline solution and mice plasma. Overall, [^{18}F]**32** is a promising PET imaging agent for cerebral Aβ plaques and merits further evaluation in human subjects.

4 Experimental Section

4.1 General remarks

All reagents used in the synthesis were commercial products and were used without further purification unless otherwise indicated. The ^{1}H-NMR spectra were obtained at 400MHz on Bruker spectrometer in $CDCl_3$ or DMSO-d_6 solutions at room temperature with TMS as an internal standard. Chemical shifts were reported as δ values relative to the internal TMS. Coupling constants were reported in hertz. Multiplicity is defined by s(singlet), d(doublet), t(triplet), and m(multiplet). Mass spectra were acquired under a Surveyor MSQ Plus(ESI) instrument. Radiochemical purity was determined by HPLC performed on a Shimadzu system SCL-20 AVP equipped with a SPD-20A UV detector($\lambda = 254$nm) and Bioscan flow count 3200 NaI/PMT γ-radiation scintillation detector. HPLC separations were achieved on a Venusil MP C18 reverse phase column(Agela Technologies, 5μm, 10mm×250mm), and elution was with a binary gradient system at a flow rate of 4.0mL/min. HPLC analyses were achieved on a Venusil MP C18 reverse phase column(Agela Technologies, 5μm, 4.6mm×250mm), and elution was with a binary gradient system at a flow rate of 1.0mL/min. Mobile phase A was water, while mobile phase B was acetonitrile. The purity of the synthesized key compounds(28-33 and 40-45) was determined using analytical HPLC and was found to be more than 96%. Fluorescent observation was performed by the Oberver Z1 (Zeiss, Germany) equipped with GFP filter set(excitation, 505nm). Normal ICR mice(5weeks, male) were used for biodistribution experiments. Transgenic mice(C57BL6, APPswe/PSEN1, 12 months old), used as an Alzheimer's model, were purchased from the Institute of Laboratory Animal Science, Chinese Academy of Medical Sciences. All protocols requiring the use of mice were approved by the Animal Care Committee of Beijing Normal University.

4.2 Chemistry

2,2'-Oxybis(ethane-2,1-diyl) Bis(4-methylbenzenesulfonate) (**18**). Intermediate **18** was synthesized according to the literature[37]. Briefly, to a solution of tosyl chloride(19.8g, 10mmol) and 2,2'-oxidiethanol(5.0g, 5mmol) in CH_2Cl_2 (200mL) was added KOH(21.1g, 40mmol). The mixture was stirred at 0℃ for 6h. After reaction, the mixture was washed with water(3 × 100mL). The organic layer was dried over anhydrous $MgSO_4$. After removal of the solvent, the residue was purified by recrystallization in anhydrous MeOH to give **18** (18.1g, 92.7%). ^{1}H NMR(400MHz, $CDCl_3$): δ 7.78 (d, J = 8.3Hz, 4H), 7.34 (d, J = 8.2Hz, 4H), 4.09 (t, J = 4.6Hz, 4H), 3.61 (t, J = 4.7Hz, 4H), 2.45 (s, 6H).

(Ethane-1,2-diylbis(oxy))bis(ethane-2,1-diyl) Bis(4-methylbenzenesulfonate) (**19**). The same reaction as described above to prepare 18 was used, and 19 was obtained as a white crystal(24.7g, 71.5%). ^{1}H NMR(400MHz, $CDCl_3$): δ 7.79 (d, J = 8.3Hz, 4H), 7.34 (d, J = 8.2Hz, 4H), 4.14 (t, J = 4.7Hz, 4H), 3.65 (t, J = 4.9Hz, 4H), 3.53 (s, 4H), 2.44 (s, 6H).

2-(2-Fluoroethoxy) ethyl 4-Methylbenzenesulfonate (**20**). Intermediate **20** was synthesized according to the literature[38]. Briefly, to a solution of **18**(2.3g, 5mmol) in dry THF(30mL) was added anhydrous TBAF(10mL, 10mmol, 1M in THF). The solution was stirred at 70℃ for 5h. After removal of solvent, the residue was purified by silica gel chromatography to give **20** (540mg, 37.1%) as a colorless oil. ^{1}H NMR(400MHz, CDCl$_3$): δ 7.81(d, J = 8.2Hz, 2H), 7.35(d, J = 8.1Hz, 2H), 4.58-4.52(m, 1H), 4.46-4.40(m, 1H), 4.21-4.16(m, 2H), 3.75-3.62(m, 4H), 2.45(s, 3H).

2-(2-(2-Fluoroethoxy) ethoxy) ethyl 4-Methylbenzenesulfonate (**21**). The same reaction as described above to prepare **20** was used, and **21** was obtained as a colorless oil (309.2mg, 20.2%). ^{1}H NMR(400MHz, CDCl$_3$): δ 7.80(d, J = 8.1Hz, 2H), 7.34(d, J = 8.1Hz, 2H), 4.59(t, J = 3.8Hz, 1H), 4.48(t, J = 3.8Hz, 1H), 4.17(t, J = 4.6Hz, 2H), 3.74(t, J = 4.2Hz, 1H), 3.61-3.72(m, 3H), 3.61(s, 4H), 2.45(s, 3H).

2-Amino-4-methoxyphenol (**22**). Intermediate **22** was synthesized according to the literature[32]. Briefly, a mixture of 4-methoxy-2-nitrophenol (3.5g, 20.7mmol) and Pd/C (10%, 0.5g) in absolute methanol (100mL) under a hydrogen atmosphere (balloon) was vigorously stirred for 24h at room temperature. The mixture was filtered and the filtrate washed with methanol(20mL). Evaporation of the solvent under reduced pressure gave **22** as brown flake crystal (2.8g, 97.3%). M_p: 133.4-135.7℃.

5-(5-Methoxybenzo[d]oxazol-2-yl) pyridin-2-amine (**23**). A solution of **22**(1.0g, 7mmol) and 5-(trifluoromethyl)pyridin-2-amine(1.7g, 10.8mmol) in NaOH(1M in water, 20mL) was stirred at 90℃ for 3h. After the mixture was cooled to room temperature, the precipitate was collected by filtration. Compound **23**(1.5g, 88.9%) was obtained as a white solid. M_p: 204.1-205.9℃. ^{1}H NMR (400MHz, DMSO-d_6): δ 8.71(d, J = 2.1Hz, 1H), 8.05(dd, J = 6.6, 2.2Hz, 1H), 7.58(d, J = 8.8Hz, 1H), 7.25(d, J = 2.3Hz, 1H), 6.92(dd, J = 6.5, 2.4Hz, 1H), 6.80(s, 2H), 6.59(d, J = 8.8Hz, 1H), 3.81(s, 3H). MS(ESI): m/z calcd for $C_{13}H_{11}N_3O_2$ 241.09; found 241.9(M + H)$^+$.

5-(5-Methoxybenzo[d]oxazol-2-yl)-N-methylpyridin-2-amine (**24**). To a mixture of **23** (480mg, 2mmol) and paraformaldehyde(240mg, 8mmol) in MeOH(20mL) was added CH$_3$ONa (1.08g, 20mmol). The mixture was stirred under reflux for 1h. After the mixture was cooled, NaBH$_4$(160mg, 4mmol) was added, and the mixture was brought to reflux again for 2h. The reaction mixture was poured onto ice-water, and the precipitate was collected by filtration to obtain **24** as a white solid (500mg, 98.0%). M_p: 178.6-179.8℃. ^{1}H NMR (400MHz, DMSO-d_6): δ 8.78(d, J = 1.9Hz, 1H), 8.04(d, J = 8.5Hz, 1H), 7.59(d, J = 8.8Hz, 1H), 7.35(d, J = 4.5Hz, 1H), 7.25(d, J = 2.3Hz, 1H), 6.91(dd, J = 6.5, 2.3Hz, 1H), 6.61(d, J = 8.9Hz, 1H), 3.81(s, 3H), 2.87(d, J = 4.7Hz, 3H). MS(ESI): m/z calcd for $C_{14}H_{13}N_3O_2$ 255.1; found 255.9(M + H)$^+$.

5-(5-Methoxybenzo[d]oxazol-2-yl)-N,N-dimethylpyridin-2-amine(**25**). To a solution of **23** (480mg, 2mmol) and paraformaldehyde (300mg, 10mmol) in acetic acid(50mL) was added NaCNBH$_3$(189mg, 6mmol) in one portion at room temperature. The resulting mixture was stirred at room temperature overnight. After neutralization with NH_4OH, water was added, and the precipitate was collected by filtration to give **25** as a white solid (481mg, 89.4%). M_p: 140.0-141.6℃. ^1H NMR (400MHz, DMSO-d_6): δ 8.84 (d, J = 2.1Hz, 1H), 8.14 (dd, J = 6.9, 2.2Hz, 1H), 7.59 (d, J = 8.8Hz, 1H), 7.26 (d, J = 2.3Hz, 1H), 6.92 (dd, J = 6.4, 2.4Hz, 1H), 6.80 (d, J = 9.1Hz, 1H), 3.81 (s, 3H), 3.14 (s, 6H). MS(ESI): m/z calcd for $C_{15}H_{15}N_3O_2$ 269.12, found 270.0 $(M + H)^+$.

2-(6-(Methylamino)pyridin-3-yl)benzo[d]oxazol-5-ol (**26**). To a solution of **24** (480mg, 1.88mmol) in CH_2Cl_2 (30mL) was added BBr_3 (10mL, 1M solution in CH_2Cl_2) dropwise at -78℃ in a liquid nitrogen-ethanol bath. The mixture was allowed to warm to room temperature and was stirred overnight. Water was added to quench the reaction. After neutralization with NH_4OH, the mixture was extracted with CH_2Cl_2. The solvent was removed, and **26** was obtained as a white solid (442mg, 97%). M_p: 237.1-238.6℃. ^1H NMR (400MHz, DMSO-d_6): δ 9.41 (s, 1H), 8.76 (d, J = 1.6Hz, 1H), 8.02 (dd, J = 7.2, 1.4Hz, 1H), 7.47 (d, J = 8.7Hz, 1H), 7.32 (d, J = 4.4Hz, 1H), 7.00 (d, J = 2.1Hz, 1H), 6.75 (dd, J = 6.4, 2.2Hz, 1H), 6.60 (d, J = 8.8Hz, 1H), 2.87 (d, J = 4.7Hz, 3H). MS(ESI): m/z calcd for $C_{13}H_{11}N_3O_2$ 241.1, found 241.8 $(M + H)^+$.

2-(6-(Dimethylamino)pyridin-3-yl)benzo[d]oxazol-5-ol (**27**). The same reaction as described above to prepare **26** was used, and **27** was obtained as a white solid (348mg, 88.5%). M_p: 285.2-286.7℃. ^1H NMR (400MHz, DMSO-d_6): δ 9.43 (s, 1H), 8.82 (d, J = 2.1Hz, 1H), 8.13 (dd, J = 9.0, 2.3Hz, 1H), 7.48 (d, J = 8.7Hz, 1H), 7.01 (d, J = 2.2Hz, 1H), 6.80 (d, J = 9.1Hz, 1H), 6.76 (dd, J = 8.7, 2.3Hz, 1H), 3.14 (s, 6H). MS(ESI): m/z calcd for $C_{14}H_{13}N_3O_2$ 255.1, found 255.9 $(M + H)^+$.

5-(5-(2-(2-Fluoroethoxy)ethoxy)benzo[d]oxazol-2-yl)-N-methylpyridin-2-amine (**28**). A solution of **26** (48mg, 0.2mmol), 2-(2-fluoroethoxy)ethyl 4-methylbenzenesulfonate (60mg, 0.2mmol), and K_2CO_3 (28mg, 0.2mmol) in DMF (6mL) was stirred at 110℃ for 3h. The aqueous portion was extracted with dichloromethane (3 × 15mL), and the organic layer was dried over anhydrous $MgSO_4$. CH_2Cl_2 was removed in vacuum. The residue was purified by column chromatography (petroleum ether/AcOEt, 1/2) to give **28** as a white solid (41mg, 62.2%). M_p: 134.8-135.6℃. ^1H NMR (400MHz, $CDCl_3$): δ 8.92 (s, 1H), 8.21 (dd, J = 7.0, 1.7Hz, 1H), 7.41 (d, J = 8.8Hz, 1H), 7.21 (d, J = 2.2Hz, 1H), 6.93 (dd, J = 6.6, 2.2Hz, 1H), 6.48 (d, J = 8.8Hz, 1H), 5.12 (d, J = 4.4Hz, 1H), 4.61 (dt, J = 47.6, 4.0Hz, 2H), 4.19 (t, J = 4.5Hz, 2H), 3.92 (t, J = 4.8Hz, 2H), 3.87 (t, J = 4.0Hz, 1H), 3.80 (t, J = 4.1Hz, 1H), 3.02 (d, J = 5.1Hz, 3H). HRMS(EI): m/z calcd for $C_{17}H_{19}N_3O_3F$ 332.1410; found 332.1412 $(M + H)^+$.

5-(5-(2-(2-(2-Fluoroethoxy)ethoxy)ethoxy)benzo[d]oxazol-2-yl)-N-methylpyridin-2-amine (**29**). The same reaction as described above to prepare **28** was used, and **29** was obtained as a white solid (30mg, 40.2%). M_p: 87.2-89.1℃. ^1H NMR (400MHz, $CDCl_3$): δ 8.92 (s, 1H), 8.21 (dd, J = 7.0, 1.8Hz, 1H), 7.40 (d, J = 8.8Hz, 1H), 7.21 (d, J = 2.2Hz, 1H), 6.92 (dd, J = 6.6, 2.2Hz, 1H), 6.49 (d, J = 8.8Hz, 1H), 5.12 (d, J = 4.4Hz, 1H), 4.57 (dt,

J = 47.7, 4.0Hz, 2H), 4.18 (t, J = 4.5Hz, 2H), 3.90 (t, J = 4.8Hz, 2H), 3.82-3.72 (m, 6H), 3.80 (t, J = 4.1Hz, 1H), 3.02 (d, J = 5.0Hz, 3H). HRMS(EI): m/z calcd for $C_{19}H_{23}N_3O_4F$ 376.1673; found 376.1674(M + H)$^+$.

5-(5-(2-(2-Fluoroethoxy)ethoxy)benzo[d]oxazol-2-yl)-N,N-dimethylpyridin-2-amine (**30**). The same reaction as described above to prepare **28** was used, and **30** was obtained as a white solid (30mg, 43.5%). M_p: 88.0-89.7℃. ^{1}H NMR(400MHz, CDCl$_3$): δ 8.98 (s, 1H), 8.20 (dd, J = 6.8, 2.2Hz, 1H), 7.40 (d, J = 8.8Hz, 1H), 7.21 (d, J = 2.3Hz, 1H), 6.92 (dd, J = 6.4, 2.4Hz, 1H), 6.59 (d, J = 9.0Hz, 1H), 4.61 (dt, J = 47.7, 4.1Hz, 2H), 4.19 (t, J = 4.5Hz, 2H), 3.92 (t, J = 4.9Hz, 2H), 3.88 (t, J = 4.2Hz, 1H), 3.80 (t, J = 4.2Hz, 1H), 3.19 (s, 6H). HRMS(EI): m/z calcd for $C_{18}H_{21}N_3O_3F$ 346.1567; found 346.1559(M + H)$^+$.

5-(5-(2-(2-(2-Fluoroethoxy)ethoxy)ethoxy)benzo[d]oxazol-2-yl)-N,N-dimethylpyridin-2-amine (**31**). The same reaction as described above to prepare **28** was used, and **31** was obtained as a white solid (40mg, 50%). M_p: 120.0-121.8℃. ^{1}H NMR(400MHz, CDCl$_3$): δ 8.98 (s, 1H), 8.19 (d, J = 8.9Hz, 1H), 7.39 (d, J = 8.8Hz, 1H), 7.27 (d, J = 0.9Hz, 1H), 6.91 (d, J = 8.6Hz, 1H), 6.59 (d, J = 9.0Hz, 1H), 4.50-4.63 (m, 2H), 4.18 (t, J = 4.1Hz, 2H), 3.90 (t, J = 4.0Hz, 2H), 3.62-3.82 (m, 6H), 3.19 (s, 6H). HRMS(EI): m/z calcd for $C_{20}H_{25}N_3O_4F$ 390.1829; found 390.1829(M + H)$^+$.

5-(5-(2-Fluoroethoxy)benzo[d]oxazol-2-yl)-N-methylpyridin-2-amine (**32**). The same reaction as described above to prepare **28** was used, and **32** was obtained as a white solid (30mg, 52.5%). M_p: 212.2-212.8℃. ^{1}H NMR(400MHz, DMSO-d_6): δ 8.78 (d, J = 1.9Hz, 1H), 8.04 (d, J = 7.5Hz, 1H), 7.60 (d, J = 8.8Hz, 1H), 7.36 (d, J = 4.5Hz, 1H), 7.30 (d, J = 2.3Hz, 1H), 6.95 (dd, J = 6.4, 2.4Hz, 1H), 6.61 (d, J = 8.9Hz, 1H), 4.77 (dt, J = 47.8, 3.5Hz, 2H), 4.33 (t, J = 3.7Hz, 1H), 4.25 (t, J = 3.8Hz, 1H), 2.87 (d, J = 4.7Hz, 3H). HRMS(EI): m/z calcd for $C_{15}H_{15}N_3O_2F$ 288.1148; found 288.1147(M + H)$^+$.

5-(5-(2-Fluoroethoxy)benzo[d]oxazol-2-yl)-N,N-dimethylpyridin-2-amine (**33**). The same reaction as described above to prepare **28** was used, and **33** was obtained as a white solid (54mg, 89.7%). M_p: 167.0-168.4℃. ^{1}H NMR(400MHz, CDCl$_3$): δ 8.98 (s, 1H), 8.20 (d, J = 8.0Hz, 1H), 7.42 (d, J = 8.4Hz, 1H), 7.21 (s, 1H), 6.93 (d, J = 7.5Hz, 1H), 6.60 (d, J = 8.4Hz, 1H), 4.85 (s, 1H), 4.73 (s, 1H), 4.30 (s, 1H), 4.23 (s, 1H), 3.20 (s, 6H). HRMS(EI): m/z calcd for $C_{16}H_{17}N_3O_2F$ 302.1305; found 302.1307(M + H)$^+$.

2-Amino-5-methoxybenzenethiol (**34**). Intermediate **34** was synthesized according to the literature[14]. Briefly, a mixture of 6-methoxybenzo[d]thiazol-2-amine (9g, 50mmol) and KOH (28g, 500mmol) in H_2O (100mL) was stirred at 120℃ overnight. After removal of the scraps by filtering, the filtrate was neutralized by acetic acid (30% in water), and the precipitate was collected by filtration to give **34** (7.75g, 88.9%) as a light yellow solid. M_p: 81.6-82.8℃.

5-(6-Methoxybenzo[d]thiazol-2-yl)pyridin-2-amine (**35**). The same reaction as described above to prepare **23** was used, and **35** was obtained (1.29g, 50.2%). M_p: 224.7-225.4℃. ^{1}H NMR(400MHz, DMSO-d_6): δ 8.58 (s, 1H), 7.98 (d, J = 8.2Hz, 1H), 7.84 (d, J =

8.8Hz,1H),7.65(s,1H),7.08(d, J =8.2Hz,1H),6.70(s,2H),6.57(d, J =8.5Hz,1H),3.84(s,3H). MS(ESI): m/z calcd for $C_{13}H_{11}N_3OS$ 257.1; found 257.5(M+H)$^+$.

5-(6-Methoxybenzo[d]thiazol-2-yl)-N-methylpyridin-2-amine (**36**). The same reaction as described above to prepare **24** was used, and **36** was obtained as a white solid(798mg,98%). M_p: 194.7-195.6℃. ^{1}H NMR(400MHz, DMSO-d_6): δ 8.64(s,1H),7.96(d, J =7.4Hz,1H),7.83(d, J =8.8Hz,1H),7.64(d, J =1.6Hz,1H),7.20(d, J =4.6Hz,1H),7.07(dd, J =6.8,2.1Hz,1H),6.57(d, J =8.8Hz,1H),3.84(s,3H),2.86(d, J =4.6Hz,3H). MS(ESI): m/z calcd for $C_{14}H_{13}N_3OS$ 271.1, found 271.8(M+H)$^+$.

5-(6-Methoxybenzo[d]thiazol-2-yl)-N,N-dimethylpyridin-2-amine(**37**). The same reaction as described above to prepare **25** was used, and **37** was obtained(0.65g,91.1%). M_p: 175.4-176.7℃. ^{1}H NMR(400MHz, CDCl$_3$):δ 8.76(d, J =2.3Hz,1H),8.13(dd, J =9.0,2.3Hz,1H),7.87(d, J =8.9Hz,1H),7.32(d, J =2.4Hz,1H),7.05(dd, J =8.9,2.4Hz,1H),6.58(d, J =9.0Hz,1H),3.88(s,3H),3.18(s,6H). MS(ESI): m/z calcd for $C_{15}H_{15}N_3OS$ 285.1; found 285.6(M+H)$^+$.

2-(6-(Methylamino)pyridin-3-yl)benzo[d]thiazol-6-ol(**38**). The same reaction as described above to prepare **26** was used, and **38** was obtained as a white solid(653mg,88.4%). M_p: 238.5-244.2℃; ^{1}H NMR(400MHz, DMSO-d_6): δ 9.72(s,1H),8.61(s,1H),7.94(d, J =8.3Hz,1H),7.74(d, J =8.6Hz,1H),7.36(s,1H),7.17(d, J =3.8Hz,1H),6.93(d, J =6.8Hz,1H),6.57(d, J =8.7Hz,1H),2.86(d, J =4.2Hz,3H). MS(ESI): m/z calcd for $C_{13}H_{11}N_3OS$ 257.1, found 257.9(M+H)$^+$.

2-(6-(Dimethylamino)pyridin-3-yl)benzo[d]thiazol-6-ol(**39**). The same reaction as described above to prepare **26** was used, and **39** was obtained as a white solid(0.61g,68.2%). M_p: 287.2-288.7℃. ^{1}H NMR(400MHz, DMSO-d_6): δ 9.74(s,1H),8.68(d, J =2.1Hz,1H),8.04(dd, J =9.0,2.2Hz,1H),7.75(d, J =8.7Hz,1H),7.36(d, J =2.1Hz,1H),6.94(dd, J =8.7,2.1Hz,1H),6.76(d, J =9.0Hz,1H),3.12(s,6H). MS(ESI): m/z calcd for $C_{14}H_{13}N_3OS$ 271.1; found 271.5(M+H)$^+$.

5-(6-(2-(2-Fluoroethoxy)ethoxy)benzo[d]thiazol-2-yl)-N-methylpyridin-2-amine(**40**). The same reaction as described above to prepare **28** was used, and **40** was obtained as a white solid(40mg,55.2%). M_p: 155.7-157.5℃. ^{1}H NMR(400MHz, CDCl$_3$): δ 8.72(d, J =2.0Hz,1H),8.10(dd, J =6.5,2.2Hz,1H),7.86(d, J =8.9Hz,1H),7.35(d, J =2.3Hz,1H),7.08(dd, J =6.5,2.4Hz,1H),6.46(d, J =8.8Hz,1H),4.98(d, J =4.6Hz,1H),4.61(dt, J =47.6,4.0Hz,2H),4.21(t, J =4.5Hz,2H),3.93(t, J =4.9Hz,2H),3.87(t, J =4.1Hz,1H),3.80(t, J =4.2Hz,1H),2.99(d, J =5.1Hz,3H). HRMS(EI): m/z calcd for $C_{17}H_{19}N_3O_2FS$ 348.1182; found 348.1179(M+H)$^+$.

5-(6-(2-(2-(2-Fluoroethoxy)ethoxy)ethoxy)benzo[d]thiazol-2-yl)-N-methylpyridin-2-amine(**41**). The same reaction as described above to prepare **28** was used, and **41** was obtained as a white solid(45mg,57.3%). M_p: 126.4-128.7℃. ^{1}H NMR(400MHz, CDCl$_3$):δ 8.71(d, J =2.0Hz,1H),8.10(dd, J =6.5,2.2Hz,1H),7.86(d, J =8.9Hz,1H),7.34(d, J =2.3Hz,1H),7.08(dd, J =6.5,2.4Hz,1H),6.46(d, J =8.8Hz,1H),4.98(d, J =4.6Hz,

1H),4.57(dt, J = 47.7, 4.0Hz, 2H),4.20(t, J = 4.6Hz, 2H),3.90(t, J = 4.9Hz, 2H),3.69-3.81(m, 6H),2.99(d, J = 5.1Hz, 3H). HRMS(EI): m/z calcd for $C_{19}H_{23}N_3O_3FS$ 392.1444; found 392.1446(M + H)$^+$.

5-(6-(2-(2-Fluoroethoxy)ethoxy)benzo[d]thiazol-2-yl)-N, N-dimethylpyridin-2-amine (**42**). The same reaction as described above to prepare **28** was used, and **42** was obtained as a white solid (50mg, 75%). M_p: 140.3-141.1℃. ^{1}H NMR(400MHz, CDCl$_3$): δ 8.76(s, 1H),8.12(dd, J = 6.6, 2.3Hz, 1H),7.86(d, J = 8.9Hz, 1H),7.34(d, J = 2.4Hz, 1H),7.07(dd, J = 6.5, 2.4Hz, 1H),6.57(d, J = 9.0Hz, 1H),4.60(dt, J = 47.7, 4.0Hz, 2H),4.21(t, J = 4.5Hz, 2H),3.93(t, J = 4.9Hz, 2H),3.87(t, J = 4.2Hz, 1H),3.80(t, J = 4.1Hz, 1H),3.10(s, 6H). HRMS(EI): m/z calcd for $C_{18}H_{21}N_3O_2FS$ 362.1339; found 362.1338(M + H)$^+$.

5-(6-(2-(2-(2-Fluoroethoxy)ethoxy)ethoxy)benzo[d]thiazol-2-yl)-N, N-dimethylpyridin-2-amine(**43**). The same reaction as described above to prepare **28** was used, and **43** was obtained as a white solid(32.4mg, 40%). M_p: 109.9-111.3℃. ^{1}H NMR(400MHz, CDCl$_3$): δ 8.76(s, 1H),8.11(dd, J = 6.6, 2.4Hz, 1H),7.85(d, J = 8.9Hz, 1H),7.34(d, J = 2.3Hz, 1H),7.07(dd, J = 6.5, 2.4Hz, 1H),6.57(d, J = 9.0Hz, 1H),4.56(dt, J = 47.7, 4.0Hz, 2H),4.20(t, J = 4.6Hz, 2H),3.90(t, J = 4.9Hz, 2H),3.71-3.81(m, 6H),3.17(s, 6H). HRMS(EI): m/z calcd for $C_{20}H_{25}N_3O_3FS$ 406.1601; found 406.1590(M + H)$^+$.

5-(6-(2-Fluoroethoxy)benzo[d]thiazol-2-yl)-N-methylpyridin-2-amine (**44**). The same reaction as described above to prepare **28** was used, and **44** was obtained as a white solid (40mg, 64.6%). M_p: 182.8-184.6℃; ^{1}H NMR(400MHz, CDCl$_3$): δ 8.72(d, J = 2.0Hz, 1H),8.11(dd, J = 6.5, 2.3Hz, 1H),7.88(d, J = 8.9Hz, 1H),7.36(d, J = 2.4Hz, 1H),7.09(dd, J = 6.4, 2.4Hz, 1H),6.47(d, J = 8.8Hz, 1H),4.80(dt, J = 47.4, 4.0Hz, 2H),4.32(t, J = 4.2Hz, 1H),4.25(t, J = 4.2Hz, 1H),3.00(d, J = 5.1Hz, 3H). HRMS(EI): m/z calcd for $C_{15}H_{15}N_3OFS$ 304.0920; found 304.0924(M + H)$^+$.

5-(6-(2-Fluoroethoxy)benzo[d]thiazol-2-yl)-N, N-dimethylpyridin-2-amine(**45**). The same reaction as described above to prepare **28** was used, and **45** was obtained as a white solid (30mg, 78.9%). M_p: 166.1-168.5℃. ^{1}H NMR(400MHz, CDCl$_3$): δ 8.69(d, J = 2.3Hz, 1H),8.05(dd, J = 6.6, 2.4Hz, 1H),7.80(d, J = 8.9Hz, 1H),7.28(d, J = 2.4Hz, 1H),7.01(dd, J = 6.4, 2.4Hz, 1H),6.51(d, J = 9.0Hz, 1H),4.72(dt, J = 47.4, 4.0Hz, 2H),4.24(t, J = 4.2Hz, 1H),4.17(t, J = 4.2Hz, 1H),3.10(s, 6H). HRMS(EI): m/z calcd for $C_{16}H_{17}N_3OFS$ 318.1076; found 318.1075(M + H)$^+$.

2-(2-(6-(Dimethylamino)pyridin-3-yl)benzo[d]oxazol-5-yloxy)ethyl 4-Methylbenzenesulfonate(**46**). To a solution of **27**(60mg, 0.24mmol) in acetone(30mL) was added ethane-1,2-diyl bis(4-methylbenzenesulfonate)(131mg, 0.35mmol), K_2CO_3(124mg, 0.9mmol), and 18-crown-6 in catalytic amount. The mixture was stirred at 75℃ for 3h. Acetone was removed in vacuum. The residue was purified by column chromatography(petroleum ether/AcOEt, 1/1) to give **46** as a white solid(38.5mg, 35.4%). M_p: 179.1-179.5℃. ^{1}H NMR(400MHz, CDCl$_3$): δ 8.97(s, 1H),8.19(d, J = 8.9Hz, 1H),7.83(d, J = 8.3Hz, 2H),7.37(d, J = 8.8Hz, 1H),

7.34(d, J = 8.1Hz, 2H), 7.06(d, J = 2.4Hz, 1H), 6.77(dd, J = 8.8, 2.5Hz, 1H), 6.60(d, J = 9.0Hz, 1H), 4.17-4.42(m, 4H), 3.21(s, 6H), 2.44(s, 3H). MS(ESI): m/z calcd for $C_{23}H_{23}N_3O_5S$ 453.14, found 454.11(M + H)$^+$.

2-(2-(2-(2-(6-(Dimethylamino)pyridin-3-yl)benzo[d]oxazol-5-yloxy)ethoxy)ethoxy)ethyl 4-Methylbenzenesulfonate(**47**). The same reaction as described above to prepare **46** was used, and **47** was obtained as a white solid (60mg, 27.7%). M_p: 86.3-87.9℃. ^{1}H NMR (400MHz, CDCl$_3$): δ 8.98(s, 1H), 8.22(d, J = 7.7Hz, 1H), 7.79(d, J = 8.2Hz, 2H), 7.40(d, J = 8.8Hz, 1H), 7.32(d, J = 8.1Hz, 2H), 7.20(d, J = 2.4Hz, 1H), 6.91(dd, J = 6.6, 2.2Hz, 1H), 6.62(d, J = 8.5Hz, 1H), 4.16(m, 4H), 3.85(t, J = 4.9Hz, 2H), 3.61-3.73(m, 6H), 3.22(s, 6H), 2.42(s, 3H). MS(ESI): m/z calcd for $C_{27}H_{31}N_3O_7S$ 541.2, found 542.0(M + H)$^+$.

2-(2-(6-(Methylamino)pyridin-3-yl)benzo[d]oxazol-5-yloxy)ethyl 4-Methylbenzenesulfonate(**48**). The same reaction as described above to prepare **46** was used, and **48** was obtained as a white solid (40mg, 30.5%). M_p: 169.7-170.7℃. ^{1}H NMR(400MHz, CDCl$_3$): δ 8.90(s, 1H), 8.22(dd, J = 8.8, 2.0Hz, 1H), 7.83(d, J = 8.2Hz, 2H), 7.38(d, J = 8.8Hz, 1H), 7.34(d, J = 8.1Hz, 2H), 7.07(d, J = 2.4Hz, 1H), 6.78(dd, J = 8.8, 2.4Hz, 1H), 6.51(d, J = 8.8Hz, 1H), 5.37(s, 1H), 4.40(t, J = 4.4Hz, 2H), 4.18(t, J = 5.0Hz, 2H), 3.02(d, J = 5.0Hz, 3H), 2.44(s, 3H). MS(ESI): m/z calcd for $C_{22}H_{21}N_3O_5S$ 439.12, found 440.09(M + H)$^+$.

2-(2-(2-(2-(6-(Methylamino)pyridin-3-yl)benzo[d]oxazol-5-yloxy)ethoxy)ethoxy)ethyl 4-Methylbenzenesulfonate(**49**). The same reaction as described above to prepare **46** was used, and **49** was obtained as a white solid(75mg, 47.8%). M_p: 91.3-92.2℃. ^{1}H NMR(400MHz, CDCl$_3$): δ 8.90(d, J = 2.0Hz, 1H), 8.16(dd, J = 6.7, 2.1Hz, 1H), 7.77(d, J = 8.2Hz, 2H), 7.37(d, J = 8.8Hz, 1H), 7.29(d, J = 8.1Hz, 2H), 7.17(d, J = 2.4Hz, 1H), 6.88(dd, J = 6.4, 2.4Hz, 1H), 6.45(d, J = 8.8Hz, 1H), 5.25(d, J = 4.8Hz, 1H), 4.11-4.17(m, 4H), 3.82(t, J = 4.9Hz, 2H), 3.59-3.70(m, 6H), 2.98(d, J = 5.1Hz, 3H), 2.39(s, 3H). MS(ESI): m/z calcd for $C_{26}H_{29}N_3O_7S$ 527.17, found 528.19(M + H)$^+$.

2-(2-(6-($tert$-Butoxycarbonyl)pyridin-3-yl)benzo[d]oxazol-5-yloxy)ethyl 4-Methylbenzenesulfonate(**50**). To a solution of **48**(40mg, 0.09mmol) in THF(30mL) was added excessive (Boc)$_2$O. The mixture was stirred at 85℃ for 28h. After solvent was removed in vacuum, the residue was purified by column chromatography(petroleum ether/AcOEt, 1/1) to give **50** as a white solid(32.7mg, 66.7%). M_p: 172.1-173.2℃. ^{1}H NMR(400MHz, CDCl$_3$): δ 9.16(s, 1H), 8.37(dd, J = 8.9, 6.6Hz, 1H), 8.02(d, J = 8.8Hz, 1H), 7.83(d, J = 8.3Hz, 2H), 7.44(d, J = 8.9Hz, 1H), 7.35(d, J = 8.0Hz, 2H), 7.12(d, J = 2.4Hz, 1H), 6.86(dd, J = 8.9, 2.5Hz, 1H), 4.19-4.43(m, 4H), 3.51(s, 3H), 2.45(s, 3H), 1.57(s, 9H). MS(ESI): m/z calcd for $C_{27}H_{29}N_3O_7S$ 539.17, found 540.18(M + H)$^+$.

2-(2-(2-(2-(6-($tert$-Butoxycarbonyl)pyridin-3-yl)benzo[d]oxazol-5-yloxy)ethoxy)ethoxy)ethyl 4-Methylbenzenesulfonate(**51**). The same reaction as described above to prepare **50** was used, and **51** was obtained as a white solid(20mg, 22.5%). M_p: 85.0-85.5℃. ^{1}H NMR

(400MHz, CDCl$_3$) : δ 9.16(s,1H), 8.37(dd, J = 6.8, 2.0Hz, 1H), 7.99(d, J = 8.9Hz, 1H), 7.79(d, J = 8.0Hz, 2H), 7.45(d, J = 8.8Hz, 1H), 7.32(d, J = 8.0Hz, 2H), 7.24(d, J = 2.0Hz, 1H), 6.98(dd, J = 6.7, 2.2Hz, 1H), 4.15-4.19(m, 4H), 3.86(t, J = 4.8Hz, 2H), 3.67-3.69(m, 4H), 3.63(t, J = 5.2Hz, 2H), 3.49(s, 3H), 2.42(s, 3H), 1.56(s, 9H). MS (ESI) : m/z calcd for C$_{31}$H$_{37}$N$_3$O$_9$S 627.23, found 628.0(M + H)$^+$.

2-(2-(2-(2-(6-(Dimethylamino) pyridin-3-yl) benzo[d] thiazol-6-yloxy) ethoxy) ethoxy) ethyl 4-Methylbenzenesulfonate(**52**). The same reaction as described above to prepare **46** was used, and **52** was obtained as a white solid(87mg, 48.7%). M_p: 107.6-108.7℃. ^{1}H NMR (400MHz, CDCl$_3$) : δ 8.76(d, J = 2.2Hz, 1H), 8.12(dd, J = 6.6, 2.4Hz, 1H), 7.85(d, J = 8.9Hz, 1H), 7.79(d, J = 8.2Hz, 2H), 7.30-7.34(m, 3H), 7.05(dd, J = 6.4, 2.5Hz, 1H), 6.58(d, J = 9.0Hz, 1H), 4.15-4.18(m, 4H), 3.86(t, J = 4.8Hz, 2H), 3.61-3.72(m, 6H), 3.17(s, 6H), 2.42(s, 3H). MS(ESI) : m/z calcd for C$_{27}$H$_{31}$N$_3$O$_6$S$_2$ 557.17, found 558.0 (M + H)$^+$.

2-(2-(6-(Methylamino) pyridin-3-yl) benzo[d] thiazol-6-yloxy) ethyl 4-methylbenzenesulfonate(**53**). The same reaction as described above to prepare **46** was used, and **53** was obtained as a white solid(55mg, 40.3%). M_p: 153.0-155.4℃. ^{1}H NMR(400MHz, CDCl$_3$) : δ 8.70(s, 1H), 8.11(dd, J = 8.8, 2.3Hz, 1H), 7.81-7.85(m, 3H), 7.33(d, J = 8.0Hz, 2H), 7.20(d, J = 2.5Hz, 1H), 6.92(dd, J = 8.9, 2.6Hz, 1H), 6.47(d, J = 8.8Hz, 1H), 5.02(d, J = 4.5Hz, 1H), 4.20-4.43(m, 4H), 3.01(d, J = 5.2Hz, 3H), 2.44(s, 3H). MS(ESI) : m/z calcd for C$_{22}$H$_{21}$N$_3$O$_4$S$_2$ 455.1, found 456.06(M + H)$^+$.

2-(2-(2-(2-(6-(Methylamino) pyridin-3-yl) benzo[d] thiazol-6-yloxy) ethoxy) ethoxy) ethyl 4-Methylbenzenesulfonate(**54**). The same reaction as described above to prepare **46** was used, and **54** was obtained as a white solid (63mg, 38.4%). M_p: 110.1-110.9℃; ^{1}H NMR (400MHz, CDCl$_3$) : δ 8.71(d, J = 2.2Hz, 1H), 8.10(dd, J = 6.4, 2.3Hz, 1H), 7.86(d, J = 8.9Hz, 1H), 7.79(d, J = 8.2Hz, 2H), 7.30-7.35(m, 3H), 7.06(dd, J = 6.4, 2.5Hz, 1H), 6.46(d, J = 8.8Hz, 1H), 4.98(d, J = 4.6Hz, 1H), 4.15-4.20(m, 4H), 3.86(t, J = 4.9Hz, 2H), 3.63-3.72(m, 4H), 3.61(t, J = 3.7Hz, 2H), 3.00(d, J = 5.1Hz, 3H), 2.42(s, 3H). MS(ESI) : m/z calcd for C$_{26}$H$_{29}$N$_3$O$_6$S$_2$ 543.15, found 543.9(M + H)$^+$.

2-(2-(6-(*tert*-Butoxycarbonyl) pyridin-3-yl) benzo[d] thiazol-6-yloxy) ethyl 4-Methylbenzenesulfonate(**55**). The same reaction as described above to prepare **50** was used, and **55** was obtained as a white solid(53mg, 79.6%). M_p: 163.4-165.4℃. ^{1}H NMR(400MHz, CDCl$_3$) : δ 8.97(d, J = 2.2Hz, 1H), 8.27(dd, J = 8.8, 2.4Hz, 1H), 7.92(dd, J = 8.8, 7.2Hz, 2H), 7.82(d, J = 8.3Hz, 2H), 7.34(d, J = 8.1Hz, 2H), 7.25(d, J = 2.6Hz, 1H), 6.98(dd, J = 9.0, 2.6Hz, 1H), 4.22-4.44(m, 4H), 3.48(s, 3H), 2.44(s, 3H), 1.56(s, 9H). MS(ESI) : m/z calcd for C$_{27}$H$_{29}$N$_3$O$_6$S$_2$ 555.15, found 556.13(M + H)$^+$.

2-(2-(2-(6-(*tert*-Butoxycarbonyl) pyridin-3-yl) benzo[d] thiazol-6-yloxy) ethoxy) ethoxy) ethyl 4-Methylbenzenesulfonate(**56**). The same reaction as described above to prepare **50** was used, and **56** was obtained as a white solid (45mg, 63.4%). M_p: 88.2-90.1℃. ^{1}H NMR(400MHz, CDCl$_3$) : δ 8.96(s, 1H), 8.25(dd, J = 6.4, 2.4Hz, 1H), 7.92(t, J = 8.7Hz,

2H), 7.79(d, J = 8.2Hz, 2H), 7.34(d, J = 2.5Hz, 1H), 7.31(d, J = 8.1Hz, 2H), 7.11(dd, J = 6.6, 2.3Hz, 1H), 4.15-4.22(m, 4H), 3.87(t, J = 4.7Hz, 2H), 3.62-3.73(m, 6H), 3.47(s, 3H), 2.42(s, 3H), 1.56(s, 9H). MS(ESI): m/z calcd for $C_{31}H_{37}N_3O_8S_2$ 643.20, found 644.18(M + H)$^+$.

4.3 Radiolabeling

[^{18}F]Fluoride trapped on a QMA cartridge was eluted with 1mL of (Kryptofix 222)/K_2CO_3 solution(13mg of Kryptofix 222 and 1.1mg of K_2CO_3 in CH_3CN/H_2O, 0.8∶0.2). The solvent was removed at 120℃ under a stream of nitrogen gas. The residue was azeotropically dried with 1mL of anhydrous acetonitrile three times at 120℃ under a stream of nitrogen gas.

For [^{18}F]**29**, [^{18}F]**32**, [^{18}F]**41**, and [^{18}F]**44**, a solution of the tosylate precursors **51**, **50**, **56**, and **55**(1.0mg) in CH_3CN(0.5mL) was added to the reaction vessel containing the ^{18}F$^-$ activity, respectively. The mixture was heated at 100℃ for 5min. After the solution was cooled to room temperature, HCl(1M aqueous solution, 0.2mL) was added and the mixture was heated at 100℃ again for 5min. An aqueous solution of $NaHCO_3$ was added to adjust to pH 8-9. Water (10mL) was added, and the mixture was passed through a Sep-Pak C18 cartridge(Waters). The cartridge was washed with 10mL of water, and the labeled compound was eluted with 2mL of acetonitrile. After the solvent was removed, the residue was dissolved in CH_3CN and subjected to HPLC for purification(Agela Technologies, 5μm, 10mm × 250mm, CH_3CN/water = 6/4; flow rate = 4mL/min). The retention times of [^{18}F]**29**, [^{18}F]**32**, [^{18}F]**41**, and [^{18}F]**44** were 7.72min, 6.36min, 5.28min, and 7.81min, respectively, in this HPLC system. The preparation took 60min, and the radiochemical yield was 20%-30%(decay not corrected). The radiochemical purity of all tracers was greater than 98%.

For [^{18}F]**31**, [^{18}F]**33**, and [^{18}F]**43**, a solution of the tosylate precursors **47**, **46**, and **52** (1.0mg) in CH_3CN(0.5mL) was added to the reaction vessel containing the ^{18}F$^-$ activity, respectively. The mixture was heated at 100℃ for 5min. Water(10mL) was added, and the mixture was passed through a Sep-Pak C18 cartridge(Waters). The cartridge was washed with 10mL of water, and the labeled compound was eluted with 2mL of acetonitrile. The eluted compound was purified by HPLC(Agela Technologies, 5μm, 10mm × 250mm). The retention times of [^{18}F]**31**, [^{18}F]**33**, and [^{18}F]**43** were 9.09min, 5.93min, and 8.23min, respectively, in this HPLC system(CH_3CN/water = 7/3; flow rate = 4mL/min). The preparation took 40min, and the radiochemical yield was 40%-50%(decay not corrected). The radiochemical purity of all tracers was greater than 98%. Specific activity, estimated by comparing the UV peak intensity of purified ^{18}F-labeled compounds with reference nonradioactive compounds, was approximately 200GBq/μmol.

4.4 Binding assay using Aβ$_{1-42}$ aggregates

Inhibition experiments were carried out in 12mm × 75mm borosilicate glass tubes according to procedures described previously with some modifications. The radioligand [^{125}I]IMPY was prepared according to procedures described previously[12]. After HPLC purification, the radiochemical purity was greater than 95%. The reaction mixture contained 100μL of aggregated Aβ$_{1-42}$ fibrils, 100μL of radioligands ([^{125}I]IMPY, 60000-100000cpm/100μL), 100μL of inhibitors (10^{-4}-10$^{-8.5}$M in ethanol), and 700μL of 0.05% bovine serum albumin solution in a final volume of 1mL. The mixture was incubated at 37℃ for 2h, and the free radioactivity was separated by vacuum filtration through Whatman GF/B filters using a Brandel Mp-48T cell harvester followed by 3×4mL washes with PBS(0.02M, pH 7.4) at room temperature. Filters with the bound [^{125}I]IMPY were counted in a γ counter (WALLAC/Wizard 1470, U.S.) with 70% efficiency. Inhibition experiments were repeated three times, and the half maximal inhibitory concentration (IC_{50}) was determined using GraphPad Prism 4.0, and the inhibition constant (K_i) was calculated using the Cheng-Prusoff equation[39]: $K_i = IC_{50}/(1 + [L]/K_d)$.

4.5 Partition coefficient determination

The determination of partition coefficients of radiofluorinated tracers was performed according to the procedure previously reported[32]. A solution of ^{18}F-labeled tracer (1.5MBq) was added to premixed suspensions containing 3.0g of n-octanol and 3.0g of PBS(0.05M, pH 7.4) in a test tube. The test tube was vortexed for 3min at room temperature, followed by centrifugation for 5min at 3000r/min. Two samples from the n-octanol (50μL) and water (500μL) layers were measured. The partition coefficient was expressed as the logarithm of the ratio of the count per gram from n-octanol versus PBS. Samples from the n-octanol layer were repartitioned until consistent partition coefficient values were obtained. The measurement was done in triplicate and repeated three times.

4.6 Biodistribution studies

The biodistribution experiments were performed in normal male mice (5 weeks, male, average weight, about 20g) and approved by the Animal Care Committee of Beijing Normal University. A saline solution containing the ^{18}F-labeled tracers (370KBq per 100μL) was injected directly into the tail. The mice ($n=5$ for each time point) were sacrificed at 2min, 10min, 30min, and 60min after injection. The organs of interest were removed and weighed, and radioactivity was measured with an automatic γ counter. The percent dose per gram of wet tissue was calculated by a comparison of the tissue counts to suitably diluted aliquots of the injected material.

4.7 In vitro autoradiography studies

Paraffin-embedded brain sections were deparaffinized with 2×20min washes in xylene; 2×5min washes in 100% ethanol, a 5min wash in 90% ethanol/H$_2$O, a 5min wash in 80% ethanol/H$_2$O, a 5min wash in 60% ethanol/H$_2$O, and a 10min wash in running tap water and then incubated in PBS(0.2M, pH 7.4) for 30min. After that, they were incubated with ^{18}F-labeled tracers (370KBq per 100μL) for 1h at room temperature. They were then washed in 40% EtOH before being rinsed with water for 1min. After drying, the sections were exposed to a phosphorus plate(PerkinElmer, U. S.)for 2h. In vitro autoradiographic images were obtained using a phosphor imaging system(Cyclone, Packard). After autoradiographic examination, the same mouse brain sections were stained by thioflavin S to confirm the presence of Aβ plaques. For the staining of thioflavin S, sections were immersed in a 0.125% thioflavin S solution containing 10% EtOH for 3min and washed in 40% EtOH. After drying, fluorescent observation was performed using the Oberver Z1(Zeiss, Germany)equipped with GFP filter set(excitation, 505nm).

4.8 Ex vivo autoradiography studies

The ex vivo evaluation was performed using a Tg(C57BL6-APP/PS1, 10 months old, male) mouse, which was used as an Alzheimer's model, and an age-matched control(C57BL6, 10 months, male). A saline solution of the labeled agent[^{18}F]**32**(19.2MBq) was injected directly into the tail vein. The mice were sacrificed by decapitation at 30min after intravenous injection. The brains were immediately removed and frozen in a liquid nitrogen bath. Sections of 20μm were cut and exposed to a phosphorus plate(PerkinElmer, U. S.)for 5h. Ex vivo film autoradiograms were thus obtained using a phosphor imaging system(Cyclone, Packard). After autoradiographic examination, the same sections were stained with thioflavin S to confirm the presence of amyloid plaques.

4.9 In vitro stability studies

The in vitro stability of[^{18}F]**32** in mouse plasma was determined by incubating 1.85 MBq purified[^{18}F]**32** with 100μL of mouse plasma at 37℃ for 2min, 10min, 30min, and 60min. Proteins were precipitated by adding 200μL of acetonitrile after centrifugation at 5000r/min for 5min at 4℃. The supernatant was collected. Approximately 0.1mL of the supernatant solution was analyzed using HPLC.

Associated Content

◎ Supporting Information

Purity of key target compounds together with HPLC chromatograms, ^{1}H NMR spectra of compounds, and HRMS data of key target compounds. This material is available free of charge via the Internet at http://pubs.acs.org.

Author Information

Corresponding Author

*For M. C. : phone, +86-10-58808891; fax, +86-10-58808891; e-mail, cmc@bnu.edu.cn.
For B. L. : phone, +86-10-58808891; e-mail, liuboli@bnu.edu.cn.

Author Contributions

These authors contributed equally.

Notes

The authors declare no competing financial interest.

Acknowledgments

The authors thank Dr. Jin Liu (College of Life Science, Beijing Normal University) for assistance in the in vitro neuropathological staining. This work was supported by National Natural Science Foundation of China (Grants 21201019, 21071023, and 30670586).

Abbreviations Used

AD, Alzheimer's disease; Aβ, β-amyloid; NFT, neurofibrillary tangle; CR, Congo Red; Th-T, thioflavin T; SPECT, single photon emission computed tomography; PET, positron emission tomography; IMPY, 2-(4'-dimethylaminophenyl)-6-imidazo[1,2-a] pyridine; PEG, polyethylene glycol; FPEG, fluoropolyethylene glycol; THF, tetrahydrofuran; BOC, butyloxycarbonyl; TBAF, tetra-n-butylammonium fluoride.

References

[1] Selkoe, D. J. The origins of Alzheimer disease: a is for amyloid. JAMA, J. Am. Med. Assoc. 2000, 283, 1615-1617.

[2] Selkoe, D. J. Alzheimer's disease: genes, proteins, and therapy. Physiol. Rev. 2001, 81, 741-766.

[3] Hardy, J. ; Selkoe, D. J. The amyloid hypothesis of Alzheimer's disease: progress and problems on the road to therapeutics. Science 2002, 297, 353-356.

[4] Hardy, J. The amyloid hypothesis for Alzheimer's disease: a critical reappraisal. J. Neurochem. 2009, 110, 1129-1134.

[5] Reitz, C. Alzheimer's disease and the amyloid cascade hypothesis: a critical review. Int. J. Alzheimer's Dis. 2012, 2012, 369808.

[6] Mathis, C. A. ; Wang, Y. ; Klunk, W. E. Imaging β-amyloid plaques and neurofibrillary tangles in the aging human brain. Curr. Pharm. Des. 2004, 10, 1469-1492.

[7] Cai, L. ; Innis, R. B. ; Pike, V. W. Radioligand development for PET imaging of β-amyloid (Aβ)-current status. Curr. Med. Chem. 2007, 14, 19-52.

[8] Sugimoto, H. Development of anti-Alzheimer's disease drug based on β-amyloid hypothesis. Yakugaku Zasshi 2010, 130, 521-526.

[9] Valotassiou, V. ; Archimandritis, S. ; Sifakis, N. ; Papatriantafyllou, J. ; Georgoulias, P. Alzheimer's disease: spect and pet tracers for β-amyloid imaging. Curr. Alzheimer Res. 2010, 7, 477-486.

[10] Herholz, K. ; Ebmeier, K. Clinical amyloid imaging in Alzheimer's disease. Lancet Neurol. 2011, 10, 667-670.

[11] Kung, M. P. ; Hou, C. ; Zhuang, Z. P. ; Zhang, B. ; Skovronsky, D. ; Trojanowski, J. Q. ; Lee, V. M. ; Kung, H. F. IMPY: an improved thioflavin-T derivative for in vivo labeling of β-amyloid plaques. Brain Res. 2002, 956, 202-210.

[12] Zhuang, Z. P. ; Kung, M. P. ; Wilson, A. ; Lee, C. W. ; Plossl, K. ; Hou, C. ; Holtzman, D. M. ; Kung, H. F. Structure-activity relationship of imidazo[1,2-a] pyridines as ligands for detecting β-amyloid

plaques in the brain. J. Med. Chem. 2003,46,237-243.

[13] Newberg, A. B. ; Wintering, N. A. ; Plossl, K. ; Hochold, J. ; Stabin, M. G. ; Watson, M. ; Skovronsky, D. ; Clark, C. M. ; Kung, M. P. ; Kung, H. F. Safety, biodistribution, and dosimetry of ^{123}I-IMPY: a novel amyloid plaque-imaging agent for the diagnosis of Alzheimer's disease. J. Nucl. Med. 2006,47,748-754.

[14] Mathis, C. A. ; Wang, Y. ; Holt, D. P. ; Huang, G. F. ; Debnath, M. L. ; Klunk, W. E. Synthesis and evaluation of ^{11}C-labeled 6-substituted 2-arylbenzothiazoles as amyloid imaging agents. J. Med. Chem. 2003,46,2740-2754.

[15] Klunk, W. E. ; Engler, H. ; Nordberg, A. ; Wang, Y. ; Blomqvist, G. ; Holt, D. P. ; Bergstrom, M. ; Savitcheva, I. ; Huang, G. F. ; Estrada, S. ; Ausen, B. ; Debnath, M. L. ; Barletta, J. ; Price, J. C. ; Sandell, J. ; Lopresti, B. J. ; Wall, A. ; Koivisto, P. ; Antoni, G. ; Mathis, C. A. ; Langstrom, B. Imaging brain amyloid in Alzheimer's disease with Pittsburgh compound-B. Ann. Neurol. 2004,55,306-319.

[16] Johnson, K. A. Amyloid imaging of Alzheimer's disease using Pittsburgh compound B. Curr. Neurol. Neurosci. Rep. 2006,6,496-503.

[17] Johnson, A. E. ; Jeppsson, F. ; Sandell, J. ; Wensbo, D. ; Neelissen, J. A. ; Jureus, A. ; Strom, P. ; Norman, H. ; Farde, L. ; Svensson, S. P. AZD2184: a radioligand for sensitive detection of β-amyloid deposits. J. Neurochem. 2009,108,1177-1186.

[18] Nyberg, S. ; Jonhagen, M. E. ; Cselenyi, Z. ; Halldin, C. ; Julin, P. ; Olsson, H. ; Freund-Levi, Y. ; Andersson, J. ; Varnas, K. ; Svensson, S. ; Farde, L. Detection of amyloid in Alzheimer's disease with positron emission tomography using [^{11}C]AZD2184. Eur. J. Nucl. Med. Mol. Imaging 2009,36,1859-1863.

[19] Ono, M. ; Wilson, A. ; Nobrega, J. ; Westaway, D. ; Verhoeff, P. ; Zhuang, Z. P. ; Kung, M. P. ; Kung, H. F. ^{11}C-labeled stilbene derivatives as Aβ-aggregate-specific PET imaging agents for Alzheimer's disease. Nucl. Med. Biol. 2003,30,565-571.

[20] Verhoeff, N. P. ; Wilson, A. A. ; Takeshita, S. ; Trop, L. ; Hussey, D. ; Singh, K. ; Kung, H. F. ; Kung, M. P. ; Houle, S. In-vivo imaging of Alzheimer disease β-amyloid with [^{11}C]SB-13 PET. Am. J. Geriatr. Psychiatry 2004,12,584-595.

[21] Koole, M. ; Lewis, D. M. ; Buckley, C. ; Nelissen, N. ; Vandenbulcke, M. ; Brooks, D. J. ; Vandenberghe, R. ; Van Laere, K. Whole-body biodistribution and radiation dosimetry of 18F-GE067: a radioligand for in vivo brain amyloid imaging. J. Nucl. Med. 2009,50,818-822.

[22] Jureus, A. ; Swahn, B. M. ; Sandell, J. ; Jeppsson, F. ; Johnson, A. E. ; Johnstrom, P. ; Neelissen, J. A. ; Sunnemark, D. ; Farde, L. ; Svensson, S. P. Characterization of AZD4694, a novel fluorinated Aβ plaque neuroimaging PET radioligand. J. Neurochem. 2010,114,784-794.

[23] Rowe, C. C. ; Ackerman, U. ; Browne, W. ; Mulligan, R. ; Pike, K. L. ; O'Keefe, G. ; Tochon-Danguy, H. ; Chan, G. ; Berlangieri, S. U. ; Jones, G. ; Dickinson-Rowe, K. L. ; Kung, H. P. ; Zhang, W. ; Kung, M. P. ; Skovronsky, D. ; Dyrks, T. ; Holl, G. ; Krause, S. ; Friebe, M. ; Lehman, L. ; Lindemann, S. ; Dinkelborg, L. M. ; Masters, C. L. ; Villemagne, V. L. Imaging of amyloid β in Alzheimer's disease with 18F-BAY94-9172, a novel PET tracer: proof of mechanism. Lancet Neurol. 2008,7,129-135.

[24] Lin, K. J. ; Hsu, W. C. ; Hsiao, I. T. ; Wey, S. P. ; Jin, L. W. ; Skovronsky, D. ; Wai, Y. Y. ; Chang, H. P. ; Lo, C. W. ; Yao, C. H. ; Yen, T. C. ; Kung, M. P. Whole-body biodistribution and brain PET imaging with [^{18}F]AV-45, a novel amyloid imaging agent-a pilot study. Nucl. Med. Biol. 2010,37,497-508.

[25] Choi, S. R. ; Golding, G. ; Zhuang, Z. ; Zhang, W. ; Lim, N. ; Hefti, F. ; Benedum, T. E. ; Kilbourn, M. R. ; Skovronsky, D. ; Kung, H. F. Preclinical properties of ^{18}F-AV-45: a PET agent for Aβ plaques in the brain. J. Nucl. Med. 2009,50,1887-1894.

[26] Wong, D. F. ; Rosenberg, P. B. ; Zhou, Y. ; Kumar, A. ; Raymont, V. ; Ravert, H. T. ; Dannals, R. F. ; Nan-

di, A. ; Brasic, J. R. ; Ye, W. ; Hilton, J. ; Lyketsos, C. ; Kung, H. F. ; Joshi, A. D. ; Skovronsky, D. M. ; Pontecorvo, M. J. In vivo imaging of amyloid deposition in Alzheimer disease using the radioligand ^{18}F-AV-45(florbetapir[corrected] F 18). J. Nucl. Med. 2010, 51, 913-920.

[27] Rowe, C. C. ; Villemagne, V. L. Brain amyloid imaging. J. Nucl. Med. 2011, 52, 1733-1740.

[28] Stephenson, K. A. ; Chandra, R. ; Zhuang, Z. P. ; Hou, C. ; Oya, S. ; Kung, M. P. ; Kung, H. F. Fluoro-pegylated(FPEG) imaging agents targeting Aβ aggregates. Bioconjugate Chem. 2007, 18, 238-246.

[29] Cheng, Y. ; Ono, M. ; Kimura, H. ; Kagawa, S. ; Nishii, R. ; Saji, H. A novel ^{18}F-labeled pyridyl benzofuran derivative for imaging of β-amyloid plaques in Alzheimer's brains. Bioorg. Med. Chem. Lett. 2010, 20, 6141-6144.

[30] Ono, M. ; Cheng, Y. ; Kimura, H. ; Cui, M. ; Kagawa, S. ; Nishii, R. ; Saji, H. Novel ^{18}F-labeled benzofuran derivatives with improved properties for positron emission tomography (PET) imaging of β-amyloid plaques in Alzheimer's brains. J. Med. Chem. 2011, 54, 2971-2979.

[31] Cheng, Y. ; Ono, M. ; Kimura, H. ; Kagawa, S. ; Nishii, R. ; Kawashima, H. ; Sajio, H. Fluorinated benzofuran derivatives for PET imaging of β-amyloid plaques in Alzheimer's disease brains. ACS Med. Chem. Lett. 2010, 1, 321-325.

[32] Cui, M. ; Ono, M. ; Kimura, H. ; Ueda, M. ; Nakamoto, Y. ; Togashi, K. ; Okamoto, Y. ; Ihara, M. ; Takahashi, R. ; Liu, B. ; Saji, H. Novel (18) F-Labeled benzoxazole derivatives as potential positron emission tomography probes for imaging of cerebral β-amyloid plaques in Alzheimer's disease. J. Med. Chem. [Online early access]. DOI: 10.1021/jm 300251n. Published Online: Jun 12, 2012.

[33] Qiao, J. X. ; Wang, T. C. ; Hu, C. ; Li, J. ; Wexler, R. R. ; Lam, P. Y. Transformation of anionically activated trifluoromethyl groups to heterocycles under mild aqueous conditions. Org. Lett. 2011, 13, 1804-1807.

[34] Cui, M. ; Ono, M. ; Kimura, H. ; Kawashima, H. ; Liu, B. L. ; Saji, H. Radioiodinated benzimidazole derivatives as single photon emission computed tomography probes for imaging of β-amyloid plaques in Alzheimer's disease. Nucl. Med. Biol. 2011, 38, 313-320.

[35] Toyama, H. ; Ye, D. ; Ichise, M. ; Liow, J. S. ; Cai, L. ; Jacobowitz, D. ; Musachio, J. L. ; Hong, J. ; Crescenzo, M. ; Tipre, D. ; Lu, J. Q. ; Zoghbi, S. ; Vines, D. C. ; Seidel, J. ; Katada, K. ; Green, M. V. ; Pike, V. W. ; Cohen, R. M. ; Innis, R. B. PET imaging of brain with the β-amyloid probe, [11C]6-OH-BTA-1, in a transgenic mouse model of Alzheimer's disease. Eur. J. Nucl. Med. Mol. Imaging 2005, 32, 593-600.

[36] Kung, H. F. ; Choi, S. R. ; Qu, W. ; Zhang, W. ; Skovronsky, D. [18]F stilbenes and styrylpyridines for PET imaging of A beta plaques in Alzheimer's disease: a miniperspective. J. Med. Chem. 2010, 53, 933-941.

[37] Bonger, K. M. ; van den Berg, R. J. ; Heitman, L. H. ; AP, I. J. ; Oosterom, J. ; Timmers, C. M. ; Overkleeft, H. S. ; van der Marel, G. A. Synthesis and evaluation of homo-bivalent GnRHR ligands. Bioorg. Med. Chem. 2007, 15, 4841-4856.

[38] Kim, D. Y. ; Kim, H. J. ; Yu, K. H. ; Min, J. J. Synthesis of [(18)F]-labeled(2-(2-fluoroethoxy)ethyl) tris (4-methoxyphenyl) phosphonium cation as a potential agent for positron emission tomography myocardial imaging. Nucl. Med. Biol. 2012, 39, 1093-1098.

[39] Cheng, Y. ; Prusoff, W. H. Relationship between the inhibition constant(K1) and the concentration of inhibitor which causes 50 per cent inhibition(I50) of an enzymatic reaction. Biochem. Pharmacol. 1973, 22, 3099-3108.

[40] Zhang, W. ; Oya, S. ; Kung, M. P. ; Hou, C. ; Maier, D. L. ; Kung, H. F. F-18 polyethyleneglycol stilbenes as PET imaging agents targeting Aβ aggregates in the brain. Nucl. Med. Biol. 2005, 32, 799-809.

^{18}F-labeled 2-phenylquinoxaline Derivatives as Potential Positron Emission Tomography Probes for *in vivo* Imaging of β-amyloid Plaques *

Abstract In continuation of our study on the 2-phenylquinoxaline scaffold as potential β-amyloid imaging probes, two [^{18}F]fluoro-pegylated 2-phenylquinoxaline derivatives, 2-(4-(2-[^{18}F]fluoroethoxy)phenyl)-*N*-methylquinoxalin-6-amine ([^{18}F]4a) and 2-(4-(2-(2-(2-[^{18}F]fluoroethoxy)ethoxy)ethoxy)phenyl)-*N*-methylquinoxalin-6-amine ([^{18}F]4b) were prepared. Both of them displayed high binding affinity to Aβ$_{1-42}$ aggregates (K_i = (10.0 ± 1.4) nM for 4a, K_i = (5.3 ± 3.2) nM for 4b). The specific and high binding of [^{18}F]4a and [^{18}F]4b to Aβ plaques was confirmed by *in vitro* autoradiography on brain sections of AD human and transgenic mice. In biodistribution in normal mice, [^{18}F]4a displayed high initial brain uptake (8.17% ID/g at 2min) and rapid washout from the brain. These preliminary results suggest [^{18}F]4a may be a potential PET imaging agent for Aβ plaques in the living human brain.

Key words Alzheimer's disease, β-amyloid plaque, 2-phenylquinoxaline, autoradiography, biodistribution

1 Introduction

Alzheimer's disease (AD) is the most common senile dementia. AD patients suffer from growing dementia and disability including cognitive decline, irreversible memory loss, disorientation, and language impairment, which are because of the progressive neurodegeneration in the brain. Till now, there is no effective therapeutic method for AD, however, the rapid growing of AD population has brought a ruin to not only the AD patients but also their families. AD is histopathologically characterized by β-amyloid (Aβ) plaques and neurofibrillary tangles, which presents in the gray matter of AD patient even before the dementia[1-3]. In addition, Aβ plaques had not been found in other kinds of dementia such as frontotemporal dementia or pure vascular dementia[4]. At this point of view, *in vivo* detection of Aβ plaques in the brain by positron emission tomography (PET) should be useful for early diagnosis of AD[5-8].

In the past decades, a number of PET imaging probes for Aβ plaques have been reported, several of which have been reported for clinical trials. The first Aβ plaques tracer for PET is 2-(4-([^{11}C]methylamino)phenyl)-6-hydroxybenzothiazol ([^{11}C]PIB, K_i = (0.87 ± 0.18) nM)[9,10], a neutral analog of thioflavin-T. After that, other ^{11}C-labeled Aβ imaging agents such as 4-*N*-[^{11}C]methylamino-4'-hydroxystilbene ([^{11}C]SB-13, K_i = (6.0 ± 1.5) nM)[11,12] and

---

* Copartner: Yu Pingrong, Cui Mengchao, Wang Xuedan, Zhang Xiaojun, Li Zijing, Yang Yanping, Jia Jianhua, Zhang Jinming, Masahiro Ono, Hideo Saji, Jia Hongmei. Reprinted from *European Journal of Medicinal Chemistry*, 2012, 57: 51-58.

2-[6-([^{11}C] methylamino) pyridin-3-yl]-1, 3-benzothiazol-6-ol ([^{11}C] AZD2184, K_i = (1.70 ± 0.54) nM)[13-15] had also been reported under clinical trials and displayed effective result in differencing AD patients with controls by measuring the brain uptake and retention. As a ^{11}C-labeled tracer, the chemical structure is retained, which completely maintained the biological property of molecular. But the short half-life of ^{11}C ($t_{1/2}$ = 20min) leads to that the supply of ^{11}C-labeled tracers will be limited to centers equipped with an on-site cyclotron, which may be the most important reason for preventing ^{11}C-labeled tracers to be more widely used. Meanwhile, ^{18}F ($t_{1/2}$ = 110min) has a longer radioactive decay half-life, which permits a more widespread application, and allowed multiple injections from a single production batch. In this case, ^{18}F may be the better radionuclide for PET imaging. Great efforts have been focused on the development of ^{18}F-labeled Aβ plaques tracers. Some of them like 4-(N-methylamino)-4'-(2-(2-(2-[^{18}F]fluoroethoxy) ethoxy) ethoxy)-stilbene ([^{18}F] BAY94-9172, florbetaben, K_i = (2.22 ± 0.54) nM)[16,17], 2-(3-[^{18}F] fluoro-4-methyaminophenyl) benzothiazol-6-ol ([^{18}F] GE-067, flutemetamol, K_i = (0.74 ± 0.38) nM)[18] had already been reported under clinical trials. In April 2012, (E)-4-(2-(6-(2-(2-(2-[^{18}F]fluoroethoxy) ethoxy) ethoxy) pyridin-3-yl) vinyl)-N-methylaniline([^{18}F] AV-45, florbetapir, K_i = (2.87 ± 0.17) nM)[19-21], had been approved by the U. S. Food and Drug Administration (FDA) as a radioactive diagnostic agent indicated for brain imaging of Aβ plaques in patients who are being evaluated for AD and other causes of cognitive impairment.

However, clinical trials for ^{18}F-labeled Aβ imaging tracers show that there is greater white matter retention compared with ^{11}C-labeled agents, and high non-specific white matter retention may limit the sensitivity of PET imaging. The mechanism of white matter retention seems to be owing to non-specific binding[22], besides further modifications in order to decrease lipophilicity are needed. In addition, almost all of the tracers evaluated in humans are thioflavin-T or stilbene derivatives. Thus, development of tracers with a new scaffold may lead a new way to improve the *in vivo* properties including higher affinity to Aβ plaques and less nonspecific binding in the white matter of the brain.

In a search for novel Aβ imaging probes, we have recently reported a lipophilic ^{125}I-labeled 2-phenylquinoxaline derivative ([^{125}I] QN-1)[23], which displayed high affinity to Aβ aggregates (K_i = (4.1 ± 0.7) nM). Film autoradiography and fluorescent staining confirmed the specific binding to Aβ plaques on postmortem AD brain sections. In biodistribution experiment, this probe exhibited high initial uptake into the brain (6.03% ID/g at 2min), but unsatisfactory rate of washing out (2.98% ID/g at 60min). This may be caused by its high lipophilicity (lg D = 4.02) due to the existence of iodine atom. To improve the pharmacokinetic profile of [^{125}I] QN-1 for using as a PET imaging agent, the iodine atom was replaced by a short fluorine end-capped polyethylene glycol chain (n = 1 or 3)[24]. Furthermore, aiming of circumventing the problem of the rapid *in vivo* N-demethylation for dimethylamino group which appeared in metabolism of tracers like 6-[^{123}I]iodo-2-(4'-dimethylamino-) phenyl-imidazo[1,2] pyridine ([^{123}I] IMPY)[25-27] and (E)-4-(2-(6-(2-(2-(2-[^{18}F] fluoroethoxy) ethoxy) ethoxy) pyridin-3-yl) vinyl)-*N*, *N*-dimethylaniline ([^{18}F]AV-19)[7], most of the radiotracers for Aβ plaques in clinical trials have a monomethylamino group. Accounting for the same reason, the dimethylamino group in [^{125}I]QN-1 was changed to monomethylamino group.

In the present study, we reported the synthesis and evaluation of novel ^{18}F-labeled 2-phenylquinoxaline probes for Aβ plaques with fluoro-pegylated side chains of different length ($n = 1$ or 3) and a monomethylamino group, expecting of better *in vivo* properties comparing with the radioiodinated ligand (Fig. 1 and Fig. 2).

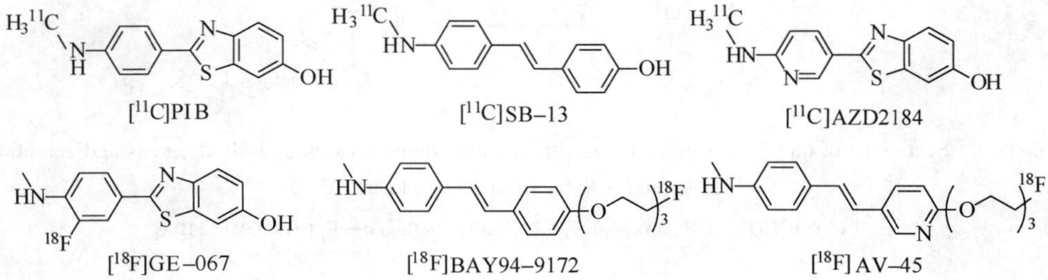

Fig. 1　Chemical structure of reported Aβ imaging probes for clinical trials

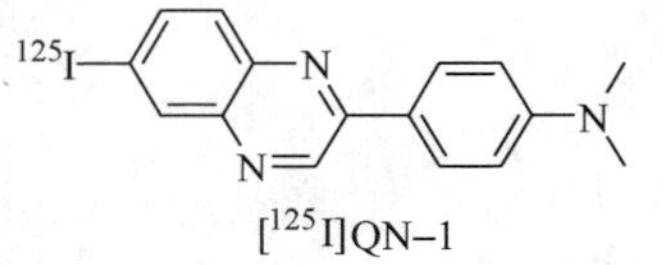

Fig. 2　Chemical structure of 2-phenylquinoxaline derivative [^{125}I]QN-1

2　Results and Discussion

2.1　Chemistry

The synthesis was shown in Scheme 1 and Scheme 2. The derivative **1** was formed by a one-pot tandem oxide condensation procedure in DMSO (yield 93.6%). The amino derivative **2** was obtained from **1** by reduction with excess hydrazine hydrate in ethanol in which Pd/C was added as catalyzer (yield 95.6%). Monomethylation of derivative **2** was achieved with paraformaldehyde, sodium methoxide and sodium borohydride to obtain derivative **3** (yield 88.7%). The corresponding fluoropegylated derivative **4a**, **4b** were prepared by **3** with K_2CO_3, 1-bromo-2-fluoroethane or 2-(2-(2-fluoroethoxy)ethoxy)ethyl 4-methylbenzenesulfonate in DMF (yield 51.8%, 35.0%). Ethane-1,2-diyl bis(4-methylbenzenesulfonate) or (ethane-1,2-diylbis(oxy))bis(ethane-2,1-diyl) bis(4-methylbenzenesulfonate) was coupled with the hydroxy group of **3** with K_2CO_3 and 18-crown-6 in acetone to obtain **5a**, **5b** (yield 27.5%, 73.6%). The tosylate precursor **6a** and **6b** were obtained by protecting the methylamino groups of **5a** and **5b** with di-tert-butyldicarbonate in THF (yield 43.1%, 52.6%).

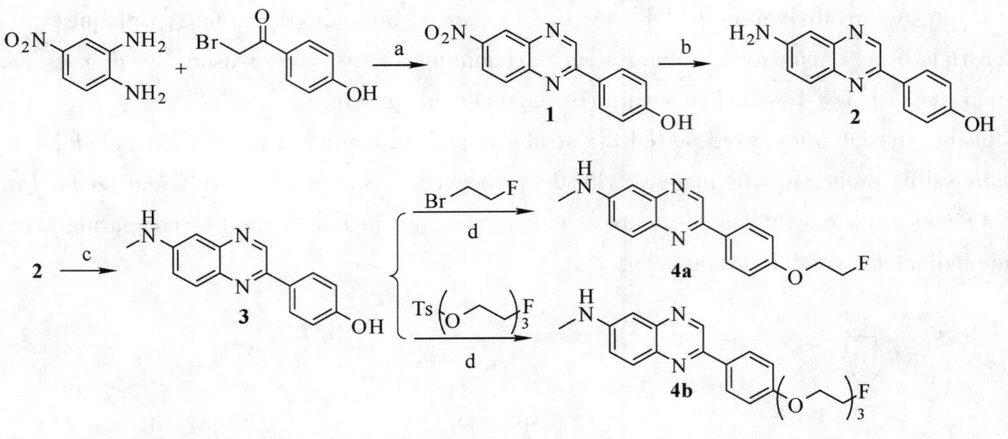

Scheme 1 Synthesize of the fluoropegylated phenylquinoxaline derivatives **4a** and **4b**. Reagents and conditions

a—DMSO, rt; b—$N_2H_4 \cdot H_2O$, Pd/C, EtOH, reflux;

c—(1)(HCHO)$_n$, CH_3ONa, reflux; (2) $NaBH_4$, reflux; d—K_2CO_3, DMF, 110℃

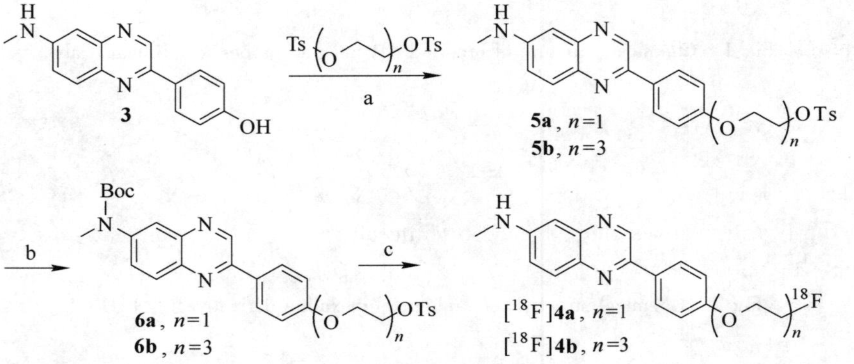

Scheme 2 Synthesize of the precursors and radiolabeling. Reagents and conditions

a—K_2CO_3, acetone, 18-crown-6, reflux; b—$(Boc)_2O$, THF;

c—(1) Kryptofix 222, K_2CO_3, $^{18}F^-$, CH_3CN, 100℃; (2) 1M HCl, 100℃

2.2 Radiolabeling

To get the radiofluorinated ligands [^{18}F]**4a**, [^{18}F]**4b**, the N-BOC-protected tosylate precursor **6a**, **6b** was mixed with [^{18}F]fluoride, potassium carbonate and Kryptofix 222 in acetonitrile under heating at 110℃ for 5min. The mixtures were treated with aqueous hydrochloric acid to remove the N-BOC-protecting group, and neutralized by sodium dicarbonate. The mixture was loaded on a Sep-Pak Plus-C18 cartridge (Waters), and the elution was concentrated and the residue was purified by HPLC. The ^{18}F-labeled [^{18}F]**4a** and [^{18}F]**4b** were prepared with an average radiochemical yield of 20% and 52% (no decay corrected), and radiochemical purity of >98%. The identity of [^{18}F]**4a** and [^{18}F]**4b** was verified by a comparison of the retention time with that of the nonradioactive compound (Fig. 3), and their specific activity was estimated at approximately 200GBq/μmol.

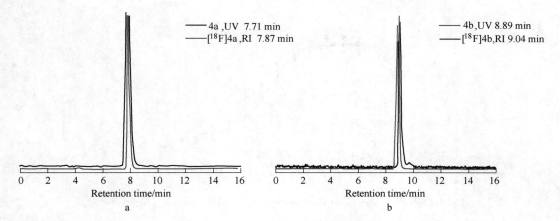

Fig. 3 HPLC profiles of **4a**, [^{18}F]**4a** (a) and **4b**, [^{18}F]**4b** (b)
(HPLC conditions: Venusil MP C18 column (Agela Technologies, 10mm×250mm), CH_3CN/H_2O = 80/20 for 4a, CH_3CN/H_2O = 70/30 for 4b, 4mL/min, UV, 254nm)

2.3 Biological evaluation

To evaluate the binding affinity of the two 2-phenylquinoxaline derivatives (**4a** and **4b**) to $A\beta_{1-42}$ aggregates, *in vitro* inhibition assay was carried out in solutions with [^{125}I]IMPY as the competing radioligand according to conventional methods[28]. The result is shown in Table 1. As expected, the two derivatives showed good binding affinity to $A\beta_{1-42}$ aggregates (K_i = (10.0 ± 1.4)nM for **4a**, K_i = (5.3 ± 3.2)nM for **4b**) comparable to the value determined under the same assay system for IMPY (K_i = (10.5 ± 1.0)nM). Compared with the iodinated tertiary *N*,*N*-dimethylamino analog QN-1 (K_i = (4.1 ± 0.7)nM), these fluorinated secondary monomethylamino analogs still kept the high affinity.

Table 1 Inhibition constants (K_i) for binding to aggregates of $A\beta_{1-42}$ versus [^{125}I]IMPY[①]

Compound	K_i/nM	Compound	K_i/nM
4a	10.0 ± 1.4	QN-1[②]	4.1 ± 0.7
4b	5.3 ± 3.2	IMPY[②]	10.5 ± 1.0

① Measured in triplicate with results given as the mean ± SD.
② Data from literature [23].

The lg*D* values (3.14 ± 0.23 for [^{18}F]**4a**, 2.79 ± 0.14 for [^{18}F]**4b** and 4.02 ± 0.12 for [^{125}I]QN-1) showed in Table 3 confirm that the importing of fluoropegylated chain is an effective path to decrease the lipophilicity of the tracers. The lipophilicity was reduced as the length of the fluoro-pegylated chain being increased, which may lower the non-specific binding in brain (Fig. 4).

In vitro autoradiography in sections of brain tissue from AD patients or Tg model mice (C57BL6, APPswe/PSEN1, 12 months old) was carried out to confirm the high binding affinity of [^{18}F]**4a** and [^{18}F]**4b** to Aβ plaques. As shown in Fig. 5a and c, specific labeling of plaques

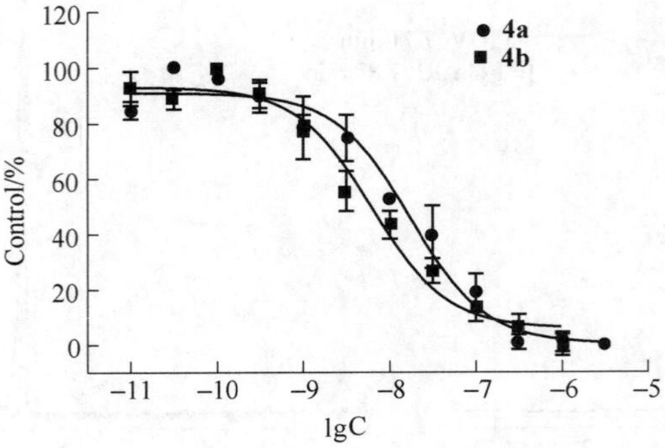

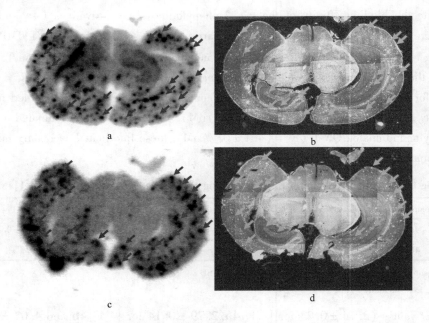

Fig. 4 Inhibition curves for the binding of $[^{125}I]$IMPY to $A\beta_{1-42}$ aggregates

was observed in the brain sections of transgenic mice. The presence and distribution of $A\beta$ plaques was consistent with the results of fluorescent staining using thioflavin-S on the same sections (Fig. 5b and d). Furthermore, intense labeling of plaques and low non-specific background were observed in the brain sections of AD patients (Fig. 6a and b). In contrast, no apparent labeling was observed in normal adult brain sections (Fig. 6c and d).

Fig. 5 *In vitro* autoradiography of $[^{18}F]$**4a** and $[^{18}F]$**4b** (a, c) on a Tg model mouse (C57BL6, APPswe/PSEN1, 12 months old, male), and the presence and distribution of plaques in the sections were confirmed by fluorescence staining using thioflavin-S on the same sections with a filter set for GFP (b, d)

Biodistribution experiments in normal male ICR mice were carried out to evaluate the ability of radiofluorinated tracers ($[^{18}F]$**4a** and $[^{18}F]$**4b**) to penetrate the blood-brain barrier (BBB) and properties of clearance from the brain. As shown in Table 2 and Table 3, $[^{18}F]$**4a** with a short ($n=1$) fluoro-pegylated side chain displayed a higher initial brain uptake (8.17% ID/g

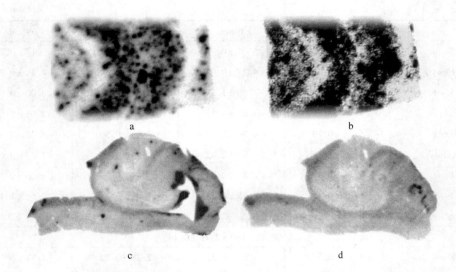

Fig. 6 *In vitro* autoradiography of [^{18}F]**4a** and [^{18}F]**4b** on AD human brain sections(a, b) and control human brain sections(c, d)

at 2min) than that of [^{18}F]**4b** with a long ($n = 3$) side chain (2.49% ID/g at 2min). The lower brain uptake for [^{18}F]**4b** may be due to more hydrogen bonds formed between the longer fluoro-pegylated chain and water or other biomoleculars *in vivo* which decrease the initial brain uptake. Compared with [^{125}I]QN-1 (6.03% ID/g at 2min) and [^{18}F]AV-45 (7.33% ID/g at 2min), the initial brain uptake of [^{18}F]**4a** is superior. High initial brain uptake and high brain$_{2min}$/brain$_{60min}$ ratio in normal mouse brain are considered to be important as *in vivo* pharmacokinetic indexes for selecting appropriate Aβ imaging tracers. As shown in Table 3, the brain$_{2min}$/brain$_{60min}$ ratio is 2.58, 3.89, 2.07, 3.90 for [^{18}F]**4a**, [^{18}F]**4b**, [^{125}I]QN-1, and [^{18}F]AV-45, respectively. The ratios of [^{18}F]**4a** and [^{18}F]**4b** are obviously higher than that of [^{125}I]QN-1, this may due to the introduction of fluoro-pegylated chains which reduce the lipophilicity of these two probes. [^{18}F]**4a** and [^{18}F]**4b** also distributed to several other organs. The liver and kidney showed an initial uptake with washout, continuous gastrointestinal accumulation of the radiotracers resulted in an intestine uptake (18.75% ID/g and 39.02% ID/g at 60min). In addition, accumulation of radioactivity in the bone, 2.46% ID/g observed already at 2min pi and steadily increased up to 6.20% ID/g during the course of the experiment, suggesting little defluorination *in vivo* of [^{18}F]**4a**, while the bone uptake of [^{18}F]**4b** remained almost constant, indicating no defluorination *in vivo*.

Table 2 Biodistribution of in ICR normal mice after iv injections of [^{18}F] tracers

(% ID/g, Mean ± SD, $n = 5$)

Organ	2min	10min	30min	60min
[^{18}F]**4a** ($\lg D = 3.14 \pm 0.23$)				
Blood	4.93 ± 0.43	4.68 ± 0.32	5.09 ± 0.30	5.24 ± 0.45

Continued 2

Organ	2min	10min	30min	60min
Brain	8.17±1.33	5.07±0.71	3.45±0.13	3.17±0.22
Heart	6.08±0.37	4.27±0.60	4.28±0.54	4.24±0.22
Liver	9.50±1.04	7.80±1.33	5.37±0.51	5.00±0.84
Spleen	4.71±0.17	3.87±0.57	3.35±0.50	3.04±0.64
Lung	6.38±0.79	4.36±0.42	3.78±0.56	3.77±0.36
Kidney	9.37±1.11	5.96±0.42	4.25±0.43	3.79±0.66
Bone	2.46±0.73	2.84±0.54	5.73±0.29	6.20±0.25
Stomach[①]	1.81±0.42	2.55±0.54	1.84±0.36	2.50±1.19
Intestine[①]	9.63±1.10	12.87±2.61	17.15±1.23	18.75±3.90
[^{18}F]**4b** ($\lg D = 2.79 \pm 0.14$)				
Blood	5.51±0.34	3.32±0.32	2.38±0.21	1.54±0.12
Brain	2.49±0.22	1.44±0.11	0.83±0.17	0.64±0.07
Heart	3.21±0.46	1.76±0.19	1.19±0.19	1.03±0.15
Liver	14.18±1.66	14.09±1.49	7.90±1.07	4.89±0.74
Spleen	2.07±0.97	1.65±0.11	1.21±0.21	0.94±0.39
Lung	4.07±0.45	2.40±0.33	1.76±0.25	1.15±0.11
Kidney	5.52±0.46	3.91±0.27	3.27±0.74	1.65±0.94
Bone	1.49±0.34	1.00±0.40	1.12±0.23	1.78±0.06
Stomach[①]	1.55±0.12	2.81±1.31	3.21±1.62	6.53±2.03
Intestine[①]	6.06±0.64	12.26±2.94	32.83±6.30	39.02±6.98

① Expressed as % ID.

Table 3 Comparison of inhibition constants (K_i) and brain kinetics between radiolabeled 2-phenylquinoxaline derivatives and [^{18}F]AV-45

Compound	K_i/nM	Brain[①]$_{2min}$	Ratio$_{2min/60min}$	$\lg D$
[^{18}F]**4a**	10.0	8.17	2.58	3.14
[^{18}F]**4b**	5.3	2.49	3.89	2.79
[^{125}I]QN-1[②]	4.1	6.03	2.07	4.02
[^{18}F]AV-45[③]	2.9	7.33	3.90	2.41

① Expressed as %ID/g.
② Data from Ref. [23].
③ Data from Ref. [19].

3 Conclusion

In conclusion, two novel 2-phenylquinoxaline derivatives containing an end-capped fluoro-pegylated chains ($n = 1, 3$) had been successfully prepared and evaluated as PET imaging tracers for Aβ plaques. These two compounds appeared to have good binding affinities to Aβ aggregates. $[^{18}F]$**4a** with a short fluoro-pegylated chain ($n = 1$) displayed a high initial brain uptake and good ratio of brain$_{2min}$/brain$_{60min}$, and the *in vitro* autoradiography studies confirmed its specific binding to Aβ plaques and low non-specific binding, which suggests $[^{18}F]$**4a** may be a potential PET imaging agent for *in vivo* detection of Aβ plaques.

4 Experimental

4.1 General information

All reagents used in the synthesis were commercial products and were used without further purification unless otherwise indicated. $^{18}F^-$ was obtained from the Chinese PLA General Hospital. The ^{1}H-NMR spectra were obtained at 400MHz on Bruker spectrometer in CDCl$_3$ or [D$_6$]-DMSO at room temperature with TMS as an internal standard. Chemical shifts were reported as δ values with respect to residual solvents. The multiplicity is defined by s (singlet), d (doublet), t (triplet), m (multiplet). Mass spectrometry was acquired under the Surveyor MSQ Plus (ESI)(Waltham, MA, USA) instrument. Reactions were monitored by TLC (precoated silica gel plate F254, Merck). Radiochemical purity was determined by HPLC performed on a Shimadzu SCL-20 AVP equipped with a Bioscan Flow Count 3200 NaI/PMT γ-radiation scintillation detector. Separations were achieved on a Venusil MP C18 column (Agela Technologies, 10μm, 10mm × 250mm) eluted with a binary gradient system at a 4.0mL/min flow rate. Mobile phase A was water while mobile phase B was acetonitrile. Fluorescent observation was performed by the LSM 510 META (Zeiss, Germany) equipped with a LP 505 filter set (excitation, 405nm; long-pass filter, 505nm). The purity of the synthesized key compounds was determined using analytical HPLC and was found to be more than 95%. ICR Mice (five weeks, 20-22g, male) were used for biodistribution experiments. All protocols requiring the use of mice were approved by the animal care committee of Beijing Normal University. Postmortem brain tissues from an autopsy-confirmed case of AD (5μm, temporal lobe) and a control subject (5μm, temporal lobe) were obtained from BioChain Institute Inc. Transgenic mice brain tissues (C57BL6, APPswe/PSEN1, 12 months old, male) were purchased from Institute of Laboratory Animal Sciences, Chinese Academy of Medical Sciences.

4.2 4-(6-nitroquinoxalin-2-yl)phenol (1)

A mixture of 2-bromo-1-(4-hydroxyphenyl)ethanone (1mmol) and 4-nitrobenzene-1,2-diamine (1mmol) was stirred in 5mL DMSO at room temperature. The reaction mixture was poured into water, and quickly filtered over a buchner funnel. The filter residue was washed with water, oven dried and purified by silica gel chromatography (petroleum ether/ethylacetate = 2∶1), to give 250mg of **1** in a yield of 93.6%. ^{1}H NMR (400MHz, [D$_6$]-DMSO) δ 10.27 (s, 1H), 9.68 (d, J = 4.5Hz, 1H), 8.81 (dd, J = 13.2, 2.5Hz, 1H), 8.50 (dd, J = 9.2, 2.6Hz, 1H), 8.29 (d, J = 8.8Hz, 2H), 8.23 (t, J = 9.0Hz, 1H), 6.98 (d, J = 8.7Hz, 2H), 5.37 (s, 1H). MS (ESI) m/z calcd for C$_{14}$H$_9$N$_3$O$_3$ 267.1, found 267.7 [M + H]$^+$.

4.3 4-(6-aminoquinoxalin-2-yl)phenol (2)

A mixture of compound **1** (2mmol) and hydrazine hydrate (8mmol) in ethanol (20mL) was refluxed over night, in which Pd/C was added as catalyzer. The reaction mixture was cooled to room temperature and then filtered. The filtrate was concentrated, and purified by silica gel chromatography (petroleum ether/ethylacetate = 1 : 1) to give 453mg of **2** in a yield of 95.6%. ^{1}H NMR (400MHz, [D_6]-DMSO) δ 9.86(s, 1H), 9.21(s, 1H), 8.10(d, J = 8.7Hz, 2H), 7.78(d, J = 9.0Hz, 1H), 7.28(dd, J = 9.0, 2.4Hz, 1H), 6.99(d, J = 2.4Hz, 1H), 6.98(s, 1H), 6.95(s, 1H), 6.03(s, 2H). MS (ESI) m/z calcd for $C_{14}H_{11}N_3O$ 237.1, found 237.7 [M+H]$^+$.

4.4 4-(6-(methylamino)quinoxalin-2-yl)phenol (3)

A solution of compound **3** (2.3mmol) in methanol (20mL) was added with sodium methoxide (4.5mmol) and paraformaldehyde (9mmol). The reaction mixture was refluxed for 2h then cooled to 0℃ in an ice bath. Sodium borohydride (9mmol) was added carefully. The mixture was refluxed again for 1h and cooled to room temperature. After solvent being removed in vacuo, water (10mL) was added, and then ethylacetate (3 × 50mL) was used for extraction. The organic phase was dried over MgSO$_4$, and concentrated latter. The residue was purified by silica gel chromatography (petroleum ether/ethylacetate = 1 : 1) to give 512mg of **3** in a yield of 88.7%. ^{1}H NMR (400MHz, [D_6]-DMSO) δ 9.80(s, 1H), 9.18(s, 1H), 8.05(d, J = 8.7Hz, 2H), 7.74(d, J = 9.1Hz, 1H), 7.25(dd, J = 9.1, 2.5Hz, 1H), 6.91(d, J = 8.6Hz, 2H), 6.74(d, J = 2.4Hz, 1H), 2.83(s, 3H). MS (ESI) m/z calcd for $C_{15}H_{13}N_3O$ 251.1, found 251.8 [M+H]$^+$.

4.5 2-(4-(2-fluoroethoxy)phenyl)-*N*-methylquinoxalin-6-amine (4a)

A mixture of compound **3** (0.35mmol) and 1-bromo-2-fluoroethane (0.5mmol) in DMF (5mL) was refluxed for 2h. After being cooled to room temperature, water (10mL) and CH$_2$Cl$_2$ (20mL) were added to the mixture. The organic layer was separated, washed with water (3 × 10mL), and dried over MgSO$_4$. The solvent was removed in vacuo, and the residue was purified by silica gel chromatography (petroleum ether/ethylacetate = 2 : 1) to give 54mg of **4a** in a yield of 51.8%. ^{1}H NMR (400MHz, CDCl$_3$) δ 9.10(s, 1H), 8.08(d, J = 8.8Hz, 2H), 7.87(d, J = 9.1Hz, 1H), 7.13(dd, J = 9.1, 2.6Hz, 1H), 7.08(d, J = 8.8Hz, 2H), 6.98(d, J = 2.5Hz, 1H), 4.89-4.85(m, 1H), 4.77-4.73(m, 1H), 4.35-4.33(m, 1H), 4.28-4.25(m, 1H), 3.00(s, 3H). ^{13}C NMR (100MHz, CDCl$_3$) δ 159.50, 149.80, 147.20, 143.75, 142.77, 137.12, 130.77, 130.02, 128.27, 122.07, 115.17, 103.34, 82.70, 81.00, 67.37, 67.17, 30.53. HRMS (ESI) m/z calcd for $C_{17}H_{16}FN_3O$ 298.1356, found 298.1360 [M+H]$^+$.

4.6 2-(4-(2-(2-(2-fluoroethoxy)ethoxy)ethoxy)phenyl)-*N*-methylquinoxalin-6-amine (4b)

The reaction described for **4a** was used, and 27mg of **4b** was obtained in a yield of 35.0%.

^{1}H NMR (400MHz, CDCl$_3$) δ 9.09(s, 1H), 8.06(d, J = 8.6 Hz, 2H), 7.86(d, J = 9.0Hz, 1H), 7.12(dd, J = 9.1, 2.1Hz, 1H), 7.06(d, J = 8.6Hz, 2H), 6.98(d, J = 1.7Hz, 1H), 4.65-4.62(m, 1H), 4.53-4.50(m, 1H), 4.24-4.20(m, 2H), 3.93-3.90(m, 2H), 3.82-3.79(m, 1H), 3.78-3.72(m, J = 12.5, 4.9Hz, 6H), 3.00(s, 3H). ^{13}C NMR (100MHz, CDCl$_3$) δ 158.85, 148.76, 146.25, 142.55, 141.65, 136.08, 129.22, 128.93, 127.12, 121.08, 114.11, 102.12, 82.96, 81.29, 69.90, 69.86, 69.55, 69.35, 68.77, 66.55, 29.49. HRMS (ESI) m/z calcd for C$_{21}$H$_{25}$FN$_3$O$_3$ 386.1880, found 386.1891 [M + H]$^+$.

4.7 2-(4-(6-(methylamino)quinoxalin-2-yl)phenoxy)ethyl 4-methylbenzenesulfonate (5a)

A mixture of compound **3** (0.4mmol), ethane-1,2-diyl bis(4-methylbenzenesulfonate) (0.6mmol) and K$_2$CO$_3$ (1.2mmol) in acetone (10mL) was refluxed for 10h, in which 18-Crown-6 was added as catalyzer. After the solvent being removed in vacuo, the residue was purified by silica gel chromatography (petroleum ether/ethylacetate = 1 : 2) to give 49.4mg of **5a** in a yield of 27.5%. ^{1}H NMR (400MHz, CDCl$_3$) δ 9.08(s, 1H), 8.04(d, J = 8.8Hz, 2H), 7.85(t, J = 8.4Hz, 3H), 7.36(d, J = 8.0Hz, 2H), 7.13(dd, J = 9.1, 2.6Hz, 1H), 6.98(d, J = 2.4Hz, 1H), 6.93(d, J = 8.8Hz, 2H), 4.42(dd, J = 5.5, 3.9Hz, 2H), 4.24(d, J = 4.9Hz, 2H), 3.00(s, 3H), 2.45(s, 3H). MS (ESI) m/z calcd for C$_{24}$H$_{23}$N$_3$O$_4$S 449.1, found 449.6 [M + H]$^+$.

4.8 2-(2-(2-(4-(6-(methylamino)quinoxalin-2-yl)phenoxy)ethoxy)ethoxy)ethyl 4-methylbenzenesulfonate (5b)

The reaction described for **5a** was used, and 210mg of **5b** was obtained in a yield of 73.6%. ^{1}H NMR (400MHz, CDCl$_3$) δ 9.09(s, 1H), 8.06(d, J = 8.7Hz, 2H), 7.85(d, J = 9.0Hz, 1H), 7.80(d, J = 8.2Hz, 2H), 7.32(d, J = 8.1Hz, 2H), 7.13(dd, J = 9.1, 2.5Hz, 1H), 7.05(d, J = 8.8Hz, 2H), 6.97(d, J = 2.4Hz, 1H), 4.21-4.15(m, J = 6.4, 5.0Hz, 4H), 3.89-3.84(m, 2H), 3.73-3.67(m, 4H), 3.66-3.62(m, J = 5.9, 3.1Hz, 2H), 2.99(s, 3H), 2.42(s, 3H). MS(ESI) m/z calcd for C$_{28}$H$_{32}$N$_3$O$_6$S 538.2, found 538.5 [M + H]$^+$.

4.9 2-(4-(6-(*tert*-butoxycarbonyl)quinoxalin-2-yl)phenoxy)ethyl 4-methylbenzenesulfonate (6a)

Compound **5a** (0.1mmol) was added to a solution of di-*tert*-butyldicarbonate (0.4mmol) and DIEA (0.2mmol) in THF, and the reaction mixture was refluxed over night. After the solvent being removed in vacuum, the residue was purified by silica gel chromatography (petroleum ether/ethylacetate = 1 : 1) to give 24mg of **6a** in a yield of 43.1%. ^{1}H NMR (400MHz, CDCl$_3$) δ 9.24(s, 1H), 8.12(d, J = 8.6Hz, 2H), 8.03(d, J = 8.9Hz, 1H), 7.90-7.76(m, 4H), 7.36(d, J = 8.0Hz, 2H), 6.96(d, J = 8.6Hz, 2H), 4.46-4.39(m, 2H), 4.29-4.21(m, 2H), 3.43(s, 3H), 2.46(s, 3H), 1.50(s, 9H). MS (ESI) m/z calcd for C$_{29}$H$_{31}$N$_3$O$_6$S 549.2, found 549.7 [M + H]$^+$.

4.10 2-(2-(2-(4-(6-(*tert*-butoxycarbonyl)quinoxalin-2-yl)phenoxy)ethoxy)ethoxy)ethyl 4-methylbenzenesulfonate (6b)

The reaction described for **6a** was used, and 67mg of **6b** was obtained in a yield of 52.6%. ^{1}H NMR (400MHz, CDCl$_3$) δ 9.25(s,1H),8.15(d, J=8.8Hz,2H),8.03(d, J=9.0Hz,1H),7.85(d, J=2.3Hz,1H),7.83-7.77(m,3H),7.32(d, J=8.0Hz,2H),7.08(d, J=8.9Hz,2H),4.22-4.16(m,4H),3.90-3.85(m,2H),3.73-3.68(m,4H),3.66-3.61(m,2H),3.43(s,3H),2.42(s,3H),1.50(s,9H). MS(ESI) m/z calcd for C$_{33}$H$_{39}$N$_3$O$_6$S 637.2, found 637.7 [M+H]$^+$.

4.11 Radiolabeling

[^{18}F]Fluoride trapped on a QMA cartridge was eluted with 1mL of Kryptofix 222/K$_2$CO$_3$ solution. The solvent was removed at 110℃ under a stream of nitrogen gas. The residue was azeotropically dried with 1mL of anhydrous acetonitrile twice at 110℃ under a stream of nitrogen gas. A solution of tosylate precursor (1mg) in 1mL CH$_3$CN was added to the reaction tube containing dried ^{18}F$^-$ activities. The mixture was heated at 110℃ under nitrogen protecting for 5min. 200μL HCl (1M) was added into the tube and the mixture was heated for another 5min at 110℃ under nitrogen protecting for deprotection, and neutralized by NaHCO$_3$ after being cooled down to room temperature. Water (5mL) was added, and the mixture was passed through a preconditioned Sep-Pak Plus-C18 cartridge (Waters), which was washed with 10mL water later. The ^{18}F-labeled compound was eluted with 2mL of acetonitrile, and purified by HPLC. The ^{18}F-labeled [^{18}F]**4a** and [^{18}F]**4b** were prepared with an average radiochemical yield of 20% and 52% (no decay corrected), and radiochemical purity of >98%.

4.12 Binding assay in vitro using Aβ$_{1-42}$ aggregates

Inhibition experiments were carried out in 12mm × 75mm borosilicate glass tubes according to procedures described previously with some modifications. The radio-ligand [^{125}I]IMPY was prepared according to procedures described previously[25], after HPLC purification, the radiochemical purity was greater than 95%. For the inhibition assay, 100μL Aβ$_{1-42}$ aggregates solution, 100μL [^{125}I]IMPY solution, 100μL BSA solution (1%) and 100μL inhibitors (10^{-4}-10$^{-9.5}$M) were added into borosilicate glass tubes. The mixture was incubated at 37℃ for 2h, and the free radioactivity were separated by vacuum filtration through Whatman GF/B filters using a Brandel Mp-48T cell harvester followed by 3 × 4mL washes with PBS (0.02M, pH 7.4) at room temperature. Filters with the bound [^{125}I]IMPY were counted in gamma counter (WALLAC/Wizard 1470, USA) with 70% efficiency. Inhibition experiments were repeated three times, and the half maximal inhibitory concentration (IC_{50}) was determined using GraphPad Prism 4.0, and the inhibition constant (K_i) was calculated using the Cheng-Prusoff equation: $K_i = IC_{50}/(1+[L]/K_d)$[29].

4.13 Autoradiography *in vitro* using brain sections of human and transgenic model mouse

Paraffin-embedded brain sections were deparaffinized with 2 × 20min washes in xylene; 2 × 5min washes in 100% ethanol; a 5min wash in 90% ethanol/H_2O; a 5min wash in 80% ethanol/H_2O; a 5min wash in 60% ethanol/H_2O and a 10min wash in running tap water, and then incubated in PBS (0.2M, pH = 7.4) for 30min. The sections were incubated with [^{18}F]**4a**, [^{18}F]**4b** (1.85MBq/200μL) for 1h at room temperature. They were then washed in 40% EtOH, before being rinsed with water for 1min. After drying, the sections were exposed to a phosphorus plate (PerkinElmer, USA) for 2h. *In vitro* autoradiographic images were obtained using a phosphor imaging system (Cyclone, Packard). After autoradiographic examination, the same mouse brain sections were stained by thioflavin-S to confirm the presence of Aβ plaques. For the staining of thioflavin S, sections were immersed in a 0.125% thioflavin-S solution containing 10% EtOH for 3min and washed in 40% EtOH. After drying, the fluorescent observation was performed by the LSM 510 META (Zeiss, Germany) equipped with a LP 505 filter set (excitation, 405nm; long-pass filter, 505nm).

4.14 Biodistribution experiments in normal mice

A saline (0.1mL, 10% EtOH) solution containing ^{18}F-labeled tracer (370kBq) was injected into the tail vein of ICR mice (five weeks, 20-22g, male). The mice ($n = 5$ for each time point) were executed by decollation at designated time points post injection. The organs of interest were removed and weighed, and radioactivity was counted with an automatic gamma counter (WALLAC/Wizard 1470, USA). The percent dose per gram (%ID/g) of wet tissue was calculated by a comparison of the tissue counts to suitably diluted aliquots of the injected material.

4.15 Partition coefficient determination

The ^{18}F-labeled tracer (740kBq) was added to a premixed suspension containing 3g of *n*-octanol and 3g of PBS (0.05M, pH = 7.4) in a test tube. The test tube was vortexed for 3min at room temperature, and centrifuged for 5min at 3000r/min. Two weighed samples from the *n*-octanol (100μL) and buffer (500μL) layers were measured. The partition coefficient was expressed as the logarithm of the ratio of the count per gram from *n*-octanol versus PBS. Samples from the *n*-octanol layer were repartitioned until consistent partition coefficient values were obtained. The measurement was done in triplicate and repeated three times.

Acknowledgments

The authors thank Dr. Jin Liu (College of Life Science, Beijing Normal University) for assistance in the *in vitro* neuropathological staining. This work was funded by NSFC (21071023).

Appendix A. Supplementary information

Supplementary data related to this article can be found at http://dx.doi.org/10.1016/j.ejmech.2012.08.31.

References

[1] J. A. Hardy, G. A. Higgins, Alzheimer's disease: the amyloid cascade hypothesis, Science 256 (1992) 184-185.

[2] D. J. Selkoe, The origins of Alzheimer disease: a is for amyloid, JAMA: the Journal of the American Medical Association 283 (2000) 1615-1617.

[3] J. Hardy, D. J. Selkoe, The amyloid hypothesis of Alzheimer's disease: progress and problems on the road to therapeutics, Science 297 (2002) 353-356.

[4] C. C. Rowe, V. L. Villemagne, Brain amyloid imaging, Journal of Nuclear Medicine 52 (2011) 1733-1740.

[5] D. J. Selkoe, Imaging Alzheimer's amyloid, Nature Biotechnology 18 (2000) 823-824.

[6] L. Cai, R. B. Innis, V. W. Pike, Radioligand development for PET imaging of beta-amyloid (Abeta)-current status, Current Medicinal Chemistry 14 (2007) 19-52.

[7] H. F. Kung, S. R. Choi, W. Qu, W. Zhang, D. Skovronsky, 18F stilbenes and styrylpyridines for PET imaging of A beta plaques in Alzheimer's disease: a miniperspective, Journal of Medicinal Chemistry 53 (2010) 933-941.

[8] C. A. Mathis, Y. Wang, W. E. Klunk, Imaging beta-amyloid plaques and neurofibrillary tangles in the aging human brain, Current Pharmaceutical Design 10 (2004) 1469-1492.

[9] C. A. Mathis, Y. Wang, D. P. Holt, G. F. Huang, M. L. Debnath, W. E. Klunk, Synthesis and evaluation of ^{11}C-labeled 6-substituted 2-arylbenzothiazoles as amyloid imaging agents, Journal of Medicinal Chemistry 46 (2003) 2740-2754.

[10] W. E. Klunk, H. Engler, A. Nordberg, Y. Wang, G. Blomqvist, D. P. Holt, M. Bergstrom, I. Savitcheva, G. F. Huang, S. Estrada, B. Ausen, M. L. Debnath, J. Barletta, J. C. Price, J. Sandell, B. J. Lopresti, A. Wall, P. Koivisto, G. Antoni, C. A. Mathis, B. Langstrom, Imaging brain amyloid in Alzheimer's disease with Pittsburgh compound-B, Annals of Neurology 55 (2004) 306-319.

[11] M. Ono, A. Wilson, J. Nobrega, D. Westaway, P. Verhoeff, Z. P. Zhuang, M. P. Kung, H. F. Kung, ^{11}C-labeled stilbene derivatives as Abeta-aggregate-specific PET imaging agents for Alzheimer's disease, Nuclear Medicine and Biology 30 (2003) 565-571.

[12] N. P. Verhoeff, A. A. Wilson, S. Takeshita, L. Trop, D. Hussey, K. Singh, H. F. Kung, M. P. Kung, S. Houle, In-vivo imaging of Alzheimer disease beta-amyloid with [^{11}C]SB-13 PET, The American Journal of Geriatric Psychiatry 12 (2004) 584-595.

[13] A. E. Johnson, F. Jeppsson, J. Sandell, D. Wensbo, J. A. Neelissen, A. Jureus, P. Strom, H. Norman, L. Farde, S. P. Svensson, AZD2184: a radioligand for sensitive detection of beta-amyloid deposits, Journal of Neurochemistry 108 (2009) 1177-1186.

[14] S. Nyberg, M. E. Jonhagen, Z. Cselenyi, C. Halldin, P. Julin, H. Olsson, Y. Freund-Levi, J. Andersson, K. Varnas, S. Svensson, L. Farde, Detection of amyloid in Alzheimer's disease with positron emission tomography using [^{11}C]AZD2184, European Journal of Nuclear Medicine and Molecular Imaging 36 (2009) 1859-1863.

[15] J. D. Andersson, K. Varnas, Z. Cselenyi, B. Gulyas, D. Wensbo, S. J. Finnema, B. M. Swahn, S. Svensson,

S. Nyberg, L. Farde, C. Halldin, Radiosynthesis of the candidate beta-amyloid radioligand [(11)C] AZD2184: positron emission tomography examination and metabolite analysis in cynomolgus monkeys, Synapse 64 (2010) 733-741.

[16] C. C. Rowe, U. Ackerman, W. Browne, R. Mulligan, K. L. Pike, G. O' Keefe, H. Tochon-Danguy, G. Chan, S. U. Berlangieri, G. Jones, K. L. Dickinson-Rowe, H. P. Kung, W. Zhang, M. P. Kung, D. Skovronsky, T. Dyrks, G. Holl, S. Krause, M. Friebe, L. Lehman, S. Lindemann, L. M. Dinkelborg, C. L. Masters, V. L. Villemagne, Imaging of amyloid beta in Alzheimer's disease with ^{18}F-BAY94-9172, a novel PET tracer: proof of mechanism, Lancet Neurology 7 (2008) 129-135.

[17] G. J. O' Keefe, T. H. Saunder, S. Ng, U. Ackerman, H. J. Tochon-Danguy, J. G. Chan, S. Gong, T. Dyrks, S. Lindemann, G. Holl, L. Dinkelborg, V. Villemagne, C. C. Rowe, Radiation dosimetry of beta-amyloid tracers ^{11}C-PiB and ^{18}F-BAY94-9172, Journal of Nuclear Medicine 50 (2009) 309-315.

[18] M. Koole, D. M. Lewis, C. Buckley, N. Nelissen, M. Vandenbulcke, D. J. Brooks, R. Vandenberghe, K. Van Laere, Whole-body biodistribution and radiation dosimetry of 18F-GE067: a radioligand for *in vivo* brain amyloid imaging, Journal of Nuclear Medicine 50 (2009) 818-822.

[19] S. R. Choi, G. Golding, Z. Zhuang, W. Zhang, N. Lim, F. Hefti, T. E. Benedum, M. R. Kilbourn, D. Skovronsky, H. F. Kung, Preclinical properties of ^{18}F-AV-45: a PET agent for Abeta plaques in the brain, Journal of Nuclear Medicine 50 (2009) 1887-1894.

[20] K. J. Lin, W. C. Hsu, I. T. Hsiao, S. P. Wey, L. W. Jin, D. Skovronsky, Y. Y. Wai, H. P. Chang, C. W. Lo, C. H. Yao, T. C. Yen, M. P. Kung, Whole-body biodistribution and brain PET imaging with [^{18}F]AV-45, a novel amyloid imaging agent-a pilot study, Nuclear Medicine and biology 37 (2010) 497-508.

[21] D. F. Wong, P. B. Rosenberg, Y. Zhou, A. Kumar, V. Raymont, H. T. Ravert, R. F. Dannals, A. Nandi, J. R. Brasic, W. Ye, J. Hilton, C. Lyketsos, H. F. Kung, A. D. Joshi, D. M. Skovronsky, M. J. Pontecorvo, *In vivo* imaging of amyloid deposition in Alzheimer disease using the radioligand ^{18}F-AV-45 (florbetapir [corrected] F 18), Journal of Nuclear Medicine 51 (2010) 913-920.

[22] A. Jureus, B. M. Swahn, J. Sandell, F. Jeppsson, A. E. Johnson, P. Johnstrom, J. A. Neelissen, D. Sunnemark, L. Farde, S. P. Svensson, Characterization of AZD4694, a novel fluorinated Abeta plaque neuroimaging PET radioligand, Journal of Neurochemistry 114 (2010) 784-794.

[23] M. Cui, M. Ono, H. Kimura, B. Liu, H. Saji, Novel quinoxaline derivatives for *in vivo* imaging of beta-amyloid plaques in the brain, Bioorganic & Medicinal Chemistry Letters 21 (2011) 4193-4196.

[24] K. A. Stephenson, R. Chandra, Z. P. Zhuang, C. Hou, S. Oya, M. P. Kung, H. F. Kung, Fluoro-pegylated (FPEG) imaging agents targeting Abeta aggregates, Bioconjugate chemistry 18 (2007) 238-246.

[25] M. P. Kung, C. Hou, Z. P. Zhuang, B. Zhang, D. Skovronsky, J. Q. Trojanowski, V. M. Lee, H. F. Kung, IMPY: an improved thioflavin-T derivative for *in vivo* labeling of beta-amyloid plaques, Brain Research 956 (2002) 202-210.

[26] Z. P. Zhuang, M. P. Kung, A. Wilson, C. W. Lee, K. Plossl, C. Hou, D. M. Holtzman, H. F. Kung, Structure-activity relationship of imidazo[1,2-a]pyridines as ligands for detecting beta-amyloid plaques in the brain, Journal of Medicinal Chemistry 46 (2003) 237-243.

[27] A. B. Newberg, N. A. Wintering, K. Plossl, J. Hochold, M. G. Stabin, M. Watson, D. Skovronsky, C. M.

Clark, M. P. Kung, H. F. Kung, Safety, biodistribution, and dosimetry of 123I-IMPY: a novel amyloid plaque-imaging agent for the diagnosis of Alzheimer's disease, Journal of Nuclear Medicine 47 (2006) 748-754.

[28] M. Cui, M. Ono, H. Kimura, B. Liu, H. Saji, Synthesis and structure-affinity relationships of novel dibenzylideneacetone derivatives as probes for beta-amyloid plaques, Journal of Medicinal Chemistry 54 (2011) 2225-2240.

[29] Y. Cheng, W. H. Prusoff, Relationship between the inhibition constant (K1) and the concentration of inhibitor which causes 50 per cent inhibition (I50) of an enzymatic reaction, Biochemical Pharmacology 22 (1973) 3099-3108.

(E)-5-styryl-1H-indole and (E)-6-styrylquinoline Derivatives Serve as Probes for β-amyloid Plaques[*]

Abstract We report the synthesis and biological evaluation of novel (E)-5-styryl-1H-indole and (E)-6-styrylquinoline derivatives as probes for imaging β-amyloid (Aβ) plaques. These derivatives showed binding affinities for $A\beta_{1-40}$ aggregates with K_i values varying from 4.1 to 288.4nM. (E)-5-(4-iodostyryl)-1H-indole (**8**) clearly stained Aβ plaques in the brain sections of Alzheimer's disease (AD) model mice (APP/PS1). Furthermore, autoradiography for [^{125}I]**8** displayed intense and specific labeling of Aβ plaques in the brain sections mentioned above with low background. In biodistribution experiments using normal mice [^{125}I]**8** showed high initial brain uptake followed by rapid washout (4.27% ID/g and 0.64% ID/g at 2min and 30min post injection, respectively). These findings suggests that [^{123}I]**8** may be a potential SPECT imaging agent for detecting Aβ plaques in AD brain.

Key words Alzheimer's disease, β-amyloid plaques, binding affinity, imaging agent, SPECT

1 Introduction

Alzheimer's disease (AD) is a kind of irreversible, progressive brain disease characterized by dementia, cognitive impairment and memory loss. Although currently the pathogenesis of AD is not completely understood, it is generally accepted that β-amyloid (Aβ) plaques is considered to be one of the biomarkers for early diagnosis of AD[1-3]. Therefore, *in vivo* imaging agent for Aβ plaques applicable for PET (positron emission tomography) or SPECT (single photon emission computed tomography) would be very useful for early diagnosis of AD and provide significant information to evaluate the efficacy of AD therapies[4,5].

To date, several radiolabeled ligands have been developed as imaging probes for Aβ plaques[6]. For example, [^{11}C]SB-13[7,8], [^{18}F]BAY94-9172[9] and [^{18}F]AV-45[10,11] derived from Congo Red (CR), [^{11}C]PIB[12,13], [^{18}F]GE-067[14] and [^{123}I]IMPY[15-17] derived from thioflavin T (ThT) (Fig. 1). However, [^{123}I]IMPY, the only SPECT tracer tested in human studies, has failed because of its low *in vivo* stability and its insufficient target-to-background ratio. In comparison with PET, SPECT is a more widely accessible and cost-effective technique in terms of routine diagnostic use. Consequently, the development of more useful imaging agents for Aβ plaques labeled with ^{123}I ($T_{1/2}$, 13h, 159keV) or ^{99m}Tc ($T_{1/2}$, 6h, 140keV) for SPECT has been a critical issue.

Previously, we successfully developed a series of novel imaging agents for β-amyloid plaques

---

[*] Copartner: Yang Yang, Jia Hongmei. Reprinted from *Molecules*, 2012, 17(4):4252-4265.

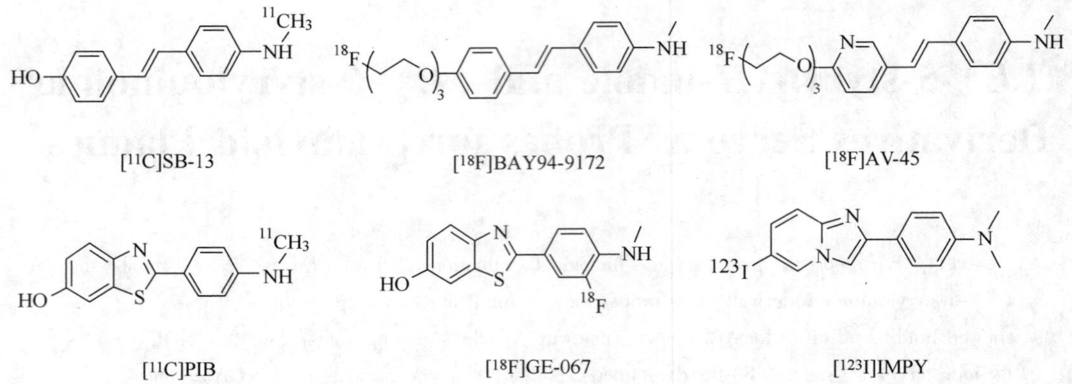

Fig. 1 Chemical structures of Aβ imaging probes for clinical study

based on the N-benzoylindole core which showed high binding affinities with K_i values in the nM range[18]. The brain uptake of these derivatives was encouraging, but their washout from the brain in normal mice appeared to be relatively slow. Qu et al. have developed indolylphenylacetylenes as potential Aβ plaques imaging agent, and the use of indolyl groups may improve the brain kinetics for β-amyloid imaging agents[19]. Recently, Watanabe et al. have developed phenylindoles for image β-amyloid in brain, these derivatives demonstrated high binding affinities to Aβ$_{1-42}$ aggregates[20]. Following these successful results, we applied highly conjugated (E)-5-styryl-1H-indole as a core structure for Aβ imaging agents to explore more useful candidates with favorable pharmacokinetics as Aβ imaging probes, and developed (E)-6-styrylquinoline derivatives for further studies (Fig. 2). Reported herein are the synthesis and biological evaluation of novel (E)-5-styryl-1H-indole and (E)-6-styrylquinoline derivatives and especially, two radioiodinated derivatives as potential SPECT tracers for imaging β-amyloid plaques in the brain.

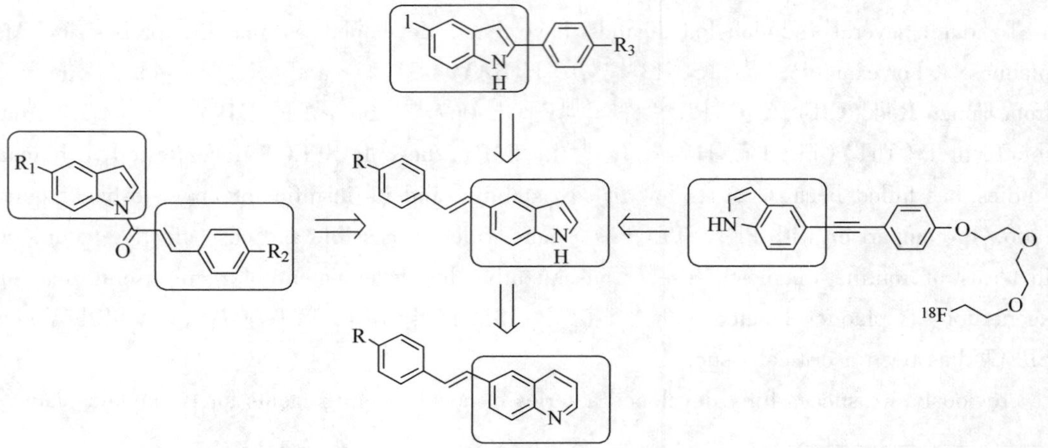

Fig. 2 Design considerations of (E)-5-styryl-1H-indole and (E)-6-styrylquinoline derivatives

2 Results and Discussion

2.1 Chemistry and radiochemistry

The synthetic route to the (E)-5-styryl-1H-indole and (E)-6-styrylquinoline derivatives is shown in Scheme 1.

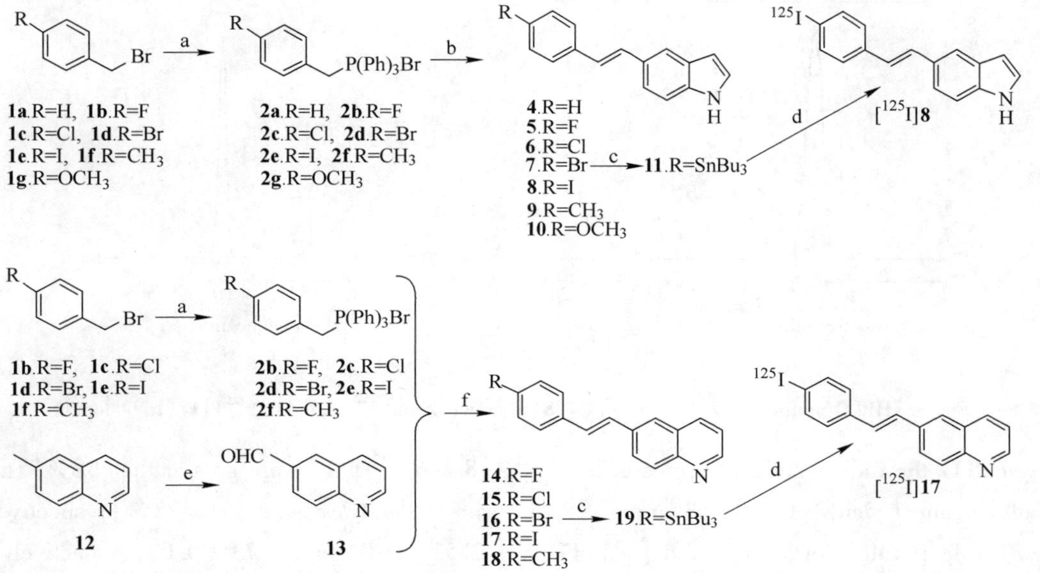

Scheme 1 Synthetic route of (E)-5-styryl-1H-indole and (E)-6-Styrylquinoline derivatives

Reagents and Conditions: a—PPh$_3$, xylene, reflux; b—1H-indole-5-carbaldehyde, CH$_3$ONa, CH$_3$OH; c—(Bu$_3$Sn)$_2$, (Ph$_3$P)$_4$Pd, toluene, reflux; d—[^{125}I]NaI, H$_2$O$_2$, HCl, rt; e—SeO$_2$, 160℃, 12h; f—CH$_3$ONa, CH$_3$OH

The key step was the base-catalyzed Wittig reaction between substituted triphenyl phosphonium ylides **2a-g** and 1H-indole-5-carbaldehyde or quinoline-6-carbaldehyde. The tributyltin derivatives **11, 19** were prepared in yields of 22.3% and 28.6%, respectively, from the bromoprecursors **7, 16** using an exchange reaction catalyzed by Pd(0). [^{125}I]**8** and [^{125}I]**17** were prepared via a iodo-destannylation reaction using hydrogen peroxide as the oxidant. The products were purified by radio-HPLC using a reverse-phase column and mobile phase consisting of acetonitrile with a flow rate of 1mL/min. In order to identify the radiotracer, the non-radioactive **8** and **17** were co-injected and co-eluted with the corresponding radioactive product, respectively. Their HPLC profiles using acetonitrile and water (90∶10 v/v) as mobile phase at a flow rate of 1mL/min are present in Fig. 3.

From Fig. 3, the retention times of non-radioactive **8** and [^{125}I]**8** were observed to be 6.45min and 6.89min, respectively. The retention times of non-radioactive **17** and [^{125}I] **17** were observed to be 19.56min and 19.94min, respectively. The differences in retention time were in good agreement with the time lag which corresponds with the volume and flow rate within the distance between the UV and radioactive detector of our HPLC system. After purification

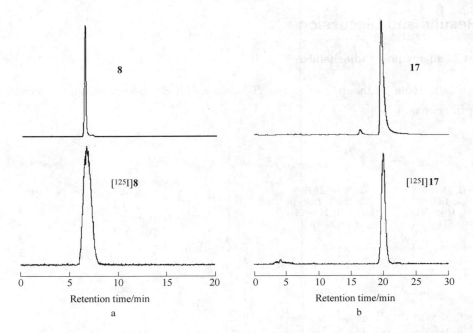

Fig. 3 HPLC profiles of **8** (a, top), [^{125}I]**8**(a, bottom) and **17**(b, top), [^{125}I]**17**(b, bottom)

by HPLC, the radiochemical purities of both [^{125}I]**8** and [^{125}I]**17** were greater than 98%. The radiochemical yields of [^{125}I]**8** and [^{125}I]**17** were 48%-67% and 61%-78%, respectively. The lg D values of [^{125}I]**8** and [^{125}I]**17** were 2.52±0.04 and 2.73±0.03, respectively, which are in the appropriate range for brain imaging agents indicative of good permeability through the blood-brain barrier (BBB).

2.2 *In vitro* binding studies using the aggregated Aβ$_{1-40}$

The affinity of (E)-5-styryl-1H-indole and (E)-6-styrylquinoline derivatives for Aβ$_{1-40}$ aggregates was determined by competition binding assay using [^{125}I]TZDM as radio-ligand. TZDM was also screened using the same competition assay for comparison. The K_i values shown in Table 1 were varied from 4.1 to 288.4 nM suggesting that all these compounds share the same binding site with ThT. The K_i value of TZDM was 4.2 nM, which is comparable to that of previously reported in the literature (K_i = 0.9 nM)[21]. (E)-5-styryl-1H-indole (**4**) without any substituents showed moderate binding affinity (K_i = 25.1 nM). Introducing a F, Cl or OCH$_3$ group at the *para*-position of the phenyl ring decreased the binding affinity (K_i = 89.3 nM, 51.5 nM and 32.4 nM for compounds **5**, **6** and **10**, respectively), while introducing a Br or CH$_3$ group at the same position increased the affinity (K_i = 16.3 nM and 15.8 nM for compounds **7** and **9**, respectively). It is noteworthy that compound **8** with a iodo group showed K_i value of 4.1 nM, which is comparable with that of TZDM. In general, (E)-5-styryl-1H-indole derivatives showed slightly better potency in binding to Aβ$_{1-40}$ aggregates than (E)-6-styrylquinoline derivatives. Since derivatives **8** and **17** with iodine at the *para*-position of the phenyl ring displayed nanomolar affinities for $A\beta_{1-40}$ aggregates, we prepared $[^{125}I]\mathbf{8}$ and $[^{125}I]\mathbf{17}$ for further evaluation as potential ligands for ^{123}I-labeled SPECT imaging agents.

Table 1 K_i values of (E)-5-styryl-1H-indole and (E)-6-styrylquinoline derivatives for $A\beta_{1-40}$ aggregates against $[^{125}I]$TZDM

Compound	$K_i^{①}$/nM	Compound	$K_i^{①}$/nM
4	25.1 ± 2.1	10	32.4 ± 1.9
5	89.3 ± 2.6	14	270.4 ± 1.5
6	51.5 ± 1.0	15	45.0 ± 1.3
7	16.3 ± 1.7	16	23.5 ± 1.3
8	4.1 ± 0.2	17	8.6 ± 1.2
9	15.8 ± 1.5	18	288.4 ± 1.3
TZDM	4.2 ± 0.4	TZDM②	0.9 ± 0.2

① Measured in triplicate with results given as the mean ± SD;
② Data from Ref. [21].

2.3 *In vitro* fluorescent staining of amyloid plaques in brain sections from transgenic mouse

To confirm the binding affinities of these derivatives for Aβ plaques in the brain, *in vitro* fluorescent staining of brain sections (8μm) from a transgenic model mouse (APP/PS1, 12 months, male) was carried out with compound **8**. As shown in Fig. 4, many fluorescence spots were observed in the brain sections of transgenic mice (Fig. 4b). The fluorescent labeling pattern was consistent with that observed with thioflavin-S (Fig. 4a). These results suggested that **8** show specific binding to Aβ plaques in the transgenic model mouse brain.

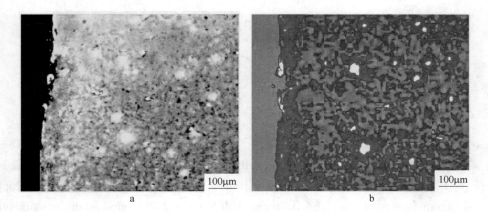

Fig. 4 The labeled plaques were confirmed by staining of the adjacent sections by thioflavin-S (a) and fluorescence staining of compound **8** on AD model mouse sections from the cortex (b)

2.4 *In vitro* labeling of brain sections from transgenic mouse by autoradiography

The results of *in vitro* autoradiography of $[^{125}I]\mathbf{8}$ in the brain sections of a transgenic model mouse (APP/PS1, 12 months, male) are shown in Fig. 5. $[^{125}I]\mathbf{8}$ showed excellent labeling of

Aβ plaques in the cortex region of the brain sections, and no remarkable accumulation of radioactivity were observed in white matter. The same sections were also stained with thioflavin-S and the localizations of Aβ plaques were in accord with the results of autoradiography. These results demonstrated that [^{125}I]**8** was specific for Aβ plaques, which were consistent with the high binding affinity of compound 8 to Aβ$_{1\text{-}40}$ aggregates.

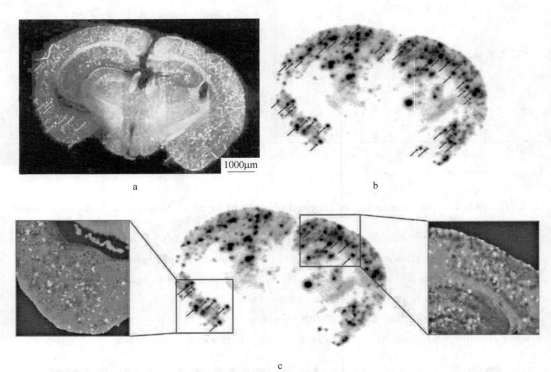

Fig. 5 The presence and distribution of plaques in the sections were confirmed with thioflavin-S staining (a, c) (red arrows); autoradiography of [^{125}I]**8** *in vitro* in Tg model mouse (APP/PS1, 12 months, male) brain sections (b)

2.5 *In vivo* biodistribution studies

In vivo biodistribution studies of [^{125}I]**8** and [^{125}I]**17** were performed in normal mice. The uptake of radiotracer in the organs of interest at different time points after intravenous administration of [^{125}I]**8** and [^{125}I]**17** is summarized in Table 2. [^{125}I]**8** showed high initial brain uptake followed by rapid clearance (4.27% ID/g and 0.28% ID/g at 2min and 60min post injection, respectively). On the other hand, [^{125}I]**17** showed relatively low brain uptake and slow washout (2.05% ID/g and 0.55% ID/g at 2min and 60min post injection, respectively). As compared with previously reported radioiodinated *N*-benzoylindole derivatives[18], radioiodinated (*E*)-5-styryl-1*H*-indole derivative [^{125}I]**8** showed greatly improved brain uptake. Because there are no plaques in normal brain, potential Aβ-specific probe should possess high brain uptake followed by fast washout in normal mice. The brain$_{2\min}$/brain$_{60\min}$ ratio has been used as an index to compare the washout rate from normal brain and select candidate tracers with appropriate kinetics *in vivo*. It was reported that [^{123}I]IMPY showed a high initial brain uptake and fast washout in normal mice (2.88% ID/organ and 0.21% ID/organ at 2min and 60min postinjection, respectively)[16]. The brain$_{2min}$/brain$_{60min}$ ratio of [^{125}I]**8** (15.3) is higher than that of [^{123}I]IMPY (13.7), indicating [^{125}I]**8** may possess suitable pharmacokinetic properties for imaging Aβ plaques in AD brain. Accordingly, [^{125}I]**8** may be comparable or even better for detecting Aβ plaques. Therefore, (*E*)-5-styryl-1*H*-indole derivative [^{125}I]**8**, with nanomolar affinity to Aβ$_{1-40}$ aggregates, excellent BBB permeability as well as fast washout from the normal brain, may be suitable for development as a novel Aβ imaging agent.

Table 2 Biodistribution in normal mice after iv injection of [^{125}I]**8** and [^{125}I]**17** (%ID/g, avg of 5 mice ± SD) and its partition coefficient (*D*)

Organ	2min	15min	30min	60min	120min	240min
[^{125}I]**8** (lg*D* = 2.52 ± 0.04)						
Blood	11.91 ± 0.62	12.29 ± 1.30	7.64 ± 0.62	4.49 ± 0.28	2.93 ± 0.30	1.50 ± 0.13
Brain	4.27 ± 0.49	1.37 ± 0.16	0.64 ± 0.11	0.28 ± 0.06	0.20 ± 0.08	0.10 ± 0.02
Heart	5.76 ± 0.38	3.70 ± 0.40	2.55 ± 0.52	1.59 ± 0.28	1.47 ± 0.48	0.65 ± 0.16
Liver	14.73 ± 0.66	10.66 ± 0.31	7.14 ± 1.13	4.45 ± 0.23	4.19 ± 0.61	3.06 ± 0.36
Spleen	4.38 ± 0.33	4.21 ± 0.26	3.44 ± 0.21	2.44 ± 0.12	1.87 ± 0.28	1.37 ± 0.09
Lung	10.76 ± 0.63	8.22 ± 0.88	5.39 ± 0.83	3.28 ± 0.10	2.30 ± 0.20	1.43 ± 0.52
Kidney	11.66 ± 1.52	14.89 ± 4.23	9.46 ± 1.95	4.65 ± 0.98	1.93 ± 0.76	1.34 ± 0.36
Stomach[①]	1.23 ± 0.56	4.82 ± 0.46	3.50 ± 0.19	1.72 ± 0.21	3.48 ± 0.79	2.31 ± 1.44
Muscle	2.64 ± 0.40	2.01 ± 0.24	1.25 ± 0.04	0.81 ± 0.12	0.61 ± 0.24	0.35 ± 0.09
[^{125}I]**17** (lg*D* = 2.73 ± 0.03)						
Blood	11.39 ± 1.56	6.31 ± 0.51	5.80 ± 0.37	3.86 ± 0.74	1.97 ± 0.35	1.38 ± 0.24
Brain	2.05 ± 0.25	1.18 ± 0.17	0.93 ± 0.13	0.55 ± 0.11	0.26 ± 0.03	0.14 ± 0.02
Heart	7.70 ± 0.86	3.80 ± 0.12	3.21 ± 0.11	2.55 ± 0.21	1.33 ± 0.06	0.80 ± 0.19
Liver	22.45 ± 1.79	9.95 ± 0.18	9.12 ± 0.53	6.97 ± 0.28	3.93 ± 0.46	2.96 ± 0.28
Spleen	5.88 ± 0.30	5.91 ± 0.58	4.66 ± 0.52	4.71 ± 0.97	2.25 ± 0.36	1.59 ± 0.17
Lung	13.56 ± 1.71	6.42 ± 0.47	5.64 ± 0.41	4.55 ± 0.49	2.01 ± 0.18	1.43 ± 0.28
Kidney	15.01 ± 1.56	7.47 ± 1.01	6.57 ± 0.50	4.72 ± 0.69	2.20 ± 0.34	1.64 ± 0.23
Stomach[①]	4.05 ± 0.09	15.84 ± 0.78	8.11 ± 1.21	6.75 ± 0.43	11.12 ± 2.48	7.03 ± 1.51
Muscle	2.78 ± 0.42	1.68 ± 0.20	2.30 ± 0.32	1.40 ± 0.34	0.72 ± 0.21	0.61 ± 0.08

① Expressed as %ID/organ.

3 Experimental

3.1 General

Unless otherwise indicated, all chemicals used in synthesis were commercial products and were used without further purification. Na^{125}I (2200Ci/mmol) was obtained from PerkinElmer Life and Analytical Sciences, USA. The double transgenic (APP/PS1) AD model mouse was obtained from Institute of Laboratory Animal Science, Chinese Academy of Medical Sciences and Comparative Medicine Center of Peking Union Medical College (Beijing, China). ^{1}H-NMR spectra were obtained on Bruker (400MHz) NMR spectrometer at room temperature with TMS as an internal standard. Chemical shifts are reported as δ values relative to internal TMS. Coupling constants are reported in hertz. The multiplicity is defined by s (singlet), d (doublet), t (triplet), and m (multiplet). Mass spectra were acquired using the Surveyor MSQ Plus (ESI) (Waltham, MA, USA) instrument. HPLC was performed on a Shimadzu SCL-10AVP system (Shimadzu Corporation, Kyoto, Japan) which consisted of a binary pump with on-line degasser, a model SPD-10AVP UV-VIS detector operating at a wavelength of 254nm, and a Packard 500TR series flow scintillation analyzer (Packard BioScience Co., Wallingford, CT, USA) with a Alltech Alltima RPC-18 column (5μm, ID = 4.6mm, length = 250mm). The samples were analyzed using acetonitrile and water (90 : 10 v/v) as mobile phase at a flow rate of 1mL/min. The sample was separated using acetonitrile as mobile phase at a flow rate of 1mL/min. All key compounds were proven by analytical HPLC analysis to show ≥95% purity (Supporting information).

3.1.1 General procedure for preparing substituted triphenyl phosphonium ylide 2 (2a-g)

The suitable 4-substituted-1-(bromomethyl) benzene **1a-g** (1mmol) and triphenylphosphine (1mmol) was heated to reflux in xylene (10mL) for about 6h. The mixture was filtered and crude materials were purified by recrystallization with toluene.

3.1.2 General procedure for preparing 4-11, 13

The appropriate substituted compounds **2a-g** (1mmol), 1H-indole-5-carbaldehyde (**3**, 1mmol), and CH$_3$ONa (1mmol) was heated to reflux in CH$_3$OH (12mL) for about 5h. The organic solvent was removed under vacuum. Crude materials were purified by column chromatography on silica gel (petroleum ether/AcOEt, 4/1).

(E)-5-Styryl-1H-indole (**4**). Yield 53.6%, ^{1}H-NMR (DMSO-d$_6$) δ: 7.73(1H, s), 7.58 (2H, d, J = 7.5Hz), 7.43-7.38 (5H, m), 7.32 (1H, d, J = 16.5Hz), 7.22 (1H, t, J = 7.2Hz), 7.12(1H, d, J = 16.4Hz), 6.44 (1H, d, J = 3.0Hz). HRMS m/z C$_{16}$H$_{13}$N found 220.1120/calcd 220.1126([M + H]$^+$). m.p.: 162-163℃.

(E)-5-(4-Fluorostyryl)-1H-indole (**5**). Yield 71.4%, ^{1}H-NMR (DMSO-d$_6$) δ: 7.72(1H, s), 7.62(2H, dd, J_1 = 8.6Hz, J_2 = 5.7Hz), 7.40 (2H, dd, J_1 = 11.4Hz, J_2 = 8.6Hz), 7.34 (1H, d, J = 3.0Hz), 7.27 (1H, d, J = 16.4Hz), 7.19 (2H, t, J = 8.8Hz), 7.12(1H, d, J = 16.4Hz), 6.44(1H, d, J = 3.0Hz). HRMS m/z C$_{16}$H$_{12}$FN found 238.0878/calcd 238.0876 ([M + H]$^+$). m.p.: 179-180℃.

(E)-5-(4-Chlorostyryl)-1H-indole (**6**). Yield 70.2%, ^{1}H-NMR (DMSO-d$_6$) δ: 7.74(1H, s), 7.60(2H, d, J = 8.5Hz), 7.44-7.38(4H, m), 7.35 (1H, d, J = 3.0Hz), 7.34(1H, d, J = 16.4Hz), 7.12(1H, d, J = 16.4Hz), 6.44 (1H, d, J = 3.0Hz). HRMS m/z C$_{16}$H$_{12}$ClN found 254.0733/calcd 254.0737([M + H]$^+$). m.p.: 205-206℃.

(E)-5-(4-Bromostyryl)-1H-indole (**7**). Yield 68.3%, ^{1}H-NMR (DMSO-d$_6$) δ: 7.74(1H, s),7.54(4H, s),7.43-7.40(2H, m),7.36(1H, d, J = 16.4Hz),7.35(1H, d, J = 3.1Hz), 7.11(1H, d, J = 16.4Hz),6.44(1H, d, J = 3.0Hz). HRMS m/z $C_{16}H_{12}BrN$ found 298.0235/calcd 298.0231 ($[M+H]^+$). m. p. : 215-216℃.

(E)-5-(4-Iodostyryl)-1H-indole (**8**). Yield 55.3%, ^{1}H-NMR (DMSO-d_6)δ : 7.73(1H, s),7.70(2H, d, J = 8.4Hz),7.43-7.39(4H, m),7.36(1H, d, J = 16.4Hz),7.34(1H, d, J = 3.1Hz),7.08(1H, d, J = 16.5Hz),6.44(1H, d, J = 2.9Hz). HRMS m/z $C_{16}H_{12}IN$ found 246.1283/calcd 246.1299($[M+H]^+$). m. p. : 211-212℃.

(E)-5-(4-Methoxystyryl)-1H-indole (**9**). Yield 67.7%, ^{1}H-NMR (DMSO-d_6)δ : 7.70(1H, s),7.47(2H, d, J = 8.0Hz),7.39(2H, dd, J_1 = 11.5Hz, J_2 = 8.6Hz),7.33(1H, d, J = 3.0Hz),7.25(1H, d, J = 16.4Hz),7.17(2H, d, J = 8.0Hz),7.08(1H, d, J = 16.4Hz),6.44(1H, d, J = 3.0Hz),2.31(3H, s). HRMS m/z $C_{17}H_{15}N$ found 234.1126/calcd 234.1126 ($[M+H]^+$). m. p. : 176-177℃.

(E)-5-(4-Methoxystyryl)-1H-indole (**10**). Yield 65.2%, ^{1}H-NMR (DMSO-d_6)δ : 7.68(1H, s),7.51(2H, d, J = 8.6Hz),7.40-7.35(2H, m),7.33(1H, d, J = 3.0Hz),7.16(1H, d, J = 16.4Hz),7.06(1H, d, J = 16.4Hz),6.93(2H, d, J = 8.5Hz),6.42(1H, d, J = 3.0Hz),3.77(3H, s). HRMS m/z $C_{17}H_{15}NO$ found 250.1241/calcd 250.1232 ($[M+H]^+$). m. p. : 163-164℃.

(E)-5-(4-(Tributylstannyl)styryl)-1H-indole (**11**). A mixture of 7 (29.8mg, 0.1mmol), bis(tributyltin)(290mg, 0.5mmol), and Pd(Ph$_3$P)$_4$(12.0mg, 0.01mmol) in toluene (15mL) was stirred at 110℃ overnight. After removing the solvent in vacuo, the crude products were purified by column chromatography (petroleum ether/AcOEt, 20/1) to give **11** as a light-yellow-colored solid with a yield of 22.3%. ^{1}H-NMR (CDCl$_3$)δ : 8.87 (1H, d, J = 2.8Hz),8.15 (1H, d, J = 8.1Hz),8.08 (1H, d, J = 8.7Hz),7.98(1H, dd, J_1 = 8.9Hz, J_2 = 1.6Hz),7.82 (1H, s),7.52(4H, dd, J_1 = 19.6Hz, J_2 = 8.5Hz),7.40(1H, dd, J_1 = 8.2Hz, J_2 = 4.2Hz), 7.28(2H, dd, J_1 = 16.4Hz, J_2 = 8.0Hz),1.64-1.55(6H, m),1.40-1.35(6H, m),1.16-1.11 (6H, m), 0.97-0.90 (9H, m). ESI-MS m/z $C_{28}H_{39}NSn$ found 510.4/calcd 509.2 ($[M+H]^+$).

Quinoline-6-carbaldehyde (**13**). Quinoline-6-carbaldehyde (**13**) was prepared from 6-methylquinoline (**12**) according to the previously reported procedure[22]. 6-Methylquinoline (**12**, 4.0g, 27.6mmol) was heated to 160℃ and selenium dioxide (2.0g, 18.4mmol) was added. The mixture was stirred for 16h, cooled to room temperature, and diluted with ethyl acetate (30mL). The solution was decanted, and the residue was extracted with ethylacetate (20mL × 2). The combined organic phase was concentrated, and the residue was purified by column chromatography on silica gel (petroleum ether/AcOEt, 4/1) to give 13 as a light gray solid (1.3g, 30%). ^{1}H-NMR (CDCl$_3$)δ : 10.21 (1H, s), 9.06 (1H, dd, J_1 = 4.2Hz, J_2 = 1.6Hz),8.38-8.35(2H, m),8.27-8.24(2H, m),7.55(1H, dd, J_1 = 8.3Hz, J_2 = 4.3Hz).

3.1.3 General procedure for preparing 14-19

The suitable substituted compounds **2b-f** (1mmol),**13** (1mmol),CH$_3$ONa (1mmol) was heated to reflux in CH$_3$OH (12mL) for about 6h. The organic solvent was removed under vacuum. Crude materials were washed by water and purified by column chromatography on silica gel (petroleum ether/AcOEt, 6/1).

(E)-6-(4-Fluorostyryl)quinoline (**14**). Yield 66.8%, ^{1}H-NMR (DMSO-d_6) δ : 8.86 (1H, dd, J_1 = 8.9Hz, J_2 = 1.4Hz), 8.35(1H, d, J = 8.0Hz), 8.11(1H, dd, J_1 = 8.8Hz, J_2 = 1.5Hz), 8.07(1H, s), 8.01(1H, d, J = 8.8Hz), 7.73(2H, dd, J_1 = 8.7Hz, J_2 = 5.7Hz), 7.54(1H, dd, J_1 = 8.2Hz, J_2 = 4.2Hz), 7.45(2H, dd, J_1 = 16.5Hz, J_2 = 7.7Hz), 7.26(1H, t, J = 8.8Hz). HRMS m/z $C_{17}H_{12}FN$ found 250.1039/calcd 250.1032([M+H]$^+$). m.p.: 120-121℃.

(E)-6-(4-Chlorostyryl)quinoline (**15**). Yield 72.3%, ^{1}H-NMR (DMSO-d_6) δ : 8.86(1H, d, J = 3.4Hz), 8.35(1H, d, J = 7.9Hz), 8.12(1H, d, J = 8.8Hz), 8.08(1H, s), 8.01(1H, d, J = 8.7Hz), 7.70(2H, d, J = 8.2Hz), 7.53(2H, dd, J_1 = 8.4 Hz, J_2 = 4.1Hz), 7.49-7.47 (4H, m). HRMS m/z $C_{17}H_{12}ClN$ found 266.0732/calcd 266.0737([M+H]$^+$). m.p.: 129-130℃.

(E)-6-(4-Bromostyryl)quinoline (**16**). Yield 73.9%, ^{1}H-NMR (DMSO-d_6) δ : 8.87(1H, dd, J_1 = 4.1Hz, J_2 = 1.5Hz), 8.36(1H, d, J = 8.0Hz), 8.14(1H, dd, J_1 = 8.8Hz, J_2 = 1.6Hz), 8.09(1H, s), 8.01(1H, d, J = 8.8Hz), 7.63(4H, dd, J_1 = 13.3Hz, J_2 = 8.8Hz), 7.54(1H, dd, J_1 = 8.1Hz, J_2 = 4.2Hz), 7.48(2H, dd, J_1 = 16.6Hz, J_2 = 13.3Hz). HRMS m/z $C_{17}H_{12}BrN$ found 310.0217/calcd 310.0231([M+H]$^+$). m.p.: 143-144℃.

(E)-6-(4-Iodostyryl)quinoline (**17**). Yield 70.2%, ^{1}H-NMR (DMSO-d_6) δ : 8.87(1H, d, J = 4.1Hz), 8.35(1H, d, J = 8.0Hz), 8.12(1H, d, J = 8.9Hz), 8.09(1H, s), 8.01(1H, d, J = 8.9Hz), 7.78(2H, d, J = 8.0Hz), 7.56-7.40(5H, m). HRMS m/z $C_{17}H_{12}IN$ found 358.0100/calcd 358.0093([M+H]$^+$). m.p.: 173-174℃.

(E)-6-(4-Methylstyryl)quinoline (**18**). Yield 71.7%, ^{1}H-NMR (DMSO-d_6) δ : 8.85(dd, J_1 = 4.0Hz, J_2 = 1.4Hz, 1H), 8.36(d, J = 8.1Hz, 1H), 8.11(dd, J_1 = 8.8Hz, J_2 = 1.6Hz, 1H), 8.06(s, 1H), 8.01(d, J = 8.8Hz, 1H), 7.57(d, J = 8.0Hz, 2H), 7.54(dd, J_1 = 8.3Hz, J_2 = 4.3Hz, 1H), 7.41(s, 2H), 7.23(d, J = 7.9Hz, 2H). HRMS m/z $C_{17}H_{12}IN$ found 246.1299/calcd 246.1283([M+H]$^+$). m.p.: 125-126℃.

(E)-6-(4-(Tributylstannyl)styryl)quinolin (**19**). The same reaction described above to prepare **11** was used, and a primrose yellow-colored solid of **19** was obtained in a yield of 28.6% from **16**. ^{1}H-NMR (CDCl$_3$) δ : 8.87(1H, d, J = 2.8Hz), 8.15(1H, d, J = 8.1Hz), 8.08(1H, d, J = 8.7Hz), 7.98(1H, dd, J_1 = 8.9Hz, J_2 = 1.6Hz), 7.82(1H, s), 7.52(4H, dd, J_1 = 19.6Hz, J_2 = 8.5Hz), 7.40(1H, dd, J_1 = 8.2Hz, J_2 = 4.2Hz), 7.28(2H, dd, J_1 = 16.4Hz, J_2 = 8.0Hz), 1.64-1.55(6H, m), 1.40-1.34(6H, m), 1.16-1.13(6H, m), 0.96-0.91(9H, m). ESI-MS m/z $C_{29}H_{39}NSn$ found 522.6/calcd 521.2([M+H]$^+$).

3.1.4 Preparation of radioiodinated ligands

The radioiodinated compounds [^{125}I]**8** and [^{125}I]**17** were prepared from the corresponding tributyltin derivatives by an iododestannylation according to the procedure described previously[21]. Briefly, H_2O_2(3%, 100μL) was added to a mixture of a tributyltin derivative (0.1mg/100μL in ethanol), sodium [^{125}I]iodide (specific activity 2200Ci/mmol), and 1M HCl (100μL) in a sealed vial. The reaction was allowed to proceed at room temperature for 15 min and then quenched by addition of saturated $NaHSO_3$ solution (50μL). The reaction mixture, after neutralization with 1M NaOH, was purified by HPLC using a Alltech Alltima RPC-18 column (250mm × 4.6mm, 5μm) and mobile phase consisting of acetonitrile with a flow rate of 1.0mL/min. Finally, the radiochemical identity of the radioiodinated ligands were verified by co-injection and co-elution with non-radioactive 8 and 17 from HPLC profiles (Alltech Alltima RPC-18 column, 250mm × 4.6mm, 5μm, CH_3CN/H_2O = 9/1 at the flow rate of 1.0mL/min). The desired fractions containing the product were collectd and evaporated to dryness and redissolved in 100% ethanol. The final products were stored at −20℃ for further studies.

3.2 Partition coefficient determination

The determination of partition coefficients of $[^{125}I]8$ and $[^{125}I]17$ was performed according to the procedure previously reported with some modifications[23]. Ligand $[^{125}I]8$ or $[^{125}I]17$ (~5μCi) was mixed with 3mL each of n-octanol and PBS (0.02M, pH 7.4) in a test tube. The test tube was vortexed for 5min at room temperature, followed by centrifugation for 10min at 3000r/min. Two weighed samples from the n-octanol (50μL) and buffer layers (400μL) were counted in a γ-counter. The partition coefficient was expressed as the logarithm of the ratio of the counts per gram from n-octanol *versus* that of PBS. Samples from the n-octanol layer were repartitioned until consistent partition coefficient values were obtained. The measurements were done in triplicate and repeated three times.

3.3 *In vitro* binding studies using the aggregated $A\beta_{1-40}$

The lyophilized white powder of β-amyloid(1-40) were purchased from AnaSpec (San Jose, CA, USA). After reconstituted by adding basic buffer (1% NH_4OH. 60-70μL) to β-amyloid (1-40)(1mg) aggregation of $A\beta_{1-40}$ was carried out by gently dissolving $A\beta_{1-40}$ (0.25mg/mL) in a PBS buffer (pH 7.4). The solution was incubated at 37℃ for 72h with gentle and constant shaking. Binding studies were carried out according to the procedure described previously with some modifications using $[^{125}I]$TZDM as the radiolabeled standard[24]. Briefly, the competition binding assays were performed by mixing $A\beta_{1-40}$ aggregates (100μL), and $[^{125}I]$TZDM (100μL) in appropriate concentration (0.02nM, diluted in 10% EtOH), test ligand (10^{-5}-10^{-10}M, 100μL) and PBS (0.02M, pH 7.4, 700μL) in a final volume of 1mL. The mixture was incubated at 37℃ for 2h. Then the bound and free radioactivities were separated by vacuum filtration through Whatman GF/B glass filters via a Brandel Mp-48T cell harvester followed by 3 × 4mL washes with PBS (0.02M, pH 7.4, 4℃) containing 10% ethanol at room temperature. Filters containing the bound ^{125}I ligand were counted in a γ-counter (WALLAC Wizard 1470, PerkinElmer Life Sciences, Waltham, MA, USA) with 75% counting efficiency. The IC_{50} values were determined using GraphPad Prism 5.0, and those for the inhibition constant (K_i) were calculated using the Cheng Prusoff equation: $K_i = IC_{50}/(1 + [L]/K_d)$[25].

3.4 *In vitro* fluorescent staining of amyloid plaques in brain sections from transgenic mouse

Paraffin-embedded brain sections of transgenic model mouse (8μm, APP/PS1, 12 months, male) were used for the *in vitro* fluorescent staining of amyloid plaques. The brain sections were deparaffinized with xylene, ethanol and distilled water. After immersion in PBS (0.02M, pH 7.4) for 30min, the brain sections were incubated with 20% ethanol solution (1μM) of compound **8** for 10min. The localization of plaques was confirmed by staining with thioflavin-S on the adjacent sections. Finally, the sections were washed with 40% ethanol and PBS (0.02M, pH 7.4). Fluorescent observation was performed by a Stereo Discovery V12 instrument (Zeiss, Oberkochen, Germany) equipped with a LP 505 filter set (excitation, 405nm).

3.5 *In vitro* labeling of brain sections from transgenic mouse by autoradiography

The brain sections mentioned above were incubated with $[^{125}I]$**8** (5μCi/100μL) for about 20min at room temperature. Then the sections were washed with saturated Li_2CO_3 in 40% EtOH for 3min and 40% EtOH for 3min, followed by rinsing with water for 30s. After drying, the ^{125}I-labeled sections were exposed to phosphorus film for 8h and then scanned with the phosphor imaging system (Cyclone, Packard) at the resolution of 600 dpi. The presence and localization of plaques were confirmed by the fluorescent staining with thioflavin-S on the same sections using a Stereo Discovery V12 (Zeiss) instrument equipped with a LP 505 filter set (excitation: 405nm).

3.6 *In vivo* biodistribution in normal mice

In vivo biodistribution studies were performed in KunMing normal mice (female, average weight **18-22g**) and in accordance with the national laws related to the care and experiments on laboratory animal. A saline solution (100μL) containing $[^{125}I]$**8** or $[^{125}I]$**17** (1μCi) was injected directly into the tail vein of mice. The mice ($n=5$ for each time point) were sacrificed at designated time points post-injection. The organs of interest were removed and weighed, and the radioactivity was counted with an automatic γ-counter (WALLAC Wizard 1470).

4 Conclusions

A series of (*E*)-5-styryl-1*H*-indole and (*E*)-6-styrylquinoline based compounds have been synthesized and evaluated as novel imaging probes for Aβ plaques. Compound **8** was found to possess nanomolar affinity for β-amyloid plaques. In autoradiography, $[^{125}I]$**8** clearly labeled amyloid plaques in the cortex region of AD model mice. Moreover, $[^{125}I]$**8** displayed high initial brain uptake and fast clearance in biodistribution studies in normal mice. The findings suggest that the (*E*)-5-styryl-1*H*-indole derivative $[^{123}I]$**8** may be a potential probe for detecting β-amyloid plaques in the AD brain.

Supplementary Materials

Supplementary materials can be accessed at: http://www.mdpi.com/1420-3049/17/4/4252/s1.

Acknowledgments

This work was funded by NSFC (20871021 and 21071023). The authors thank Mengchao Cui (College of Chemistry, Beijing Normal University, Beijing, China) for his kindness in providing the paraffin-embedded brain sections of AD model mice. The authors also thank Jin Liu (College of Life Science, Beijing Normal University) for providing the Zeiss Oberver Z1 equipment.

References

[1] Selkoe, D. J. The origins of Alzheimer disease: Aβ is for amyloid. J. Am. Med. Assoc. 2000, 283, 1615-1617.

[2] Hardy, J. ; Selkoe, D. J. The amyloid hypothesis of Alzheimer's disease: Progress and problems on the road to therapeutics. Science 2002, 297, 353-356.

[3] Hardy, J. A. ; Higgins, G. A. Alzheimer's disease: The amyloid cascade hypothesis. Science 1992, 256, 184-185.

[4] Nordberg, A. PET imaging of amyloid in Alzheimer's disease. Lancet Neurol. 2004, 3, 519-527.

[5] Cai, L. S. ; Innis, R. B. ; Pike, V. W. Radioligand development for PET Imaging of β-amyloid (Aβ)-current status. Curr. Med. Chem. 2007, 14, 19-52.

[6] Mathis, C. A. ; Wang, Y. ; Klunk, W. E. Imaging β-amyloid plaques and neurofibrillary tangles in the aging human brain. Curr. Pharm. Des. 2004, 10, 1469-1492.

[7] Ono, M. ; Wilson, A. ; Nobrega, J. ; Westaway, D. ; Verhoeff, P. ; Zhuang, Z. P. ; Kung, M. P. ; Kung, H. F. ^{11}C-labeled stilbene derivatives as Aβ-aggregate-specific PET imaging agents for Alzheimer's disease. Nucl. Med. Biol. 2003, 30, 565-571.

[8] Verhoeff, N. P. ; Wilson, A. A. ; Takeshita, S. ; Trop, L. ; Hussey, D. ; Singh, K. ; Kung, H. F. ; Kung, M. P. ; Houle, S. In vivo imaging of Alzheimer disease β-amyloid with [^{11}C]SB-13 PET. Am. J. Geriatr. Psychiatry 2004, 12, 584-595.

[9] Rowe, C. C. ; Ackerman, U. ; Browne, W. ; Mulligan, R. ; Pike, K. L. ; O'Keefe, G. ; Tochon-Danguy, H. ; Chan, G. ; Berlangieri, S. U. ; Jones, G. ; et al. Imaging of amyloid β inAlzheimer's disease with ^{18}F-BAY94-9172, a novel PET tracer: Proof of mechanism. Lancet Neurol. 2008, 7, 129-135.

[10] Choi, S. R. ; Golding, G. ; Zhuang, Z. P. ; Zhang, W. ; Lim, N. ; Hefti, F. ; Benedum, T. E. ; Kilbourn, M. R. ; Skovronsky, D. ; Kung, H. F. Preclinical properties of ^{18}F-AV-45: A PET agent for Aβ plaques in the brain. J. Nucl. Med. 2009, 50, 1887-1894.

[11] Kung, H. F. ; Choi, S. R. ; Qu, W. C. ; Zhang, W. ; Skovronsky, D. ^{18}F Stilbenes and styrylpyridines for PET imaging of Aβ plaques in Alzheimer's disease: A miniperspective. J. Med. Chem. 2010, 53, 933-941.

[12] Mathis, C. A. ; Wang, Y. M. ; Holt, D. P. ; Huang, G. F. ; Debnath, M. L. ; Klunk, W. E. Synthesis and evaluation of ^{11}C-labeled 6-substituted 2-arylbenzothiazoles as amyloid imaging agents. J. Med. Chem. 2003, 46, 2740-2754.

[13] Klunk, W. E. ; Engler, H. ; Nordberg, A. ; Wang, Y. M. ; Blomqvist, G. ; Holt, D. P. ; Bergström, M. ; Savitcheva, I. ; Huang, G. F. ; Estrada, S. ; et al. Imaging brain amyloid in Alzheimer's disease with Pittsburgh Compound-B. Ann. Neurol. 2004, 55, 306-319.

[14] Koole, M. ; Lewis, D. M. ; Buckley, C. ; Nelissen, N. ; Vandenbulcke, M. ; Brooks, D. J. ; Vandenberghe, R. ; Laere, K. V. Whole-body biodistribution and radiation dosimetry of ^{18}F-GE067: A radioligand for

in vivo brain amyloid imaging. J. Nucl. Med. 2009, 50, 818-822.

[15] Kung, M. P. ; Hou, C. ; Zhuang, Z. P. ; Zhang, B. ; Skovronsky, D. ; Trojanowski, J. Q. ; Lee, V. M. ; Kung, H. F. IMPY: An improved thioflavin-T derivative for *in vivo* labeling of β-amyloid plaques. Brain Res. 2002, 956, 202-210.

[16] Zhuang, Z. P. ; Kung, M. P. ; Wilson, A. ; Lee, C. W. ; Plössl, K. ; Hou, C. ; Holtzman, D. M. ; Kung, H. F. Structure-activity relationship of imidazo[1,2-a]pyridines as ligands for detecting β-amyloid plaques in the brain. J. Med. Chem. 2003, 46, 237-243.

[17] Newberg, A. B. ; Wintering, N. A. ; Plössl, K. ; Hochold, J. ; Stabin, M. G. ; Watson, M. ; Skovronsky, D. ; Clark, C. M. ; Kung, M. P. ; Kung, H. F. Safety, biodistribution and dosimetry of ^{123}I-IMPY: A novel amyloid plague-imaging agent for the diagnosis of Alzheimer's disease. J. Nucl. Med. 2006, 47, 748-754.

[18] Yang, Y. ; Duan, X. H. ; Deng, J. Y. ; Bing, J. ; Jia, H. M. ; Liu, B. L. Novel imaging agents for β-amyloid plaque based on the *N*-benzoylindole core. Bioorg. Med. Chem. Lett. 2011, 21, 5594-5597.

[19] Qu, W. C. ; Choi, S. R. ; Hou, C. ; Zhuang, Z. P. ; Oya, S. ; Zhang, W. ; Kung, M. P. ; Manchandra, R. ; Skovronsky, D. M. ; Kung, H. F. Synthesis and evaluation of indolinyl-and indolylphenylacetylenes as PET imaging agents for β-amyloid plaques. Bioorg. Med. Chem. Lett. 2008, 18, 4823-4827.

[20] Watanabe, H. ; Ono, M. ; Haratake, M. ; Kobashi, N. ; Saji, H. ; Nakayama, M. Synthesis and characterization of novel phenylindoles as potential probes for imaging of β-amyloid plaques in the brain. Bioorg. Med. Chem. 2010, 18, 4740-4746.

[21] Zhuang, Z. P. ; Kung, M. P. ; Hou, C. ; Skovronsky, D. M. ; Gur, T. L. ; Plössl, K. ; Trojanowski, J. Q. ; Lee, V. M. ; Kung, H. F. Radioiodinated styrylbenzenes and thioflavins as probes for amyloid aggregates. J. Med. Chem. 2001, 44, 1905-1914.

[22] Mikhail, K. ; Dmitry, P. ; Denis, L. ; Dmitry, K. Synthesis and practical use of 1*H*-1,2,3-benzotriazole-5-carboxaldehyde for reductive amination. Synth. Commun. 2005, 35, 2587-2595.

[23] Wu, C. Y. ; Wei, J. J. ; Gao, K. Q. ; Wang, Y. M. Dibenzothiazoles as novel amyloid-imaging agents. Bioorg. Med. Chem. 2007, 15, 2789-2796.

[24] Klunk, W. E. ; Wang, Y. M. ; Huang, G. F. ; Debnath, M. L. ; Holt, D. P. ; Shao, L. ; Hamilton, R. L. ; Ikonomovic, M. D. ; DeKosky, S. T. ; Mathis, C. A. The binding of 2-(4'-methylaminophenyl)benzothiazole to postmortem brain homogenates is dominated by the amyloid component. J. Neurosci. 2003, 23, 2086-2092.

[25] Cheng, Y. C. ; William, H. P. Relationship between the inhibition constant (K_i) and the concentration of inhibitor which causes 50 per cent inhibition (IC_{50}) of an enzymatic reaction. Biochem. Pharmacol. 1973, 22, 3099-3108.

Synthesis and Preliminary Evaluation of ^{18}F-labeled Pyridaben Analogues for Myocardial Perfusion Imaging with PET[*]

Abstract In this study the ^{18}F-labeled pyridaben analogs 2-tertbutyl-4-chloro-5-(4-(2-^{18}F-fluoroethoxy))benzyloxy-2H-pyridazin-3-one (^{18}F-FP1OP) and 2-tertbutyl-4-chloro-5-(4-(2-(2-(2-^{18}F-fluoroethoxy)ethoxy)ethoxy))benzyloxy-2H-pyridazin-3-one (^{18}F-FP3OP) were synthesized, characterized, and evaluated as potential myocardial perfusion imaging (MPI) agents with PET. Methods: The tosylate labeling precursors of 2-tert-butyl-4-chloro-5-(4-(2-tosyloxy-ethoxy))-benzyloxy-2H-pyridazin-3-one (OTs-P1OP), 2-tert-butyl-4-chloro-5-(4-(2-(2-(2-tosyloxy-ethoxy)ethoxy)ethoxy))-benzyloxy-2H-pyridazin-3-one (OTs-P3OP), and the corresponding nonradioactive compounds (^{19}F-FP1OP and ^{19}F-FP3OP) were synthesized and characterized by infrared, ^{1}H nuclear magnetic resonance, ^{13}C nuclear magnetic resonance, and mass spectrometry analysis. ^{18}F-FP1OP and ^{18}F-FP3OP were obtained by 1-step nucleophilic substitution of tosyl with ^{18}F and evaluated as MPI agents *in vitro* (physicochemical properties, stability), ex vivo (autoradiography), and *in vivo* (toxicity and biodistribution in normal mice; cardiac PET in healthy Chinese mini swine and in acute myocardial infarction and chronic myocardial ischemia models). Results: The total radiosynthesis time of both tracers, including final high-pressure liquid chromatography purification, was about 70-90min. Typical decay-corrected radiochemical yields were about 50%, and the radiochemical purities were more than 98% after purification. ^{18}F-FP1OP had lower hydrophilicity and higher water stability than that of ^{18}F-FP3OP. In biodistribution studies, both ^{18}F-FP1OP and ^{18}F-FP3OP had high heart uptake (31.13 ± 6.24 and 31.10 ± 3.72 percentage injected dose per gram at 2min after injection, respectively) and high heart-to-liver, heart-to-lung, and heart-to-blood ratios at all time points after injection. Further autoradiography evaluation of ^{18}F-FP1OP showed that the heart uptake could be blocked effectively by rotenone or nonradioactive ^{19}F-FP1OP. Clear cardiac PET images of ^{18}F-FP1OP were obtained in healthy Chinese mini swine at 2min, 15min, 30min, 60min, and 120min after injection, and the uptake of perfusion deficit areas was much lower than in normal tissue in both acute myocardial infarction and chronic myocardial ischemia models. Conclusion: The ^{18}F-labeled pyridaben analogs reported in this study have high heart uptake and low background uptake in both the mouse model and the Chinese mini swine model. The tracer with the shorter radiolabeling side chain (^{18}F-FP1OP) has better stability, faster clearance from the major organs, and a higher heart-to-liver ratio than the other tracer (^{18}F-FP3OP). On the basis of the promising biologic properties, this mitochondrial complex I-targeted tracer (^{18}F-FP1OP) is worthy to be developed as an MPI agent and to be compared with the other PET MPI agents in the future.

Key words MPI, MC-I, 18F, animal imaging, radiopharmaceuticals

---

[*] Copartner: Mou Tiantian, Zhao Zuoquan, Fang Wei, Peng Cheng, Guo Feng, Ma Yunchuan, Zhang Xianzhong. Reprinted from *The Journal of Nuclear Medicine*, 2012, 53(3):472-479.

Myocardial perfusion imaging (MPI) is a significant method of noninvasive measurement in the diagnosis and prognosis of coronary artery disease. Though ^{99m}Tc-sestamibi has been the gold standard of MPI for decades in nuclear medicine[1], it still has several weaknesses compared with PET[2]. Currently, since ^{18}F has an appropriate physical half-life, good biocompatibility, a suitable atomic radius, and high resolution with PET technology[3], some ^{18}F-labeled lipophilic cations (such as ^{18}F-TPP[4,5], ^{18}F-FBnTP[1,6], and ^{18}F-FERhB[7]) and analogs of mitochondrial complex I (MC-I) inhibitors (such as ^{18}F-FDHR[8], ^{18}F-RP1003/04/05[9], and 2-tert-butyl-4-chloro-5-[4-(2-^{18}F-fluoroethoxymethyl)-benzyloxy]-2H-pyridazin-3-one (BMS-747158-02)[10]) have been reported as potential MPI agents. Most of those compounds have better image quality and a better relationship to true myocardial blood flow than ^{99m}Tc-sestamibi[11,12]. Up to now, BMS-747158-02 (also known by the nonproprietary name *flurpiridaz*) has been the best evaluated ^{18}F-labeled MPI agent. It has high (94%) first-pass extraction fraction[12], its high myocardial uptake is proportional to myocardial blood flow, and it has entered clinical development (phase 3)[11]. The promising properties of BMS-747158-02 will ensure that cardiac PET will be the standard for evaluation of myocardial perfusion in coming years.

Currently, clinical use of approved PET agents for MPI is limited by the inherent properties of radioisotopes, the degree of flow alteration, or the requirement for an on-site cyclotron. The development of an ideal ^{18}F-labeled myocardial perfusion tracer that can be produced and widely distributed by a central cyclotron facility remains a challenge. Mitochondria take up about 20%-30% of the myocardial intracellular volume in the heart[13]. Therefore, the development of mitochondrion-targeted ^{18}F-labeled analogs of MC-I inhibitors as potential MPI agents is of great interest. Several classes of MC-I inhibitors have been reported, such as fenazaquin, S-chromone, tebufenpyrad, and pyridaben; all have the same binding site of the MC-I enzyme as rotenone[14,15]. Our laboratories have been particularly interested in analogs of pyridaben. Recently, we reported a pyridaben analog 2-tert-butyl-5-[4-2-(2-^{18}F-fluroethoxy)ethoxy]benzyloxy-4-chloro-2H-pyridazin-3-one (^{18}F-FP2OP) as a potential MPI agent with promising properties[16]. It has high heart uptake and good heart-to-nontarget ratios. However, ^{18}F-FP2OP is not very stable in water, obviously limiting its application. Accordingly, 2 new tracers with different lengths of radiolabeling side chain were developed to improve stability while retaining the promising properties. Herein, we report the synthesis and evaluation of 2-tertbutyl-4-chloro-5-(4-(2-^{18}F-fluoroethoxy))benzyloxy-2H-pyridazin-3-one (^{18}F-FP1OP) and 2-tertbutyl-4-chloro-5-(4-(2-(2-(2-^{18}F-fluoroethoxy)ethoxy)ethoxy))benzyloxy-2H-pyridazin-3-one (^{18}F-FP3OP) as potential MPI agents.

1 Materials and Methods

1.1 Materials

^{18}F-F$^-$ was obtained from the PET Center of Xuanwu Hospital. No-carrier-added ^{18}F-F$^-$ was trapped on a QMA cartridge (Waters) and eluted with 0.3mL of K_2CO_3 solution (10mg/mL in

H_2O) combined with 1mL of Kryptofix2.2.2. solution (Sigma-Aldrich) (13mg/mL in acetonitrile)[17,18]. Rotenone with 95% purity was purchased from Sigma-Aldrich. Other reagents and solvents were purchased from commercial suppliers. Paper electrophoresis experiments were performed using 0.025M phosphate buffer (pH 7.4) and No.1 filter paper (Xinhua) at 150V for 180min. Reversed-phase high-pressure liquid chromatography (HPLC) was performed on a system with an LC-20AT pump (Shimadzu) and a B-FC-320 flow counter (Bioscan). The C-18 reverse-phase semipreparative HPLC column (10mm × 250mm, 5-μm particle size (Venusil MP-C18; Agela Technologies Inc.)) was eluted at a flow rate of 5mL/min. A Labgen 7 homogenizer was used (Cole-Parmer Instruments). ^1H nuclear magnetic resonance spectra were recorded on a 400-MHz spectrometer (Bruker), and ^{13}C nuclear magnetic resonance spectra were recorded on a 100-MHz spectrometer (Bruker). Chemical shifts are reported in δ (ppm) values. Infrared spectra were measured on a Nicolet 360 Avatar instrument (Thermo) using a potassium bromide disk, scanning from 400 to 4000cm^{-1}. Mass spectra were recorded using an Apex IV FTM instrument (Brucker).

Kunming mice (18-20g) were obtained from the Animal Center of Peking University. Healthy Chinese mini swine and cardiac ischemia models (about 15kg) were obtained from the Animal Center of Fu Wai Hospital. All biodistribution studies were performed under a protocol approved by Beijing Administration Office of Laboratory Animals.

1.2 Chemistry

As shown in Supplemental Fig.1 (supplemental materials are available online only at http://jnm.snmjournals.org), the tosylate precursors of 2-tert-butyl-4-chloro-5-(4-(2-tosyloxy-ethoxy))-benzyloxy-2H-pyridazin-3-one (OTs-P1OP), 2-tert-butyl-4-chloro-5-(4-(2-(2-tosyloxy-ethoxy)ethoxy)ethoxy))-benzyloxy-2H-pyridazin-3-one (OTs-P3OP), and the corresponding nonradioactive reference compounds (^{19}F-FP1OP and ^{19}F-FP3OP) were synthesized and characterized according to a procedure published previously[14,16]. Details of the synthesis and the characterization data are available in the online supplemental data.

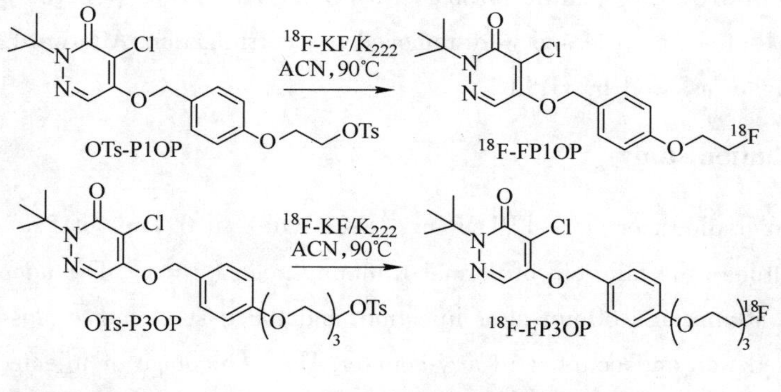

Fig. 1 Radiolabeling route of ^{18}F-FP1OP and ^{18}F-FP3OP
(ACN = acetonitrile; K_{222} = Kryptofix2.2.2.; KF = potassium fluoride)

1.3 Radiochemistry

The labeling procedure of ^{18}F-FP1OP and ^{18}F-FP3OP, shown in Fig. 1, is similar to that we published previously, with slight modification[16]. Briefly, after the solvent of the ^{18}F-F$^-$ eluate was evaporated under a stream of nitrogen at 110℃, 1.5mg of tosylate precursor dissolved in anhydrous acetonitrile (1.5mg/mL) were added. After 30min of stirring at 90℃ and then cooling to room temperature, the reaction mixture was injected onto a semi-HPLC column for purification. The column was eluted with water (solvent A) and acetonitrile (solvent B) at a flow rate of 5.0mL/min. The gradient was 95% A from 0 to 5min, 95%-60% A from 5.01 to 8min, 60%-35% A from 8.01 to 19min, and 0% A from 19.01 to 30min. The desired product was collected from the HPLC column, and the solvent was evaporated using a rotary evaporator. The product was redissolved in 5% ethanol solution and filtered through a 0.22-μm Millipore filter.

The final radiochemical purity was determined by reinjection of the product onto a radio-HPLC column, and the radioactive fraction was collected and measured in a dose calibrator for specific activity calculation. The mass of the product was calculated by comparing the area under the ultraviolet curve at 254nm with that of the standard reference (Supplemental Fig. 2).

1.4 Physicochemical properties studies

A previously published procedure[16,19] was used to measure the octanol-to-water partition coefficient and to perform the paper electrophoresis experiment. The partition coefficient value, expressed as log P, was measured in 1-octanol and phosphate buffer (0.025M, pH 7.4). For paper electrophoresis experiments, after development in phosphate buffer (0.025M, pH 7.4) the radioactivity distribution on the strip was determined using a 1470 Wizard automatic γ-counter (Perkin-Elmer).

In vitro stability was tested by following a published procedure[16,20]. Briefly, the radiotracer was incubated in water at room temperature for 1h, in 80% ethanol solution at room temperature for 4h, and in 0.5mL of murine plasma at 37℃ for 1h. Plasma proteins were precipitated by adding 100μL of acetonitrile and were removed by centrifugation. Afterward, the radiochemical purity was again assayed by HPLC.

1.5 Biodistribution study

About 185kBq of radiotracer (^{18}F-FP1OP or ^{18}F-FP3OP) in 0.1mL of 5% ethanol solution were injected through the tail vein of normal Kunming mice ($n=5$). The mice were sacrificed at 2min, 15min, 30min, and 60min after injection, and the tissues and organs of interest were collected, weighed wet, and counted in a γ-counter. The percentage of injected dose per gram (%ID/g) for each sample was calculated by comparing its activity with an appropriate standard of injected dose. The values were expressed as mean ± SD.

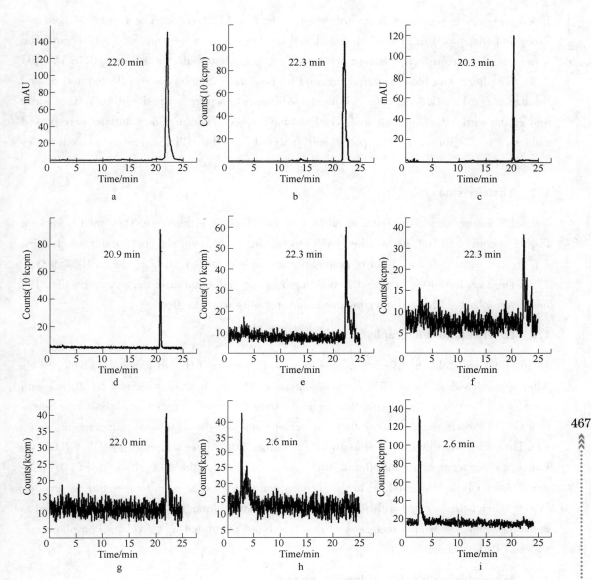

Fig. 2 HPLC chromatograms of compounds ^{19}F-FP1OP(a), ^{18}F-FP1OP(b), ^{19}F-FP3OP(c), and ^{18}F-FP3OP(d) and profiles of metabolic stability study of ^{18}F-FP1OP(soluble fractions were collected in heart at 2min(e), in blood at 2min(f), in heart at 30min(g), in blood at 30min(h), and in urine at 30min(i))

Nonradioactive compounds of ^{19}F-FP1OP and ^{19}F-FP3OP were measured with ultraviolet detector, and radioactive compounds were measured radiometrically

1.6 Metabolic stability

Normal Kunming mice were intravenously injected with 3.7MBq of the ^{18}F-FP1OP solution. The animals were sacrificed and dissected at 2 and 30min after injection. Blood, urine, and heart were collected and treated using a previously published procedure[16,21]. Briefly, the blood sample was immediately centrifuged for 5 min at 13200 r/min. The heart was homogenized and suspended in 1 mL of methanol and then centrifuged for 5 min at 13000 r/min. The urine sample was directly diluted with 1 mL of phosphate-buffered saline (0.025 M, pH 7.4). The heart and blood supernatants and diluted urine samples were collected and passed through a Sep-Pak C18 cartridge (Waters). All cartridges were washed with 0.5 mL of water and eluted with 0.5 mL of methanol. The combined aqueous and organic solutions were passed through a 0.22-μm filter (Millipore) and analyzed by radio-HPLC using the procedure described above.

1.7 Toxicity study

Normal Kunming mice were fed a standard diet for 48h before treatment. They then were kept fasting overnight (12h) and weighed before treatment[22]. A group of 5 male and 5 female mice was treated with each dose level of nonradioactive ^{19}F-FP1OP (1.0142 mg/kg, 1.1064 mg/kg, 1.2171 mg/kg, 1.3830 mg/kg, and 1.6043 mg/kg) or saline through tail vein injection. The mice were observed for 7 d after injection, and mortality was recorded.

1.8 Ex vivo autoradiography

Heart slides were obtained from a normal Sprague Dawley rat (Harlan Laboratories, Inc.)[16]. After being dehydrated with 30% sucrose solution at 4℃, the heart was sectioned (30μm) with a cryostat (CM1900; Leica) and thaw-mounted onto Superfrost microscope slides (Erie Scientific Co.). Frozen sections were fixed in acetone and air-dried at room temperature before use. The slides were pretreated with 100μL of saline, 20μM rotenone, or 20μM ^{19}F-FP1OP solution at room temperature for 30 min. After removal of that solution, 100μL of ^{18}F-FP1OP solution (about 14 kBq) were added to the sections and incubated at room temperature for another 30 min. After being washed with water and 40% ethanol solution, the sections were subjected to autoradiography for 10 min. Images were developed and quantified in a Cyclone Phosphoimager system (Perkin-Elmer Inc.).

1.9 Whole-body PET/CT of healthy animal

Animal procedures were performed following National Institutes of Health guidelines and were approved by Fu Wai Hospital, Chinese Academy of Medical Science.

For the whole-body PET/CT study, a healthy Chinese mini swine was anesthetized by a mixture of ketamine (25 mg/kg) and diazepam (1.1 mg/kg). Anesthesia was supplemented as needed. The animal was placed prone on the PET/CT bed, and a venous catheter was established for radiotracer injection. About 55 MBq of ^{18}F-FP1OP in 2mL of 5% ethanol solution were injected intravenously.

Whole-body imaging was performed at 2 min, 15 min, 30 min, 60 min, and 120 min after injection. PET data were acquired on a PET/CT system (Biograph 64; Siemens Healthcare). Whole-body scanning involved 4 bed positions, with each scanned for 2 min. Regions of interest were drawn on the myocardium, liver, and lung. Standardized uptake value was calculated as [mean region-of-interest count (cps/pixel) × body weight(kg)]/[injected dose(mCi) × calibration factor(cps/pixel)]. The myocardium-to-liver and myocardium-to-lung standardized uptake value ratios at each time point were evaluated.

1.10 Cardiac PET/CT scans of disease models

For the model of acute myocardial infarction, a Chinese mini swine was anesthetized with intravenous injection of sodium pentobarbital (25-35mg/kg), and additional sodium pentobarbital was used via intravenous injection to maintain anesthesia, when needed. Heart rate and electrocardiography were recorded throughout. The animal underwent lateral thoracotomy at the level of the fifth left intercostal space, and the heart was suspended in a pericardial cradle. The left anterior descending coronary artery was isolated to the section about 1.5cm after the first major diagonal branch. A silk suture was placed around the artery and ligated to block the coronary flow. Acute myocardial infarction induced by ligation was conformed by electrocardiography. A dose of about 55MBq of ^{18}F-FP1OP was injected at 60min after the thoracic incision had been closed.

The chronic myocardial ischemia model was created in a Chinese mini swine by ligation of the left anterior descending coronary artery at about 1.5cm after the first major diagonal branch, using an amaroid constrictor, to produce progressive vessel occlusion and ischemia. Three weeks later, when a critical stenosis (>80%) was confirmed by coronary angiography (Supplemental Fig. 3), a dose of about 55MBq of ^{18}F-FP1OP was injected before cardiac PET/CT.

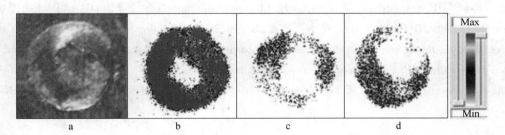

Fig. 3 Autoradiograms of heart sections from normal Sprague Dawley rat
(a is photograph of heart section before autoradiography)
Sections were inclubated with 100μL of ^{18}F-FP1OP solution(about 14kBq) for 30min at room temperature after pretreatment with 100μL of saline(b), 20μM rotenone(c), or 20μM ^{19}F-FP1OP(d)

The cardiac PET scans were performed at 2min, 15min, 30min, 60min, and 120min after injection. After a low-dose CT scan for attenuation correction (tube voltage, 120kV; tube current, 50 mAs with CareDose4D technique (Siemens); collimation, 24mm × 1.2mm; rotation time, 0.5s; pitch, 1.2; slice width, 3mm), PET was performed to evaluate myocardial uptake. Using list mode with electrocardiography gating, the scan time was 10min. The images were reconstructed with iterative ordered-subsets expectation maximization (2 iterations and 8 subsets). The matrix size was 2.0mm × 2.0mm with 128 × 128 pixels, and the slice thickness was 3.125mm.

2 Results

2.1 Chemistry

The precursors (OTs-P1OP and OTs-P3OP) and the corresponding nonradioactive references (^{19}F-FP1OP and ^{19}F-FP3OP) were synthesized and characterized by ^{1}H, ^{13}C, and ^{19}F nuclear magnetic resonance; electrospray ionization-mass spectrometry; infrared analysis; and elemental analysis. The synthesis route and all analysis data are shown in Supplemental Fig. 1.

As shown in Fig. 2, HPLC analysis of ^{19}F-FP1OP and ^{19}F-FP3OP showed that their retention times were 22.0 and 20.3min, respectively (Fig. 2a and 2c). The delayed retention time of ^{19}F-FP1OP indicated that the compound with the shorter radiolabeling chain has lower polarity and hydrophilicity. Their chemical purity was calculated as more than 98% from the HPLC chromatogram ($\lambda = 254$nm), suggesting they are acceptable reference standards for the corresponding radioactive tracers.

2.2 Radiochemistry

The radiolabeling route is shown in Fig. 1. Starting from ^{18}F-F$^-$ in a Kryptofix 2.2.2./K$_2$CO$_3$ solution, the total reaction time, including final HPLC purification, was 70-90min. The overall decay-corrected radiochemical yields were 58% ± 7.4% and 47% ± 6.8% for ^{18}F-FP1OP and ^{18}F-FP3OP, respectively. The radio-HPLC retention times of ^{18}F-FP1OP and ^{18}F-FP3OP were 22.3 and 20.9min, respectively (Fig. 2b and 2d) and were highly consistent with the corresponding nonradioactive references (Fig. 2a and 2c). The radiochemical purities calculated from radio-HPLC chromatograms of both tracers were more than 98% after purification. The specific activities of both tracers were estimated by HPLC analysis to be about 30GBq/μmol.

2.3 Physicochemical properties

According to the results of the octanol-to-water partition coefficient and paper electrophoresis, ^{18}F-FP1OP and ^{18}F-FP3OP are lipophilic and neutral compounds. Their partition coefficients (log P) were 3.07 ± 0.13 and 1.25 ± 0.00 ($n = 3$), respectively. The shorter radiolabeling chain of ^{18}F-FP1OP leads to higher lipophilicity. This result is consistent with that of HPLC analysis.

The HPLC profiles of stability studies are shown in Supplemental Fig. 4. After storage in water at room temperature for 1h, about 85% intact product of ^{18}F-FP1OP and 24% intact product of ^{18}F-FP3OP were eluted from HPLC. For the radiotracer incubated in murine plasma at 37℃ for 1h, about 80% of ^{18}F-FP1OP was intact. When stocked in 80% ethanol solution,

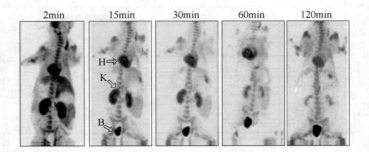

Fig. 4 Whole-body planar images of healthy Chinese mini swine. Images were obtained with 55MBq of ^{18}F-FP1OP in 5% ethanol solution at 2min, 15min, 30min, 60min, and 120min after injection
(B = urinary bladder; H = heart; K = kidney)

the tracers are stable during at least a 4-h period (Supplemental Fig. 4). Therefore, the HPLC-purified tracers are best stocked in 80% ethanol solution first. The stability studies show that the longer radiolabeling chain leads to worse stability in water (after storage in water for 1h, about 85% of ^{18}F-FP1OP, 78% of ^{18}F-FP2OP, and 24% of ^{18}F-FP3OP are stable). Therefore, in the studies performed afterward, the most stable ^{18}F-FP1OP was selected for further evaluation.

2.4 Biodistribution in mice

The biologic distribution results in mice are shown in Table 1. Both ^{18}F-FP1OP and ^{18}F-FP3OP had high initial heart uptake ((31.13 ± 6.24)% ID/g and (31.10 ± 3.72)% ID/g at 2min after injection, respectively). Meanwhile, uptake in most other tissues was low, especially in liver and lung, resulting in the high heart-to-nontarget ratios for both tracers. For ^{18}F-FP1OP, the heart-to-liver, heart-to-lung, and heart-to-blood ratios were 16.12, 9.99, and 16.63, respectively, at 15min after injection. For ^{18}F-FP3OP, the respective ratios were 4.91, 21.28, and 48.77.

As described in Table 1, ^{18}F-FP1OP had high initial heart uptake and cleared quickly; about 64% of the tracer was eliminated from myocardium during the first hour after injection ((11.32 ± 1.01)% ID/g at 60min after injection). Compared with ^{18}F-FP1OP, ^{18}F-FP3OP had much slower clearance from myocardium, obviously higher liver uptake, and lower lung uptake at 2min after injection, leading to the lower heart-to-liver and higher heart-to-lung ratios of ^{18}F-FP3OP than of ^{18}F-FP1OP. Both tracers had high initial kidney uptake, suggesting the tracers were excreted mainly via the renal system.

The reason for the slight increase of blood uptake of both tracers over time is not clear. The high uptake in muscle of both tracers may due to the high mitochondrial expression. All other organs and tissues had low uptake, and no obvious change was observed during the 60min after injection.

Table 1 Biodistribution results of ^{18}F-FP1OP and ^{18}F-FP3OP in normal mice

Tissue	Compound	Postinjection time/min			
		2	15	30	60
Heart	^{18}F-FP1OP	31.13 ± 6.24	29.00 ± 1.49	19.25 ± 1.67	11.32 ± 1.01
	^{18}F-FP3OP	31.10 ± 3.72	33.52 ± 5.03	29.85 ± 3.89	22.06 ± 3.91
Liver	^{18}F-FP1OP	2.72 ± 0.33	1.86 ± 0.14	1.77 ± 0.20	2.14 ± 0.29
	^{18}F-FP3OP	2.71 ± 0.93	6.83 ± 0.84	6.97 ± 0.91	6.77 ± 0.55
Spleen	^{18}F-FP1OP	3.45 ± 0.98	2.49 ± 0.49	2.41 ± 0.22	2.34 ± 0.17
	^{18}F-FP3OP	1.41 ± 0.44	2.69 ± 0.28	2.79 ± 0.45	2.45 ± 0.21
Lung	^{18}F-FP1OP	2.98 ± 0.28	2.92 ± 0.42	2.53 ± 0.20	2.49 ± 0.37
	^{18}F-FP3OP	4.65 ± 1.60	1.58 ± 0.33	1.78 ± 0.51	1.90 ± 0.26
Muscle	^{18}F-FP1OP	7.45 ± 1.38	9.56 ± 2.21	7.27 ± 0.66	6.06 ± 1.53
	^{18}F-FP3OP	5.55 ± 2.51	6.96 ± 0.86	7.87 ± 0.58	6.85 ± 0.86
Bone	^{18}F-FP1OP	2.23 ± 0.22	3.56 ± 0.53	4.13 ± 0.54	5.28 ± 2.29
	^{18}F-FP3OP	1.04 ± 0.29	1.76 ± 0.32	2.19 ± 0.30	2.34 ± 0.61
Kidney	^{18}F-FP1OP	23.69 ± 3.16	16.31 ± 2.76	10.50 ± 0.75	6.63 ± 0.68
	^{18}F-FP3OP	18.94 ± 4.89	17.19 ± 3.04	15.48 ± 1.99	12.60 ± 1.56
Blood	^{18}F-FP1OP	1.06 ± 0.07	1.60 ± 0.18	2.12 ± 0.18	2.51 ± 0.33
	^{18}F-FP3OP	0.82 ± 0.05	0.69 ± 0.10	0.89 ± 0.16	1.07 ± 0.12
Heart/liver	^{18}F-FP1OP	12.53	16.12	10.87	5.29
	^{18}F-FP3OP	11.48	4.91	4.28	3.26
Heart/lung	^{18}F-FP1OP	8.58	9.99	7.62	4.55
	^{18}F-FP3OP	6.69	21.28	16.75	11.61
Heart/blood	^{18}F-FP1OP	32.36	16.63	9.09	4.50
	^{18}F-FP3OP	37.9	48.77	33.57	20.61

Note: Data are %ID/g ± SD ($n = 5$).

2.5 Metabolic stability

The metabolic stability of ^{18}F-FP1OP was determined in the heart, blood, and urine of Kunming mice at 2 min and 30 min after injection. More than 95% of the total ^{18}F was eluted at 22.3 min from HPLC (Fig. 2e and 2f), suggesting that the radioactivity in myocardium and blood at 2 min after injection was intact ^{18}F-FP1OP. At 30 min after injection, the tracer was still intact in myocardium (Fig. 2g) but was metabolized in blood and urine and eluted quickly from HPLC (Fig. 2h and 2i). The peak of ^{18}F-FP1OP had almost disappeared and a new metabolite appeared at about 2.6 min, suggesting the tracer undergoes considerable degradation in both blood and urine at 30 min after injection. No defluorination of ^{18}F-FP1OP was observed in vivo from HPLC analysis. This finding was confirmed by the low bone uptake found in the biodistribution study (Table 1).

2.6 Toxicity study

Mortality data are shown in Supplemental Table 1. According to the Bliss method[23], the calculated lethal dose of 50% for the nonradioactive reference of ^{19}F-FP1OP was 1.2869mg/kg (95% confidence limits, 1.1818-1.4015mg/kg). On the basis of the toxicity result and the high specific activity (about 30GBq/μmol), this HPLC-purified tracer solution should be safe for injection.

2.7 Ex vivo autoradiography

Ex vivo autoradiography studies were performed on frozen sections of heart from a normal Sprague Dawley rat. As shown in Fig. 3, ^{18}F-FP1OP had high uptake in myocardium (Fig. 3b) and could be blocked both by preincubation with rotenone (a known MC-I inhibitor with high affinity, Fig. 3c) and preincubation with ^{19}F-FP1OP (Fig. 3d). After calculation of the net radioactivity intensity, about 74% and 62% of the radioactivity were found to have been blocked by rotenone and ^{19}F-FP1OP, respectively. The results of autoradiography suggest that ^{19}F-FP1OP has the same binding site as, and comparable affinity to, rotenone. This is potent evidence that the ^{18}F-FP1OP was taken up by myocardium through the MC-I enzyme and remained in mitochondria.

2.8 Whole-body PET/CT of healthy animal

The representative whole-body images of ^{18}F-FP1OP in a healthy Chinese mini swine are shown in Fig. 4. The heart could clearly be seen from 2 to 60min after injection and was still visible at 120min after injection, although heart uptake had obviously decreased. Radioactivity accumulated strongly in the kidney and cleared quickly to the bladder. Other organs and tissues had low background uptake. The heart-to-liver and heart-to-lung standardized uptake value ratios were, respectively, 1.83 and 4.53 at 2min after injection, 2.33 and 6.25 at 15min after injection, 2.73 and 7.39 at 30min after injection, 3.03 and 8.77 at 60min after injection, and 2.00 and 7.21 at 120min after injection. The heart-to-liver and heart-to-lung ratios were highest at 60min after injection, leading to a best PET acquisition time of 15-60min after injection. In contrast, in the biodistribution study the highest heart-to-liver and heart-to-lung ratios were reached at only 15min after injection; this difference may due to the different animal models used.

2.9 Cardiac PET/CT

The representative cardiac PET images of ^{18}F-FP1OP in healthy Chinese mini swine and the acute myocardial infarction model are shown in Fig. 5. High heart activity was observed at all time points. Clearance agreed well with the biodistribution studies in mice. The outline of the myocardium was clear and uptake in background organs (liver and lung) low at all time points. In comparison with normal myocardium, the infarct region (apical and anterior wall) was apparent on ^{18}F-FP1OP cardiac PET in the acute myocardial infarction model (Fig. 5).

Representative cardiac PET images of the chronic myocardial ischemia model are shown in Fig. 6. The perfusion deficit area (anterior wall) was clearly detected at 2-60min after injection but showed a slightly redistribution of ^{18}F-FP1OP at 120min after injection. Future experiments are needed to confirm the redistribution properties of ^{18}F-FP1OP.

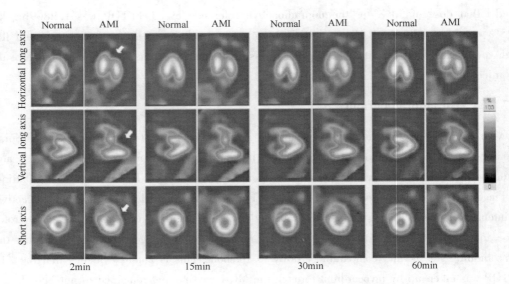

Fig. 5 Cardiac PET images of healthy Chinese mini swine (normal) and
Chinese mini swine with acute myocardial infarction (AMI)
(Images were obtained with 55MBq of ^{18}F-FP1OP in 5% ethanol solution at 2min, 15min, 30min, and 60min after injection.
Arrows indicate perfusion defect sites of infarction areas (apical and anterior walls))

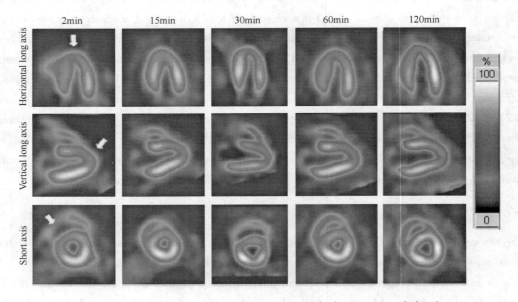

Fig. 6 Cardiac PET images of Chinese mini swine with chronic myocardial ischemia
(Images were obtained with 55MBq of ^{18}F-FP1OP in 5% ethanol solution at 2min, 15min, 30min, 60min,
and 120min after injection. Arrows indicate sites of perfusion deficit regions (anterior wall).
Redistribution was observed at 120min after injection)

3 Discussion

Pyridaben possesses a hydrophobic heterocyclic headpiece, a side chain of *p*-tert-butylphenyl moiety, and a heteroatom-containing linker[14]. In this study, we found that the side chain had a significant impact on the stability and pharmacokinetic properties of pyridaben-based radiotracers. According to the physicochemical properties study, the compound with the shorter polyethylene glycol chain had higher lipophilicity and better stability. For example, ^{18}F-FP1OP, which had the shortest polyethylene glycol chain, displayed the highest lipophilicity and the best stability. Furthermore, changes of the radiolabeling chain could also affect biodistribution (Supplemental Fig. 5). The biodistribution comparison of ^{18}F-FP1OP, ^{18}F-FP2OP[16], and ^{18}F-FP3OP showed that ^{18}F-FP1OP had the lowest liver uptake and the fastest clearance rate from the liver (Supplemental Fig. 5). ^{18}F-FP1OP had the best stability, the highest heart-to-liver ratio at all time points, and acceptable heart-to-lung and heart-to-blood ratios—all promising properties making it superior to the other 2 tracers.

^{18}F-FP1OP had higher initial heart uptake and lower initial liver uptake than that of BMS-747158-02, making it possible to acquire PET data earlier after injection. This feature will be of great benefit to patients by shortening their waiting time significantly. In addition, ^{18}F-FP1OP cleared from myocardium much more quickly than did BMS-747158-02; this property could definitely lower the irradiation dose and make the tracer a good choice for stress and rest imaging in a 1-d protocol.

Because inhibiting MC-I activity might lead to death of the animal, ex vivo autoradiography was done instead to confirm the specific binding of ^{18}F-FP1OP to MC-I enzyme. The toxicity of the tracer as a pyridaben analog should be considered seriously. In this study, the 50% lethal dose of ^{19}F-FP1OP was determined in mice by tail vein injection as 1.2869mg/kg. On the basis of the reported specific activity (about 30GBq/μmol), if 370MBq of tracer were to be injected in a 70-kg human, the mass of ^{19}F-FP1OP would be about 4.37μg, and the safety factor is calculated as more than 20000 times the lethal dose of 50%. Therefore, we believe the radiotracer is safe enough for PET application. Acute or single-dose toxicity and other special safety studies will be done in the future.

Conclusion

The 2 lipophilic and neutral tracers ^{18}F-FP1OP and ^{18}F-FP3OP were successfully prepared with high radiochemical yield (≈50%) and radiochemical purity (>98%). Both had high initial heart uptake and good heart-to-nontarget ratios. The better stability and faster clearance from myocardium of ^{18}F-FP1OP make it worthy to be evaluated as an MPI agent, and the high specific activity makes it a safe injection. The heart of a healthy Chinese mini swine could be imaged clearly at 2-60min after injection by cardiac PET. The infarction region could be detected clearly in both acute myocardial infarction and chronic myocardial ischemia models of Chinese mini swine, suggesting the potential usefulness of ^{18}F-FP1OP as an MPI agent.

Disclosure Statement

The costs of publication of this article were defrayed in part by the payment of page charges.

Therefore, and solely to indicate this fact, this article is hereby marked "advertisement" in accordance with 18 USC section 1734.

Acknowledgments

We thank Huihui Jing, Wenjiang Yang, and Wenyan Guo for their generous help. This project was sponsored by the National Natural Science Foundation of China (20871020) and Beijing Natural Science Foundation (2092018) and supported partially by Fundamental Research Funds for the Central Universities and the Scientific Research Foundation for the Returned Overseas Chinese Scholars, State Education Ministry. No other potential conflict of interest relevant to this article was reported.

References

[1] Madar I, Ravert H T, Du Y, et al. Characterization of uptake of the new PET imaging compound ^{18}F-fluorobenzyl triphenyl phosphonium in dog myocardium. J Nucl Med. 2006;47:1359-1366.

[2] Le Guludec D, Lautamäki R, Knuuti J, Bax J, Bengel F. Present and future of clinical cardiovascular PET imaging in Europe: a position statement by the European Council of Nuclear Cardiology (ECNC). Eur J Nucl Med Mol Imaging. 2008;35:1709-1724.

[3] Kopka K, Schober O, Wagner S. ^{18}F-labelled cardiac PET tracers: selected probes for the molecular imaging of transporters, receptors and proteases. Basic Res Cardiol. 2008;103:131-143.

[4] Cheng Z, Subbarayan M, Chen X, Gambhir S S. Synthesis of (4-[^{18}F]fluorophenyl) triphenylphosphonium as a potential imaging agent for mitochondrial dysfunction. J Labelled Comp Radiopharm. 2005;48:131-137.

[5] Shoup T M, Elmaleh D, Brownell A L, Zhu A, Guerrero J, Fischman A. Evaluation of (4-[^{18}F]fluorophenyl) triphenylphosphonium ion: a potential myocardial blood flow agent for PET. Mol Imaging Biol. 2011;13:511-517.

[6] Ravert H T, Madar I, Dannals R F. Radiosynthesis of 3-[^{18}F]fluoropropyl and 4-[^{18}F]fluorobenzyl triarylphosphonium ions. J Labelled Comp Radiopharm. 2004;47:469-476.

[7] Heinrich T K, Gottumukkala V, Snay E, et al. Synthesis of fluorine-18 labeled rhodamine B: a potential PET myocardial perfusion imaging agent. Appl Radiat Isot. 2010;68:96-100.

[8] Marshall R C, Powers-Risius P, Reutter B W, et al. Kinetic analysis of ^{18}F-fluorodihydrorotenone as a deposited myocardial flow tracer: comparison to ^{201}Tl. J Nucl Med. 2004;45:1950-1959.

[9] Yu M, Guaraldi M, Kagan M, et al. Assessment of ^{18}F-labeled mitochondrial complex I inhibitors as PET myocardial perfusion imaging agents in rats, rabbits, and primates. Eur J Nucl Med Mol Imaging. 2009;36:63-72.

[10] Yu M, Guaraldi M, Mistry M, et al. BMS-747 158-02: a novel PET myocardial perfusion imaging agent. J Nucl Cardiol. 2007;14:789-798.

[11] Nekolla S G, Saraste A. Novel F-18-labeled PET myocardial perfusion tracers: bench to bedside. Curr Cardiol Rep. 2011;13:145-150.

[12] Huisman M C, Higuchi T, Reder S, et al. Initial characterization of an ^{18}F-labeled myocardial perfusion tracer. J Nucl Med. 2008;49:630-636.

[13] Kronauge J F, Chiu M L, Cone J S, et al. Comparison of neutral and cationic myocardial perfusion agents: characteristics of accumulation in cultured cells. Int J Rad Appl Instrum B. 1992;19:141-148.

[14] Purohit A, Radeke H, Azure M, et al. Synthesis and biological evaluation of pyridazinone analogues as potential cardiac positron emission tomography tracers. J Med Chem. 2008;51:2954-2970.

[15] Okun J G, Lummen P, Brandt U. Three classes of inhibitors share a common binding domain in mitochondrial complex I(NADH:ubiquinone oxidoreductase). J Biol Chem. 1999;274:2625-2630.

[16] Mou T, Jing H, Yang W, et al. Preparation and biodistribution of [^{18}F]FP2OP as myocardial perfusion imaging agent for positron emission tomography. Bioorg Med Chem. 2010;18:1312-1320.

[17] Yang W, Mou T, Peng C, et al. Fluorine-18 labeled galactosyl-neoglycoalbumin for imaging the hepatic asialoglycoprotein receptor. Bioorg Med Chem. 2009;17:7510-7516.

[18] Zhang X, Cai W, Cao F, et al. ^{18}F-labeled bombesin analogs for targeting GRP receptor-expressing prostate cancer. J Nucl Med. 2006;47:492-501.

[19] Zhang X, Zhou P, Liu J, et al. Preparation and biodistribution of ^{99m}Tc-tricarbonyl complex with 4-[(2-methoxyphenyl)piperazin-1-yl]-dithioformate as a potential 5-HT1A receptor imaging agent. Appl Radiat Isot. 2007;65:287-292.

[20] Mou T, Yang W, Peng C, Zhang X, Ma Y. [^{18}F]-labeled 2-methoxyphenylpiperazine derivative as a potential brain positron emission tomography imaging agent. Appl Radiat Isot. 2009;67:2013-2018.

[21] Yang W, Mou T, Shao G, Wang F, Zhang X, Liu B. Copolymer-based hepatocyte asialoglycoprotein receptor targeting agent for SPECT. J Nucl Med. 2011;52:978-985.

[22] Tubaro A, Del Favero G, Beltramo D, et al. Acute oral toxicity in mice of a new palytoxin analog: 42-hydroxy-palytoxin. Toxicon. 2011;57:755-763.

[23] Zhou H J. Statistical Methods for Biological Tests. Beijing, China: People's Medical Publishing House; 1988:1214.

Radioiodinated Benzyloxybenzene Derivatives: a Class of Flexible Ligands Target to β-amyloid Plaques in Alzheimer's Brains*

Abstract Benzyloxybenzene, as a novel flexible scaffold without rigid planarity, was synthesized and evaluated as ligand toward Aβ plaques. The binding site calculated for these flexible ligands was the hydrophobic Val18_Phe20 channel on the flat surface of Aβ fiber. Structure-activity relationship analysis generated a common trend that binding affinities declined significantly from para-substituted ligands to ortho-substituted ones, which was also quantitatively illustrated by 3D-QSAR modeling. Autoradiography in vitro further confirmed the high affinities of radioiodinated ligands [^{125}I]**4**, [^{125}I]**24** and [^{125}I]**22** (K_i = 24.3nM, 49.4nM, and 17.6nM, respectively). In biodistribution, [^{125}I]**4** exhibited high initial uptake and rapid washout property in the brain with brain$_{2min}$/brain$_{60min}$ ratio of 16.3. The excellent in vitro and in vivo biostability of [^{125}I]**4** enhanced its potential for clinical application in SPECT imaging of Aβ plaques. This approach could also allow the design of a new generation of Aβ targeting ligands without rigid and planar framework.

⑤ Supporting Information

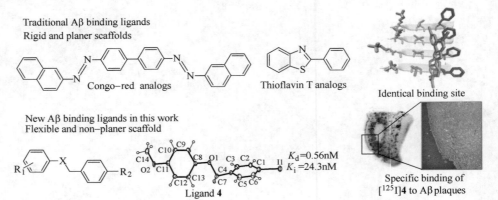

1 Introduction

As the world's population ages, we confront a looming global prevalence of Alzheimer's disease (AD), which represents a daunting, worldwide challenge for society and healthcare providers[1]. Today, the early diagnosis of AD in its preclinical phase is at the forefront of biomedical research. The amyloid cascade hypothesis, which posits that the deposition of extracellular β-amyloid (Aβ) plaques in the brain plays a causative role in the pathogeny of AD, has domina-

---

* Copartner: Yang Yanping, Cui Mengchao, Zhang Xiaoyang, Dai Jiapei, Zhang Zhiyong, Lin Chunping, Guo Yuzhi. Reprinted from *J. Med. Chem*, 2014, 57: 6030-6042.

ted research for the past two decades[2]. Therefore, it's of paramount clinical value to precisely detect the Aβ plaques at preclinical AD with noninvasive techniques such as positron emission tomography (PET) or single photon emission computed tomography (SPECT). It is expected to provide clues to the underlying disease pathology and aid in earlier evaluation of new disease modifying treatments.

Early diagnosis of AD has become a hot spot, with an enormous number of published studies describing in vivo or ex vivo attempts to image Aβ plaques with various radiolabeled probes. Initial inspirations toward Aβ imaging came from two categories of historic amyloid binding dyes used in neuropathology, Congo red (CR) and thioflavin T (ThT)(Fig. 1). However, these dyes failed to penetrate the blood-brain barrier (BBB) because of their bulky and ionic natures[3,4]. Subsequent approaches focused on structural modifications of these dyes resulted in a great diversity of radiolabeled Aβ probes for PET or SPECT imaging. The chemical scaffolds derived from ThT including benzothiazole[5,6], benzoxazole[7], benzothiophene[8], benzofuran[9], imidazopyridine[10], benzoimidazole[11], and so on. Stilbene[12,13] and its analogues such as biphenyl[14], biphenylalkyne[15], and diphenyl-1, 3, 4-oxadiazole[16] were also well studied. In addition, another series of frameworks generated from plant pigments like curcumin[17], flavone[18], chalcone[19], and aurone[20] were radiolabeled and biologically investigated. Among them, several PET imaging agents targeting to Aβ plaques have achieved wonderful progress. [^{11}C]PIB[5] has been studied in human from 2002, and almost 10000 PET scans have been conducted worldwide. Better yet, [^{18}F]AV-45 (florbetapir)[13,21], [^{18}F]GE-067 (flutemetamol)[6], and [^{18}F]BAY 94-9172 (florbetaben)[22] have been approved by the U.S. FDA in the last two years. In contrast, the approach in developing a selective SPECT agent to quantify Aβ plaques in living human brain has intensively lagged behind. Preliminary clinical data of [^{123}I]IMPY (Fig. 1), the only SPECT tracer tested in humans, showed excellent binding property, high brain penetration and fast washout kinetics[10]. Unfortunately, its signal-to-noise ratio for plaque labeling was not robust in AD and normal subjects, thus preventing its translation to clinical use. The in vivo instability and rapid metabolism were deemed as main causes of the low contrast and poor image quality[23].

From the foregoing overview of these Aβ binding probes, we can easily draw a conclusion that the chemical diversity of their scaffolds has been confined to small variations of the traditional dyes[24]. Most, if not all, of them contain two aromatic moieties and retain the highly rigid and planar scaffold with π-conjugated system, which are widely believed as vital points of the specific binding between probes and Aβ fibers[25]. This common feature gave strong impulse to our bold attempts to break the conjugated backbone and make "out of box" innovation. Herein, we describe the preliminary characterization of a series of benzyloxybenzene derivatives without highly conjugated structure to screen novel Aβ probes with favorable binding affinities and brain pharmacokinetics. This study could probably expand the chemical library of Aβ probes and provide more impetus for designing novel tracers off from the traditional thought pattern that the conjugated and planer framework is essential. Computational simulations including three-dimensional quantitative structure-activity relationship (3D-QSAR) modeling and molecular docking were also carried out to further clarify the interaction mechanism between benzyloxybenzene derivatives and Aβ fibers.

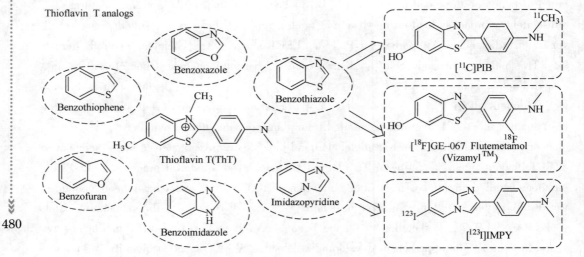

Fig. 1 Chemical structures of dyes originally utilized for Aβ staining and scaffolds explored for PET/SPECT imaging of Aβ plaques in AD patients

2 Results and Discussion

2.1 Chemistry

The synthesis of the benzyloxybenzene derivatives is outlined in Scheme 1. Twenty-two iodinated ligands and three tributyltin precursors were successfully produced with common methods

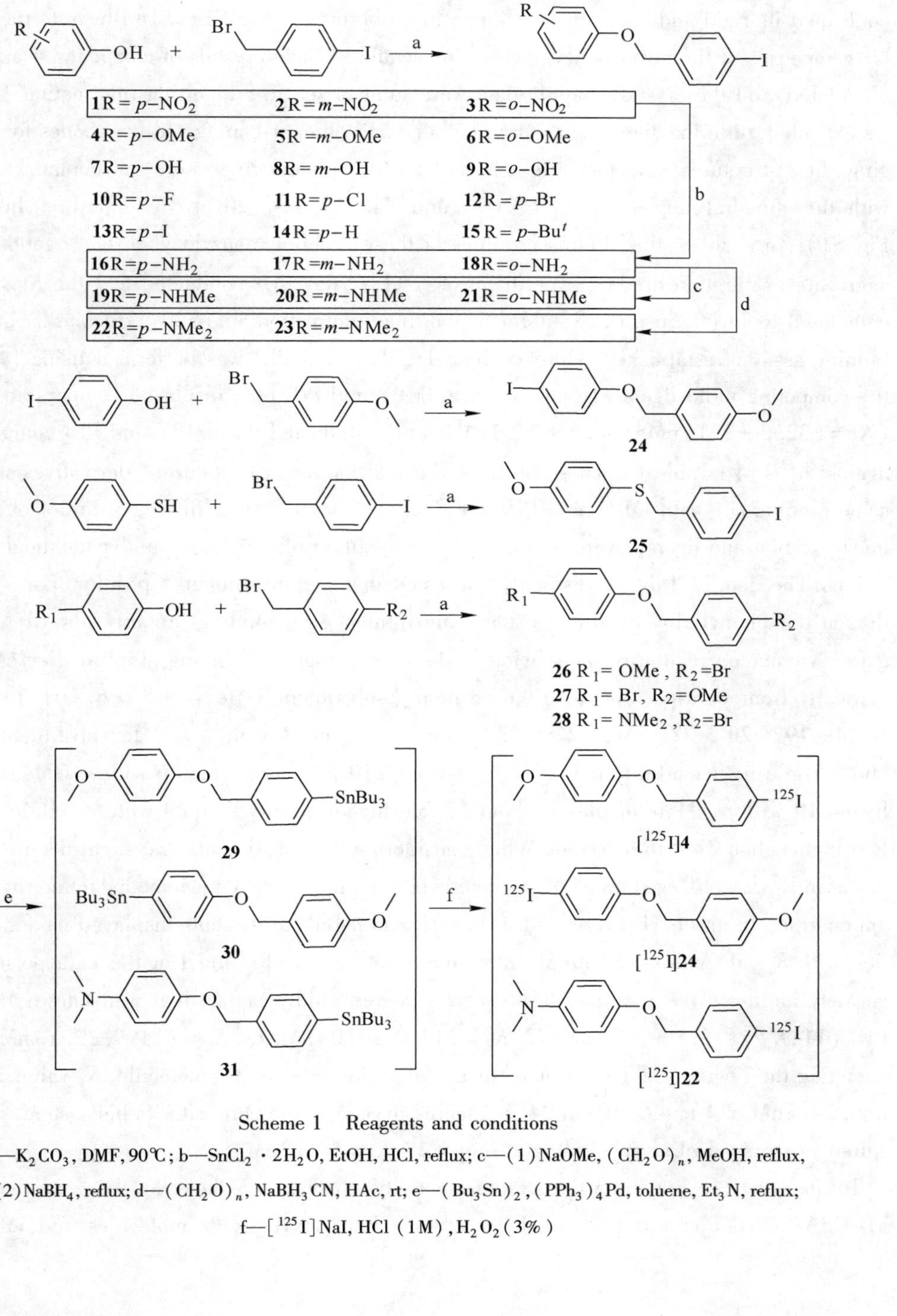
1 R = p–NO2
2 R = m–NO2
3 R = o–NO2
4 R = p–OMe
5 R = m–OMe
6 R = o–OMe
7 R = p–OH
8 R = m–OH
9 R = o–OH
10 R = p–F
11 R = p–Cl
12 R = p–Br
13 R = p–I
14 R = p–H
15 R = p–Bu^t
16 R = p–NH2
17 R = m–NH2
18 R = o–NH2
19 R = p–NHMe
20 R = m–NHMe
21 R = o–NHMe
22 R = p–NMe2
23 R = m–NMe2
24
25
26 R1 = OMe, R2 = Br
27 R1 = Br, R2 = OMe
28 R1 = NMe2, R2 = Br
29
30
31
[125I]4
[125I]24
[125I]22

Scheme 1 Reagents and conditions

a—K_2CO_3, DMF, 90℃; b—$SnCl_2 \cdot 2H_2O$, EtOH, HCl, reflux; c—(1) NaOMe, $(CH_2O)_n$, MeOH, reflux, (2) $NaBH_4$, reflux; d—$(CH_2O)_n$, $NaBH_3CN$, HAc, rt; e—$(Bu_3Sn)_2$, $(PPh_3)_4Pd$, toluene, Et_3N, reflux; f—[^{125}I] NaI, HCl (1M), H_2O_2 (3%)

and verified with ^{1}H/^{13}C NMR and MS (EI). Purities of key compounds were proved to be higher than 98% by high-performance liquid chromatography (HPLC).

2.2 Structure-activity relationships

As analyzed above, almost all the previously reported Aβ probes were flat molecules. The crystal of stilbene scaffold, one of the most studied scaffolds for developing Aβ imaging agents, visually confirmed its rigid and flat natures (Supporting Information Fig. S1e). To illustrate the molecular geometries of the novel benzyloxybenzene scaffold, single crystals of compound **4** and **6** were elucidated, and the crystallographic data were summarized in Supporting Information Table S1. As expected, differing significantly from stilbene and other π-conjugated backbones for Aβ imaging, their structures were more flexible and the two phenyl rings were not coplanar any more with the dihedral angles up to 60.02° and 74.81°, respectively (Supporting Information Fig. S1). To evaluate the binding affinities of these different benzyloxybenzene ligands to Aβ$_{42}$ aggregates, saturation binding experiment of [^{125}I]**4**, was first conducted and the K_d value was calculated to be (0.56 ± 0.18) nM (Supporting Information Fig. S2). Afterward, competition binding assay of established reference ligands IMPY and PIB was performed using [^{125}I]**4** as the competing radio-ligand. To our surprise, IMPY and PIB both displayed comparable affinities (K_i = (32.2 ± 2.1) nM; (38.8 ± 2.6) nM) which indicated that IMPY and PIB competed well against [^{125}I]**4** (Table 1). These results indicated that benzyloxybenzene derivatives shared the same binding site with IMPY and PIB, the thioflavin binding site. Binding affinities of iodinated benzyloxybenzene ligands were evaluated by competition binding assay under identical test conditions. The data in Table 1 distinctly suggested that the substituents' position, para/meta/ortho, on the phenyl ring exhibited astonishing influence on binding property toward Aβ aggregates. A common regularity was obviously discovered that the binding affinities decreased dramatically from para-position to ortho-position (substituent OMe, **4** > **5** > **6**; OH, **7** > **8** > **9**; NHMe, **19** > **20** > **21**; NMe$_2$, **22** > **23**). Notably, ligand **4** with a p-OMe substituent, showed about 5-fold higher affinity than **5** with: m-OMe, 210-fold higher than **6** with: o-OMe, similarly, ligand **19** with p-NHMe displayed about 12-fold higher affinity than **20** with m-NHMe and 438-fold higher than **21** with o-NHMe. When considering the substituents, the slight decline in binding affinity was observed from N,N-dimethylamino, methoxy, N-monomethylamino, hydroxyl to amino group. Benzyloxybenzene with p-F, p-H, and p-Bu^t substituents displayed moderate affinities, with K_i values around 100nM. Introducing halogen Cl, Br, and I at the para position staggeringly improved the binding affinities to a degree mildly better than well-known IMPY and PIB (**11**, K_i = (18.8 ± 2.2) nM; **12**, K_i = (12.0 ± 1.0) nM; **13**, K_i = (21.9 ± 2.1) nM). When changing the position of iodine atom to the opposite side of the molecule, K_i value increased from 24.3nM of **4** to 49.4nM of **24**. Replacing oxygen in the core with its bioisostere, sulfur, resulted in strong decline in binding affinity (**25**, K_i = 530.2nM).

To quantitatively illustrate the structure-activity correlation of benzyloxybenzene derivatives, 3D-QSAR modeling was performed. Twenty-one benzyloxybenzene molecules and IMPY were first geometry optimized at the B3LYP/6-31G and 3-21G level in the water phase. The optimized structures of compound **4** and **6** conformed well to the geometries of their X-ray crystals in root-mean-square deviations (RMSD) of 0.020nm and 0.007nm, respectively. The small RMSD values guaranteed the credibility of using the optimized structures for the QSAR and molecualr docking calculations. 3D-QSAR studies successfully yielded two statistically reliable models, including comparative molecular field analysis (CoMFA: r^2, 0.947; q^2, 0.606) and comparative molecular similarity indices analysis (CoMSIA: r^2, 0.988; q^2, 0.777), which were able to predict binding abilities of novel derivatives accurately. CoMFA and CoMSIA analysis results were summarized in Fig. 2c, and it showed that the electrostatic field with the highest contributions (53.4% in CoMFA; 38.2% in CoMSIA) was the dominant factor for Aβ binding affinities. The actual and predicted pK_i values obtained from CoMFA and CoMSIA models were graphed in parts d and e of Fig. 2, respectively. Ideally, all of the data points appeared on or close to the diagonal, which sufficiently demonstrated that the calculated pK_i were in good agreement with the experimental data. The deviations of the calculated pK_i values from the corresponding actual data in both models were all smaller than 0.5 log unit. To graphically visualize the CoMFA/CoMSIA results, the steric, electrostatic, hydrophobic, hydrogen-bond donor, and hydrogen-bond acceptor fields were aligned with the conformation of ligand **4** and presented as 3D coefficient contour plots in Fig. 2f-i. The CoMFA steric contours (Fig. 2f) show a favored green area at the para-position of the phenyl ring indicating that the bulky substituents are desirable. All the ortho-substituted ligands (**6**, **9**, **18** and **21**), which displayed shocking degradation of binding affinities, exhibited significant encroachment in the disfavored yellow contours. In Fig. 2g, the blue counter around the para-position and a small blue region near the ortho-position suggest that more positively charged substituents at the position may enhance the binding abilities. On the contrary, two red areas illustrate that more negatively charged groups are favorable at that position. A yellow-colored polyhedron close to the para-position (Fig. 2h) indicates that adding hydrophobic substituents will be propitious to improve binding affinities. Three large white polyhedrons represent regions where decreased hydrophobic interaction is helpful in increasing binding abilities. In Fig. 2i, large cyan regions represent areas where hydrogen-bond donor groups on the ligand are predicted to enhance binding, while the purple area is predicted to disfavor binding. A small region in red was hydrogen-bond acceptors unfavorable. All the structural insights acquired from 3D-QSAR contour maps were in coincidence with the experimental affinity data and reasonably instructive for future drug design.

Table 1 Inhibition constants for the binding of [^{125}I]4 to Aβ$_{42}$ aggregates[①]

Scaffold

Crystal structure of **4**

Continued 1

Compd	R_1	X	R_2	K_i/nM	pK_i
4	p-OMe	O	I	24.3 ± 6.8	7.61
5	m-OMe	O	I	130.6 ± 4.8	6.88
6	o-OMe	O	I	5164.8 ± 410.3	5.29
7	p-OH	O	I	113.4 ± 23.8	6.95
8	m-OH	O	I	385.6 ± 114.2	6.41
9	o-OH	O	I	2831.2 ± 517.0	5.55
10	p-F	O	I	107.1 ± 15.4	6.97
11	p-Cl	O	I	18.8 ± 2.2	7.73
12	p-Br	O	I	12.0 ± 1.0	7.92
13	p-I	O	I	21.9 ± 2.1	7.66
14	p-H	O	I	79.4 ± 5.2	7.10
15	p-Bu^t	O	I	117.6 ± 17.7	6.93
16	p-NH$_2$	O	I	409.2 ± 45.0	6.39
17	m-NH$_2$	O	I	1534.7 ± 159.7	5.81
18	o-NH$_2$	O	I	1028.0 ± 49.0	5.99
19	p-NHMe	O	I	48.2 ± 4.3	7.32
20	m-NHMe	O	I	593.6 ± 78.2	6.23
21	o-NHMe	O	I	20662 ± 2653	4.68
22	p-NMe$_2$	O	I	17.6 ± 1.6	7.75
23	m-NMe$_2$	O	I	894.2 ± 88.0	6.05
24	I	O	OMe	49.4 ± 3.0	7.31
25	p-OMe	S	I	530.2 ± 109.4	6.28
IMPY				32.2 ± 2.1	7.49
PIB				38.8 ± 2.6	7.41

① Measured in triplicate with values given as the mean ± SD.

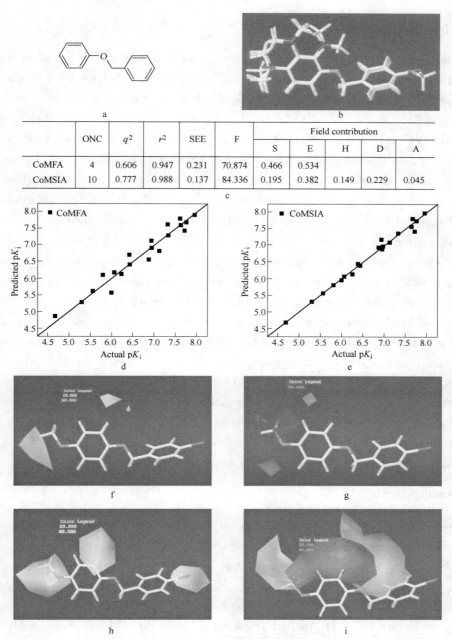

Fig. 2　Construction of good predictive CoMFA and CoMSIA 3D-QSAR models

a—Common core selected for database alignment; b—Superposition of compounds 4-24; c—Statistical parameters of CoMFA and CoMSIA models. ONC: optimum number of components. q^2: leave one out (LOO) cross-validated correlation coefficient. r^2: nonvalidated correlation coefficient. SEE: standard error of estimate. F: F-test value. S, E, H, D, A: relative contributions of steric, electrostatic, hydrophobic, and hydrogen-bond donor and acceptor fields, respectively; d, e—Correlation between actual and predicted pK_i of 3D-QSAR models (d, CoMFA model; e, CoMSIA model); f-i—Contour maps of CoMFA (f, g) and CoMSIA (h, i) models (Scattered green areas are sterically favored, yellow areas are unfavorable. Red regions are negative potential favored, while blue areas are unfavorable. Yellow areas are regions where hydrophobic groups are desirable, white areas are undesirable. Large cyan regions represent areas where hydrogen donors on the ligand are predicted to enhance binding, while purple areas are predicted to disfavor binding. Regions in magenta are hydrogen acceptor favored, and regions in red are unfavorable. The most promising compound 4 was superposed as the reference molecule in these maps. These contour maps generated depict regions with scaled coefficients greater than 80% (favored) or less than 20% (disfavored))

2.3 Molecular docking

To further understand the binding nature of these novel benzyloxybenzene ligands to Aβ fiber, docking simulations were also carried out. Several previously reported articles, which investigated molecular interactions of Aβ binding ligands with fiber, revealed that the hydrophobic cleft formed between Val18 and Phe20 was the most probable and feasible binding site[26]. Thus resides 16-KLVFFA-21 on the surface of Aβ fiber (PDB 2LMO) were selected for docking with grids centered on this region. Computational docking results revealed that all the benzyloxybenzene analogues, similar to IMPY, inserted into the hydrophobic Val18_Phe20 channel with long molecular axes oriented longitudinally to the fiber axis (Fig. 3a-c). This observation sufficiently verified the results from inhibition binding assay that benzyloxybenzene ligands shared the same binding site with IMPY. Analyzing the binding energy (ΔG) of top-ranking conformations for each analogue gained a qualitative linear correction with experimental pK_i in high R^2 value of 0.875 (Fig. 3d). After packing into the binding patch, all the flexible benzyloxybenzene analogues with substituents at the para-position tended to be locked into a near-flat conformation. For instance, the dihedral angle between the two phenyl planes in ligand **4** decreased from 60.02° to 40.40° after docking. The near-flat characteristic of these ligands favored their impaction into the hydrophobic channel with low binding energies, which exactly justified their high binding affinities. On the contrary, ligand **6** and **21** with ortho-substituents held greater degrees of nonplanarity in the top-ranking poses and were harder to be accommodated in the shallow binding channel, correspondingly resulting in high binding energies and low binding affinities fit the binding pocket. This binding mode generated by docking simulations was in good agreement with the experimental data and 3D-QSAR models.

2.4 Radiochemistry

Iodinated ligands **4** and **22** with high binding affinities were selected for radiolabeling and further biological evaluations. To evaluate the impact of iodine's position on biological properties, ligand **24** with iodine atom on the opposite side of the backbone was also pitched on. Radiochemical synthesis of [^{125}I]**4**, [^{125}I]**24**, and [^{125}I]**22** was reproducibly achieved via an iododestannylation reaction in high radiochemical yields of 86.2%, 94.9%, and 92.9%, respectively. After purification by HPLC, the radiochemical purity of these radiotracers was higher than 95%. The radiochemical identities of ^{125}I-labeled tracers were verified by comparison of the retention times with that of the nonradioactive compounds (Supporting Information Table S2, Fig. S3).

2.5 In vitro autoradiography

We further confirmed the specific binding of [^{125}I]**4**, [^{125}I]**24** and [^{125}I]**22** to Aβ plaques with in vitro autoradiography on brain sections from AD patients and Tg model mice. As shown in Fig. 4, marked labeling of Aβ plaques was observed on sections of AD patients and Tg mice

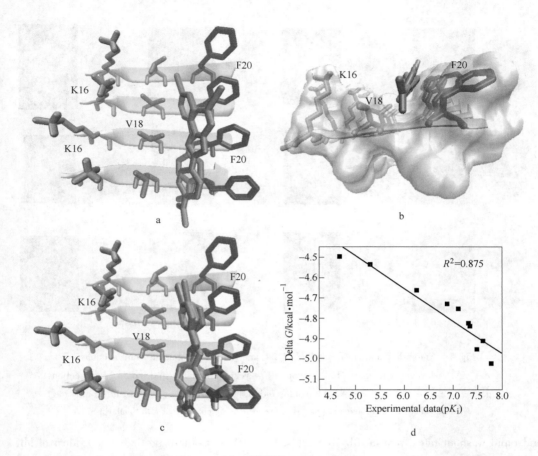

Fig. 3 Computational docking models of IMPY and benzyloxybenzene analogues on Aβ fiber

a—IMPY and ligand 4 bound along the hydrophobic grooves formed between Val18 and Phe20 running longitudinally to the long axis of the β-sheet; b—Identical binding model viewed down the fiber axis with molecular surface of the protein represented by transparent material; c—Lowest energy docked conformations of nine benzyloxybenzene derivatives; d—ΔG correlated well with experimental pK_i in reasonably high R^2 value of 0.875

brain, and the location of accumulated radioactivity was in conformity with the position of Aβ plaques stained with thioflavin-S, a conventional dye usually used for plaque staining (Fig. 4a, e, i and c, g, k), while the healthy human and wild-type mice brain showed no such concentration of radioactivity (Fig. 4b, f, j and d, h, l).

2.6 In vivo biodistribution

To assess BBB penetration and washout properties of [^{125}I]**4**, [^{125}I]**24**, and [^{125}I]**22**, partition coefficient determination and biodistribution assays in normal ICR mice were implemented. The log D values determined for [^{125}I]**4**, [^{125}I]**24**, and [^{125}I]**22** were 4.00 ± 0.08, 4.16 ± 0.29, and 2.89 ± 0.09, respectively, indicating that the three ^{125}I-labeled tracers have appropriate lipophilicity to penetrate BBB (Supporting Information Table S3). As shown in Fig. 5, all the three ^{125}I-labeled ligands exhibited high initial brain uptakes (>4% ID/g at 2min postinjection)

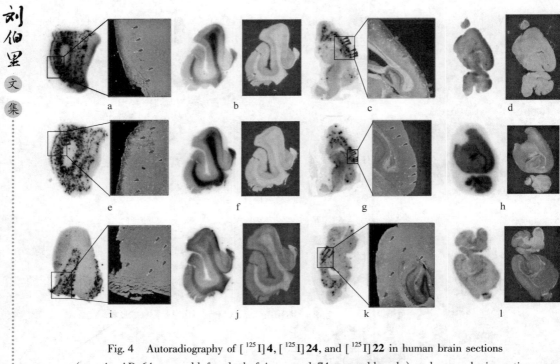

Fig. 4　Autoradiography of [^{125}I]**4**, [^{125}I]**24**, and [^{125}I]**22** in human brain sections
(a, e, i—AD, 64 years old, female; b, f, j—normal, 74 years old, male) and mouse brain sections
(c, g, k—Tg mouse, APPswe/PSEN1, 11 months old; d, h, l—wild-type, C57BL6, 11 months old)
(The presence and distribution of plaques in the sections were confirmed with thioflavin-S)

and rapid washout rates from healthy brain (<0.5% ID/g at 60min postinjection). Most of all, [^{125}I]**4** provided the highest brain uptake with 6.18% ID/g at 2min and fast clearance from normal brain with brain$_{2min}$/brain$_{60min}$ ratio of 16.3, which were superior to the results of [^{125}I]IMPY gained under the same procedure. Moreover, [^{125}I]IMPY displayed slender deiodination, while no increase of accumulated radioactivity in thyroids was observed over time in the biodistribu-

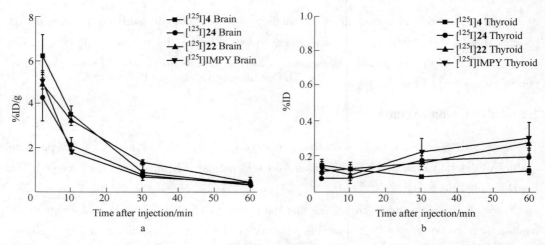

Fig. 5　Comparison of brain and thyroid uptake of [^{125}I]**4**, [^{125}I]**24**, [^{125}I]**22**, and [^{125}I]IMPY in normal ICR mice

tion experiment of [^{125}I]**4**. The excellent pharmacokinetics of the three ^{125}I-labeled ligands strongly demonstrated that the novel chemical scaffold was qualified for imaging Aβ plaques in the brain, and [^{125}I]**4** with optimal properties was selected for further investigation.

2.7 In vitro biostability of [^{125}I]4

When incubated in mouse plasma at 37 ℃, [^{125}I]IMPY underwent severe chemical degradation (Supporting Information Fig. S4c). At 60min, only half of radioactivity remained as parent tracer and three other metabolized radioactive substances with shorter retention times were detected. Conversely, [^{125}I]**4** displayed excellent in vitro biostability when incubated in mouse plasma and liver homogenate under the same condition with 100% of radioactivity existed as the intact form (Supporting Information Fig. S4a, 4b). The in vitro experiments in both plasma and liver homogenate sufficiently demonstrated high biostability of [^{125}I]**4**, which is an indispensable criterion for drug delivery.

2.8 In vivo biostability of [^{125}I]4

The in vivo metabolic stability of [^{125}I]**4** was evaluated by HPLC analysis of each organ sample of normal ICR mice obtained at different postinjection time points (Fig. 6). Two polar radioactive metabolites were detected. In brain, only one metabolic product, metabolite 2, was observed. The percentage of the parent tracer was 95.0% at 2min, and almost 40% of the radioactivity remained to be the intact form at 60min. In plasma, the fraction of unchanged [^{125}I]**4** decreased rapidly from 71.5% at 2min to 10.6% at 60min, whereas the major metabolite 1 increased to 85.3% at 60min. A small amount of metabolite 2 was also found in plasma. In the liver at 2min, 42.4% parent tracer was presented and the rest radioactivity was partitioned 42.9% as metabolite 1 and 14.7% as metabolite 2. From 30 to 60min, almost no parent tracer was discovered, and the radioactivity presented mainly as metabolite 1. In urine sample, radioactivity was detected only after 10min and more than three-quarters existed as metabolite 1. Compared with other tissues, the amount of radioactivity in feces was too low to be detected by HPLC. This preliminary metabolism study illustrated that [^{125}I]**4** was converted to two radiochemical forms. Metabolite 2 was predominately generated in brain, while metabolite 1 was mainly yielded in liver and cannot cross the BBB.

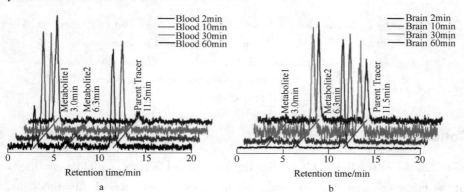

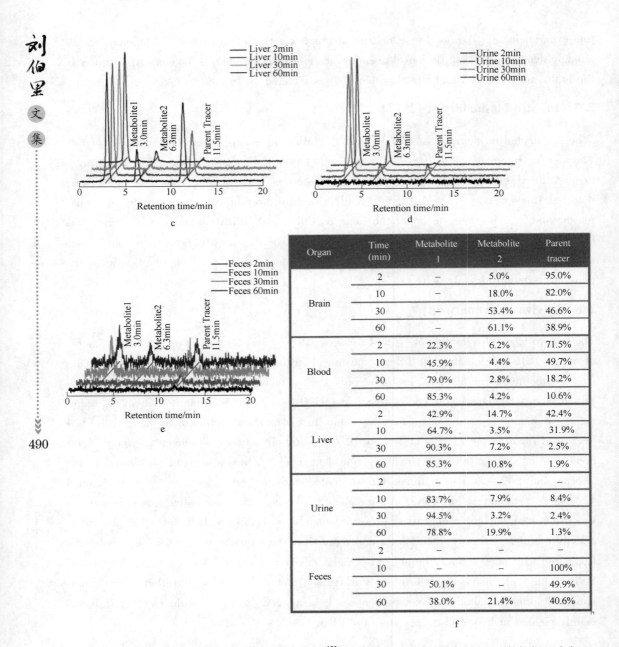

Fig. 6 HPLC profiles for in vivo biostabilities of $[^{125}I]$**4** in ICR mice blood (a), brain (b), liver (c), urine (d) and feces (e), and Percentages of the parent tracer, metabolite 1 and metabolite 2 in each organ at 2min, 10min, 30min, and 60min time points (f)
(HPLC conditions: Venusil MP C18 column (Agela Technologies, 5μm, 4.6mm×250mm), $CH_3CN/H_2O = 80\%/20\%$, 1mL/min)

3 Conclusion

Different from the conventional Aβ binding ligands, a novel flexible scaffold, benzyloxybenzene, without the big π-conjugated system and rigid planarity was discovered. Amazingly, benzyloxybenzene derivatives competed well against IMPY and exhibited various binding affinities to $A\beta_{42}$ aggregates, with K_i values differing from 12. 0nM to larger than 20000nM. Molecular docking successfully determined that this series of flexible ligands and IMPY shared the same hydrophobic Val18_Phe20 channel on the flat spine of Aβ fiber. Together with 3D-QSAR studies, structure-activity relationships were clarified. Severe decline in binding affinities was found from para-substituted ligands to ortho-substituted ones. Subsequently, [^{125}I]**4**, [^{125}I]**24**, and [^{125}I]**22** with high binding affinities (K_i = 24. 3nM, 49. 4nM, and 17. 6nM, respectively) were reproducibly synthesized in radiochemical yields higher than 85%. Autoradiography in vitro showed that the ^{125}I-labeled ligands specifically labeled Aβ plaques on sections of AD patients and Tg mice brain. In biodistribution, the ^{125}I-labeled ligands displayed excellent in vivo pharmacokinetics. Especially, [^{125}I]**4** exhibited the highest brain uptake with 6. 18% ID/g at 2min postinjection and rapid clearance property with brain$_{2min}$/brain$_{60min}$ ratio of 16. 3. Contrary to IMPY, no deiodination of [^{125}I]**4** was observed. More importantly, [^{125}I]**4** displayed good in vitro biostability in mouse plasma and liver homogenate (100% intact form at 60min) and considerably overcame the instability of [^{125}I]IMPY which was a limiting factor for its clinical use. In metabolism studies, two polar radioactive metabolites of [^{125}I]**4** were detected. In the brain, [^{125}I]**4** showed agreeable in vivo biostability with almost 40% of the radioactivity remained to be the parent tracer until 60min.

In conclusion, this innovative finding could offer new insights for imaging of Aβ plaques in Alzheimer's brains. The rigid and planer scaffold with big π-conjugated system was not indispensable for Aβ probes, and the binding pocket of Aβ fiber exhibited considerable tolerance to a certain degree of distortion of small molecule ligands. This would provide new hints for developing Aβ targeting ligands with flexible framework to enable improved localization and quantitative imaging of Aβ plaques. In addition, we believe that [^{125}I]**4** with optimal qualities could potentially be translated into clinical practice for SPECT imaging of Aβ plaques in AD brains.

4 Experimental Section

4.1 General remarks

All reagents used for chemical synthesis were commercial products and were used without further purification. [^{125}I]NaI was purchased from PerkinElmer. ^{1}H NMR (400MHz) and ^{13}C NMR (100MHz) spectra were acquired on Bruker Avance III NMR spectrometers in CDCl$_3$ or DMSO-d_6 solutions at room temperature with trimethylsilyl (TMS) as an internal standard. Mass spectra were acquired with a GCT CA127 Micronass UK instrument. X-ray crystallography data were collected on a Bruker Smart APEX II diffractometer (Bruker Co., Germany). HPLC was performed on a Shimadzu SCL-20 AVP (which was equipped with a Bioscan Flow Count 3200 NaI/PMT γ-radiation scintillation detector and a SPD-20A UV detector, λ = 254nm) and a Venusil MP C18 reverse phase column (Agela Technologies, 5μm, 4. 6mm ×

250mm) eluted with a binary gradient system (acetonitrile : water = 80% : 20%) at a 1.0mL/min flow rate. Fluorescent observation was performed on the Axio Observer Z1 inverted fluorescence microscope (Zeiss, Germany) equipped with a DAPI filter set (excitation, 405nm). Normal ICR mice (5 weeks, male) were used for biodistribution experiments. Human brain sections of an autopsy-confirmed AD (64 years old, female), and a control subject (74 years old, male) were obtained from Chinese Brain Bank Center. Transgenic mice (APPswe/PSEN1, male, 11 months old) and wild-type mice (C57BL6, male, 11 months old) were purchased from the Institute of Laboratory Animal Science, Chinese Academy of Medical Sciences. All protocols requiring the use of mice were approved by the animal care committee of Beijing Normal University.

1-Iodo-4-((4-nitrophenoxy)methyl)benzene (**1**). To a solution of 4-nitrophenol (695.6mg, 5.0mmol) and 1-(bromomethyl)-4-iodobenzene (1.48g, 5.0mmol) in anhydrous DMF (5mL) was added K_2CO_3 (1.38g, 10.0mmol). The resulting mixture was stirred at 90℃ for 2h; after cooling to room temperature, 50mL of water was added and a white precipitate was formed. The precipitate was collected by filtration, washed with 50mL of water, and recrystallized from methanol to give **1** as a white solid (1.69g, 95.2%). HPLC: 7.51min, 99.8%; mp: 147.4-148.1℃. ^{1}H NMR (400MHz, $CDCl_3$) δ 8.21(d, J = 9.2Hz, 2H), 7.75(d, J = 8.3Hz, 2H), 7.18(d, J = 8.3Hz, 2H), 7.01(d, J = 9.2Hz, 2H), 5.10(s, 2H). ^{13}C NMR (101MHz, $CDCl_3$) δ 163.37, 141.85, 137.91, 135.19, 129.24, 125.94, 114.84, 94.10, 69.97. MS (EI): m/z calcd for $C_{13}H_{10}INO_3$ 355; found 355 M^+.

1-(4-Iodobenzyloxy)-3-nitrobenzene (**2**). The procedure described above for the preparation of **1** was employed to afford **2** as a white solid (345.5mg, 97.3%). HPLC: 8.53 min, 99.9%; mp: 75.2-75.9℃. ^{1}H NMR (400MHz, $CDCl_3$) δ 7.85(d, J = 8.0Hz, 1H), 7.80(s, 1H), 7.74(d, J = 8.2Hz, 2H), 7.45(t, J = 8.2Hz, 1H), 7.30-7.26(m, 1H), 7.19(d, J = 8.1Hz, 2H), 5.09(s, 2H). ^{13}C NMR (101MHz, $CDCl_3$) δ 158.91, 149.19, 137.84, 135.41, 130.06, 129.29, 121.84, 116.18, 109.21, 93.99, 69.88. MS (EI): m/z calcd for $C_{13}H_{10}INO_3$ 355; found 355 M^+.

1-(4-Iodobenzyloxy)-2-nitrobenzene (**3**). The procedure described above for the preparation of **1** was employed to afford **3** as a yellow solid (238.9mg, 67.3%). HPLC: 6.03min, 99.5%; mp: 70.6-71.5℃. ^{1}H NMR (400MHz, $CDCl_3$) δ 7.87(dd, J = 8.1, 1.4Hz, 1H), 7.73(d, J = 8.2Hz, 2H), 7.56-7.46(m, 1H), 7.22(d, J = 8.1Hz, 2H), 7.09-7.04(m, 2H), 5.18(s, 2H). ^{13}C NMR (101MHz, $CDCl_3$) δ 151.62, 140.23, 137.80, 135.27, 134.07, 128.78, 125.73, 120.90, 115.03, 93.80, 70.45.

1-Iodo-4-((4-methoxyphenoxy)methyl)benzene (**4**). The procedure described above for the preparation of **1** was employed to afford **4** as a white solid (255.3mg, 75.1%). HPLC: 8.28min, 99.5%; mp: 128.9-130.2℃. ^{1}H NMR (400MHz, $CDCl_3$) δ 7.71(d, J = 8.2Hz, 2H), 7.17(d, J = 8.2Hz, 2H), 6.89(d, J = 9.2Hz, 2H), 6.83(d, J = 9.2Hz, 2H), 4.96(s, 2H), 3.77(s, 3H). ^{13}C NMR (101MHz, $CDCl_3$) δ 154.18, 152.68, 137.63, 137.08, 129.25, 115.92, 114.72, 93.30, 70.07, 55.72. HRMS (EI): m/z calcd for $C_{14}H_{13}IO_2$ 339.9960;

found 339. 9966 M$^+$.

1-(4-Iodobenzyloxy)-3-methoxybenzene (**5**). The procedure described above for the preparation of **1** was employed to afford **5** as a white solid (211.7mg, 62.2%). HPLC: 8.68min, 98.5%; mp: 77.4-78.7℃. ^{1}H NMR (400MHz, CDCl$_3$) δ 7.71 (d, J = 8.1Hz, 2H), 7.21-7.17 (m, 3H), 6.57-6.50 (m, 2H), 4.99 (s, 2H), 3.79 (s, 3H). ^{13}C NMR (101MHz, CDCl$_3$) δ 160.87, 159.75, 137.65, 136.71, 129.94, 129.26, 106.93, 106.73, 101.44, 93.42, 69.30, 55.27. MS (EI): m/z calcd for C$_{14}$H$_{13}$IO$_2$ 340; found 340 M$^+$.

1-(4-Iodobenzyloxy)-2-methoxybenzene (**6**). The procedure described above for the preparation of **1** was employed to afford **6** as a white solid (288.1mg, 84.7%). HPLC: 6.63min, 98.7%; mp: 110.4-110.8℃. ^{1}H NMR (400MHz, CDCl$_3$) δ 7.69 (d, J = 8.1Hz, 2H), 7.19 (d, J = 8.0Hz, 2H), 6.97-6.91 (m, 2H), 6.85 (d, J = 3.8Hz, 2H), 5.10 (s, 2H), 3.89 (s, 3H). ^{13}C NMR (101MHz, CDCl$_3$) δ 149.78, 147.89, 137.58, 137.00, 129.11, 121.76, 120.78, 114.41, 112.03, 93.25, 70.41, 55.92. MS (EI): m/z calcd for C$_{14}$H$_{13}$IO$_2$ 340; found 340 M$^+$.

4-(4-Iodobenzyloxy)phenol (**7**). To a solution of hydroquinone (110.1mg, 1.0mmol) and 1-(bromomethyl)-4-iodobenzene (296.9mg, 1.0mmol) in anhydrous DMF (5mL) was added K$_2$CO$_3$ (276.4mg, 2.0mmol). The resulting mixture was stirred at 90℃ for 2h, after cooling to room temperature, 50mL of water was added and extracted by CH$_2$Cl$_2$ (3 × 10mL). Combined organic layers were dried over MgSO$_4$, filtered, and concentrated under a vacuum. The crude mixture was purified by silica gel chromatography (petroleum ether/AcOEt = 4/1) to give **7** as a white solid (49.3mg, 15.1%). HPLC: 3.75min, 99.3%; mp: 152.2-153.7℃. ^{1}H NMR (400MHz, CDCl$_3$) δ 7.71 (d, J = 8.3Hz, 2H), 7.17 (d, J = 8.2Hz, 2H), 6.85-6.81 (m, 2H), 6.78-6.74 (m, 2H), 4.95 (s, 2H), 4.41 (s, 1H). ^{13}C NMR (101MHz, CDCl$_3$) δ 152.75, 149.83, 137.65, 137.00, 129.27, 116.09, 115.91, 93.35, 70.11. MS (EI): m/z calcd for C$_{13}$H$_{11}$IO$_2$ 326; found 326 M$^+$ (different from the procedure in ref 45, 4-(4-iodobenzyloxy)phenol was gained though a SN2 reaction between hydroquinone and 1-(bromomethyl)-4-iodobenzene).

3-(4-Iodobenzyloxy)phenol (**8**). The procedure described above for the preparation of **7** was employed to afford 8 as a white solid (93.1mg, 28.5%). HPLC: 3.95min, 98.6%; mp: 109.9-110.6℃. ^{1}H NMR (400MHz, CDCl$_3$) δ 7.71 (d, J = 7.9Hz, 2H), 7.17 (d, J = 8.2Hz, 2H), 7.16-7.09 (m, 1H), 6.54 (d, J = 8.6Hz, 1H), 6.46-6.44 (m, 2H), 4.98 (s, 2H), 4.74 (s, 1H). ^{13}C NMR (101MHz, CDCl$_3$) δ 159.87, 156.65, 137.67, 136.59, 130.24, 129.23, 108.32, 107.39, 102.55, 93.45, 69.36. MS (EI): m/z calcd for C$_{13}$H$_{11}$IO$_2$ 326; found 326 M$^+$.

2-(4-Iodobenzyloxy)phenol (**9**). The procedure described above for the preparation of **7** was employed to afford **9** as yellow oil (109.8mg, 33.7%). HPLC: 4.25min, 99.9%; mp: 62.1-63.4℃. ^{1}H NMR (400MHz, CDCl$_3$) δ 7.74 (d, J = 8.2Hz, 2H), 7.17 (d, J = 8.0Hz, 2H), 6.98-6.95 (m, 1H), 6.93-6.88 (m, 2H), 6.85-6.81 (m, 1H), 5.62 (s, 1H), 5.06 (s, 2H). ^{13}C NMR (101MHz, CDCl$_3$) δ 145.91, 145.56, 137.88, 136.09, 129.54, 122.13, 120.20, 114.98, 112.30,

93.99, 70.44. MS (EI): m/z calcd for $C_{13}H_{11}IO_2$ 326; found 326 M$^+$.

1-Fluoro-4-(4-iodobenzyloxy)benzene (**10**). The procedure described above for the preparation of **1** was employed to afford **10** as a white solid (253.1mg, 77.1%). HPLC: 8.92min, 98.0%; mp: 62.3-62.8℃. ^{1}H NMR (400MHz, CDCl$_3$) δ 7.72(d, J = 8.2Hz, 2H), 7.17(d, J = 8.2Hz, 2H), 7.00-6.95 (m, 2H), 6.91-6.85 (m, 2H), 4.97 (s, 2H). ^{13}C NMR (101MHz, CDCl$_3$) δ 157.49 (d, J = 238.9Hz, 1C), 154.61 (d, J = 2.1Hz, 1C), 137.71, 136.58, 129.23, 115.95 (d, J = 12.3Hz, 2C), 115.88 (d, J = 18.8Hz, 2C), 93.51, 70.00. MS (EI): m/z calcd for $C_{13}H_{10}FIO$ 328; found 328 M$^+$.

1-Chloro-4-(4-iodobenzyloxy)benzene (**11**). The procedure described above for the preparation of **1** was employed to afford **11** as a white solid (136.2mg, 79.0%). HPLC: 12.86min, 99.6%; mp: 107.5-108.1℃. ^{1}H NMR (400MHz, CDCl$_3$) δ 7.72(d, J = 8.2Hz, 2H), 7.24 (d, J = 8.9Hz, 2H), 7.16 (d, J = 8.2Hz, 2H), 6.87 (d, J = 8.9Hz, 2H), 4.98 (s, 2H). ^{13}C NMR (101MHz, CDCl$_3$) δ 157.06, 137.72, 136.30, 129.40, 129.19, 126.08, 116.15, 93.59, 69.60. MS (EI): m/z calcd for $C_{13}H_{10}ClIO$ 344; found 344 M$^+$.

1-Bromo-4-(4-iodobenzyloxy)benzene (**12**). The procedure described above for the preparation of **1** was employed to afford **12** as a white solid (302.5mg, 77.8%). HPLC: 14.61min, 99.3%; mp: 122.6-123.5℃. ^{1}H NMR (400MHz, CDCl$_3$) δ 7.71(d, J = 8.2Hz, 2H), 7.37 (d, J = 9.0Hz, 2H), 7.16 (d, J = 8.2Hz, 2H), 6.82 (d, J = 8.9Hz, 2H), 4.97 (s, 2H). ^{13}C NMR (101MHz, CDCl$_3$) δ 157.58, 137.74, 136.26, 132.35, 129.20, 116.68, 113.39, 93.62, 69.53. MS (EI): m/z calcd for $C_{13}H_{10}BrIO$ 388; found 388 M$^+$ (compared with the method in Ref. [46], the reaction was carried out in DMF instead of acetone and gave 1-bromo-4-(4-iodobenzyloxy)benzene in a higher yield of 77.8%).

1-Iodo-4-(4-iodobenzyloxy)benzene (**13**). The procedure described above for the preparation of **1** was employed to afford **13** as a white solid (436.0mg, 89.3%). HPLC: 17.49min, 98.2%; mp: 135.0-135.9℃. ^{1}H NMR (400MHz, CDCl$_3$) δ 7.71(d, J = 8.2Hz, 2H), 7.56 (d, J = 8.7Hz, 2H), 7.15 (d, J = 8.1Hz, 2H), 6.72 (d, J = 8.8Hz, 2H), 4.97 (s, 2H). ^{13}C NMR (101MHz, CDCl$_3$) δ 158.33, 138.29, 137.72, 136.21, 129.17, 117.25, 93.60, 83.31, 69.37. MS (EI): m/z calcd for $C_{13}H_{10}I_2O$ 436; found 436 M$^+$.

1-Iodo-4-(phenoxymethyl)benzene (**14**). The procedure described above for the preparation of **1** was employed to afford **14** as a white solid (310.1mg, 72.1%). HPLC: 9.59min, 99.2%; mp: 96.7-97.6℃. ^{1}H NMR (400MHz, CDCl$_3$) δ 7.72(d, J = 8.2Hz, 2H), 7.32-7.27(m, 2H), 7.19 (d, J = 8.1Hz, 2H), 6.99-6.94 (m, 3H), 5.02 (s, 2H). ^{13}C NMR (101MHz, CDCl$_3$) δ 158.49, 137.64, 136.81, 129.51, 129.22, 121.15, 114.84, 93.37, 69.20. MS (EI): m/z calcd for $C_{13}H_{11}IO$ 310; found 310 M$^+$ (different from the procedure in Ref. [47], 1-iodo-4-(phenoxymethyl)benzene was gained though a SN2 reaction between phenol and 1-(bromomethyl)-4-iodobenzene).

1-Tert-Butyl-4-(4-iodobenzyloxy)benzene (**15**). The procedure described above for the preparation of **1** was employed to afford **15** as a white solid (366.2mg, 87.4%). HPLC:

27.14min, 98.3%; mp: 91.9-93.0℃. ^{1}H NMR (400MHz, CDCl$_3$) δ 7.71 (d, J = 8.2Hz, 2H), 7.31 (d, J = 8.8Hz, 2H), 7.18 (d, J = 8.2Hz, 2H), 6.89 (d, J = 8.8Hz, 2H), 4.99 (s, 2H), 1.30 (s, 9H). ^{13}C NMR (101MHz, CDCl$_3$) δ 156.28, 143.85, 137.62, 137.04, 129.23, 126.28, 114.26, 93.29, 69.30, 34.07, 31.50. MS (EI): m/z calcd for C$_{17}$H$_{19}$IO 366; found 366 M$^+$.

4-(4-Iodobenzyloxy)aniline (**16**). To a solution of **1** (1.42g, 4.0mmol) and SnCl$_2$·2H$_2$O (1.66g, 8.0mmol) in EtOH (25mL) was added concentrated HCl (2mL). The resulting mixture was stirred at 80℃ for 2h; after cooling to room temperature, 1M NaOH (30mL) was added and extracted by ethyl acetate (3 × 10mL). Combined organic layers were dried over MgSO$_4$, filtered, and concentrated under a vacuum. The crude mixture was purified by silica gel chromatography (petroleum ether/AcOEt = 2/1) to give **16** as a pink solid (643.3mg, 49.5%). HPLC: 4.16min, 98.4%; mp: 138.6-140.0℃. ^{1}H NMR (400MHz, CDCl$_3$) δ 7.70 (d, J = 8.2Hz, 2H), 7.16 (d, J = 7.9Hz, 2H), 6.78 (d, J = 8.7Hz, 2H), 6.64 (d, J = 8.8Hz, 2H), 4.93 (s, 2H), 3.44 (s, 2H). ^{13}C NMR (101MHz, DMSO-d_6) δ 149.39, 142.65, 137.65, 137.02, 129.66, 115.75, 114.84, 93.42, 69.11. MS (EI): m/z calcd for C$_{13}$H$_{12}$INO 325; found 325 M$^+$.

3-(4-Iodobenzyloxy)aniline (**17**). The procedure described above for the preparation of **16** was employed to afford **17** as a white solid (695.5mg, 73.1%). HPLC: 4.50min, 99.4%; mp: 153.9-154.8℃. ^{1}H NMR (400MHz, CDCl$_3$) δ 7.70 (d, J = 8.1Hz, 2H), 7.17 (d, J = 8.0Hz, 2H), 7.07 (t, J = 8.0Hz, 1H), 6.40-6.34 (m, 3H), 4.97 (s, 2H). ^{13}C NMR (101MHz, DMSO-d_6) δ 158.95, 143.26, 137.13, 136.95, 129.97, 129.69, 110.51, 106.88, 104.24, 93.73, 68.31. MS (EI): m/z calcd for C$_{13}$H$_{12}$INO 325; found 325 M$^+$.

2-(4-Iodobenzyloxy)aniline (**18**). The procedure described above for the preparation of **16** was employed to afford **18** as a white solid (99.3mg, 11.4%). HPLC: 4.88min, 99.1%; mp: 99.1-100.1℃; ^{1}H NMR (400MHz, CDCl$_3$) δ 7.71 (d, J = 8.3Hz, 2H), 7.19 (d, J = 8.3Hz, 2H), 6.85-6.78 (m, 3H), 6.75-6.71 (m, 1H), 5.03 (s, 2H). ^{13}C NMR (101MHz, CDCl$_3$) δ 146.14, 137.66, 136.90, 136.48, 129.34, 121.73, 118.39, 115.30, 112.14, 93.46, 69.73. MS (EI): m/z calcd for C$_{13}$H$_{12}$INO 325; found 325 M$^+$.

4-(4-Iodobenzyloxy)-N-methylaniline (**19**). A solution of NaOCH$_3$ (54.0mg, 1.0mmol) in methanol (5mL) was added to a mixture of **16** (162.6mg, 0.5mmol) and paraformaldehyde (60.0mg, 2.0mmol) in methanol (30mL) dropwise. The resulting mixture was stirred under reflux for 2h. After cooling to room temperature, NaBH$_4$ (75.6mg, 2.0mmol) was added and the solution was stirred under reflux for 2h. The solvent was evaporated under a vacuum. Then 1M NaOH (50mL) was added and a white precipitate was formed. The precipitate was collected by filtration, washed with 50mL water, and recrystallized from methanol to give **19** as a pink solid (152.4mg, 89.9%). HPLC: 6.34min, 99.2%; mp: 93.9-95.2℃. ^{1}H NMR (400MHz, CDCl$_3$) δ 7.70 (d, J = 8.2Hz, 2H), 7.17 (d, J = 8.1Hz, 2H), 6.84 (d, J = 8.7Hz, 2H), 6.58 (d, J = 8.5Hz, 2H), 4.94 (s, 2H), 2.81 (s, 3H). ^{13}C NMR (101MHz, CDCl$_3$) δ 150.89, 144.11, 137.55, 137.42, 129.28, 116.23, 116.21, 113.50, 93.16, 70.26, 70.24, 31.49. MS

(EI): m/z calcd for $C_{14}H_{14}INO$ 339; found 339 M^+.

3-(4-Iodobenzyloxy)-*N*-methylaniline (**20**). A solution of $NaOCH_3$ (54.0mg, 1.0mL) in methanol (5mL) was added to a mixture of **17** (162.6mg, 0.5mmol) and paraformaldehyde (60.0mg, 2.0mmol) in methanol (30mL) dropwise. The resulting mixture was stirred under reflux for 2h. After cooling to room temperature, $NaBH_4$ (75.6mg, 2.0mmol) was added and the solution was stirred under reflux for 2h. The solvent was evaporated under a vacuum. Then 1M NaOH (50mL) was added and extracted by CH_2Cl_2 (3 × 10mL). Combined organic layers were dried over $MgSO_4$, filtered, and concentrated under a vacuum. The crude mixture was purified by silica gel chromatography (petroleum ether/AcOEt = 4/1) to give **20** as a colorless oil (71.0mg, 41.8%). HPLC: 6.60min, 99.9%; mp: 45.3-46.6℃. ^{1}H NMR (400MHz, $CDCl_3$) δ 7.71(d, J = 8.1Hz, 2H), 7.19(d, J = 8.0Hz, 2H), 7.09(t, J = 8.0Hz, 1H), 6.32(d, J = 8.0Hz, 1H), 6.27(d, J = 7.9Hz, 1H), 6.24(s, 1H), 4.99(s, 2H), 2.83(s, 3H). ^{13}C NMR (101MHz, $CDCl_3$) δ 159.84, 150.77, 137.59, 137.09, 129.94, 129.25, 106.16, 103.06, 99.24, 93.26, 69.11, 30.68. MS (EI): m/z calcd for $C_{14}H_{14}INO$ 339; found 339 M^+.

2-(4-Iodobenzyloxy)-*N*-methylaniline (**21**). To a solution of **18** (50.0mg, 0.15mmol) and K_2CO_3 (41.5mg, 0.30mmol) in acetone (10mL) was added CH_3I (32.6mg, 0.23). The resulting mixture was stirred at room temperature for 10h. Then 50mL water was added and extracted by CH_2Cl_2 (3 × 10mL). Combined organic layers were dried over $MgSO_4$, filtered, and concentrated under a vacuum. The crude mixture was purified by silica gel chromatography (petroleum ether/AcOEt = 4/1) to give **21** as a yellow oil (34.2mg, 67.2%). HPLC: 8.19min, 98.1%. ^{1}H NMR (400MHz, $CDCl_3$) δ 7.71 (d, J = 8.3Hz, 2H), 7.17 (d, J = 8.4Hz, 2H), 6.92(td, J = 7.7, 1.3Hz, 1H), 6.78(d, J = 7.9Hz, 1H), 6.66-6.60(m, 2H), 5.00(s, 2H), 4.24(s, 1H), 2.85(s, 3H). ^{13}C NMR (101MHz, $CDCl_3$) δ 145.77, 139.59, 137.66, 136.88, 129.43, 121.96, 116.15, 110.88, 109.60, 93.48, 69.72, 30.31. MS (EI): m/z calcd for $C_{14}H_{14}INO$ 339; found 339 M^+.

4-(4-Iodobenzyloxy)-*N*,*N*-dimethylaniline (**22**). To a solution of **16** (162.6mg, 0.5mmol) and paraformaldehyde (150.0mg, 5.0mmol) in acetic acid (20mL) was added $NaBH_3CN$ (157.0mg, 2.5). The resulting mixture was stirred at room temperature for 24h. Then 1M NaOH (20mL) was added and a white precipitate was formed. The precipitate was collected by filtration, washed with 50mL water, and recrystallized from methanol to give **22** as a white solid (168.4mg, 95.4%). HPLC: 10.63min, 98.4%; mp: 128.2-129.3℃. ^{1}H NMR (400MHz, $CDCl_3$) δ 7.69(d, J = 8.2Hz, 2H), 7.17(d, J = 8.0Hz, 2H), 6.88(d, J = 8.9Hz, 2H), 6.82-6.68(s, 2H), 4.95(s, 2H), 2.87(s, 6H). ^{13}C NMR (101MHz, $CDCl_3$) δ 150.92, 145.99, 137.59, 137.41, 129.28, 115.95, 114.77, 93.19, 70.16, 41.73. HRMS (EI): m/z calcd for $C_{15}H_{16}INO$ 353.0277; found 353.0282 M^+.

3-(4-Iodobenzyloxy)-*N*,*N*-dimethylaniline (**23**). The procedure described above for the preparation of **22** was employed to afford **23** as a white solid (171.3mg, 97.1%). HPLC: 11.30min, 99.9%; mp: 68.8-70.1℃. ^{1}H NMR (400MHz, $CDCl_3$) δ 7.71 (d, J = 8.2Hz, 2H), 7.19(d, J = 8.2Hz, 2H), 7.15(t, J = 8.1Hz, 1H), 6.43-6.30(m, 3H), 5.00(s, 2H),

2.94(s, 6H). ^{13}C NMR (101MHz, CDCl$_3$) δ 159.68, 152.02, 137.62, 137.18, 129.77, 129.31, 106.15, 102.16, 100.03, 93.26, 69.20, 40.55. MS (EI): m/z calcd for C$_{15}$H$_{16}$INO 353; found 353 M$^+$.

1-Iodo-4-(4-methoxybenzyloxy)benzene (**24**). The procedure described above for the preparation of **1** was employed to afford **24** as a white solid (906.3mg, 88.9%). HPLC: 8.55min, 99.7%; mp: 130.3-131.5℃. ^{1}H NMR (400MHz, CDCl$_3$) δ 7.55(d, J = 9.0Hz, 2H), 7.33(d, J = 8.7Hz, 2H), 6.91(d, J = 8.7Hz, 2H), 6.74(d, J = 8.9Hz, 2H), 4.95(s, 2H), 3.81(s, 3H). ^{13}C NMR (101MHz, CDCl$_3$) δ 159.59, 158.71, 138.22, 129.19, 128.55, 117.36, 114.08, 82.95, 69.91, 55.31. HRMS (EI): m/z calcd for C$_{14}$H$_{13}$IO$_2$ 339.9960; found 339.9964 M$^+$ (compared with the method in Ref. [48], the reaction was carried out in DMF without the use of tetra-n-butylammonium iodide and gave 1-iodo-4-(4-methoxybenzyloxy)benzene in a comparable yield).

(4-Iodobenzyl)(4-methoxyphenyl)sulfane (**25**). The procedure described above for the preparation of **1** was employed to afford **25** as a white solid (712.4mg, 94.7%). mp: 87.9-88.5℃. ^{1}H NMR (400MHz, CDCl$_3$) δ 7.56(d, J = 8.3Hz, 2H), 7.23(d, J = 8.8Hz, 2H), 6.90(d, J = 8.3Hz, 2H), 6.79(d, J = 8.8Hz, 2H), 3.89(s, 2H), 3.78(s, 3H). ^{13}C NMR (101MHz, CDCl$_3$) δ 159.40, 137.99, 137.39, 134.38, 130.81, 125.39, 114.51, 92.32, 55.30, 40.76. MS (EI): m/z calcd for C$_{14}$H$_{13}$IOS 356; found 356 M$^+$.

1-Bromo-4-((4-methoxyphenoxy)methyl)benzene (**26**). The procedure described above for the preparation of **1** was employed to afford **26** as a white solid (2.93g, 93.2%). mp: 105.3-106.7℃. ^{1}H NMR (400MHz, CDCl$_3$) δ 7.51(d, J = 8.3Hz, 2H), 7.30(d, J = 8.3Hz, 2H), 6.89(d, J = 9.2Hz, 2H), 6.83(d, J = 9.2Hz, 2H), 4.97(s, 2H), 3.77(s, 3H). ^{13}C NMR (101MHz, CDCl$_3$) δ 154.15, 152.65, 136.38, 131.65, 129.05, 121.74, 115.89, 114.70, 69.97, 55.70. MS (EI): m/z calcd for C$_{14}$H$_{13}$BrO$_2$ 292; found 292 M$^+$.

1-Bromo-4-(4-methoxybenzyloxy)benzene (**27**). The procedure described above for the preparation of **1** was employed to afford **27** as a white solid (681.4 mg, 77.8%); mp: 122.1-122.9℃. ^{1}H NMR (400MHz, CDCl$_3$) δ 7.37(d, J = 9.0Hz, 2H), 7.33(d, J = 8.7Hz, 2H), 6.91(d, J = 8.7Hz, 2H), 6.84(d, J = 9.0Hz, 2H), 4.96(s, 2H), 3.82(s, 3H). ^{13}C NMR (101MHz, CDCl$_3$) δ 159.57, 157.93, 132.25, 129.19, 128.57, 116.74, 114.06, 113.03, 70.04, 55.30. MS (EI): m/z calcd for C$_{14}$H$_{13}$BrO$_2$ 292; found 292 M$^+$ (compared with the method in Ref. [49], 1-bromo-4-(4-methoxybenzyloxy)benzene was prepared though a common SN2 reaction between 4-bromophenol and 1-(chloromethyl)-4-methoxybenzene in a comparable yield).

4-(4-Bromobenzyloxy)-N,N-dimethylaniline (**28**). The procedure described above for the preparation of **1** was employed to afford **28** as a white solid (303.5mg, 99.1%); mp: 125.8-127.1℃. ^{1}H NMR (400MHz, CDCl$_3$) δ 7.52-7.48(m, 2H), 7.32-7.28(m, 2H), 6.92-6.85(m, 2H), 6.76(s, 2H), 4.96(s, 2H), 2.88(s, 6H). ^{13}C NMR (101MHz, DMSO-d_6) δ 149.92, 145.65, 137.14, 131.20, 129.56, 120.60, 115.62, 114.09, 68.84, 41.00. MS (EI):

m/z calcd for $C_{15}H_{16}BrNO$ 305; found 305 M$^+$.

Tributyl(4-((4-methoxyphenoxy)methyl)phenyl)stannane (**29**). A mixture of **25** (146.6mg, 0.5mmol), (Bu$_3$Sn)$_2$ (580.1mg, 1.0mmol), and (Ph$_3$P)$_4$Pd (57.8mg, 0.05mmol) in toluene (10mL) was stirred under reflux overnight. The mixture was concentrated under a vacuum and purified by silica gel chromatography (petroleum ether/AcOEt = 15/1) to give **29** as a colorless oil (89.5mg, 35.6%). ^{1}H NMR (400MHz, CDCl$_3$) δ 7.49(d, J = 7.9Hz, 2H), 7.39(d, J = 7.8Hz, 2H), 6.93(d, J = 9.1Hz, 2H), 6.85(d, J = 9.1Hz, 2H), 5.00(s, 2H), 3.78(s, 3H), 1.59-1.51(m, 6H), 1.39-1.29(m, 6H), 1.15-0.97(m, 6H), 0.90(t, J = 7.3Hz, 9H). ^{13}C NMR (101MHz, CDCl$_3$) δ 153.96, 153.11, 141.68, 136.90, 136.67, 127.08, 115.83, 114.65, 70.83, 55.72, 29.09, 27.38, 13.67, 9.60. MS (EI): m/z calcd for $C_{26}H_{40}O_2Sn$ 504; found 504 M$^+$.

Tributyl(4-(4-methoxybenzyloxy)phenyl)stannane (**30**). The procedure described above for the preparation of **29** was employed to afford **30** as a colorless oil (75.0mg, 29.8%). ^{1}H NMR (400MHz, CDCl$_3$) δ 7.40-7.32 (m, 3H), 7.31-7.27 (m, 1H), 6.99-6.95 (m, 2H), 6.91 (d, J = 8.4Hz, 2H), 4.98(s, 2H), 3.82(s, 3H), 1.69-1.60(m, 6H), 1.41-1.30(m, 12H), 0.92(t, J = 7.3Hz, 9H). ^{13}C NMR (101MHz, CDCl$_3$) δ 153.98, 153.12, 141.68, 136.91, 136.66, 127.07, 115.85, 114.67, 70.85, 55.73, 29.09, 27.38, 13.66, 9.60. MS (EI): m/z calcd for $C_{26}H_{40}O_2Sn$ 504; found 504 M$^+$.

N,N-Dimethyl-4-((4-(tributylstannyl)benzyl)oxy)aniline (**31**). The procedure described above for the preparation of **29** was employed to afford **31** as a colorless oil (56.2mg, 21.8%). ^{1}H NMR (400MHz, CDCl$_3$) δ 7.47(d, J = 7.9Hz, 2H), 7.38(d, J = 7.9Hz, 2H), 6.92(d, J = 9.1Hz, 2H), 6.74(d, J = 8.9Hz, 2H), 4.98(s, 2H), 2.87(s, 6H), 1.58-1.50(m, 6H), 1.38-1.28(m, 6H), 1.11-0.97(m, 6H), 0.88(t, J = 7.3Hz, 9H). ^{13}C NMR (101MHz, CDCl$_3$) δ 151.32, 145.96, 141.49, 137.24, 136.63, 127.10, 115.84, 114.76, 70.90, 41.74, 29.09, 27.38, 13.66, 9.60. MS (EI): m/z calcd for $C_{27}H_{43}NOSn$ 517; found 517 M$^+$.

4.2 X-ray crystallography

50mg of ligand **4** and **6** were dissolved in 3mL of ethyl acetate, respectively. Upon slow evaporation at room temperature, colorless crystals suitable for an X-ray diffraction analysis were yielded. The X-ray single-crystal structures were determined on a Bruker Smart APEX II CCD area-a-detector diffractometer at 100(2) or 150(2) K with graphite monochromated Mo Kα radiation (λ = 0.071073nm). Absorption correction was performed by the SADABS program[27]. All structures were solved using SHELXL-97[28] program and refined with full-matrix least-squares on F^2 method. All the hydrogen atoms were geometrically fixed using the riding model.

4.3 Radiolabeling and bioevaluation

The radio-iodinated ligands **4, 24** and **22** were prepared from the corresponding tributyltin derivatives by iododestannylation reaction through previously reported method[29].

Bioevaluation including binding assay in vitro using Aβ aggregates, autoradiography in vitro using brain sections from AD patients and transgenic model mice, biodistribution studies, and partition coefficient determination were all conducted according to previously reported methods[30].

K_d determination for [^{125}I]**4**: Peptides Aβ$_{42}$ were purchased from Osaka Peptide Institute (Osaka, Japan). Aggregation was performed according to the procedure described previously[30]. Saturation binding assay was set up by adding 100 μL of Aβ$_{42}$ aggregates (10 nM in the final assay mixture) into a mixture containing 100 μL of different concentrations of [^{125}I]**4** (final concentrations vary from 0.2 to 1.2 nM), 100 μL of EtOH, and a corresponding amount of 0.1% bovine serum albumin in a final volume of 1 mL. While for nonspecific binding group, an additional 100 μL of cold **4** was added in a final concentration of 1 μM. After incubation for 3 h at room temperature, the bound and free radioactive fractions were separated by vacuum filtration through glass fiber filters (Whatman GF/B) using a Mp-48T cell harvester (Brandel, Gaithersburg, MD). The radioactivity of filters containing the bound ^{125}I-ligand was measured in an automatic γ-counter (WALLAC/Wizard1470, USA) with 70% efficiency. The dissociation constant K_d for [^{125}I]4 was determined by analysis of the saturation binding curves of three independent assays with GraphPad Prism 5.0.

In vitro biostability in mouse plasma and liver homogenate: [^{125}I]4(30 μCi, in 30 μL of ethanol) was added to 300 μL of mouse plasma and liver homogenate separately and then incubated at 37℃ for 2 min, 10 min, 30 min, and 60 min. After that, 300 μL of acetonitrile was added to each sample followed by filtration using nylon syringe filters (13 mm × 0.22 μm, Troody, China) to remove the denatured proteins. The filtrates were analyzed by HPLC. Biostability of [^{125}I]IMPY in plasma was conducted in the same way.

In vivo biostability of [^{125}I]4: Four ICR mice were injected [^{125}I]4 (250 μCi, 200 μL) via tail vein and sacrificed at 2 min, 10 min, 30 min, and 60 min postinjection. Then 2 mL blood sample was collected and centrifuged at 3000 r/min for 3 min to separate plasma. Urine was also harvested. Brain, liver, and feces were obtained and homogenized in physiological saline. Subsequently, 200 μL of each sample was removed and mixed with 300 μL of acetonitrile, and then samples were analyzed with the identify method as described for in vitro biostability.

Computational Methods

Ligands **4** and **6** with X-ray crystal structural data were directly minimized to the nearest ground states using hybrid density functional technique B3LYP[31,32] with basis set 3-21G[33] for iodine and 6-31G[34] for other atoms in Gaussian 09[35] with tight convergence. The RMSD values between the optimized structures and the geometries of corresponding X-ray crystallographic structures were calculated using VMD 1.9.1 software[36] after alignment. The remaining 19 molecules were constructed with GaussView 5.0 and geometry optimized with the same method.

CoMFA and CoMSIA analysis was carried out to generate good predictive QSAR models using SYBYL-X 1.1 software[37] according to method reported previously with some modifications[38].

Gasteiger-Hückel charges were assigned to all the QM-optimized molecules. Structural alignment was performed by using most promising compound 4 as a template and benzyloxybenzene backbone (Fig. 2a) as a common structure. The aligned molecules are shown in Fig. 2b. The pK_i values of 21 benzyloxybenzene ligands in Table 1, which covered a range of 3 log units, were used as dependent variables. These pK_i values were imported into a molecular spreadsheet, and then the Tripos standard CoMFA/CoMSIA fields were added to the spreadsheet. Compared with CoMFA, three additional descriptors (hydrophobic and hydrogen-bond donor and acceptor) were also defined in CoMSIA method. Default parameters were used for constructing 3D-QSAR models. Partial leastsquare (PLS) methods were utilized to construct 3D-QSAR models. The PLS algorithmic rule with leave-one-out cross-validation method was applied to gain optimum number of components, cross-validated coefficient q^2 and to evaluate the statistical significance of each model. Column filter value was set to 1.0kcal/mol for all cross-validated PLS analyses. Conventional correlation coefficient (r^2), and its standard error (SEE), F-test value (F) were further computed by PLS analysis with no validation method.

AutoDock4.0[39-41] was employed to conduct the docking simulations on a solid-state NMR-derived model of an Aβ fiber (PDB 2LMO)[42]. All hydrogen atoms were added and Gasteiger charges[43] were assigned to the receptor using AutoDock Tools (ADT)[44]. QM-optimized IMPY and several benzyloxybenzene analogues with various binding affinities were served as the input ligands. Nonpolar hydrogen atoms were merged and all torsions were set to be rotatable during docking. Residues 16-KLVFFA-21 of the Aβ$_{40}$ fiber were chosen for docking, and each grid was centered on this site. Grid maps were generated using AutoGrid 4.0 with 3.4 ×6.2 × 8nm^3 dimensions and a grid spacing of 0.0375nm, which was large enough to encompass the entire chosen site[39]. Lamarckian genetic algorithm was applied for searching the lowest energy binding conformations with the default parameters from AutoDock4.0[39]. The binding energy is a value representing the fitness of the ligand with the protein, which integrates a set of state variables including values describing the translation, orientation, and conformation of the ligand. It is evaluated using the energy function. Then 100 GA ran with a population size of 300, a maximum of 2.5 ×10^7 energy evaluations, and a maximum of 27000 generations. The resulting 100 solutions from docking were clustered into groups with RMS deviations lower than 0.10nm, and the lowest-energy/top-ranked docked pose was extracted as the best conformation for each ligand. We confirmed that the best conformation for each ligand was located in the identical position of Aβ$_{40}$ fiber with similar orientation.

Associated Content

Supporting Information

Additional tables, figures, NMR/MS spectra, and CIF files. This material is available free of charge via the Internet at http://pubs.acs.org.

Author Information

Corresponding Author

*Phone/Fax: +86-10-58808891. E-mail: cmc@ bnu. edu. cn.

Author Contributions

Conceived and designed the experiments: M. Cui and B. Liu. Performed the experiments: Y. Yang, M. Cui and X. Zhang. Analyzed the data: Y. Yang, X. Zhang. Contributed reagents/materials/analysis tools: M. Cui, B. Liu, J. Dai, Z. Zhang, C. Lin and Y. Guo. Wrote the paper: Y. Yang, M. Cui and B. Liu.

Notes

The authors declare no competing financial interest.

Acknowledgments

This work was funded by the National Natural Science Foundation of China (No. 21201019) and the Fundamental Research Funds for the Central Universities (No. 2012LYB19). We thank Dr. Jin Liu (College of Life Science, Beijing Normal University) for assistance in the in vitro neuropathologueical staining, and Prof. Huabei Zhang, Prof. Yan Wang, and Dr. Shiyuan Hu (College of Chemistry, Beijing Normal University) for assistance in the computional experiments.

Abbreviations Used

AD, Alzheimer's disease; Aβ, β-amyloid; PET, positron emission tomography; SPECT, single photon emission computed tomography; CR, Congo red; ThT, thioflavin T; BBB, blood-brain barrier; PIB, 2-(4-(methylamino)phenyl)benzo[d]thiazol-6-ol; U. S. FDA, United States Food and Drug Administration; IMPY, 4-(6-iodoimidazo[1,2-a]pyridin-2-yl)-N,N-dimethylaniline; B3LYP, 3-parameter hybrid Becke exchange/Lee-Yang-Parr correlation functional; RMSD, root-mean-square deviations; 3D-QSAR, three-dimensional quantitative structure-activity relationship; CoMFA, comparative molecular field analysis; CoMSIA, comparative molecular similarity indices analysis; NMR, nuclear magnetic resonance; J, coupling constant (in NMR spectrometry); MS (EI), mass spectrometry (electron ionization); HRMS, high-resolution mass spectrometry; HPLC, high-performance liquid chromatography; DMF, N,N-dimethylformamide; QM, quantum mechanics; PLS, partial least-square; ADT, autodock tools; GA, genetic algorithm.

References

[1] Brookmeyer,R.;Johnson,E.;Ziegler-Graham,K.;Arrighi,H. M. Forecasting the global burden of Alzhei-

mer's disease. Alzheimer's Dementia 2007,3,186-191.

[2] Hardy,J. ;Selkoe,D. J. The amyloid hypothesis of Alzheimer's disease:progress and problems on the road to therapeutics. Science 2002,297,353-356.

[3] Han,H. ; Cho,C. G. ; Lansbury, P. T. Technetium complexes for the quantitation of brain amyloid. J. Am. Chem. Soc. 1996,118,4506-4507.

[4] Raimon,S. ;Joan,E. Pinacyanol as effective probe of fibrillar beta-amyloid peptide:comparative study with Congo Red. Biopolymers 2003,72,455-463.

[5] Kadir,A. ;Marutle, A. ;Gonzalez,D. ;Schöll,M. ;Almkvist,O. ;Mousavi,M. ;Mustafiz,T. ;Darreh-Shori, T. ;Nennesmo,I. ;Nordberg, A. Positron emission tomography imaging and clinical progression in relation to molecular pathologuey in the first Pittsburgh Compound B positron emission tomography patient with Alzheimer's disease. Brain 2011,134,301-317.

[6] Koole,M. ;Lewis,D. M. ;Buckley,C. ;Nelissen,N. ;Vandenbulcke,M. ;Brooks,D. J. ;Vandenberghe,R. ; Van Laere,K. Whole-body biodistribution and radiation dosimetry of ^{18}F-GE067:a radioligand for in vivo brain amyloid imaging. J. Nucl. Med. 2009,50,818-822.

[7] Okamura, N. ; Shiga, Y. ; Furumoto, S. ; Tashiro, M. ; Tsuboi, Y. ; Furukawa, K. ; Yanai, K. ; Iwata, R. ; Arai,H. ;Kudo,Y. ; Itoyama, Y. ; Dohura, K. In vivo detection of prion amyloid plaques using [^{11}C]BF-227 PET. Eur. J. Nucl. Med. Mol. Imaging 2010,37,934-941.

[8] Chang, Y. S. ; Jeong, J. M. ; Lee, Y. S. ; Kim, H. W. ; Ganesha, R. B. ; Kim, Y. J. ; Lee, D. S. ; Chung, J. -K. ;Lee,M. C. Synthesis and evaluation of benzothiophene derivatives as ligands for imaging β-amyloid plaques in Alzheimer's disease. Nucl. Med. Biol. 2006,33,811-820.

[9] Juréus,A. ;Swahn, B. M. ; Sandell, J. ; Jeppsson, F. ; Johnson, A. E. ; Johnström, P. ; Neelissen, J. A. M. ; Sunnemark,D. ;Farde,L. ;Svensson, S. P. S. Characterization of AZD4694, a novel fluorinated Aβ plaque neuroimaging PET radioligand. J. Neurochem. 2010,114,784-794.

[10] Kung, M. P. ; Hou, C. ; Zhuang, Z. P. ; Zhang, B. ; Skovronsky, D. ; Trojanowski, J. Q. ; Lee, V. M. Y. ; Kung,H. F. IMPY:an improved thioflavin-T derivative for in vivo labeling of β-amyloid plaques. Brain Res. 2002,956,202-210.

[11] Cui,M. ;Ono,M. ;Kimura,H. ;Kawashima,H. ;Liu,B. L. ;Saji,H. Radioiodinated benzimidazole derivatives as single photon emission computed tomography probes for imaging of β-amyloid plaques in Alzheimer's disease. Nucl. Med. Biol. 2011,38,313-320.

[12] Ono, M. ; Wilson, A. ; Nobrega, J. ; Westaway, D. ; Verhoeff, P. ; Zhuang, Z. P. ; Kung, M. P. ; Kung, H. F. ^{11}C-labeled stilbene derivatives as Aβ-aggregate-specific PET imaging agents for Alzheimer's disease. Nucl. Med. Biol. 2003,30,565-571.

[13] Choi,S. ;Golding,G. ;Zhuang,Z. ;Zhang,W. ;Lim,N. ;Hefti,F. ;Benedum,T. ;Kilbourn,M. ;Skovronsky, D. ; Kung, H. Preclinical properties of ^{18}F-AV-45: a PET agent for abeta plaques in the brain. J. Nucl. Med. 2009,50,1887-1894.

[14] Zhuang,Z. P. ;Kung,M. P. ;Hou,C. ;Ploessl,K. ;Kung,H. F. Biphenyls labeled with technetium 99m for imaging β-amyloid plaques in the brain. Nucl. Med. Biol. 2005,32,171-184.

[15] Wey,S. P. ;Weng,C. C. ;Lin,K. J. ;Yao,C. H. ;Yen,T. C. ;Kung,H. F. ;Skovronsky,D. ;Kung,M. P. Validation of an ^{18}F-labeled biphenylalkyne as a positron emission tomography imaging agent for β-amy-

loid plaques. Nucl. Med. Biol. 2009,36,411-417.

[16] Watanabe, H. ; Ono, M. ; Ikeoka, R. ; Haratake, M. ; Saji, H. ; Nakayama, M. Synthesis and biologueical evaluation of radioiodinated 2,5-diphenyl-1,3,4-oxadiazoles for detecting β-amyloid plaques in the brain. Bioorg. Med. Chem. 2009,17,6402-6406.

[17] Ryu, E. K. ; Choe, Y. S. ; Lee, K. H. ; Choi, Y. ; Kim, B. T. Curcumin and dehydrozingerone derivatives: synthesis, radiolabeling, and evaluation for β-amyloid plaque imaging. J. Med. Chem. 2006, 49, 6111-6119.

[18] Ono, M. ; Yoshida, N. ; Ishibashi, K. ; Haratake, M. ; Arano, Y. ; Mori, H. ; Nakayama, M. Radioiodinated flavones for in vivo imaging of β-amyloid plaques in the brain. J. Med. Chem. 2005,48,7253-7260.

[19] Ono, M. ; Watanabe, R. ; Kawashima, H. ; Cheng, Y. ; Kimura, H. ; Watanabe, H. ; Haratake, M. ; Saji, H. ; Nakayama, M. Fluoro-pegylated chalcones as positron emission tomography probes for in vivo imaging of β-amyloid plaques in Alzheimer's disease. J. Med. Chem. 2009,52,6394-6401.

[20] Ono, M. ; Maya, Y. ; Haratake, M. ; Ito, K. ; Mori, H. ; Nakayama, M. Aurones serve as probes of β-amyloid plaques in Alzheimer's disease. Biochem. Biophys. Res. Commun. 2007,361,116-121.

[21] Camus, V. ; Payoux, P. ; Barré, L. ; Desgranges, B. ; Voisin, T. ; Tauber, C. ; Joie, R. ; Tafani, M. ; Hommet, C. ; Chételat, G. ; Mondon, K. ; Sayette, V. ; Cottier, J. P. ; Beaufils, E. ; Ribeiro, M. J. ; Gissot, V. ; Vierron, E. ; Vercouillie, J. ; Vellas, B. ; Eustache, F. ; Guilloteau, D. Using PET with ^{18}F-AV-45 (florbetapir) to quantify brain amyloid load in a clinical environment. Eur. J. Nucl. Med. Mol. Imaging 2012,39, 621-631.

[22] Barthel, H. ; Gertz, H. ; Dresel, S. ; Peters, O. ; Bartenstein, P. ; Buerger, K. ; Hiemeyer, F. ; Wittemer-Rump, S. ; Seibyl, J. ; Reininger, C. ; Sabri, O. Cerebral amyloid-beta PET with florbetaben(^{18}F) in patients with Alzheimer's disease and healthy controls: a multicentre phase 2 diagnostic study. Lancet Neurol. 2011,10,424-435.

[23] Newberg, A. B. ; Wintering, N. A. ; Plössl, K. ; Hochold, J. ; Stabin, M. G. ; Watson, M. ; Skovronsky, D. ; Clark, C. M. ; Kung, M. P. ; Kung, H. F. Safety, biodistribution, and dosimetry of ^{123}I-IMPY: a novel amyloid plaque-imaging agent for the diagnosis of Alzheimer's disease. J. Nucl. Med. 2006,47,748-754.

[24] Eckroat, T. J. ; Mayhoub, A. S. ; Garneau-Tsodikova, S. Amyloid-β probes: review of structure-activity and brain-kinetics relationships. Beilstein J. Org. Chem. 2013,9,1012-1044.

[25] Cui, M. Past and recent progress of molecularimaging probes for β-amyloid plaques in the brain. Curr. Med. Chem. 2014,21,82-112.

[26] Cook, N. P. ; Ozbil, M. ; Katsampes, C. ; Prabhakar, R. ; Martí, A. A. Unraveling the photoluminescence response of light-switching ruthenium(II) complexes bound to amyloid-β. J. Am. Chem. Soc. 2013,135, 10810-10816.

[27] Sheldrick, G. M. SADABS, Program for Empirical Absorption Correction of Area Detector Data; University of Gottingen: Gottingen, Germany, 1996.

[28] Sheldrick, G. M. SHELXL-97 program for the Refinement of Crystal Structure from Diffraction Data; University of Gottingen: Gottingen, Germany, 1997.

[29] Cui, M. ; Ono, M. ; Kimura, H. ; Liu, B. ; Saji, H. Synthesis and structure-affinity relationships of novel dibenzylideneacetone derivatives as probes for β-amyloid plaques. J. Med. Chem. 2011,54,2225-2240.

[30] Yang, Y. ; Cui, M. ; Jin, B. ; Wang, X. ; Li, Z. ; Yu, P. ; Jia, J. ; Fu, H. ; Jia, H. ; Liu, B. ^{99m}Tc-labeled dibenzylideneacetone derivatives as potential SPECT probes for in vivo imaging of β-amyloid plaque. Eur. J. Med. Chem. 2013, 64, 90-98.

[31] Becke, A. D. Density-functional thermochemistry. III. The role of exact exchange. J. Chem. Phys. 1993, 98, 5648-5652.

[32] Stephens, P. J. ; Devlin, F. J. ; Chabalowski, C. F. ; Frisch, M. J. Ab initio calculation of vibrational absorption and circular dichroism spectra using density functional force fields. J. Chem. Phys. 1994, 98, 11623-11627.

[33] Dobbs, K. D. ; Hehre, W. J. Molecular orbital theory of the properties of inorganic and organometallic compounds. 6. Extended basis sets for second-row transition metals. J. Comput. Chem. 1987, 8, 880-893.

[34] Hehre, W. J. ; Ditchfield, R. ; Pople, J. A. Self-consistent molecular orbital methods. XII. Further extensions of Gaussian-type basis sets for use in molecular orbital studies of organic molecules. J. Chem. Phys. 1972, 56, 2257-2261.

[35] Frisch, M. J. ; Trucks, G. W. ; Schlegel, H. B. ; Scuseria, G. E. ; Robb, M. A. ; Cheeseman, J. R. ; Scalmani, G. ; Barone, V. ; Mennucci, B. ; Petersson, G. A. ; Nakatsuji, H. ; Caricato, M. ; Li, X. ; Hratchian, H. P. ; Izmaylov, A. F. ; Bloino, J. ; Zheng, G. ; Sonnenberg, J. L. ; Hada, M. ; Ehara, M. ; Toyota, K. ; Fukuda, R. ; Hasegawa, J. ; Ishida, M. ; Nakajima, T. ; Honda, Y. ; Kitao, O. ; Nakai, H. ; Vreven, T. ; Montgomery, J. A. , Jr. ; Peralta, J. E. ; Ogliaro, F. ; Bearpark, M. ; Heyd, J. J. ; Brothers, E. ; Kudin, K. N. ; Staroverov, V. N. ; Keith, T. ; Kobayashi, R. ; Normand, J. ; Raghavachari, K. ; Rendell, A. ; Burant, J. C. ; Iyengar, S. S. ; Tomasi, J. ; Cossi, M. ; Rega, N. ; Millam, J. M. ; Klene, M. ; Knox, J. E. ; Cross, J. B. ; Bakken, V. ; Adamo, C. ; Jaramillo, J. ; Gomperts, R. ; Stratmann, R. E. ; Yazyev, O. ; Austin, A. J. ; Cammi, R. ; Pomelli, C. ; Ochterski, J. W. ; Martin, R. L. ; Morokuma, K. ; Zakrzewski, V. G. ; Voth, G. A. ; Salvador, P. ; Dannenberg, J. J. ; Dapprich, S. ; Daniels, A. D. ; Farkas, O. ; Foresman, J. B. ; Ortiz, J. V. ; Cioslowski, J. ; Fox, D. J. Gaussian 09, Revision B. 01; Gaussian, Inc. : Wallingford CT, 2010.

[36] Humphrey, W. ; Dalke, A. ; Schulten, K. VMD: Visual molecular dynamics. J. Mol. Graph. 1996, 14, 33-38.

[37] Sybyl-X, version 1. 1; Tripos Inc. : St. Louis, MO, 2010.

[38] Zeng, H. ; Zhang, H. Combined 3D-QSAR modeling and molecular docking study on 1, 4-dihydroindeno [1, 2-c] pyrazoles as VEGFR-2 kinase inhibitors. J. Mol. Graphics Model. 2010, 29, 54-71.

[39] Morris, G. M. ; Goodsell, D. S. ; Halliday, R. S. ; Huey, R. ; Hart, W. E. ; Belew, R. K. ; Olson, A. J. Automated docking using a lamarckian genetic algorithm and an empirical binding free energy function. J. Comput. Chem. 1998, 19, 1639-1662.

[40] Huey, R. ; Morris, G. M. ; Olson, A. J. ; Goodsell, D. S. A semiempirical free energy force field with charge-based desolvation. J. Comput. Chem. 2007, 28, 1145-1152.

[41] Huey, R. ; Goodsell, D. S. ; Morris, G. M. ; Olson, A. J. Grid-based hydrogen bond potentials with improved directionality. Lett. Drug Des. Discovery 2004, 1, 178-183.

[42] Petkova, A. T. ; Yau, W. M. ; Tycko, R. Experimental constraints on quaternary structure in alzheimer's β-amyloid fibrils. Biochemistry 2005, 45, 498-512.

[43] Gasteiger, J. ; Marsili, M. Iterative partial equalization of orbital electronegativity—a rapid access to atom-

ic charges. Tetrahedron 1980,36,3219-3228.

[44] Morris, G. M. ; Huey, R. ; Lindstrom, W. ; Sanner, M. F. ; Belew, R. K. ; Goodsell, D. S. ; Olson, A. J. AutoDock4 and AutoDockTools4: automated docking with selective receptor flexibility. J. Comput. Chem. 2009,30,2785-2791.

[45] Raghavan, S. ; Rajender, A. Target alcohol/phenol release by cyclative cleavage using glycine as a safety catch linker. Chem. Commun. 2002,1572-1573.

[46] Abe, Y. ; Aoki, T. ; Jia, H. ; Hadano, S. ; Namikoshi, T. ; Kakihana, Y. ; Liu, L. ; Zang, Y. ; Teraguchi, M. ; Kaneko, T. Chiral Teleinduction in Asymmetric Polymerization of 3,5-Bis(hydroxymethyl)phenylacetylene Having a Chiral Group via a Very Long and Rigid Spacer at 4-Position. Chem. Lett. 2012,41,244-246.

[47] Kimball, F. S. ; Romero, F. A. ; Ezzili, C. ; Garfunkle, J. ; Rayl, T. J. ; Hochstatter, D. G. ; Hwang, I. ; Bogeret, D. L. Optimization of α-Ketooxazole Inhibitors of Fatty Acid Amide Hydrolase. J. Med. Chem. 2008,51,937-947.

[48] Ma, L. ; Morgan, J. C. ; Stancill, W. E. ; Allen, W. E. Phosphorylated 1,6-diphenyl-1,3,5-hexatriene. Bioorg. Med. Chem. Lett. 2004,14,1075-1078.

[49] Oriyama, T. ; Noda, K. ; Yatabe, K. Highly Efficient and Convenient Methods for the Direct Conversion of Aryl Silyl Ethers and Aryl Acetates into Aryl Alkyl Ethers. Synlett 1997,710-703.

其他

放射性同位素半衰期测定中可能遇见的问题

自从 Rutherford 和 Soddy 等提出衰变指数定律以来，人们就利用半对数曲线的线性关系来测定放射性核素的半衰期。长期以来，这种线性规律一直被认为是没有例外的普遍规律。本文首次得到了直链衰变体系、支链衰变体系以及在实际测量中可能出现表观半衰期的普遍关系式，其中包括肖伦等人对二次衰变体系用图解法首次得到的条件。

1 直链衰变体系

众所周知，纯放射性物质的衰变，服从简单的指数定律。对于二次衰变链体系：

$$A \xrightarrow{\lambda_1} B \xrightarrow{\lambda_2} C$$

$$N_2\lambda_2 = \frac{N_1^0 \lambda_1 \lambda_2}{\lambda_2 - \lambda_1}(e^{-\lambda_1 t} - e^{-\lambda_2 t}) \tag{1}$$

式中，λ_1 和 λ_2 分别为母体 A 和子体 B 的衰变常数。体系的总放射性为：

$$N\lambda = N_1\lambda_1 + N_2\lambda_2 = N_1^0 \lambda_1 e^{-\lambda_1 t} + \frac{N_1^0 \lambda_1 \lambda_2}{\lambda_2 - \lambda_1}(e^{-\lambda_1 t} - e^{-\lambda_2 t}) \tag{2}$$

肖伦等用图解法得到，当 λ_2 为定值时，λ_1 由 ∞ 变到 0，体系总放射性在半对数图上的变化如图 1 所示，由图中可以看出，当 $\lambda_1 = 2\lambda_2$ 时，体系总放射性随时间变化，在半对数图上是一直线。

将式（2）重排后，由式（3）不难看出，当总放射性在半对数图上呈现线性时，只有在

$$1 + \frac{\lambda_2}{\lambda_2 - \lambda_1} = 0$$

时，才有可能。由此同样得到 $\lambda_1 = 2\lambda_2$ 的结论。

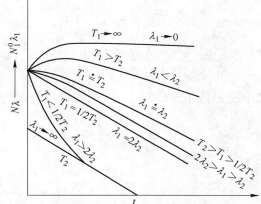

图 1 二次衰变直链体系的总放射性随时间的变化

$$N\lambda = N_1^0 \lambda_1 e^{-\lambda_1 t}\left(1 + \frac{\lambda_2}{\lambda_2 - \lambda_1}\right) - \frac{N_1^0 \lambda_1 \lambda_2}{\lambda_2 - \lambda_1}e^{-\lambda_2 t} \tag{3}$$

---

* 本文合作者：聂镕。原发表于《科学通报》，1979(10)：447~449。

作者进一步对三次、四次……n 次体系进行了分析，对于 n 次直链衰变体系：

$$A \xrightarrow{\lambda_1} B \xrightarrow{\lambda_2} \cdots \xrightarrow{\lambda_{n-1}} M \xrightarrow{\lambda_n} N$$

体系的总放射性为：

$$N\lambda = N_1\lambda_1 + N_2\lambda_2 + \cdots + N_n\lambda_n \tag{4}$$

根据 Bateman 方程式，可得到各代子体在 t 时的量，代入式（4），经整理后得到：

$$\begin{aligned}
N\lambda = & \left[1 + \frac{\lambda_2}{\lambda_2 - \lambda_1} + \frac{\lambda_2\lambda_3}{(\lambda_2 - \lambda_1)(\lambda_3 - \lambda_1)} + \cdots + \frac{\lambda_2\lambda_3\cdots\lambda_n}{(\lambda_2 - \lambda_1)(\lambda_3 - \lambda_1)\cdots(\lambda_n - \lambda_1)}\right] \times \\
& N_1^0\lambda_1 e^{-\lambda_1 t} + \left[1 + \frac{\lambda_3}{\lambda_3 - \lambda_2} + \frac{\lambda_3\lambda_4}{(\lambda_3 - \lambda_2)(\lambda_4 - \lambda_2)} + \cdots + \right. \\
& \left. \frac{\lambda_3\lambda_4\cdots\lambda_n}{(\lambda_3 - \lambda_2)(\lambda_4 - \lambda_2)\cdots(\lambda_n - \lambda_2)}\right] \times \frac{N_1^0\lambda_1\lambda_2}{\lambda_1 - \lambda_2}e^{-\lambda_2 t} + \\
& \left[1 + \frac{\lambda_4}{\lambda_4 - \lambda_3} + \frac{\lambda_4\lambda_5}{(\lambda_4 - \lambda_3)(\lambda_5 - \lambda_3)} + \cdots + \frac{\lambda_4\lambda_5\cdots\lambda_n}{(\lambda_4 - \lambda_3)(\lambda_5 - \lambda_3)\cdots(\lambda_n - \lambda_3)}\right] \times \\
& \frac{N_1^0\lambda_1\lambda_2\lambda_3}{(\lambda_1 - \lambda_3)(\lambda_2 - \lambda_3)}e^{-\lambda_3 t} + \cdots + \left[1 + \frac{\lambda_n}{\lambda_n - \lambda_{n-1}}\right] \times \\
& \frac{N_1^0\lambda_1\lambda_2\lambda_3\cdots\lambda_{n-1}}{(\lambda_1 - \lambda_{n-1})(\lambda_2 - \lambda_{n-1})\cdots(\lambda_{n-2} - \lambda_{n-1})}e^{-\lambda_{n-1} t} + \\
& \frac{N_1^0\lambda_1\lambda_2\cdots\lambda_n}{(\lambda_1 - \lambda_n)(\lambda_2 - \lambda_n)\cdots(\lambda_{n-1} - \lambda_n)}e^{-\lambda_n t}
\end{aligned} \tag{5}$$

n 次衰变链体系总放射性在半对数图上呈现线性的条件为：

$$\begin{cases}
1 + \dfrac{\lambda_2}{\lambda_2 - \lambda_1} + \dfrac{\lambda_2\lambda_3}{(\lambda_2 - \lambda_1)(\lambda_3 - \lambda_1)} + \cdots + \dfrac{\lambda_2\lambda_3\cdots\lambda_n}{(\lambda_2 - \lambda_1)(\lambda_3 - \lambda_1)\cdots(\lambda_n - \lambda_1)} = 0 \\
1 + \dfrac{\lambda_3}{\lambda_3 - \lambda_2} + \dfrac{\lambda_3\lambda_4}{(\lambda_3 - \lambda_2)(\lambda_4 - \lambda_2)} + \cdots + \dfrac{\lambda_3\lambda_4\cdots\lambda_n}{(\lambda_3 - \lambda_2)(\lambda_4 - \lambda_2)\cdots(\lambda_n - \lambda_2)} = 0 \\
1 + \dfrac{\lambda_4}{\lambda_4 - \lambda_3} + \dfrac{\lambda_4\lambda_5}{(\lambda_4 - \lambda_3)(\lambda_5 - \lambda_3)} + \cdots + \dfrac{\lambda_4\lambda_5\cdots\lambda_n}{(\lambda_4 - \lambda_3)(\lambda_5 - \lambda_3)\cdots(\lambda_n - \lambda_3)} = 0 \\
\quad\vdots \\
1 + \dfrac{\lambda_n}{\lambda_n - \lambda_{n-1}} = 0
\end{cases}$$

上式的解为：

$$\begin{cases} \lambda_{n-1} = 2\lambda_n \\ \lambda_{n-2} = \dfrac{3}{2}\lambda_{n-1} \\ \lambda_{n-3} = \dfrac{4}{3}\lambda_{n-2} \\ \lambda_{n-4} = \dfrac{5}{4}\lambda_{n-3} \\ \quad \vdots \\ \lambda_1 = \dfrac{n}{n-1}\lambda_2 \end{cases} \qquad (6)$$

所以当 $\lambda_1, \lambda_2, \lambda_3, \ldots, \lambda_n$ 的数值满足式（6）时，刚纯化的具有 n 次衰变链体系的总放射性随时间的变化，在半对数图上呈线性关系。兹将 $n=4$ 的结果作图，如图2所示。

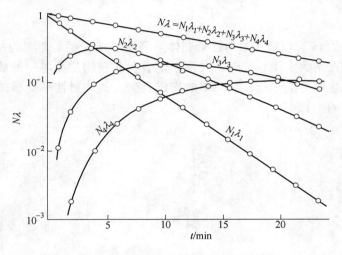

图2　四次衰变直链体系的总放射性随时间的变化

2 支链衰变体系

按盖伯尔（Garber）1968年所得的分支衰变体系的一般方程式，没有得到呈现线性的条件。

对于简单的分支衰变：

可以得到呈现线性的条件为：

$$\lambda_1 = (2-u)\lambda_2 \qquad (7)$$

当 $u=0$ 时，上式即为 $\lambda_1 = 2\lambda_2$，即二次直链衰变体系，其总放射性呈现线性的条件。

3 在实际测量中的应用

设 η_1 和 η_2 分别为母体和子体的探测效率，则只有当 $\lambda_2 = \dfrac{\eta_1}{\eta_1 + \eta_2}\lambda_1$ 时，体系总放射性在实际测量时才能在半对数坐标上呈线性关系。不难看出，当 $\eta_1 \neq \eta_2$ 时，母子体的衰变常数在相当大的范围内变化，在实际测量中均可呈现线性关系，出现表观半衰期。同理，对于简单分支衰变，可得以下关系：

$$\lambda_2 = \frac{\eta_1}{\eta_2 + \eta_1 - \eta_2 u}\lambda_1$$

满足上述关系时，在实际测量中可能呈现线性关系。

4 结论

式（6）和式（7）反映了放射性衰变体系可能呈现表观半衰期的一般条件。刚纯化的放射性物质按指数定律衰变时，并不一定表明此核物质为单一的放化纯，其中可能有多代子体在生长，也可能存在简单的支链衰变，其放射性平衡后的表观半衰期均为最后一代子体的半衰期。

（n, γ）核反冲法浓集^{24}Na 的研究*

为了浓集（n, γ）核反应所生成的放射性核素，可以利用 Szilard-Chalmers 效应[1~7]。对于碱金属由（n, γ）核反应生成的放射性核素，因其所处的价态和状态一般与靶子物中的稳定同位素几乎相同，因此利用上述效应很难达到浓集目的。有关碱金属的浓集工作报道很少[8,9,16]。但利用非均相体系的浓集工作有所开展[10~17]。本文在袁志熙[8]工作的基础上，探讨了固-固反冲的不同浓集条件，研究了以 Na_2CO_3 为反冲相，以 MgO、SiO_2 和 Al_2O_3 为接受相，以不同的比例、不同的颗粒半径、不同的洗涤剂对^{24}Na 的产额和浓集系数的影响。并根据^{24}Na 在接受相中的平均产额，近似计算了^{24}Na 的反冲射程[10]，测定了^{24}Na 在固液两相间的半交换期，初步探索了用颗粒反冲法浓集^{24}Na 的实验条件。

1 实验结果和讨论

1.1 试剂及测量仪器

本工作选用 Na_2CO_3 为反冲相，MgO、SiO_2 和 Al_2O_3 为接受相。以 Na_2CO_3-MgO、Na_2CO_3-SiO_2 和 Na_2CO_3-Al_2O_3 三种体系作为靶子物。所用 Na_2CO_3 为保证试剂，SiO_2、盐酸和 40% HF 为分析纯试剂，Al_2O_3 和 MgO 为德国标准试剂。

上述靶子物在反应堆内均经过单独的空白照射，在测量的误差范围内，没有发现放射性 γ 杂质的干扰。

试验中所用的全部试剂，除 Na_2CO_3 外，均经过稳定钠杂质的测定，并对实验结果作了校正。钠含量的测定采用 Lange-6 型火焰光度计，测量的相对误差为 1.5%~2%。反冲相和接受相颗粒半径的测定，采用 MLI 生物显微镜，放大倍数为 1500 倍。γ 计数装置符合相对测量的各项要求，放射性测量的标准误差为 2%。

1.2 靶子的制备

实验所用的 Na_2CO_3-MgO, Na_2CO_3-Al_2O_3 和 Na_2CO_3-SiO_2 靶子物，在混合前均单独于玛瑙研钵中研磨 60min。然后在红外灯照射下分别混合研磨 10min, 50min, 100min 和 180min。用 MLI 显微镜分别测定 Na_2CO_3、MgO、Al_2O_3 和 SiO_2 的平均颗粒半径。将不同比例的靶子混合物密封在石英安瓿瓶中，在反应堆内照射。根据靶子物中 Na_2CO_3 的含量和需要的强度，照射时间在 0.5~5h 范围内。靶子物从反应堆取出后，冷却一定时间，进行化学处理。

---

* 本文合作者：华英圣。原发表于《北京师范大学学报（自然科学版）》，1979(2)：29~36。

1.3 靶子物的化学处理

(1) Na_2CO_3-MgO 靶子物的化学处理。称取 40mg 或 100mg 的照射样品，用蒸馏水或乙醇作为洗涤剂，在离心试管中溶解洗涤，洗涤剂每次用量为 3mL，根据实验的不同要求，分别洗涤二到五次。洗涤后均经过离心沉降，取出上层清液，冲稀到一定体积，取样测其放射性强度（a）。

未溶的 MgO，用 9N HCl 3mL 溶解，冲稀到一定体积，取样测其放射性强度（b）。由测得的结果，可求出接受相中 ^{24}Na 的产额。然后分别测量洗涤液（c）和 MgO 溶解液（d）中稳定钠的含量，并求出相应的浓集系数。

(2) Na_2CO_3-SiO_2 靶子物的化学处理。处理顺序同（1）。将洗涤后未溶解的 SiO_2 转入铂坩埚中，用 6mL 40% HF 加热使其溶解。蒸干后再用 4mL 9N HCl 处理。将溶解液冲稀到一定体积，取样测其放射性强度（b）。用上述相同的方法，求出产额和浓集系数。

(3) Na_2CO_3-Al_2O_3 靶子物的化学处理。处理顺序同（1）。将洗涤后未溶解的 Al_2O_3 转入铂坩埚中，用 8mL 40% HF 加热使其溶解。蒸浓后再加入 9mL 浓盐酸，以除去过量的 HF。蒸干后，用 6mL 9N HCl 使沉淀溶解，将溶解液冲稀到一定体积，取样测其放射性强度（b），按照上述相同方法，求出产额和浓集系数。

1.4 结果及讨论

1.4.1 不同接受体和不同混合比对 ^{24}Na 产额和浓集系数的影响

(1) 选用 MgO 为接受体，以不同比例与 Na_2CO_3 混合，MgO 与 Na_2CO_3 的平均颗粒半径为 1.98μm 和 2.12μm❶。靶子物的质量为 100mg，用蒸馏水洗涤四次，所得产额和浓集系数如图 1 所示。

图 1 表明，不同混合比对浓集系数有较大影响，当 Na_2CO_3 和 MgO 比例为 1∶2 时，浓集系数呈现一个极大值。但不同混合比对产额的影响不大，除体系的混合比为 1∶1 时产额略低以外，其他混合比的产额均为 3% 左右。因此，从浓集系数和产额两方面来考虑，Na_2CO_3 与 MgO 的混合比采用 1∶2 是适宜的。

当体系的比例为 1∶1 时，由于 Na_2CO_3 的量较多，Na_2CO_3 颗粒周围不能充分和 MgO 相接触，因此一部分反冲 ^{24}Na 不能进入 MgO 接受相，而使产额降低。另一方面混合靶子

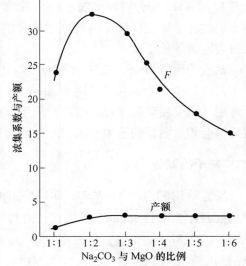

图 1 Na_2CO_3 与 MgO 的比例变化对浓集系数和产额的影响

---

❶ 颗粒半径的平均值是指 100 次测定的平均值。

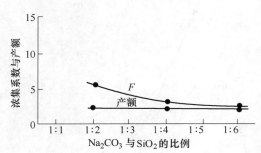

图 2　Na_2CO_3 与 SiO_2 的比例变化
对浓集系数与产额的影响

在溶解和洗涤过程中，由于 Na_2CO_3 的含量较大，溶液中稳定钠离子和 MgO 相中 ^{24}Na 的同位素交换速度也随之加大，这也是使产额降低的可能原因。

（2）选用 SiO_2 为接受体，以不同的比例与 Na_2CO_3 混合，Na_2CO_3 与 SiO_2 的平均颗粒半径分别为 $2.12\mu m$ 和 $1.69\mu m$。靶子物的质量为 40mg，洗涤一次。所得产额和浓集系数如图 2 所示。

图 2 表明：不同的混合比对产额的影响不大，一般在 2% 左右。但浓集系数随 SiO_2 比例的增加而减少。

（3）选用 Al_2O_3 为接受体，以不同的比例与 Na_2CO_3 混合，Al_2O_3 与 Na_2CO_3 的平均颗粒半径分别为 $1.16\mu m$ 和 $1.66\mu m$，靶子物的质量为 100mg，洗涤四次。所得产额和浓集系数如图 3 所示。

图 3 表明：当体系的混合比为 1∶5 时，产额有一个最大值，但浓集系数变化很小，一般在 1.5～2。

从上述三种结果可以看出：用 Al_2O_3 为接受体时，产额虽然最高可达 53% 左右，但浓集系数较低。当用 MgO 为接受体时，其产额虽较 Al_2O_3 为低，但浓集系数较高，从浓集的观点来考虑，是有意义的。用 SiO_2 为接受相时，浓集系数和产额均不理想。从化学处理方法来比较，MgO 相较 Al_2O_3 和 SiO_2 易于处理。

1.4.2　不同洗涤剂对浓集 ^{24}Na 的影响

为了提高产额，降低靶子物在溶解和洗涤过程中的同位素交换速度，采用不同浓度的乙醇溶液作为洗涤剂。

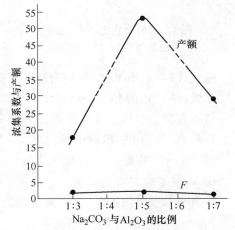

图 3　Na_2CO_3 与 Al_2O_3 的比例变化
对浓集系数与产额的影响

以 MgO 为接受相，MgO 和 Na_2CO_3 的平均颗粒半径为 $1.98\mu m$ 和 $2.12\mu m$，靶子物质量为 40mg，用不同浓度的乙醇溶液分别洗涤四次，所得产额和浓集系数如图 4 所示。

图 4 表明：用乙醇溶液作洗涤剂时，产额随乙醇浓度的增加而上升，但浓集系数却下降。这意味着随着乙醇浓度的增加，MgO 相中稳定钠离子的含量降低不多，洗涤效果不好。图 5 表明：用水和不同浓度乙醇溶液洗涤 MgO 接收相后其中尚存留的稳定钠含量。

图 5 中，所用靶子混合物为 MgO 和 Na_2CO_3，混合比为 5∶1，平均颗粒半径分别为 $1.98\mu m$ 和 $2.12\mu m$。

曲线（1）是用二次蒸馏水作为洗涤剂。

曲线（2）乙醇和水的体积比为 1∶9。

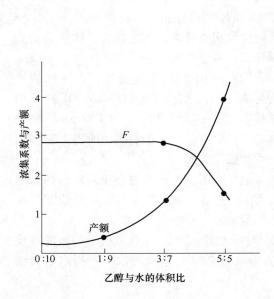

图 4　乙醇与水的体积比变化对浓集系数与产额的影响

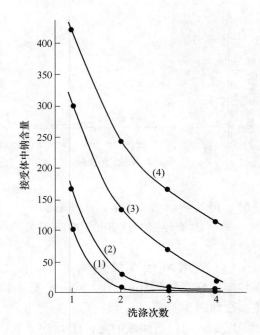

图 5　洗涤次数对接受体中钠含量的影响

曲线（3）乙醇和水的体积比为 3∶7。
曲线（4）乙醇和水的体积比为 5∶5。
结果表明，水为上述体系的良好洗涤剂。

1.4.3　Na_2CO_3 颗粒半径对浓集 ^{24}Na 的影响

由（n,γ）核反应所生成的 ^{24}Na，反冲射程很小，为了提高产额，进一步观察了 Na_2CO_3 颗粒半径对它的影响。

以 MgO 接受体为代表，所用靶子物的质量为 100mg，Na_2CO_3 和 MgO 的比例为 1∶5，用蒸馏水洗涤四次。实验结果如图 6 所示。

图 6 表明：产额随 Na_2CO_3 颗粒半径的减少而增大，但浓集系数下降较快。这可能是因为 Na_2CO_3 和 MgO 二相混合研磨的时间越长，Na_2CO_3 的颗粒半径虽不断减小，但黏合在 MgO 相表面的 Na_2CO_3 似不易洗涤下来，以致在洗涤次数相同的情况下，研磨时间越长，颗粒半径越小，但浓集系数反而呈现下降的情况。

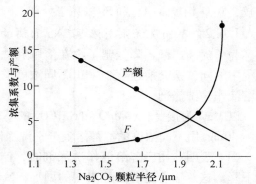

图 6　Na_2CO_3 颗粒半径对浓集系数与产额的影响

1.4.4　MgO 接受体中 ^{24}Na 与溶液中稳定钠之间的同位素交换

考虑到在溶解和洗涤过程中存在同位素交换过程，为了提高产额，有必要研究它们的交换规律。

假定两相间的同位素交换过程如下：

$$^{24}\text{Na}_{(\text{MgO})} + {}^{23}\text{Na}_{(溶液)} \rightleftharpoons {}^{23}\text{Na}_{(\text{MgO})} + {}^{24}\text{Na}_{(溶液)}$$

设 x 为在 t 时间内固-液相之间交换的原子数。a 表示 $t=0$ 时，MgO 中 ^{24}Na 的原子数，b 表示所加入的溶液中稳定钠的原子数，R 表示交换速度常数。

考虑到 ^{24}Na 的反冲射程很小，假定 ^{24}Na 一般均反冲在 MgO 表面的晶格内，作为一级近似，可以忽略 ^{24}Na 在 MgO 相中的扩散过程。因此根据二级动力学方程，得到：

$$\frac{\mathrm{d}x}{\mathrm{d}t} = R(a-x)(b-x)$$

由于 $b \gg x$, $b-x \approx b$, 所以：

$$\frac{\mathrm{d}x}{\mathrm{d}t} = Rb(a-x) \tag{1}$$

将式（1）改写成如下形式，并进行积分：

$$\frac{\mathrm{d}x}{a-x} = Rb\mathrm{d}t \tag{2}$$

$$\ln(a-x) = -Rbt + c \quad c = \ln a$$

$$\frac{a-x}{a} = \mathrm{e}^{-Rbt} \quad 或 \quad 1-\frac{x}{a} = \mathrm{e}^{-Rbt} \tag{3}$$

式（3）中，$\frac{x}{a} = F$ 即交换度：

$$1 - F = \mathrm{e}^{-Rbt} \tag{4}$$

式（4）表明，MgO 相中放射性钠与溶液中稳定钠之间的非均相同位素交换服从指数定律。

取 100mg 经过照射的靶子物，Na_2CO_3 与 MgO 的比例为 1∶5。放入离心试管中，用重蒸馏水洗涤六次。将含有 ^{24}Na 的 MgO 接受相与稳定 Na_2CO_3 溶液（100mg/3mL）等时间交换，离心沉降后，测定 Na_2CO_3 溶液中的放射性 ^{24}Na, 结果如图 7 所示。

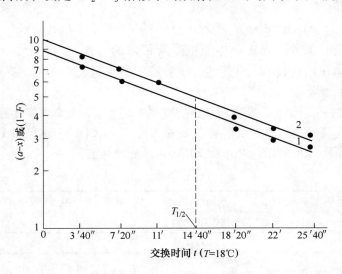

图 7　Na_2CO_3 溶液中的放射性 ^{24}Na

图 7 中直线 1 和 2 分别为 $(a-x)$、$(1-F)$ 对 t 作图的结果，它表明 MgO 相中 ^{24}Na

与溶液中稳定钠之间的非均相同位素交换服从指数规律。一般来说非均相交换不遵守简单的指数规律，但在某些特定条件下，两相间的交换速度近似地只取决于两相界面上的交换过程，可以忽略扩散过程，则同样可以服从简单的指数规律。从图 7 可以求得 ^{24}Na 的半交换期为 15min。为了提高产额，在洗涤和操作过程中，尽可能缩短时间是必要的。

1.4.5 放射性 ^{24}Na 反冲射程的近似计算

^{24}Na 在接受相中的产额 R 和反冲相的颗粒半径 r，反冲原子在该相中的射程 P 有如下关系：

$$R = \frac{\frac{1}{4} \times \frac{4}{3}\pi[r^3 - (r-P)^3]}{\frac{4}{3}\pi r^3} \tag{5}$$

如果 $P \ll r$，

$$P = \frac{4}{3}Rr \tag{6}$$

利用式（6），计算了 ^{24}Na 在 Na_2CO_3 中的反冲射程。

（1）以 Al_2O_3 为接受体时，其平均产额 $R = 3.0\%$，Na_2CO_3 颗粒的平均半径为 2.12μm，^{24}Na 的平均射程 $P = \frac{4}{3}Rr = 0.084$μm。

（2）以 MgO 为接受体时，其平均产额 $R = 4.0\%$，Na_2CO_3 颗粒的平均半径为 2.00μm，^{24}Na 的平均射程 $P = \frac{4}{3}Rr = 0.10$μm。

在计算平均射程时，选择产额低的实验值比较可靠。因为根据厚靶产额的计算可以看到，最大产额不应超过 25%。考虑到 ^{24}Na 的射程较 Na_2CO_3 颗粒半径小得多，因此凡实验所得产额大于 25% 的产额均为表观产额，这主要是由接受相中 Na_2CO_3 的洗涤不够完全所引起的。

1.4.6 MgO、Al_2O_3 和 SiO_2 对钠（稳定钠 + ^{24}Na）吸附的影响

在测定产额时，考虑到接受体的吸附所带来的误差。为此分别对 MgO、Al_2O_3 和 SiO_2 三种接受体在颗粒大小、放射性 ^{24}Na 和稳定钠含量完全相同的条件下，测量了由于吸附所引起的误差。

（1）SiO_2 对钠的吸附。吸附所用的 SiO_2 的质量为 30mg，颗粒半径为 1.69μm。所用的吸附溶液为靶子物的第一次溶解洗涤液（其中 Na_2CO_3 与 SiO_2 的比例为 1∶4），实验结果如表 1 所示（原始溶液强度为 23610（每分钟计数））。

表 1 SiO_2 对钠吸附的影响

洗涤次数	SiO_2 相中强度/cpm	吸附/%	洗涤次数	SiO_2 相中强度/cpm	吸附/%
1	111	0.46	4	2	0.0084
2	6	0.043			

（2）Al_2O_3 对钠的吸附。所取 Al_2O_3 的质量为 83mg，颗粒半径为 1.16μm 的 Na_2CO_3 与 Al_2O_3 的比例为 1∶5，所用的吸附溶液为靶子物第一次溶解洗涤液，实验结果如下：原始溶液强度为 208641，洗涤 4 次后 Al_2O_3 相中强度为 69，吸附 0.033%。

(3) MgO 对钠的吸附。所取 MgO 的质量为 83mg，颗粒半径为 1.98μm，Na_2CO_3 与 MgO 的比例为 1:5，所用的吸附溶液为靶子物的第一次溶解洗涤液，实验结果如表 2 所示（原始溶液强度为 21970）。

表 2 MgO 对钠吸附的影响

洗涤次数	MgO 相中强度/cpm	吸附/%	洗涤次数	MgO 相中强度/cpm	吸附/%
1	105	0.47	2	65	0.29

上述结果表明：MgO、Al_2O_3 和 SiO_2 经二次洗涤后，接受相中的放射性强度已接近本底。事实上接受相一般均经过四次洗涤，吸附对产额造成的影响很小。

1.4.7 ^{24}Na 半衰期的鉴定

对浓集所得的样品进行了半衰期鉴定，计数均经过死时间的校正，用最小二乘法求得的半衰期为 14.62h（文献值 14.97h）。结果如图 8 所示。

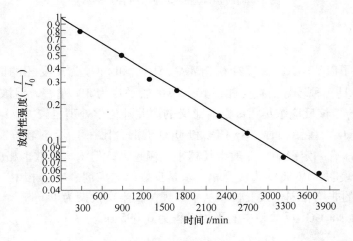

图 8 ^{24}Na 半衰期的测试结果

1.4.8 γ 辐照对 Al_2O_3 表面及对 MgO 溶解速度的影响

实验表明，Al_2O_3 经不同剂量的 γ 射线辐照后，用水洗涤，发现接受体中的钠含量随剂量的增加而加大。这意味着，在洗涤条件相同的情况下，Al_2O_3 表面不易洗涤干净。应该指出，对于 SiO_2、MgO 接受体，没有观察到上述现象。实验结果如图 9 所示。

此外还观察到经 γ 射线辐照后的 MgO 样品，在盐酸中的溶解速度变慢，实验结果如表 3 所示。

表 3 经 γ 射线辐照后的 MgO 样品在盐酸中的溶解情况

γ 辐射剂量/rad	3mL 4N HCl 溶解所需时间/s	γ 辐射剂量/rad	3mL 4N HCl 溶解所需时间/s
0	38	599.0×10^4	50
199.0×10^4	43	824.0×10^4	49
396.0×10^4	50		

关于反冲原子在固相中的慢化问题以及 γ 辐照所产生的滞留问题，未作进一步的研究。

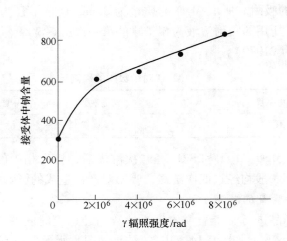

图 9 辐照强度对接受体中钠含量的影响

2 结论

(1) 研究了以 Na_2CO_3 为反冲相，MgO、Al_2O_3 和 SiO_2 为接受相的固-固反冲效应。结果表明 MgO 是一种较好的接受相，其产额虽然平均为3%左右，但浓集系数可达30左右，从 (n, γ) 核反应浓集 ^{24}Na 是有意义的。此外化学处理也较 Al_2O_3 和 SiO_2 简单。

(2) 以 MgO 为接受相时，Na_2CO_3 和 MgO 的混合比在 1∶2 时较好。反冲相的颗粒半径对浓集 ^{24}Na 有一定影响，一般半径越小产额越大，但浓集系数也相应降低。

(3) 实验表明，以 MgO 为接受相，水是良好的洗涤剂。MgO 相中 ^{24}Na 与溶液中稳定钠离子之间的非均相同位素交换服从指数规律，半交换期为 15min。

(4) ^{24}Na 在 Na_2CO_3 反冲相中的平均射程为 0.095μm。

参 考 文 献

[1] Proceedings of the Symposium on the Chemical Effects of Nuclear Transformations, I. A. E. A., Vienna, 1961.

[2] 王榕树：原子能科学技术，6，432(1963).

[3] Proceedings of the Symposium on the Chemical Effects of Nuclear transformations, I. A. E. A., Vienna, 1965.

[4] Harbottle, G：Ann, Rev, Nuclear Sci., 15, 89(1965).

[5] Stoecklin, G: "Chemistry of Hot Atoms", Verlag Chemie, Weinheim, 284(1969).

[6] Maddock, A. G: International Review of Science, Series one, Inorganic Chemistry Vol 8, 213(1972).

[7] Maddock, A. G: International Review of Science, Series two, Inorganic Chemistry Vol 8, 237(1975).

[8] 袁志熙：原子能科学技术，6，432(1960).

[9] Pertessis, M: Radiochim Acta; 1, 58(1968).

[10] Sue. P et al; C. R. 240, 2226 (1955), 240, 2415(1955).

[11] Несмеянов, A. H; Радиохимия, 6, 694(1959).

[12] Colonomos, and Parker, W; Radiochim Acta, 12, 163(1969).

[13] Parker, W. and Alarcon J. P; Radiochem, Radioanal Lett, 3, 223(1970).

[14] Reichold, P. and Wolb, P; Radiochim Acta, 15, 1, (1971).

[15] Collins, K. E et al; Radiochem, Radioanal, Lett, 11, 303, (1972).

[16] Lai, P. P and Rees, L. V. C; Faraday Transactions I 72, 1809(1976); 72, 1818(1976); 72, 1827(1976).

[17] 李旺长、王通、傅克坚、刘元方：化学学报，31，359(1965).

WilzBach 气曝法标记 H³-三尖杉酯碱[*]

1 前言

三尖杉酯碱是一种从植物中提取的天然碱[1]，其结构式为[2,3]：

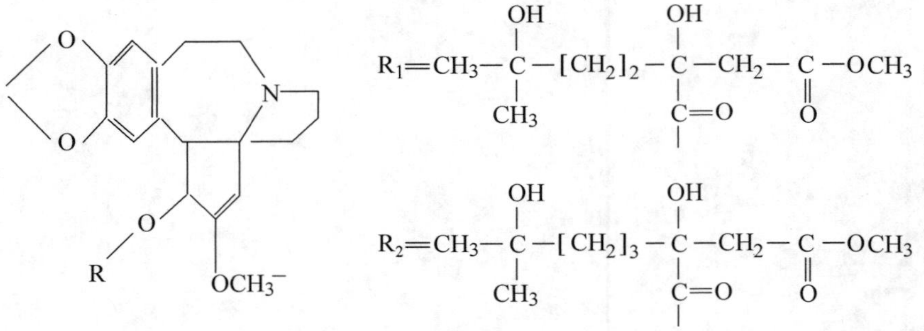

其中支链 R 有 R_1 和 R_2 两种。这种碱经临床证明对白血病有一定疗效，目前已引起国内外的重视。从结构上可以看出，它容易水解成母核和支链两部分。为了探讨三尖杉酯碱在动物体内的药理代谢及分布，我们采用了放电曝射法标记 H^3，结果表明有 86% 的 H^3 集中标记在母核上。H^3 放电曝射法标记三尖杉酯碱到目前为止，国内外还未见报道。

2 H³-三尖杉酯碱的制备

称取 20mg 三尖杉酯碱，放入曝射瓶内的镍舟中，滴入几滴甲醇，使样品能够均匀展开在镍舟的内壁上，待甲醇挥发后，曝射瓶接到加氚的真空系统上，抽真空至 5×10^{-3} mm(Hg)，然后在曝射瓶内通入 42mm(Hg) 的氚 4.3Ci，在超声波的作用下每天放电 10min(高频放电 1000V)，共八天。回收剩余氚气。用甲醇洗掉样品上的不稳定氚，每次 3.0mL 共四次。将氚化样品用 4.0mL 甲醇溶解。在 β 液体闪烁计数器上测量粗样品的比度为 25mc/mL，放射性总强度为 100mc。

3 H³-三尖杉酯碱的分析鉴定

（1）样品的纯化：用氧化铝薄板层析法纯化，收集 $R_f = 0.56$ 处的氧化铝，用甲醇溶解，吸取上层清液，反复三次，作为样品的提纯液，总体积 11.3mL，放射性比度 1.0mc/mL。

---

[*] 本文合作者：潘邵华、张连水、仪明光。原发表于《北京师范大学学报（自然科学版）》，1979(1)：59~61。

（2）H^3-三尖杉酯碱的水解，为了研究三尖杉酯碱母核和支链的分布，在几滴提纯液内加入 1~2mL 0.5M 乙醇钠溶液，放置 12h 进行水解。然后用薄板层析法在氧化铝干板上测定 H^3 的标记分布，测量结果如图 2 所示（氧化铝干板所用条件与样品纯化相同）：氧化铝 16~200 目，120℃烘 2h 发展剂，氯仿：甲醇 = 5:0.2。显色剂为碘铋酸钾、R_f =0.6 处的母核位置上出现明显的标记峰，所占标记 = 88%。在母核峰值的上面，R_f = 0.77 处出现一个较小的峰，可能是支链的标记峰，但占的标记很少。

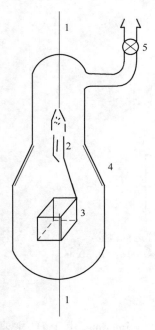

图 1
1—钨棒电极；2—放电区；
3—镍舟；4—磨口；5—二通活门

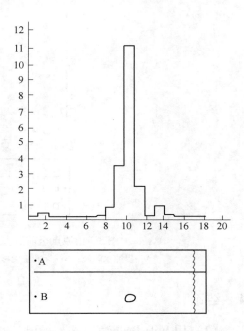

图 2 由原点至前沿，每隔 1cm
刮取一个样，然后进行测量
A—提纯样品；B—标准母核样

在进行氧化铝干板鉴定的同时，我们还在碱板上进行了 H^3 标记分布的测定，碱板的制法如下：称取 12~13g 硅胶用 1N NaOH 34mL 调成糊状铺板，每块约 1g 硅胶，烘干使用。乙醚：丙酮 = 2:1 为展开剂，碘铋酸钾显色，测量结果如图 3 所示。结果表明母核标记占总强度的 86%，从上图可以看出 H^3 标记在母核上的峰值与标准母核显色点一致。在标准三尖杉显色点处没有出现峰值，说明水解是完全的。

实验表明 86% 的 H^3 标记在母核上，几次实验结果是重复的。为了进一步验证其可靠性，又用萃取法加以对照，在酸性条件下用氯仿萃取支链，母核留在水相，达到分离的目的，然后分别测量水相和有机相强度。

方法：在分液漏斗中加入 5% 乙酸 15mL、氯仿 15mL 和少许三尖杉提纯水解液，摇动 10min，萃取后有机相分别用 15mL 5% 和 15mL 5% Na_2CO_3 各洗涤一次，以防止支链对母核测定的影响。然后各取水相和有机相 1mL 测定。测定结果为：支链强度 1%，母核强度 99%。

萃取测定结果表明，与碱板、氧化铝干板层析结果基本一致。由于萃取测量误差较小，所以结果比碱板、干板层析好。

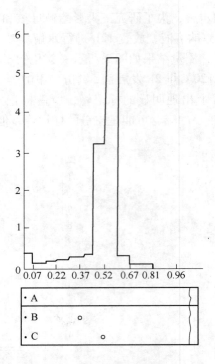

图 3　由原点至前沿，每隔 1cm 刮取一个样进行测量
A—提纯水解液；B—标准三尖杉酯碱，$R_f = 0.31$；C—标准母核，$R_f = 0.48$

4　结论

（1）用 Wilz Bach 气曝法，标记三尖杉酯碱产品比度为 5mc/mg，由于 H^3 主要集中在母核上，可以满足实际工作的需要。母核上的标记率为 86%。

（2）H^3 在母核上的标记位置目前尚难确定，关于用人工半合成的方法，把 H^3 标记在支链上的工作，正在继续进行。

本实验所用三尖杉酯碱是由中国科学院药物研究所提供的。在此对他们的大力协助表示衷心感谢。

参 考 文 献

[1] 李述文，中草药通讯，3，43(1976).
[2] 陆永宁，中草药通讯，4，50(1974).
[3] Paudler. W. W，J, org. Chem, 38，2110(1973).
[4] Mikolajczak, K. L, Tetrahedron, 28，1995(1972).

H³-L 门冬酰胺的标记*

1 前言

H³-L 门冬酰胺作为蛋白质掺入的前体，在白血病细胞动力学的研究中有一定的作用。关于 H³-L 门冬氨酸的标记，特别是 H³-L 门冬酰胺的标记，报道极少。

关于 H³-L.D 门冬氨酸，M. A. Каломийцев，Ц. Д. Гамкрелидзе 曾用反冲法[1]，高桥忠男等用气曝法、化学合成法及溶液中同位素交换法[2]进行了研究。

本实验以二溴琥珀酸为前体，通过正压加氚、化学合成、生物发酵得到 H³-L 门冬酰胺。反应式如下：

$$\begin{array}{c}\text{COOH}\\\text{CH-Br}\\\text{CH-Br}\\\text{COOH}\end{array} \xrightarrow{CH_3OH+KOH} \begin{array}{c}\text{COOH}\\\text{C}\\\text{\|\|\|}\\\text{C}\\\text{COOH}\end{array} \xrightarrow[Pd/C]{H^3} \begin{array}{c}\text{COOH}\\\text{T-C}\\\text{\|\|}\\\text{T-C}\\\text{COOH}\end{array} \xrightarrow[200℃]{封闭} \begin{array}{c}\text{COOH}\\\text{C-T}\\\text{\|\|}\\\text{C-T}\\\text{HOOC}\end{array}$$

$$\xrightarrow{发酵}\begin{array}{c}\text{COOH}\\\text{HCT}\\\text{H}_2\text{NCT}\\\text{HOOC}\end{array} \xrightarrow[酯化]{CH_3OH,H_2SO_4}\begin{array}{c}\text{COOCH}_3\\\text{HCT}\\\text{H}_2\text{NCT}\\\text{HOOC}\end{array}\xrightarrow{NH_3}\begin{array}{c}\text{CONH}_2\\\text{HCT}\\\text{H}_2\text{NCT}\\\text{HOOC}\end{array}$$

2 实验结果

2.1 丁炔二酸[3]

10g α,β-二溴丁二酸，12g 氢氧化钾及 75mL(95%) 甲醇，用水蒸气共同回流。得到的混合盐溶于含有一定量 H_2SO_4 的水中，产生白色结晶沉淀。抽滤，晶体溶于一定量水中（水中含一定量 H_2SO_4). 用乙醚萃取，得到白色丁炔二酸晶体。

2.2 H³-顺丁烯二酸

在图 1 所示装置中通氚。在 8.3mL 反应瓶内加入丁炔二酸 53mg，5% Pd/C 13.8mg（军事医学科学院制备），乙酸乙酯 0.5mL，液氮冷冻。

系统抽真空至 5×10^{-3} mm(Hg)，关闭真空系统，加热铀粉瓶，放氚。利用活性炭冷指"D"将 8.3mL(8.3Ci)氚转至反应瓶"C"中，使"C"中压力为 745mm(Hg). 解冻，

---

* 本文合作者：包华影、丁绍凤、孟昭兴、仪明光。原发表于《北京师范大学学报（自然科学版）》，1979（2）：75~78, 90。

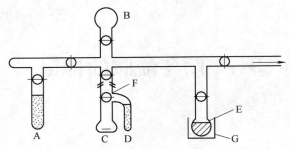

图 1 氚化装置示意图
A, D—活性炭冷指；B—氚气；C—反应瓶；
E—贮氚铀粉瓶；F—三通活塞；G—电炉
（标有"圆中加一横"处均为三通活塞）

吸氚，电磁搅拌 2h 后停止反应。残余氚气回收到铀粉瓶内。

反应液滤除催化剂，滤液加入顺丁烯二酸载体 680mg 后，转入具有喇叭口形状的安瓿瓶中，抽干乙醇，得到淡黄色固体。

2.3 H^3-反丁烯二酸

上述安瓿瓶由细处封管，在 200℃加热 1h。少量乙醇使样品溶解，经活性炭处理后抽干乙醇，用水重结晶。然后用升华法精制。

2.4 H^3-L 门冬氨酸

（1）培养液的制备。升华精制后的 H^3-反丁烯二酸以 116mg：12mg 比例与 $MgSO_4$ 混合，加入 1mL H_2O，用 NH_4OH 调节 pH 值至 8.5。

（2）AS1.881 固定化细胞（由微生物所提供）的处理。将固定化细胞放入布氏漏斗中，用水反复洗涤，湿称重后使用。

（3）发酵。培养液中加入 AS1.881 固定化细胞，在 37℃恒温下静置培养 1h。

（4）发酵后处理。滤出固定化细胞，并用少量水冲洗细胞。

滤液用 HCl 调节 pH 值至 2.8，析出 H^3-L 门冬氨酸（如无结晶析出时，可先在 100℃下浓缩）晶体。

2.5 H^3-L 门冬酰胺

在 1.9mL 甲醇和 0.04mL 浓硫酸中分批加入 H^3-L 门冬氨酸 69.5mg，在 27～28℃恒温下酯化 42h。

酯化液通氨气 25min，放置 6 天。离心除去 $(NH_4)_2SO_4$，上清液排氨后进行板层分离[4]。

以硅胶（G）加少量羧甲基纤维素钠铺板，硅胶用量为 $8g/15×20cm^2$。以乙二胺：丙酮：正丁醇：水(1.2：6：6：3)体系作为推进剂，上行 10～12cm。用 2%～2.5%茚三酮（含 3%醋酸）丙酮溶液显色，确定 H^3-L 门冬酰胺位置，R_f 值为 $R_{门冬氨酸}=0.2$，$R_{门冬酰胺}=0.4$。

将门冬酰胺部分刮下，用水洗脱，离心除去硅胶，蒸干水后加 14mL 0.9% NaCl 水溶液。

2.6 分析鉴定

利用 FJ-353G₁ 型双道液体闪烁计数器测量 H³-L 门冬酰胺的放射性强度。比度为 0.5mc/mg。

薄层层析鉴定放化纯度大于 95%。

3 讨论

3.1 加氚压力的影响

加氚时，压力增加有利于加氚反应的进行，但是加氚反应系统一般在大于一个大气压时容易造成泄漏，因此在适当减压情况下加氚较为适宜。

我们分别进行了高于一个大气压（参看图 2）和略低于一个大气压（参看图 3）时的加氢试验。实验表明：用自制 Lindlar-Pd 催化剂，一般经 1.5h 即可达到平衡，加氢量为理论值的 1/3，实验结果平行。

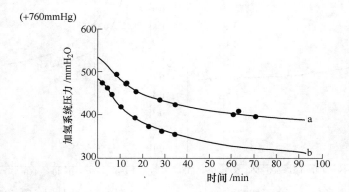

图 2 吸氢量与反应时间的关系

（催化剂：Lindlar-Pd 5mg，乙炔二羧酸 114mg，乙酸乙酯 0.5mL）

a—起始压力 760mmHg + 540mmH₂O； b—起始压力 760mmHg + 485mmH₂O

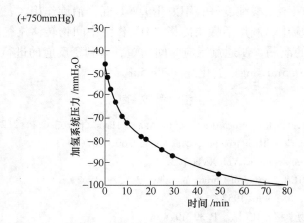

图 3 吸氢量与反应时间的关系

（催化剂：Lindlar-Pd 5mg，乙炔二羧酸 20mg，乙酸乙酯 0.5mL）

在减压情况下，加氢平衡时间和加氢量与上述情况大致相同。一般经1h后，已接近平衡量，如再延长0.5h，加氢量增加不大。

3.2 加氢催化剂的影响

H. Lindlar[5]指出：使用白金、钯这样的强催化剂，可还原到琥珀酸止，因此顺丁烯二酸的收率低于30%。使用比白金、钯弱一些的Pt-C催化剂，产率可达60%。使用Lindlar-Pd催化剂，收率最好。

本实验采用5% Pd/C，与Lindlar-Pd未作比较。

3.3 关于构型

曾对比做过化学合成法与生物发酵法对构型的影响。利用偏振光旋光仪测定$[\alpha]_D^{20}$。发酵法制备的L-门冬氨酸$[\alpha]_D^{20}$为+26.6°（图4），与标准L-门冬氨酸试剂（$[\alpha]_D^{20}$ +25.2°，图5）很好地获得一致。

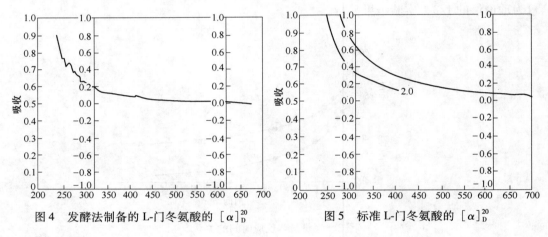

图4 发酵法制备的L-门冬氨酸的$[\alpha]_D^{20}$　　图5 标准L-门冬氨酸的$[\alpha]_D^{20}$

4 结论

（1）采用AS1.881固定化细胞培养制备L-门冬氨酸，与化学合成法相比，其操作简便，所得产物L-门冬氨酸纯度好，不发生D、L混合消旋作用。

（2）在通氚反应中，利用活性炭冷指"D"作为中间物转移氚气，正压加氚，避免整个管道系统氚气的泄漏，使通氚反应时间缩短，有利于反应的进行。

（3）产品比度0.5mL/mg。放化纯度79.5%。

参 考 文 献

[1] М. А. Каломийцев, Ц. Д. Гамкрелидзе. Сообщ. Акад, Наук. Груз. С. С. Р. 42(2),311～314(1966).

[2] 高桥忠男等, Proc. Conf. Radioisotopes, 5th. No. 3, 260～261(1963).

[3] R. C. Fuson, Organic syntheses Vol XVIII p. 3.

[4] E. Von ARX, NEHER. Journal of Chromatography 12, 329(1963).

[5] H. Lindlar; Helv., 35, 446(1952).

[6] 孟广震等, 微生物学报, 18(1), P. 39(1978).

[7] E. A. Evans, Tritium and its Compounds(1974).

用中子活化分析法测定 ^{238}U/^{235}U 同位素丰度比[*]

摘要 本文论述了用中子活化分析法测定含微量铀的样品中^{238}U/^{235}U同位素丰度比的原理及方法。样品在反应堆中接受短时间照射后，用 Ge(Li) 探头或高纯锗探头-多道能谱分析仪-计算机系统测量射线的能谱，可以分辨出^{238}U 和^{235}U 的许多监测峰。利用这两种监测峰计数之比与这两种同位素丰度比成正比的关系，分析铀的同位素丰度比，在^{235}U 丰度为 0.6% ~ 18% 范围时精密度为 1% ~ 2%，在贫化铀和 18% ~ 60% 丰度^{235}U 时，精密度为 2% ~ 3%。

关键词 中子活化分析 ^{238}U/^{235}U 同位素丰度比 微量铀 γ射线和 X 射线的测量

1 引言

含铀样品经反应堆照射后，^{239}Np 对各种裂变产物衰变强度的比值和样品中^{238}U/^{235}U 同位素丰度比值成正比关系[1,2,4,5]。利用这种关系可以分析^{235}U 丰度低于 10% 的样品中铀的同位素成分[1~6]。把样品密封在镉盒中照射，还可以使分析范围扩大到丰度达 93%[7] 的样品。

本工作用原子能研究所重水堆照射一系列不同丰度的标准样品，用 Ge(Li) 探头测量 80 ~ 2000keV 范围内的能谱，分析^{239}Np 的三个 γ 峰和 8 种不同半衰期（9 ~ 200h）的裂变产物的 γ 峰，获得良好的正比关系。另外，还用高纯锗探头测量 10 ~ 200keV 范围内的能谱，分析了^{239}Np 的一个 γ 峰与两种裂变产物的 γ 峰，获得同样良好的正比关系，同时，还显示^{239}Np 的 β$^-$衰变子体激发态的 Pu 与若干种裂变产物元素的特征 X 射线峰的计数比和^{238}U/^{235}U 丰度比也存在着正比关系。

针对丰度范围不同的样品采取不同的冷却时间，选择不同的监测峰，在^{235}U 丰度从 0.6% ~ 60% 的范围内都获得了良好正比关系，平行样品的相对偏差都不大于 ±0.5%。

2 原理

在粒子束流照射下，样品内任何一种元素的两种同位素 a、b 若同时发生核反应，其产物 a′、b′在停止照射时的放射性测量计数可由下式计算：

$$A_{a'} = \Phi \cdot f_1(\overline{\sigma}_a) \cdot \frac{W}{M} \cdot \theta_a \cdot 6.023 \cdot 10^{23} \cdot \eta_{a'} \cdot \varepsilon_{a'}$$

$$A_{b'} = \Phi \cdot f_2(\overline{\sigma}_b) \cdot \frac{W}{M} \cdot \theta_b \cdot 6.023 \cdot 10^{23} \cdot \eta_{b'} \cdot \varepsilon_{b'}$$

式中，Φ 为束流通量；$f(\overline{\sigma})$ 为有关核反应截面、衰变常数、分支比以及照射时间的函数式；W 为样品中该元素的含量；M 为相对原子质量（当同位素成分相差不大时，相

---

[*] 本文合作者：张存和。原发表于《核化学与放射化学》，1982,4(3)：167~173。

对原子质量的差异可忽略不计);θ 为百分比丰度;η 为几何探测效率;ε 为能量探测效率。用同一个探头测量时,$\eta_{a'} = \eta_{b'}$,两式相比可得:

$$\frac{A_{a'}}{A_{b'}} = k \cdot \frac{\theta_a}{\theta_b}$$

$$k = \frac{f_1(\overline{\sigma}_a)}{f_2(\overline{\sigma}_b)} \cdot \frac{\varepsilon_{a'}}{\varepsilon_{b'}}$$

对于用同样能量的粒子束流所照射的若干份样品,$f(\overline{\sigma})$ 值不变,K 值相等。即每个样品两种监测射线的相对计数比与同位素丰度比成正比。如果其中一个样品的同位素丰度已知,其余样品的丰度比就可算出。

中子照射铀的主要核反应为:

$$^{238}U(n,\gamma)^{239}U \xrightarrow{2.35\min} {}^{239}Np \xrightarrow{56.4h} {}^{239}Pu$$

$$^{235}U(n,f)F.P.(裂变产物)$$

用高分辨率的半导体探头-多道能谱仪-计算机系统测量中子照射后的铀样品的能谱,可以分辨出许多 ^{239}Np、激发态的 Pu 以及裂变产物的能峰,并精确给出这些能峰的净峰计数。则:

$$\frac{A_{^{239}Np}}{A_{F.P.}} = k \cdot \frac{\theta_{238}}{\theta_{235}}$$

$$k = \frac{\sigma_{238俘获}}{\sigma_{235裂变} \cdot Y_{F.P.}} \cdot \frac{\varepsilon_{^{239}Np}}{\varepsilon_{F.P.}}$$

式中,A 为 γ 峰净峰计数;Y 为裂变产额。

^{239}Np 的 β^- 衰变产物 Pu 处于核激发态,在极短时间内退激,同时以一定的分支比发射 γ 射线-内转换电子,并伴随发射 Pu 的特征 X 射线。这种 X 射线的强度与 ^{239}Np 原子核数成正比,表观半衰期与 ^{239}Np 相同。因此,Pu 的特征 X 射线峰可以作为 ^{238}U 的监测峰。许多裂变产物在先 β^- 后 γ 射线-内转换电子衰变过程中也都伴随发射子体核所属元素的特征 X 射线,例如:

$$^{143}Ce \xrightarrow[33h]{\beta^-, \gamma-内转换} {}^{143}Pr$$

过程中就发射 Pr 的特征 X 射线,其强度与 ^{143}Ce 原子数成正比,表观半衰期也为 33h,因此可以作为 ^{235}U 的监测峰。即:

$$\frac{A_{X(Pu)}}{A_{X(F.P.)}} = k' \cdot \frac{\theta_{238}}{\theta_{235}}$$

3 实验

(1) 制靶。由于 ^{238}U 的俘获截面与 ^{235}U 的裂变截面都很大,因此这一方法具有很高的灵敏度。一般可取 10～100μg 铀直接放在高纯铝箔上，或配制成含铀浓度为 1～4μg 铀/mg 溶液的稀硝酸溶液，在高纯铝箔上加一小滴（约 10mg 溶液，含铀 10～40μg）。烘干后折叠成小方箔，若干份样品包在一起送交照射。

（2）照射。制好的靶放在铝照射罐中，在反应堆的反射层孔道接受照射。在通量为 10^{13} n/(cm^2·s) 的情况下照射 20min～1h，就可获得宜于测量的强度。对于不同丰度的样品，选择不同镉比值的孔道照射可以得到更好的效果。

（3）测量。为了使 ^{239}U 全部衰变成 ^{239}Np，同时尽量减小短寿命裂变产物的强度，并且使一些作为监测核素的母子体达到瞬变平衡（否则很难计算停止照射时的强度），照射后至少应冷却 20h。分析贫化铀及天然铀时，冷却时间以 20～24h 为宜，分析高丰度铀时，冷却时间应延长到 50～70h。

用半导体探头-4096 道能谱仪-计算机系统测量每个样品的能谱。调整样品与探头的距离，使死时间为 7%～10%，测量 0.5h，一些主要能峰净峰计数测量误差可达 0.3%。打印机给出每个峰的净峰计数、本底计数及其误差，绘图机绘出能谱图。

图 1 为 Ge(Li) 探头测得的一份能谱图，图 2 为高纯锗探头测得的一份能谱图。

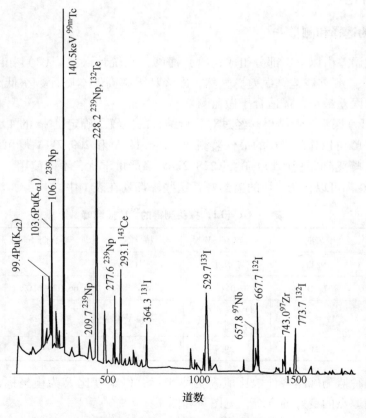

图 1　Ge（Li）探头所测能谱图
（样品丰度比 ^{238}U/^{235}U = 4.673；冷却时间 64h）

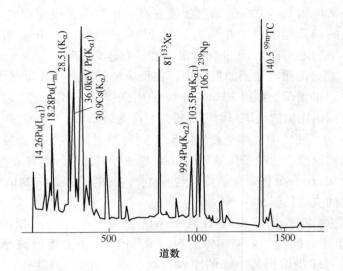

图2 高纯锗探头测得能谱图
（样品丰度比 $^{238}U/^{235}U = 3.671$；冷却时间65h）

4 结果及讨论

4.1 监测峰的选择和测量

净峰计数误差由以下三部分组成：（1）净峰计数统计误差；（2）扣除本底计数的计算误差；（3）解重峰运算误差。显然，净峰计数率高、本底计数率低、不受重叠峰干扰的峰计数误差最小，最适合于做监测峰。

（1）Ge(Li)探头测量结果。经过对分析结果的比较，确定 ^{239}Np 的4个主要γ峰当中，277.6keV 峰可以作为 ^{238}U 的最佳监测峰，106.1keV 和 209.7keV 峰可作为辅助监测峰，228.2keV 峰受到裂变产物 ^{132}Te 的 228.2keV 峰严重干扰，不能使用。有8种裂变产物的10个γ峰都可以作为 ^{235}U 的监测峰，这些峰都列在表1中。

表1 Ge(Li)探头测得的 ^{235}U 监测峰

核 素	半衰期/h	能量/keV	核 素	半衰期/h	能量/keV
^{135}Xe	9.17	249.1	^{143}Ce	33	293.1
^{91}Sr-^{91}Y	9.7~0.85	556.0（^{91}Y）	^{99}Mo-^{99m}Tc	66.02~6.02	140.5
^{97}Zr-^{97}Nb	17~1.22	743.0（^{97}Zr），657.8（^{97}Nb）	^{132}Te-^{132}I	78~2.28	667.7（^{132}I），773.7（^{132}I）
^{133}I	21	529.7	^{131}I	193	364.3

以上述两种监测峰净峰计数比值 K 为纵坐标，以丰度比 R 为横坐标，所绘的实验图线都是通过原点的良好的直线，见图3和图4。

（2）高纯锗探头测量结果。高纯锗探头在 10~200keV 范围内比 Ge(Li) 探头有更高的分辨率（见图1和图2），在所测能量范围内也有很好的监测峰。经过分析，确定

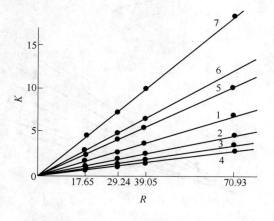
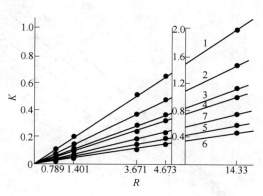

图 3　Ge(Li)探头测得的两监测峰 K 值
与 R 值的线性关系

1—$A_{277.6keV(^{239}Np)}/A_{293.1keV(^{143}Ce)}$；
2—$A_{277.6}/A_{140.5(^{99m}Tc)}$；3—$A_{277.6}/A_{743.0(^{97}Zr)}$；
4—$A_{277.6}/A_{529.7(^{133}I)}$；5—$A_{277.6}\times 0.25/A_{364.6(^{131}I)}$；
6—$A_{277.6}\times 4/A_{657.8(^{97}Nb)}$；7—$A_{277.6}/A_{667.6(^{131}I)}$

图 4　Ge(Li)探头测得的两监测峰 K 值
与 R 值的线性关系

1—$A_{277.6keV(^{239}Np)}\times 0.25/A_{364.3keV(^{131}I)}$；
2—$A_{277.6}\times 0.5/A_{647.4(^{132}I)}$；3—$A_{277.6}/A_{293.1(^{143}Ce)}$；
4—$A_{277.6}\times 2/A_{529.7(^{133}I)}$；5—$A_{277.6}/A_{743.0(^{97}Zr)}$；
6—$A_{277.6}/A_{657.8(^{97}Nb)}$；7—$A_{277.6}/A_{140.5(^{99m}Tc)}$

106.1 keV（^{239}Np 的 γ 峰）、103.6 keV（Pu 的 $K_{\alpha 1}$ X 射线峰）、99.4 keV（Pu 的 $K_{\alpha 2}$ X 射线峰）都可以作为 ^{238}U 的监测峰。81 keV（^{133}Xe 的 γ 射线峰，半衰期 127h）和 140.5 keV（^{99m}Tc 的 γ 峰）可以作为 ^{235}U 的最佳监测峰，表 2 中列出的三种裂变产物元素的特征 X 射线峰可以作为 ^{235}U 的辅助监测峰。

表 2　高纯锗探头测得的裂变产物元素的 X 射线峰

能量/keV	元素	X 射线系	母体
28.5	I	$K_{\alpha 1}$ 和 $K_{\alpha 2}$	^{131m}Te→^{131}I (30h) ^{132}Te→^{132}I (78h)
30.9	Cs	$K_{\alpha 1}$ 和 $K_{\alpha 2}$	^{135}Xe→^{135}Cs (9.17h) ^{133}Xe→^{133}Cs (127h)
36.0	Pr	$K_{\alpha 1}$	^{143}Ce→^{143}Pr (33h)

由于这些 X 射线峰在一定程度上都受到邻近峰的重叠，经过解重峰程序所给出的计数误差稍大些。

以上述两种监测峰净峰计数比值 K 为纵坐标、以丰度比 R 为横坐标，所绘实验图线也都是通过原点的良好直线，见图 5 和图 6。

4.2　中子能量的影响

由于 $\sigma_{235俘获}/\sigma_{235裂变}$ 的比值随中子能量的变化有很大改变，因此，中子能量略微不均匀也会引起不同样品的 k 值有明显的差异。只有把各份样品紧紧包在一起，尽量减小总体积（小于 1 cm^3），并且在照射时不停地旋转照射罐，才能确保 k 值相同。

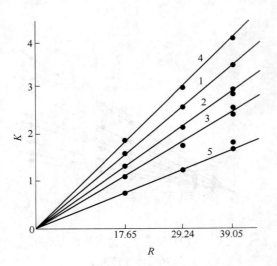

图 5　高纯锗探头测得两监测峰 K 值
与 R 值的线性关系

1—$A_{99.4keV(239Pu)} \times 0.25/A_{31.0keV(Cs)}$；
2—$A_{99.4}/A_{28.5(I)}$；3—$A_{99.4} \times 0.7/A_{36.0(Pr)}$；
4—$A_{106.0(239Np)}/A_{140.5(99Tc)}$；5—$A_{196.0} \times 0.3/A_{28.5(I)}$

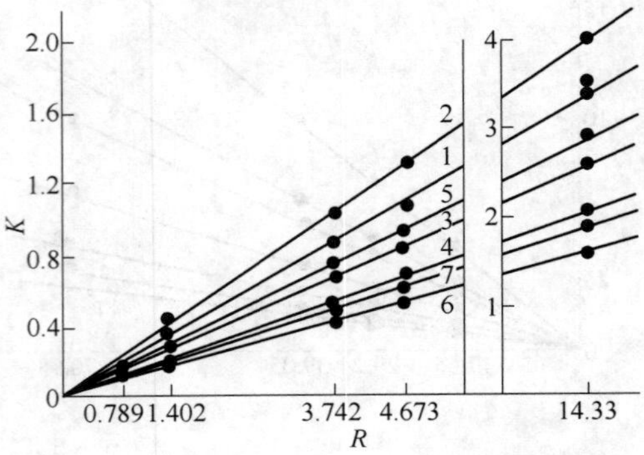

图 6　高纯锗探头测得两监测峰 K 值
与 R 值的线性关系

1—$A_{99.4keV(239Pu)} \times 4/A_{28.5keV(I)}$；
2—$A_{99.4} \times 4/A_{36.0(Pr)}$；3—$A_{99.4} \times 0.5/A_{31.0(Cs)}$；
4—$A_{106.0(239Np)} \times 0.8/A_{140.5(99Tc)}$；
5—$A_{106.0} \times 0.3/A_{31.0(Cs)}$；
6—$A_{106.0}/A_{28.5(I)}$；7—$A_{106.0}/A_{36.0(Pr)}$

利用 k 值对中子能谱的敏感性可以更好地分析不同丰度范围的样品。用充分慢化了的热中子照射贫化铀，可以减小 k 值，相对提高裂变产物对 ^{239}Np 的强度。用超热中子成分较多、热中子成分较少的中子流照射丰度较高的样品，可以提高 k 值，相对提高 ^{239}Np 对裂变产物的强度。掌握适当的冷却时间，可以使两种监测峰的计数误差都降到 0.3%。

4.3　干扰核反应

（1）中子能量大于 1MeV 后，^{238}U 开始裂变，3～5MeV 时 ^{238}U 裂变截面约为 5bar。^{235}U 热中子裂变截面为 580bar；因此，分析高丰度样品时，中子束流中有少量大于 1MeV 的快中子成分并不会引起可觉察的误差，分析贫化铀和天然铀时，则不允许有大于 1MeV 的快中子成分。

（2）^{234}U 的丰度极低，^{234}U 俘获中子产生 ^{235}U 的影响可以忽略不计。^{234}U 的裂变截面为 0.65bar，其裂变影响也可忽略不计。

（3）不允许样品中含有其他可裂变核素。

4.4　方法精度

由于仅比较同一探头测得的一个样品中两种射线的相对计数比，因此样品形状、测量位置、死时间校正等因素不会引起误差。在中子束流能谱不均匀所引起的误差可以忽略不计的条件下，在适当镉比值的孔道中照射样品，冷却适当的时间，使 ^{239}Np 峰计数率约为裂片峰的 1.5 倍时，可以得到最佳测量精密度约为 0.4%。

在我们的工作中，分别把两种不同丰度的标准试剂溶解成一定浓度的稀硝酸溶液，用严格的重量法称重后互相混合，配成一系列中间丰度的标准样品溶液，每种样品取 2~3 份制靶，检查分析结果的重现性，平行样品的相对偏差列在表 3 中。

表 3　平行样品的相对偏差

$^{238}U/^{235}U$①	相对偏差/%	$^{238}U/^{235}U$①	相对偏差/%
146.1	±0.07	14.33	±0.48
118.6	±0.31	4.673	±0.25
70.93	±0.11	3.671	±0.28
39.05	±0.36	1.401	±0.50
29.24	±0.04	0.789	±0.50
17.65	±0.12		

① $^{238}U/^{235}U$ 值为配制标准的计算结果。

对于 ^{235}U 从 0.4%~60% 范围内的标准系列，考查两监测峰计数比与丰度比的正比关系，所获得的精密度为 1%~3%。

在配制标准的工作中得到原子能研究所严叔衡、苏树新、杨景霞、孙淑英等同志的大力帮助，在测量能谱的工作中得到地质科学研究院物理探矿研究所张玉君、赵美卓等同志的大力帮助，在此衷心致谢。

参 考 文 献

[1] G. W. Leddicotte et al., TID-7531(1975).
[2] A. Kjelberg et al., Radiochimica Acta, 5, 104(1966).
[3] 吉田博之等, Radioisotopies, 19, 13(1970).
[4] F. T. Bunus, Radiochimica Acta, 15, 118(1971).
[5] M. Mantel et al., J. Radioanal. Chem., 2, 395(1969).
[6] R. Ganapathy et al., J. Radioanal. Chem., 44, 199(1978).
[7] Wen-deh Lu et al., J. Inorg. Nucl. Chem., 36, 2433(1974).

Determination of $^{238}U/^{235}U$ Isotopic Ratio by Neutron Activation Analysis

Zhang Cunhe, Liu Boli

(Beijing Normal University, Division of Radiochemistry, Beijing, 100875, China)

Abstract　A method is described for the determination of the isotopic composition of uranium by activation analysis. After irradiation of samples containing a small quantity of uranium in reactor, the energy spectrum of ^{239}Np formed from ^{238}U and those of fission products formed from ^{235}U are measured with high resolving power semiconductor-computer system. The ratio of the peak intensities is proportional to the $^{238}U/^{235}U$ ratio in the sample.

A series of standard samples with ^{235}U abundance ranging from 0.6% to 60% were irradiated in the heavy water reactor of the Institute of Atomic Energy at a neutron flux of 3×10^{13} n cm^{-2} · s^{-1}. The intensity ratios of the three ^{239}Np photo peaks and nine fission product peaks were plotted against the ^{238}U/^{235}U ratio of the samples respectively. Straight lines of zero intercept were obtained in all cases using Ge(Li) detector. High purity Ge detector were used to determine the X-ray characteristic peaks of ^{239}Pu and three fission products. Straight lines were also obtained when the X-ray characteristic intensity ratios of the ^{239}Pu and those of some of the fission products were plotted against the ^{238}U/^{235}U ratio.

These methods can be used for the analysis of samples containing 1-100 μg uranium with a precision of 1%-3% for samples with ^{235}U abundance ranging from 0.6% to 55%.

Key words neutron activation analysis, abundance ratio of ^{238}U/^{235}U, trace of uranium, measurement of γ-ray and X-ray

关于不同卤原子之间的交换反应[*]

1 前言

当前在核医学临床诊断中,缺中子放射性卤素 ^{18}F、^{34m}Cl、^{77}Br、^{123}I 起着十分重要的作用[1,2]。这些"有机"核素不仅可以用来置换几乎所有有机化合物中的氢原子,从而制备各种放射性药物,而且它们的核性质既宜于进行体内的研究,又便于体外的探测。因此自20世纪60年代以来,各种卤素的放射性药物已广泛用于临床诊断和人体代谢的研究[3~6]。特别像重卤素 ^{211}At,由于它是 α-衰变核素,有可能用于辐射治疗,近年来也重新引起各国的关注[7~9]。

为了制备各种标有放射性卤素的药物,曾发展了多种卤原子的标记方法,其中为了满足短寿命卤素标记的需要,又发展了若干快速的标记方法,如衰变诱导交换[10,11]、熔融交换[12]、酶促交换[13]和其他交换法[14],在上述标记方法中同位素交换和卤原子之间的交换反应占有重要地位。利用卤原子之间的碘、溴交换,溴、氟交换,碘、砹交换反应,曾制备了一系列高比度、无载体的放射性药物[15],积累了大量资料,其中有关同位素交换反应的机构及动力学的报道较多[16]。但迄今为止,尚未见到卤原子之间交换反应动力学的详细讨论[17]。本文首次得到了一个描述卤原子交换反应的普遍动力学方程。其中在特定条件下,包含了一般同位素交换反应的指数定律。上述方程还可以用来帮助判别一级交换反应和二级交换反应的机构。

2 一般性的讨论

考虑交换反应

$$AX^* + BY \rightleftharpoons BX^* + AY \tag{1}$$

设 AX^* 和 AY 分别为标记以放射性同位素的和未被标记的两个无机卤化物(X 和 Y 为卤素),如在某冠醚介质中的标以 I-123 的 NaI 和稳定的 NaBr;BX^* 和 BY 为相应的有机卤化物,如 ω-I-硬脂酸和 ω-Br-硬脂酸;$[M_1]_t$,$[M_2]_t$,$[M_3]_t$ 和 $[M_4]_t$ 分别为在时间 t 时的 AX^*、BY、BX^* 和 AY 的浓度,在 $t=0$ 时,$[M_3]_0 = [M_4]_0 = 0$。

2.1 二级反应

若该交换反应属二级反应,则 k_1 和 k_2 分别为正向和反向反应的速率常数,反应平衡常数 $K = \dfrac{k_1}{k_2}$。因此:

---

[*] 本文合作者:冯锡璋、国毓智。原发于《北京师范大学学报》,1983(3):63~69。

$$\frac{d[M_3]_t}{dt} = k_1[M_1]_t[M_2]_t - k_2[M_3]_t[M_4]_t \tag{2}$$

由于 $[M_3]_t = [M_4]_t$，$[M_1]_t = [M_1]_0 - [M_3]_t$ 和 $[M_2]_t = [M_2]_0 - [M_4]_t$，因此：

$$\frac{d[M_3]_t}{dt} = k_1([M_1]_0 - [M_3]_t)([M_2]_0 - [M_3]_t) - k_2[M_3]_t^2$$

$$= k_1[M_1]_0[M_2]_0 - k_1([M_1]_0 + [M_2]_0)[M_3]_t + (k_1 - k_2)[M_3]_t^2 \tag{3}$$

若设 $[M_3]_t = x(t)$，$[M_1]_0 + [M_2]_0 = a$ 和 $[M_1]_0[M_2]_0 = b$，则：

$$\frac{dx}{dt} = k_1 b - k_1 a x + (k_1 - k_2)x^2 = (k_1 - k_2)(f - x)(g - x)$$

这里

$$f = \frac{1}{2}\left[\left(\frac{k_1 a}{k_1 - k_2}\right) + \sqrt{\left(\frac{k_1 a}{k_1 - k_2}\right)^2 - \frac{4k_1 b}{k_1 - k_2}}\right]$$

$$g = \frac{1}{2}\left[\left(\frac{k_1 a}{k_1 - k_2}\right) - \sqrt{\left(\frac{k_1 a}{k_1 - k_2}\right)^2 - \frac{4k_1 b}{k_1 - k_2}}\right]$$

因此

$$\frac{dx}{(f-x)(g-x)} = \frac{1}{f-g} \times \left(\frac{-dx}{f-x} - \frac{-dx}{g-x}\right) = (k_1 - k_2)dt$$

$$\frac{1}{f-g}[\ln(f-x) - \ln(g-x)]\Big|_{x(0)=0}^{x(t)} = (k_1 - k_2)t\Big|_0^t$$

$$\ln\left(\frac{f-x}{f}\right) - \ln\left(\frac{g-x}{g}\right) = (k_1 - k_2)(f-g)t \tag{4}$$

并进一步可以得出：

$$x(t) = g + (f-g)\left\{1 - \frac{f}{g}\exp[(k_1 - k_2)(f-g)t]\right\}^{-1} \tag{5}$$

若 $k_1 = k_2$，$K = 1$，则式(5)可简化为：

$$x(t) = \frac{b}{a}(1 - e^{-k_1 at}) \tag{6}$$

$$x(\infty) = 平衡值 = \frac{b}{a}$$

$$-\ln\left[1 - \frac{x(t)}{x(\infty)}\right] = -\ln(1 - F) = k_1 a t \tag{7}$$

因此，$[-\ln(1-F)]$ 与 t 呈线性关系，其斜率为 $k_1 a$。若 $k_1 > k_2$，$K > 1$，则 $f > g > 0$，$x(\infty) = g$，当 K 接近于 1 时：

$$x(\infty) = g \approx \frac{b}{a}\left(1 + \frac{b}{a^2} \times \frac{K-1}{K}\right) \tag{8}$$

$$f \approx \frac{Ka}{K-1} - \frac{b}{a} - \frac{b^2(K-1)}{a^3 K} \gg x(\infty) > x(t) \tag{9}$$

$$f - g \approx \frac{Ka}{K-1} - \frac{2b}{a} - \frac{2b^2(K-1)}{a^3 K} \tag{10}$$

直接用式(4)

$$-\ln\left(1 - \frac{x}{g}\right) = -\ln(1-F) = (k_1 - k_2)(f-g)t - \ln\left(1 - \frac{x}{f}\right) \tag{11}$$

即
$$-\ln(1-F) + \ln\left(1 - \frac{x}{f}\right) \approx -\ln(1-F) - \frac{x}{f} - \frac{x^2}{2f^2}$$

$$\approx -\ln(1-F) - \frac{x(K-1)}{aK} \approx (k_1 - k_2)(f-g)t$$

$$\approx \left[k_1 a - \frac{2b}{a}(k_1 - k_2)\right]t \tag{12}$$

若将 $-\ln(1-F)$ 减去 cx 后对 t 作图而得一直线,则:

$$c = \frac{K-1}{Ka} = \frac{k_1 - k_2}{k_1 a}, \text{斜率 } m \text{ 为 } k_1 a - \frac{2b}{a}(k_1 - k_2)$$

$$k_1 = \frac{m}{a - 2bc} \tag{13}$$

$$k_2 = (1 - ac)k_1 = \frac{m(1-ac)}{a - 2bc} \tag{14}$$

若 $k_1 < k_2, K < 1$,则 $f > 0, g < 0, |g| > f, x(\infty) = f$,当 K 接近于 1 时:

$$x(\infty) = f \approx \frac{b}{a}\left(1 + \frac{b}{a^2} \times \frac{K-1}{K}\right) \tag{15}$$

$$g \approx -\frac{Ka}{1-K} - \frac{b}{a} + \frac{b^2(1-K)}{Ka^3} \tag{16}$$

$$f - g \approx \frac{Ka}{1-K} + \frac{2b}{a} - \frac{2b^2(1-K)}{Ka^3}, |g| \gg x(\infty) > x(t)$$

$$-\ln\left(1 - \frac{x}{f}\right) = -\ln(1-F) = (k_1 - k_2)(f-g)t - \ln\left(1 - \frac{x}{g}\right) \tag{17}$$

即
$$-\ln(1-F) + \ln\left(1 - \frac{x}{g}\right) = (k_2 - k_1)(f-g)t$$

$$\approx [-\ln(1-F)] - \frac{x}{g} - \frac{x^2}{2g^2}$$

$$\approx -\ln(1-F) + \frac{x(1-K)}{Ka}$$

$$\approx \left[k_1 a + \frac{2b}{a}(k_2 - k_1)\right]t \tag{18}$$

若将 $-\ln(1-F)$ 加上 $c'x$ 后对 t 作图而得一直线,则 $c' = \frac{1-K}{Ka} = \frac{k_2 - k_1}{k_1 a}$,斜率 m' 为

$$k_1 a + \frac{2b}{a}(k_2 - k_1)$$

$$k_1 = \frac{m'}{a + 2bc'} \tag{19}$$

$$k_2 = (1 + ac')k_1 = \frac{m'(1 + ac')}{a + 2bc'} \tag{20}$$

更精确并适用于任何 K 值的方法是：在等待较长的时间以达到交换反应的基本平衡后，测定 $x(t\to\infty)$，它接近于为 g（当 $k_1 > k_2$ 时）或 f（当 $k_1 < k_2$ 时）。若 $x(\infty) > \frac{b}{a}$，则 $k_1 > k_2$；若 $x(\infty) < \frac{b}{a}$，则 $k_1 < k_2$。由 $x(t\to\infty)$ 可以算出 $\frac{k_1}{k_2}$，并由此计算出 f（当 $k_1 > k_2$ 时）或 g（当 $k_1 < k_2$ 时），再由 $x(t)$ 值求得 $\frac{x(t)}{g} = F(t)$（当 $k_1 > k_2$ 时）与 $\frac{x(t)}{f}$，或 $\frac{x(t)}{f} = F(t)$（当 $k_1 < k_2$ 时）与 $\frac{x(t)}{g}$，并进一步求得 $-\ln[1 - F(t)]$ 值与 $\ln\left[1 - \frac{x(t)}{f}\right]$ 或 $\ln\left[1 - \frac{x(t)}{g}\right]$ 值。当 $k_1 > k_2$ 时，将 $-\ln[1 - F(t)] + \ln\left[1 - \frac{x(t)}{f}\right]$ 对 t 作图得直线，其斜率 $m = (k_1 - k_2)(f - g)$。由 m 和 $\frac{k_1}{k_2}$ 可分别得出 k_1 与 k_2。当 $k_1 < k_2$ 时，将 $-\ln[1 - F(t)] + \ln\left[1 - \frac{x(t)}{g}\right]$ 对 t 作图得直线，其斜率 $m' = (k_2 - k_1)(f - g)$。由 m' 和 $\frac{k_1}{k_2}$ 可分别得出 k_1 与 k_2。这里应该指出，若 $[M_1]_0 \ll [M_2]_0$，如 $[M_1]_0 = 0.001[M_2]_0$，则当 $\frac{k_2}{2} < k_1 < 2k_2$ 时，$-\ln[1 - F(t)]$ 对 t 作图为（或很近于为）一直线，其斜率为（或很近于为）$k_1 a$，当 $[M_1]_0 \approx [M_2]_0$ 和 $k_1 \neq k_2$ 时，$-\ln[1 - F(t)]$ 对 t 作图，则为通过原点的曲线，如图 1 所示。

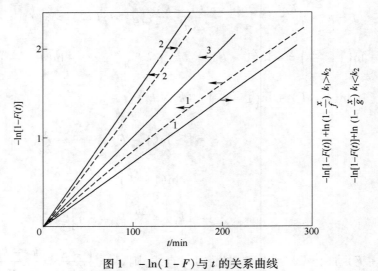

图 1　$-\ln(1 - F)$ 与 t 的关系曲线
($k_1 a = 0.01 \text{min}^{-1}$)

1—$k_1 = 2k_2$；2—$k_1 = \frac{k_2}{2}$；3—$k_1 = k_2$，$[M_1]_0 = [M_2]_0$

对相同卤素原子之间的交换，如：

$$AX^* + BX \underset{k}{\overset{k}{\rightleftharpoons}} AX + BX^* \tag{21}$$

或

$$AY^* + BY \underset{k'}{\overset{k'}{\rightleftharpoons}} AY + BY^* \tag{22}$$

则

$$-\ln(1 - F) = kat \tag{23}$$

或

$$-\ln(1 - F) = k'at \tag{24}$$

但是，k、k'、k_1 和 k_2 都不一定相等。

2.2 一级反应

若该反应属一级反应，反应历程可设想为：

(1) $BY \xrightarrow{k_1} B^+ + Y^-$ 慢、决速步

(2) $BX^* \xrightarrow{k_2} B^+ + X^{*-}$ 慢、决速步

(3) $B^+ + X^{*-} \xrightarrow{k_3} BX^*$ 快速反应

(4) $B^+ + Y^- \xrightarrow{k_4} BY$ 快速反应

则

$$\frac{dx}{dt} = k_1[BY]_t \frac{k_3[AX^*]_t}{k_3[AX^*]_t + k_4[AY]_t} - k_2[BX^*]_t \frac{k_4[AY]_t}{k_3[AX^*]_t + k_4[AY]_t}$$

$$= [k_3([M_1]_0 - x) + k_4 x]^{-1}[k_1 k_3 b - k_1 k_3 a x + (k_1 k_3 - k_2 k_4)x^2]$$

$$= [k_3[M_1]_0 - (k_3 - k_4)x]^{-1}[k_1 k_3 b - k_1 k_3 a x + (k_1 k_3 - k_2 k_4)x^2] \tag{25}$$

设 $k_3[M_1]_0 = \alpha$，$k_3 - k_4 = \beta$，则：

$$\frac{\alpha - \beta x}{(p-x)(q-x)} dx = (k_1 k_3 - k_2 k_4) dt \tag{26}$$

$$p = \frac{1}{2}\left[\frac{k_1 k_3 a}{k_1 k_3 - k_2 k_4} + \sqrt{\left(\frac{k_1 k_3 a}{k_1 k_3 - k_2 k_4}\right)^2 - \frac{4 k_1 k_3 b}{k_1 k_3 - k_2 k_4}}\right]$$

$$q = \frac{1}{2}\left[\frac{k_1 k_3 a}{k_1 k_3 - k_2 k_4} - \sqrt{\left(\frac{k_1 k_3 a}{k_1 k_3 - k_2 k_4}\right)^2 - \frac{4 k_1 k_3 b}{k_1 k_3 - k_2 k_4}}\right]$$

因此：

$$\left(\frac{\alpha - \beta p}{p - q}\right)\left(\frac{-dx}{p - x}\right) - \left(\frac{\alpha - \beta q}{p - q}\right)\left(\frac{-dx}{q - x}\right) = (k_1 k_3 - k_2 k_4) dt$$

$$\left(\frac{\alpha - \beta p}{p - q}\right)\ln\left(1 - \frac{x}{p}\right) - \left(\frac{\alpha - \beta q}{p - q}\right)\ln\left(1 - \frac{x}{q}\right) = (k_1 k_3 - k_2 k_4) t$$

若 $k_1 k_3 > k_2 k_4$，则 $p > q > 0$，$x(\infty) = q$，并

$$[-\ln(1-F)] + \frac{\alpha - \beta p}{\alpha - \beta q}\ln\left(1 - \frac{x}{p}\right) = \frac{p - q}{\alpha - \beta q}(k_1 k_3 - k_2 k_4) t \tag{27}$$

若 $k_1k_3 < k_2k_4$，则 $p > 0$，$q < 0$，$|q| > p$，$x(\infty) = p$，并

$$[-\ln(1-F)] + \frac{\alpha - \beta q}{\alpha - \beta p}\ln\left(1 - \frac{x}{q}\right) = (k_2k_4 - k_1k_3)\left(\frac{p-q}{\alpha - \beta p}\right)t \tag{28}$$

式（27）和式（28）虽与式（11）和式（17）相似，但较复杂。首先需考虑 k_1 和 k_2 的分别测定。以 k_1 为例，用式（22）的交换反应。简单的推导（一级反应）得：

$$-\ln(1-F) = \frac{k_1 a}{[M_1]_0} t \tag{29}$$

从式（29）和式（24）可以看出，一级反应与二级反应的差别在于它们的 $-\ln(1-F)$ 对 t 的直线的斜率不一样，前者为 $\frac{k_1 a}{[M_1]_0}$，与 $\frac{a}{[M_1]_0}$ 成正比，后者则为 $k'a$，与 a 成正比。因此，改变 $[M_1]_0$ 与 $[M_2]_0$ 之间的比值，可以区别一级和二级反应。若属一级反应，则可由式（29）得出 k_1。类似地，用式（21）的交换反应可以得出 k_2。再用处理式（11）和式（17）以得出 k_1 和 k_2 的方法处理式（27）和式（28），则有可能得出 k_3 和 k_4 之比值。

若 $k_3 \gg |k_3 - k_4|$，则式（25）可简化为：

$$\frac{dx}{dt} = (k_3[M_1]_0)^{-1}[k_1k_3 b - k_1k_3 ax + (k_1k_3 - k_2k_4)x^2] \tag{30}$$

将式（3）中的 k_1 改为 k_1k_3，k_2 改为 k_2k_4 和 dt 改为 $\frac{dt}{k_3[M_1]_0}$，即可转化为式（30）。因此，若 k_1 和 k_2 为已知，则可得出 k_3 和 k_4 之比值。

3 结论

从以上分析可以看到，用示踪原子法和改变反应物的浓度的比值，可以区分相同卤素原子之间的交换反应属一级反应或二级反应，并测定其决速步的速率常数。不同卤素原子之间的交换反应，也可用同样的方法来确定其为一级或二级反应。若为二级反应，则还可得出其正向和反向反应的速率常数 k_1 和 k_2；若为一级反应，则在从相同卤素原子之间的交换反应得出两个决速步的速率常数后，可进一步得出两个快速反应的速率常数之比值 $\frac{k_3}{k_4}$。

参 考 文 献

[1] R. M. Lambrecht, et al., Radiopharmaceutical and Labelled Compounds, Vol. 1, 1973, 275.
[2] C. P. Madhusudhan, et al., J. Radioanal. Chem., 53(1979), 299.
[3] G. J. Stocklin, Int. J. Appl. Radiat. Isot., 28(1977), 131.
[4] A. J. Palmer, ibic., 28(1977), 53.
[5] C. D. Robinson, ibid., 28(1977), 149.
[6] J. F. Harwig, et al., ibid., 28(1977), 157.
[7] W. M. Gerard, et al., J. Label. Comp. Radiopharm., 18(1982), 799.
[8] C. A. Aijeet, et al., Int. J. Appl. Radiat. Isot., 26(1975), 25.

[9] I. Brown, Int. J. Appl. Radiat. Isot., 33(1982), 75.
[10] M. J. Welch, J. Am. Chem. Soc., 92(1970), 408.
[11] R. M. Lambrecht, et al., J. Nucl. Med., 13(1972), 266.
[12] H. Elias, Proc. 9th Japan Conf. on Radioisotopes, Tokyo, 1968, 538.
[13] R. M. Lambrecht, J. Nucl. Med., 15(1974), 863.
[14] 刘伯里等，核化学与放射化学，3(1980), 242.
[15] G. J. Stocklin, J. Label. Comp. Radiopharm., 17(1980), 353.
[16] E. A. Evans, et al., Radiotracer Techniques and Applications, Vol. 1, (1977), 405.
[17] 国毓智、刘伯里，核化学与放射化学，4 (1983), 1.

Discussion on the Exchange Reactions between Different Halogen Atoms

Feng Xizhang, Liu Boli, Guo Yuzhi

Abstract In this paper, the exchange reactions between different halogen atoms are generally discussed. A general chemical kinetic equation describing these reactions is derived. The exponential expression for the general isotopic exchange reaction is included as a special case. This equation may also be used to distinguish between the first order and the second order exchange reactions.

Synthesis, Characterization, and Biodistribution of [^{113m}In] TE-BAT: a New Myocardial Imaging Agent[*]

Abstract In order to develop a new myocardial perfusion agent, new lipid-soluble complexes containing a net charge of +1 were evaluated. Synthesis, radiolabeling, characterization, and biodistribution of a unique indium complex, [^{113m}In] TE-BAT (tetraethyl-bis-aminoethanethiol), are described. The complex formation between In^{3+} and TE-BAT ligand is rapid, simple, and of high yield ($\geqslant 95\%$). This process is amenable to kit formulation. The complex has a net charge of +1 and an In/ligand ratio of 1 : 1. Biodistribution in mice shows higher heart uptake and longer retention as compared to ^{201}Tl. This complex, when labeled with ^{111}In, shows promise as a possible tracer for myocardial perfusion imaging.

Several new technetium-99m (^{99m}Tc) agents for regional myocardial perfusion imaging have been reported[1-8]. These agents are potential substitutes for thallium-201 (^{201}Tl) as the major radiopharmaceutical for routine nuclear medicine application. The ^{99m}Tc-labeled isonitriles, initially developed by Jones and Davison, have reached the final stage of clinical trial[1]. The first agent of this series was [^{99m}Tc] TBI (tbutylisonitrile), which showed very high myocardial uptake and retention reflecting regional perfusion. However, the initial lung uptake and the subsequent high liver retention of [^{99m}Tc] TBI clearly indicates the need for further research and development[2]. A new generation of isonitriles, [^{99m}Tc] MIBI (2-methoxyisobutylnitrile) and CPI (2-carboxypropanylisonitrile) which show improved liver and lung washout, and the same high myocardial uptake and prolonged retention was reported[3-5]. The [^{99m}Tc] MIBI is currently under phase II clinical trial. New boronic acid adducts of vicinyl dioximes, "cage" complexes of ^{99m}Tc, (BATO, Squibb, SQ-30217) have been reported[6]. Initial clinical study has also indicated that they may be useful as myocardial perfusion tracers[7]. Another group of Tc. (Arene)$_2$+ compounds, which is a "sandwich" complex, was reported[8]. The clinical evaluation of these agents in humans has not yet been reported; nevertheless, initial studies in animals showed very high myocardial uptake and prolonged retention.

Recent advances in chemistry of ^{99}Tc complexes based on N$_2$S$_2$ ligands has dramatically enhanced our ability to predict the chemical structure of the final ^{99m}Tc complexes. This series of ligands forms strong complexes with (Tc =O)$^{+3}$. The X-ray crystallography studies of several N$_2$S$_2$ complexes developed by ourselves and others[9-19] has confirmed the (Tc =O)$^{3+}$ chemical state and the pyramidal core structure. Indium-111 (^{111}In) is a radionuclide with a $T_{1/2} = 2.8$

---

[*] Copartner: H. F. Kung, Y. T. Jin, L. Zhu, M. Meng. Reprinted from *The Journal of Nuclear Medicine*, 1989, 30 (3): 367-373.

days and gamma rays of 172keV and 247keV, which are suitable for nuclear medicine imaging studies. Currently, the radionuclide [^{111}In]oxine is being used for white blood cell[20] and monoclonal antibody labeling. In order to investigate further the radiochemistry of indium, we have initiated a study using the N_2S_2 ligand, tetraethylbis-(aminoethanethiol)(TE-BAT), for complexing In^{3+}. This paper presents our data on synthesis, radiolabeling, characterization, and biodistribution of this unique In complex. For convenience in this study, ^{113m}In eluted from a tin-113m indium-113m(^{113m}Sn-^{113m}In) generator was employed as the tracer. However, for imaging studies, ^{111}In is more suitable.

1 Materials and Methods

1.1 General

The preparation of TE-BAT was achieved by a method reported previously[10]. The only difference is that lithium aluminum hydride was employed for the last reduction step of diimine intermediate[19]. The dimercapto hydrochloride salt of TE-BAT was precipitated and used for this study. Indium-113m was obtained by eluting a ^{113}Sn-^{113m}In generator (Institute of Atomic Energy, Beijing, China) with 0.1N HCl.

1.2 Radiolabeling

No-carrier-added [^{113m}In]chloride(1mCi/mL) eluted with 0.1N HCl was added to a test tube containing the BAT-TE ligand(2mg) in 1mL of water. In order to maintain the pH at 4-5, a simultaneous addition of a solution of 5% NaOH in the reaction mixture is needed. The mixture was vortexed and kept in a water bath at 80℃ for 0.5h. The percent labeling yield was measured by thin layer chromatography(TLC)(Silica gel plate, developing solvent: acetone, R_f = 0.6). The radiochemical purity usually was over 96%. This mateial was used directly for animal studies. The effect of acidity and reaction time on the formation of this complex was determined by the same TLC technique.

1.3 Characterization of In-TE-BAT complex

1.3.1 Determination of composition

The composition of the complex, [^{113m}In]TE-BAT, was determined by a pH titration method (Radiometer, PH64). The formation of this complex follows the equation:

$$\text{TE-BAT} + ^{113m}In^{3+} \rightleftharpoons [^{113m}\text{In-TE-BAT}]^+ + 2H^+$$

When [^{113m}In]TE-BAT is formed, two equivalents of [H^+] are released and can be titrated by a standardized sodium hydroxide solution(0.01N)(Fig. 1). The titration is performed under two different conditions(with or without [In^{3+}]): solution(A) containing 1.0mg(0.391mM) of TE-BAT in 7mL of 1mM HCl solution and solution(B) containing the same amount of ligand, 1.0mg of TE-BAT in 3mL of 1mM HCl, and 4mL of In(NO$_3$)$_3$ solution(1.9mg in 50mL of 1mM HCl, 1.26×10^{-4}mM). Both of the solutions contain 0.1N of NaCl(the same ionic strength). Based on the difference between titration curves A and B, the formation function can be calculated[21]:

$$\bar{n} = \frac{[^{113m}\text{In-TE-BAT}]}{[T_M]}$$

where, $[T_M]$ = total concentration of In^{3+}. At the endpoint of titration the formation function $\bar{n}$ approaches unity, if the In/ligand ratio is equal to one.

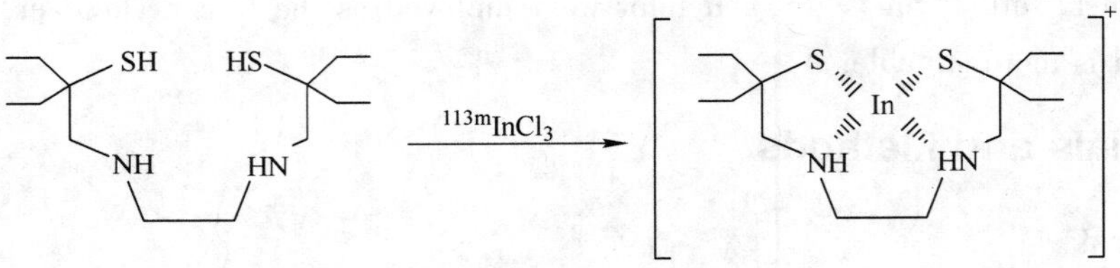

Fig. 1 Chemical equation for the formation of $[^{113m}\text{In}]$TE-BAT

1.3.2 Determination of net charge

Determination of net charge of this complex was achieved by the ion exchange method[21]. Ion exchange resin(cation, 10mg/each experiment) was placed in a test tube together with a solution of $[^{113m}\text{In}]$TE-BAT(5mL, at pH 0.9-2.3). The mixture was shaken for 1h. The resin and the solution were separated. The residual radioactivity in the solution was measured and the distribution coefficient(D) was calculated by counts in resin/counts in solution.

$$[^{113m}\text{In-TE-BAT}]^{x+} + \text{RH}_x \rightleftharpoons \text{R-}[^{113m}\text{In-TE-BAT}]^+ + x\text{H}^+$$

RH: Cation exchange resin

The equilibrium constant = K

$$K = \frac{\text{R-}[^{113m}\text{In-TE-BAT}]^+[\text{H}^+]^x}{[^{113m}\text{In-TE-BAT}]^{x+}_{aq}[\text{RH}_x]}$$

Distribution Coefficient = D

$$D = \frac{\text{R-}[^{113m}\text{In-TE-BAT}]^+}{[^{113m}\text{In-TE-BAT}]^{x+}_{aq}}$$

$$\lg D = \lg K + \lg[\text{RH}_x] + x\text{pH} \quad \lg D = x\text{pH} + C$$

The relationship between $\lg D$ and pH is a straight line and the slope, x, is equal to the net charge of the complex.

1.4 Biodistribution in mice

Biodistribution of $[^{113m}\text{In}]$TE-BAT was studied in male mice(18-22g) which were allowed access to food and water adlib. Saline solution containing $[^{113m}\text{In}]$TE-BAT in a volume of 0.1mL was injected directly into the tail vein. Mice were killed(at various time points, 2min to 1h, postinjection) by cardioc excision under ether anesthesia. The organs of interest were removed and counted using a well gamma counter. Percent dose per organ was calculated by comparison of tissue counts to suitably diluted aliquots of injected material. Total activities of blood and muscle were calculated assuming that they are 7% and 40% of total body weight, respectively.

2 Results

2.1 Characterization of [^{113m}In]TE-BAT

Effects of acidity. It is well known that In(OH)$_3$ is the predominant form when the pH of the reaction is higher than 5. The formation of the complex was evaluated at various pHs to determine the optimum conditions for labeling. The results shown in Fig. 2 suggest that the labeling yield reaches a plateau between pH 4 and 5. At higher pH, precipitation of the ligand, owing to its limited solubility in water, is observed (Fig. 2).

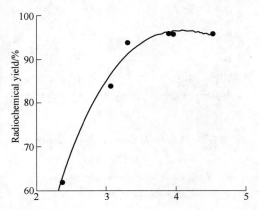

Fig. 2 Effects of pH on the formation of [^{113m}In]TE-BAT
(The optimum pH range is between 4 and 5)

2.1.1 Reaction time

At pH 4, the labeling yield was evaluated at various reaction times. The formation of the complex reaches a plateau at 20min (Fig. 3). Prolonged heating appears to have no significant effect on the labeling yield.

2.1.2 Determination of composition of In-TE-BAT

As indicated in Fig. 1, the formation of no-carrier-added ^{113m}In-TE-BAT produces two hydrogen ions. After the complexation, the pH of the reaction solution decreases. The decrease of pH is stoichiometrically proportional to the formation of the complex. This change can be measured by using acid-base titration techniques. The titration curves for the TE-BAT ligand at the same concentration with(B) and without(A) the presence of indium metal ion are represented in Fig. 4. From this figure the concentration of [H$^+$] can be calculated. The ionic strength of the solutions, under which curves A and B are generated, is the same, except that solution B contained In^{3+} (1.00 × 10^{-4}). At the same pH value, curves A and B show that a different volume of sodium hydroxide is consumed. The difference is a reflection of complex formation, and can be employed to calculate the concentration of the complex. Based on the titration curves and the stoichiometric relationship of hydrogen ion release and complex formation, the formation function(n) can be calculated. The relationship of formation function and pH is represented in Fig. 5. This figure clearly indicates that the composition of the complex is 1 : 1, confirming the structure shown in Fig. 1.

2.1.3 Determination of the net charge of the complex

Using the ion exchange method, to determine the distribution constant(D) between resin and aqueous solution at various pHs, the net charge of the complex can be determined based on the following equation:

$$\lg D = x\mathrm{pH}_{aq} + C$$

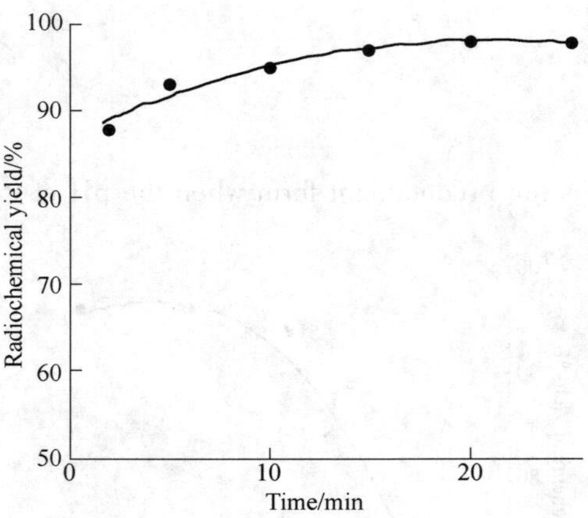

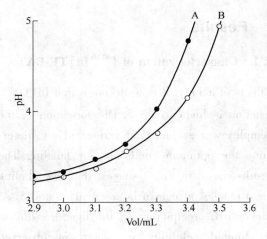

Fig. 3 Effects of time on the formation of [^{113m}In]TE-BAT

(The formation of the complex reaches a plateau at 20min)

Fig. 4 The titration curves of the ligand: TE-BAT, with(B) and without(A) the presence of indium metal ion(1.00×10^{-4}M)

(The difference between these two curves at the same pH value is stoichiometrically proportional to the complex formation)

From Fig. 6 the net charge, x, is determined to be 1.17. It is most likely that the net charge of this complex is +1. This is one more piece of evidence suggesting that the chemical structure in Fig. 1 is correct.

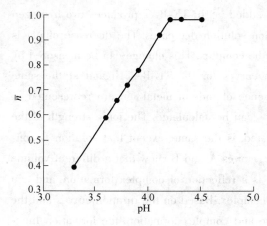
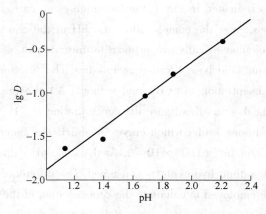

Fig. 5 The relationship of the formation function ($\bar{n}$) and pH

(This figure indicates that the composition of the complex is 1 : 1, comfirming the chemical structure shown in Fig. 1)

Fig. 6 The relationship of the distribution constant(D) between resin and aqueous solution at various pHs

(The net charge of the complex can be determined based on the slope of this straight line)

2.2 Biodistribution in mice

After an i. v. injection of [^{113m}In]TE-BAT in mice, a significant heart uptake(32.9% dose/g) at 2min(i. v.) was observed. The heart uptake dropped to 22.5% dose/g at 15min and 10.1%

dose/g at 1h(Table 1 and Fig. 7). The uptake at these time points is higher than those reported for thallium-201(^{201}Tl) and technetium-99m(^{99m}Tc) TBI[1,21]. The heart to lung, and heart to blood ratios for this complex are comparable or superior to those reported for ^{201}Tl and [^{99m}Tc] TBI. There is significant uptake in liver and lung which washes out with time. The ratios of heart/blood, heart/lung and heart/liver are reported in Table 1.

Table 1 Biodistribution of [^{113m}In]TE-BAT[①]

Organ	Time(postinjection)					
	2min	5min	10min	20min	30min	60min
Heart	32.93 ± 8.07	32.14 ± 5.12	26.94 ± 3.80	21.34 ± 3.90	18.04 ± 2.48	10.13 ± 2.26
Blood	4.82 ± 1.53	2.33 ± 0.27	1.44 ± 0.22	1.08 ± 0.14	1.02 ± 0.17	0.81 ± 0.23
Lung	48.38 ± 9.64	27.21 ± 6.05	16.02 ± 3.23	9.17 ± 2.68	8.48 ± 2.98	5.82 ± 1.20
Liver	14.41 ± 2.03	12.13 ± 2.21	10.32 ± 2.81	6.52 ± 1.11	6.03 ± 1.99	3.97 ± 0.26
H/Blood	6.19	13.85	17.50	21.27	15.73	12.13
H/Lung	0.50	1.00	1.64	2.25	2.05	1.72
H/Liver	2.19	2.80	2.62	2.52	2.91	2.43

① Mean percent dose per gram ± s. d. (six mice).

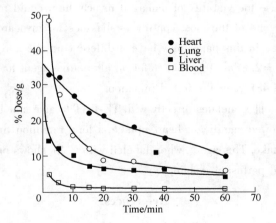

Fig. 7 Biodistribution of[^{113m}In]TE-BAT in mice at different time points after i. v. injection

3 Discussion

The complex formation between In and TE-BAT ligand is very rapid, simple and occurs in high yield(≥95%). The high labeling efficiency and excellent purity of this labeling reaction means that it requires no further purification before being used in animal studies. It is possible that this process is amenable for kit formulation.

The labeling reaction is pH sensitive; the optimum pH range is between 4-5. This pH can be easily maintained by the addition of buffer solution and is, therefore, easily adaptable for a simple one step reaction. The net charge of the no-carrier-added [^{113m}In] TE-BAT is determined to be +1. In view of the fact that almost all of the myocardial perfusion imaging agents reported are +1 charge molecules, it is not surprising that [^{113m}In] TE-BAT, with the same net charge, also displays good heart uptake and retention. In mice, this agent displays fast myocardial uptake and rapid blood and lung washout; at 20min postinjection the heart/blood and heart/lung ratios reach 21 and 2.25, respectively. At 1h postinjection the heart uptake still remains high: 10.13% dose/g. The heart uptake is comparable to that reported for ^{201}Tl and [^{99m}Tc]TBI[23]. The biologic behavior of [^{113m}In] TE-BAT clearly suggests that this agent is potentially useful for myocardial perfusion imaging. It is necessary to determine the chemical structure by preparing "cold"In-TE-BAT complexes. In addition, further studies in primates and humans are needed, especially the "redistribution"of this agent in myocardial tissue, to fully characterize the physiological properties.

Technetium-99m-labeled myocardial perfusion agents are currently being developed, which could potentially replace the ^{201}Tl, the agent being used in the clinics at present. However, despite the superior physical characteristics of the ^{99m}Tc isotope(gamma ray 140keV, $T_{1/2}$ =6h), the biologic behavior(no redistribution) of ^{99m}Tc isonitriles is different from ^{201}Tl (with delayed redistribution). The lack of redistribution for the ^{99m}Tc agents is being perceived as being less useful than ^{201}Tl, because the viability of damaged myocardium could not be studied effectively[24]. This is probably one of the most controversial issues for myocardial perfusion imaging. The In-complex reported in this paper may have a different uptake and retention pattern when labeled with ^{111}In($T_{1/2}$ =2.8days); it may offer an alternative agent for evaluation of "redistribution" at 24h or even 48h after the initial injection.

In conclusion, ^{113m}In(Ⅲ) chelates directly with TE-BAT to give a +1 charged complex. Biodistribution in mice showed significant heart uptake, a long retention time, high heart to blood ratios and low liver uptake. This agent, when labeled with ^{111}In, shows promise as a possible radiotracer for myocardial perfusion imaging.

References

[1] Jones A G, Abrams M J, Davison A, et al. Biological studies of a new class of technetium complexes: the hexakis(alkylisonitrile)-technetium(Ⅰ) cations. Int J Nucl Med Biol 1984;11:225.

[2] Holman B L, Campbell C A, Lister-James J, et al. Effect of reperfusion and hyperemia on the myocardial distribution of technetium-99m t-butylisonitrile. J Nucl Med 1986;27:1172-1177.

[3] Holman B L, Sporn V, Jones A G, et al. Myocardial imaging with technetium-99m CPI: initial experience in human. J Nucl Med 1987; 28:13-18.

[4] McKusick K, Holman B L, Jones A G, et al. Comparison of three Tc-99m isonitriles for detection of ischemic heart disease in human [Abstract]. J Nucl Med 1986; 27:878.

[5] Sia S T B, Holman B L. Dynamic myocardial imaging in ischemic heart disease: use of technetium-99m isonitriles. Am J Cardiac Imaging 1987; 1:125.

[6] Nunn A D, Treher E N, Feld T. Boronic acid adducts of technetium oxime complexes(BATOS) a new class

of neutral complexes with myocardial imaging capabilities [Abstract]. J Nucl Med 1986; 27:893.

[7] Coleman R E, Maturi M, Nunn A D, et al. Imaging of myocardial perfusion with Tc-99m SQ30217: dog and human studies [Abstract]. J Nucl Med 1986;27:894.

[8] Wester D W, Nosco D L, Coveney J R, et al. New Tc-99m myocardial agent with low plasma binding and fast blood clearance [Abstract]. J Nucl Med 1986;27:894.

[9] Lever S Z, Burns H D, Kervitzky T M, et al. Design, preparation and biodistribution of a technetium-99m triaminodithiol complex to access regional cerebral blood flow. J Nucl Med 1985;26:1287-1294.

[10] Kung H F, Molnar M, Billings J, Wicks R, Blau M. Synthesis and biodistribution of neutral lipid-soluble Tc-99m complexes which cross the blood brain barrier. J Nucl Med 1984; 25:326-332.

[11] Kung H F, Yu C C, Billings J, Molnar M, Blau M. Synthesis of new bisaminoethanethiol(BAT) derivatives: possible ligands for Tc-99m brain imaging agents. J Med Chem 1985;28:1280-1284.

[12] Efange S M N, Kung H F, Billings J, Guo Y Z, Blau M. Tc-99m Bis(aminoethanethiol)(BAT) complexes with amine sidechains—potential brain perfusion imaging agents for SPECT. J Nucl Med 1987; 28: 1012-1019.

[13] Chiolellis E, Varvarigou A D, Maina T H, et al. Comparative evaluation of ^{99m}Tc-labeled aminothiols as possible brain perfusion imaging agents. Nucl Med Biol 1988;15:215-223.

[14] Scheffel U, Goldfarb H W, Lever S Z, Gungon R L, Burns H D, Wagner, Jr, H N. Comparison of technetium-99m aminoalkyl diaminodithiol analogs as potential brain blood flow imaging agents. J Nucl Med 1988; 29:73-82.

[15] Efange S M N, Kung H F, Billings J, Blau M. Synthesis and biodistribution of ^{99m}Tc-labeled piperidinyl bis (aminoethanethiol) complexes: potential brain imaging agents for single photon emission computed tomography. J Med Chem 1988;31:1043.

[16] Kung H F, Guo Y Z, Yu C C, Billings J, Subramanyam V, Calabrese J. New brain perfusion imaging agents based on Tc-99m bis-Aminoethanethiol(BAT) complexes: stereoisomers and biodistribution. J Med Chem: in press.

[17] Kasina S, Fritzberg A R, Johnson D L, Eshima D. Tissue distribution of technetium-99m-diamide-dimercaptide complexes and potential use as renal radiopharmaceuticals. J Med Chem 1986; 29:1933.

[18] Davison A, Jones A G, Orvig C, et al. A new class of oxotechnetium(+5) chelate complexes containing a $TcON_2S_2$ Core. Inorgan Chem 1981; 20:1632.

[19] Lever S Z. Correction: design, preparation, and biodistribution of a technetium-99m triaminedithiol complex to assess regional cerebral blood flow. J Nucl Med 1987;28:1064-1065.

[20] Green M A, Huffman J C. The molecular structure of Indium oxine. J Nucl Med 1988;29:417-420.

[21] Hindman J C, Sullivan J C. Principles and methods for the study of metal complex ion equilibria. In: Martell A E, ed. Coordination chemistry. Vol. 1. New York: Van Nostrand Reinhold;1971:419.

[22] de Kieviet, W. Technetium radiopharmaceuticals: chemical characterization and tissue distribution of Tc-glucoheptonate using Tc-99m and carrier Tc-99. J Nucl Med 1981;22:703-709.

[23] Burns H D, Dannals R F, Woud J, et al. Radiotracers for studying the cholinergic system. In: Richard P. Spencer, ed. Radiopharmaceuticals structure—activity relationships. New York: Grune & Stratton; 1981:573-594.

[24] Gutman J, Berman D S, Freeman M, et al. Time to completed redistribution of thallium-201 in exercise myocardial scintigraphy: relationship to the degree of coronary artery stenosis. Am Heart J 1983; 106:989.

A New Myocardial Imaging Agent: Synthesis, Characterization, and Biodistribution of Gallium-68-BAT-TECH[*]

Abstract In order to develop a new myocardial perfusion agent for positron emission tomography (PET), a new lipid-soluble gallium complex was evaluated. Synthesis, radiolabeling, characterization, and biodistribution of a unique gallium complex, [^{67}Ga]BAT-TECH (bis-aminoethanethiol-tetraethyl-cyclohexyl), are described. The complex formation between Ga^{3+} and BAT-TECH ligand is simple, rapid, and of high yield($\geqslant 95\%$). This process is amenable to kit formulation. The complex has a net charge of +1 and a Ga/ligand ratio of 1:1. Biodistribution in rats shows high uptake in the heart as well as in the liver. When [^{68}Ga]BAT-TECH was injected into a monkey, the heart and liver are clearly delineated by PET imaging, suggesting that this complex may be a possible tracer for myocardial perfusion imaging.

Generator-based radiopharmaceuticals may provide a useful and effective way of positron emission tomography (PET) imaging without an on-site cyclotron. The germanium-68/gallium-68 generator is commonly used in PET facilities as a source of positron radionuclide for physics experiments, and it is also suitable for preparing radiopharmaceuticals. The physical half-life of the parent, ^{68}Ge, is 287 days, which means that the generator is useful for about one year. The half-life of the daughter, ^{68}Ga, is 68min, which is convenient for multi-step chemical preparation. A large number of ^{68}Ga complexes have been reported[1-11]. However, there are only a few ^{68}Ga radiopharmaceuticals currently being used in humans. Development of lipid-soluble gallium complexes for imaging the brain and heart has not been successful. A series of lipidsoluble gallium complexes potentially useful for myocardial imaging has been reported[3,4]. Unfortunately, these agents behave neither as freely diffusible tracers nor as microspheres; therefore, they are not useful as myocardial perfusion agents. Other types of neutral and highly lipid-soluble gallium complexes designed for brain perfusion imaging have been reported[6,11]. These complexes showed little brain uptake, which suggests that lipid-solubility is not the sole requirement for molecules to penetrate the intact blood-brain barrier.

Despite its short half-life(75s), rubidium-82, produced by a strontium-82/rubidium-82 generator, is useful for myocardial perfusion imaging[12,13]. It has now been approved for routine clinical use. The generator-produced agent can support clinical cardiac PET imaging without an on-site cyclotron. A comparable ^{68}Ga compound with a half-life of 68min may provide significant improvements for PET myocardial imaging.

---

[*] Copartner: H. F. Kung, D. Mankoff, M. P. Kung, J. J. Billings, L. Francesconi, A. Alavi. Reprinted from *The Journal of Nuclear Medicine*, 1990, 31(10):1635-1640.

Another potentially useful positron-generator is zinc-62/copper-62[14-17] ($T_{1/2}$ is 9h and 9min for parent and daughter radionuclides, respectively). Several recent reports indicate that this generator may also be feasible for routine clinical use[15-17]. Since the parent half-life is relatively short(9h), the generator is only useful for one to two days. Nonetheless, the clinical potential of a series of copper(Ⅱ)(bisthiosemicarbazone) complexes, specifically Cu(PTSM)(Fig. 1), as myocardial and cerebral perfusion tracers has been demonstrated[18,19]. The Cu(PTSM) is based on an N_2S_2 ligand and is a neutral and lipid-soluble compound. After an i. v. injection, the compound passes through the cell membrane, including the intact bloodbrain barrier. Apparently, the compound decomposes intracellularly after interacting with sulfhydryl groups[20]. The regional distribution is a reflection of regional perfusion, a property consistent with "chemical microspheres". Therefore, this agent in combination with the $^{62}Zn/^{62}Cu$ generator may provide a convenient source of radiopharmaceuticals for measuring regional blood perfusion of the brain and heart. However, ^{68}Ga-labeled compounds may offer some advantages because the longer half-lives of the parent and daughter may greatly enhance the clinical potential as PET radiopharmaceuticals.

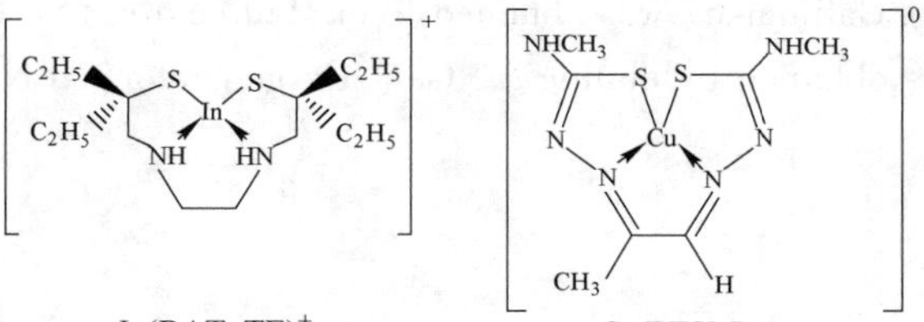

Fig. 1 Chemical structure of In(BAT-TE)$^+$ and Cu(PTSM)

Recent advances in technetium-99 chemistry of complexes based on N_2S_2(bisaminoethanethiol, BAT) ligands have dramatically enhanced our ability to predict the chemical structure of the final ^{99m}Tc complexes. This series of ligands forms strong complexes with(Tc=O)$^{3+}$ [21-33]. The X-ray crystallography studies of several N_2S_2 complexes developed by us and others have confirmed the(Tc=O)$^{3+}$ chemical state and the pyramidal core structure[21,30,31]. We have extended the use of the BAT ligands to investigate the radiochemistry of indium, a plus three cation[34]. We have initiated a study using the N_2S_2 ligand, bis-(aminoethanethiol) tetraethyl (BAT-TE), for complexing In^{3+}. The result suggested that a lipid-soluble and plus one charged In(BAT-TE)$^+$ was formed(Fig. 1) and that it may be useful as a myocardial perfusion imaging agent. In this paper, we turn our attention to synthesis, radiolabeling, characterization, and biodistribution of a similar gallium complex, Ga(BAT-TECH)$^+$(bis-aminoethanethiol-tetraethyl-cyclohexyl)(Fig. 2). For convenience, [^{67}Ga]gallium citrate from commercial sources was employed as the tracer in this paper. However, for imaging studies, ^{68}Ga is the radionuclide suitable for PET imaging.

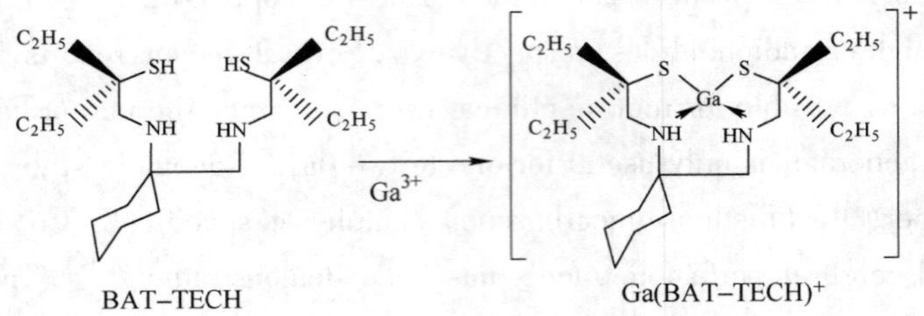

Fig. 2 Chemical equation for the formation of Ga(BAT-TECH)$^+$

1 Materials and Methods

1.1 General

The preparation of BAT-TECH was achieved by a method reported previously[23]. The only difference is that lithium aluminum hydride was employed for the final reduction step of the diimine intermediate[24,28,30]. The dimercapto hy-drochloride salt of BAT-TECH was precipitated and used for this study. Gallium-67 was obtained from Mallinckrodt(St. Louis, MO) as gallium citrate. Gallium-68 was obtained by eluting a ^{68}Ge/^{68}Ga generator(NEN/DuPont, N. Billerica, MA) with 0.1N HCl.

1.2 Radiolabeling

No-carrier-added ^{67}Ga-citrate(1mCi/mL) was added to a test tube containing the BAT-TECH ligand(1mg) in 0.5mL of water and adjusting the pH to 3.1 ± 0.1 by the dropwise addition of a solution of 5% NaOH or 1N HCl. The mixture was vortexed and kept in a heating block at 75℃ for 0.5h. The percent labeling yield was measured by thin-layer chromatography(silica gel plate, developing solvent: acetone: aceticacid 3 : 1, V/V, R =0.1 and 0.7 for Ga-citrate and Ga-BAT-TECH, respectively). The radiochemical purity is usually over 96%. This material was used directly for animal studies. The effects of pH, temperature, and ligand concentration on the formation of this complex was determined by the same TLC technique. For charge determination experiments, the ^{67}Ga-BAT-TECH complex was purified on preparative silica gel plates(developed by the same solvent system). The desired fraction was scraped from the plates and redissolved in water. The solution was centrifuged and the supernatant containing the ^{68}Ga-BAT-TECH complex was used(radiochemical purity >99%).

For a monkey imaging study, ^{68}Ga was eluted from a ^{68}Ge/^{68}Ga generator and extracted in a 6N HCl solution with ether(3 ×1.5mL)[35]. The combined extract was dried under a stream of nitrogen. To this residue, BAT-TECH ligand(3mg/mL, pH 3.1) was added. The mixture was heated in a heating block at 75℃ for 15min. After filtration through a 0.22-micron filter, the material was assayed and injected into a monkey. The whole preparation was accomplished in 40min(yield 40%, purity >98%).

1.3 Characterization of [^{67}Ga]BAT-TECH complex

The same methods as those reported previously for characterization of In(BAT-TE)$^+$ were also employed for identifying the Ga(BAT-TECH)$^+$ complex[34,36].

1.3.1 Determination of composition

The composition of the complex was determined by a pH titration method(Orion, pH meter 611). The formation of this complex follows the equation:

$$\text{BAT-TECH} + \text{Ga}^{3+} \rightleftharpoons (\text{Ga-BAT-TECH})^+ + 2\text{H}^+$$

When [Ga]BAT-TECH is formed, two equivalents of [H$^+$] are released and can be titrated by a standardized sodium hydroxide solution(0.01N). The titration is performed under two different conditions: solution(A) containing 1.0mg(0.391mM) of BAT-TECH in 7mL of 1mM HCl solution and solution(B) containing the same amount of ligand, 1.0mg of BAT-TECH in 3mL of 1mM HCl, and 4mL of Ga(NO$_3$)$_3$ solution(1.9mg in 50mL of 1mM HCl, 0.168mM). Both of the solutions contain 0.1N NaCl(the same ionic strength). Based on the difference between titration curves A and B, the formation function can be calculated[34,36]:

$$\bar{n} = \frac{(\text{Ga-BAT-TECH})^+}{[T_M]}$$

where, [T_M] = total concentration of Ga^{3+}.

At the end point of titration the formation function n approaches unity if the Ga/ligand ratio is equal to one.

1.3.2 Determination of net charge

Determination of net charge of this complex was achieved by the ion exchange method[34,37]. Ion exchange resin(strong cation R-SO$_3$H, 10mg/each experiment) was placed in a test tube with a solution of radioactive(no carrier-added) [^{67}Ga]BAT-TECH(5mL, at pH 0.9-2.3). The mixture was shaken for 1h. The resin(RH$_x$) and the solution were separated. The residual radioactivity in the solution was measured and the distribution coefficient(D) was calculated by counts in resin/counts in solution.

$$[^*\text{Ga-BAT-TECH}]_{aq}^{x+} + \text{RH}_x \rightleftharpoons \text{R-}[^*\text{Ga-BAT-TECH}] + x\text{H}^+$$

where RH = cation exchange resin.

The equilibrium constant = K:

$$K = \frac{\text{R-}[^*\text{Ga-BAT-TECH}][\text{H}^+]^x}{[^*\text{Ga-BAT-TECH}]_{aq}^{x+}[\text{RH}_x]}$$

Distribution coefficient = D:

$$D = \frac{\text{R-}[^*\text{Ga-BAT-TECH}]}{[^*\text{Ga-BAT-TECH}]_{aq}^{+x}}$$

$$\lg D = \lg K + \lg[\text{RH}_x] + x\text{pH} \Longrightarrow \lg D = x\text{pH} + C$$

The relationship between $\lg D$ and pH is a straight line and the slope, x, is equal to the net charge of the complex.

1.4 Biodistribution in rats

Biodistribution of [^{67}Ga]BAT-TECH was studied in male Sprague-Dawley rats(200-250g), which were allowed access to food and water ad lib. Saline solution containing [^{67}Ga]BAT-TECH in a volume of 0.2mL was injected directly into a femoral vein. Rats were killed at 2min, 30min, and 60min postinjection by cardiac excision under ether anesthesia. The organs of interest were removed and counted using a well-type gamma counter. Percent dose per organ was calculated by comparison of tissue counts to suitably diluted aliquots of injected material. Total activities of blood and muscle were calculated assuming that they are 7% and 40% of total body weight, respectively. The % dose/gram of each organ can be calculated by dividing the % dose/organ by the mean organ weight(i.e., average 200g rat: heart, 0.85g; brain, 1.65g; blood, 18g; liver, 9g; kidney, 1.9g; lungs, 1.6g). Each time point consists of a group of three rats.

1.5 Imaging study in a monkey

A monkey(cynomologous, male, 10lb) was sedated with ketamine(50mg i.m.) and then anesthetized with nembutal(0.2mL, 65mg/mL, additional amount was used as needed). The monkey was positioned in the PENN-PET[38] tomograph and the scan started at 7min after an i.v. injection of [^{68}Ga]BAT-TECH (424μCi/3mL of saline). The monkey was scanned for 15min and a total of 5.8 million counts were collected. Data were reconstructed in 45 overlapping 8-mm thick slices using filtered backprojection with a Hanning filter. In this preliminary study, no attenuation correction was performed. Slice spacing was 2mm, yielding image data on a 2mm×2mm×2mm grid suitable for displaying transverse sections.

2 Results

2.1 Characterization of [^{67}Ga]BAT-TECH

2.1.1 Effects of acidity, temperature, and ligand concentration

The formation of the complex was evaluated at various pHs to determine the optimum conditions for labeling. The results shown in Fig.3 suggest that the labeling yield reaches a plateau at pH 3-5. At a more basic pH, precipitation of the ligand, due to the limited solubility in water, is observed. The reaction temperature is also an important factor controlling the rate of complex formation, however, as shown in Fig.4, when the reaction temperature is above 40℃, the labeling yield appears to be constant at >93%. The concentration of the ligand in the reaction mixture also affects the labeling yield. When the concentration is above 3mg/mL, the labeling yield is >97% (Fig.5).

Determination of composition of [Ga]BAT-TECH. As indicated in Fig.2, the formation of no-carrieradded [Ga]BAT-TECH produces two hydrogen ions. Due to the release of these two hydrogen ions, the pH of the reaction solution will decrease. This change can be measured by

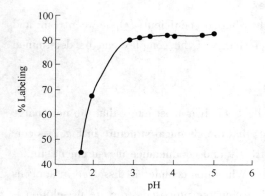

Fig. 3 Effects of pH on the formation of Ga(BAT-TECH)$^+$
(The optimum pH range is between 3 and 5)

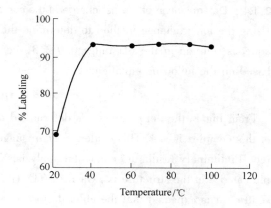

Fig. 4 Effects of temperature on the formation of Ga(BAT-TECH)$^+$
(The formation of the complex reaches a plateau above 40℃)

using acid-base titration techniques. The titration curves for BAT-TECH ligand at the same concentration (0.391mM) with and without the presence of the gallium metal ion (0.148mM) are presented in Fig. 6. From this figure, the [H$^+$] can be calulated. The ionic strength of the solutions for generating curves A and B is the same. At an equal pH value, curves A and B show that a different volume of sodium hydroxide is consumed. This is due to the hydrogen ion which is released during the interaction of Ga^{3+} with the ligand. The difference is a reflection of complex formation, and can be employed to calculate the concentration of the complex. Based on the titration curves and the stoichiometric relationship of hydrogen ion release and complex formation, the formation function (n) can be calculated. The relationship of formation function and pH is presented in Fig. 7. This figure clearly indicates that the composition of the complex is 1 : 1, confirming the proposed structure shown in Fig. 2.

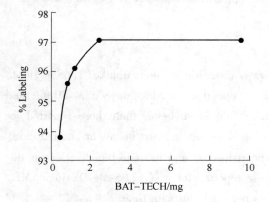

Fig. 5 Effects of ligand concentration on the formation of Ga(BAT-TECH)$^+$
(The formation of the complex reaches 97% at 3mg/mL)

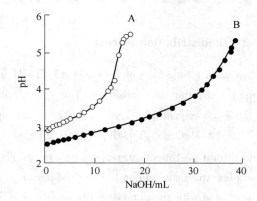

Fig. 6 The titration curves of the ligand: BAT-TECH, with (B) and without (A) the presence of gallium metal ion (0.148mM)
(The difference between these two curves at the same pH value is proportional to the extent of the complex formation)

2.1.2 Determination of the net charge of the complex

Using the ion exchange method to determine the distribution coefficient(D) between resin and aqueous solution in the pH range 0.9-2.3, the net charge of the complex can be determined based on the following equation:

$$\lg D = x\mathrm{pH}_{aq} + C$$

From Fig. 8, the net charge, x, is determined to be 1.17. It is most likely that the net charge of this complex is +1. This evidence again suggests that the chemical structure in Fig. 2 is correct. Preliminary results on elemental analysis, NMR, IR, and conductance measurement studies indicated that the structure is [Ga-BAT-TECH.Cl]. When this complex is dissolved in aqueous solution it is expected that the chloride ion is ionized and the proposed structure in solution is correct(Kung, unpublished data).

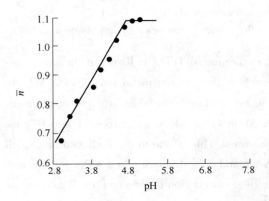

Fig. 7 The relationship of the formation function(n) and pH
(This figure indicates that the composition of the complex is 1∶1, confirming the chemical structure shown in Fig. 1)

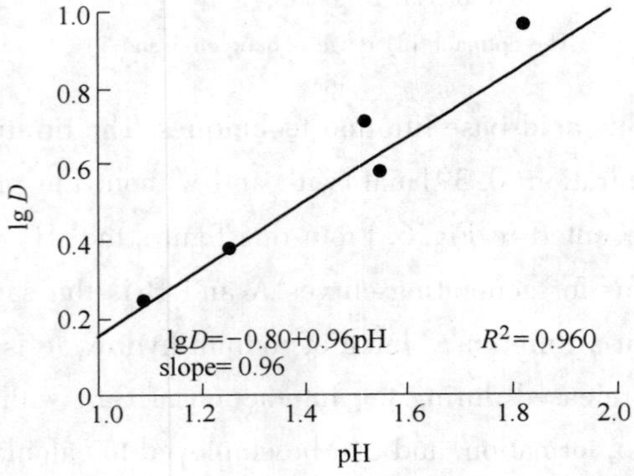

Fig. 8 The relationship of the distribution coefficient (D) between resin and aqueous solution at various pHs
(The net charge(+1) of the complex is determined based on the slope($x=1$) of this straight line)

2.2 Biodistribution in rats

After an i.v. injection of [^{67}Ga]BAT-TECH in rats, a significant heart uptake(1.68% dose/organ) at 2min was observed. The heart uptake dropped to 0.52% dose/g at 30min and 0.26% dose/g at 1h(Table 1). The heart uptake values are better than those reported for [^{68}Ga](5-MeOSal)$_3$TAME(0.97, 0.23 and 0.14% dose/whole heart in rats at 1min, 30min, and 60min postinjection, respectively)[3,4]. The heart-to-lung and heart-to-blood ratios for this complex are also comparable to or superior to those reported for [^{68}Ga](5-Me-OSal)$_3$TAME. There is significant uptake in the liver which does not wash out with time.

2.3 Imaging study in a monkey

After an i.v. injection of [^{68}Ga]BAT-TECH(424μCi in 3mL saline), the compound quickly localized in the heart and liver. Images taken with the PENN-PET at 7min postinjection clearly show that the agent is localized in the heart (Fig. 9). In all views, the myocardial cavity is clearly delineated, indicating an acceptable heart/blood tracer concentration ratio.

Table 1 Biodistribution of [^{67}Ga]BAT-TECH in rats after intravenous injection

(% dose/organ)

Organ	2min	30min	60min
Blood	10.18 ± 0.30	3.58 ± 0.08	4.54 ± 1.10
Heart	1.68 ± 0.12	0.52 ± 0.08	0.26 ± 0.02
Muscle	13.89 ± 3.21	21.14 ± 2.18	10.79 ± 1.85
Lung	2.07 ± 0.07	0.46 ± 0.09	0.37 ± 0.009
Kidney	6.94 ± 0.31	2.00 ± 0.10	1.06 ± 0.14
Spleen	0.50 ± 0.06	0.15 ± 0.009	0.11 ± 0.001
Liver	21.52 ± 1.11	33.54 ± 4.42	46.41 ± 2.39
Skin	5.44 ± 1.65	7.56 ± 1.60	5.78 ± 0.92
Brain	0.02 ± 0.004	0.01 ± 0.001	0.01 ± 0.002

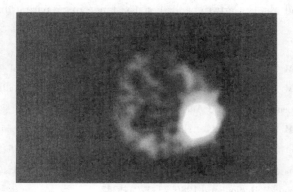

Fig. 9 PET images of the chest of a monkey (transverse and sagittal views) after an i. v. injection of ^{68}Ga(BAT-TECH)$^+$ (0.42mCi)

3 Discussion

The complex formation between Ga^{3+} and BAT-TECH ligand is very rapid, simple, and occurs in high yield (≥95%). The high labeling efficiency and excellent purity of this labeling reaction yields a product that requires no further purification before animal study. It is possible that this process is amenable for kit formulation.

The labeling reaction is pH sensitive, the optimum pH range is 3-5. This pH can be easily maintained by the addition of buffer solution and is, therefore, easily adaptable for a simple one-step reaction. The Ga^{3+} and BAT-TECH apparently form a 1∶1 complex with release of two hydrogen ions and the net charge of the nocarrier-added [^{67}Ga]BAT-TECH is probably +1. In view of the fact that the corresponding In(BAT-TE) complex showed a net charge of +1, it is not surprising that [^{67}Ga]BAT-TECH may have the same net charge. The [^{67}Ga]BAT-TECH$^+$

also displays good heart uptake and retention. In rats, this agent displays fast myocardial uptake and rapid blood and lung washout. The biologic behavior of [^{67}Ga]BAT-TECH suggests that this agent is potentially useful for myocardial perfusion imaging. Further studies examining the structure and chemistry of "cold" Ga-BAT-TECH are in progress. An examination of the quantitative relationship of tracer distribution and the regional blood flow of this agent, or agents in this series, will also be necessary before a successful agent can be developed for clinical use.

In conclusion, ^{67}Ga or [^{68}Ga]BAT-TECH can be readily prepared by direct complexation of [^{67}Ga]gallium citrate or [^{68}Ga]GaCl$_3$, respectively, with BAT-TECH. Biodistribution in rats and a monkey showed significant heart uptake. When labeled with ^{68}Ga, this agent, or related complexes in this series, may be useful as possible radiotracers for myocardial perfusion imaging for PET.

Acknowledgments

This work is partially supported by a grant (NS-15908) awarded by National Institute of Health and a grant from DOE (DE-AC02-80EV10402). The authors thank Dr. Gerd Muehllehner for helpful discussions and Ms. C. Cartwright for her assistance in preparing this manuscript.

References

[1] Green M A, Welch M J. Gallium radiopharmaceutical chemistry. Nucl Med Biol 1989;16:435-448.

[2] Green M A, Welch M J. Synthesis and crystallographic characterization of a gallium salicylaldimine complex of radiopharmaceutical interest. J Am Chem Soc 1984;106:3689.

[3] Green M A, Welch M J, Mathias C J, et al. Gallium-68 1,1,1-tris(5-methoxysalicylaldimino-methyl)ethane: a potential tracer for evaluation of myocardial blood flow. J Nucl Med 1985;26:170-180.

[4] Green M A. Synthesis and biodistribution of a series of lipophilic gallium-67 tris(salicylaldimine) complexes. J Labeled Compounds Radiopharm 1986;23:1221-1222.

[5] Hawkins R A, Phelps M E, Huang S C, et al. A kinetic evaluation of blood-brain barrier permeability in human brain tumors with (Ga-68)-EDTA and positron computed tomography. J Cereb Blood Flow Metab 1984;4:504-515.

[6] Mathias C J, Sun Y, Welch M J, et al. Targeting radiopharmaceuticals: comparative biodistribution studies of gallium and indium complexes of multidentate ligands. Nucl Med Biol, Int J Radiat Appl Instrum Part B 1988;15:69-81.

[7] Mintun M A, Dennis D R, Welch M J, et al. Measurements of pulmonary vascular permeability with positron emission tomography and Ga-68 transferring. J Nucl Med 1987;28:1704-1716.

[8] Moore D A, Motekaitis R J, Martell A E, et al. A new aminothiol ligand for radiopharmaceutical use with indium and gallium[Abstract]. J Nucl Med 1989;30:922.

[9] Nelson W O, Rettig S J, Orvig C. Aluminum and gallium complexes of l-ethyl-3-hydroxy-2-methyl-4-pyridnone: a new exoclatharate matrix. Inorg Chem 1989;28:3153-3157.

[10] Reger D L, Knox S J, Lebioda L. Dihydrobis(pyrazolyl)borate complexes of gallium. X-ray crystal structure of [H$_2$B(pz)$_2$]$_2$GaCl(pz = Pyrazolyl Ring). Inorg Chem 1989;28:3092-3093.

[11] Moerlein S M, Welch M J, Raymond K N. Use of tricatecholamine legends to alter the biodistribution of gallium-67. J Nucl Med 1982;23:501-506.

[12] Goldstein R A, Mullani N A, Wong W H, et al. Positron imaging of myocardial infarction with rubidium-82. J Nucl Med 1986;27:1824-1829.

[13] Gould K L, Goldstein R A, Mullani N A. Economic analysis of clinical positron emission tomography of the heart with rubidium-82. J Nucl Med 1989;30:707-717.

[14] Robinson G D. Generator systems for positron emitters. In: Reivich M, Alavi A, eds. Positron emission tomography. New York: AR Liss;1985:81-102.

[15] Robinson G D, Zielinski F W, Lee A W. Zn-62/Cu-62 generator: a convenient source of copper-62 radiopharmaceuticals. Int J Appl Radiat Isotopes 1980;31:111-116.

[16] Thakur M L, Nunn A D. Preparation of carrier-free zinc-62 for medical use. Radiochem Radioanal Letters 1969;2:301-306.

[17] Ueda N, Nakamoto S, Tanaka Y, et al. Production of Zn-62 and development of Zn-62/Cu-62 generator system [Abstract]. J Nucl Med 1983;24:P124.

[18] Green M A, Klippenstein D L, Tennison J R. Copper(Ⅱ)bis(thiosemicarbazone) complexes as potential tracers for evaluation of cerebral and myocardial blood flow with PET. J Nucl Med 1989;29:1549-1557.

[19] Green M A. A potential copper radiopharmaceutical for imaging the heart and brain: copper-labeled pyruvaldehyde bis(N4-methylthiosemicarbazone). Nucl Med Biol, Int J Radiat Appl Instrum Part B 1989;14:59-61.

[20] Baerga I D, Maickel R P, Green M A. Subcellular distribution of tissue radiocopper following intravenous administration of[Cu-62]-Cu(PTSM)[Abstract]. J Nucl Med 1989;30:920.

[21] Davison A, Jones A G, Orvig C, et al. A new class of oxotechnetium(+5) chelate complexes containing a $TcON_2S_2$ Core. Inorg Chem 1981;20:1632.

[22] Kasina S, Fritzberg A R, Johnson D L, Eshima D. Tissue distribution of technetium-99m-diamide-dimercaptide complexes and potential use as renal radiopharmaceuticals. J Med Chem 1986;29:1933.

[23] Kung H F, Molnar M, Billings J, Wicks R, Blau M. Synthesis and biodistribution of neutral lipid-soluble Tc-99m complexes which cross the blood-brain barrier. J Nucl Med 1984;25:326-332.

[24] Kung H F, Yu C C, Billings J, Molnar M, Blau M. Synthesis of new bis-aminoethanethiol(BAT) derivatives: possible ligands for Tc-99m brain imaging agents. J Med Chem 1985;28:1280-1284.

[25] Efange S M N, Kung H F, Billings J, Guo Y Z, Blau M. Tc-99m Bis(aminoethanethiol)(BAT) complexes with amine sidechains—potential brain perfusion imaging agents for SPECT. J Nucl Med 1987;28:1012-1019.

[26] Chiolellis E, Varvarigou A D, Maina T H, et al. Comparative evaluation of ^{99m}Tc-labeled aminothiols as possible brain perfusion imaging agents Nucl Med Biol 1988;15:215-223.

[27] Scheffel U, Goldfarb H W, Lever S Z, Gungon R L, Burns H D, Wagner Jr H N. Comparison of technetium-99m aminoalkyl diaminodithiol analogs as potential brain blood flow imaging agents. J Nucl Med 1988;29:73-82.

[28] Lever S Z. Correction: design, preparation, and biodistribution of a technetium-99m triaminedithiol complex to assess regional cerebral blood flow. J Nucl Med 1987;28:1064-1065.

[29] Lever S Z, Burns H D, Kervitzky T M, et al. Design, preparation, and biodistribution of a technetium-99m triaminedithiol complex to access regional cerebral blood flow. J Nucl Med 1985;26:1287-1294.

[30] Kung H F, Guo Y Z, Yu C C, Billings J, Subramanyam V, Calabrese J. New brain perfusion imaging agents based on Tc-99m bis-aminoethanethiol(BAT) complexes: stereoisomers and biodistribution. J Med Chem 1989;32:433-437.

[31] Mach R H, Kung H F, Guo Y Z, Yu C C, Subramanyam V, Calabrese J. Synthesis, characterization, and biodistribution of neutral and lipid-soluble ^{99m}Tc-PAT-HM and ^{99m}Tc-TMR for brain imaging. Intl J Nucl Med Biol 1989;16:828-837.

[32] Walovitch R C, Hill T C, Garrity S T, et al. Characterization of technetium-99m-L,L-ECD for brain perfusion imaging, part 1: pharmacology of technetium-99m-ECD in nonhuman primates. J Nucl Med 1989;30:1892-1901.

[33] Léveillé J, Demonceau G, De Roo M, et al. Characterization of technetium-99m-L,L-ECD for brain perfusion imaging, part 2: biodistribution and brain imaging in humans. J Nucl Med 1989;30:1902-1910.

[34] Liu B L, Kung H F, Jin Y T, Zhu L, Meng M. A new myocardial imaging agent: synthesis, characterization, and biodistribution of [^{113m}In]TE-BAT. J Nucl Med 1989;30:367-373.

[35] Yano Y, Budinger T F, Ebbe S N, et al. Gallium-67 lipophilic complexes for labeling platelets. J Nucl Med 1985;26:1429-1437.

[36] Hindman J C, Sullivan. Principles and methods for study of the metal complex ion equilibria. In: Martell AE, ed. Coordination chemistry, Volume 1. New York: Van Nostrand Reihold;1971:419.

[37] de Kieviet W. Technetium radiopharmaceuticals: chemical characterization and tissue distribution of Tc-glucoheptonate using Tc-99m and carrier Tc-99. J Nucl Med 1981;22:703-709.

[38] Muehllehner G, Karp J S, Mankoff D A, Beerbohm I D, Ordonez C E. Design and performance of a new positron tomograph. IEEE Trans Nucl Sci. 1988;35:670-674.

Carbon-11 Labeled Stilbene Derivatives from Natural Products for the Imaging of Aβ Plaques in the Brain*

Abstract Four stilbene derivatives from natural products were screened as novel β-amyloid (Aβ) imaging ligands. *In vitro* binding assay showed that the methylated ligand, (E)-1-methoxy-4-styrylbenzene(**8**) displayed high binding affinity to $A\beta_{1\text{-}42}$ aggregates ($K_i = 19.5$ nM). Moreover, the ^{11}C-labeled ligand, [^{11}C]**8** was prepared through an O-methylation reaction using [^{11}C]CH$_3$OTf. *In vitro* autoradiography with sections of transgenic mouse brain also confirmed the high and specific binding of [^{11}C]**8** to Aβ plaques. *In vivo* biodistribution experiments in normal mice indicated that [^{11}C]**8** displayed high initial uptake ((9.41 ± 0.51)% ID/g at 5min post-injection) into and rapid washout from the brain, with a brain$_{5\text{min}}$/brain$_{30\text{min}}$ ratio of 6.63. These preliminary results suggest that [^{11}C]**8** may be served as a novel Aβ imaging probe for PET.

Key words Alzheimer disease, β-amyloid plaque, autoradiography, biodistribution, carbon-11 PET imaging

1 Introduction

Alzheimer's disease (AD) is the most common form of "dementia" among older people, which causes problems of memory, thinking and behavior. To date, no therapeutics are available to cure the disease although some drugs may help keep symptoms from getting worse for a limited period. Unfortunately, a definite diagnosis of AD is very difficult in clinic, only possible or probable AD can be routinely diagnosed. Despite decades of research, the mechanism that triggers the onset of AD still remains unclear. Based on the amyloid cascade hypothesis, the accumulation of miss folded β-amyloid (Aβ) plaques and neurofibrillary tangles (NFTs) in the brain may play an important role in the development of the disease[1,2]. Therefore, *in vivo* imaging and quantification of Aβ plaque burden in the brain is thought to be a valuable tool for the early diagnosis and monitoring new therapy approaches of AD[3,4].

Based on the chemical structure of Congo red (CR) and thioflavin T (Th-T) (commonly used fluorescent dyes for detection of Aβ plaques), a number of radiolabeled ligands with high affinity and specificity for Aβ plaques have been developed for positron emission tomography (PET) or single photon emission computed tomography (SPECT) during the last decade. Some of them have been tested in humans (Fig. 1), such as [^{11}C]-2-(4-(methylamino)phenyl)-6-

---

* Copartner: Cui Mengchao, Tang Ruikun, Zhang Jinming, Zhang Xiaojun, Li Zijing, Jia Hongmei. Reprinted from *Radiochim. Acta*, 2014, 102(1-2):185-192.

hydroxybenzothiazol ([^{11}C]PIB), a Th-T derivative, which is the most extensively studied PET tracer for Aβ plaques. PET studies indicate that AD and control cases can be clearly distinguished by [^{11}C]PIB[5,6]. After that, Koole et al.[7] reported a ^{18}F-labeled PIB derivative, [^{18}F]-2-(3-fluoro-4-methyaminophenyl)-benzothiazol-6-ol ([^{18}F]GE-067). [^{18}F]GE-067 displayed an uptake pattern similar to [^{11}C]PIB, and has been just approved by the U. S. Food and Drug Administration (FDA) last year in October. Recently, Andersson et al.[8,9] reported a close analogue of PIB, [^{11}C]-2-[6-(methylamino)pyridin-3-yl]-1,3-benzothiazol-6-ol ([^{11}C]AZD2184), by displacing the phenyl ring with a pyridyl ring. After this modification, [^{11}C]AZD2184 displayed an apparent lower degree of non-specific binding in white matter. In addition, the stilbene scaffold, which is derived from CR, has also been selected for developing Aβ imaging agents. [^{11}C]-4-N-methylamino-4'-hydroxystilbene ([^{11}C]SB-13) is the first reported ^{11}C-labeled stilbene derivative. Initial PET imaging *in vivo* with [^{11}C]SB-13 demonstrated potential usefulness in detecting Aβ plaques in the human brain[10,11]. However, the ^{18}F-based Aβ imaging agents are more available for routine clinical use and will allow longitudinal studies across multiple centers. To date, two ^{18}F-labeled stilbene derivatives were developed for human studies. [^{18}F]-4-(N-methylamino)-4'-(2-(2-(2-fluoroethoxy)ethoxy)ethoxy)stilbene ([^{18}F]BAY94-9172) is now under phase III trial in Europe, and is initiated by the Bayer company[12]. [^{18}F]-(E)-4-(2-(6-(2-(2-(2-fluoroethoxy)ethoxy)ethoxy)pyridin-3-yl)vinyl)-N-methylani-line([^{18}F]AV-45), which is the pyridyl derivative of [^{18}F]BAY94-9172, has been approved by FDA in 2012[13,14].

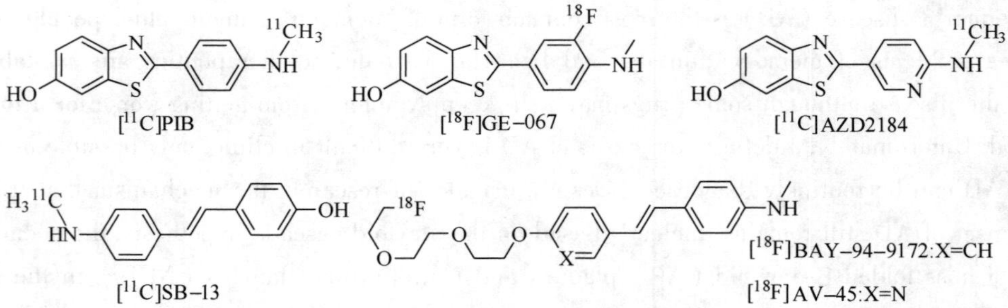

Fig. 1 Structures of Aβ imaging agents for PET that have been evaluated in human subjects

Beside the artificial fluorescent dyes, the extensive natural products library is another good source for the screening of new Aβ imaging ligands. Several types of stilbene derivatives were isolated from plants and have been shown to possess a variety of interesting biological properties[15-17]. Resveratrol (Fig. 2) is the most well studied polyphenol derivative bearing the stilbene moiety and is known to exhibit a variety of promising biological activities[18-20]. Some research indicated that resveratrol inhibits Aβ$_{1-42}$ fibril formation and reduces the cytotoxicity, which implies that resveratrol may serve as a potential medicine for the treatment and reducing the risk of AD[21]. Gester et al.[22] reported the first ^{18}F-labeled resveratrol derivative (**1**) by replacing the 4′-OH with ^{18}F in order to study the radiopharmacological properties of this polyphenol derivative *in vivo*. [^{18}F]**1** displayed extensive uptake and metabolism in the liver and kidney of normal Wistar rats. However, the brain uptake of [^{18}F]**1** was relatively lower (0.33% ID/g at 5min post-injection). More recently, Lee et al.[23] reported that some fluorinated resveratrol derivatives (**1-4**) displayed high binding affinity to $A\beta_{1-42}$ aggregates. [^{18}F]**4** showed the highest binding affinity ($K_i = 0.49$nM), but the *in vivo* pharmacokinetics in normal mice brain was slow (3.26% ID/g at 2min and 2.09% ID/g at 30min post-injection). In addition, [^{18}F]**4** displayed a significant bone uptake (9.73% ID/g at 60min post-injection) of radioactivity due to the metabolic defluorination.

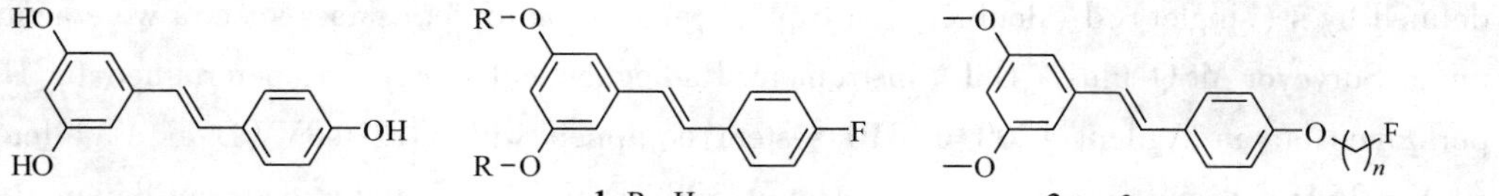

Fig. 2 Structure of resveratrol and it's fluorinated derivatives

In addition, stilbene glucoside derivatives were considered to be significant active components from *polygonum multiflorum thumb* (PMT), which is a traditional Chinese herb. Some research indicated that stilbene glucosides isolated from PMT can reverse the down-regulation of brain-drived neurtrophic factor (BDNF) expression in hippocampus CA 1 in rats induced by $A\beta_{1-40}$[24].

In the present study, four stilbene derivatives (Polydatin(**5**), isolated from *polygonum cuspidatum*; Rhapontin (**6**), isolated from *rhubarb* roots; (E)-4-styrylphenol(**7**), isolated from *pinus excelas*; (E)-1-methoxy-4-styrylbenzene(**8**), also isolated from *pinus excelas*) from our natural products library (Fig. 3) were selected to study as potential $A\beta$ imaging agents. Selected ligand, **8**, was radiolabeled with ^{11}C and evaluated as PET radioligand for $A\beta$ plaque imaging.

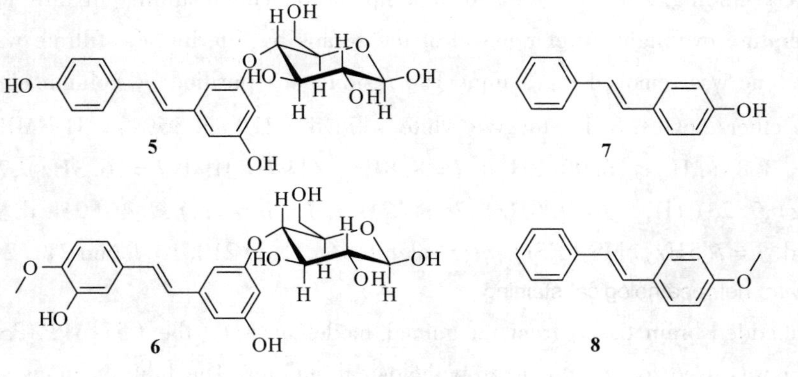

Fig. 3 Selected stilbene derivatives from natural products

2 Experimental

2.1 General information

All reagents used in the synthesis were commercial products and were used without further purification unless otherwise indicated. Polydatin, Rhapontin and (E)-4-styrylphenol were obtained from National Institute for the Control of Pharmaceutical and Biological Products (purity >99%). The ^{1}H-NMR spectra were obtained at 400MHz on Bruker spectrometer in CDCl$_3$ solutions at room temperature with TMS as an internal standard. Chemical shifts were reported as δ values relative to the internal TMS. Coupling constants were reported in Hertz. Multiplicity is defined by s (singlet), d (doublet), t (triplet), and m (multiplet). Mass spectra were acquired under Surveyor MSQ Plus (ESI) instrument. Radiochemical purity was determined by HPLC performed on an Agilent 1200 G1310 system equipped with G1314BVWD as UV detector (UV = 254nm) and a Bioscan Flow Count 3200 NaI/PMT γ-radiation scintillation detector. HPLC analysis was achieved on an Agilent Eclipse XDB – C18 reverse phase column (Agilent Technologies, 5μm, 4.6mm × 150mm) eluted with a binary gradient system at a flow rate of 1.0mL/min. Mobile phase A was deionized water while mobile phase B was acetonitrile (CH$_3$CN/H$_2$O = 65/35). Fluorescent observation was performed by the Oberver Z1 (Zeiss, Germany) equipped with GFP filter set (excitation, 505nm). Normal ICR mice (five weeks, male) were used for biodistribution experiments. Transgenic mice (C57BL6, APPswe/PSEN1, male, 12 months old), used as an Alzheimer's model, were purchased from the Institute of Laboratory Animal Science, Chinese Academy of Medical Sciences. All protocols requiring the use of mice were approved by the animal care committee of Beijing Normal University.

2.2 Chemistry

2.2.1 Synthesis of (E)-1-methoxy-4-styrylbenzene (8)

To a solution of **7** (544mg, 2.8mmol) and CH$_3$I (800mg, 5.6mmol) in acetone (50mL) was added K$_2$CO$_3$ (386mg, 2.8mmol) at room temperature. The resulting mixture was stirred at room temperature overnight. After removal of the scraps by filtering, the filtrate was dried over MgSO$_4$. Acetone was removed in vacuum. The residue was purified by column chromatography (petroleum ether/AcOEt, 8/1) to give white solid **8** (210mg, 36%). ^{1}H-NMR (400MHz, CDCl$_3$) δ: 3.83(3H, s), 6.90(2H, d, J = 8.8Hz), 6.97(1H, d, J = 16.3Hz), 7.07(1H, d, J = 16.8Hz), 7.23(1H, t, J = 7.6Hz), 7.34(2H, t, J = 7.6 Hz), 7.46(2H, d, J = 8.8Hz), 7.49(2H, d, J = 7.5Hz). MS (ESI): m/z calcd for C$_{15}$H$_{14}$O 210.10, found 211.2 (M + H)$^+$.

2.2.2 *In vitro* neuropathological staining

Paraffin-embedded brain tissue from an animal model of AD (the C57-APP/PS1 mouse, 12 months old) was used for *in vitro* neuropathological staining. The brain sections were deparaffinized with 2 × 20min washes in xylene; 2 × 5min washes in 100% ethanol; 5min washes in

90% ethanol/H_2O; 5min washes in 80% ethanol/H_2O; a 5min wash in 60% ethanol/H_2O; and a 10min wash under running tap water, and then incubated in PBS (0.2M, pH = 7.4) for 30min. Next they were incubated with a 10% ethanol solution (1μM) of **7** and **8** for 10min. The location of plaques was confirmed by staining with Thioflavin-S (1μM) in adjacent sections.

2.2.3 Binding assay using Aβ_{1-42} aggregates

Inhibition experiments were carried out in 12 × 75mm borosilicate glass tubes according to procedures described previously with some modifications[25]. The radio-ligand [^{125}I]TZDM was prepared according to procedures described previously[26], after HPLC purification, the radiochemical purity was greater than 95%. The reaction mixture contained 100μL of aggregated Aβ_{1-42} fibrils, 100μL of radioligands ([^{125}I]TZDM, 60000-100000cpm/100μL), 100μL of inhibitors (10^{-4}-$10^{-8.5}$M in ethanol) and 700μL of 0.05% bovine serum albumin solution in a final volume of 1mL. The mixture was incubated at 37℃ for 2h, and the free radioactivity were separated by vacuum filtration through Whatman GF/B filters using a Brandel Mp-48T cell harvester followed by 3 × 4mL washes with PBS (0.02M, pH 7.4) at room temperature. Filters with the bound [^{125}I]TZDM were counted in gamma counter (WALLAC/Wizard 1470, USA) with 70% efficiency. Inhibition experiments were repeated three times, and the half maximal inhibitory concentration (IC_{50}) was determined using GraphPad Prism 4.0, and the inhibition constant (K_i) was calculated using the Cheng-Prusoff equation[27]: $K_i = IC_{50}/(1 + [L]/K_d)$.

2.3 Radiochemistry

2.3.1 Production of [^{11}C]CH_3I and [^{11}C]CH_3OTf

[^{11}C]CH_3I and [^{11}C]CH_3OTf were produced according to the method reported previously[28] with some modifications. [^{11}C]CO_2 was produced in-target via the $^{14}N(p, \alpha)^{11}C$ reaction on a N_2-O_2(99.5/0.5, v/v) gaseous system, by irradiation with 20 MeV proton beam at 40tA in a Sumitomo HM-20S (Osaka, Japan) cyclotron for 10min. At the end of bombardment, the [^{11}C]CO_2 was collected through a stainless steel loop cooled with liquid nitrogen. The frozen [^{11}C]CO_2 was warmed up to -20℃ and transferred by a stream of nitrogen gas at 15mL/min to the reactor containing 150μL of lithium aluminium hydride solution (1.0M) in THF at room temperature. Then THF was removed by a stream of nitrogen gas at 40mL/min under 170℃. After cooling, 200μL of 57% HI was added and the produced [^{11}C]CH_3I was carried by a stream of nitrogen gas, and subsequently passed over a silver triflate/C online with the gas/solid exchange reaction at 210℃ yielding the [^{11}C]CH_3OTf.

2.3.2 Preparation of [^{11}C]8

Gaseous [^{11}C]CH_3OTf was delivered into the reaction tube containing compound **7** (1.5mg) in DMSO (0.3mL) with 10μL of NaOH (5M). The methylation reaction was completed in five minutes at room temperature. After the reaction, the solution was injected into the semipreparative HPLC. The fraction that contained the product was evaporated to dryness by a stream of nitrogen gas and dissolved into saline. The radiochemical identity of [^{11}C]**8** was verified by co-injection with nonradioactive compound **8**.

2.3.3　*In vitro* autoradiography studies

Paraffin-embedded mouse brain section (C57-APP/PS1, 12 months old, male) was deparaffinized using the same protocol employed for the neuropathological staining assay. The section was incubated with ^{11}C-labeled ligand (3.7MBq per 100μL) for 1h at room temperature, then washed in 40% EtOH, before being rinsed with water for 1min. After drying, the section was exposed to a phosphorus plate (PerkinElmer, USA) for 2h. *In vitro* autoradiographic image was obtained using a phosphor imaging system (Cyclone, Packard). After autoradiographic examination, the same mouse brain section was stained by thioflavin-S to confirm the presence of Aβ plaques. For the staining of thioflavin-S, sections were immersed in a 0.125% thioflavin-S solution containing 10% EtOH for 3min and washed in 40% EtOH. After drying, the fluorescent observation was performed by the Oberver Z1 (Zeiss, Germany) equipped with GFP filter set (excitation, 505nm).

2.3.4　Partition coefficient determination

The determination of partition coefficients of ^{11}C-labeled ligand was done according to the procedure previously reported[25]. A solution of ^{11}C-labeled tracer (37MBq) was added to a premixed suspension containing 3.0g *n*-octanol and 3.0g PBS (0.05M, pH = 7.4) in a test tube. The test tube was vortexed for 3min at room temperature, followed by centrifugation for 5min at 3000r/min. Two samples from the *n*-octanol (50μL) and water (500μL) layers were measured. The partition coefficient was expressed as the logarithm of the ratio of the count per gram from *n*-octanol *vs*. PBS. Samples from the *n*-octanol layer were repartitioned until consistent partition coefficient values were obtained. The measurement was done in triplicate and repeated three times.

2.3.5　Biodistribution studies

The biodistribution experiments were performed in normal male mice (5 weeks, male, average weight, about 20g) and approved by the animal care committee of Beijing Normal University. A saline solution containing the ^{11}C-labeled ligand (9.25MBq per 100μL) was injected directly into the tail. The mice ($n = 5$ for each time point) were sacrificed at 2min, 5min, 10min, 20min and 30min post injection. The organs of interest were removed and weighed, radioactivity was measured with an automatic gamma counter. The percent dose per gram of wet tissue was calculated by a comparison of the tissue counts to suitably diluted aliquots of the injected material.

3　Results and Discussion

3.1　Chemistry

(E)-1-methoxy-4-styrylbenzene (**8**) was obtained by methylation of the hydroxy group of **7** using methyl iodide in 36% yield (see Scheme 1).

3.2　*In vitro* neuropathological staining

To investigate the specific binding of these stilbene derivatives to Aβ plaques, *in vitro* neuro-

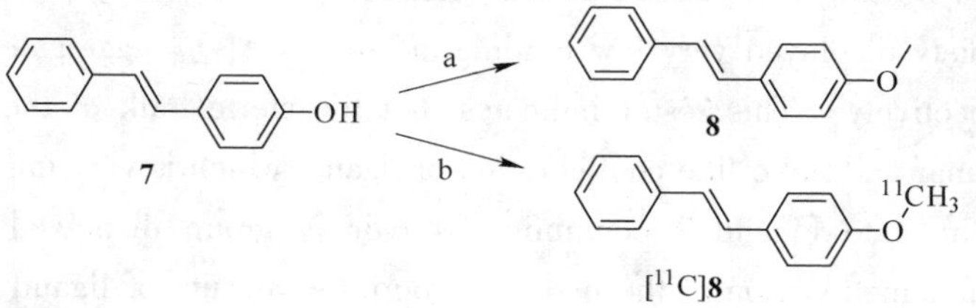

Scheme 1 Reagents and conditions

a—CH_3I, K_2CO_3, acetone, r. t. ; b—[^{11}C]-Triflate-CH_3, DMSO, NaOH(5M), r. t.

pathological staining was performed on the brain sections of a transgenic model mouse (C57BL6, APPswe/PSEN1, 12 months old). As shown in Fig. 4, ligand 7 and 8 distinctively stained Aβ plaques on the brain sections with low background (Fig. 4a and c). The similar pattern of Aβ plaques was consistent with that stained with thioflavin-S using the adjacent sections (Fig. 4b and d). The staining of ligands 5 and 6 (data not shown) was very weak due to their low binding affinity to $Aβ_{1-42}$ aggregates.

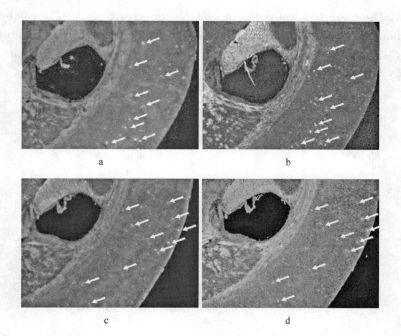

Fig. 4 Fluorescence staining of **7** (a) and **8** (c) on sections of brain tissue from a Tg model mouse (C57BL6, APP/PS1, 12 months old, male) with a filter set for GFP and the presence of plaques was confirmed by staining of the adjacent section with thioflavin S (b, d)

3.3 *In vitro* binding studies

The affinities of these stilbene derivatives (**5-8**) for $Aβ_{1-42}$ aggregates were determined by competing binding assays using [^{125}I]TZDM as the competing radioligand. Ligands **5** and **6** containing a glucose moiety displayed very low binding affinity to $A\beta_{1-42}$ aggregates ($K_i > 10000$nM and 7355nM, respectively). This result indicates that the steric bulk of the glucose fragment may distort the planar and flake-like configuration of ligand, which is very important for entering into the $A\beta$ binding site. Ligand **7** containing a hydroxy group displayed moderate affinity ($K_i = 384$nM). After methylation of the hydroxy group, the affinity of ligand **8** was greatly improved ($K_i = 19.5$nM). This finding was similar to that reported for the resveratrol derivatives[23].

3.4 Radiochemistry

Due to the high binding of **8** to $A\beta_{1-42}$ aggregates ($K_i = 19.5$nM), compound **8** was selected for ^{11}C-labelling. [^{11}C]**8** was labeled by direct O-methylation using [^{11}C]CH$_3$OTf in the presence of NaOH (5M) as base. Briefly, [^{11}C]CH$_3$OTf was trapped at room temperature in a glass vessel containing 1.5mg of **7** and 10μL NaOH (5M) in DMSO (0.3mL), and the mixture was reacted at room temperature for 5min with a radiochemical yield of 60% (Scheme 1). The radiochemical purity of [^{11}C]**8** was greater than 97% after purification by high performance liquid chromatography (HPLC). The radiochemical identity of ^{11}C-labeled ligand was verified by co-injection with nonradioactive compound **8** from HPLC profiles (Fig. 5).

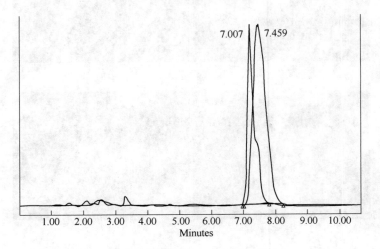

Fig. 5 HPLC profiles of **8** and [^{11}C]**8**
(HPLC conditions: Agilent Eclipse XDB-C18 reverse phase column
(Agilent Technologies. 5μm, 4.6mm × 150mm), CH$_3$CN/H$_2$O = 65/35, 1mL/min,
UV, 254 nm, **8**t_R(UV) = 7.01min, [^{11}C]**8**t_R(RI) = 7.46min)

3.5 *In vitro* autoradiography studies

On the basis of the *in vitro* binding data for these stilbene derivatives, ligand **8**, which exhibits the highest affinity, was selected for ^{11}C labeling and further biological evaluations. *In vitro* autoradiography was carried out in brain sections of transgenic model mouse (C57BL6, APPswe/PSEN1, 12 months old) using ^{11}C-labeled ligand [^{11}C]**8**. As shown in Fig. 6, specific and intense labeling of plaques was observed in the brain sections of transgenic mice, with numerous hot spots in the cortex region, as well as minimal background noise. The distribution of Aβ plaques was consistent with the results of autoradiography with thioflavin-S.

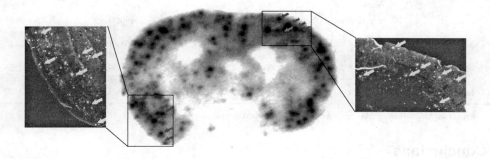

Fig. 6 *In vitro* autoradiography of [^{11}F]**8** on the brain section of Tg model mouse
(C57BL6, APPswe/PSEN1, 12 months old, male)
(The presence and distribution of plaques in the section were confirmed by fluorescence
staining using thioflavin-S on the same section with a filter set for GFP)

3.6 *In vivo* biodistribution studies

The ability to penetrate the intact blood-brain barrier (BBB) is one of the key qualifications for brain imaging agents. The lg$D$ value of [^{11}C]**8** obtained under the experimental condition (measured by a partition between *n*-octanol and pH 7.4 phosphate buffer) was 3.02, which is desirable for penetrating the BBB. To evaluate the pharmacokinetic properties of [^{11}C]**8**, *in vivo* biodistribution experiments were performed in normal ICR mice. As shown in Table 1, [^{11}C]**8** displayed high BBB penetration with the initial brain uptakewith (9.41 ± 0.51) % ID/g at 5min post-injection. Since there are no Aβ plaques in the brain of normal mice, the high brain uptake was subsequently followed by a rapid washout with (1.42 ± 0.09) % ID/g at 30min. The brain$_{5min}$/brain$_{30min}$ ratio as an index to compare washout rate is used for determining radioactivity pharmacokinetics *in vivo*. Compared with the ^{18}F-labeled resveratrol derivative [^{18}F]**4** (1.56)[23] and [^{11}C]SB-13 (2.74)[10], the ^{11}C-labeled ligand [^{11}C]**8** in this study had superior brain washout rate (6.63). In addition, compared with the brain uptake, the blood background levels during the experiment were lower, resulting in high brain-to-blood ratios (3.31, 4.44, 2.40, 1.66 and 1.61 at 2min, 5min, 10min, 20min and 30min post-injection, respectively), which is good for reducing nonspecific binding. The high initial uptakes in the liver and the kidneys were followed by a moderate washout, indicating that [^{11}C]**8** was excreted predominantly by the hepatobiliary and excretory systems.

Table 1 Biodistribution in normal ICR mice after i. v. injection of [^{11}C]8[①]

Organ	2min	5min	10min	20min	30min
Blood	2.26 ± 0.40	2.12 ± 0.04	1.02 ± 0.03	1.03 ± 0.08	0.88 ± 0.05
Brain	7.47 ± 0.35	9.41 ± 0.51	2.45 ± 0.16	1.71 ± 0.27	1.42 ± 0.09
Heart	6.14 ± 0.60	5.07 ± 0.29	1.30 ± 0.11	1.42 ± 0.22	1.40 ± 0.03
Liver	9.14 ± 0.41	12.84 ± 0.65	7.40 ± 0.87	6.52 ± 0.98	5.63 ± 0.21
Lung	10.77 ± 1.19	8.99 ± 0.24	2.83 ± 0.22	2.49 ± 0.49	2.85 ± 0.31
Kidney	9.55 ± 0.45	8.34 ± 1.22	3.87 ± 0.33	2.76 ± 0.43	2.48 ± 0.15
Spleen	3.86 ± 0.18	4.77 ± 0.90	1.45 ± 0.07	1.59 ± 0.31	1.88 ± 0.28
Muscle	3.21 ± 0.54	3.16 ± 0.04	1.04 ± 0.05	0.95 ± 0.23	1.00 ± 0.27

① Expressed as % dose/g. Averages for 5 mice ± standard deviation.

4 Conclusions

In summary, we have screened four natural products with stilbene moiety as novel amyloid imaging agent. After methylation of the hydroxy group, compound 8 displayed high affinity for Aβ$_{1-42}$ aggregates (K_i values in nM range). The ^{11}C-labeled ligand [^{11}C]8 showed specific labeling of Aβ plaques in sections of brain tissue from transgenic model mouse (C57BL6, APPswe/PSEN1, 12 months old). In addition, [^{11}C]8 displayed high initial uptake into and rapid washout from the brain in normal mice with a brain$_{5min}$/brain$_{30min}$ ratio of 6.63, which is highly desirable for Aβ imaging agents. These results suggest that [^{11}C]8 derived from natural products could be potentially useful for the imaging of Aβ plaques in living subjects. Furthermore, we could also screen potential therapeutic drugs for AD from natural product library.

Acknowledgement

The authors thank Dr. Jin Liu (College of Life Science, Beijing Normal University) for assistance in the *in vitro* neuropathological staining. This work was supported by National Natural Science Foundation of China (20871021, 21201019 and 21071023).

References

[1] Hardy, J., Selkoe, D. J.: The amyloid hypothesis of Alzheimer's disease: progress and problems on the road to therapeutics. Science 297, 353 (2002).

[2] Reitz, C.: Alzheimer's disease and the amyloid cascade hypothesis: a critical review. Int. J. Alzheimers Dis. 2012, 369808 (2012).

[3] Mathis, C. A., Wang, Y., Klunk, W. E.: Imaging beta-amyloid plaques and neurofibrillary tangles in the aging human brain. Curr. Pharm. Des. 10, 1469 (2004).

[4] Cai, L., Innis, R. B., Pike, V. W.: Radioligand development for PET imaging of beta-amyloid (Abeta)-current status. Curr. Med. Chem. 14, 19 (2007).

[5] Mathis, C. A., Wang, Y., Holt, D. P., Huang, G. F., Debnath, M. L., Klunk, W. E.: Synthesis and

evaluation of ^{11}C-labeled 6-substituted 2-arylbenzothiazoles as amyloid imaging agents. J. Med. Chem. 46, 2740 (2003).

[6] Klunk, W. E. , Engler, H. , Nordberg, A. , Wang, Y. , Blomqvist, G. , Holt, D. P. , Bergström, M. , Savitcheva, I. , Huang, G. F. , Estrada, S. , Ausén, B. , Debnath, M. L. , Barletta, J. , Price, J. C. , Sandell, J. , Lopresti, B. J. , Wall, A. , Koivisto, P. , Antoni, G. , Mathis, C. A. , Långström, B. : Imaging brain amyloid in Alzheimer's disease with Pittsburgh Compound-B. Ann. Neurol. 55, 306 (2004).

[7] Koole, M. , Lewis, D. M. , Buckley, C. , Nelissen, N. , Vandenbulcke, M. , Brooks, D. J. , Vandenberghe, R. , Van Laere, K. : Whole-body biodistribution and radiation dosimetry of ^{18}F-GE067: a radioligand for *in vivo* brain amyloid imaging. J. Nucl. Med. 50, 818 (2009).

[8] Johnson, A. E. , Jeppsson, F. , Sandell, J. , Wensbo, D. , Neelissen, J. A. , Jureus, A. , Strom, P. , Norman, H. , Farde, L. , Svensson, S. P. : AZD2184: a radioligand for sensitive detection of beta-amyloid deposits. J. Neurochem. 108, 1177 (2009).

[9] Nyberg, S. , Jonhagen, M. E. , Cselenyi, Z. , Halldin, C. , Julin, P. , Olsson, H. , Freund-Levi, Y. , Andersson, J. , Varnas, K. , Svensson, S. , Farde, L. : Detection of amyloid in Alzheimer's disease with positron emission tomography using [^{11}C]AZD2184. Eur. J. Nucl. Med. Mol. Imaging 36, 1859 (2009).

[10] Ono, M. , Wilson, A. , Nobrega, J. , Westaway, D. , Verhoeff, P. , Zhuang, Z. P. , Kung, M. P. , Kung, H. F. : ^{11}C-labeled stilbene derivatives as Abeta-aggregate-specific PET imaging agents for Alzheimer's disease. Nucl. Med. Biol. 30, 565 (2003).

[11] Verhoeff, N. P. , Wilson, A. A. , Takeshita, S. , Trop, L. , Hussey, D. , Singh, K. , Kung, H. F. , Kung, M. P. , Houle, S. : *In vivo* imaging of Alzheimer disease beta-amyloid with [^{11}C]SB-13 PET. Am. J. Geriatr. Psychiatry 12, 584 (2004).

[12] Rowe, C. C. , Ackerman, U. , Browne, W. , Mulligan, R. , Pike, K. L. , O'Keefe, G. , Tochon-Danguy, H. , Chan, G. , Berlangieri, S. U. , Jones, G. , Dickinson-Rowe, K. L. , Kung, H. P. , Zhang, W. , Kung, M. P. , Skovronsky, D. , Dyrks, T. , Holl, G. , Krause, S. , Friebe, M. , Lehman, L. , Lindemann, S. , Dinkelborg, L. M. , Masters, C. L. , Villemagne, V. L. : Imaging of amyloid beta in Alzheimer's disease with ^{18}F-BAY94-9172, a novel PET tracer: proof of mechanism. Lancet Neurol 7, 129 (2008).

[13] Lin, K. J. , Hsu, W. C. , Hsiao, I. T. , Wey, S. P. , Jin, L. W. , Skovronsky, D. , Wai, Y. Y. , Chang, H. P. , Lo, C. W. , Yao, C. H. , Yen, T. C. , Kung, M. P. : Whole-body biodistribution and brain PET imaging with [^{18}F]AV-45, a novel amyloid imaging agent-a pilot study. Nucl. Med. Biol. 37, 497 (2010).

[14] Choi, S. R. , Golding, G. , Zhuang, Z. , Zhang, W. , Lim, N. , Hefti, F. , Benedum, T. E. , Kilbourn, M. R. , Skovronsky, D. , Kung, H. F. : Preclinical properties of ^{18}F-AV-45: a PET agent for Abeta plaques in the brain. J. Nucl. Med. 50, 1887 (2009).

[15] Dai, J. R. , Hallock, Y. F. , Cardellina, J. H. 2nd. , Boyd, M. R. : HIV-inhibitory and cytotoxic oligostilbenes from the leaves of Hopea malibato. J. Nat. Prod. 61, 351 (1998).

[16] Condori, J. , Sivakumar, G. , Hubstenberger, J. , Dolan, M. C. , Sobolev, V. S. , Medina-Bolivar, F. : Induced biosynthesis of resveratrol and the prenylated stilbenoids arachidin-1 and arachidin-3 in hairy root cultures of peanut: Effects of culture medium and growth stage. Plant Physiol. Biochem. 48, 310 (2010).

[17] Heide, L. : Prenyl transfer to aromatic substrates: genetics and enzymology. Curr. Opin. Chem. Biol. 13, 171 (2009).

[18] Fremont, L. : Biological effects of resveratrol. Life Sci. 66, 663 (2000).

[19] Fremont, L. , Belguendouz, L. , Delpal, S. : Antioxidant activity of resveratrol and alcohol-free wine polyphenols related to LDL oxidation and polyunsaturated fatty acids. Life Sci. 64, 2511 (1999).

[20] Howitz, K. T. , Bitterman, K. J. , Cohen, H. Y. , Lamming, D. W. , Lavu, S. , Wood, J. G. , Zipkin, R. E. ,

Chung, P., Kisielewski, A., Zhang, L. L., Scherer, B., Sinclair, D. A.: Small molecule activators of sirtuins extend Saccharomyces cerevisiae lifespan. Nature 425, 191 (2003).

[21] Feng, Y., Wang, X. P., Yang, S. G., Wang, Y. J., Zhang, X., Du, X. T., Sun, X. X., Zhao, M., Huang, L., Liu, R. T.: Resveratrol inhibits beta-amyloid oligomeric cytotoxicity but does not prevent oligomer formation. Neurotoxicology 30, 986 (2009).

[22] Gester, S., Wuest, F., Pawelke, B., Bergmann, R., Pietzsch, J.: Synthesis and biodistribution of an ^{18}F-labelled resveratrol derivative for small animal positron emission tomography. Amino Acids 29, 415 (2005).

[23] Lee, I., Choe, Y. S., Choi, J. Y., Lee, K. H., Kim, B. T.: Synthesis and evaluation of ^{18}F-labeled styryltriazole and resveratrol derivatives for beta-amyloid plaque imaging. J. Med. Chem. 55, 883 (2012).

[24] Qiu, G., Wu, X. Q., Luo, X. G.: Effect of polygonum multiflorum thunb on BDNF expression in rat hippocampus induced by amyloid beta-protein (Abeta) 1-40. J. Cent. South Univ. (Med. Sci.) 31, 194 (2006).

[25] Cui, M. C., Li, Z. J., Tang, R. K., Liu, B. L.: Synthesis and evaluation of novel benzothiazole derivatives based on the bithiophene structure as potential radiotracers for beta-amyloid plaques in Alzheimer's disease. Bioorg. Med. Chem. 18, 2777 (2010).

[26] Zhuang, Z. P., Kung, M. P., Hou, C., Skovronsky, D. M., Gur, T. L., Plossl, K., Trojanowski, J. Q., Lee, V. M., Kung, H. F.: Radioiodinated styrylbenzenes and thioflavins as probes for amyloid aggregates. J. Med. Chem. 44, 1905 (2001).

[27] Cheng, Y., Prusoff, W. H.: Relationship between the inhibition constant (K1) and the concentration of inhibitor which causes 50 per cent inhibition (I50) of an enzymatic reaction. Biochem. Pharmacol 22, 3099 (1973).

[28] Zhang, J., Zhang, X., Li, Y., Tian, J.: Simple, rapid and reliable preparation of $[^{11}C]$-(+)-α-DTBZ of high quality for routine applications. Molecules 17, 6697 (2012).

Evaluation of Molecules Based on the Electron Donor-acceptor Architecture as Near-infrared β-amyloidal-targeting Probes*

Abstract A novel class of near-infrared molecules based on the donor-acceptor architecture were synthesized and evaluated as Aβ imaging probes. *In vivo* imaging studies suggested that MCAAD-3 could penetrate the blood-brain barrier and label Aβ plaques in the brains of transgenic mice. Computational studies could reproduce the experimental trends well.

To date the diagnosis and treatment of Alzheimer's disease(AD), a disease that increases in prevalence in old age, are difficult despite intensive research. The amyloid hypothesis, which suggests that β-amyloid (Aβ) is causative in the pathogenesis of AD, is widely accepted and has become the pivotal focus of current AD research[1]. The asymptomatic detection of Aβ plaques in the brain *in vivo* is critical for obtanining valid diagnostic and therapeutic strategies for AD; hence, several imaging techniques, such as positron emission tomography (PET)[2], single photon emission computed tomography (SPECT)[3], optical imaging[4], and magnetic resonance imaging (MRI)[5] have been exploited for this purpose.

Recent advances in optical imaging technology have driven the development of fluorescent probes into a rapidly expanding era[6]. In particular, fluorescence molecular tomography (FMT) imaging has gained adequate attention in the realm of localization and quantification in deep tissues, where it is used necessarily, with probes fluorescing in the near-infrared (NIR) spectral range, which provide sufficient light propagation into tissue and avoid auto-fluorescence from biological matter[7]. Research in the field of developing fluorescent probes targeting Aβ in the brain has been deficient, although some probes, including AOI987[8], 2C40[9], NIAD-4[10], CRANAD-2[11] and BAP-1[12] have been reported. AOI987 and 2C40 are charged molecules which have poor capability of blood-brain barrier (BBB) penetration. NIAD-4 was not optimal for NIR imaging with its emission maximum less than 650nm and CRANAD-2 had limited application in NIR imaging due to the low brain egress. BAP-1 had high QY and high affinity for Aβ aggregates. However, the narrow Stokes shift has lowered its application *in vivo*.

To fill this gap, we attempted to develop an effective *in vivo* NIR imaging dye for Aβ by merging the following empirical requirements and rationales. (1) We employed a push-pull architecture with an electron-rich donor and an electron-deficient acceptor as end groups bridged by a π-conjugated system as the backbone. A benzene ring and a polyenic chain were selected as

---

* Copartner: Fu Hualong, Cui Mengchao, Tu Peiyu, Pan Ziwei. Reprinted from *Chem. Commun.*, 2014, 50: 11875-11878.

the "π-electron bridge", and the HOMO-LUMO gap could be narrowed by elongating the polyenes[13], leading the spectra of the resulting molecule to fall in the "NIR window" (650-900nm), an ideal region for imaging living biosome[7b]. (2) The solubility, lipophilicity, molecular weight, and molecular geometry of the resulting probes can be tailored by rationally varying the donor and acceptor moieties[6a, 14]. A N, N'-dimethylamino group is an ideal electron donating moiety and possesses good furtherance for Aβ binding affinity[15]. The four π-electron acceptors we chose are presented in Fig. 1. (3) The probes should be "turned on" concomitant with significant changes in their fluorescence properties (i. e., emission wavelength and fluorescence intensity) upon interacting with Aβ plaques[11]. Recently, our group reported the NIR probe DANIR 2c (Fig. 1), which has high affinity for Aβ aggregates and performed excellently for *in vivo* imaging of Aβ plaques in the brains of transgenic (Tg) mice; however, its emission wavelength was only 625nm when bound to Aβ aggregates[16]. Here, we reported the synthesis, characterization, and biological assessment of these electron donor-acceptor molecules as novel NIR imaging probes targeting Aβ plaques in AD; computational studies were also performed and conformed well with the experimental results.

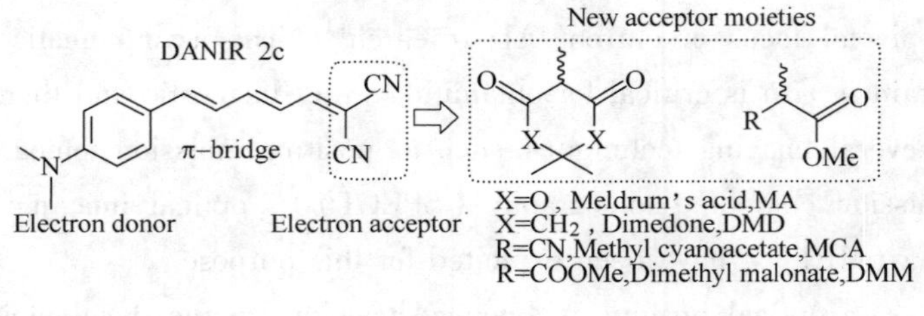

Fig. 1 Chemical structures of DANIR 2c and appropriate acceptor moieties

The synthesis of the probes is presented in Scheme 1. The purity of the compounds was determined to be higher than 96% by HPLC (see Fig. S1 in the ESI†). Subsequently, the absorption and fluorescent properties of the probes were assessed in solvents of different polarities (see Spectroscopic Measurements in the ESI†). As expected, by lengthening the polyenic chains, the probes underwent an excited state charge transfer (CT) and striking solvent dependency ac-

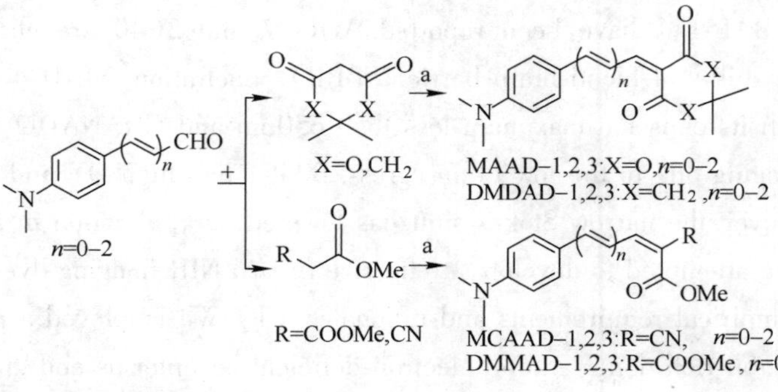

Scheme 1 Synthesis of the probes

a—For MAAD-1,2 and MCAAD/DMMAD series, methanol, piperidine, r. t. ;
for MAAD-3 and DMDAD-2,3, methanol, r. t or 100 ℃; for DMDAD-1, no solvent, 150 ℃

companied by a larger dipole moment. As a result, the emission of the four probes with the longest polyenic chains, MAAD-3, DMDAD-3, MCAAD-3 and DMMAD-3 red-shifted to the "NIR window" in PBS with λ_{em} ranging from 685 to 725 nm; concurrently, these four NIR fluorochromes (NIRFs) displayed optimal Stokes shifts larger than 69nm in PBS (Table 1, Fig. S6-S9 and Table S1 in the ESI†).

Table 1 Selected spectroscopic and binding data for the synthesized NIRFs

Probes	$\lambda_{em1}^{①}$/nm	λ_{em2}/nm	$\phi^{②}$/%	Fold[③]	$K_i^{④}$/nM
MAAD-3	704	674	4.71/0.048	15	354.3 ± 43.5
DMDAD-3	725	694	2.68/0.033	7	645.2 ± 77.2
MCAAD-3	685	654	1.23/0.25	26	106.0 ± 29.8
DMMAD-3	687	642	0.10/0.068	8	652.6 ± 143.0

① The emission measured in PBS and upon binding to $A\beta_{1-42}$ aggregates is presented as λ_{em1} and λ_{em2}, respectively.
② Fluorescence quantum yields were measured in dichloromethane/PBS, respectively.
③ The fold increase in the fluorescence intensity upon binding to $A\beta_{1-42}$.
④ Measured in triplicate with results given as the mean ± SD.

We first tested the fluorescence properties of the NIRFs in PBS upon interaction with $A\beta_{1-42}$ aggregates or bovine serum albumin (BSA). As shown in Fig. 2c, f, i and l, a strong increase in fluorescence intensity (7-to 26-fold) was observed when the NIRFs bound to the $A\beta_{1-42}$ aggregates in PBS. Conversely, weak fluorescence intensity was observed for the NIRFs alone or in combination with BSA in PBS. The emission wavelength of three of the NIRFs (except DMMAD-3) still fell in the NIR range and manifested significant blue shifts (>30nm, Table 1) even after binding with the $A\beta_{1-42}$ aggregates, indicating that the NIRFs likely intercalated into the hydrophobic pocket of the aggregated amyloid fibrils[11,16,17]. This result suggested that our NIRFs could be "turned on" upon interacting with $A\beta_{1-42}$ aggregates with an increase in fluorescence intensity and a blue shift in the emission wavelength.

Next, to evaluate the binding selectivity of our NIRFs for Aβ plaques, in vitro neuropathological fluorescence staining was carried out on slices of brain tissue from a double Tg mouse (APPswe/PSEN1, 11 months old, male) with Aβ plaques. We observed high-contrast fluorescent spots on the brain sections from the Tg mouse when stained with the NIRFs, which was highly consistent with the staining pattern of the adjacent brain sections stained with Thioflavin-S (Th-S) (Fig. 2). As anticipated, no spots were observed on the brain slices from the wild-type mouse (see Fig. S12 in the ESI†). These results indicated that our NIRFs are capable of selectively binding to Aβ plaques.

Furthermore, to quantify the binding affinity to $A\beta_{1-42}$ aggregates, we measured the K_i values of all of the probes by competitive binding assays using [^{125}I]IMPY as the competing radioligand (see Table 1 and Table S2, Fig. S13 in the ESI†). The probes with the shortest polyenic chains (MAAD-1, DMDAD-1, MCAAD-1 and DMMAD-1) had the poorest affinities (K_i > 2700nM); while the other probes with longer polyenic chains had moderate affinities. Among

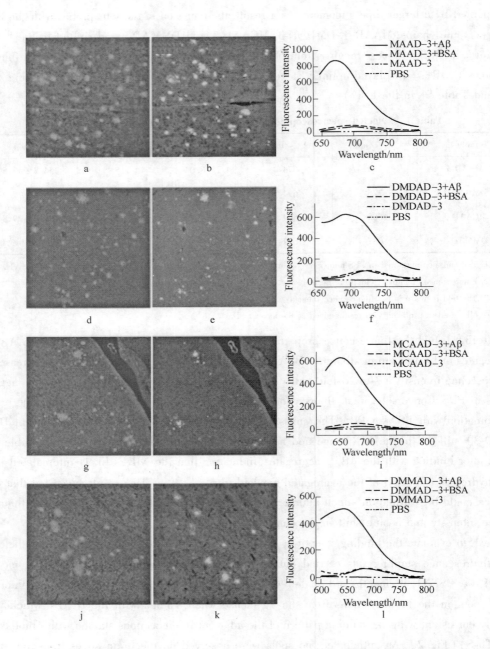

Fig. 2 Brain sections from Tg mice stained with MAAD-3(a), DMDAD-3(d), MCAAD-3(g), and DMMAD-3(j), and the emission wavelengths after the NIRFs bound to Aβ$_{1-42}$ aggregates or BSA (c, f, i, and l, respectively), and the adjacent brain sections were stained with Th-S (b, e, h, and k, respectively)

them, MCAAD-3 displayed the highest affinity of 106.0nM, a decreased affinity compared to that of DANIR 2c, which was probably caused by the interruption of the planarized geometry of DANIR 2c when the dicyanomethylene group was replaced by the methyl cyanoacetate group[16,18]. This result was consistent with the increase in fluorescence intensity observed for these probes upon interaction with the Aβ$_{1-42}$ aggregates. Overall, the binding data indicated that increased affinity could be obtained by lengthening the polyenic chain, and that further modifications were needed to improve the binding affinities.

To further validate the capability of our NIRFs in *in vivo* imaging, BBB penetrating determination, *in vivo* NIR imaging and *ex vivo* fluorescence staining were conducted. The results of the BBB penetrating test showed that MCAAD-3 and DMMAD-3 could readily penetrate the BBB with a high initial brain uptake and fast brain egress (Table S5 and Fig. S14 in the ESI†). By virtue of the valid BBB penetration and high affinity to Aβ aggregates, MCAAD-3 was selected for *in vivo* imaging on Tg mice and age-matched wild-type (WT) ones. As shown in Fig. 3a, the fluorescence signals of MCAAD-3 in the brains of Tg mice diminished considerably more slowly than in those of WT ones. Furthermore, the brain kinetic curves showed that the differences of fluorescence signals between Tg and WT groups could be observed at the earliest time point of 30min (Fig. 3b and Fig. S15 in the ESI†). Finally, the binding of MCAAD-3 to Aβ plaques *in vivo* was confirmed by *ex vivo* histology. A higher number of Aβ plaques were observed in the cortex, hippocampus and cerebellum regions of brain slices from Tg mice and this result was further confirmed by staining the same section with Th-S (Fig. 4c and d and Fig. S16 in the ESI†). These results demonstrated that MCAAD-3 could penetrate BBB and bind to Aβ plaques *in vivo*.

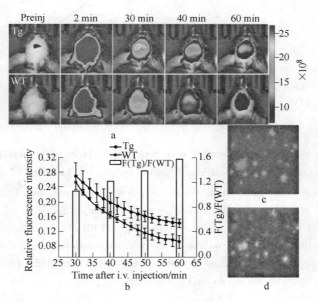

Fig. 3 *In vivo* NIR imaging images of Tg and WT mice at representative time points before and after i. v. injection of MCAAD-3(a), brain kinetic curves of MCAAD-3(left Y axis) and the values of F(Tg)/F(WT) at selected time points (right Y axis)(b), and *Ex vivo* histology results of brain slices from Tg mouse stained with MCAAD-3 and Th-S(c,d)

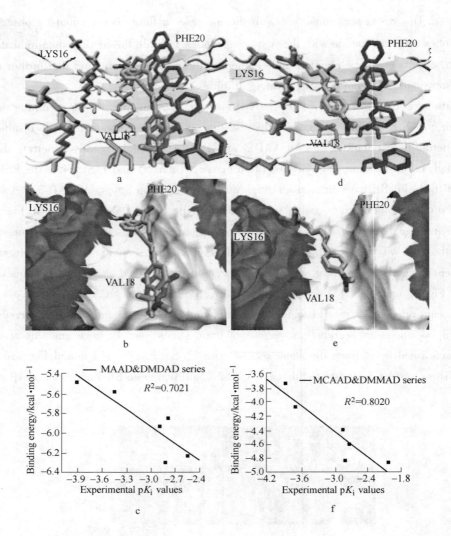

Fig. 4 The lowest-energy/top-ranked dock conformations of the NIRFs on Aβ fibrils and the correlations between the calculated binding energies and experimental pK_i values.

The residues of Aβ fibrils are vividly presented in the licorice style (a and d) and as a molecule surface representation (b and e)

(a and b, d and e are merged dock conformations for MAAD-3/DMDAD-3 and MCAAD-3/DMMAD-3, respectively. c and f show the qualitative linear relationships between the calulated binding energy at the "IMPY site" and the experimental pK_i values of the MAAD/DMDAD series and the MCAAD/DMMAD series, respectively)

To gain insight into the binding nature of these NIRFs to Aβ fibrils, molecular docking simulations were performed. By recapitulating the docking results, two major binding sites were found. One was identified as the binding site of IMPY (named the "IMPY site", see Fig. S18 in the ESI†), which was a hydrophobic cleft running longitudinally to the fibrillar axis formed by VAL18 and PHE20[19]. Another site was formed by LYS16, VAL18, and PHE20 and was a wider channel, which was named the "non-IMPY site". The lowest-energy/top-ranked docking poses are presented in Fig. 4 and Fig. S19 (ESI†). Bound probes of the MAAD/DMDAD series were aligned into the "IMPY site", which was anchored by hydrophobic interactions and ran along the longitudinal fibrillar spine. Likewise, molecules of the MCAAD/DMMAD series occupied a "non-IMPY site" that was dominated by hydrogen bonds between the hydrogen atoms of a polar residue LYS16 and the oxygen atoms of the electron acceptor moieties (Fig. S21 in the ESI†) and hydrophobic interactions between the ligands and the nonpolar residues VAL18 and PHE20. A good correlation coefficient ($R^2 = 0.7021, 0.8020$) between the docking scores and the experimental pK_i values was obtained after analysing the binding energies of the docking poses located at the "IMPY site" for all the probes (Fig. 4c and f). Furthermore, with the extension of the conjugated double bonds of each analogue, the hydrophobic interactions between the probes and the Aβ fibrils strengthened, the binding energy diminished and the binding affinity increased. Overall, the experimental studies and computational simulations were highly consistent. A more detailed discussion about the docking results can be found in the ESI†.

In summary, a series of probes with the electron donor-acceptor architecture for the detection of Aβ plaques have been designed, synthesized and characterized using spectroscopic and biological methods, and the binding characteristics of these probes were further investigated by computational methods. Four probes with the longest conjugated double bond systems displayed moderate binding affinity to Aβ$_{1-42}$ aggregates and rational optical properties upon association with Aβ$_{1-42}$ aggregates. MCAAD-3, a probe possessed efficient BBB penetration and high affinity for Aβ aggregates, performed excellent *in vivo* imaging, and was an optimal NIR probe for the detection of Aβ plaques both *in vitro* and *in vivo*. Furthermore, computational studies provided insights into the binding sites of our probes on the Aβ fibrils. This research opens promising avenues to a new generation of Aβ-targeting molecules for diagnostic applications.

This work was funded by the NSFC (No. 21201019). The authors thank Dr Jin Liu (College of Life Science, Beijing Normal University) for assistance in the *in vitro* neuropathological staining.

References

[1] (a) J. Hardy and D. J. Selkoe, Science, 2002, 297, 353; (b) D. J. Selkoe, JAMA, J. Am. Med. Assoc., 2000, 283, 1615; (c) D. J. Selkoe, Physiol. Rev., 2001, 81, 741.

[2] (a) S. R. Choi, G. Golding, Z. Zhuang, W. Zhang, N. Lim, F. Hefti, T. E. Benedum, M. R. Kilbourn, D. Skovronsky and H. F. Kung, J. Nucl. Med., 2009, 50, 1887; (b) C. A. Mathis, Y. Wang, D. P. Holt, G. F. Huang, M. L. Debnath and W. E. Klunk, J. Med. Chem., 2003, 46, 2740.

[3] A. Newberg, N. Wintering, C. Clark, K. Ploessl, D. Skovronsky, J. Seibyl, M. P. Kung and H. Kung, J. Nucl. Med., 2006, 47, 78P.

[4] B. J. Bacskai, S. T. Kajdasz, R. H. Christie, C. Carter, D. Games, P. Seubert, D. Schenk and B. T. Hyman, Nat. Med., 2001, 7, 369.

[5] M. Higuchi, N. Iwata, Y. Matsuba, K. Sato, K. Sasamoto and T. C. Saido, Nat. Neurosci., 2005, 8, 527.

[6] (a) N. Karton-Lifshin, L. Albertazzi, M. Bendikov, P. S. Baran and D. Shabat, J. Am. Chem. Soc., 2012, 134,

20412; (b) S. Lee, K. Park, K. Kim, K. Choi and I. C. Kwon, Chem. Commun. ,2008,4250.

[7] (a) F. Stuker, J. Ripoll and M. Rudin, Mol. Pharmaceutics,2011,3,229; (b) R. Weissleder, Nat. Biotechnol. ,2001,19,316.

[8] M. Hintersteiner, A. Enz, P. Frey, A. L. Jaton, W. Kinzy, R. Kneuer, U. Neumann, M. Rudin, M. Staufenbiel, M. Stoeckli, K. H. Wiederhold and H. U. Gremlich, Nat. Biotechnol. ,2005,23,577.

[9] Q. Li, J. S. Lee, C. Ha, C. B. Park, G. Yang, W. B. Gan and Y. T. Chang, Angew. Chem. ,2004,43,6331.

[10] E. E. Nesterov, J. Skoch, B. T. Hyman, W. E. Klunk, B. J. Bacskai and T. M. Swager, Angew. Chem. ,2005,44,5452.

[11] C. Ran, X. Xu, S. B. Raymond, B. J. Ferrara, K. Neal, B. J. Bacskai, Z. Medarova and A. Moore, J. Am. Chem. Soc. ,2009,131,15257.

[12] M. Ono, H. Watanabe, H. Kimura and H. Saji, ACS Chem. Neurosci. ,2012,3,319.

[13] (a) V. Alain, S. Rédoglia, M. Blanchard-Desce, S. Lebus, K. Lukaszuk, R. Wortmann, U. Gubler, C. Bosshard and P. Günter, Chem. Phys. ,1999,245,51; (b) L. Yao, S. Zhang, R. Wang, W. Li, F. Shen, B. Yang and Y. Ma, Angew. Chem. ,2014,53,2119.

[14] S. Ellinger, K. R. Graham, P. Shi, R. T. Farley, T. T. Steckler, R. N. Brookins, P. Taranekar, J. Mei, L. A. Padilha, T. R. Ensley, H. Hu, S. Webster, D. J. Hagan, E. W. Van Stryland, K. S. Schanze and J. R. Reynolds, Chem. Mater. ,2011,23,3805.

[15] (a) L. Cai, R. B. Innis and V. W. Pike, Curr. Med. Chem. ,2007,14,19; (b) M. Cui, M. Ono, H. Kimura, H. Kawashima, B. L. Liu and H. Saji, Nucl. Med. Biol. ,2011,38,313.

[16] M. Cui, M. Ono, H. Watanabe, H. Kimura, B. Liu and H. Saji, J. Am. Chem. Soc. ,2014,136,3388.

[17] (a) Principles of Fluorescence Spectroscopy, ed. J. Lakowicz, Springer, US, 2006, pp. 1-26; (b) A. Jacobson, A. Petric, D. Hogenkamp, A. Sinur and J. R. Barrio, J. Am. Chem. Soc. ,1996,118,5572.

[18] A. Petric, S. A. Johnson, H. V. Pham, Y. Li, S. Ceh, A. Golobic, E. D. Agdeppa, G. Timbol, J. Liu, G. Keum, N. Satyamurthy, V. Kepe, K. N. Houk and J. R. Barrio, Proc. Natl. Acad. Sci. U. S. A. ,2012,109,16492-16497.

[19] (a) N. P. Cook, M. Ozbil, C. Katsampes, R. Prabhakar and A. A. Marti, J. Am. Chem. Soc. ,2013,135,10810; (b) C. Wu, J. Scott and J. E. Shea, Biophys. J. ,2012,103,550.

Radiopharmaceuticals in China: Current Status and Prospects*

Abstract The review provides an overview of the current status of radiopharmaceuticals in China for *in vivo* clinical use and also describes some important advances in the past three decades. Development of the diagnostic and therapeutic radiopharmaceuticals as well as basic research on radiopharmaceutical chemistry are being introduced. The radiotracers developed in China include: (1) Brain perfusion imaging agents and CNS radiotracers for β-amyloid plaques, σ_1 receptors, and dopamine D_2 or D_4 receptors; (2) ^{99m}Tc- and ^{18}F-labeled myocardial perfusion imaging agents; (3) tumor imaging agents including integrin-targeting radiotracer, novel sentinel lymph node imaging agents, hypoxia imaging agents, ^{99m}Tc-labeled glucose derivatives, σ_2 receptor imaging agents, folate receptor imaging agents, and potential radiotracers for imaging of human telomerase reverse transcriptase expression; (4) Potential infection imaging agents; (5) Potential asialoglycoprotein receptor imaging agents; (6) Other imaging agents. Moreover, some prospects of research and development of radiopharmaceuticals in the near future are discussed.

Key words radiopharmaceuticals, current status, important advances in recent years, potential radiotracers, future prospects

1 Introduction

The research work and production of medical radionuclides and radiopharmaceuticals in China began in 1958. After more than 50 years' development, a comparatively complete production and management system has been built up and is well equipped. Reactor produced radionuclides and radiopharmaceuticals for medical applications have been the subject of concentrated effort in the past years. Now, three reactors are running for production of medical radionuclides. Another two reactors will be put in use for production of medical radionuclides in one or two years. Among the above five reactors, two are situated in Beijing, and others are in the Southwest China. Research has also been reported for the production of ^{99}Mo using a solution reactor. The most promising reactor-produced radionuclides for SPECT and therapy in China are ^{99}Mo-^{99m}Tc, ^{131}I, ^{125}I, ^{198}Au, ^{90}Sr-^{90}Y, ^{186}Re, ^{153}Sm, etc. The latest effort in radiopharmaceutical research has been focused on the development of cyclotron-produced radionuclides and their labeled compounds, which will bloom in the future. Up to date, there are four cyclotrons used in production of ^{67}Ga, ^{111}In, ^{201}Tl, ^{123}I, ^{103}Pd, ^{68}Ge, ^{62}Zn, ^{57}Co and ^{18}F, etc. A cyclotron, called Cyclone-30 (built in cooperation with IBA of Belgium) is installed in China Institute of Atomic Energy (Beijing)[1]. Another Cyclone-30 is installed in Shanghai Institute of Applied Physics

---

* Copartner: Jia Hongmei. Reprinted from *Radiochim. Acta*, 2014, 102(1-2) :53-67.

(Chinese Academy of Sciences). The third machine, a CS-30 cyclotron (TCC company, USA) is installed in Sichuan University (Chengdu, Sichuan Province) [2]. The fourth one, CS-22 cyclotron, is installed in Beijing Normal University (Beijing). ^{67}Ga, ^{111}In, and ^{201}Tl are routinely produced for preparation of related radiopharmaceuticals. However, the main product is [^{18}F] FDG. Small amounts of [^{13}N]NH$_3$, [^{15}O]H$_2$O, [^{18}F]FDOPA, etc. are also used in some hospitals. It should be pointed out that Beijing Atom High Tech Co., Ltd. (BAHC) of China Institute of Atomic Energy is the major producer and supplier of medical radionuclides and radiopharmaceuticals in China. Currently, BAHC has established two radiopharmaceutical production systems based on using reactors and cyclotrons, respectively.

Currently, Chinese Isotope Society (CIS), Chinese Nuclear and Radiochemistry Society (CNRS), and Chinese Society of Nuclear Medicine (CSNM) have established radiopharmaceutical professional groups. They, together with China Isotope and Radiation Corporation, have organized a series of national symposia on labeled compounds and radiopharmaceuticals in the past three decades. The latest 11th National Symposium was held in 2011. Moreover, the Chinese Society for Molecular Imaging (CSMI) was founded in June 2011. The CSNM, CNRS, and CIS have established their official journals as the Chinese Journal of Nuclear Medicine and Molecular Imaging, Journal of Nuclear and Radiochemistry, and Journal of Isotopes, respectively. Significant progress has also been achieved in the past decades in the basic research on radiopharmaceutical chemistry and its applications. The main research institutes and universities in the radiopharmaceutical area in China are as follows:

(1) China Institute of Atomic Energy, Beijing;

(2) Chinese Academy of Medical Sciences, Beijing;

(3) Shanghai Institute of Applied Physics (Chinese Academy of Sciences), Shanghai;

(4) Institute of High Energy Physics (Chinese Academy of Sciences), Beijing;

(5) China Academy of Engineering Physics, Sichuan Province;

(6) Jiangsu Institute of Nuclear Medicine, Jiangsu Province;

(7) Peking University, Beijing;

(8) Fudan University, Shanghai;

(9) Xiamen University, Fujian Province;

(10) Capital Medical University, Beijing;

(11) Sichuan University, Sichuan Province;

(12) Suzhou University, Jiangsu Province;

(13) Beijing Normal University, Beijing; and many Departments of Nuclear Medicine and PET Centers in the hospitals.

It is well known that radiopharmaceuticals could serve as effective diagnostic and therapeutic tools for many human diseases. Moreover, they allow the assessment of metabolism and functions by providing quick, non-invasive and real-time visualization of physiological and pathological processes in the living humans at the molecular level together with PET (positron emission tomography) and SPECT (single photon emission computed tomography) imaging modalities.

Radionuclide-labeled molecular probes could provide new concepts, new methods and new approaches for truly early diagnosis and therapy, and possible pathways for preventive medicine, translational medicine and personalized medicine. The present review provides an overview of current status of radiopharmaceuticals in China for *in vivo* use. In addition, we discuss some prospects of research and development on radiopharmaceuticals in the near future.

2 Current Status of Radiopharmaceuticals in Clinical Trials

Nuclear medicine departments and laboratories have been set up since 1950s in many large hospitals and a few medium-sized hospitals. Most of the large hospitals in China are equipped with SPECT which are used for radiopharmaceutical studies and for routine examination. Since PET possesses several technical advantages over SPECT, the number of PET facilities has been increasing very rapidly in China in the past few years. Over the last twenty years, the radiopharmaceutical sciences in China have been developing and advancing together with the rapid economic growth. The first PET center in China was built in 1995, but now the number of PET centers is increasing rapidly. Till January 2012, there were 162 sets of PET and PET/CT. There were 605 sets of SPECT, SPECT/CT, coincidence SPECT and gamma-camera in the whole nation[3]. Following sixteen radiopharmaceuticals are listed in the Pharmacopoeia of People's Republic of China (2010):

Sodium phosphate[^{32}P] Oral Solution, Sodium phosphate [^{32}P] Injection, Colloidal chromium phosphate[^{32}P] Injection, Sodium chromate [^{51}Cr] Injection, Gallium citrate [^{67}Ga] Injection, Sodium pertechnetate [^{99m}Tc] Injection, Technetium methylendiphosphonate [^{99m}Tc] Injection, Technetium etifenin [^{99m}Tc] Injection, Technetium phytate[^{99m}Tc] Injection, Technetium pentetate[^{99m}Tc] Injection, Technetium pyrophosphate [^{99m}Tc] Injection. Technetium albumin aggregated [^{99m}Tc] Injection, Sodium iodohippurate [^{131}I] Injection, Sodium iodide [^{131}I] Capsules, Sodium iodide [^{131}I] Oral Solution, and Xenon [^{133}Xe]Injection.

The examples of diagnostic and therapeutic radionuclides and radiopharmaceuticals approved by China Food and Drug Administration (CFDA) are presented in Table 1 and Table 2, respectively. Currently, 11 kinds of ^{99m}Tc-labeled radiopharmaceuticals including sodium pertechnetate [^{99m}Tc] Injection (^{99m}TcO$_4^-$) approved by CFDA have been widely used in clinic. Besides the ten associated kits for preparation of ^{99m}Tc-radiopharmaceuticals in Table 1, there are another two kits approved by CFDA commercially available in China. One is the kit of sodium pyrophosphate (PYP) and stannous chloride for injection. Another is the kit of sodium gluceptate and stannous chloride for injection. Most of them are based on the methods reported in the international literature. Among ^{99m}Tc-radiopharmaceuticals, ^{99m}Tc-labeled NOET is in phase 3 clinical trial for myocardial perfusion imaging. More than 20 kinds of products including ^{99m}Tc-radiopharmaceuticals and associated kits are routinely available and their use in diagnosis is steadily increasing. For the production of ^{99m}Tc, two types of ^{99m}Tc-generators have been manufactured from fission products and ^{98}Mo(n,γ) ^{99}Mo nuclear reaction (gel type). The sold amount of ^{99}Mo-^{99m}Tc generators by BAHC has increased with time since 1990. For example, the radioactivity of 7013 Ci of ^{99}Mo for the production of ^{99}Mo-^{99m}Tc generators was supplied by BAHC in 2010. In 2011 and 2012, the supplied amount was increased to 7282 Ci and 8456 Ci, respectively. No doubt, ^{99m}Tc-imaging agents will continue to play an important role in the next decade in China.

Table 1 Example of diagnostic radionuclides and radiopharmaceuticals approved by CFDA[①]

Name	Applications
Sodium Pertechnetate [^{99m}Tc] Injection (^{99m}TcO$_4^-$)	Clinical diagnosis and preparation of ^{99m}Tc-radiopharmaceuticals
Technetium [^{99m}Tc] Bicisate Injection (^{99m}Tc-ECD)	Brain scintigraphy to delineate focal perfusion abnormalities
Technetium [^{99m}Tc] Methoxyisobutyl Isonitrile Injection (^{99m}Tc-MIBI)	Myocardial perfusion studies
Technetium [^{99m}Tc] Tetrofosmin Injection (^{99m}Tc-tetrofosmin)	Myocardial perfusion studies
Technetium [^{99m}Tc] Methylenediphosphonate Injection (^{99m}Tc-MDP)	Diagnosis of bone disease
Technetium [^{99m}Tc] Phytate Injection (^{99m}Tc-phytate)	Liver, spleen and bone marrow scintigraphy
Technetium [^{99m}Tc] Etifenin Injection (^{99m}Tc-etifenin)	Hepatobiliary scintigraphy
Technetium [^{99m}Tc] Pentetate Injection (^{99m}Tc-DTPA)	Renal studies, cerebral scintigraphy
Technetium [^{99m}Tc] Dimercaptosuccinate Injection (^{99m}Tc-DMSA)	Renal scintigraphy
Technetium [^{99m}Tc] Albumin Aggregated Injection	Lung perfusion scintigraphy
Technetium [^{99m}Tc] N, N'-Ethylenedicysteine Injection (^{99m}Tc-EC)	Examination of renal function
Sodium iodohippurate [^{131}I] injection	Diagnosis of kidney function
Fludeoxyglucose [^{18}F] Injection ([^{18}F]FDG)	Diagnosis of tumors and diseases in brain and heart by studying the status of glucose metabolism

① Data from http://www.sfda.gov.cn.

Table 2 Examples of therapeutic radionuclides and radiopharmaceuticals approved by CFDA[①]

Name	Applications
Sodium Phosphate [^{32}P] Oral Solution	Treatment of hypercythemia, neuropathic deaf, neuropathic dermatitis, etc
Colloidal Chromium Phosphate [^{32}P] Injection	Inhibition of thoracic ascites and assistant therapy of some malignant cancers
Strontium [^{89}Sr] Chloride Injection	Alleviation treatment of bone metastasis
Technetium [^{99}Tc] Methylenediphosphonate Injection ([^{99}Tc]MDP, Yunke)	Treatment of adjuvant arthritis
Sodium iodide [^{125}I] Solution	Preparation of labeled compounds
Iodine [^{125}I] Brachy Therapy Source	Possible treatment of tumors in head, neck, lung, liver, pancreas and prostate (early stages) with an applicator under the TPS system
Sodium Iodide [^{131}I] Oral Solution	Diagnosis and therapy of thyroid disease, preparation of labeled compounds
Sodium Iodide [^{131}I] Capsules	Therapy of thyroid disease
Iodine [^{131}I] Tumor Necrosis Therapy Monoclonal Antibody Injection	Therapy of lung cancer
Iodine [^{131}I] Metuximab Injection	Therapy of liver cancer
Skin Test Preparation for Iodine [^{131}I] Metuximab	Skin test for iodine [^{131}I] Metuximab
Samarium [^{153}Sm] lexidronam injection	Therapy of bone metastasis

① Data from http://www.sfda.gov.cn.

Besides the radiopharmaceuticals, synthetic facilities such as the synthetic modules of [^{18}F] FDG, and ^{11}C-labeled radiotracers are also commercially available in China. The above home-made modules are quite useful for development of new PET tracers in China. In addition, a dedicated PET for breast cancer detection with cylindrical geometry named PEMi has been developed. The initial performance results reflect a promising potential of this system for early breast cancer detection.

3 Important Advances in the Past Three Decades

3.1 Brain imaging agents

3.1.1 Brain perfusion imaging agents

It is well known that ^{99m}Tc-d, l-HMPAO and ^{99m}Tc-L, L-ECD have been approved by the U. S. Food and Drug Administration (FDA) for brain perfusion imaging agents. However, ^{99m}Tc-d, l-HMPAO is not stable *in vitro*. SPECT imaging of ^{99m}Tc-L, L-ECD may not accurately reflect rCBF at high flow rate. Therefore, investigation of brain perfusion imaging agents with suitable properties is still an interesting area in the past decade. A number of new ligands were designed and synthesized for this purpose[4-7]. For example, ^{99m}Tc-MPBDA showed higher stability than ^{99m}Tc-d, l-HMPAO and better retention ability than ^{99m}Tc-L, L-ECD[4-6]. [^{99m}TcN(IPEDTC)$_2$] also possessed good stability *in vitro*. Biodistribution studies in mice showed that this radiotracer exhibited high uptake and good retention in the brain with the brain/blood ratio 2.34 and 3.22 at 30min and 60min postinjection (p. i.), respectively. These results suggested that [^{99m}TcN(IPEDTC)$_2$] could be a potential brain perfusion imaging agent[7].

3.1.2 CNS imaging agents

3.1.2.1 Potential radiotracers for β-amyloid plaques in Alzheimer's disease

The population of China is now more than 1.3 billion, out of which about 9% are over the age of 65. Alzheimer's disease (AD) is the most common form of dementia. It affects up to 10% of people over the age of 65 and 30%-35% or more of those over the age of 85 years. AD is becoming an extensive health problem with the ever-increasing aging population in China. At present there are no effective treatments to reverse or halt the progression of the disease. The best hope might be to detect the disease at an early stage and initiate treatment before the brain damage is widespread. Although the etiology of AD has not been definitively established, the amyloid cascade hypothesis is most prevailing to explain the pathogenesis of AD, which postulates that the deposition of the Aβ peptide in the brain is a crucial step that ultimately leads to AD. Therefore, the development of novel imaging agents specifically targeting Aβ plaques may lead to early diagnosis of AD and monitoring of the effectiveness of novel therapies for this devastating disease.

More recently, [^{18}F]AV-45 (FLORBETAPIR F-18, brand-named AMYVID) and [^{18}F]GE-067 (FLUTEMETAMOL F-18, brand-named VIZAMYL) have been approved by FDA. Several other PET imaging probes for Aβ plaques such as [^{11}C]PIB and [^{18}F]BAY94-9172 (florbetaben) have been reported for clinical trials. In China, great efforts have been invested for the investigation of novel promising Aβ imaging agents. Novel radiotracers including ^{11}C-labeled derivatives, ^{18}F-labeled 2-phenylquinoxaline[8], 2-phenylindole[9], 2-pyridinylbenzoxazole and 2-pyridinylbenzothiazole derivatives[10], ^{99m}Tc-labeled benzothiazole[11,12], phenylbenzoxazole derivatives[13], 6-dialkylamino-2-naphthylethylidene derivatives[14], dibenzylideneacetone derivatives[15], and cyclopentadienyl tricarbonyl complexes mimicking chalcone[16], ^{125}I-labeled anilinophthalimide[17], N-benzoylindole[18], (E)-5-styryl-2, 2'-bithiophene derivatives[19], (E)-5-styryl-1H-indole and (E)-6-styrylquinoline derivatives[20] have been reported in the past decade. Among them, 5-(5-(2-[^{18}F]fluoroethoxy)benzo[d]oxazol-2-yl)-N-methylpyrmethylpyridin-2-amine is a promising Aβ imaging agent for PET and deserves further evaluation in human subjects[10]. It possesses high binding affinities for Aβ$_{1-42}$ aggregates with K_i value of (8.0 ± 3.2) nM. This tracer displayed high initial uptake with (7.23 ± 0.04)%ID/g at 2min p. i. Moreover, it displayed a brain$_{2min}$/brain$_{60min}$ ratio of 4.66, which is highly desirable for Aβ imaging agents. Target specific binding of this tracer to Aβ plaques was validated by *ex vivo* autoradiographic experiment with transgenic model mouse.

3.1.2.2 Potential radiotracers for sigma-1 receptors imaging

Sigma-1 (σ_1) receptors represent a distinct class of intracellular membrane proteins. It has been reported that σ_1 receptors regulate several targets, including various ion channels, G-protein-coupled receptors, lipids, and other signaling proteins by a direct protein-protein interaction. More and more evidence suggests that the σ_1 receptors are linked to a number of human diseases, including brain disorders, tumors as well as heart failure. Development of specific radiotracers for *in vivo* imaging of σ_1 receptors may provide useful diagnostic tools for investigation of their pathophysiology.

Many potential radiotracers for imaging σ_1 receptor expression with PET and SPECT have been reported in the past few years. But until now, only a few tracers such as [^{11}C]SA4503, [^{18}F]FPS, and [^{123}I]TPCNE have been evaluated in human studies. However, [^{18}F]FPS and [^{123}I]TPCNE were found to have irreversible kinetics. The *in vivo* [^{11}C]SA4503, as the first useful PET radiotracer, imaging studies showed that the density of σ_1 receptors was decreased in the brain of Parkinson's disease (PD) and AD patients. In order to develop radiotracers as σ_1 receptor imaging agents with suitable properties, ^{125}I-, ^{18}F-and ^{99m}Tc-labeled 4-benzylpiperazine[21-23] and spirocyclic piperidine derivatives[23-25] have been developed in China in cooperation with some foreign institutes. Among these, 1'-(4-(2-[^{18}F]fluoroethoxy)benzyl)-3H-spiro[2-benzofuran-1,4'-piperidine] was evaluated as a promising imaging agent for investigation of the σ_1 receptors in humans[25]. It exhibited low nanomolar affinity for σ_1 receptors and high subtype selectivity. Biodistribution studies in ICR mice indicated that this tracer displayed excellent initial brain uptake (8.07%ID/g at 2min), and slow washout. Pretreatment of animals with SA4503 resulted in a significant reduction in radiotracer uptake in organs known to contain σ_1 receptors. *Ex vivo* autoradiography in Sprague-Dawley rat demonstrated high accumulation of the radiotracer in brain areas known to express high levels of σ_1 receptors. Micro PET imaging and blocking studies confirmed the specific binding to σ_1 receptors *in vivo*. It also needs to be pointed out that ^{99m}Tc-labeled 4-benzylpiperazine complex has potential to be developed as ^{99m}Tc-labeled σ_1 receptor imaging agent. It possesses low nanomolar range affinity for σ_1 receptors, high brain uptake (3.25% ID/g at 15min) and specific binding to σ_1 receptors[23].

3.1.2.3 Potential radiotracers for dopamine D_2 or D_4 receptors imaging

For the CNS imaging agents, several radiotracers targeting other receptors have also been investigated in China. For example, Shen et al. reported the synthesis of ^{125}I-labeled dopamine D_2 receptor antagonists amino-substituted benzamide compounds[26,27]. The brain uptake of [^{125}I] AIBZM is relatively low due to its lower lipophilicity as compared to IBZM, IBF, epidepride and ioxipride. But this tracer possesses some advantages such as higher specific binding and lower unspecific binding than [^{123}I] IBZM. In addition, ^{18}F-labeled 3-{[4-(4-fluorobenzyl)] piperazin-1-yl} methyl-1H-pyrrolo[2,3-b]-pyridine[28], 3-(4-fluorobenzyl)-8-hydroxy-1,2,3,4-tetrahydrochromeno[3,4-c]pyridin-5-one (FHTP) and 3-(4-fluorobenzyl)-8,9-dimethoxy-1,2,3,4-tetrahydrochromenol[3,4-c]pyridin-5-one (FDTP)[29,30] were reported. Among them, the K_i value of FHTP for cloned human dopamine $D_{4.2}$ receptor was determined to be 2.9nM and FHDP displayed a 2000-fold D_4-selectivity over the D_{2long} subtype. Biodistribution, blocking distribution and metabolism studies in rats demonstrated specific distribution of [^{18}F]FHTP in brain regions, suggesting that [^{18}F]FHTP may be a suitable PET imaging agent for *in vivo* studies of the dopamine D_4 receptors.

3.2 Myocardial perfusion imaging agents

Coronary artery disease (CAD) is the main cause of death and remains a major health problem worldwide. Myocardial perfusion imaging is a well established noninvasive method for diagnosing CAD. Recently, Lin et al.[31] have provided a detailed review on the development of various radiotracers for myocardial perfusion imaging, including SPECT perfusion imaging agents and PET perfusion imaging agents and introduced novel basic research data related to this area. In addition, the current status in myocardial perfusion imaging agents and the future directions were discussed in that review. Therefore, we only give a brief introduction of this topic here.

3.2.1 ^{99m}Tc labeled myocardial perfusion imaging agents

In order to decrease the liphophilicity and increase the heart/liver ratio, [^{99m}Tc(CO)$_3$(CNR)$_3$]$^+$ complex was designed and prepared[32-34]. The results showed that the heart/liver ratio of [^{99m}Tc(CO)$_3$(MIBI)$_3$]$^+$ complex was much higher than that of the [^{99m}Tc(MIBI)$_6$]$^+$ complex. The contrastive imaging of [^{99m}Tc(CO)$_3$(MIBI)$_3$]$^+$ and [^{99m}Tc(MIBI)$_6$]$^+$ in dog showed better myocardium perfusion imaging achieved by the former complex[35]. Moreover, investigation of different synthetic conditions and the inter-transformation of Tc-CO-MIBI complexes was also performed[36]. In order to explain this phenomenon, the mechanism of the water substitution reactions and the carbonyl ligand exchanges by MIBI was proposed and investigated with Gaussian software[37]. The energy barriers for the water substitution for fac-[^{99m}Tc(CO)$_3$(H$_2$O)$_3$]$^+$ are less than 20kcal/mol, while the energy barriers for CO substitution reaction for

^{99m}Tc(CO)$_3$(MIBI)$_3$ are more than 30kcal/mol without any catalyst. Therefore it can be concluded that [^{99m}Tc(CO)$_3$(MIBI)$_3$]$^+$ was easily formed, while [^{99m}Tc(CO)$_x$(MIBI)$_{6-x}$]$^+$ (x = 2,1,0) complexes were not obtained under acidic condition. Later, the corresponding carrier-added [^{99}Tc-CO-MIBI]$^+$ complexes were synthesized and the structures were confirmed by LC-MS analyses. The proposed mechanism of water exchange with MIBI on fac-[^{99m}Tc(CO)$_3$(H$_2$O)$_3$]$^+$ could explain the complex formation with [^{99m}Tc/^{99}Tc(CO)$_3$(MIBI)$_3$]$^+$ under acidic condition. The CO exchange mechanism on [^{99m}Tc/^{99}Tc(CO)$_3$(MIBI)$_3$]$^+$ by CNR with OH$^-$ ion as a catalyst needs to be investigated further.

3.2.2 ^{18}F-labeled myocardial perfusion imaging agents

It is well known that PET possesses several technical advantages over SPECT for cardiac imaging, including better spatial and temporal resolution, robust attenuation correction and listmode data acquisition. Nowadays obesity prevalence increases and poses a challenge for image quality of CAD. Therefore, it is necessary to develop PET radiotracers for myocardial perfusion imaging (MPI) in order to improve diagnostic accuracy, reader confidence and versatility. Mitochondria take up about 20%-30% of the myocardial intracellular volume in the heart. Therefore, development of mitochondrion-targeted ^{18}F-labeled analogs of MC-I inhibitors as potential MPI agents is of great interest.

Recently, analogs of mitochondrial complex I (MC-I) inhibitors (such as [^{18}F]FDHR, [^{18}F]RP1003/04/05, and 2-tert-butyl-4-chloro-5-[4-(2-[^{18}F]-fluoroethoxymethyl)-benzyloxy]-2H-pyridazin-3-one(BMS-747158-02)) have been reported as potential MPI agents. Up to now, BMS-747158-02 (also known by the nonproprietary name flurpiridaz) has been the best evaluated ^{18}F-labeled MPI agent, which has entered clinical development (phase 3)[38]. The promising properties of BMS 747158-02 will ensure that cardiac PET will serve the standard for evaluation of myocardial perfusion in coming years.

Several classes of MC-I inhibitors have been reported, such as fenazaquin, S-chromone, tebufenpyrad, and pyridaben. All have the same binding site of the MC-I enzyme as rotenone. Recently, Mou et al. reported three pyridaben analogs ([^{18}F]FP1OP, [^{18}F]FP2OP, and [^{18}F]FP3OP)[39,40]. Among these, [^{18}F]FP1OP seems to be a potential MPI agent[40]. It showed high heart uptake and low background uptake in both the mouse model and the Chinese mini swine model. It possessed lower initial liver uptake than BMS-747158-02, which made it possible to perform imaging at earlier time with minimum injection dose. The fast myocardium clearance of [^{18}F]FP1OP could lower the radiation dose and make it a good choice for stress and rest imaging in one-day protocol. On the basis of the promising biological properties, [^{18}F]FP1OP warrants further investigation as an MPI agent.

3.3 Tumor imaging agents

3.3.1 Integrin-targeting radiotracer ^{99m}Tc-3PRGD2

Integrins have been identified as an important family of transmembrane receptors which were found to play essential roles in angiogenesis and tumor metastasis. Among this family, the integrin $\alpha_v\beta_3$ receptor was found to express preference on various types of tumor cells and activated endothelial cells of tumor angiogenesis with very little influence on the quiescent vessel cells and other normal cells. Therefore, the integrin $\alpha_v\beta_3$ receptor is becoming a valuable target for diagnosis and treatment of malignant tumors. Since tripeptide sequence of arginine-glycine-aspartic acid (RGD) showed specific binding to the integrin $\alpha_v\beta_3$ receptor, a variety of radiolabeled RGD-based peptides have been developed in the past few years. Up to date, [^{18}F]galacto-RGD and [^{18}F]AH111585 have been well investigated in clinical trials. ^{99m}Tc-NC100692, a ^{99m}Tc-labeled cyclic peptide that contains a monomer RGD tripeptide sequence, has proved to be feasible for detection of breast and lung cancers in human beings. In China, a series of ^{99m}Tc-labeled RGD dimeric peptides with different linkers have been studied[41-48]. Among these, ^{99m}Tc-3PRGD2 represents the best one investigated until now. It was designed as a dimeric cyclic RGD peptide that inserted 2 PEG4 between the 2 RGD motifs, which was expected to possess significantly increased binding affinity compared with a cyclic RGD monomer. It exhibited high tumor uptake in mouse models, nonhuman primates and patients. In addition, its synthetic procedure is simple, efficient, and reproducible, allowing a kit formulation and the easy availability for routine clinical use. Thus, ^{99m}Tc-3PRGD2 has recently been translated into clinical trials in China[47].

3.3.2 Novel sentinel lymph node imaging agents

Currently, novel sentinel lymph node (SLN) imaging agent, ^{99m}Tc-Rituximab[49-53], initially developed in China, has been used in clinic. ^{99m}Tc-Rituximab, prepared by introducing ^{99m}Tc into 2-mercaptoethanol modified Rituximab with a yield of 95%, can show SLNs status in both mice and patients. Dynamic SLNs mapping results in mice showed clear SLNs imaging from 10min to 16h p.i. with no detectable secondary lymph nodes. The uptake of SLNs was 3.0% to 3.5%. In patients, both the sensitivity and accuracy of sentinel lymph nodes biopsy (SLNB) were 100%. ^{99m}Tc-Rituximab can also be prepared from 2-iminothiolane (2-IT) modified Rituximab. Dynamic SLN mapping results showed clear imaging of SLNs from 30min to 24h p.i. with no higher order nodes visualization. In biodistribution studies, the uptake in SLNs was 4.49% at 24h. Nearly no uptakes were seen in secondary and tertiary lymph nodes. When ^{99m}Tc-Rituximab together with patent blue was used for SLNB, the sensitivity, the specificity and accuracy of SLNB were 97% (30/31), 100% (51/51) and 99% (81/82), respectively. The false negative rate was only 3% (1/30). The negative and positive predictive values reached 98% (51/52) and 100% (31/31), respectively. More recently, the freeze-dried kit containing 2-mercaptoethanol modified rituximab for ^{99m}Tc-labeling was developed. Biodistribution studies of ^{99m}Tc-Rituximab prepared from the kit in normal rats *via* front pad injection showed that the accumulation in the sentinel lymph node increased from 1h (0.93% ID) to 4h (2.14% ID). The ratio of the uptake in the sentinel lymph node to that in the injected site kept no significant change with 14.8 at 4h and 13.03 at 18h. SPECT imaging in rat demonstrated clear accumulation in the sentinel lymph node with no secondary lymph node imaging at 30min to 18h p.i. The kit for the synthesis of ^{99m}Tc-Rituximab is available for its clinical applications in China.

3.3.3 Potential hypoxia imaging agents

Hypoxia is a common characteristic of many solid tumors. Hypoxia in tumors decreases the efficacy of radiotherapy and chemotherapy of cancer. Researchers have made lots of efforts to develop hypoxia markers for noninvasive hypoxia imaging. [^{18}F]FMISO is considered to be the gold standard for tracing hypoxia with PET. Recently, Li and Chu[54] have provided an overview of the radionuclide labeled hypoxia-imaging agents developed in the past decade. In that review, the radionuclide labeled bio-reducible compounds including nitroimidazoles and other redox-sensitive compounds such as Tc-BnAO and Cu-ATSM are summarized. Hypoxia-imaging agents used to detect hypoxia with endogenous hypoxia markers such as carbonic anhydrase IX (CA IX) and hypoxia-inducible factor-1 (HIF-1) were introduced. In addition, the effect of a second redox center on the selectivity, multi-modality molecular hypoxia-imaging agents, as well as the problems related to the hypoxia marker development were discussed.

In China, great efforts have been devoted to the investigation of novel promising hypoxia marker. The reported novel radiotracers include ^{131}I-labeled nitroimidazole analogues[55] and N-[4-(benzothiazol-2-yl)phenyl]-11-(2-nitroimidazole-1-yl) undecanamide (2NUBTA)[56], ^{99m}Tc-labeled nitroimidazole derivatives[57-60], nitrobenzoimidazole and nitrotriazole derivatives[61], as well as ^{99m}Tc-BnAO derivatives[62]. Among them, ^{99m}Tc-N4IPA showed accumulation in S180 tumor and slow clearance from it. The tumor-to-tissue uptake ratios increase with time, suggesting that ^{99m}Tc-N4IPA would be a marker for imaging tumor hypoxia[57]. Moreover, effects of a second nitroimidazole or nitrotriazole redox centre on the accumulation of a hypoxia marker were explored[63-65]. The findings indicated that a second nitroimidazole redox centre with appropriate reduction potential might play an important role in the hypoxic accumulation. The compounds containing multi-redox centres are worthy of further investigations and may serve as potential hypoxia markers in the future.

3.3.4 ^{99m}Tc-labeled glucose derivatives

To seek novel ^{99m}Tc-labeled glucose derivatives as tumor imaging agents, Zhang et al. recently reported the synthesis of deoxyglucose dithiocarbamate (DGDTC) and ^{99m}Tc-labeled ^{99m}TcN-DGDTC[66] and ^{99m}TcO-DGDTC[67] complexes for tumor imaging. The ^{99m}TcN-DGDTC complex was prepared in high yield through a ligand-exchange reaction, which can be easily used for the preparation of a radiopharmaceutical through a freeze-dried kit formulation. Biodistribution studies in mice showed that the complex accumulated in the tumor with good retention. However, the tumor uptake of the ^{99m}TcN-DGDTC ((1.16 ± 0.57)%ID/g) is lower than that of ^{99m}Tc-DTPA-DG ((1.59 ± 0.04)%ID/g) and [^{18}F]FDG ((1.42 ± 0.12)%ID/g). ^{99m}TcO-DGDTC was very stable at room temperature over a period of 6h. *In vitro* cell studies showed that there was an increase in the uptake of ^{99m}TcO-DGDTC as a function of incubation time and the cellular uptake of ^{99m}TcO-DGDTC was possibly mediated by a way of D-glucose mechanism. The biodistribution of ^{99m}TcO-DGDTC in S180 tumor bearing mice showed that the complex accumulated in the tumor with good uptake and excellent retention. Compared with other reported ^{99m}Tc-labeled glucose derivatives, ^{99m}TcO-DGDTC showed the highest tumor uptake and good tumor/muscle ratios. SPECT image studies showed that there was a visible accumulation in tumor sites, suggesting that ^{99m}TcO-DGDTC would be a promising candidate for tumor imaging.

3.3.5 Sigma-2 receptor tumor imaging agents

Progress in cancer biology has revealed that both σ_1 and σ_2 receptors are over-expressed in many types of cancer. The density of σ_2 receptor is significantly higher in proliferating tumor cells than that in quiescent tumor cells. Therefore, the expression of σ_2 receptor protein is assumed to be a suitable biomarker of cellular proliferation in solid tumors. Accordingly, novel σ_2 receptor radiotracers were investigated for tumor imaging.

First, a novel ^{99m}Tc-labeled complex, [N-[2-((2-oxo-2-(4-(3-phenylpropyl) piperazin-1-yl)ethyl)(2-mercaptoethyl) amino) acetyl]-2-aminoethanethiolato] technetium (V) oxide (PPPE-MAMA'-^{99m}TcO) and the corresponding rhenium complex (PPPE-MAMA'-ReO) have been designed and prepared based on the integrated approach[68]. The radiotracer of PPPE-MAMA'-^{99m}TcO possessed moderate affinity for σ receptors and retained certain tumor uptake in MCF-7 human breast tumor bearing mice. Later, [(Cp-R)^{99m}Tc(CO)$_3$] core was applied to design novel ^{99m}Tc-labeled receptor-targeting radiotracers with retention of high affinity and selectivity[69,70]. A novel ^{99m}Tc-labeled 4-(4-cyclohexylpiperazine-1-yl)-butan-1-one-1-cyclopentadienyltricarbonyl technetium based on the lead compound PB28 was synthesized and evaluated as a potential SPECT tracer for imaging of σ_2 receptors in tumors. The corresponding Re complex showed relatively high affinity towards σ_2 receptors with K_i value of (64.4 ± 18.5) nM and moderate selectivity vs. σ_1 receptors ($K_i\sigma_1/K_i\sigma_2$ ratio was 12.5). Biodistribution studies in mice revealed comparably high initial brain uptake of radiotracer and slow washout. Administration of haloperidol 5min prior to injection of the radiotracer significantly reduced the radiotracer uptake in brain, heart, lung, and spleen by 40%-50% at 2h p.i. Moreover, this radiotracer showed high uptake in C6 glioma cell lines (8.6%) after incubation for 1h. Blocking with haloperidol significantly reduced the cell uptake. Preliminary blocking study in C6-brain-tumor bearing rats showed that this radiotracer was bound to σ receptors in the brain-tumor specifically. Moreover, another ^{99m}Tc-labeled complex based on the lead compound RHM-1 possessed nanomolar affinity and moderate selectivity for σ_2 receptors[70]. These results are encouraging for further exploration of ^{99m}Tc-labeled probes for σ_2 receptor tumor imaging *in vivo*.

3.3.6 Potential radiotracers for folate receptor imaging agents

Folate receptor (FR) is overexpressed in many types of human tumors. Folate-based targeting systems provide an effective means of selectively delivering therapeutic or imaging agents to tumors. ^{99m}Tc labeled folate conjugates[71,72] and folate derivatives[73] have been investigated in China. Among them, ^{99m}Tc(HYNIC-NHHN-FA)(tricine/TPPTS) displayed high KB cell binding. Biodistribution in KB tumor bearing athymic nude mice showed that this radiotracer had high uptake in FR-positive tumor ((9.79 ± 1.66)%ID/g at 4h p.i.), and the blocking studies confirmed the specific accumulation of the radiotracer *in vivo*. Moreover, ^{99m}Tc-dendrimer poly(amido)-amine folic acid conjugates have also been reported[74,75]. The ^{99m}Tc-labeled PE-

Gylated dendrimer PAMAM folic acid conjugate (^{99m}Tc-G5-Ac-pegFA-DTPA) showed much higher uptake in KB cancer cells. Biodistribution in KB tumor bearing nude mice exhibited specific accumulation of the radiotracer in tumors, which was in agreement with the results of micro-SPECT imaging. These findings suggest that PEGylation of PAMAM dendrimer folic acid conjugate could improve the tumor targeting[75].

3.3.7 Potential radiotracers for imaging of human telomerase reverse transcriptase expression

The expression of human telomerase reverse transcriptase (hTERT) is present in most malignant cells (more than 85%) while it is undetectable in most normal cells. Visualization of hTERT expression using radionuclide labeled tracers can provide important diagnostic information and prognostic value in malignant tumors. Recently, ^{131}I-and ^{99m}Tc-labeled antisense oligonucleotide (^{99m}Tc-MAG$_3$-ASON) targeting hTERT mRNA have been reported[76,77]. The radiotracer preserved the capacity to bind living hTERT-expressing cells specifically and to inhibit the expression of hTERT mRNA significantly as well as ASON. In nude mice bearing hTERT-expressing MCF-7 xenografts, tumor uptake of ^{99m}Tc-MAG$_3$-ASON was significantly higher than that of ^{99m}Tc-MAG$_3$-SON. The hTERT-expressing xenografts were clearly imaged at 4-8h noninvasively after injection of ^{99m}Tc-MAG$_3$-ASON, whereas the xenografts were not imaged at any time after injection of ^{99m}Tc-MAG$_3$-SON. The above findings suggested that ^{99m}Tc-MAG$_3$-ASON can be used as a potential candidate for visualization of hTERT expression in carcinomas. Later, ^{99m}Tc-labeled duplex siRNA (^{99m}Tc-hTERT siRNA) was developed[78]. In HepG2 tumor-bearing mice, significantly higher accumulation of radiotracer in tumors and a higher tumor-to-blood ratio were observed compared to that of control siRNA. Scintigraphy of ^{99m}Tc-hTERT siRNA showed clear tumor images at 0.5h, 1h, 3h, and 6h p.i. Ratios of uptake in tumor to that in contralateral region of hTERT-targeted siRNA were significantly higher than those of control siRNA at each time point. Therefore, ^{99m}Tc-hTERT siRNA could be used for noninvasive visualization of siRNA delivery *in vivo*.

3.4 Infection imaging agents

Infection is an important problem that needs accurate and prompt diagnosis for early management to avoid serious complications. Currently, radiolabeled leukocytes are still the optimum radiopharmaceutical agents used in the diagnosis of focal bacterial infection and inflammation. However, this technique is time-consuming, needs a sterile environment, and has risk associated with handling of potentially contaminated blood. Thus, there is still great interest in the development of new radiopharmaceuticals for infection imaging.

Zhang et al. reported the synthesis and evaluation of lipophilic and neutral complexes as potential bacteria-specific infection imaging agents[79-81]. The bacterial binding assay showed that ^{99m}TcN-NFXDTC had a good binding affinity. Biodistribution results in bacterially infected mice and in turpentine-induced abscesses mice indicated that ^{99m}TcN-NFXDTC was able to discriminate infection and sterile inflammation. It showed a higher uptake at the sites of infection and better abscess/blood and abscess/muscle ratios than those of ^{99m}Tc-ciprofloxacin and ^{99m}TcN-

CPFXDTC. SPECT image studies showed a visible accumulation at infection sites, suggesting that it would be a promising candidate for bacterial infection imaging.

3.5 Potential asialoglycoprotein receptor imaging agents

Asialoglycoprotein receptors (ASGP-Rs) are well known to exist in the mammalian liver, situated on the surface of hepatocyte membrane, and participate in the hepatic metabolism of serum proteins. Quantitative imaging of ASGP-Rs could estimate the function of the liver, a unique way to noninvasively diagnose disease. ASGP receptor imaging agent can assess the anatomy and function of the liver, help the early diagnosis of hepatic diseases and accurate evaluation of functional status. Noninvasive quantification of the liver uptake of radiolabeled ASGP can provide a unique way to estimate the liver function especially in terms of spatial distribution of hepatic function. In China, novel radiotracers including ^{18}F-labeled galactosyl-neoglycoalbumin[82] and galactosylated-chitosan[83], ^{99m}Tc-labeled DMP-NGA[84], and copolymer[85] have been reported recently. Among them, copolymer-based ^{99m}Tc[P(VLA-co-VNI)](tricine)$_2$ is a potential hepatic ASGP receptor imaging agent and warrants further evaluation. It displayed high liver uptake with (125.33 ± 10.99) %ID/g at 10min p.i. in mice. The accumulation of tracer in liver could be blocked significantly by preinjection of free neogalactosylalbumin or P(VLA-co-VNI). SPECT images with high quality were obtained at 15min, 30min, 60min, and 120min after injection of the radiotracer. Significant radioactivity defect was observed in the liver cancer model. The promising biological properties of ^{99m}Tc[P(VLA-co-VNI)](tricine)$_2$ afford potential applications for the assessment of hepatocyte function in the future.

3.6 Other imaging agents

Besides the radiopharmaceuticals mentioned above, some radiotracers targeting other organs including ^{99m}Tc-BPHA for renal function[86], ^{99m}Tc-labeled fatty acid[87,88], ^{99m}Tc-labeled RGD-BBN peptide[89], ^{99m}Tc-labeled radiotracers for apoptosis imaging[90-92], ^{99m}Tc-labeled zoledronic acid derivatives for bone imaging agents[93-99] have been reported. In the past two decades, ^{131}I-radiopharmaceuticals have also been investigated. However, most of the ^{131}I-radiopharmaceuticals for diagnosis have been or will be replaced by ^{123}I- or ^{99m}Tc-radiopharmaceuticals in the future.

3.7 Therapeutic radiopharmaceuticals

Several therapeutic radionuclides and radiopharmaceuticals are being manufactured in China according to standard methods. Some of them have already been in clinical use or have been put into trials, e.g., [^{125}I]octreotide, [^{125}I]UdR, ^{125}I and ^{103}Pd seeds, ^{90}Y and ^{32}P glass microspheres, [^{153}Sm]EDTMP, [^{186}Re]HEDP, etc. Many low-energy β^- emitters, α emitters, as well as Auger and conversion electron emitters are ideal for therapy of diseases such as cancer, the rheumatoid arthritis and the palliation of pain associated with metastatic bone cancer. The large market requirements will greatly promote the researches of therapeutic radiopharmaceuticals.

^{177}Lu, a low-energy β$^-$ emitter, is a therapeutic radionuclide suitable for the treatment of small and metastatic tumors. ^{177}Lu-labeled radiopharmaceuticals have recently become of great interest. ^{177}Lu-labeled EDTMP and DOTMP[100-102], cyclic RDGfK dimer[103], DOTA/DTPA-Bz-Cys-RGD dimer[104], DOTA-Bz-RGD dimer and DOTA-Bz-PEG4-RGD dimer[105], DOTA-Bz-Cys-RGD tetramer and DOTA-RGD tetramer[106], DTPA-BIS-Biotin[107,108], as well as knottin peptides[109] have been reported within China.

Regarding the α-emitters, ^{211}At is attractive for endoradiotherapeutic applications. Liu et al. summarized the recent progress of ^{211}At-labeled radiopharmaceuticals[110]. ^{211}At-labeled astatomethyl-19-norcholest-5(10)-en-3β-ol[111], 6-^{211}At-cholesterol[112], IgG[113], monoclonal antibody (McAb) and its Fab fragment[114,115], protein[116,117], insulin[118], and amidobisphophonates[119] have been investigated.

3.8 Basic research on radiopharmaceutical chemistry

Together with the development of novel radiopharmaceuticals, basic research on radiopharmaceutical chemistry including labeling methods and theoretical studies has also been going on in China. For example, direct and indirect labeling methods of McAbs with radionuclides including ^{90}Y, ^{99m}Tc, ^{111}In, ^{111}Ag, ^{169}Yb, ^{188}Re and ^{199}Au were investigated[120]. In addition, some rapid isotope exchange methods including isotopic exchange by hydrothermal melt method[121], in the pure crown ether medium[121,122] and solid phase[123] have been developed. A general isotopic exchange law was derived for I exchanging with Br, Cl and At[124]. The homogenous exponential isotopic exchange law is only its special case.

In addition, some basic technetium chemistry research has also been carried out. Quantitative study of the structure-stability relationship of Tc complexes with the cone packing model was investigated[125] and a general rule of stability of ^{99m}Tc-complexes was established. Ligand exchange reaction results, monitored by HPLC, showed that actual direction of the ligand exchange reaction was consistent with the predicted order of SAS (Solid Angle Factor Sum)[126]. Structure-activity relationship of ^{99m}Tc-radiopharmaceuticals for brain, myocardium and receptors were also investigated[127,128]. Moreover, reaction mechanism of technetium complexes was studied[129,130]. These fundamental researches were very useful in the design of new ^{99m}Tc-radiopharmaceuticals. Together with technetium chemistry, considerable progress has also been achieved in Rechemistry.

4 Future Trends

From the international point of view, considerable progress has been achieved in the area of radiopharmaceuticals in the past decades. More and more new concepts emerged with the radiopharmaceutical developments. For example, a combination of PET and internal radiotherapy was shown by the Juelich group using the radionuclide pair ^{86}Y/^{90}Y[131]. Srivastava further discussed this concept about the dual-purpose theragnostic radionuclides or radionuclide pairs with emission suitable for both pretherapy low-dose imaging and higher-dose therapy in the same patient[132]. Interesting theragnostic pairs are ^{124}I/^{131}I, ^{44}Sc/^{47}Sc, ^{64}Cu/^{67}Cu, ^{86}Y/^{90}Y, and ^{83}Sr/^{89}Sr. The methods of production of the corresponding positron emitters have been recently reviewed[133]. A low-dose molecular imaging (SPECT/CT, PET/CT) could provide the necessary information on biodistribution, dosimetry, the limits on critical organ or tissues, and the maximum tolerated dose (MTD), etc. It would be safe and appropriate to follow up with the dose ranging experiment to allow higher-dose molecular therapy in the same patient with the greatest effectiveness using the same radiopharmaceuticals. Based on the current status of radiopharmaceuticals in China discussed above, important future trends might include the following aspects. (1) Production of medical radionuclides using both reactors and accelerators; (2) Investigations on the basic radiopharmaceutical chemistry; (3) Development of receptor-based imaging agents; (4) Development of multi-modality imaging probes, etc.

In order to meet the basic requirements of the development of radiopharmaceuticals both for diagnosis and therapy, the production of the most important medical radionuclides, including ^{99}Mo (^{99m}Tc), ^{131}I, $^{188/186}$Re and ^{123}I, need to be emphasized in China. In addition, novel therapeutic radionuclides including low-energy β$^-$ emitters and α-particle emitters as well as Auger and conversion electron emitters also need to be paid more attentions[134].

In order to accelerate the development of novel radiotracers, investigations on basic radiopharmaceutical chemistry should be emphasized. Efforts related to the basic chemistry of the very important medical radionuclides, such as the basic fluorination chemistry and technetium chemistry, need to be enhanced. As to the basic fluorination chemistry, efforts might be directed toward finding new ^{18}F-fluorination methods such as click reaction and microfluidic technique. For the ^{99m}Tc-radiopharmaceuticals, future trends might go in the directions of adding new targets (new specific biochemical processes as targets), finding new targeting molecules, developing new cores (basic technetium chemistry), and selecting new proper linker between the technetium core and targeting molecule[135]. For example, ^{99m}Tc is an Auger emitter and could be applied in therapy if it could be delivered to the cell nucleus. The cell nucleus or the DNA of malignant cells might be an important future target. The recently developed small *fac*-[^{99m}TcO$_3$]$^+$ core might have a high potential for Tc radiopharmaceutical applications[136]. Moreover, the recent broadened concept of "multifunctional ligands" is also useful to develop novel ^{99m}Tc-radiopharmaceuticals with suitable biological properties.

For the purpose of promoting the development of radiopharmaceuticals in China, many new efforts will be made including researches and development of CNS receptor binding radiopharmaceuticals with high affinity and selectivity, small peptide imaging agents, and gene expression imaging agents. In order to facilitate and accelerate the discovery of new molecular probes in the radiopharmaceutical development process, the investigators in the radiopharmaceutical area will take advantage of the recent developments in the field of combinational chemistry, computer-aided drug design, click chemistry and micro-flow reaction. We expect that all our efforts in the radiopharmaceutical area will improve and enhance the quality of human life and health care in China.

Acknowledgment

The authors are grateful to Prof. Luo Zhi-Fu and Beijing Atom High Tech Co., Ltd. for providing the data on ^{99}Mo-^{99m}Tc generators. All the authors cited in this paper are highly appreciated. This work was supported by the National Natural Science Foundation of China (No. 21071023).

Abbreviations

BPHA: N, N'-bis(2-aminoethyl)propanediamine hexaacetic acid; CPFXDTC: ciprofloxacin dithiocarbamate; HMPAO: hexamethylpropyleneamine oxime. IPEDTC: N-isopentyl dithiocarbamato; MPBDA: N-(2-mercapto-propyl)-1,2-benzenediamine; NFXDTC: norfloxacin dithiocarbamate; P (VLA-co-VNI): Poly (N-p-vinylbenzyl-[O-b-D-galactopyranosyl-(1/4)-D-gluconamide]-co-N-p-vinylbenzyl-6-[2-(4-dimethylamino)benzaldehydehydrazono]nicotinate).

References

[1] Cui, H., Wang, G., Zhang, H.: Target system of cyclone-30 accelerator and its target preparation. Atom. Energ. Sci. Technol. (Chinese) 30, 46(1996).

[2] Liu, N., Yang, Y., Jin, J., Lin, R., Cao, Y., Liao, J., Liao, X.: Preparation of radioactive isotopes by CS-30 cyclotron and their applications. J. Isot. (Chinese) 25, 189(2012).

[3] Chinese Society of Nuclear Medicine: Survey on the status of nuclear medicine in China in 2012. Chin. J. Nucl. Med. (Chinese) 32, 357(2012).

[4] Miao, Y., Liu, B.: Synthesis of new N_2S ligands, preparation of ^{99m}Tc complexes and their preliminary biodistribution in mice. J. Labelled Compds. Radiopharm. 42, 629(1999).

[5] Miao, Y., Liu, B.: A new potential cerebral perfusion imaging agent. SPECT study in monkey. Quarterly J. Nucl. Med. 42, 41(1998).

[6] Wang, R., Zhang, C., Zhu, S., Miao, Y., Tang, Z., Liu, B.: Preparation and animal studies of a novel potential cerebral perfusion imaging agent. Chin. J. Nucl. Med. (Chinese) 21, 118(2001).

[7] Zhang, J., Wang, X., Li, C.: Synthesis and biodistribution of a new ^{99m}Tc nitrido complex for cerebral imaging. Nucl. Med. Biol. 29, 665(2002).

[8] Yu, P., Cui, M., Wang, X., Zhang, X., Li, Z., Yang, Y., Jia, J., Zhang, J., Ono, M., Saji, H., Jia, H., Liu, B.: ^{18}F-Labeled 2-phenylquinoxaline derivatives as potential positron emission tomography probes for *in vivo* imaging of β-amyloid plaques. Eur. J. Med. Chem. 57, 51(2012).

[9] Fu, H., Yu, L., Cui, M., Zhang, J., Zhang, X., Li, Z., Wang, X., Jia, J., Yang, Y., Yu, P., Jia, H., Liu, B.: Synthesis and biological evaluation of ^{18}F-labled 2-phenylindole derivatives as PET imaging probes for β-amyloid plaques. Bioorg. Med. Chem. 21, 3708(2013).

[10] Cui, M., Wang, X., Yu, P., Zhang, J., Li, Z., Zhang, X., Yang, Y., Ono, M., Jia, H., Saji, H., Liu, B.: Synthesis and evaluation of novel ^{18}F labeled 2-pyridinylbenzoxazole and 2-pyridinylbenzothiazole derivatives as ligands for positron emission tomography(PET) imaging of β-amyloid plaques. J. Med. Chem. 55, 9283(2012).

[11] Chen, X., Yu, P., Zhang L., Liu, B.: Synthesis and biological evaluation of ^{99m}Tc, Re-monoamine-monoamide conjugated to 2-(4-aminophenyl)benzothiazole as potential probes for β-amyloid plaques in the

brain. Bioorg. Med. Chem. Lett. 18,1442(2008).

[12] Cui,M. C. ,Li,Z. J. ,Tang,R. K. ,Liu,B. L. :Synthesis and evaluation of novel benzothiazole derivatives based on the bithiophene structure as potential radiotracers for β-amyloid plaques in Alzheimer's disease. Bioorg. Med. Chem. 18,2777(2010).

[13] Wang,X. ,Cui,M. ,Yu,P. ,Li,Z. ,Yang,Y. ,Jia,H. ,Liu,B. :Synthesis and biological evaluation of novel technetium-99m labeled phenylbenzoxazole derivatives as potential imaging probes for β-amyloid plaques in brain. Bioorg. Med. Chem. Lett. 22,4327(2012).

[14] Cui, M. C. , Tang, R. K. , Li, Z. J. , Ren, H. Y. , Liu, B. L. : Tc-99m-and Re-labeled 6-dialkylamino-2-naphthylethylidene derivatives as imaging probes for β-amyloid plaques. Bioorg. Med. Chem. Lett. 21, 1064(2011).

[15] Yang,Y. ,Cui,M. ,Jin,B. ,Wang,X. ,Li,Z. ,Yu,P. ,Jia,J. ,Fu,H. ,Jia,H. ,Liu,B. :^{99m}Tc-labeled dibenzylideneacetone derivatives as potential SPECT probes for in vivo imaging of β-amyloid plaque. Eur. J. Med. Chem. 64,90(2013).

[16] Li,Z. ,Cui,M. ,Dai,J. ,Wang,X. ,Yu,P. ,Yang,Y. ,Jia,J. ,Fu,H. ,Ono,M. ,Jia,H. ,Saji,H. ,Liu, B. :Novel cyclopentadienyl tricarbonyl complexes of ^{99m}Tc mimicking chalcone as potential single-photon emission computed tomography imaging probes for β-amyloid plaques in Brain. J. Med. Chem. 56, 471 (2013).

[17] Duan,X. H. ,Qiao,J. P. ,Yang,Y. ,Cui,M. C. ,Zhou,J. N. ,Liu,B. L. :Novel anilinophthalimide derivatives as potential probes for β-amyloid plaque in the brain. Bioorg. Med. Chem. 18,1337(2010).

[18] Yang,Y. ,Duan,X. H. ,Deng,J. Y. ,Jin,B. ,Jia,H. M. ,Liu,B. L. : Novel imaging agents for β-amyloid plaque based on the N-benzoylindole core. Bioorg. Med. Chem. Lett. 21,5594(2011).

[19] Cui,M. C. ,Li,Z. J. ,Tang,R. K. ,Jia,H. M. ,Liu,B. L. :Novel(E)-5-styryl-2,2'-bithiophene derivatives as ligands for β-amyloid plaques. Eur. J. Med. Chem. 46,2908(2011).

[20] Yang,Y. ,Jia,H. M. ,Liu,B. L. : (E)-5-styryl-1H-indole and (E)-6-styrylquinoline derivatives serve as probes for β-amyloid plaques. Molecules 17,4252(2012).

[21] Li, Z. J. , Ren, H. Y. , Cui, M. C. , Deuther-Conrad, W. , Tang, R. K. , Steinbach, J. , Brust, P. , Liu, B. L. ,Jia,H. M. :Synthesis and biological evaluation of novel 4-benzylpiperazine ligands for sigma-1 receptor imaging. Bioorg. Med. Chem. 19,2911(2011).

[22] Wang,X. ,Li,Y. ,Deuther-Conrad,W. ,Xie,F. ,Chen,X. ,Cui,M. C. ,Zhang,X. J. ,Zhang,J. M. ,Steinbach,J. ,Brust,P. ,Liu,B. L. ,Jia,H. M. :Synthesis and biological evaluation of ^{18}F labeled fluoro-oligo-ethoxylated 4-benzylpiperazine derivatives for sigma-1 receptor imaging. Bioorg. Med. Chem. 21, 215 (2013).

[23] Wang,X. ,Li Y. ,Deuther-Conrad,W. ,Li,D. ,Cui,M. ,Steinbach,J. ,Brust,P. ,Liu,B. ,Jia,H. :Synthesis and biological evaluation of two novel ^{99m}Tc cyclopentadienyl tricarbonyl complexes for sigma-1 receptor imaging. J. Labelled Compd. Radiopharm. 56,S77(2013).

[24] Chen, R. Q. , Li, Y. , Zhang, Q. Y. , Jia, H. M. , Deuther-Conrad, W. , Schepmann, D. , Steinbach, J. , Brust,P. ,Wünsch,B. ,Liu,B. L. :Synthesis and biological evaluation of a radioiodinated spiropiperidine ligand as a potential σ_1 receptor imaging agent. J. Labelled Compd. Radiopharm. 53,569(2010).

[25] Li,Y. ,Wang,X. ,Zhang,J. ,Deuther-Conrad,W. ,Xie,F. ,Zhang,X. ,Liu,J. ,Qiao,J. ,Cui,M. ,Steinbach,J. ,Brust,P. ,Liu,B. ,Jia,H. :Synthesis and evaluation of novel ^{18}F-labeled spirocyclic piperidine derivatives as σ_1 receptor ligands for positron emission tomography imaging. J. Med. Chem. 56, 3478 (2013).

[26] Shen,M. ,Gong,J. ,Wang,Q. ,Zhu,T. ,Liu,B. ,Meng,Z. ,Tang,Z. :Synthesis of dopamine D_2 receptor

antagonists amino-substituted benzamide compounds. Nucl. Tech. (Chinese)18,651(1995).

[27] Li,H. F. ,Gildehaus,F. J. ,Dresel,S. ,Patt,J. T. ,Shen,M. H. ,Zhu,T. ,Liu,B. L. ,Tang,Z. G. ,Tatsch,K. ,Hahn,K. :Comparison of in vivo dopamine D_2 receptor binding of ^{123}I-AIBZM and ^{123}I-IBZM in rat brain. Nucl. Med. Biol. 28,383(2001).

[28] Hai-Bin,T. ,Duan-Zhi,Y. ,Jun-Ling,L. ,Lan,Z. ,Cun-Fu,Z. ,Yong-Xian,W. ,Wei,Z. :3-1{4-(4-[^{18}F]fluorobenzyl)lpiperazin-1-yl}methyl-1H-pyrrolo[2,3-b] pyridine:A potential dopamine D_4 receptor imaging agent. Radiochim. Acta 91,241(2003).

[29] Li,G. C. ,Yin,D. Z. ,Wang,M. W. ,Cheng,D. F. ,Wang,Y. X. :Syntheses of two potential dopamine D_4 receptor radioligands:F-18 labelled chromeno [3,4-c] pyridin-5-ones. Radiochim. Acta 94,119(2006).

[30] Li,G. C. ,Yin,D. Z. ,Cheng,D. F. ,Zheng,M. Q. ,Han,Y. J. ,Cai,H. C. ,Xia,J. Y. ,Liang,S. ,Xu,W. B. ,Wang,Y. X. :In vitro and in vivo evaluation of [^{18}F]FHTP as a potential dopamine D_4 receptor PET imaging agent. J. Radioanal. Nucl. Chem. 280,15(2009).

[31] Lin,X. ,Zhang,J. ,Wang,X. ,Tang,Z. ,Zhang,X. ,Lu,J. :Development of radiolabeled compounds for myocardial perfusion imaging. Curr. Pharm. Design 18,1041(2012).

[32] Jiang,Y. ,Liu,B. :A novel ^{99m}Tc labeled [^{99m}Tc(CO)$_3$(MIBI)$_3$]$^+$ complex as myocardial imaging agent. Chin. Sci. Bull. (Chinese)46,727(2001).

[33] Hao,G. Y. ,Zang,J. Y. ,Zhu,L. ,Guo,Y. Z. ,Liu,B. L. :Synthesis, separation and biodistribution of ^{99m}Tc-CO-MIBI complex. J. Labelled Compd. Radiopharm. 47,513(2004).

[34] Hao,G. ,Zang,J. ,Liu,B. :Preparation and biodistribution of novel ^{99m}Tc(CO)$_3$-CNR complexes for myocardial imaging. J. Labelled Compd. Radiopharm. 50,13(2007).

[35] Wang,J. ,Liu,B. ,Mi,H. ,Jiang,Y. ,Tang,Z. ,Gui,H. :Comparative study of pharmacology of a new myocardial imaging agent [^{99}Tc^m(CO)$_3$(MIBI)$_3$]$^+$ and ^{99}Tc^m-MIBI. Chin. J. Nucl. Med. (Chinese)22,231 (2002).

[36] Chen,X. ,Guo,Y. ,Zhang,Q. ,Hao,G. ,Jia,H. ,Liu,B. :Preparation and biological evaluation of ^{99m}Tc-CO-MIBI as myocardial perfusion imaging agent. J. Organomet. Chem. 693,1822(2008).

[37] Yu,L. H. ,Fang,D. C. ,Ren,H. Y. ,Jia,H. M. ,Liu,B. L. :Ligand exchange mechanism of fac-[^{99m}Tc(CO)$_3$(H$_2$O)$_3$]$^+$ complex for ^{99m}Tc-CO-MIBI radiopharmaceuticals. Nucl. Med. Biol. 37, 704 (2010).

[38] Yu,M. ,Nekolla,S. G. ,Schwaiger,M. ,Robinson,S. P. :The next generation of cardiac positron emission tomography imaging agents:discovery of flurpiridaz F-18 for detection of coronary disease. Semin. Nucl. Med. 41,305(2011).

[39] Mou,T. ,Jing,H. ,Yang,W. ,Fang,W. ,Peng,C. ,Guo,F. ,Zhang,X. ,Pang,Y. ,Ma,Y. :Preparation and biodistribution of [^{18}F]FP2OP as myocardial perfusion imaging agent for positron emission tomography. Bioorg. Med. Chem. 18,1312(2010).

[40] Mou,T. ,Zhao,Z. ,Fang,W. ,Peng,C. ,Guo,F. ,Liu,B. ,Ma,Y. ,Zhang,X. :Synthesis and preliminary evaluation of ^{18}F-labeled pyridaben analogues for myocardial perfusion imaging with PET. J. Nucl. Med. 53,472(2012).

[41] Jia,B. ,Shi,J. ,Yang,Z. ,Xu,B. ,Liu,Z. ,Zhao,H. ,Liu,S. ,Wang,F. :^{99m}Tc-labeled cyclic RGDfK dimer:initial evaluation for SPECT imaging of glioma integrin $\alpha_v\beta_3$ expression. Bioconjugate Chem. 17, 1069(2006).

[42] Wang,L. ,Shi,J. ,Kim,Y. ,Zhai,S. ,Jia,B. ,Zhao,H. ,Liu,Z. ,Wang,F. ,Chen,X. ,Liu,S. :Improving tumor-targeting capability and pharmacokinetics of ^{99m}Tc-labeled cyclic RGD dimers with PEG4 linkers. Mol. Pharm. 6,231(2009).

[43] Shi, J., Kim, Y. S., Chakraborty, S., Jia, B., Wang, F., Liu, S.: 2-Mercaptoacetylglycylglycyl (MAG2) as a bifunctional chelator for ^{99m}Tc-labeling of cyclic RGD dimers: effect of technetium chelate on tumor uptake and pharmacokinetics. Bioconjugate Chem. 20, 1559 (2009).

[44] Liu, Z., Jia, B., Shi, J., Jin, X., Zhao, H., Li, F., Liu, S., Wang, F.: Tumor uptake of the RGD dimeric probe ^{99m}Tc-G3-2P4-RGD2 is correlated with integrin $\alpha_v\beta_3$ expressed on both tumor cells and neovasculature. Bioconjugate Chem. 21, 548 (2010).

[45] Jia, B., Liu, Z., Zhu, Z., Shi, J., Jin, X., Zhao, H., Li, F., Liu, S., Wang, F.: Blood clearance kinetics, biodistribution and radiation dosimetry of a kit-formulated integrin $\alpha_v\beta_3$-selective radiotracer ^{99m}Tc-3PRGD2 in non-human primates. Mol. Imaging Bio. 13, 730 (2011).

[46] Ma, Q., Ji, B., Jia, B., Gao, S., Ji, T., Wang, X., Han, Z., Zhao, G.: Differential diagnosis of solitary pulmonary nodules using ^{99m}Tc-3P4-RGD2 scintigraphy. Eur. J. Nucl. Med. Mol. Imaging 38, 2145 (2011).

[47] Zhu, Z., Miao, W., Li, Q., Dai, H., Ma, Q., Wang, F., Yang, A., Jia, B., Jing, X., Liu, S., Shi, J., Liu, Z., Zhao, Z., Wang, F., Li, F.: ^{99m}Tc-3PRGD2 for integrin receptor imaging of lung cancer: A multicenter study. J. Nucl. Med. 53, 716 (2012).

[48] Li, Y., Liu, Z., Dong, C., He, P., Liu, X., Zhu, Z., Jia, B., Li, F., Wang, F.: Noninvasive detection of human-induced pluripotent stem cell (hiPSC)-derived teratoma with an integrin-targeting agent ^{99m}Tc-3PRGD2. Mol Imaging Biol. 15, 58 (2013).

[49] Wang, X., Lin, B., Yang, Z., Ouyang, T., Li, J., Xu, B., Zhang, Y., Zhang, M.: Preliminary study on a new sentinel lymphoscintigraphy agent ^{99m}Tc-Rituximab for breast patient. Chin. J. Oncol. (Chinese) 28, 200 (2006).

[50] Li, J., Ouyang, T., Wang, X., Wang, T., Xie, Y., Fan, Z., Lin, B., Yang, Z., Lin, B.: Preliminary study of new imaging agent, ^{99m}Tc-Rituximab, for sentinel lymph node biopsy of primary breast cancer. Chin. J. Surg. (Chinese) 44, 600 (2006).

[51] Wang, X., Yang, Z., Lin, B., Xu, B., Zhang, Y., Zhang, M.: The preparation and localization study of a novel sentinel lymphoscintigraphy agent ^{99m}Tc-IT-Rituximab. Chin. J. Nucl. Med. (Chinese) 26, 226 (2006).

[52] Li, N., Lin, B., Ouyang, T., Yang, Z., Xu, X., Zhang, Y., Ma, Y.: The results of sentinel lymph node imaging and biopsy with a novel lymphoscintigraphy agent ^{99m}Tc-Rituximab in 467 breast cancer patients. Chin. J. Nucl. Med. (Chinese) 29, 3 (2009).

[53] Wang, X., Wang, R., Yang, Z., Lin, B., Xu, B., Zhang, Y., Zhang, M.: Sentinel lymph node imaging with a novel radiotracer: preparation of ^{99m}Tc-Rituximab and experimental study. Chin. J. Med. Imaging Technol. (Chinese) 22, 139 (2006).

[54] Li, Z., Chu, T.: Recent advances on radionuclide labeled hypoxia-imaging agents. Curr. Pharm. Design 18, 1084 (2012).

[55] Li, Z., Chu, T., Liu, X.: Wang, X. Synthesis and *in vitro* and *in vivo* evaluation of three radioiodinated nitroimidazole analogues as tumor hypoxia markers. Nucl. Med. Biol. 32, 225 (2005).

[56] Chu, T., Li, Z., Wang, X.: Synthesis and biological evaluation of radioiodinated 2NUBTA as a cerebral ischemia marker. Bioorg. Med. Chem. Lett. 19, 658 (2009).

[57] Chu, T., Hu, S., Wei, B., Wang, Y., Liu, X., Wang, X.: Synthesis and biological results of the technetium-99m-labeled 4-nitroimidazole for imaging tumor hypoxia, Bioorg. Med. Chem. Lett. 14, 747 (2004).

[58] Chu, T., Li, R., Hu, S., Wang, Y., Liu, X., Wang, X.: Synthesis and biodistribution of the ^{99m}Tc nitrido complex with 2-(4-nitro-1H-imidazolyl) ethyl dithiocarbamate (NIET). J. Radioanal. Nucl. Chem. 261,

199(2004).

[59] Chu,T. ,Li,R. ,Hu,S. ,Liu,X. ,Wang,X. : Preparation and biodistribution of technetium-99m-labeled 1-(2-nitroimidazole-1-yl)-propanhydroxyiminoamide (N2IPA) as a tumor hypoxia marker. Nucl. Med. Biol. 31,199(2004).

[60] Chu,T. ,Xu,H. ,Yang,Z. ,Wang,X. : Synthesis and *in vitro* evaluation of three ^{99m}Tc-labeled hydroxamamide-based ligands as markers for hypoxic cells. Appl. Radiat. Isot. 67,590(2009).

[61] Zhang,Y. ,Chu,T. ,Gao,X. ,Liu,X. ,Yang Z. ,Guo,Z. ,Wang,X. : Synthesis and preliminary biological evaluation of the ^{99m}Tc labeled nitrobenzoimidazole and nitrotriazole as tumor hypoxia markers. Bioorg. Med. Chem. Lett. 16,1831(2006).

[62] Sun,X. ,Chu,T. ,Wang,X. : Preliminary studies of ^{99m}Tc-BnAO and its analogues: synthesis, radiolabeling and *in vitro* cell uptake. Nucl. Med. Biol. 37,117(2010).

[63] Huang,H. ,Zhou,H. ,Li,Z. ,Wang,X. : Effect of a second nitroimidazole redox centre on the accumulation of a hypoxia marker: Synthesis and *in vitro* evaluation of ^{99m}Tc-labeled bisnitroimidazole propylene amine oxime complexes. Bioorg. Med. Chem. Lett. 22,172(2012).

[64] Huang,H. ,Mei,L. ,Chu,T. : Synthesis, Radiolabeling and biological evaluation of propylene amine oxime complexes containing nitrotriazoles as hypoxia markers. Molecules 17,6808(2012).

[65] Mei,L. ,Wang,Y. ,Chu,T. : ^{99m}Tc/Re complexes bearing bisnitroimidazole or mononitroimidazole as potential bioreductive markers for tumor: Synthesis, physicochemical characterization and biological evaluation. Eur. J. Med. Chem. 58,50(2012).

[66] Zhang,J. ,Ren,J. ,Lin,X. ,Wang,X. : Synthesis and biological evaluation of a novel ^{99m}Tc nitrido radiopharmaceutical with deoxyglucose dithiocarbamate, showing tumor uptake. Bioorg. Med. Chem. Lett. 19,2752(2009).

[67] Lin,X. ,Jin,Z. ,Ren,J. ,Pang,Y. ,Zhang,W. ,Huo,J. ,Wang,X. ,Zhang,J. ,Zhang,Y. : Synthesis and biodistribution of a new ^{99m}Tc-oxo complex with deoxyglucose dithiocarbamate for tumor imaging. Chem. Biol. Drug Des. 79,239(2012).

[68] Fan,C. ,Jia,H. ,Deuther-Conrad,W. ,Brust,P. ,Steinbach,J. ,Liu,B. : Novel ^{99m}Tc labeled σ receptor ligand as a potential tumor imaging agent. Sci. China Ser. B-Chem. 49,169(2006).

[69] Chen,X. ,Cui,M. C. ,Deuther-Conrad,W. ,Tu,Y. F. ,Ma,T. ,Xie,Y. ,Jia,B. ,Li,Y. ,Xie,F. ,Wang,X. ,Steinbach,J. ,Brust,P. ,Liu,B. L. ,Jia,H. M. : Synthesis and biological evaluation of a novel ^{99m}Tc cyclopentadienyl tricarbonyl complex ([(Cp-R)^{99m}Tc(CO)$_3$]) for sigma-2 receptor tumor imaging. Bioorg. Med. Chem. Lett. 22,6352(2012).

[70] Li,D. ,Wang,X. ,Deuther-Conrad,W. ,Chen,X. ,Cui,M. ,Steinbach,J. ,Brust,P. ,Liu,B. ,Jia,H. : Synthesis and preliminary evaluation of novel [(Cp-R)M(CO)$_3$](M = Re,^{99m}Tc) complexes as potent sigma-2 receptor ligands. J. Labelled Compd. Radiopharm. 56,S383(2013).

[71] Lu,J. ,Pang,Y. ,Xie,F. ,Guo,H. ,Li,Y. ,Yang,Z. ,Wang,X. : Synthesis and *in vitro/in vivo* evaluation of ^{99m}Tc-labeled folate conjugates for folate receptor imaging. Nucl. Med. Biol. 38,557(2011).

[72] Liu,L. ,Wang,S. ,Li,F. ,Teng,B. ,Wang,K. ,Qiu,F. : Synthesis and animal imaging of ^{99m}Tc-hydrazinonicotinamide-folate as a new folate receptor-targeted tumor imaging agent. Acta Acad. Med. Sin. (Chinese)28,786(2006).

[73] Guo,H. ,Xie,F. ,Zhu,M. ,Li,Y. ,Yang,Z. ,Wang,X. ,Lu,J. : The synthesis of pteroyl-lys conjugates and its application as technetium-99m labeled radiotracer for folate receptor-positive tumor targeting. Bioorg. Med. Chem. Lett. 21,2025(2011).

[74] Zhang,Y. ,Sun,Y. ,Xu,X. ,Zhu,H. ,Huang,L. ,Zhang,X. ,Qi,Y. ,Shen,Y. M. : Radiosynthesis and

micro-SPECT imaging of ^{99m}Tc-dendrimer poly(amido)-amine folic acid conjugate. Bioorg. Med. Chem. Lett. 20,927(2010).

[75] Zhang,Y. ,Sun,Y. ,Xu,X. ,Zhang,X. ,Zhu,H. ,Huang,L. ,Qi,Y. ,Shen,Y. M. : Synthesis,biodistribution, and microsingle photon emission computed tomography(SPECT) imaging study of technetium-99m labeled PEGylated dendrimer poly(amidoamine) (PAMAM)-folic acid conjugates. J. Med. Chem. 53, 3262(2010).

[76] Wang,R. F. ,Shen,J. ,Qiu,F. ,Zhang,C. L. : Study on biodistribution and imaging of radioiodinated antisense oligonucleotides in nude mice bearing human lymphoma. J. Radioanal. Nucl. Chem. 273, 19 (2007).

[77] Liu,M. ,Wang,R. F. ,Zhang,C. L. ,Yan,P. ,Yu,M. M. ,Di,L. J. ,Liu,H. J. ,Guo,F. Q. : Noninvasive imaging of human telomerase reverse transcriptase(hTERT) messenger RNA with ^{99m}Tc-radiolabeled antisense probes in malignant tumors. J. Nucl. Med. 48,2028(2007).

[78] Kang,L. ,Wang,R. F. ,Yan,P. ,Liu,M. ,Zhang,C. L. ,Yu,M. M. ,Cui,Y. G. ,Xu,X. J. : Noninvasive visualization of RNA delivery with ^{99m}Tc-radiolabeled small-interference RNA in tumor xenografts. J. Nucl. Med. 51,978(2010).

[79] Zhang,J. ,Guo,H. ,Zhang,S. ,Lin,Y. ,Wang,X. : Synthesis and biodistribution of a novel ^{99m}TcN complex of ciprofloxacin dithiocarbamate as a potential agent for infection imaging. Bioorg. Med. Chem. Lett. 18,5168(2008).

[80] Zhang,J. ,Zhang,S. ,Guo,H. ,Wang,X. : Synthesis and biological evaluation of a novel ^{99m}Tc(CO)$_3$ complex of ciprofloxacin dithiocarbamate as a potential agent to target infection. Bioorg. Med. Chem. Lett. 20,3781(2010).

[81] Zhang,S. ,Zhang,W. ,Wangch,Y. ,Jin,Z. ,Wang,X. ,Zhang,J. ,Zhang,Y. : Synthesis and biodistribution of a novel ^{99m}TcN complex of norfloxacin dithiocarbamate as a potential agent for bacterial infection imaging. Bioconjugate Chem. 22,369(2011).

[82] Yang,W. ,Mou,T. ,Peng,C. ,Wu,Z. ,Zhang,X. ,Li,F. ,Ma,Y. : Fluorine-18 labeled galactosyl-neoglycoalbumin for imaging the hepatic asialoglycoprotein receptor. Bioorg. Med. Chem. 17,7510(2009).

[83] Yang,W. ,Mou,T. ,Guo,W. ,Jing,H. ,Peng,C. ,Zhang,X. ,Ma,Y. ,Liu,B. : Fluorine-18 labeled galactosylated chitosan for asialoglycoprotein-receptor-mediated hepatocyte imaging. Bioorg. Med. Chem. Lett. 20,4840(2010).

[84] Yang,W. ,Mou,T. ,Zhang,X. ,Wang,X. : Synthesis and biological evaluation of ^{99m}Tc-DMP-NGA as a novel hepatic asialoglycoprotein receptor imaging agent. Appl. Radiat. Isot. 68,105(2010).

[85] Yang,W. J. ,Mou,T. T. ,Shao,G. Q. ,Wang,F. ,Zhang,X. Z. ,Liu,B. L. : Copolymer-based hepatocyte asialoglycoprotein receptor targeting agent for SPECT. J. Nucl. Med. 52,978(2011).

[86] Liu, G. , Liu, B. : Synthesis of a new polyaminopolycarboxybic acid (BPHA) and labeling with ^{99m}Tc. J. Labelled Compd. Radiopharm. XLI,97(1998).

[87] Chu,T. ,Zhang,Y. ,Liu,X. ,Wang,Y. ,Hu,S. ,Wang,X. : Synthesis and biodistribution of ^{99m}Tc-carbonyltechnetium-labeled fatty acids. Appl. Radiat. Isot. 60,845(2004).

[88] Liang,J. ,Hu,J. ,Chen,B. ,Luo,L. ,Li,H. ,Shen,L. ,Luo,Z. : Preparation and biological evaluation of ^{99}Tc^m-labelled fatty acids. Nucl. Sci. Tech. 18,159(2007).

[89] Liu,Z. ,Huang,J. ,Dong,C. ,Cui,L. ,Jin,X. ,Jia,B. ,Zhu,Z. ,Li,F. ,Wang,F. : ^{99m}Tc-labeled RGD-BBN peptide for small-animal SPECT/CT of lung carcinoma. Mol. Pharm. 9,1409(2012).

[90] Luo,Q. ,Zhang,Z. ,Wang,F. ,Lu,H. ,Guo,Y. ,Zhu,R. : Preparation, in vitro and in vivo evaluation of ^{99m}Tc-Annexin B1: a novel radioligand for apoptosis imaging. Biochem. Biophys. Res. Comm. 335, 1102

(2005).

[91] Luo,Q. Y. ,Wang,F. ,Zhang,Z. Y. ,Zhang,Y. ,Lu,H. K. ,Sun,S. H. ,Zhu,R. S.:Preparation and bio-evaluation of ^{99m}Tc-HYNIC-annexin B1 as a novel radioligand for apoptosis imaging. Apoptosis 13,600(2008).

[92] Ye,F. ,Fang,W. ,Wang,F. ,Hua,Z. C. ,Wang,Z. ,Yang,X.:Evaluation of adenosine preconditioning with ^{99m}Tc-His$_{10}$-annexin V in a porcine model of myocardium ischemia and reperfusion injury:preliminary study. Nucl. Med. Biol. 38,567(2011).

[93] Wang,H. ,Luo,S. ,Xie,M. ,Liu,X. ,Feng,Y. ,Chen,Z. ,Ye,W.:The new bone-imaging agent:preparation and biodistribution of ^{99m}Tc-ZL. Nucl. Tech. (Chinese)29,438(2006).

[94] Lin,J. ,Luo,S. ,Chen,C. ,Qiu,L. ,Wang,Y. ,Cheng,W. ,Ye,W. ,Xia Y.:Preparation and preclinical pharmacological study on a novel bone imaging agent ^{99m}Tc-EMIDP. Appl. Radiat. Isot. 68,1616(2010).

[95] Wang,Y. ,Luo,S. ,Lin,J. ,Qiu,L. ,Cheng,W. ,Zhai,H. ,Nan,B. ,Ye,W. ,Xia,Y.:Animal studies of ^{99m}Tc-i-PIDP:A new bone imaging agent. Appl. Radiat. Isot. 69,1169(2011).

[96] Qiu,L. ,Cheng,W. ,Lin,J. ,Luo,S. ,Xue,L. ,Pan,J.:Synthesis and biological evaluation of novel ^{99m}Tc-labelled bisphosphonates as superior bone imaging agents. Molecules 16,6165(2011).

[97] Qiu,L. ,Xue,L. ,Chen,Y. ,Lin,J. ,Nan,B. ,Cheng,W. ,Luo,S.:Preparation and *in vivo* biological investigation of ^{99m}Tc-HEIDP as a novel radioligand for bone scanning. J. Labelled Compd. Radiopharm. 55,307(2012).

[98] Lin,J. ,Qiu,L. ,Cheng,W. ,Luo,S. ,Xue,L. ,Zhang,S.:Development of superior bone scintigraphic agent from a series of ^{99m}Tc-labeled zoledronic acid derivatives. Appl. Radiat. Isot. 70,848(2012).

[99] Qiu,L. ,Lin,J. ,Luo,S. ,Wang,Y. ,Cheng,W. ,Zhang,S.:A novel ^{99m}Tc-labeled dimethyl-substituted zoledronic acid(DMIDP)with improved bone imaging efficiency. Radiochim. Acta100,463(2012).

[100] Deng,X. ,Li,H. ,Ye,Z. ,Guo,H. ,Li,F. ,Luo,Z.:Radiolabeling and biodistribution of ^{177}Lu-EDTMP and ^{177}Lu-DOTMP. J. Isot. (Chinese)22,71(2009).

[101] Li,H. ,Liang,J. ,Xiang,X. ,Deng,X. ,Zheng,D. ,Luo,H. ,Chen,Y. ,Liu,H. ,Lu,J. ,Luo,Z.:Biodistribution and imaging study on ^{177}Lu-EDTMP prepared by kit method. J. Isot. (Chinese)23,65(2010).

[102] Liu,Y. ,Yin,G. ,Fu,B. ,Ye,Z. ,Wang,X.:Establishment of analysis method for the radiochemical purity of ^{177}Lu-EDTMP. J. Isot. (Chinese)23,163(2010).

[103] Shi,J. ,Liu,Z. ,Jia,B. ,Yu,Z. ,Zhao,H. ,Wang,F.:Potential therapeutic radiotracers:preparation,biodistribution and metabolic characteristics of ^{177}Lu-labeled cyclic RGDfK dimer. Amino Acids 39,111(2010).

[104] Sheng,F. ,He,W. ,Liu,Z. ,Zhao,H. ,Jia,B. ,Wang,F.:Preparation of ^{177}Lu-DOTA/DTPA-Bz-Cys-RGD dimer and biodistribution evaluation in normal mice. J. Nucl. Radiochem. (Chinese) 30,70(2008).

[105] Shi,J. ,Yu,Z. ,Jia,B. ,Zhao,H. ,Wang,F.:Preparation and evaluation of ^{177}Lu-DOTA-Bz-RGD dimer and ^{177}Lu-DOTA-Bz-PEG4-RGD dimer. J. Isot. (Chinese)20,214(2007).

[106] Zhang,Y. ,He,W. ,Jia,B. ,Wang,F.:Comparison of ^{90}Y/^{177}Lu labeled DOTA-Bz-RGD tetramer and DOTA-RGD tetramer. J. Nucl. Radiochem. (Chinese)30,93(2008).

[107] Deng,X. ,Du,J. ,Luo,Z.:Preparation of ^{177}Lu-DTPA-BIS-BIOTIN and biodistribution evaluation in mornal mice. J. Nucl. Radiochem. (Chinese)32,184(2010).

[108] Deng,X. ,Du,J. ,Zhai,S. ,Shen,Y. ,Luo,Z.:^{177}Lu-DTPA-BIS-BIOTIN binding of octreotide-dextran-avidinated PANC-1 cell lines *in vitro*. J. Isot. (Chinese)24,146(2011).

[109] Jiang,L. ,Miao,Z. ,Kimura R. H. ,Liu,H. ,Cochran,J. R. ,Culter,C. S. ,Bao,A. ,Li,P. ,Cheng,Z.:

Preliminary evaluation of ^{177}Lu-labeled knottin peptides for integrin receptor-targeted radionuclide therapy. Eur. J. Nucl. Med. Mol. Imaging 38,613(2011).

[110] Liu,N. ,Yang,Y. ,Liao,J. ,Jin,J. :New progress of ^{211}At labeled radiopharmaceuticals recently. J. Isot. (Chinese)24(suppl.),36(2011).

[111] Liu,B. ,Jin,Y. ,Liu,Z. ,Luo,C. ,Kojima,M. ,Maeda,M. :Halogen exchange using crown ethers:synthesis and preliminary biodistribution of 6-[^{211}At]-astatomethyl-19-norcholest-5(10)-en-3β-ol. Int. J. Appl. Radiat. Isot. 36,561(1985).

[112] Liu,B. ,Jin,Y. ,Li,T. :Synthesis,labelling and thermal stability of the new compound. 6-^{211}At-cholesterol. J. Beijing Normal University(Natural Sci. ,Chinese),57(1985).

[113] Liu,N. ,Jin,J. N. ,Mo,S. W. ,Chen,H. L. ,Yu,Y. P. :Preparation and premilinary evaluation of astatine-211 labeled IgG via DTPA anhydride. J. Radioanal. Nucl. Chem. 227,187(1998).

[114] Liu,N. ,Jin,J. ,Zhang,S. ,Mo,S. ,Yang,Y. ,Wang,J. ,Zhou,M. :At-211 labeling of a monoclonal antibody and its Fab fragment:Cytotoxicity on human gastric cancer cells and biodistribution in nude mice with tumor xenografts. J. Radioanal. Nucl. Chem. 247,129(2001).

[115] Liu,N. ,Jin,J. N. ,Zhang,S. Y. ,Luo,D. Y. ,Wang,J. A. ,Zou,M. L. ,Luo,L. ,Wang,F. Y. :Astatine-211 labeling of a monoclonal-antibody and its Fab fragment-Synthesis, immunoreactivity. J. Labelled Compd. Radiopharm. 36,1105(1995).

[116] Yi,C. H. ,Jin,J. ,Zhang,S. Y. ,Wang,K. T. ,Zhang,D. Y. ,Zhou,M. L. :Astatine-211 Labeled proteins and their stability in vivo. J. Radioanal. Nucl. Chem. 129,377(1989).

[117] Yang,Y. ,Lin,R. ,Liu,N. ,Liao,J. ,Wei,M. ,Jin,J. :Astatine-211 labeling of protein using TCP as a bi-functional linker:synthesis and preliminary evaluation *in vivo* and *in vitro*. J. Radioanal. Nucl. Chem. 288,71(2011).

[118] Liu,N. ,Yang,Y. ,Zan,L. ,Liao,J. ,Jin,J. :Astatine-211 labeling of insulin:Synthesis and preliminary evaluation in vivo and *in vitro*. J. Radioanal. Nucl. Chem. 272,85(2007).

[119] Yang,Y. ,Liu,N. ,Liao,J. ,Pu,M. ,Liu,Y. ,Wei,M. ,Jin,J. :Preparation and preliminary evaluation of ^{211}At-labeled amidobisphophonates. J. Radioanal. Nucl. Chem. 283,329(2010).

[120] Xiangyun,W. ,Yonghui,W. ,Yi,W. ,Yuanfang,L. :Radiopharmaceutical chemistry in Peking University (PKU). J. Nucl. Radiochem. Sci. 1,15(2000).

[121] Liu,B. ,Jin,Y. ,Li,T. :Heterogeneous isotope exchange reaction. Hydrothermal melt method for isotope labeling. J. Nucl. Radiochem. (Chinese)7,29(1985).

[122] Liu,B. ,Chen,S. :Effect of metal ions on halogen labeling in a crown ether medium. J. Nucl. Radiochem. (Chinese)7,155(1985).

[123] Qi,P. ,Liu,B. :A study on solid phase isotopic exchange in the presence of ammonium sulfate. Nucl. Tech. (Chinese)9,58(1986).

[124] Feng,X. ,Liu,B. ,Guo,Y. :Discussion on the exchange reaction between different halogen atoms. J. Beijing Normal University(Natural Sci. ,Chinese),3,63(1983).

[125] Wei,Y. ,Liu,B. :Study of technetium chemistry,I. The general rule of stability for structures of technetium compounds. J. Nucl. Radiochem. (Chinese)10,65(1988).

[126] Kung,H. F. ,Liu,B. ,Wei,Y. ,Pan,S. :Quantitative study of the structure-stability relationship of technetium oxide [Tc^{V}O(Ⅲ)] complexes. Appl. Radiat. Isot. 41,773(1990).

[127] Song,W. ,Xue,L. ,Chen,W. ,Wang,C. ,Jia,H. ,Liu,B. :The relationship between relative hydration free energies of ^{99m}TcON$_2$S$_2$ complexes and their brain uptakes. Chem. Res. Chin. Univ. (English)19, 340(2003).

[128] Jia, H., Ma, X., Wang, C., Liu, B.: Solvent effects on brain uptake of isomers of ^{99m}Tc-brain radiopharmaceuticals. Chin. Sci. Bull. (English) 47, 1786(2002).

[129] Wang, X. Y., Wang, Y., Liu, X. Q., Chu, T. W., Hu, S. W., Wei, X. H., Liu, B. L.: The structure, energy and stability of components formed in the preparation of fac-$[^{99m}Tc(CO)_3(H_2O)_3]^+$. Phy. Chem. Phys. 5, 456(2003).

[130] Jia, H. M., Fang, D. C., Feng, Y., Zhang, J. Y., Fan, W. B., Zhu. L.: The interconversion mechanism between TcO^{3+} and TcO_2^+ core of ^{99m}Tc labeled amine-oxime (AO) complexes. Theor. Chem. Acc. 121, 271(2008).

[131] Herzog, H., Roesch, F., Stoecklin, G., Lueders, C., Qaim, S. M., Feinendegen, L. E.: Measurement of pharmacokinetics of yttrium-86 radiopharmaceuticals with PET and radiation dose calculation of analogous yttrium-90 radiotherapeutics. J. Nucl. Med. 34, 2222(1993).

[132] Srivastava, S. C.: Paving the way to personalized medicine: production of some theragnostic radionuclides at Brookhaven National laboratory. Radiochim. Acta 99, 635(2011).

[133] Qaim, S. M.: Development of novel positron emitters for medical applications: nuclear and radiochemical aspects. Radiochim. Acta 99, 611(2011).

[134] Qaim, S. M.: The present and future of medical radionuclide production. Radiochim. Acta 100, 635 (2012).

[135] Alberto, R.: Chapter 17 Future trends in the development of technetium radiopharmaceuticals. International Atomic Energy Agency. Technetium-99m Radiopharmaceuticals: Status and Trends, IAEA, Radioisotopes and Radiopharmaceuticals series No. 1. Vienna, 347(2009).

[136] Braband, H., Tooyama, Y., Fox, T., Simms, R., Forbes, J., Valliant, J. F., Alberto, R.: fac-$[TcO_3(tacn)]^+$: a versatile precursor for the labelling of pharmacophores, amino acids and carbohydrates through a new ligand-centred labelling strategy. Chem. Eur. J. 17, 12967(2011).